W9-CRL-470

Superconducting Magnet Systems

by

H. Brechna

With 396 Figures and 46 Tables

Springer-Verlag New York · Heidelberg · Berlin
J. F. Bergmann Verlag · München
1973

PHYSICS

Technische Physik in Einzeldarstellungen

Founded by W. Meissner
Edited by F. X. Eder

Volume 18

Dr. sc. tech. H. Brechna: Eidgenössische Technische Hochschule Zürich (ETHT); Institut für Experimentelle Kernphysik Karlsruhe (IEKP); Stanford Linear Accelerator Center (SLAC), Stanford University, California

ISBN 0-387-06103-7 Springer-Verlag New York · Heidelberg · Berlin
ISBN 3-540-06103-7 Springer-Verlag Berlin · Heidelberg · New York

This work is subject to copyright. All rights are reserved, whether the whole or part of the material is concerned, specifically those of translation, reprinting, re-use of illustrations, broadcasting, reproduction by photocopying machine or similar means, and storage in data banks.

Under § 54 of the German Copyright Law where copies are made for other than private use, a fee is payable to the publisher, the amount of the fee to be determined by agreement with the publisher. © by Springer-Verlag, Berlin · Heidelberg 1973, Printed in Germany. Library of Congress Catalog Card Number 72-96051.

The use of registered names, trademarks, etc. in this publication does not imply, even in the absence of a specific statement, that such names are exempt from the relevant protective laws and regulations and therefore free for general use.

Typesetting and printing: Gebr. Parcus KG, München.
Binding: Konrad Triltsch, Würzburg.

QC 760
.7
B731
PHYSICS

Preface

The renaissance of magnet technology started in the early 1950s with the establishment of high-energy accelerators. About a decade later in 1961, or fifty years after the discovery of superconductivity, high-field superconducting laboratory magnets became a reality. Conventional electromagnets, which are still the major beam-handling and experimental devices used in laboratories, operate at zero efficiency. To generate high magnetic fields in a useful volume, considerable amounts of power are needed. Superconducting d.c. magnets do not require any power at all.

It is somewhat depressing to note that, sixty years after the first superconductor was tested, the experimental d.c. superconducting magnet is still the only large-scale equipment operated in laboratories. Although there has been considerable activity in the area of superconductivity, superconductors are used on quite a modest scale in electronic and quantum devices, in medicine and biology, and in physical experiments where high magnetic fields are essential. It is only recently that Type II superconductors have been introduced in power engineering (power generation, storage and transport) to replace pulsed accelerator magnets; for fast and economical transportation vehicles (levitated trains) where superconductors may ultimately replace the wheel; to make new means of energy generation economically feasible, such as in magnetohydrodynamics and in fusion reactors; and for high-efficiency electric motors. High-field superconducting magnets are being proposed for desalination of seawater, for magnetic separation in the mining industry, for cleaning polluted water, and for sewage treatment.

Man is not accustomed to handling low-temperature materials and his natural fear has inhibited the large-scale application of superconducting technology. The technological obstacles in the way of the practical application of superconductivity are no more formidable than those overcome in many other sophisticated but well-established technologies. Refrigerators and liquefiers are commercially available over a wide energy spectrum and the transport of cryogens is no more complicated than that of conventional coolants. Superconducting materials are available to carry thousands of amperes at high magnetic fields and the existing technology already provides sufficient "know-how" to replace conventional magnets with superconducting magnets which can operate reliably without close supervision.

The published literature on the properties of cryogenic materials and the design and operation of superconducting magnet systems is vast but scattered. Data on the physical and mechanical properties of materials are available from various sources, but it was necessary to collect them

11113

and arrange them systematically. Information on methods of generating magnetic fields, handling cryogenic problems, and controlling magneto-mechanical forces, and the experience gained in designing and operating magnet systems — all this had to be brought together in one place. Progress in superconductivity is rapid, and the volume of work performed over the entire field of high-field magnets is enormeous. During the time this book was being written, many new results were reported; most of this new information is incorporated. However, the book had to be kept to a reasonable size, so that not all the data collected could be included. For example, the lists of references at the end of each chapter are certainly not complete. However, the book will be of value to the physicist and engineer working in this field and to the student who wants to learn about this technology.

Throughout the book the "mks" system of units is used. The nomenclature is in accordance with normal practice, thus H_{c1} and H_{c2} are used for the critical fields, instead of B_{c1} and B_{c2}, though the latter may be more appropriate.

I would like to express my gratitude to the late Francis Bitter of MIT from whom I obtained most of my knowledge on high field magnets; to W. K. H. Panofsky who supported and encouraged the superconducting work at SLAC, and to F. X. Eder for his continuous encouragement and patience and for many discussions during the time this book was being written. It is also a pleasure to acknowledge the help of many colleagues from ANL, BNL, CERN, IEKP, LBL, NAL, NBS, NML, ORNL, RHEL, and Saclay and to manufacturers such as AIRCO, IGE, IMI, Siemens, Supercon Inc., and VAC for providing specimen, samples and test data. I would like also to thank Mrs. V. Smoyer, and Mrs. A. Schwartz of SLAC for typing the manuscript and preparing the drawings.

Dietlikon, Switzerland, January 1973

H. Brechna

Contents

1. Methods of Magnetic Field Generation

1.1 High Magnetic Field Laboratories

Since the discovery of type II superconductors in 1961, research activity in discovery and improvement of new type II superconductors, in utilizing properties of these materials in magnets, electric and electronic devices, electrical machines, in their use in the generation, storage and transfer of energy has been quite extensive. For these purposes a number of laboratories were established in the sixtieth to develope new types of d.c. and pulsed magnets, to study material properties at high magnetic fields at room, elevated and cryogenic temperatures. It was possible to generate magnetic fields as low as 10^{-14} T and as high as 10^3 T. From a large number of magnetic field research establishments we name only a few:

Two national laboratories are primarily associated with the development of high field magnets:

(a) The Francis Bitter National Magnet Laboratory at Cambridge, Massachusetts, which utilizes an 8 MW d.c. power capacity and has the only 25 T, d.c. water-cooled magnet in operation in the world. The laboratory was completed in 1963.

(b) The German National Magnet Laboratory at Braunschweig, which has a capacity of 5 MW at 500 volts. This laboratory was scheduled to be completed by the end of 1970; it is expected that magnetic fields up to 20 T can be generated for continuous duty operation.

A large number of high magnetic field laboratories are in existence throughout the world. Most of them have been built for quite specific or narrow scientific objectives in many disciplines of physics, but there is no single source or a directory listing and describing their capabilities. There are, however, a number of laboratories whose broad interests in high magnetic fields for research at least partially overlap those of the two above mentioned magnet laboratories. We name the following:

1. Royal Radar Establishment, Malvern, England

At present, this is probably the best equipped high field laboratory in England, with six magnet stations and a total of 5.5 megawatts of power. Silicon rectifier equipments provide 2.5 megawatts continuously, and a bank of storage batteries provide 3 megawatts on a 15% duty cycle. A hybrid system would produce 20.5 T continuously or 25.5 T 15% of the time.

2. Nijmegen, Netherlands

This laboratory has a 6 MW solid-state power supply and a 15.0 T water-cooled magnet under construction. At 6 MW, a hybrid magnet may generate a field of more than 26.0 T.

3. Australian National University, Canberra

This high field magnet laboratory has an unusual power supply in the form of a homopolar machine. It can deliver 500 megajoules of electrical energy at a peak rate of 300 megawatts. Typically, however, a 5 MW, 100 second long pulse is used for experiments at fields of 15.0 T. A 30 T multi-second pulsed magnet, which can be powered up to 30 MW, is in the final design stages. The amount of research which can be performed is unfortunately limited by the low-duty cycle, as is in most pulse magnet systems.

4. McGill University, Montreal

This laboratory is built around the concept of cryogenic magnets which are cooled to about 15 K. At this temperature, the resistivity of pure aluminum has decreased to a fraction of its room temperature value and there is a trade-off between the small power supply needed to produce high fields and the refrigeration system needed for intermittent operations. It is expected that fields of 25 T can be obtained in magnets which combine cryogenic and superconducting coils.

5. Lebedev Institute, Moscow

This laboratory has recently been improved and now has installed 9 MW of d.c. power. Water-cooled magnets up to 17.6 T are used routinely. It is understood that a 20 T magnet is being constructed. A hybrid system would provide fields up to 29.3 T.

6. Other United States Laboratories

Four laboratories in the United States have research facilities which partially overlap those of the Francis Bitter National Magnet Laboratory:

	Power in megawatts	Potential hybrid field intensity
Oak Ridge National Laboratory	7	27.5 T
University of Pennsylvania	6	26.3 T
Naval Research Laboratory	3	21.5 T
Graduate Research Center of the Southwest	1.7	19.5 T

7. Grenoble, France

The establishment of a new laboratory at Grenoble has been underway for some time. It is believed to be patterned after the Francis Bitter National Magnet Laboratory, with a 10 MW power supply.

Notes:

1. The Bell Telephone Laboratories have recently closed down their water-cooled magnet facility because superconducting magnets of equal (10.0 T) field strength were more economical to operate. Their scientists use facilities of other laboratories for experiments in the 15.0 T and higher field ranges.

2. We have not tried to be exhaustive in listing all the laboratories which can only produce fields of about 10.0 T. They are located in Austria, Poland, the Netherlands, United Kingdom, and Japan.

Before going into detail, we give in the following introductory sections, a short survey of various methods of field generation, and then illustrate the present role of superconducting magnets among other families of magnets still being used in many areas of physics.

1.2 Conventional Continuous Duty Magnets with and without Iron

Detailed data about the first pioneering attempts to produce high magnetic fields are given by DeKlerk [1]. From this book, we mention a few interesting achievements:

In 1911, Du Bois [1] introduced a new type of laboratory magnet, utilizing iron yokes and poles. The largest magnet built had a weight of 1400 kg. With a power consumption of 14 kW, this magnet produced a field of 5.5 T in a one-millimeter gap width between pole pieces of 6 mm diameter.

Weiss [2] added to these achievements several important improvements, such as using oil as a coolant. The coils were immersed in the oil bath and thus increasing the power density in coils, compared to Du Bois' air-cooled conductors. Later, Weiss built the first hollow conductor, pumping water through current-carrying tubes. This design is still the most efficient method of designing high power density electro-magnets. The largest Weiss magnet was built by de Haas in 1929 in the Kamerlingh Onnes Laboratory [3] and is still in existence. The magnet yoke is 1.5 m wide and weights about 1400 kg. The weight of each pole piece is 900 kg. Using a power of 80 kW, a field of 2.4 T was produced in a gap of 6 cm width between pole faces of 10 cm diameter. With pole pieces of 1.8 mm diameter at the gap, spaced 1.2 mm apart, a field of 6.7 T was generated!

In 1931, Dreyfus [4] introduced a new design principal, tapering the pole pieces 57° which allowed a drastic reduction of all magnet dimensions and consequently the iron weight. The Uppsala magnet [5], which was built according to Dreyfus' calculations, has 37,000 kg weight. Energized from a 14 kW power supply, it generated a magnetic field of 4.4 T in a 4 cm gap between pole pieces, having 6 cm diameter at the gap.

These types of laboratory magnets are still widely used in many areas of physics. Modifications of the basic Dreyfus principle have been to improve the magnet performance by reinforcing coil structures, adding mobility to the poles to achieve a variable gap width for different experiments, by utilizing remote-controlled motor-driven poles, introducing direct cooled and new hollow conductor configurations, and improved water and water vapour repellant conductor insulations based on glass fiber tapes impregnated under vacuum in suitable thermosettings, other laminats (H-films, Nylons, Teflons), or oxyde surfaces. Great improvements are made in field computation, field shaping, and field screening methods.

Effective use of the ferromagnetic return paths in the magnet, as well as achieving high power density coils, was first reported by Bitter [6] in 1936. Bitter used disc-shaped conductors and helical coil design, which were cooled axially. He was able to produce the first 10.0 T electromagnets for continuous d.c. operation using a power of $\sim$10 MW produced from a d.c. motor-generator set. Bitter's basic ideas are still used in many laboratories.

Although in Chapter II a detailed study of magnetic field calculations and a theoretical analysis of magnet optimization are given, it should be mentioned that Fabry [7] was first to link the power consumption of electromagnets to the produced magnetic field in the magnetic center of solenoids. Bitter extended Fabry's relation and developed curves of constant "Fabry" or "geometry" factors for various air-core type magnets.

Since 1945, interest in new types of electromagnets to be used with high-energy accelerators has increased markedly. According to the type of performance, field configuration, and duty cycle, high-energy magnets did undergo gross changes. The performance of ferromagnetic materials, field shaping by shimming pole faces, new methods of field calculations, improving magnet lifetime under severe environmental conditions (such as high irradiation doses), were explored in great detail.

From this time, called the period of giant water-cooled electromagnets, we recall only a few magnets. They are shown in Table 1.1 without giving details.

Another group comprises "iron-core" (i.e., the working field volume is contained in a gap between poles) and iron-bound magnets (i.e., the ferromagnetic material is employed only as a flux return path) for momentum analysis, spectroscopy, around bubble-, spark-, and wire-chambers, or for other special purposes such as storage of particles. A few examples of these magnets are given in Table 1.2.

The Tables 1.1 and 1.2 can be extended to a great length. But the few selected examples will illustrate the importance of electromagnets for high-energy physics experiments.

About 1945, the electron synchrotron was conceived by McMillan [8] and Veksler [9], independently. In 1949, the first 320 MeV, LRL electron accelerator came into operation. The synchrocyclotron as a proton accelerator requires a solid core; they have been successful in the 100 to 700 MeV range. At relativistic energies, the weight and cost of these magnets increase with the third power of the linear magnet dimensions. For several GeV energies, the magnet cost would become exorbitant. The obvious method to reduce cost would be to build a ring magnet covering only a narrow band. Such a fixed-orbit radius requires the synchrotron principle of acceleration, using a pulsed magnetic field: The

Table 1.1. *Giants magnets for particle acceleration*

Institution	Particles	Energy (MeV)	Year of completion	Gap height (m)	Gap length or Pole diameter (m)	Cooling mode (coils)	Peak field (T)	Total weight (10^3 kg)	Power (kW)
CERN Synchrocyclotron	Protons	600	1955	0.36—0.45	5	water	1.8	2500	875
Chicago Synchrocyclotron	Protons	460	1952	0.407	4.32	water	1.86	2120	650
LRL[a] Bevatron	Protons	6,200	1954	0.305	35.4	water	1.6	9100	3600
LRL 184″ Cyclotron	Protons Deuterons	730 460	1946	0.356	4.7	water	2.33	3700	780
Univ. of Brit. Columbia Isochron Cyclotron (TRIUMF)	H^-Mesons	200—500	1972	0.5	8.13	water	0.3 ... 0.6	2900	1940

[a] consists of four magnets

magnetic field is increased with time as the proton gains energy to maintain constant orbit. The proposal of a ring magnet was made by Oliphant [10] in 1943. In 1947, a theoretical analysis was published by Gooden, Jensen, and Symonds [11].

A new principle of magnetic focusing for accelerators, called alternating gradient (AG) or strong focusing, was introduced in 1952. It led to the building of a series of accelerators capable of much higher energies than were economically practical with the other types of machines. The AG focusing originated at Brookhaven, and its concept was published by Courant, Livingston, and Snyder [12, 13].

In the AG system, the particles pass alternately inward and outward. The focusing and the bending can be accomplished in a series of combined-function magnets, or focusing lenses and bending magnets can be separated, such as in separate-function machines. The choice relies heavily upon economical principles.

In Table 1.3, only a few examples of synchrotrons are given. The list can be extended by mentioning the currently being constructed 300 GeV proton synchrotron (NAL) at Weston, Illinois (separate-function machine), the various storage rings at CERN, at DESY, at Frascati, and at SLAC, and the synchrotron being built at CERN II. We refer to the extensive literature, specifically to the books by Livingston and Blewett [14] and by J. Livingood [15] on particles accelerators.

The use of water-cooled magnets for high-energy physics experiments, having usable field volumes of one m^3 or more for flux densities in excess of 3 T, is economically unsound due to the enormous power requirement. Ferromagnetic return paths, even at high flux densities (> 2 T), are useful for shielding the field from experimental equipments around the magnet, for enhancement of the field ($< 10\%$) in the magnet aperture due to the coil ampere-turns, and balancing forces between the coils. Ironbound magnets are popular, and we deal with this type of magnet in more detail in Section II.

1.3 Pulsed Magnets

To produce high fields in excess of 10 T in small volumes, d.c. magnets require considerable power which is actually wasted in the form of Joule's heating and must be removed by the coolant. If the experiment can be performed for a short time by synchronizing it to the maximum field, then the magnetic field can be adjusted to the duration of the experiment and does not have to persist over a long time, such as is the case in a direct current (d.c.) iron magnet. In magnets where the field is "pulsed" within a short time, the required energy is stored externally at a slow rate and discharged fast, according to the circuit impedance. The mean power requirement of such magnets is low (depending on the duty cycle) compared to d.c. magnets; however, the peak or maximum power can be, according to the pulse duration and the field energy, extremely high.

We may distinguish between three modes of pulsed fields:

(a) *Long Pulses* in the order of several seconds to minutes with a flat top, such as is commonly used in synchrotrons. The pulsed peak field is

Table 1.2. *Experimental magnets*

Institution	Type	Year of completion	Gap height or useful bore width (m)	Gap length or pole (bore) diameter (m)	Peak field (T)	Total weight (10^3 kg)	Power (MW)
CERN g-2 Bending magnet	Iron-core	1960	0.15	7	1.8	96	~0.6
CERN H_2 Bubble chamber magnet	Iron-bound	1964	~1	2.1	1.7	700	6
LRL (72″) Bubble chamber magnet	Iron-bound	1959	1.06	2 (?)	1.8	123	2.53
SLAC 2 m Spark chamber	Iron-core	1967	1	2	1.5	410	5.8
SLAC 20 GeV Spectrometer	Mixed[a]	1967	—	—	—	282	4.1

[a] The SLAC spectrometer is composed of four bending magnets, four quadrupoles, and two sextupoles rigidly mounted on a 36.5×10^3 kg heavy frame.

Table 1.3. *Combined-function accelerators (Proton synchrotrons)*

Institution	Year of completion or status	Energy (GeV)	Magnet (Accelerator ring) Max. field (T)	Field index n	Power (max.) (MW)	Diameter (m)	Aperture h (cm)	Aperture w (cm)	Total weight ($\times 10^3$ kg)	Number of magnets	Length of each magnet (m)
Argonne (ZGS)	1963	12.7	2.15	0	110 (10)[a]	55	13	87	4,320	8	16.3
Brookhaven (AGS)	1960	33	1.3	357	30 (2.4)[a]	257	6.3	13.3	4,000	240	144×2.28 96×1.90
CERN (CPS)	1959	28	1.4	288	32 (1.6)[a]	200	7	14.6	2,850	100	4.76
Rutherford (NIMROD)	1963	7	1.4	0.6	160 (2.5)[a]	54	20	100	6,600	8	14.751
Saclay (SATURNE)	1958	3	1.49	0.6	24 (1)[a]	22	10.5	36	1,500	4	13.225
Serpukov (IHE)	1968	70	1.2	422	100 (?)	472	11.5	17	18,800	120	9—10
		Combined-function accelerators (Electron synchrotrons)									
Cambridge (CEA)	1962	6.28	0.76	91	1.0[a]	75	3	14	310	48	4
Daresbury (NINA)	1966	4 (5.3)	0.64 (0.85)	46.17 (47.17)	8.0 (1.9)[a]	70.2	6.1 (7.6)	13 (9)	470	40	3.265
Hamburg (DESY)	1964	6.25	0.62	70	1.1[a]	50.4	7	12	585	48	4.85
Yerevan	1967	6	0.792	115	1.6[a]	70	4.5	12	385	48	3.2

[a] Mean power

generally less than 2.0 T in large magnets, but may reach more than 10.0 T in composite system magnets.

(b) *Short Pulses* ranging between micro- and milliseconds. Peak fields achieved range between 10 T to 10^2 T. The field of 10^2 T is also the limit between destructive and non-destructive experiments.

(c) *Ultrashort Pulses* ranging between 1 μsec and 0.1 μsec. Peak fields in the range of 100 T — 1400 T have been produced by utilizing explosive methods. The coil, as well as the experimental setup within the bore, is destroyed at each pulse.

Pulsed fields (> 10 T) represent high-energy densities, and thus high pressures on the current-carrying conductor are exerted by them. To illustrate the relationship between energy density, linear current density, and magnet pressure, we refer to Table 1.4 as a guideline.

Table 1.4. *Relationship between magnetic fields, energy density, and magnetic pressure*

Magnetic field (T)	Field intensity[a] ($A \cdot cm^{-1}$)	Magnetic pressure (kp/cm^2)	Energy density[b] ($Joule/cm^3$)
10^{-4}	8×10^{-1}[b]	4×10^{-8}[c]	4×10^{-9}[c]
10^{-2}	$8 \times 10^{+1}$	4×10^{-4}	4×10^{-5}
10^{0}	8×10^{3}	4	4×10^{-1}
10^{2}	8×10^{5}	4×10^{4}	4×10^{3}
10^{3}	8×10^{6}	4×10^{6}	4×10^{5}

[a] Also called linear current density
[b] Exact number is $1/0.4\,\pi = 0.7958$
[c] Exact numbers are 39.79 and 3.979, respectively

In the megagauss field region, the energy density and the magnetic pressure become so high that the exerted stress levels on the conductor exceed the mechanical strength of the materials known. We illustrate the stress limitation with respect to strength of the various materials being utilized or proposed for pulsed coils in Fig. 1.1. At fields above ~40 T, copper alloys having high mechanical strength but also (unfortunately) high resistivity, are utilized in magnets. At fields of 100 T or higher, single turn coils based on steel, titanium, or high-strength beryllium are sometimes preferred.

In using ultra-high field pulsed magnets, we have to consider three effects which are summarized us follows:

(a) The pulsed field produced in a certain experimental area within the coil bore leaks out through the conductor and insulation. The flux penetration into the conductor is a function of the finite resistivity of the material and the Alfven velocity (material displacement rate), expressed by $v_A = B/2\,(\mu_0 \cdot \delta)^{1/2}$. This effect will set a lower v_A limit to attain a certain desirable field. At a field of 10^2 T and a density of 10^4 kg per m^3, v_A is in the order of 10^3 m/sec for metallic conductors.

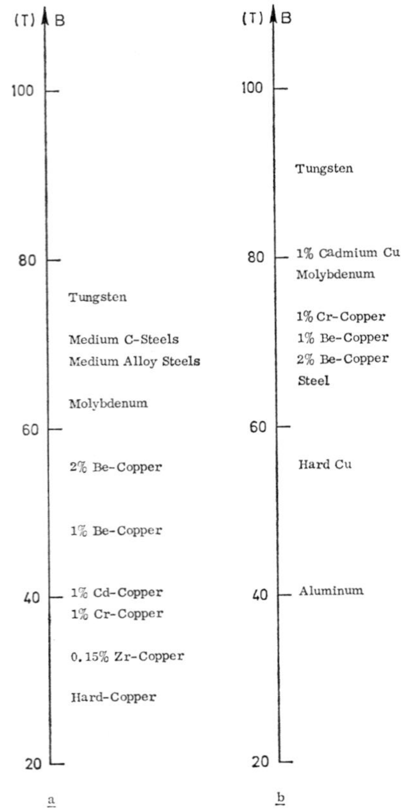

Fig. 1.1.a Magnetic field expressed in terms of mechanical yielding. (The strain basis is the tensile strength of metals and alloys at 300 K)
b Magnetic field expressed in terms of the melting point of metals and alloys. Heat conduction during the pulse is neglected

(b) The kinetic energy supplied to the conductor by the magnetic forces must be large enough to overcome the mechanical stresses during the field rise time. This implies that the rising of the field must be sufficiently fast in order that the peak field is reached prior to the time when damage is caused to the conductor.

(c) The field penetrates into the conductor due to the finite resistivity of the conductor material and causes Joule's heating. The current density in "the field penetration layer" can be so high that the conductor may not only liquefy but may even evaporate. In this case, the material relaxation time must be longerthau the field rise time. A field of "hundred" corresponds to a tensile strength in the order of 4×10^4 kp/cm^2 and a temperature of 3000 K, neglecting heat" conduction during the pulse time.

Detailed analysis of these phenomena are given by Furth [16], Lewin and Smith [17].

Pulsed magnetic fields were pioneered between 1924 and 1927 by Kapitza [18], who already had generated 50 T in a bore with a diameter of 0.1 cm, and who also had discussed generation of 100 T fields. Since this time, fields up to 1500 T were generated through flux compression by means of explosions. In Table 1.5, we give a review of pulsed magnets which have been built and operated.

Up to fields of 70 T, magnets are energized by fast-discharging capacitor banks and are repeatedly usable. Above this limit, specifically in the 100 T region, part of the coil and the experimental arrangements are destroyed. However, the importance of generating 100 T fields for solid states and high-energy physics has spurred several laboratories [19] to develop highly sophisticated techniques to accumulate considerable energy densities (10^{11} watts/cm^3) during the short periods of 0.1—10 μsec by means of explosives. The chemical energy of the explosive is converted through a fast-moving "liner" into electromagnetic energy by the principle of magnetic field compression.

Table 1.5. *Comparison of various pulsed magnets*

Designer	Pulse time (μsec)	Maximum field (T)	Useful field volume (cm^3)	Cooling	Magnetic energy (kJ)
Kapitza [18]	10^4	32.0	2	air	5
De Haas [21]	10^4	25.0	0.5	H_2	?
Olsen [22]	2.5×10^3	13.0	0.19	N_2	0.9
Birdsall, Furth [23]	300	20.7	227	?	68
Foner, Kolm [24]	120	75.0	0.24	air	9
Morpurgo [25]	1.3×10^4	20.0	100	oil	300
Kolb [26]	4.5×10^3	20.0	1500	?	285
Brechna *et al.* [27]	5×10^4	15.0	90	lN_2	15
Furth, Waniek [28]	10[a]	160.0	0.08	lN_2	1.5
Forster, Martin [29]	2[a]	250.0	0.3	air	7.5
Di Gregorio *et al.* [30]	11[b]	550.0	0.4[d]	air	200[c]
Fowler *et al.* [31]	10.5[d]	1430.0	0.54[d]	air	30[c]
Cnare [20]	18[b]	210.0	5[d]	air	136

[a] Quarter period
[b] Compression time using implosives
[c] Capacitor bank energy
[d] Final diameter or linear diameter

An initial "seed" field is produced in a coil in a period less than 1 millisecond. The energy source exclusively cousist of high-voltage capacitor banks with spark-gap switches. The detonator is fired when the peak field is reached. The seed flux is trapped within the cylinder inside the detonator, and the flux $\varphi = \pi r^2 B$ is compressed. The magnetic field is amplified according to $\lambda r_0^2/r^2$, where r is the inner radius of the liner and r_0 the initial radius. $\lambda = \varphi/\varphi_0$ is the flux loss coefficient. If B_0 is the seed field, then a final field amplification of $B/B_0 = 100$ or higher may be achieved with some efforts, which means that with a seed field of 10 T, a maximum field of 1000 T is obtainable.

For many high-energy physics experiments, a new method of flux compression by imploding a metal foil is proposed by Cnare [20]. The metallic liner is imploded by means of a capacitor bank and acts as a power transformer. Cnare's arrangement consists essentially of a single-turn coil which simultaneously provides the initial field and the driving force for the acceleration of the foil. Field amplitudes achieved with this method run up to 400 T. The vaporization of the seamless foil by currents induced during acceleration sets an upper limit to the speed of the foil and thus the upper field limit.

Means of Energy Storage

The most economical energy storage depends on the total energy to be stored, the duration of the pulse, or the rise time in which the peak field is reached (current rise time), and the duty cycle (Fig. 1.2).

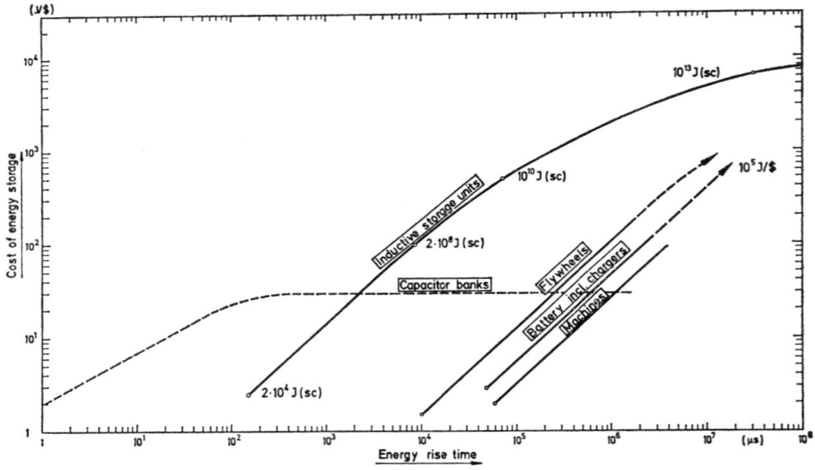

Fig. 1.2. Methods of energy storage: For energy charging up to 10^3 μs capacitors are most economical. In the range of 10^3—10^9 μs inductive coils are utilized in combination with controlled thermonuclear reactors [32]

Energy can be stored by means of storage batteries (lead acid cells, dry cells). Typically, a 12-volt cell can store an energy of $\sim 4 \times 10^6$ Joules, but only a fraction of the total stored energy can be dissipated within a short period of less than one second. In the form of mechanical energy (i.e., in rotating motors and generators, and homopolar machines, for increased inertia, flywheels are utilized), pulses of long duration (several seconds) are produced which are used in AGS-type synchrotrons. Inductors (iron core inductor circuits, conventional and superconducting coils) are frequently utilized for energy storage, energy dissipation, and energy distribution in synchrotrons.

For long pulses, it is sufficient to use a combination of synchronous motors and d.c. generators with high overload capabilities, e.g., 50% overload for 15 minutes, 100% for 10 seconds, 300% for 5 seconds, etc. If the overload capability of the system is adequate and the motor generator set is carefully coordinated with the utility system, operation even without flywheels is possible.

In the case of the Francis National Bitter Magnet Laboratory, the interconnected utility system has approximately 100 MW of power generating capability, i.e., about three times the utilized maximum pulse power. Therefore, it is necessary to employ large flywheels for the two motor-generator sets. Each of these motor-generator sets consists of one synchronous motor and two d.c. generators (each with 2 MW continuous power), a flywheel (with 6 m diameter and a weight of 76×10^3 kg), and a running-up motor. Pulses of 32 MW (4 times normal power) can be provided for 2 seconds [33]. For such slow pulses, a conventional magnet design can be used if mechanical forces and heat generation during the pulse are appropriately considered.

The National University in Canberra, Australia, uses a power supply for pulsed magnet field operation having a homo-polar machine with high inertia. It is designed to store 500 megajoules of kinetic energy. A pulsed magnet of 30 T using 300 MW of electrical power is being constructed. A 15 T magnet with a 100 second pulse time is presently available.

For experiments requiring high magnetic fields, pulse times in the order of milliseconds are provided. Thus, in most cases, high-current, low-voltage sources must be excluded. High voltage capacitor banks are convenient energy sources; they are versatile and can be synchronized precisely.

In some of the existing laboratories, elaborate capacitor banks are installed, such as the Institute for Plasma Physics at Garching, Germany, providing the energy of 2.7 MJ at a maximum charging voltage of 40 kV. Other examgles are the Los Alamos 600 kJ, the Livermore 300 kJ, the Frascati 200 kJ, the MIT 100 kJ, the CERN 300 kJ, and SLAC 300 kJ facilities. The peak voltages in these laboratories range between 3—30 kV. For switching capacitor banks, ignitrons or spark gaps are generally used. For long discharge times, ignitrons are advantageous because of their small internal resistance. However, their current-carrying capacity is limited; over-rating for short periods is not possible. Frequent prefiring of ignitrons is a drawback, particularly when one ignitron fires and the others follow.

Spark gap switches are most useful for fast pulses; they are less susceptible to prefiring. Due to their high internal resistance, they are ideal for short-time pulse work. For long pulses, part of the capacitor energy is dissipated into the spark gap.

As an example, we mention the experimental arrangement at the Los Alamos Laboratory, "SCYLLA IV." The magnet consists of a single loop with an inner diameter of 11 cm, and has a length of 100 cm. The energy source, a 600 kJ capacitor bank, has a peak voltage of 50 kV.

Pulsed peak fields of 10 T can be produced in the bore. The field rise time is 4 μsec, the field decay 20 μsec. One additional capacitor bank (3 MJ, 20 kV) can be used to increase the pulse duration ("crowbar action") and to increase the magnetic field. After reaching 10 T, the magnetic field is increased to 17.5 T in 25 μsec, and the pulse duration is extended to 60 μsec.

1.4 Cryogenic Magnets

The interest in cryogenic magnets is based on the reduction of the resistivity of the conductor material at cryogenic temperatures. Hence, at low temperatures, higher fields at lower Joule's losses per unit weight of coil material can be generated. For larger coils, however, (magnetic field energies larger than 10^6 Joules) considerable refrigeration is required if the magnet has to operate at continuous duty. New ultra-pure metals, such as copper and aluminum, are commercially available, where the resistivity from room (300 K) to liquid helium temperature (4.2 K) is reduced by a factor of more than 4×10^4. Generally, ultra-pure metals have poor mechanical strength; and special reinforcements will be required to cope with the Lorentz forces encountered in high-field magnets (see chapter 2, section 4).

The power reduction in a magnet is proportional to the material resistivity. At the same power, the central field is proportional to the square root of the resistivity. Compared to coils of equal geometry producing the same field, the reduction of power will be proportional to ϱ_c/ϱ_T, with ϱ_c the resistivity at cryogenic, and ϱ_T at room or elevated temperature.

Using oxygenization methods, have been achived with copper in single crystals r values of 40,000 to 53,000 ($r = \varrho_{300\,K}/\varrho_{4.2\,K}$). Thus, it would be possible to utilize these materials for coil design. But there are two drawbacks: High purity metals are exceedingly soft and exhibit yield strength of less than 200 kp/cm². High purity copper exhibit magneto-resistance effects which, at high magnetic fields, increases the resistivity markedly [34].

The resistivity of aluminum is increased with the field to a certain saturation value (at 4.2 K) and does not increase further when this value is exceeded [35]. However, Al drawn into a tape shows size effect. In Section 6, we will treat these effects extensively. In Table 1.6 we give the resistivity of a few metals being used as coil winding. Copper and aluminum alloys have a relatively small r value. Although they exhibit high mechanical strength, they are unattractive for low temperature magnet use. As indicated in Fig. 1.1, they are the only useful materials if fields in excess of 20 T are required. Copper-zirconium alloys show, a r value of about 50, at 4.2 K. Their magneto-resistance effect is small. Thus they can be utilized in cryogenic magnets having long pulse duration. To make full use of the low resistance properties of pure metals, coils are designed as edge-cooled sandwich-type pancakes. The conductor (e.g., aluminum with $r \simeq 10{,}000$) is produced as strips or tapes and is sandwiched with a

Table 1.6. *Resistivity of copper, copper alloys, aluminum, Nb (60%) Ti, and Nb_3Sn*

Material	ϱ (0; 4.2 K) (Ohm · m)	ϱ (0; 21 K) (Ohm · m)	ϱ (0; 78 K) (Ohm · m)	ϱ (0; 300 K) (Ohm · m)
Copper ETP (annealed)	1.3×10^{-10}	1.5×10^{-10}	0.3×10^{-8}	1.71×10^{-8}
Copper OFHC (annealed)	1.6×10^{-10}	1.8×10^{-10}	0.26×10^{-8}	1.72×10^{-8}
Copper, pure (99.995%)	1.98×10^{-11}	0.98×10^{-8}	0.24×10^{-8}	1.65×10^{-8}
Beryllium copper[a] (2.5% Be)	4.8×10^{-8}	4.92×10^{-8}	5.123×10^{-8}	7.24×10^{-8}
Beryllium copper[a] (1.0% Be)	1.84×10^{-8}	1.93×10^{-8}	2.06×10^{-8}	3.72×10^{-8}
Cr copper[b] (1% Cr)	3.8×10^{-9}	4.5×10^{-9}	0.9×10^{-8}	2.3×10^{-8}
Zr copper[c] (0.15% Zr)	1.35×10^{-9}	1.8×10^{-9}	0.56×10^{-8}	1.9×10^{-8}
Aluminum, pure (99.998%)	2.0×10^{-11}	2.8×10^{-11}	3.66×10^{-9}	2.55×10^{-8}
Aluminum, commercial electric grade (99.95%)	1.01×10^{-9}	1.1×10^{-9}	4.0×10^{-9}	2.55×10^{-8}
Sodium (99.99%)	1.36×10^{-10}	2.87×10^{-10}	1.03×10^{-9}	4.29×10^{-8}
Nb(60%)Ti	— $< 10^{-17}$	2.6×10^{-7}	3.4×10^{-7}	7.3×10^{-7}
Nb_3Sn	— $< 10^{-17}$	2.7×10^{-7}	2.9×10^{-7}	4.0×10^{-7}

[a] At 4.2 K, ϱ is unchanged within 1% up to 6 T transverse field.
[b] Samples as received. When samples were annealed at 500 °C for 10 hours in an argon atmosphere, $\varrho_{300\,K}/\varrho_{4.2\,K} \cong 10$ is achieved.
[c] Samples as received. When annealed at 500 °C for 10 hours in argon atmosphere, $\varrho_{300\,K}/\varrho_{4.2\,K} \cong 20 \ldots 50$.

high-strength metal tape backed by an insulation. If each pancake is wound such that pretension is applied to the high-strength tape material, then a large portion of the magneto-mechanical stresses is taken by the reinforcing type. Using the reinforcement, some active coil cross-sectional area is lost, but this lost fraction can be limited to less than 20%. The added safety guarantees a high number of pulses without coil damage. Sandwiched conductor configurations, using edge-cooling methods, have been proposed [36] and utilized by many magnet designers.

With reduction in temperature and increase in magnet size (load), the refrigeration cost and installed power will obviously increase accordingly. Thus an optimum system, including magnet, power supply, refrigerator, and safety circuits must rely on the choice of conductor material, the cryogenic temperature at which the magnet must operates, and the refrigerator type, mode of operation, and duty cycle.

Obviously, the cost optimum corresponds in essence to the conductor resistivity and thus to the ohmic losses for a selected current density. These losses are the sum of the actual electric losses in the magnet and the power requirement of the refrigeration system to provide the net refrigeration. The so-called "loss parameter" of a magnet system, which is a good indicator for the choice of temperature and material, is proportional to $1/r$ and depends on the type, size, and Carnot efficiency of the refrigerator (Section 6).

Although cryogenic magnets found a serious competitor in superconducting coils, there are several areas where cryogenic magnets are still being implied:

(a) For pulsed work with short and long pulse duration.

(b) For small magnets where the Lorentz forces and thus conductor stresses can be controlled effectively.

(c) Precooled pulsed magnets, where the heat transfer to the coolant during the pulse duration is curtailed.

(d) Hybrid type d.c. or pulsed magnets, where the inner coil section may be cooled to an optimum cryogenic temperature, the outer coil may be superconducting.

In case (a), low-duty cycle magnets are attractive because the size of the refrigerator can be kept small. The cryogen can be accumulated in the time period between successive pulses and pressurized immediately prior to the pulse into the coil container. In the time interval between pulses, the coil is kept at operating temperature, evaporating only a moderate amount of cryogen in the container.

The economical aspects of low-duty cycle magnets are discussed by Adair [37] and Taylor [38]. A number of low-duty cycle, liquid hydrogen-cooled coils up to 8 T are built by Laquer [39], Purcell [40], and Arp [41]. A liquid nitrogen-cooled pulsed flux concentrator producing fields up to 15 T was reported by Brechna [27], and a 10 T pulsed liquid neon-cooled magnet was designed by Laurence [42].

In case (b), the coil operates under a condition which utilizes the heat transfer to the coolant. Under atmospheric conditions, the heat flux is limited to about 0.5 W/cm^2 in free boiling liquid hydrogen. Forced convection may increase this limit to about 1 W/cm^2. With liquid helium near the nucleate boiling region, a heat flux of ~ 0.7 W/cm^2 can be removed. Using forced cooling schemes, a heat flux of about 2 W/cm^2 can be removed with liquid nitrogen. Higher heat transfer values than in the case of free boiling cryogens may be achieved using supercritical liquids, where the coolant is generally forced through an open structured coil and circulated between the magnet and the refrigerator system in a continuous closed loop.

1.5 Superconducting Coils

Since the discovery of type II superconductors having high current densities at elevated magnetic fields, the interest in using them in coils has been steadily increasing. Type II superconductors were tested by Schubnikov [43] in 1937. Schœnberg [44] gave a summary of early experiments with type II superconductors in 1952. A break-through in the development of type II superconductors came when Kuntzler et al. [45] demonstrated that Nb_3Sn conductors can carry current densities orders of magnitudes higher than copper conductors at fields in excess of 10 T when cooled below 18.0 K. Since 1961, several new alloys have been discovered [Nb (25%) Zr, Nb (50) Ti, V_3Ga ...] which exhibit high current densities and fields ranging between 0.8—30 T. In Table 1.7, we give some pertinent data of type II superconductors which are of interest in magnet design.

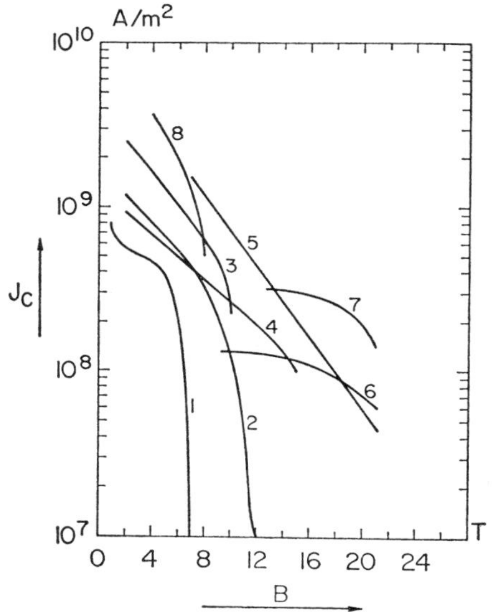

Fig. 1.3. Short sample performance of common type II superconductors at 4.2 K
1. Nb(25%) Zr 2. Nb(48) Ti 3. Nb(50) Ti 4. Nb_3Sn (Vapor deposition) 5. Nb_3Sn (Diffusion process) 6. V_3Ga (Wire) 7. V_3Ga (Ribbon) 8. $Nb_{(x)}Ti$

The theory (based on negative surface energy) explaining the behaviour of type II superconductors was derived by Abrikosov [46] in 1957, who based his work on the Ginsburg-Landau theory. It was shown by Gorter [47], however, that the current-carrying capacity of type II superconductors where fluxoids can move freely is small; only when the material is cold-worked, defects, lattice strain, and imperfections introduced in the bulk material, can high current densities be expected. Bean's [48] and London's [49] critical state model (the basic experimental work of Kim [50] which links the critical current to the penetrated flux in the superconductor) and the fluxcreep model of Anderson [51] have contributed to the knowledge and understanding of the behaviour of superconducting alloys. Primarily, type II superconductors (Fig. 1.3) were used for d.c. applications and an expressive number of small and large-size magnets were built in the time period of 1961—1968. An abbreviated list of larger superconducting magnets is given in Table 1.8, listing only the main parameters. However, theoretical and experimental work of several researchers such as Hart [52], Wipf [53], Hancox [54], and Smith [55] indicate that type II superconductors may very well be suitable for a.c. applications, specifically for accelerator magnets with low-duty cycles. We may also mention the work of Fairbank, Schwettman [56], and Wilson [57] on microwave properties of niobium which, utilized in cavities, has made feasible the building of high-energy linear accelerators.

Table 1.7. *Some properties of technical type II superconductors*

Material	Critical temperature (K)	Critical field (T) at 0 K	Critical field (T) at 4.2 K	Comments
Nb	9.24 ± 0.1	0.6—0.7	0.1—0.5	Tapes, wires, ductile
$Mo_{0.5}$—$Re_{0.5}$	12.6	2.7	1.5	
$Pb_{0.5}$—$Bi_{0.5}$	8.4	3.0[a] 1.6[b]	1.4	
Nb_3Sn	18.05	35.0[a]	20.0	Tape form, brittle
$V_{2.95}Ga$	14.5	50.0[a] 35.0[b]	21.1 (d.c.)	Tape form, brittle
$Nb_{0.75}Zr_{0.25}$	10.8	12.0[a] 7.7[b]	>7.0[c] >8.2[d]	Tape and wire, ductile
$Nb_{0.6}Ti_{0.4}$	10.6	17.6	11.8	Tape and wire, ductile
$Nb_3Al_{0.8}Ge_{0.4}$	20.7 ± 0.2	43	40.5	Presently available in short length tapes
Nb_3Al	18.67	31.9	30.8 ± 1	
Nb N	16	?	>21	
V_3Si	16.9	30.0[a] 15.6[b]	14	Tape

[a] Linear extrapolation to 0 K
[b] Parabolic extrapolation to 0 K
[c] Field perpendicular to superconductor
[d] Field parallel to superconductor

At present, the superconductor occupies an important place in many areas of physics, chemistry, medicine, and biology. Specifically for solid-state physics experiments, NMR, MHD, beam transport, bubble- and spark-chamber magnets for polarized target work, the superconducting magnets have replaced high-field conventional magnets in many labora-

Table 1.8. *Comparison of existing superconducting magnets*

Installation	Magnet type	Maximum field (T)	Useful field volume (m^3)	Bore diameter or gap height (m)	Coil weight (kg)	Field energy (kJ)
ANL	Solenoid	4.5	13×10^{-3}	0.28	363	365
LRL–Livermore	Minimum field	3.8[a]	68×10^{-3}	0.25	90	~120
NASA–Cleveland	Solenoid	13.6 (15.2[e])	5×10^{-3}	0.15	410	2.5×10^{3}
RCA	Solenoid	150	1.8×10^{-4}	0.38	80	800
Avco–Everett	MHD-Saddle	3.7—4.6	9.4×10^{-2}	0.305	2,170[b]	4.6×10^{3}
BNL	Helmholtz	4.5	1.27×10^{-2}	0.20	360	900
SLAC	Helmholtz	7.0	3.2×10^{-2}	0.30	1,600	6.2×10^{3}
CERN	Helmholtz	6.0	7×10^{-2}	0.40	800[b]	2.05×10^{3}
CERN/BEBC	Helmholtz	3.5	~20	4.72	106,000	800×10^{3}
LRL–Livermore	Solenoid	0.86	1.4	1.00	1,630	10^{3}
BNL	Tapered solenoid	~10.00	2.2×10^{-4}	0.38—0.66	?	~20
Saclay	Helmholtz	4.0	4.5×10^{-6}[f]	0.2	11[b]	400
Siemens–IEKP	Quadrupole	3.2[b] (3.4[g])	11.3×10^{-3}	0.12[d]	220	80
NASA–Cleveland	Multipole solenoids	7.5	3.5×10^{-1}	0.51	5.812×10^{3}	$\sim 16 \times 10^{3}$
RHEL	Bending	4.0	$\sim 3 \times 10^{-2}$	0.17	300[b]	1.1×10^{3}
ANL	Helmholtz[c] bubble chamber	1.80	~17	4.58	45,500[b]	80×10^{3}
BNL	Helmholtz	3.0	6.5	2.40	15,000[b]	60×10^{3}

[a] Field at conductor
[b] Weight of composite conductor only
[c] Iron-bound construction (1.45×10^{6} kg iron yoke)
[d] Warm bore diameter and iron-bound constructions
[e] Operated at 2 K
[f] Homogeneous field region ($\Delta B/B = 10^{-4}$) for polarized target experiments
[g] Maximum field at coil ends

tories. Thanks to the fact that operational costs of these magnets are considerably smaller than room-temperature magnets (see Section 6), and due to many new technological achievements in composite conductors, the investment costs (per GeV-energy) have become comparable to water-cooled magnets. Thus, serious attempts are being made to introduce superconducting pulse magnets in high energy accelerators, while superconducting beam transport magnets have been in operation for some time. The understanding of the stability criteria and magnet reliability has also removed the original fear of many scientists opposing the use of superconducting magnets as part of their experimental facilities.

After the first few superconducting magnets based on coppercladded Nb (25%) Zr conductor were built and tested, it was found (1964) that the current density of the superconductor was considerably less, when wound into a coil, compared to the short sample performance of the superconductor. The consequences of this discovery were that the optimum choice of the operational coil current was in jeopardy, and that due to the conductor degradation in the coil, more superconductor would

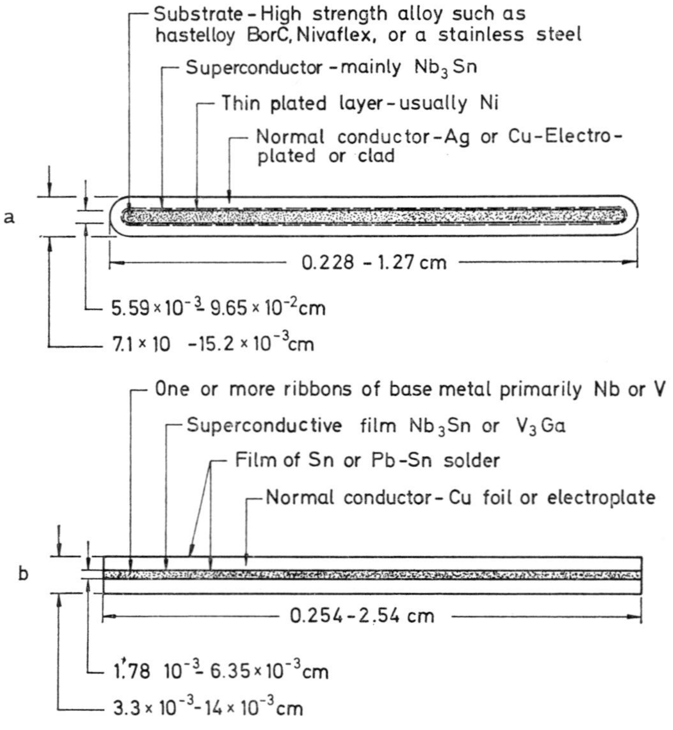

Fig. 1.4a and b. Nb_3Sn or V_3Ga tapes
a Vapor-deposition-process-Nb_3Sn ribbon, b Diffusion process-Nb_3Sn or V_3Ga ribbon

be required. Superconductor degradation led to the discovery of the cryostatic stability criterion by Stekly [58], and recently (1968) of the dynamic stability criterion. It was found that the superconductor must be in close thermomechanical contact with a high electrical conductivity metal in order to provide an alternate current path in case the superconductor becomes normal, and to increase the thermal diffusivity of the composite conductor.

Multistrand cables composed of twisted superconducting and copper filaments impregnated in indium or silber-tin alloy were introduced in early 1965. Composite conductors, based on fine Nb_xTi filaments (0.1 mm diameter) embedded in a copper matrix, became commercially available in 1967 (Fig. 4.4.1a). New tape configurations were developed for Nb_3Sn with copper backing strips for stabilization and stainless steel tapes for strength (Fig. 1.4). Multicore conductors with very fine Nb_xTi strands (down to 5 μm diameter) twisted in a copper or copper-alloy matrix transposed in braids or cables were developed in 1968 for pulsed field applications (Fig. 4.4.1b,c). Individual strands of a cable or braid consist of hundreds of fine twisted superconducting filaments embedded in copper or cupro-nickel alloy. These strands are insulated and then transposed. Compacted to a final shape, the cable can have a high degree of cross-sectional accuracy. Such a composite cable can carry up to 5000 A, and is wound into a layer or double layer. It is also common to use a suitable low temperature thermosetting for impregnation of such double layers. This type of coil construction yields high mechanical tolerances. Field homogeneities (about 10^{-3}) in the useful aperture is not uncommon. The superconducting alloy can be "stabilized" by bonding it to a metal of high electrical conductivity in good thermal contact to a coolant such as liquid helium. When the stabilized conductor is subjected to a slowly increasing transport current, no ohmic voltage will appear across the coil terminals until the superconductor critical current, Q_2, (see Fig. 1.5) is reached. Any excess current will flow through the "normal" low electrical resistivity material in the proximity of the superconductor (current sharing region). Due to the resistive behaviour of the normal conductor, the Joule heating leads to a temperature rise over the conductor, which reduces the critical current density in the superconductor (Fig. 1.6a, b). Further increase in current leads to more heating of the conductor, and the proportional amount of current through the normal metal is raised. In the early phase of the current-sharing between superconductor and normal metal (substrate), the voltage current characteristics are reversible, but eventually, as the current continues to increase, the conductor becomes thermally unstable. The transition from nucleate to film boiling occurs in the early phase of current sharing in the interface between conductor and coolant. The superconductor, now heated above its transition temperature, becomes normal (take-off current, T_2). Its resistivity in normal condition, being much higher than that of the normal metal leads to the condition where practically all the current flows through the normal metal. If the transport current is again decreased, a transition back to nucleate boiling will eventually occur, and the voltage returns to the original reversible state (recovery current, R_2).

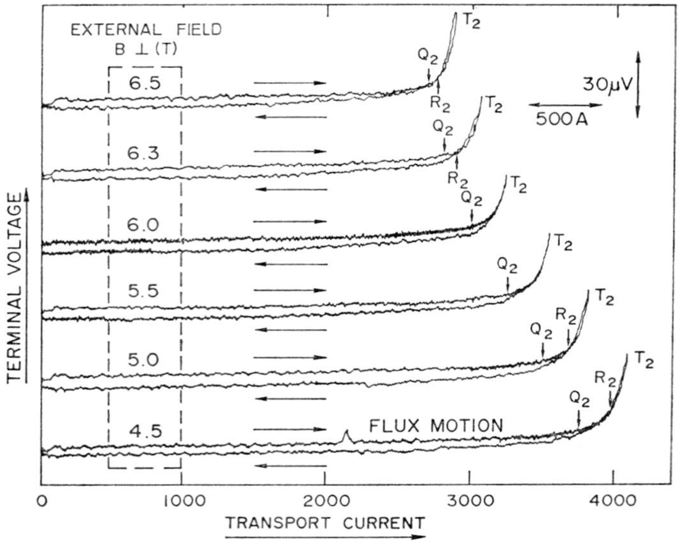

Fig. 1.5. Current-voltage characteristics of a stabilized direct current composite conductor
Q_2 critical current, T_2 Take-off current, R_2 Recovery current
The critical current value corresponds to the resistivity of 10^{-12} Ohm · cm in the composite

When a "composite" conductor which consists of superconducting wires, "filaments", and a normal metal matrix (copper of high electrical conductivity is the most common metal being used) is driven normaly by an internal or external disturbance, the current will flow through the normal metal. The composite conductor must be designed such that when a disturbance of short duration (say less than one second) has passed, the superconductor may recover its original superconducting state. This is possible when the cross-sectional areas of the superconductor and the normal metal are matched, the critical heat flux to the cryogen is not exceeded, and the cryogen has free access to the heated area. Otherwise the conductor must be intrinsically stable, which would even allow the pancakes to be potted in suitable thermosettings.

The proper design of such conductors or coils depends on the "stability criterion", a term which must be clarified due to its great importance in the design of superconducting magnets. We distinguish between three types of stabilization:

(a) *Cryostatic Stabilization*

This depends on the provision of an alternative low electrical resistance path for the current when the conductor is driven normal. Stabilization may be achieved in either the steady-state or the transient case. Basically, it is understood that a normal region in the conductor does not propagate along the conductor. Cryostatic stability criteria have been

a

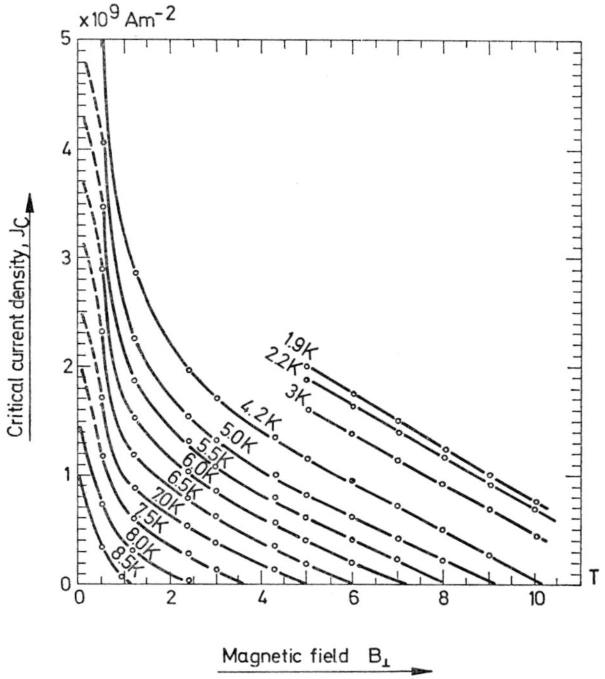

b

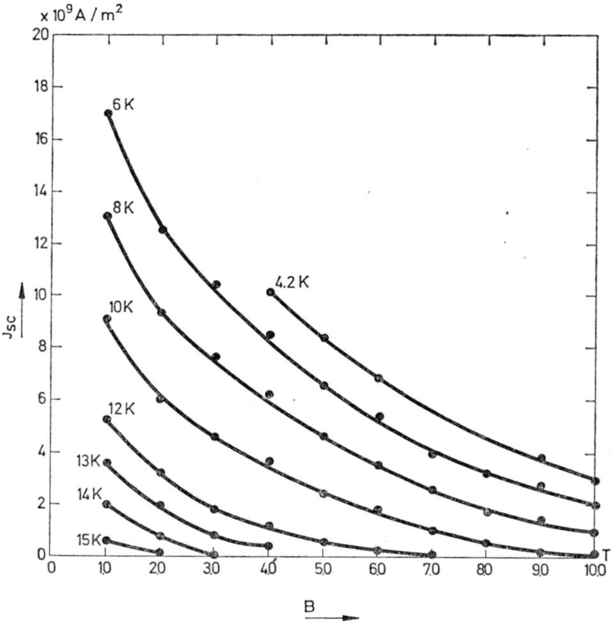

Fig. 1.6. Temperature dependence of critical current density vs. applied transverse magnetic field for *a* $Nb_{(x)}Ti$. (IMI) *b* Nb_3Sn (I.G.E)

treated by many authors, but due to simplified assumptions of actual happenings in the conductor, the obtained results are conservative. Cryostatic stability is treated in detail in Chapter 4.6.

(*b*) *Enthalpy Stabilization*

The thermal capacity of the conductor must be sufficient enough to prevent an adiabatic fluxjump from driving the superconductor normal. As copper or aluminum have low thermal capacities at cryogenic temperatures below the transition temperature of most type II superconductors, lead [59] which has a relatively high thermal capacity, and sintered open porous copper [60] to utilize the high heat capacity of helium gas, have been suggested as the conductor matrix. The *intrinsic stabilization* can be regarded as an important special case of enthalpy stabilization, in which the superconductor is divided into many fine filaments. The size of individual filaments is such that the enthalpy of the superconductor itself is sufficient for stabilization [61].

(*c*) *Dynamic stabilization*

is obtained by the addition of sufficient normal material to the superconductor to magnetically damp any fluxjump so that the energy released is removed by thermal conduction. The growth of possible instabilities is prevented.

The understanding of the above stabilization methods and their application in magnets against unstable flux motion have contributed to the rapid development of superconducting magnets in recent years.

It may be appropriate to interject a few commonly used definitions which will be encountered throughout this book:

(a) *Critical Current.* At a given temperature and magnetic field, the maximum current which a superconductor is able to carry.

(b) *Critical Field.* The maximum field at a given temperature at which a superconductor exhibits zero resistance to the flow of an electric current.

(c) *Take-Off or Break-Away Current.* The maximum current that a composite conductor can carry before the voltage gradient dU/dI approaches infinity. This current is equivalent to the critical current of a bare superconductor, and represents the value at which the superconductor becomes normal in a composite conductor in the current-sharing region.

(e) *Recovery Current.* The maximum current that a composite conductor can carry with zero resistance after the excursion into the current-sharing region.

(f) *Fluxjump.* The magnetic flux can diffuse through the superconductor much faster than can the heat generated by the moving flux; thus, the conductor can heat up if a sudden field change occurs. As the critical current density is reduced with increasing temperature, the field penetration (according to Bean's critical state model [48], the penetration depth is inversely proportional to the critical current density) in the conductor is increased, which generates additional heat, etc. The thermal-magnetic-feedback can lead, under certain conditions, to a thermal runaway, a fluxjump. The fluxjump in magnet coils is measured by placing pick-up coils close to conductors, or in short sample tests by measuring the voltage transients in the U-I diagram (Fig. 1.5).

(g) *Flux Motion.* As the field applied to a high field superconductor is increased, flux penetrates the conductor. The flux penetration (even in absence of fluxjumps) is not smooth, but is a random process accompanied by local fluctuation of flux density about the critical state profile. The flux motion leads to a coupled process of diffusion of flux and diffusion of heat.

(h) *Quench.* The status where the superconductor looses its superconducting state and becomes normal.

Referring to Table 4.2.1, we can see that the thermal diffusivity of type II superconductors is small compared to the magnetic diffusivity (in the superconducting state), which is a function of the field change dB/dt in the superconductor. To increase the thermal diffusivity, as seen above, a normal metal is added to the superconductor (composite conductor), or the filament diameter is reduced such that the superconductor becomes intrinsically stable. Figures 1.7 and 1.8 illustrate two types of stabilization. In Fig. 1.7, the addition of normal metal and improving cooling lead to enthalpy stabilization; in Fig. 1.8, the small filament diameter makes the superconductor intrinsically stable.

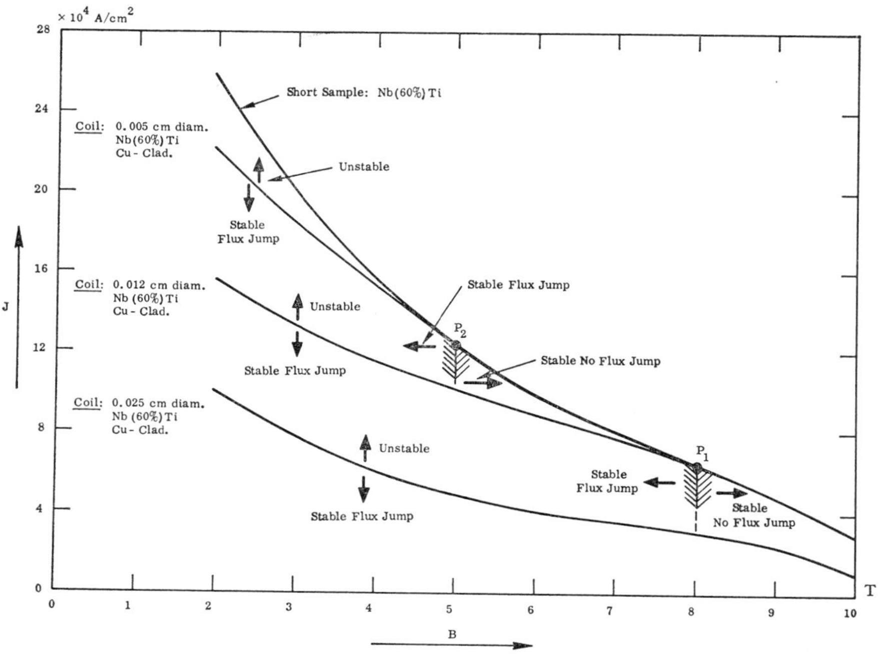

Fig. 1.7. Generalized current-field stability diagram illustrating the effect of the addition of normal metal to the superconductor

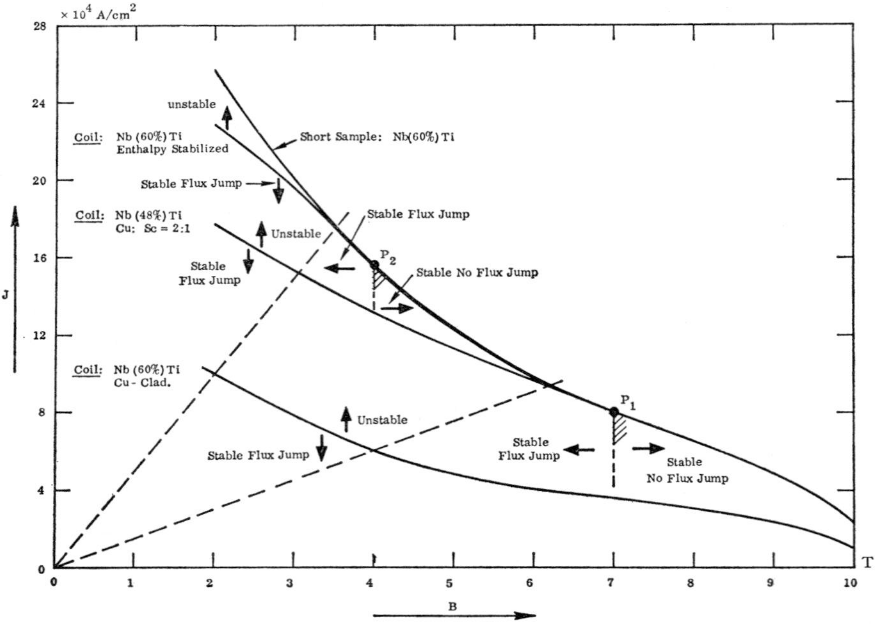

Fig. 1.8. Generalized current-field stability diagram illustrating the effect of the reduction of filament diameter and twisting

Even in intrinsically stable conductors, it is advisable to provide some stabilizing materials surrounding the filaments for emergency cases. Usually, in this case, the conductor consists of a 1 : 1 or 1.5 : 1 ratio of copper to superconductor. Stable conductors, and consequently stable magnets, are preferred for the following reasons:

(a) No coil degradation effects are encountered compared to short sample performance. This means that one is able to design the magnet based on the short-sample critical current density J_c of the superconductor, or at some values lower than J_c for added safety.

(b) When J_c is reached, the transition from superconducting to normal condition is not abrupt, but gradual. The terminal voltage will rise slowly enough to enable the operator to take preventive measures either to reduce the current until full recovery to superconducting condition is reached, or to disconnect the magnet from the current source and dissipate the free field energy into external resistors.

(c) The overall current density in the coil can be chosen appropriately higher than in a partially stabilized or nonstabilized conductor, and to take full advantage of the intrinsic stable performance of the superconductor. Optimum values of overall current densities (λJ) in magnets are chosen according to the size and the maximum operating field to which the conductor is exposed:

Experimental magnets with field energies < 1 MJ

$$B_{\max} = 10\ \text{T}; \quad \lambda J \cong 2.5 \times 10^4\ \text{A/cm}^2,$$

Beam transport magnets

$$B_{\max} \cong 6\ \text{T}; \quad \lambda J \cong 2 \quad \times 10^4\ \text{A/cm}^2,$$

Experimental magnets with field energies > 1 MJ

$$B_{\max} = 7\ \text{T}; \quad \lambda J \leqq 10^4\ \text{A/cm}^2,$$

Experimental magnets with field energies > 10 MJ

$$B_{\max} = 7\ \text{T}; \quad \lambda J \leqq 5 \times 10^3\ \text{A/cm}^2.$$

Even for large superconducting magnets, overall current densities in superconducting coils are at least one order of magnitude higher than in conventional water-cooled magnets. The cooling method utilized in superconducting magnets is commonly pool-boiling, where the composite conductor is immersed in a helium bath. He I and He II are used. Forced helium-cooling, by pumping supercritical helium through internal conductor channels [62, 63] (hollow composite conductor), or through coolant channels between layers or pancakes, convection cooling through the thermosetting in impregnated double layers in intrinsically stable conductors, is generally utilized.

References Chapter 1

[1] Du Bois, H.: Ann. d. Physik **42,** 903 (1913).
[2] Weiss, P.: J. de Phys. **6,** 353 (1907).
[3] De Haas, W. J.: Physica, Ned. T. Natuurk. 23, 113 (1932).
[4] Dryfus, L.: Elektrotechnik und Maschinenbau **53,** 205 (1935).
[5] Dryfus, L.: Arch. f. Elektrotechn. **25,** 392 (1931).
[6] Bitter, F.: Rev. Sci. Instr. I, **1,** 479; II, **7,** 482; III, **8,** 318; IV, **10,** 373 (1936—1939).
[7] Fabry, C.: L'Eclairage Electrique **XVII,** 43, p. 133 (1898).
[8] McMillen, E. M.: Phys. Rev. **68,** 143—144 (1945).
[9] Veksler, V. I.: Comptes Rendus de l'académie des Sciences de l'URSS **54,** 329—331 (1944).
[10] Oliphant, M. L., Gooden, J. S., Hide, G. S.: Proc. Phys. Soc. **59,** 666 (1947).
[11] Gooden, J. S., Jensen, H. H., Symonds, J. L.: Proc. Phys. Soc. **59,** 667 (1947).
[12] Courant, E. D., Livingston, M. S., Snyder, H. S.: Phys. Rev. **88,** 1190 to 1196 (1952).
[13] Courant, E. D., Snyder, H. S.: Annals of Physics **3,** 1—48 (1958).
[14] Blewett, J. P., Livingston, M. St.: Particle Accelerators. New York: McGraw Hill 1962.
[15] Livingood, J. J.: Principles of Cyclic Particle Accelerators. New York/ London: D. v. Norstrand 1961.
[16] Furth, H. P., Levine, M. A., Waniek, R. W.: Rev. Sci. Instr. **39,** 949 (1957).
[17] Lewin, J. D., Smith, P. F.: Rev. Sci. Instr. **46,** 541 (1964).
[18] Ter Haar, D.: The Collected Papers of P. Kapitza **1** (1964) **2** (1965).

[19] Herlach, F.: Rep. Progr. Phys. **31,** 1. Inst. of Phys. and Physical Soc. (1968).
[20] Cnare, E. C., Neilson, F. W.: Exploding Wires. **1,** 83 (1959).
[21] De Haas, W. J.: Nature **158,** 271 (1946).
[22] Olsen, J. L.: Helvetia Physica Acta **26,** 798 (1953).
[23] Birdsall, D., Furth, H. P.: Rev. Sci. Instr. **30,** 600 (1959).
[24] Foner, S., Kolm, H.: Rev. Sci. Instr. **28,** 799 (1957).
[25] Morpurgo, M., Hoffmann, L., Gibson, W. M.: CERN 60—27.
[26] Kolb, A. C.: "UN Conf. on Peaceful Uses of Atomic Energy." 2nd Conf. **31,** 328 (1958).
[27] Brechna, H., Hill, D. A., Bailey, B. M.: Rev. Sci. Instrum. **36,** 1529 (1965).
[28] Furth, H. P., Waniek, R. W.: Rev. Sci. Instr. **27,** 195 (1956).
[29] Forster, D. W., Martin, J. C.: Les Champs Magnetiques Intenses. Paris: C.N.R.S. 1967.
[30] Di Gregorio, C., Herlach, F., Knoepfel, H.: Proc. Conf. on Megagauss Magnetic Field Generation by Explosives and Related Experiments (1965).
[31] Fowler, C. M., Caird, R. S., Garn, W. B., Thomson, D. B.: High magnetic Fields, p. 269. Cambridge, Mass.: MIT Press 1962.
[32] Brechna, H., Arendt, F., Heinz, W.: Proc. 4th Int. Conf. on Magnet Technology BNL, 1972,
[33] Stevenson, D. T.: High Magnetic Fields, p. 318. Cambridge, Mass.: MIT Press 1962.
[34] Bogner, G., Franksen, H., Parsch, C. P.: Z. für angew. Physik **25,** 182 (1968).
[35] Chiang, Y. N., Emerenko, V. V., Shevchenko, O. G.: Sov. Phys. J.E.T.P. **30,** 1040 (1970).
[36] Appleton, A. D., Cowhig, T. P., Caldwell, J.: Proc. 2nd Int. Conf. on Magnet Technology, p. 553. Oxford 1967.
[37] Adair, T. W., Squire, C. F., Utley, H. B.: Rev. Sci. Instr. **31,** 416 (1960).
[38] Taylor, C. E., Post, R. F.: High Magnetic Fields, p. 101. Cambridge, Mass.: MIT Press 1961.
[39] Laquer, H. L., Hammel, E. F.: Rev. Sci. Instr. **28,** 875 (1957).
[40] Purcell, J. R., Payne, E. G.: Rev. Sci. Instr. **34,** 893 (1963).
[41] Arp, V. D.: Proc. First Int. Conf. on Magnet Technology, p. 625 (1965).
[42] Laurence, J. C., Brown, G. V., Geist, J., Zeitz, K.: High Magnetic Fields, p. 170. Cambridge, Mass.: MIT Press 1961.
[43] Shubnikov, L. W., Kotkevich, W. I., Scheplelev, J. D., Riabinin, J. N.: Phys. Z. Sowjet **10,** 165 (1937); JETP USSR **7,** 221 (1937).
[44] Schoenberg, D.: "Superconductivity." Cambridge University Press 1952.
[45] Kunzler, J. E., Buehler, E., Hsu, F. S. L., Wernick, J. H.: Phys. Rev. Letters **6,** 89 (1961).
[46] Abrikosov, A. A.: Sov. Phys. J.E.T.P. **5,** 1174 (1957).
[47] Gorter, C. J.: Phys. Letters **1,** 69; **2,** 26 (1962).
[48] Bean, C. P.: Phys. Rev. Letters **8,** 250 (1962).
[49] London, H.: Phys. Letters **6,** 162 (1963).
[50] Kim, Y. B., Hempstead, C. F., Strnad, A. R.: Phys. Rev. A **139,** 1163 (1955).
[51] Anderson, P. W.: Phys. Rev. Letters **9,** 309 (1962).

[52] Hart, H. R.: Proc. 1968 Summer Study on Supercond. Devices, p. 571. Brookhaven 1968.
[53] Wipf, S. L.: ibid, p. 511.
[54] Hancox, R.: Proc. I.E.E. **113,** 1221 (1966).
[55] Smith, P. F., Wilson, M. N., Lewin, J. D.: Proc. 1968 Summer Study on Supercond. Devices, p. 839. Brookhaven 1968.
[56] Schwettman, H. A. et al.: IEEE Trans. Nucl. Sci. NS **14,** 336 (1967).
[57] Wilson, P. B., Schwettman, H. A.: ibid NS **12,** 1045 (1965).
[58] Steckly, Z. J. J., Zar, J. L.: Trans. IEE **12,** 367 (1965).
[59] Hancox, R.: Proc. Intermag. Conf. Washington D.C. 1968.
[60] Brechna, H., Garwin, E. L.: SLAC proposal (1967).
[61] Smith, P. F., Wilson, M. N., Walters, C. R., Lewin, J. D.: Proc. 1968 Summer Study on Supercond. Devices, p. 913. Brookhaven 1968.
[62] Brechna, H.: ANL Rep. **7192,** 29 (1966).
[63] Morpurgo, M.: CERN NP. Int. Rep. 67-15 (1967).

2. Magnetic Field Calculations

2.1 Magnets without Ferromagnetic Yokes

2.1.1 Magnetic Fields due to Current Elements

The magnetic field due to a distributed current is found by superposition of fields generated by an infinite number of line currents which, together, constitute the distributed current. For a current filament with a positive current of magnitude I located at Z, the magnetic field at a point Z_0 has the value:

$$B = \frac{\mu_0 I}{2\pi |Z - Z_0|}. \tag{2.1.1}$$

The field direction perpendicular to the vector $(Z_0 - Z_1)$ is given by:

$$j \cdot \frac{Z_0 - Z}{|Z_0 - Z|}, \quad \text{with} \quad j = (-1)^{\frac{1}{2}}.$$

Thus, the magnetic field due to a line current is expressed by:

$$B = \frac{\mu_0 I}{2\pi |Z - Z_0|} \cdot \frac{j\,(Z_0 - Z)}{|Z_0 - Z|} = \frac{\mu_0 I}{2\pi} \cdot \frac{j\,(Z_0 - Z)}{(Z_0 - Z)\,(Z_0^* - Z^*)}, \tag{2.1.2}$$

$$B = \frac{\mu_0 I}{2\pi} \cdot \frac{j}{(Z_0^* - Z^*)}. \tag{2.1.3}$$

Z^* and Z_0^* are complex conjugates of Z and Z_0, respectively.

Similarly, we have:

$$\boxed{B^* = \frac{j \mu_0 I}{2\pi\,(Z - Z_0)}.} \tag{2.1.4}$$

Extending Eq. (2.1.4) to a distributed current of arbitrary cross section but uniform current density λJ, we get by integration

$$B^* = j \frac{\lambda J}{2\pi} \mu_0 \int \frac{1}{Z - Z_0}\, dX\, dY. \tag{2.1.5}$$

The integral (2.1.5) is of the form $\int F(Z)\, dX\, dY$, where $F(Z)$ is analytic and has a single value over the integration region, but with a singularity at $Z = Z_0$. The contour integration performed yields:

$$B^* = \frac{\lambda J}{4\pi} \mu_0 \oint_c \frac{Z^* - Z_0^*}{Z - Z_0}\, dZ = \frac{\lambda J}{4\pi} \mu_0 \oint_c \frac{Z^*}{Z - Z_0}\, dZ + \frac{\lambda J \mu_0 \cdot Z_0}{2j} \tag{2.1.6}$$

with:

$$B^* = B_x - jB_y$$
$$Z = X + jY; \quad Z^* = X - jY.$$

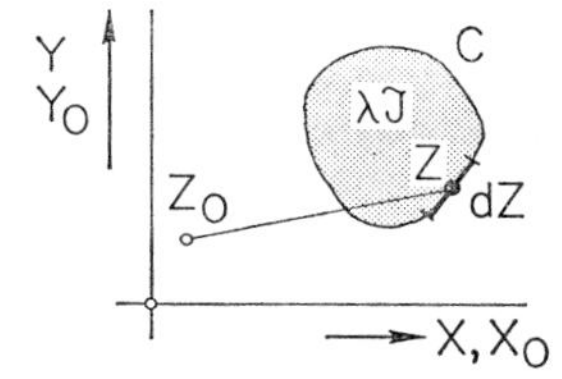

Fig. 2.1.1a.

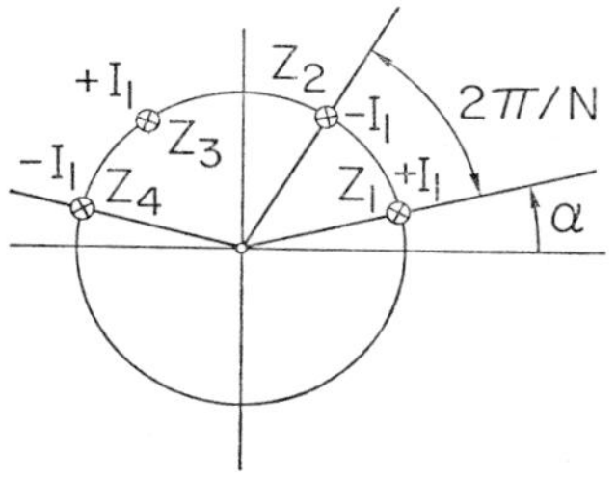

Fig. 2.1.1b.

This equation is valid inside and outside the coil region (Fig. 2.1.1.a). The field gradient is obtained by differentiating Eq. (2.1.6):

$$G^* = \frac{\partial B^*}{\partial Z_0} = \frac{\partial B_X}{\partial X_0} - j\frac{\partial B_Y}{\partial X_0},$$

$$G^* = \frac{\lambda J}{4\pi}\mu_0 \oint\limits_c \frac{Z^*}{(Z - Z_0)^2} dZ. \tag{2.1.7}$$

The arbitrary shape of the distributed current can be approximated by a polygon consisting of n straight lines, each having an equation of the form

$$X = a_k Y + b_k,$$

where a and b are complex constants:

$$a_k = \frac{X_{k+1} - X_k}{Y_{k+1} - Y_k}; \quad b_k = \frac{X_k \cdot Y_{k+1} - Y_k \cdot X_{k+1}}{Y_{k+1} - Y_k}.$$

The field at the point Z_0 is thus:

$$B^* = \frac{\lambda J}{2\pi}\mu_0 \cdot \sum_{k=1}^{N}\left[\frac{j}{j + a_k}(a_k Y_0 + b_k - X_0)\ln\left(\frac{Z_{k+1} - Z_0}{Z_k - Z}\right)\right] \tag{2.1.8}$$

and the field gradient:

$$G^* = \frac{\lambda J}{4\pi}\mu_0 \cdot \sum_{k=1}^{N}\left(\frac{a_k - j}{a_k + j}\right)\ln\left(\frac{Z_{k+1} - Z_0}{Z_k - Z_0}\right). \tag{2.1.9}$$

The field components B_X and B_Y can be derived from a vector potential $\boldsymbol{A}$ or a scalar potential V:

$$B_x = \partial \boldsymbol{A}/\partial Y = -\partial V/\partial X$$
$$B_y = \partial \boldsymbol{A}/\partial X = -\partial V/\partial Y.$$

The equations involving the scalar potential V are valid only in the region free of current-carrying conductors. The equations involving the vector potential $\boldsymbol{A}$ are valid everywhere, including coil areas in space. We write for the complex potential:

$$\boldsymbol{P}(Z) = \boldsymbol{A} + jV,$$

where $\boldsymbol{P}(Z)$ is analytic. $\boldsymbol{B}^*(Z)$ can be obtained from $P(Z)$ through the relation:

$$\boldsymbol{B}^*(Z) = j \cdot \frac{d\boldsymbol{P}(Z)}{dZ}. \tag{2.1.10}$$

Therefore, we can expand $P(Z)$ in a Taylor Series of the complex variable Z_0:

$$P(Z_0) = c_0 + c_1 Z_0 + c_2 Z_0^2 + \cdots + c_n Z_0^n = \sum_{n=0}^{\infty} c_n Z_0^n. \tag{2.1.11}$$

Thus, the field is obtained from $P(Z_0)$ through:

$$B^*(Z_0) = j \sum_{n=1}^{\infty} (n c_n Z_0^{n-1}) = \sum_{n=1}^{\infty} (d_n Z_0^{n-1}). \tag{2.1.12}$$

c_n and d_n are complex constants.

The series (2.1.11) converges only within a circle extending to the nearest coil boundary. We rewrite (2.1.11) for such a region:

$$B^*(Z_0) = \frac{\lambda J}{4\pi} \mu_0 \oint \frac{Z^*}{Z - Z_0} dZ = \frac{\lambda J}{4\pi} \mu_0 \oint \frac{Z^*}{Z(1 - Z_0/Z)} dZ$$

$$= \frac{\lambda J}{4\pi} \mu_0 \oint \frac{Z^*}{Z} \sum_{n=1}^{\infty} \left(\frac{Z_0^{n-1}}{Z^{n-1}}\right) dZ \tag{2.1.13}$$

for $|Z_0/Z| < 1$.

The two equations (2.1.11) and (2.1.12) lead to:

$$\sum_{n=1}^{\infty} (d_n Z_0^{n-1}) = \frac{\lambda J}{4\pi} \mu_0 \oint \frac{Z^*}{Z} \sum_{n=1}^{\infty} \frac{Z_0^{n-1}}{Z^{n-1}} dZ$$

where we can obtain the complex constant as:

$$d_n = \frac{\lambda J}{4\pi} \mu_0 \oint \frac{Z^*}{Z^n} dZ. \tag{2.1.14}$$

In this expression, d_n is the multipole field coefficient.

By comparing Eq. (2.1.8) with Eq. (2.1.12) and Eq. (2.1.13), we obtain the special values of d_n:

For $n = 1$:

$$d_1 = \frac{\lambda J}{2\pi} \mu_0 \sum_{k=1}^{N} \left[\left(\frac{j}{a_k + j}\right) b_k \ln\left(\frac{Z_{k+1}}{Z_k}\right)\right];$$

If $a \to \infty$:

$$d_1 = \frac{\lambda J}{2\pi} \mu_0 \sum_{k=1}^{N} \left(-j Y_k \cdot \ln\left(\frac{Z_{k+1}}{Z_k}\right)\right);$$

For $n \geqq 3$:

$$d_n = \frac{\lambda J}{2\pi} \mu_0 \sum_{k=1}^{N} \frac{j}{a_k + j} \left[\frac{a_k - j}{2j} \cdot \frac{Z_{k+1}^{2-n} - Z_k^{2-n}}{2-n} - b_k \frac{Z_{k+1}^{1-n} - Z_k^{1-n}}{1-n}\right];$$

With $a \to \infty$:

$$d_n = \frac{\lambda J}{2\pi} \mu_0 \sum_{k=1}^{N} j \left[\frac{Z_{k-1}^{2-n} - Z_k^{2-n}}{2-n} - Y_k \frac{Z_{k+1}^{1-n} - Z_k^{1-n}}{1-n}\right].$$

To express $\boldsymbol{B}$ in terms of radial and azimuthal field components, B_r and B_φ, the complex expression for the fields reads:

$$\boldsymbol{B}(r,\varphi) = B_r + jB_\varphi\,,$$

where it may be noted that $\boldsymbol{B}$ is related to B by

$$\boldsymbol{B} = B \cdot e^{-j\varphi}.$$

Since, for a symmetric 2 N-pole magnet, the relation

$$\boldsymbol{B}\left(r,\varphi + \frac{\pi}{N}\right) = -\boldsymbol{B}(r,\varphi)$$

must hold, one obtains from Eq. (2.1.10) and Eq. (2.1.11):

$$c_n \cdot e^{jn\pi/N} = -c_n, \quad \text{with} \quad n = 1, 2, 3 \ldots$$

Letting $c_0 \equiv 0$, all c_n values are zero unless $n = N(2m+1)$, with $m = 0, 1, 2 \ldots$, and we obtain the complex potential in the form:

$$P(Z) = \sum_{m=0}^{\infty} c_{N(2m+1)} \cdot Z_0^{N(2m+1)}. \tag{2.1.15}$$

$c_N Z_0^N$ is the fundamental harmonic. Due to symmetry, no odd multiples appear. If each pole of a 2 N-pole magnet has a symmetry axis and the vertex of one pole is on the X axis, then:

$$\boldsymbol{B}(r,\varphi) = \boldsymbol{B}(r,-\varphi)\,.$$

From Eq. (2.1.10) and Eq. (2.1.15), it follows that all coefficients in Eq. (2.1.15) are imaginary. By introducing the real quantity $p_N = -j\,c_N$, one obtains from Eq. (2.1.15)

$$P(Z) = j \sum_{m=0}^{\infty} P_{N(2m+1)} \cdot Z_0^{N(2m+1)}.$$

For a symmetrical (unperturbed) 2 N-pole magnet, consisting of identical poles at a scalar potential of ± 1, one has:

$$P_0(Z) = j \cdot Z_0^N\,. \tag{2.1.16}$$

The equation describing the reference pole is expressed by:

$$I_m\,[P_0(Z)] = Re\,[(Z_0^N)] = r^N \cos(N\varphi) \equiv 1\,.$$

Thus:

$$Z_0^N(\varphi) = e^{jN\varphi}/\cos(N\varphi)\,, \quad Z_0(\varphi) = e^{j\varphi}/[\cos(N\varphi)]^{1/N}.$$

In the Z_0^N plane, the reference pole is a straight line (pole of an ideal magnet). $N\varphi$ is the angle between the straight lines connecting the origin with the vertex of the pole (Fig. 2.1.2).

Remarks. Multipole coefficients are the coefficients of a Taylor expansion on Z_0. The convergence radius of the expansion is the distance from the origin to the next point of singularity (i.e., coil aperture). If P_N is set as unity, then, according to Eq. (2.1.16), the restriction of the convergence radius does not apply to $P_0(Z) = jZ_0^N$. This statement is particularly true in ideal (unperturbed) coil systems.

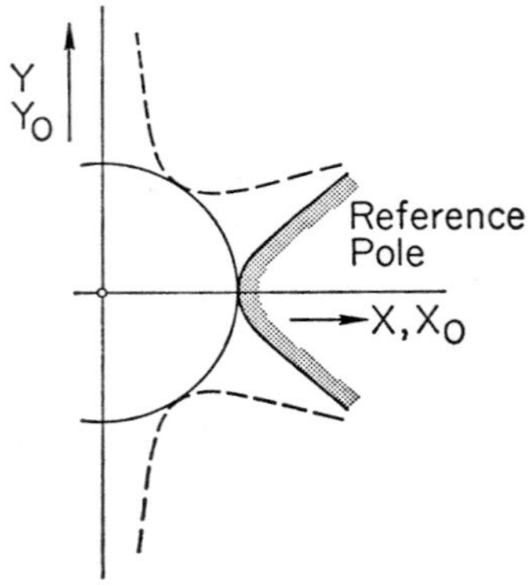

Fig. 2.1.2.

2.1.2 Fields due to Filamentary Current Lines

The field due to a current line as shown in Fig. 2.1.3a was given in Eq. (2.1.4):

$$B^* = j\frac{\mu_0 I}{2\pi}\cdot\frac{1}{Z - Z_0}.$$

The magnetic field generated by two current lines, where the currents flow in the same direction, is given by:

$$B^*(Z) = j\frac{\mu_0 I}{2\pi}\cdot\left[\frac{1}{Z_1 - Z_0} + \frac{1}{Z_2 - Z_0}\right]$$

with Z_1 and Z_2 the positions of the line currents.
If the currents flow in opposite directions, the field is represented by:

$$B^*(Z) = j\frac{\mu_0 I}{2\pi}\cdot\left[\frac{1}{Z_1 - Z_0} - \frac{1}{Z_2 - Z_0}\right].$$

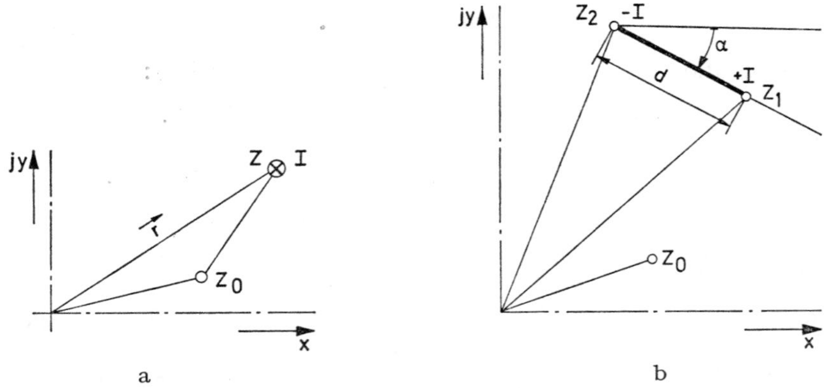

Fig. 2.1.3a and b. Arrangement of line currents in a complex space

Using the notation: $Z_1 - Z_2 = d \cdot e^{j\alpha}$,
with d the distance between the current filaments (Fig. 2.1.3b), we get:

$$B^*(Z) = j\,\frac{\mu_0 I}{2\pi}\cdot\frac{d \cdot e^{j\alpha}}{(Z_1 - Z_0)(Z_2 - Z_0)}.$$

If $Z_1 \gg d$:

$$\boxed{B^*(Z) = j\,\frac{\mu_0 I}{2\pi}\cdot\frac{d \cdot e^{j\alpha}}{(Z_1 - Z_0)^2},}$$

which is the expression for a normal complex field generated by a pair of line currents (doublet) at a point Z_0.

If N parallel current filaments are distributed on a cylinder and spaced over the angles $2\pi/N$, with N the number of poles, we have:

$$B^*(Z) = j\,\frac{I}{2\pi}\,\mu_0 \sum_{n=1}^{N} \frac{1}{Z - Z_0} = \sum_{n=1}^{\infty} (c_n Z_0^{n-1}) = j\,\frac{I}{2\pi}\,\mu_0 \sum_{n-1}^{N} \frac{1}{Z\left(1 - \frac{Z_0}{Z}\right)}$$

$$B^*(Z) = j\,\frac{I}{2\pi} \sum_{1}^{N} \left(\frac{1}{Z} \sum_{1}^{\infty} \frac{Z_0^{n-1}}{Z^{n-1}} \right) \tag{2.1.17}$$

as given in Eq. (2.1.13).

$n = 1$ corresponds to a "dipole" with $N = 2$
$n = 2$ corresponds to a "quadrupole" with $N = 4$

Generally: $N = 2m$.

The complex coefficients are of the form:

$$c_n = j\,\frac{I}{2\pi}\,\mu_0 \sum_{m=1}^{2n} \left(\frac{(-1)^{n-1}}{Z^n} \right). \tag{2.1.18}$$

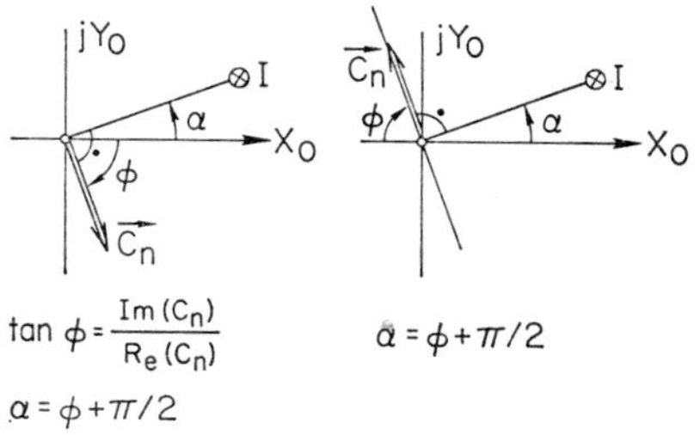

Fig. 2.1.4.

To cancel a multipole, a current I must be placed such that the vector $\boldsymbol{c}_n$ is perpendicular to the radius vector $\boldsymbol{r}$ Fig. 2.1.4). The sign of I is given by the sign of the real part of c_n. The magnitude of I is given by:

$$I = Re \left(\frac{c_n}{\frac{j}{2\pi}\,\mu_0 \sum_{m=1}^{2n} \frac{(-1)^{m-1}}{Z^n}} \right). \tag{2.1.19}$$

Field calculations by means of complex variables have been treated by Beth [1], Yourd [2], and Halbach [3].

2.1.3 Field Corrections

We express the magnetic field generated by a number of filamentary currents I_k at the points Z_k located on the circumference of a circle:

$$B^* = j \cdot \frac{\mu_0}{2\pi} \sum_{k=1}^{N} \left(\frac{I_k}{Z_k - Z_0} \right) = \frac{j\mu_0}{2\pi} \sum_{k=1}^{N} \left[\frac{I_k}{Z_k} \left(1 - \frac{Z_0}{Z_k} \right)^{-1} \right],$$

$$B^* = \frac{j\mu_0}{2\pi} \sum_{k=1}^{N} \left[\frac{I_k}{Z_k} \sum_{n=1}^{\infty} \frac{Z_0^{n-1}}{Z_k^{n-1}} \right].$$

For $|Z_0/Z_k| < 1$:

$$B^* = \frac{j\mu_0}{2\pi} \sum_{n=1}^{\infty} \left[\sum_{k=1}^{N_1} \left(\left(\frac{I_k}{Z_k^n} \right) Z_0^{n-1} \right) \right]. \tag{2.1.20}$$

and comparison with Eq. (2.1.17) yields

$$c_n = \frac{j}{2\pi} \mu_0 \sum_{k=1}^{N} \left(\frac{I_k}{Z_k^n} \right).$$

For values of $|Z_0| > |Z_k|$, we can derive the expression for the field from:

$$Z^* = -\frac{j\mu_0}{2\pi} \cdot \sum_{n=1}^{\infty} \left[\sum_{k=1}^{N} \left(I_k \, (Z_k)^{n-1} \frac{1}{(Z_0)^n} \right) \right] \tag{2.1.21}$$

for variable currents, and

$$Z^* = -\frac{j\mu_0}{2\pi} \cdot I \sum_{n=1}^{\infty} \left[\sum_{k=1}^{N} \left((-1)^{k-1} \, (Z_k)^{n-1} \frac{1}{(Z_0)^n} \right) \right] \tag{2.1.22}$$

for a constant current.

We may also write:

$$B^* = \sum_{n=1}^{\infty} \left(\frac{c_n}{(Z_0)^{n-1}} \right)$$

with:

$$c_n = \frac{jI}{2\pi} \mu_0 \sum_{k=1}^{N} \cdot \sum_{n=1}^{\infty} \left((-1)^{k-1} \frac{1}{Z^{n-2}} \right)$$

for all values of $n \geq 2$.

2.1.4 Applications

2.1.4.1 Circular Current Filament

The magnetic field of a circular current filament of constant radius a is calculated using the method of complex variables (Fig. 2.1.5a):

$$Z = a\,(\cos\varphi + j\sin\varphi) = a \cdot e^{j\varphi} = a\psi\,; \quad dZ = a\,d\psi\,; \tag{2.1.23}$$

$$Z^* = a\,e^{-j\varphi} = a/\psi\,.$$

According to Eq. (2.1.5), using the Cauchy integral,

$$F(z) = j \cdot \frac{\lambda J}{4\pi} \oint_c \frac{Z^*}{Z - Z_0},$$

we get for our cylinder:

$$F(z) = j \frac{\lambda J}{4\pi} \oint_c \frac{\frac{a}{\psi} \cdot a\, d\psi}{(a\psi - Z_0)}$$

$$= j \frac{\lambda J}{4\pi} a \oint_c \frac{d\psi}{\psi\left(\psi - \frac{Z_0}{a}\right)} = j \frac{\lambda J}{4\pi} a \oint_c f(\psi)\, d\psi. \qquad (2.1.24)$$

$f(\psi)$ has two singular poles at $\psi = 0$ and $\psi = Z_0/a$. Integrating Eq. (2.1.24) by parts and using the methods of residues, we obtain:

$$F(z) = 0, \qquad \text{for} \quad Z = Z_{in}$$

$$F(z) = \frac{\lambda J}{2} \cdot \frac{a^2}{Z_0}, \qquad \text{for} \quad Z = Z_{out},$$

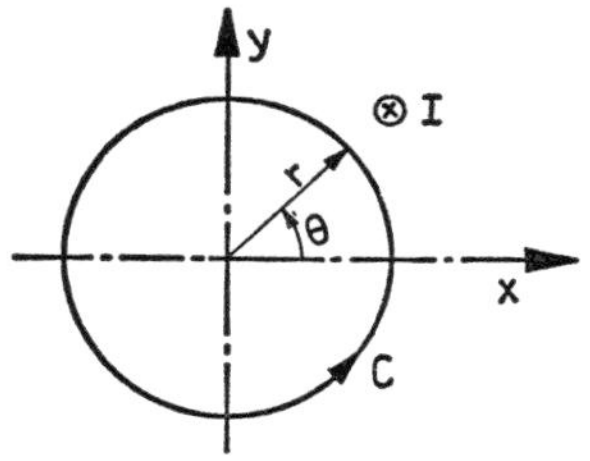

Fig. 2.1.5a. Circular current filament

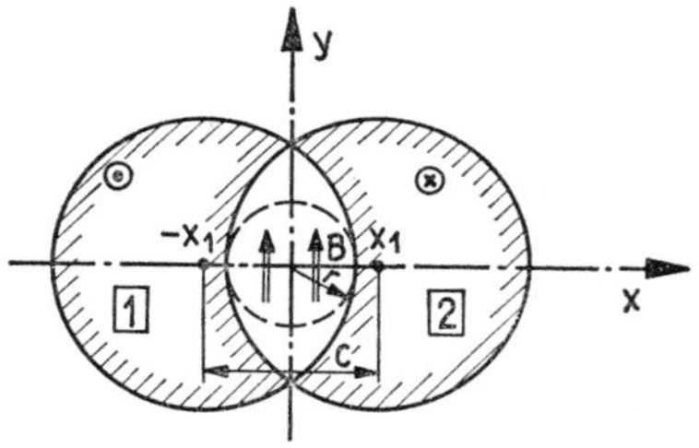

Fig. 2.1.5b. Intersecting circles

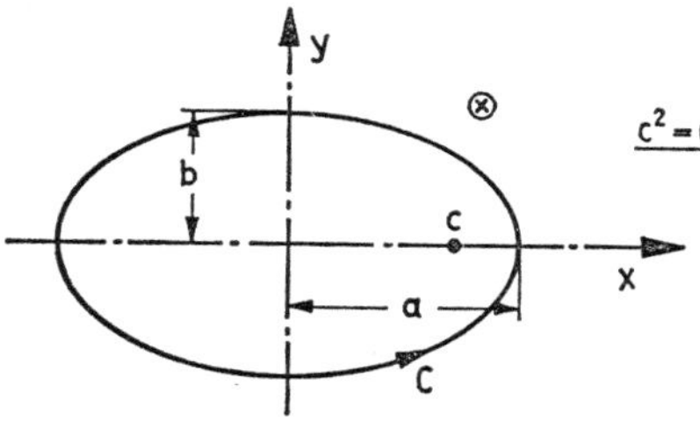

Fig. 2.1.5c. Elliptical current filament

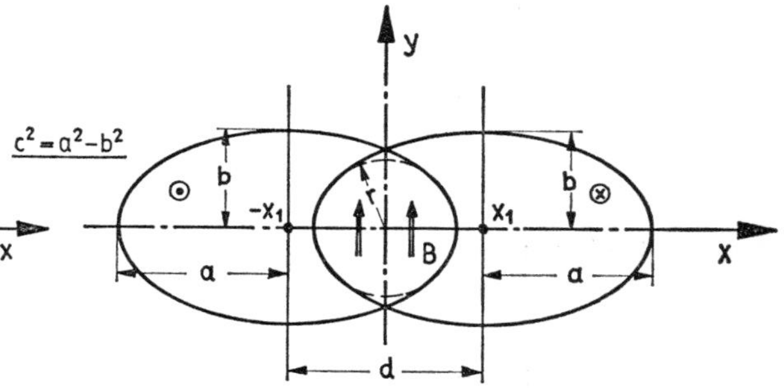

Fig. 2.1.5d. Intersecting ellipses

with Z_{in} indicating a point inside, and Z_{out} a point outside the cylindrical conductor.

From Maxwell's equation, we obtain:

$$\frac{\partial H_y}{\partial x} - \frac{\partial H_x}{\partial y} = \lambda J; \qquad \frac{\partial B_x}{\partial x} + \frac{\partial B_y}{\partial y} = 0$$

$$H(x, y) = H_y + j H_x = \frac{B(x, y)}{\mu_0}$$

$H(x, y)$ is analytic. Writing $F(z) = u + jv$
where:

$$u = H_y - \frac{\lambda J}{2} \cdot X\,; \qquad v = H_x + \frac{\lambda J}{2} \cdot Y$$

we obtain:

$$F(z) = H_y + jH_x - \frac{\lambda J}{2}(X - Y) = H_y + jH_x - \frac{\lambda J}{2} Z^* .$$

Thus, for our example:

$$H_{\text{in}} = F(z) + \frac{\lambda J}{2} Z^* = \left(\frac{\lambda J}{2}\right) \cdot Z^* \qquad (2.1.25)$$

$$B_{\text{in}} = \mu_0 \cdot \frac{\lambda J}{2}(X - jY) .$$

In the same way:

$$\boxed{B_{\text{out}} = \mu_0 \cdot \frac{\lambda J}{2} \cdot \frac{a^2}{Z_0} .} \qquad (2.1.26)$$

2.1.4.2 Elliptical Conductor

The magnetic field distribution of an elliptical conductor (Fig. 2.1.5c), with the axis $2a$, $2b$, carrying a uniform current density of $J = \text{const}$, is obtained in the same way as for example 1.

$$Z = a \cos \theta + jb \sin \theta$$

with

$$0 \leqq \theta \leqq 2\pi\,, \quad r = \frac{a+b}{2}\,, \quad \Delta = \frac{a-b}{2}\,,$$

$$c^2 = a^2 - b^2 = 4r\Delta\,; \quad \psi = e^{j\theta}.$$

We obtain:

$$Z = r \cdot \psi + \Delta \cdot \psi^{-1} .$$

The Cauchy integral for this case,

$$F(Z) = jJ \oint_c \frac{Z^* \cdot dZ}{z - Z}\,,$$

leads to:

$$F(Z) = jJ \oint_c \frac{(r + \Delta \cdot \psi^2)(r \cdot \psi^2 - \Delta)}{\psi^2 (r\psi^2 - Z\psi + \Delta)} d\psi\,,$$

which yields:

$$F(Z) = -2\pi J \cdot \frac{a-b}{a+b} \cdot Z .$$

Now from:

$$H_{\text{in}}(X, Y) = F(Z) + 2\pi J \cdot Z^*$$

for: $Z = Z_{\text{in}}$.

After inserting $F(Z)$ and rearranging:

$$H_{\text{in}}(X, Y) = \frac{4\pi J}{a+b}(bX - jaY)\,,$$

$$H_{\text{out}}(X, Y) = F_{\text{out}}(Z)\,,$$

which, leads to:

$$H_{\text{out}} = 4\,\pi J \frac{ab}{Z + \sqrt{Z^2 - c^2}},$$

a solution obtained by Beth [4]. On the conductor surface:

$$H_{\text{in}} = H_{\text{out}} = \frac{4\,\pi J}{a + b} \cdot ab \cdot e^{-j\theta}.$$

2.1.4.3 Dipole Field

The field due to a dipole coil composed of two intersecting circles of equal radii (Fig. 2.1.5b) having a constant current density distribution is given by:

$$B_{\text{in},1} = (\lambda J) \frac{\mu_0}{2} [(X + X_1) - jY],$$

$$B_{\text{in},2} = -(\lambda J) \frac{\mu_0}{2} [(X - X_1) - jY],$$

where $B_{\text{in},1}$ and $B_{\text{in},2}$ denote the field due to each current circle. The total field inside the aperture is obtained from superposition of the partial contributions,

$$B_{\text{in}} = \mu_0 (\lambda J \cdot X_1) = \text{constant}, \tag{2.1.27}$$

or with the notation of Fig. 2.1.5b,

$$B_{\text{in}} = \frac{\mu_0}{2} (\lambda J) \cdot c,$$

which means that within the coil aperture, the field is constant.

The field due to an elliptical conductor (Fig. 2.1.5c) is derived similarly to the cylindrical conductor:

$$B_{\text{in}} = \frac{\mu_0 \cdot (\lambda J)}{(a + b)} [bX - jaY], \tag{2.1.28}$$

$$B_{\text{out}} = \mu_0 (\lambda J) \cdot \frac{ab}{Z + (Z^2 - c^2)^{1/2}}. \tag{2.1.29}$$

The superposition of two ellipses yields the dipole field:

$$B_{\text{in}} = \frac{\mu_0 (\lambda J)}{2\,(a + b)} \cdot \left(\frac{bd}{a + b}\right), \tag{2.1.30}$$

which indicates that the field inside the coil aperture is constant (Fig. 2.1.5d).

It may be pointed out that in practice it is impossible to wind dipole coils which have the shapes ideally obtained by superposition of circles or ellipses. The ideal coil configuration is, in general, approximated by a number of current blocks, such as shown in Fig. 2.1.6.

To demonstrate the effect of current density deviations from the ideal shape, we approximate the intersecting circles by a polygon with 18 straight sections per coil quadrant. At an overall current density of $\lambda J = 10^8$ A/m², $a = 0.198$ m, $c = 0.096$ m, the central field is 6 T. For the ideal coil configuration, lines of constant $\frac{\Delta B}{B\,(0,0)} = \pm\, 10^{-3}$ and $\pm\, 5 \times 10^{-3}$ (constant error contours) are given in Fig. 2.1.7.

The practical case of truncated coils is shown in Fig. 2.1.8. The field contour lines may be compared directly to those of the ideal case.

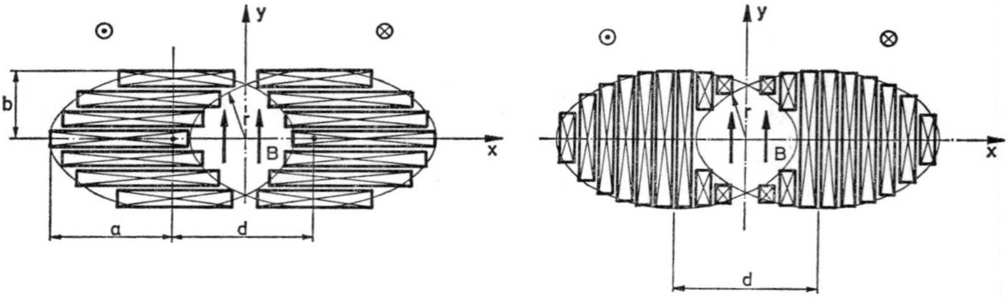

a b

Fig. 2.1.6a and b. Coil arrangements approximating intersecting ellipses

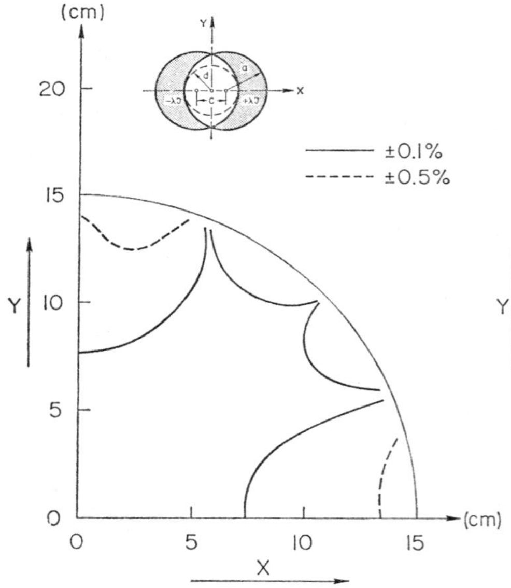

Fig. 2.1.7

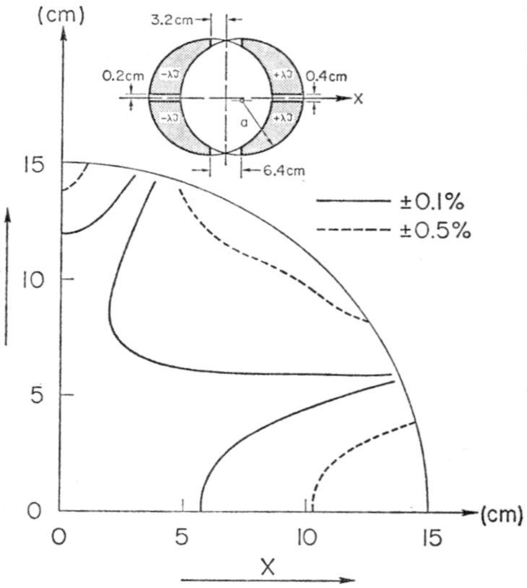

Fig. 2.1.8

Fig. 2.1.7. Lines of constant field within the aperture of a coil produced from two intersecting circles

Fig. 2.1.8. Lines of constant field within the aperture of a truncated coil simulating a practical design. The coil apertures and Ampere turns of the coils according to Figs. 2.1.7 and 2.1.8 are identical

2.1.4.4 Quadrupole Field

A quadrupole field is obtained by superimposing two ellipses one of which is rotated by 90° with respect to the other, as illustrated in Fig. 2.1.9. The field inside the ellipse I is:

$$B_{\text{in},\text{I}} = \mu_0 \cdot \frac{(\lambda J)}{(a+b)} (bX - jaY)$$

and that inside the ellipse II:

$$B_{\text{in},\text{II}} = -\mu_0 \frac{(\lambda J)}{(a+b)} (aX - jbY) .$$

Thus, the total field within the aperture:

$$B = B_{\text{in},\text{I}} + B_{\text{in},\text{II}} = \mu_0 \frac{(\lambda J)}{a+b} \cdot [(b-a)\, X + j\, (b-a)\, Y] . \qquad (2.1.31)$$

To obtain the field gradient, we separate Eq. (2.1.31) into real and imaginary components. Differentiating these components with respect to x and y, we obtain:

$$G = \frac{Re\,(B_{\text{in}})}{X} = \frac{I_m\,(B_{\text{in}})}{Y} = \mu_0 (\lambda J) \cdot \frac{b-a}{b+a} = \text{const} . \qquad (2.1.32)$$

The field gradient within the aperture of an ideal quadrupole coil is constant. For the particular case of a quadrupole given in Fig. 2.1.10, the field gradient is $G = 40$ T/m, while the maximum field at the con-

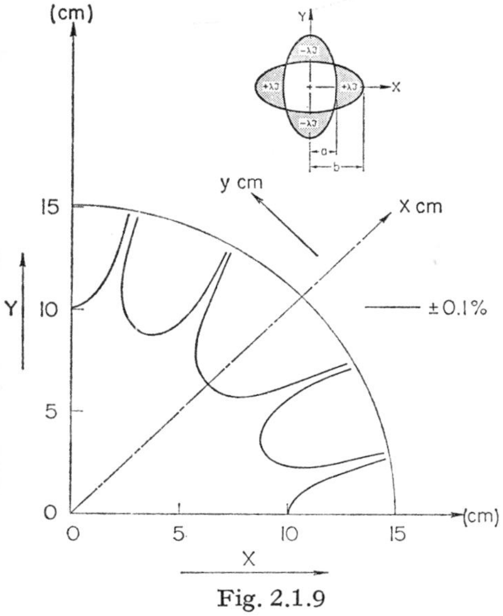

Fig. 2.1.9

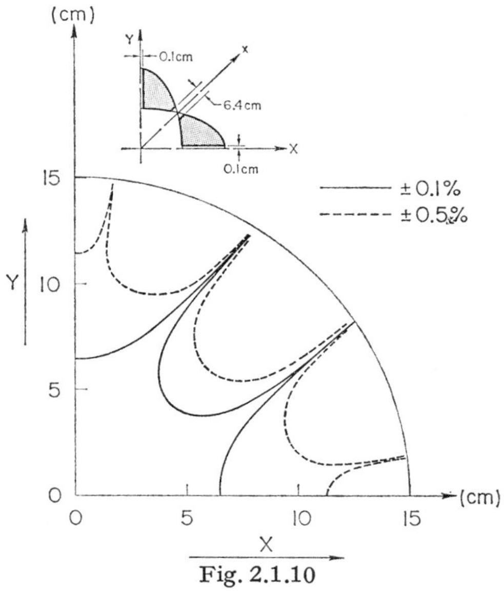

Fig. 2.1.10

Fig. 2.1.9. Lines of constant field gradient within the aperture of an ideal quadrupole generated by two intersecting and rotated ellipses

Fig. 2.1.10. Lines of constant field gradient in the aperture of a truncated quadrupole coil. The coil aperture is identical to (Fig. 2.1.9)

ductor is $B_Y = 6$ T. The corresponding current density in the coil must be $\lambda J = 10^8$ A/m², the major radius $b = 0.29$ m, and the minor radius $a = 0.15$ m.

For calculation purposes, the coil contour is approximated by 12 straight current lines such that Eq. (2.1.9) is applicable. The selection of the number of straight lines was performed such that the error due to

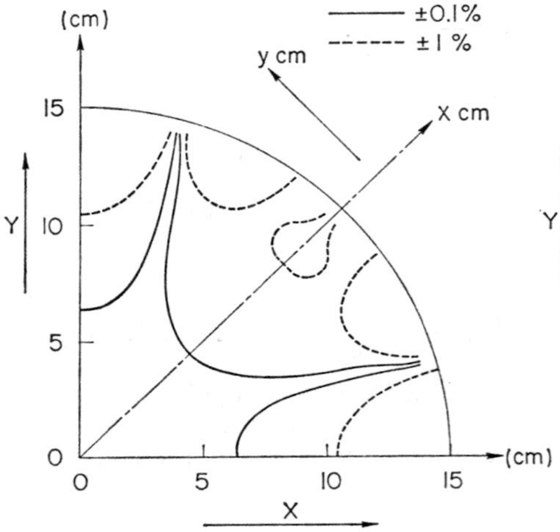

Fig. 2.1.11

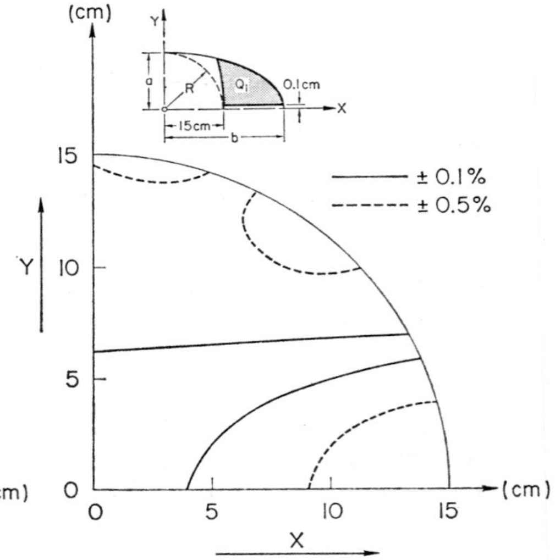

Fig. 2.1.12

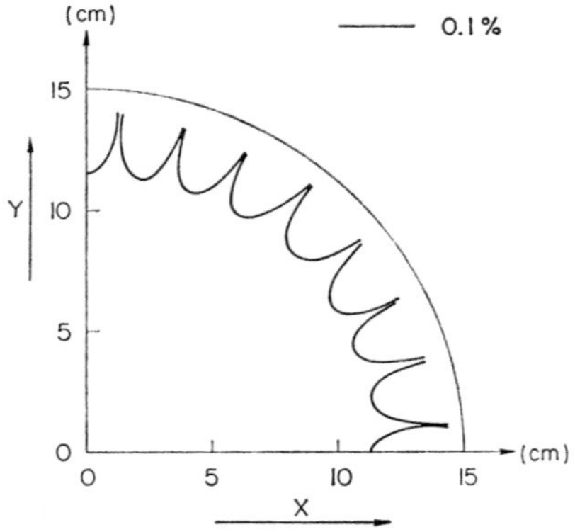

Fig. 2.1.13

Fig. 2.1.11. Lines of constant field gradient in the aperture of a quadrupole according to Fig. 2.1.10 but with corrective current filaments parallel to the axis located at the coil inner surface

Fig. 2.1.12. Lines of constant field gradient within the aperture of a quadruple with one quadrant shifted in the y direction

Fig. 2.1.13. Effect of field corrective current filaments on the field gradient of a real quadrupole the simplification (polygon contour instead of intersecting ellipses) is smaller than 10^{-3} (Fig. 2.1.11). To illustrate the effect of conductor perturbations, coil manufacturing tolerances, etc., we assume that only one coil segment (out of four) is shifted in the Y direction by a distance of 10^{-3} m. The effect of this translation is given in Fig. 2.1.12. The region where the field gradient is changed to 0.1% is now reduced considerably, compared to the ideal case given in Fig. 2.1.11. We correct for the perturbation effects by placing a 12-pole conductor configuration at the circumference of the inscribed aperture circle, Fig. 2.1.13, using the field correction method given in Chapter 2.1.2. The current through each conductor is 2.6×10^3 A, which corresponds to 0.33% of the total coil mmf. The effect of current bars placed parallel to the z-axis is obvious.

Energizing these superconducting bars placed around the coil aperture parallel to the z-axis is accomplished by means of separate power supplies, connecting the bars in series or parallel to the main coil, or connecting all the ends on each side such that the bars are connected in parallel and are short-circuited (squirrel cage). Currents are induced in the bars by a transformer effect when the field is raised in the main coil. These currents will persist.

2.1.5 Magnetic Field Calculation by Means of Current Sheets

In this section, the application of the "current sheet theorem" [5] on field calculation is given. Referring to Fig. 2.1.14b, we consider a cylinder located such that its axis is perpendicular to the z-plane. A current filament located in the cylinder surface has a complex potential given by [6]:

$$W(Z) = -2 I \mu_0 \ln \left(\frac{1}{Z - z}\right) + \text{const}\,, \qquad (2.1.33)$$

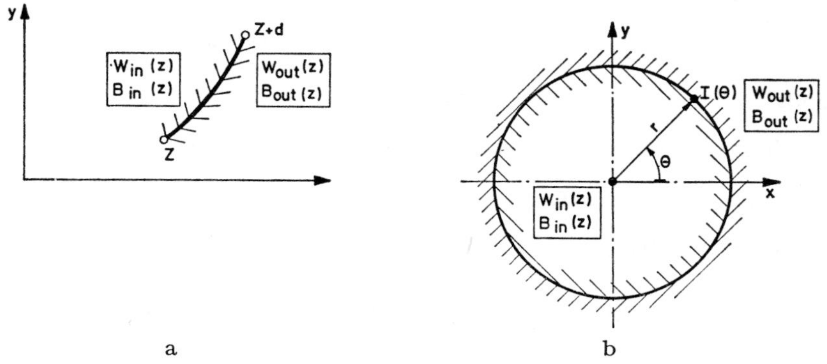

Fig. 2.1.14a and b. Arrangement of current sheets

where the minus sign corresponds to the direction of the current flow. The field is derived by differentiating Eq. (2.1.33):

$$B(Z) = \frac{dW(Z)}{dZ} = B_Y + jB_X = \mu_0 \cdot \frac{2I}{Z - z}\,. \qquad (2.1.34)$$

The Cauchy integral, taken over an arbitrary path, yields:

$$\oint B\,(Z)\,dZ = \mu_0 \cdot j \cdot I = \frac{\mu_0}{2}\,j \cdot (2\,I)\,, \tag{2.1.35}$$

where $(2\,I)$ is the so-called *current residue.*

Now, looking at a number of current filaments along the cylinder $Z = z$, and noting that ΔI is the total current in the intervall between Z and $Z + \Delta Z$, then the field obtained from Eq. (2.1.35) is written:

$$\boxed{B_{\text{out}}(z) - B_{\text{in}}(z) = \mu_0 \cdot j\,\frac{dI}{dz}.} \tag{2.1.36}$$

Equation (2.1.36) is called the "current sheet theorem". $B_{\text{out}}(z)$ and $B_{\text{in}}(z)$ are the limits of an analytic function. The current gradient dI/dz is written symbolically for the limit of $\Delta I/\Delta z$ as $\Delta z \to 0$ for any value of z. Integrating Eq. (2.1.36), we obtain the potential function:

$$W_{\text{out}}(z) - W_{\text{in}}(z) = \mu_0 j I + \text{const}\,. \tag{2.1.37}$$

Example I: Circular cylinder

According to Fig. 2.1.14b, the cylinder with the axis perpendicular to the z-plane cuts a circle of radius r with the equation

$$Z = z = z\,(\theta) = r \cdot e^{j\theta}, \quad (\text{with} \quad r = \text{const})$$

from the plane, where θ is the angular parameter. Assuming the algebraic sum of all currents in the cylinder is zero, then we may write:

$$I\,(\theta) = I\,(\theta + 2\,\pi)$$

and thus, we can express the current in a Fourier series in the form:

$$I\,(\theta) = \sum_{n=1}^{\infty} c_n \sin\,(n\theta + \theta_n) \tag{2.1.38}$$

with c_n being real.

As $\qquad \sin\,(n\theta + \theta_n) = \dfrac{1}{2\,j}\,[e^{j\,(n\theta+\theta_n)} - e^{-j\,(n\theta+\theta_n)}]$

and $\qquad z^n = r^n \cdot e^{jn\theta}\,,$

one has: $\qquad \sin\,(n\theta + \theta_n) = \dfrac{1}{2\,j}\,[e^{j\theta_n} \cdot e^{jn\theta} - e^{-j\theta_n} \cdot e^{-jn\theta}]\,.$

Thus, from Eq. (2.1.37):

$$W_{\text{out}}(z) - W_{\text{in}}(z) = \frac{\mu_0}{2} \sum_{n=1}^{\infty} c_n \cdot \left[e^{j\theta_n}\left(\frac{z}{r}\right) - e^{-j\theta_n}\left(\frac{r}{z}\right)^n\right] + \text{const}\,.$$

In order to make $W\,(Z)$ finite for both $|Z| \to 0$ and $|Z| \to \infty$, the potentials must be:

$$\left.\begin{aligned} W_{\text{out}}(Z) &= -\frac{\mu_0}{2} \sum_{n=1}^{\infty} \left(c_n \cdot e^{-j\theta_n} \cdot \left(\frac{r}{Z}\right)^n\right) + \text{const} \\ W_{\text{in}}\,(Z) &= -\frac{\mu_0}{2} \sum_{n=1}^{\infty} \left(c_n \cdot e^{j\theta_n} \cdot \left(\frac{Z}{r}\right)^n\right) + \text{const} \end{aligned}\right\}. \tag{2.1.39}$$

The fields are obtained simply by differentiating $W(Z)$ with respect to Z:

$$\left.\begin{aligned} B_{\text{out}}(Z) &= \frac{\mu_0}{2} \sum_{n=1}^{\infty} \left(n c_n r^n \cdot e^{-j\theta_n} \cdot Z^{-(n+1)}\right) \\ B_{\text{in}}(Z) &= \frac{\mu_0}{2} \sum_{n=1}^{\infty} \left(n c_n \cdot r^{-n} \cdot e^{j\theta_n} \cdot Z^{(n-1)}\right) \end{aligned}\right\} . \tag{2.1.40}$$

For any required field configuration within the cylinder, we can obtain from $B_{\text{in}}(Z)$ the current function $I(\theta)$ in the cylinder.

We explain this by calculating a few examples:

Example II: Dipole field

The field inside the cylinder must be constant, i.e.:

$$B_{\text{in}}(Z) = B_Y + jB_X = \text{const} = B_0 .$$

This gives, from Eq. (2.1.40):

$$-\frac{\mu_0}{2} \cdot n \cdot c_n \cdot r^{-n} \cdot e^{j\theta_n} \cdot Z^{n-1} = B_0 \quad (\text{for } n = 1) ,$$

$$-\frac{\mu_0}{2} \cdot n \cdot c_n \cdot r^{-n} \cdot e^{j\theta_n} \cdot Z^{n-1} = 0 \quad (\text{for } n > 1) .$$

As $\theta_n = 0$ for $n = 1, 2, 3 \ldots,$

we obtain the constants: $c_1 = -\dfrac{2 B_0 \cdot r}{\mu_0}$, $c_n = 0$ (for $n > 1$) .

Thus the current distribution inside the cylinder (Eq. 2.1.38) is obtained from:

$$\boxed{I(\theta) = -\frac{2 B_0 \cdot r}{\mu_0} \sin(\theta) .} \tag{2.1.41}$$

Example III: Quadrupole field

The field inside the cylinder of radius r has a constant gradient. Thus:

$$B_{\text{in}}(Z) = B_Y + jB_X = G(X + jY) = GZ .$$

We obtain:

$$-\frac{\mu_0}{2} n c_n r^{-n} e^{j\theta_n} = 0 \qquad \text{for } n = 1 ,$$

$$-\frac{\mu_0}{2} n c_n r^{-n} e^{j\theta_n} = G \qquad \text{for } n = 2 ,$$

$$-\frac{\mu_0}{2} n c_n r^{-n} e^{j\theta_n} = 0 \qquad \text{for } n > 2 .$$

Again, $\theta_n = 0$ for $n = 1, 2, \ldots$

Thus: $c_1 = 0$, $c_2 = -\dfrac{1}{\mu_0} \cdot G \cdot r^2$, $c_n = 0$, (for $n > 2$) .

The current distribution is, therefore:

$$\boxed{I(\theta) = -\frac{1}{\mu_0} G r^2 \cdot \sin(2\theta) .} \tag{2.1.42}$$

Example IV: Elliptical cylinder

From

$$Z = z(\theta) = a \cos \theta + j b \sin \theta$$

the combined function coil (dipole and quadrupole), the current distribution on the circumference must be according to:

$$I(\theta) = -\frac{1}{\mu_0}\left[(a+b) B_0 \sin \theta + \frac{1}{2}(a+b)^2 \cdot G \sin 2\theta\right]. \tag{2.1.43}$$

2.1.6 Magnetic Field of Cylindrical Coils

Coils with axial symmetry generally produce magnetic fields whose constant error contours are approximately spherical [7, 8]. As for beam transport coils, more restricted regions may be of interest, for example, around the symmetry axis, on a preferred plane, or around an axis of symmetry. Eventually, the problem is to evaluate

$$\boldsymbol{B} = \frac{\mu_0}{4\pi} \int\limits_{\substack{\text{source}\\ \text{volume}}} \boldsymbol{I} \cdot \frac{d\boldsymbol{l} \times \boldsymbol{r}}{|\boldsymbol{r}|^3}, \tag{2.1.44}$$

or in differential form:

$$d\boldsymbol{B} = \frac{\mu_0}{4\pi} \cdot I \cdot \frac{d\boldsymbol{l} \times \boldsymbol{r}}{r^3} = \frac{\mu_0}{4\pi} \cdot I \cdot \frac{dl}{r^2} \sin \theta , \tag{2.1.45}$$

where θ indicates the angle between $d\boldsymbol{l}$ and $\boldsymbol{r}$; $d\boldsymbol{B}$ is normal to the plane containing the two vectors, $d\boldsymbol{l}$ and $\boldsymbol{r}$.

The analytical problem represented by Eq. (2.1.44) to obtain general field points in space is quite a task. It is time-consuming and difficult. Two approximation methods are common to reduce this difficulty:

(a) Expansion of the scalar magnetic potential in spherical harmonics [7].

(b) Use of filamentary sources of simple symmetry. Parameters are adjusted to minimize the error field over the region of interest. The magnetic fields due to each filamentary source are calculated and superposed. This method of using filamentary sources to calculate the field of arbitrarily-shaped coils led to Eq. (2.1.1).

As both methods are treated extensively in literature [8, 9], we confine our calculations to superconducting magnets only.

2.1.6.1 Use of Spherical Harmonics, Axially-Symmetric Coils

Referring to Fig. 2.1.15a, which illustrates an axially-symmetric coil of symmetric cross section, we select two current loops. The axial magnetic field of such a loop arrangement is given by:

$$B_z(R,\theta) = \mu_0 \lambda J \cdot \frac{\sin^2\alpha}{c} \sum_{m=0}^{\infty} \left(\frac{R}{c}\right)^{2m} \cdot P'_{2m+1}(\cos\alpha) \cdot P_{2m}(\cos\theta) \quad (2.1.46)$$

$$(R < c)$$

$P_{2m}(\cos\theta)$ are Légendre polynomials.

The axial field at P, generated by the entire coil, is obtained by integration of Eq. (2.1.46) over the boundaries of the coil area:

$$\boxed{B_z = \mu_0 \sum_{m=0}^{\infty} \frac{\sin^2\alpha}{c^{2m+1}} \cdot R^{2m} \cdot P'_{2m+1}(\cos\alpha) \cdot \lambda J\,(c,\alpha) \cdot P_{2m}(\cos\theta)\, da\,.} \quad (2.1.47)$$

Specifically, the axial field component on the symmetry axis of a pair of axially-symmetric current loops is given by:

$$B_z(0,z) = \mu_0 \lambda J \frac{\sin^2\alpha}{c} \sum_{m=0}^{\infty} \left(\frac{z}{c}\right)^{2m} P'_{2m+1}(\cos\alpha), \quad \text{for } (R < c)\,. \quad (2.1.48)$$

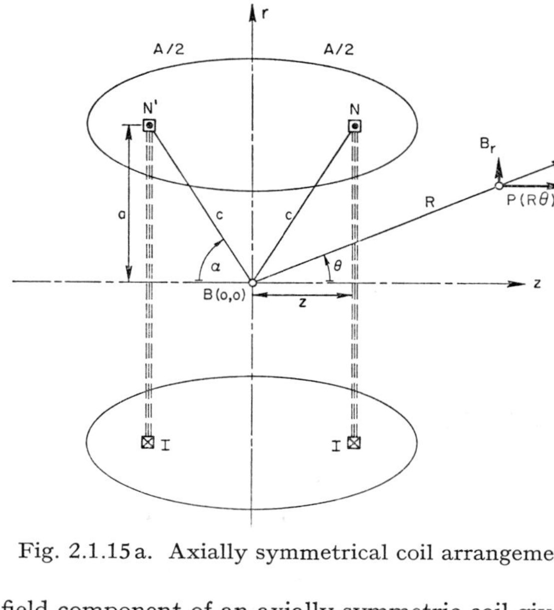

Fig. 2.1.15 a. Axially symmetrical coil arrangement

The axial field component of an axially-symmetric coil gives, accordingly (Fig. 2.1.15b):

$$B_z(0,z) = \mu_0 \int_{-b}^{+b} dz \int_{a_1}^{a_2} \left(\frac{\sin^2\theta}{c}\right) \cdot \lambda J\,(r,z) \sum_{m+0}^{\infty} \left(\frac{z}{c}\right)^{2m} P'_{2m+1}(\cos\theta)\, da\,. \quad (2.1.49)$$

Assuming constant overall current density in the coil, the field at the center of the coil system ($z = 0$) is obtained from Eq. (2.1.49):

$$B_z(0,0) = \mu_0 \lambda J \int_{-b}^{+b} dz \int_{a_1}^{a_2} \frac{\sin^2\theta}{c}\, da\,, \tag{2.1.50}$$

$$B_z(0,0) = \mu_0 \lambda J \int_{-b}^{+b} dz \int_{a_1}^{a_2} \frac{a^2}{(a^2 + s^2)^{\frac{3}{2}}}\, da\,. \tag{2.1.51}$$

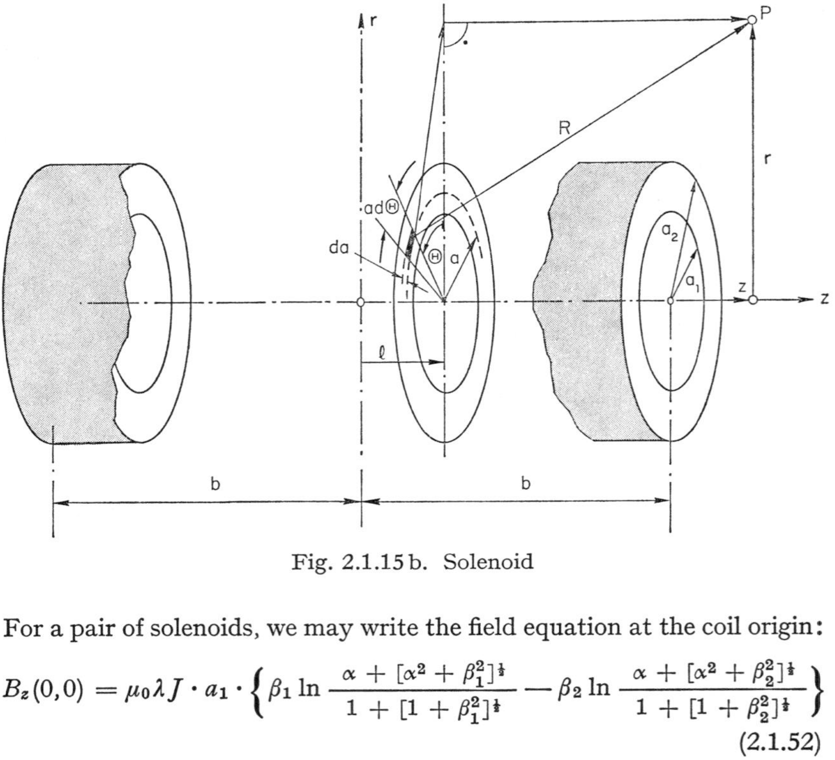

Fig. 2.1.15 b. Solenoid

For a pair of solenoids, we may write the field equation at the coil origin:

$$B_z(0,0) = \mu_0 \lambda J \cdot a_1 \cdot \left\{ \beta_1 \ln \frac{\alpha + [\alpha^2 + \beta_1^2]^{\frac{1}{2}}}{1 + [1 + \beta_1^2]^{\frac{1}{2}}} - \beta_2 \ln \frac{\alpha + [\alpha^2 + \beta_2^2]^{\frac{1}{2}}}{1 + [1 + \beta_2^2]^{\frac{1}{2}}} \right\} \tag{2.1.52}$$

with $\qquad \beta_1 = \dfrac{2\,b + g}{a_1}, \quad \beta_2 = \dfrac{g}{a_1},$

which is written in abbreviated form for the axial field at the origin:

$$B_z(0,0) = B_1 - B_2 = \mu_0 \lambda J a_1 [F_1(0) - F_2(0)] = \mu_0 \lambda J a_1 F(0)\,.$$

The axial field component at any point in space is given by:

$$B_z(0,\zeta) = \mu_0 \frac{\lambda J a_1}{2} \cdot [F_1(0,\zeta) - F_2(0,\zeta)] = \mu_0 \frac{\lambda J a_1}{2} \cdot F(\zeta) \tag{2.1.53}$$

with:

$$F(\zeta) = (\beta_j + \zeta) \ln \frac{\alpha + [\alpha^2 + (\beta_j + \zeta)^2]^{\frac{1}{2}}}{1 + [1 + (\beta_j + \zeta)^2]^{\frac{1}{2}}} + (\beta_j - \zeta) \ln \frac{\alpha + [\alpha^2 + (\beta_j - \zeta)^2]^{\frac{1}{2}}}{1 + [1 + (\beta_j - \zeta)^2]^{\frac{1}{2}}}$$

and $\zeta = z/a_1$, as the reduced axial distance, and a_1 the inner coil radius.

The axial field can be expanded in the reduced form about an origin on the axis of the plane of symmetry. All odd terms of the expansion vanish, due to symmetry, and we may write for the two field components:

$$B_z(R, \theta) = \sum_{n=0}^{\infty} \frac{1}{(2n)!} \cdot B_z^{(2n)}(0,0)\, R^{2n} P_{2n}(\cos\theta)\,,$$

$$B_r(R, \theta) = \sum_{n=0}^{\infty} \frac{1}{(2n)!} \cdot B_z^{(2n)}(0,0)\, R^{2n} P'_{2n}(\cos\theta)\,. \qquad (2.1.54)$$

Comparing Eq. (2.1.53) and (2.1.54), we may define a homogeneity factor:

$$\varkappa = \frac{F(\xi, \zeta)}{F(0)} = \sum_{n=0}^{\infty} \varepsilon_{2n} \left(\frac{r}{a_1}\right)^{2n} \cdot P_{2n}(\cos\theta) \qquad (2.1.55)$$

with $\left(\frac{r}{a_1}\right)^{2n}$ as a reduced radius to a field point, and $\xi = r/a_1$, the reduced radial distance. ε_{2n} is the 2-nth error coefficient of the system. $\varkappa_{\max}$ is obtained from:

$$\varkappa_{\max} = \frac{1}{B_0} \left[(B_z(a,z))^2 + (B_r(a,z))^2\right]^{1/2}_{\max}.$$

The error coefficients are defined as:

$$(\varepsilon_{2n})_j = \frac{1}{(2n)!\, F_j(0)} \frac{\partial^{(2n)} [F_j(\alpha, \beta_j, \zeta)]}{\partial [\zeta^{(2n)}]} \qquad (2.1.56)$$

with $j = 1$, corresponding to the reduced coil system, and $j = 2$ to the gap (Eq. 2.1.52).

For a pair of solenoids, we obtain:

$$(\varepsilon_{2n})_j = \frac{F_1(0) \cdot (\varepsilon_{2n})_1 - F_2(0) \cdot (\varepsilon_{2n})_2}{F_1(0) - F_2(0)}, \qquad (2.1.57)$$

with: (ε_{2n}) = second-order error coefficient for the system,

$(\varepsilon_{2n})_1$ = second-order error coefficient for the coil of length $(4b + 2g)$,

$(\varepsilon_{2n})_2$ = second-order error coefficient for the gap of axial length $(2g)$.

The fie'd homogeneity along the symmetry axis and in the plane of symmetry is expressed by the relations:

$$\eta_\zeta = \varkappa - 1 = \sum_{n=1}^{\infty} \varepsilon_{2n} \cdot \zeta^{2n} \tag{2.1.58}$$

and

$$\eta_\zeta = \sum_{n=1}^{\infty} (-1)^n \frac{1}{2n} \varepsilon_{2n} \cdot \zeta^{2n}. \tag{2.1.59}$$

By using this simple method, coils up to 8th order can be calculated.

The volume of the coil system is given by: $V = a_1^3 \cdot v$, with the reduced volume of the system written as: $v = 2\pi(\beta_1 - \beta_2)(\alpha^2 - 1)$. The constraints of volume minimization for a certain central field and a required field homogeneity are obtained from the relation: $\partial F(0)/\partial\alpha = 0$ for fixed β_2 and v. β_2 is either specified by the experimental requirements or by specifying the axial homogeneity over a portion of the gap.

As a basic variable for all homogeneity calculations, we used the ratio:

$$D = \frac{B_0 \cdot 10^{-2}}{\mu_0 \lambda J \cdot g} = \frac{F(0)}{\beta_2}. \tag{2.1.60}$$

Most desired quantities given in Figs. 2.1.16 to 2.1.21 are plotted against D for convenience.

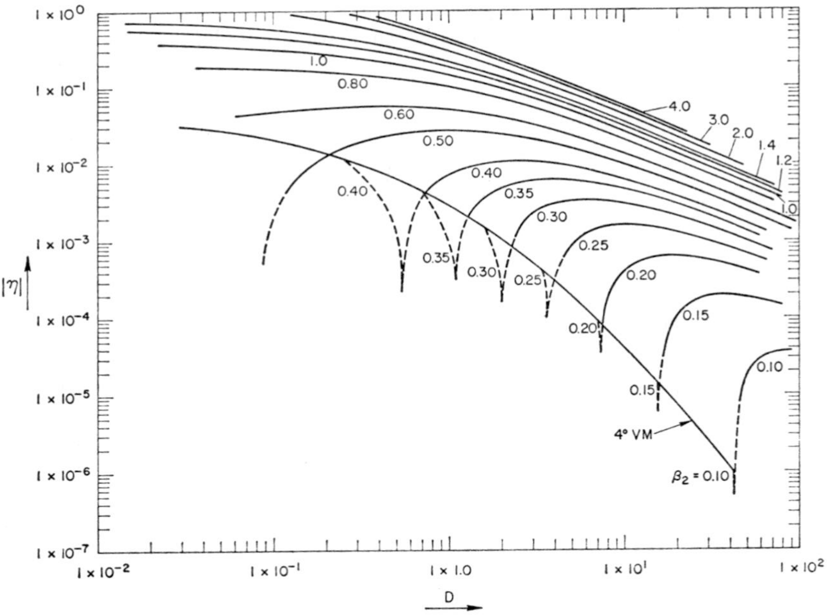

Fig. 2.1.16.

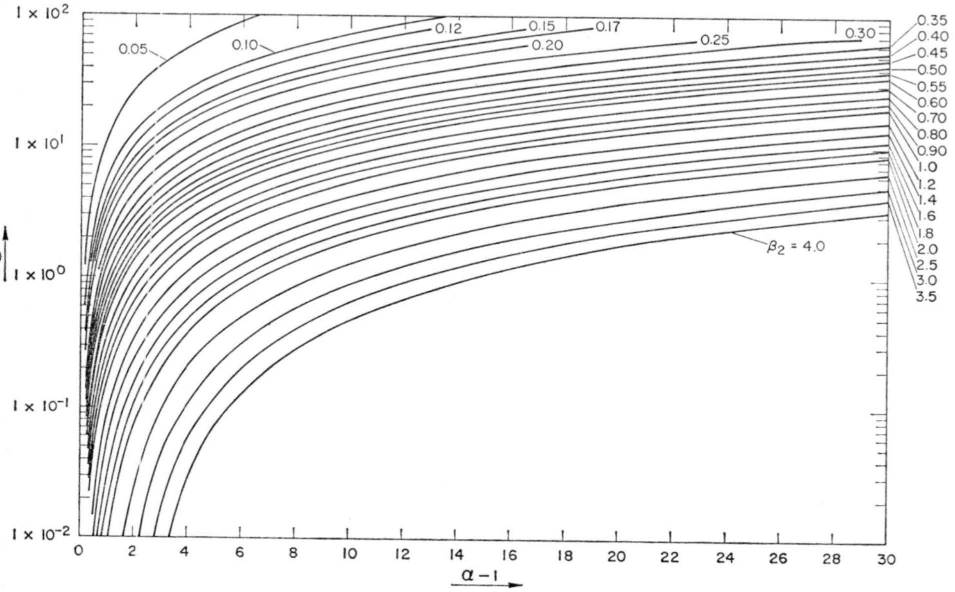

Fig. 2.1.17.

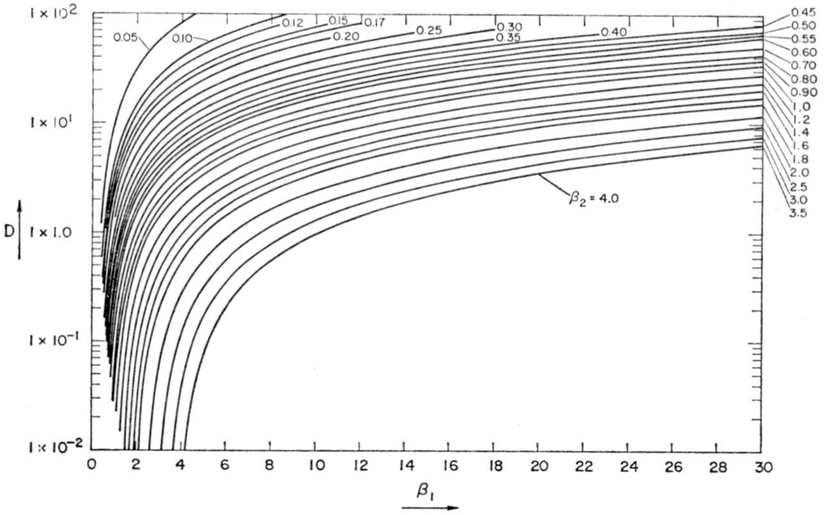

Fig. 2.1.18.

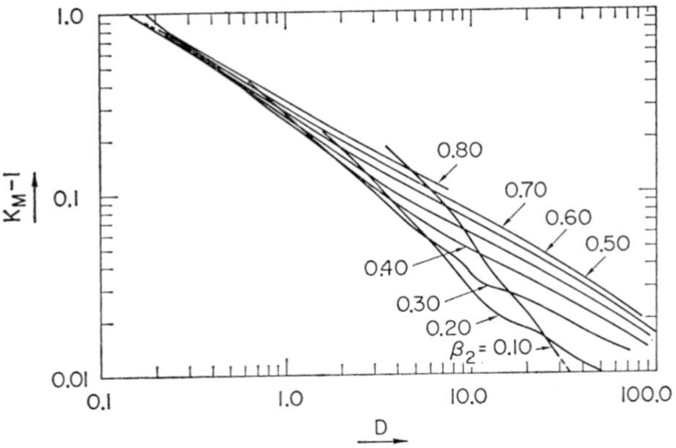

Fig. 2.1.19.

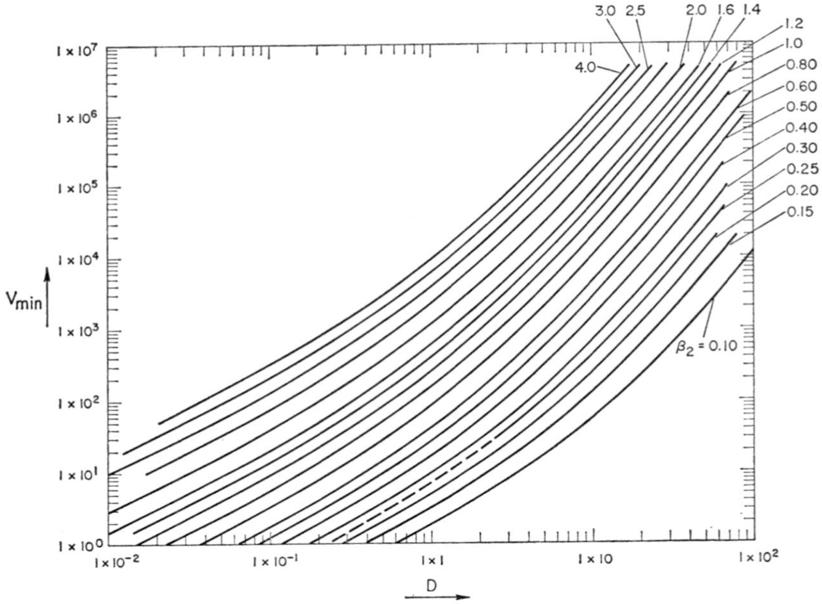

Fig. 2.1.20.

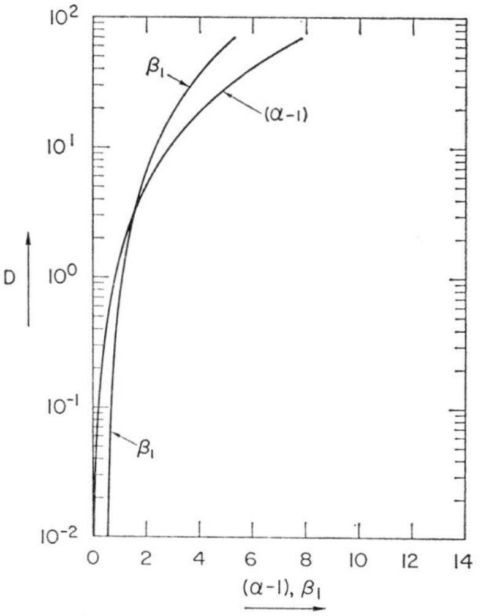

Fig. 2.1.21. D vs. $(\alpha - 1)$ and β_1 for a 4th order corrected solenoid

Examples

(*a*) *Pair of Concentric Coils*

We calculate the field homogeneity for a pair of coils with the following dimensions: $a_1 = 6.17 \times 10^{-2}$ m, $2b = 7.3612 \times 10^{-2}$ m, $a_2 = 1.383 \times 10^{-1}$ m, $g = 5.286 \times 10^{-2}$ m.

The axial field component was expressed for the plane of symmetry in terms of zonal harmonics of even power:

$$B_z = B_z(0,0) \cdot \left[1 + \varepsilon_2 \left(\frac{r}{r_0}\right)^2 \cdot P_2(\cos\theta) + \varepsilon_4 \left(\frac{r}{r_0}\right)^4 \cdot P_4(\cos\theta) + \cdots\right]. \tag{2.1.61}$$

The two error coefficients are expressed, according to Garrett [9], by:

$$\varepsilon_2 = \frac{r_0^2}{2} \frac{\sum_{n=1}^{4} \frac{1}{z_n}(1 - \sin^3\alpha_n)}{\sum_{n=1}^{4} z_n \cdot \ln[a_n + \sqrt{a_n^2 + Z_n^2}]}; \tag{2.1.62}$$

$$\varepsilon_4 = \frac{r_0^4}{4!} \frac{\sum_{n=1}^{4} z_n^{1/3} \cdot \{2 - \sin^3\alpha_n [15\cos^4\alpha_n + 3\cos^2\alpha_n + 2]\}}{\sum_{n=1}^{4} z_n \cdot \ln\left[a_n + \sqrt{a_n^2 + z_n^2}\right]}. \tag{2.1.63}$$

The summation must be taken over the four corners with alternate signs, giving positive sign to the corner nearest origin (Fig. 2.1.22).

With the data given for this coil, we get:

$$B_z = B_z(0,0)\left[1 + 4.445\times10^{-3}\left(\frac{r}{r_0}\right)^2 P_2(\cos\theta) - 1.05758\left(\frac{r}{r_0}\right)^4 P_4(\cos\theta)\ldots\right].$$

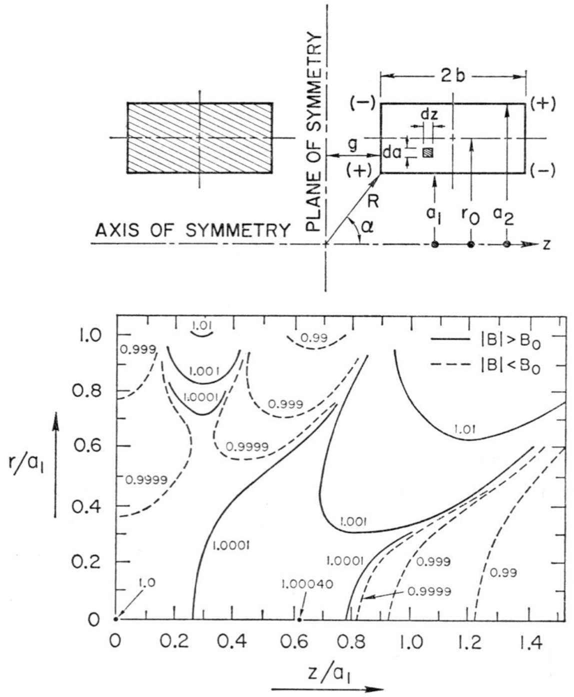

Fig. 2.1.22. Lines of constant field value within the bore of a split coil arrangement

Table 2.1 gives field uniformity in parts per million of the central field.

Table 2.1. *Field uniformity for a split-coil arrangement according to Fig. 2.1.22*

	Distance	$\theta = 0°$	15°	30°	45°	60°	75°	90°
$\frac{r}{r_0} =$	2.5×10^{-2}	+2.4	+2.2	+1.7	+0.9	—0.2	—1.2	—1.5
	5×10^{-2}	+4.5	+5.5	+6.8	+5.5	+0.5	—5.4	—8.0
	7.5×10^{-2}	—8.4	—0.4	+15	+20	+65	—15	—25
	10×10^{-2}	—61	—32	+25	+54	+75	—33	—62
	15×10^{-2}	—435	—277	+50	+243	+142	—117	—251

(b) Single Sixth-Order Coil

We design a coil having a central field of 5 T, with an inner radius of 2.5×10^{-2} m. The overall current density in a superconducting composite conductor at this field is $\sim 2.6 \times 10^8$ A/m². The overall current density in the coil is approximately 70% of this value, or

$$\lambda J = 1.84 \times 10^8 \text{ A/m}^2$$

with the notations:

$$K = \frac{B\,(0,0)}{\lambda j} \times 10^8 = \frac{5}{1.84 \times 10^8} \times 10^8 = 2.715\,,$$

$$A = \frac{a_1 \times 10^2}{K} = \frac{2.5}{2.75} = 0.92\,.$$

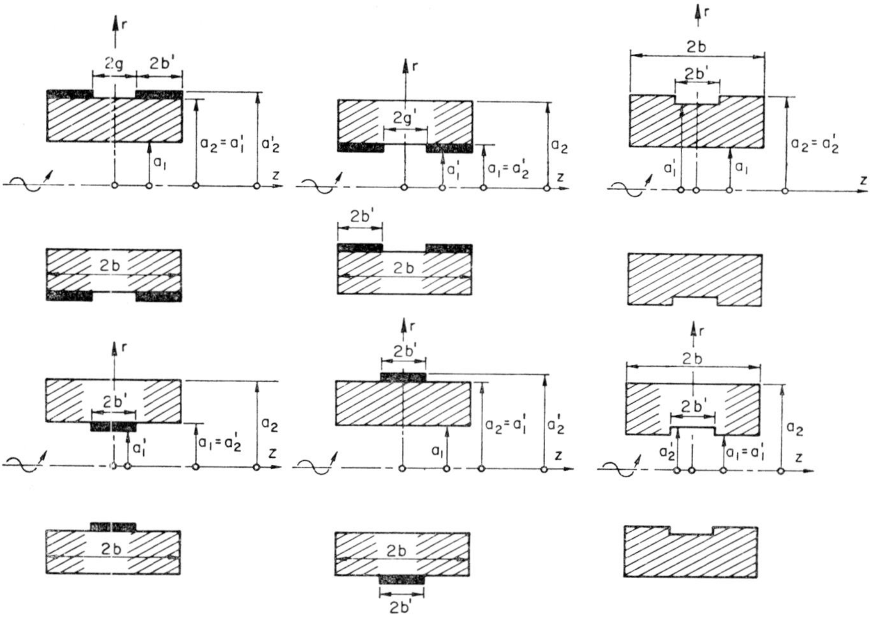

Fig. 2.1.23. Solenoid geometries to improve the field homogeneity within the useful part of the bore

Referring to Fig. 2.1.23, where we provide two additional coils at the outer coil periphery and use the notation $(a_1', a_2', \alpha', \beta)$ for the correcting coil, we may write for the axial field component in a field point $P\,(r, \theta)$:

$$B_z(r, \theta) = \mu_0 \lambda J a_2 \cdot \varepsilon_0 \left[1 + \frac{\varepsilon_6}{\varepsilon_0}\left(\frac{z}{a_2}\right)^6 + \frac{\varepsilon_8}{\varepsilon_0}\left(\frac{z}{a_2}\right)^8 + \cdots\right].$$

Note that by using proper dimensions for the auxiliary coils, which have the same current density as the main coil, the second and fourth error coefficients could be eliminated. A detailed treatment is given by

Girard and Sauzade [10]. All major parameters for the sixth-order coil are given in Fig. 2.1.24, where we obtain all coil dimensions for $A = 0.92$:

$$\begin{aligned} \beta &= 4.57 \Rightarrow b = 11.42\times 10^{-2}\ \mathrm{m} \\ \alpha &= 2.2 \Rightarrow a_2 = 5.25\times 10^{-2}\ \mathrm{m} \\ \alpha' &= 1.136 \Rightarrow a_1' = 4.62\times 10^{-2}\ \mathrm{m} \\ \beta' &= 1.552 \Rightarrow b_1' = 7.18\times 10^{-2}\ \mathrm{m} \end{aligned}$$

$$\frac{\varepsilon_6}{\varepsilon_0} = -9.7\times 10^{-4}, \qquad \frac{\varepsilon_8}{\varepsilon_0} = -5.5\times 10^{-4}.$$

Thus:

$$B_z(r,\theta) = 5\,(1 - 1.13\cdot 10^{-5} z^6 + 2.321\cdot 10^{-7} z^8 - \cdots)$$

with z-dimensions in (m)!

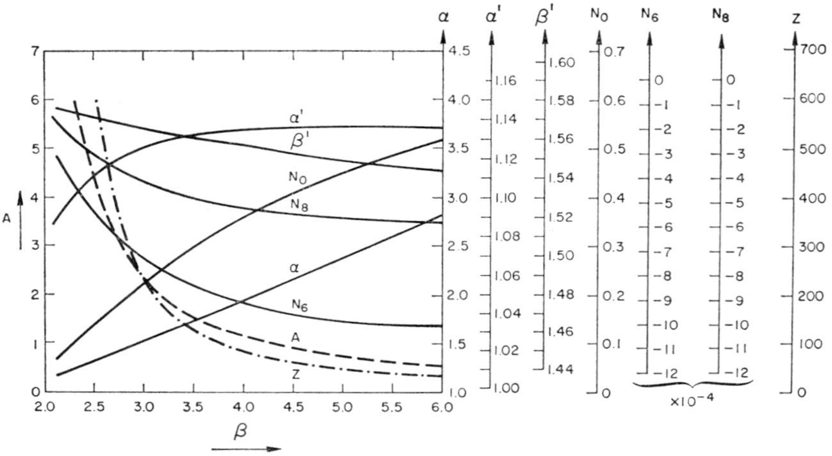

Fig. 2.1.24. To calculate the geometry of a sixth order coil. For known A and β, the dimensions of the solenoid with endcorrections according to Fig. 2.1.23 are determined

(c) *Fourth-Order Volume Minimized Coil*

To illustrate the use of the curves in Figs. 2.1.16 to 2.1.21, the field homogeneity of a coil with the following specifications is calculated:

$B\,(0,0) = 6$ T, central field, $\lambda J = 10^8$ A/m² overall current density

which is adequate for a Nb_xTi and copper composite conductor:

$$g = 5\times 10^{-2}\ \mathrm{m} \quad \text{(half axial gap width).}$$

According to Eq. (2.1.60), we obtain $D = 0.956$. Referring to Fig. 2.1.16, we get, for a fourth-order coil, $\beta_2 = 0.335$, which gives $a_1 = 15\times 10^{-2}$ m. The field homogeneity for $\zeta = \beta_2$ is, according to Fig. 2.1.16, equal to $|\eta| = 3.1\times 10^{-3}$; for $\zeta = 0.5\,\beta_2$ (Fig. 2.1.16), yields $|\eta| = 2\times 10^{-4}$.

The coi outer diameter is obtained from Fig. 2.1.17, with $\alpha = 1.8$, $a_2 = 27\times 10^{-2}$ m. The axial length of one coil section is given by Fig. 2.1.18, with $\beta_1 = 1.1$ leading to $2\,b = (\beta_1 - \beta_2)\,a_1 = 10\times 10^{-2}$ m.

The maximum field at the conductor is obtained from Fig. 2.1.19, with $K_m = 1.28$ to be $B_{\max} = 7.74$ T. The minimum reduced volume is given in Fig. 2.1.20: $v_{\min} = 12$, which gives: $V_{\min} = 12 \times (15 \times 10^{-2})^3 \cong 4.05 \times 10^{-2}$ m^3.

Finally, we have illustrated in Fig. 2.1.21, as a check to the above curves, the $(\alpha - 1)$ and (β_1) curves vs. D for a fourth-order split-coil arrangement. We obtain $\alpha = 1.8$; $\beta_1 = 1.1$, as above.

2.1.7 Magnetic Field of Non-Cylindrical Coils

For many applications, axially-symmetric coils are unsuitable, e.g., beam transport dipole and quadrupole coils, wire- and spark-chamber magnets. Solenoids would have larger diameters and certain regions not being used. Rectangularly-shaped dipole coil configurations shown in Fig. 2.1.25 with straight, circular, and bent heads, are commonly used. Field calculations for this type of coil are given by Harris [14], Lyddane [15] and Grant [16]. A computer code, MAFCO [17], is developed for general three-dimensional current elements. Only field components along the axis are given below.

Straight-head coil of filamentary conductor (Fig. 2.1.25)

$$B_y = \frac{\mu_0 I}{\pi} \frac{\sin \varphi}{R} \left[\frac{c}{\sqrt{1 + c^2}} \left(1 + \frac{1}{c^2 + \cos^2 \varphi}\right) + \frac{d}{\sqrt{1 + d^2}} \cdot \left(1 + \frac{1}{d^2 + \cos^2 \varphi}\right)\right]. \tag{2.1.64}$$

Circular-head coil (Fig. 2.1.26)

$$B_y = \frac{\mu_0 I}{\pi} \cdot \frac{\sin \varphi_1}{R} \cdot \left[\frac{c}{\sqrt{1 + c^2}} \left(\frac{2 + c^2}{1 + c^2}\right) + \frac{d}{\sqrt{1 + d^2}} \left(\frac{2 + d^2}{1 + d^2}\right)\right]. \tag{2.1.65}$$

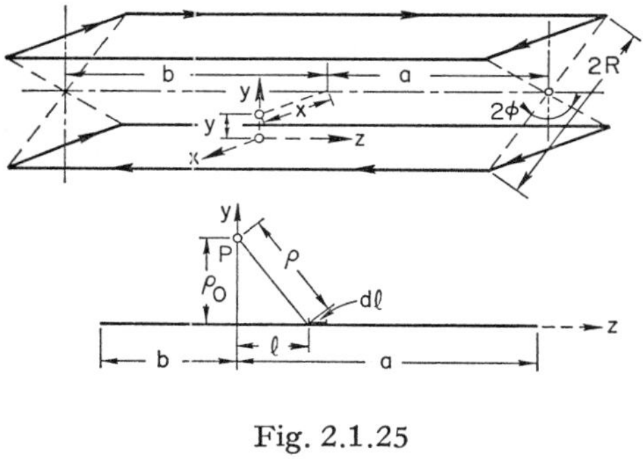

Fig. 2.1.25

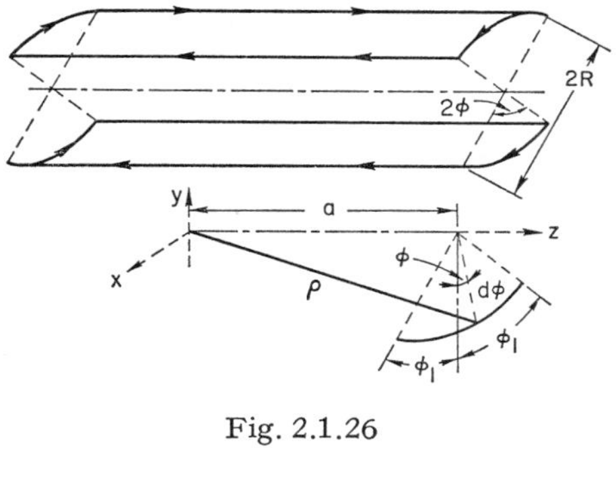

Fig. 2.1.26

Fig. 2.1.25. Rectangular coil geometries
Fig. 2.1.26. Coil geometries with circular end sections

Bent-head coil (Fig. 2.1.27)

$$B_y = \frac{\mu_0 I}{\pi}$$
$$\cdot \frac{\sin\varphi}{R}\left\{\frac{c}{\sqrt{1+c^2}}\cdot\left[1+\frac{1}{\sqrt{1+c^2}\sqrt{(1+\varepsilon)^2+c^2+\sin^2\varphi}\left(1+\frac{2\varepsilon+\varepsilon^2}{1+c^2}\right)}\right]\right.$$
$$\left.+\frac{d}{\sqrt{1+d^2}}\left[1+\frac{1}{\sqrt{1+d^2}\sqrt{(1+\varepsilon)^2+d^2+\sin^2\varphi}\left(1+\frac{2\varepsilon+\varepsilon^2}{1+c^2}\right)}\right]\right\} \quad (2.1.66)$$

with: $c = \frac{a}{R}, \quad d = \frac{b}{R}, \quad \varepsilon = \frac{C}{R} - 1.$

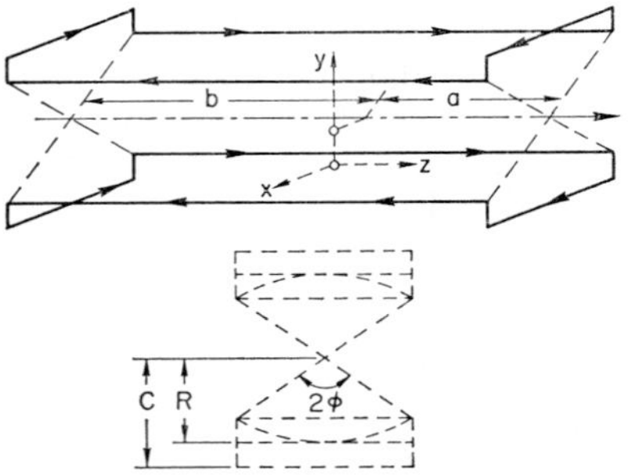

Fig. 2.1.27. Coil geometrie with bend end sections

For an infinitely long coil system, shown in Fig. 2.1.28, Lyddane and Ruark [16] obtain the following equations:

$$B_y = \frac{2\mu_0 I}{\pi}\cdot\frac{A}{\varrho_0^2}\left\{1+\frac{f}{3}\frac{a^2-b^2}{\varrho_0^2}+\frac{g}{15}\frac{3a^4-10a^2b^2+3b^2}{\varrho_0^4}+\cdots\right.$$
$$\left.+\left(\frac{r}{\varrho_0}\right)^2\cos 2\theta\left[\frac{f+2g(a^2-b^2)}{\varrho_0^2}+\cdots\right]+\left(\frac{r}{\varrho_0}\right)^4\cos 4\theta\,[g+\cdots]\right\}; \quad (2.1.67)$$

$$B_x = \frac{2\mu_0 I}{\pi}\cdot\frac{A}{\varrho_0^2}\cdot\left\{\left(\frac{r}{\varrho_0}\right)^2\sin 2\theta f - 2\left(\frac{a^2-b^2}{\varrho_0^2}\right)+\cdots\right.$$
$$\left.+\left(\frac{r}{\varrho_0}\right)^4\cos 4\theta\,[g+\cdots]+\cdots\right\} \quad (2.1.68)$$

with: $f = A^2 - 3B^2/\varrho_0^2 = \cos^2\varphi - 3\sin^2\varphi\,,$

$$g = A^4 - 10A^2B^2 + 5B^4/\varrho_0^4 = \cos^4\varphi - 10\cos^2\varphi\sin^2\varphi + 5\sin^4\varphi\,.$$

The best field uniformity is obtained by making $f = 0$, which means $A^2 - 3B^2 = 0$, or $\varphi = 30°$. For this value, $g = -1$.

The field components become:

$$B_y \cong \frac{1.5\,\mu_0 I}{\pi}\cdot\frac{1}{A}\left[1-\left(\frac{r}{\varrho_0}\right)^4\cos 4\theta\right]; \qquad B_x \cong \frac{1.5\,\mu_0 I}{\pi}\cdot\left(\frac{1}{A}\frac{r}{\varrho_0}\right)^4\sin 4\theta\,. \quad (2.1.69)$$

The neglected terms in Eq. (2.1.67) and (2.1.68) contain powers of $\left(\frac{a}{\varrho_0}\right)^4$ and $\left(\frac{b}{\varrho_0}\right)^4$. Both equations are adequate for homogeneity calculations better than 1%. To obtain higher accuracy, Grant and Strandberg [17] have developed a computer program obtaining fields for rectangular-shaped coils, as illustrated in Fig. 2.1.29a, error contours obtained by Grant are given in Fig. 2.1.29b.

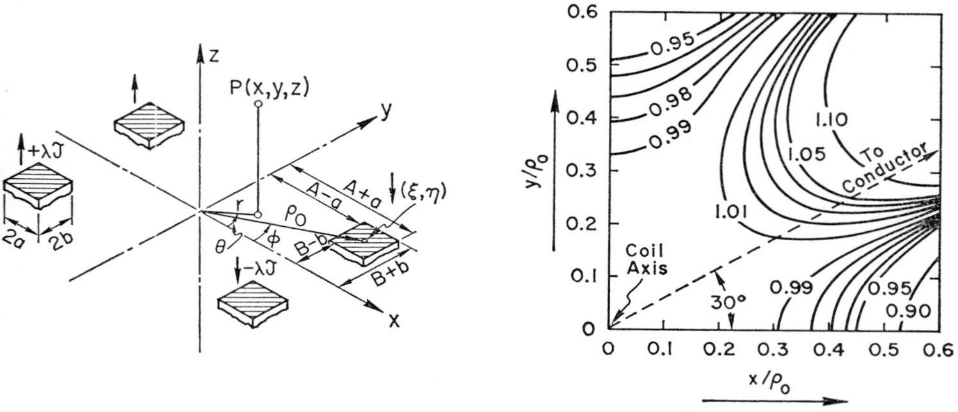

a b

Fig. 2.1.28. Field distribution due to four rectangular parallel current bars

2.1.8 Fields Produced by Means of Distributed Parallel Conductors

In this section, the equation of radial and angular field components of multipole fields are derived, where the magnetic field is generated by a chosen distribution of parallel conductors.

The calculation procedure is to express the vector potential as a function of a single current-carrying conductor, located in each of a $2p$ region over an angle α. We treat two cases: Current density in the conductor is assumed to be constant and its extension over each region is such that a desired multipole is obtained, or the current density distribution is sinusoidal and its region extends over $\pi/2p$.

2.1.8.1 Multipole Coils with Circular Aperture [18]

We consider a coil section ($a\,da\,d\varphi$) having a uniform current density λJ. The magnetic field, due to this current element (Fig. 2.1.30), located at the point $P(a,\varphi)$ generates a field $dB(r,\theta)$ at a field point $Q(r,\theta)$.

The magnetic field at $Q(r,\theta)$ is written in the usual form:

$$dB(r,\theta) = \mu_0 \frac{\lambda J}{2\pi} \cdot \frac{a\,da\,d\varphi}{R} \tag{2.1.70}$$

with: $$R^2 = a^2 + r^2 - 2ar\cos(\varphi - \theta).$$

a

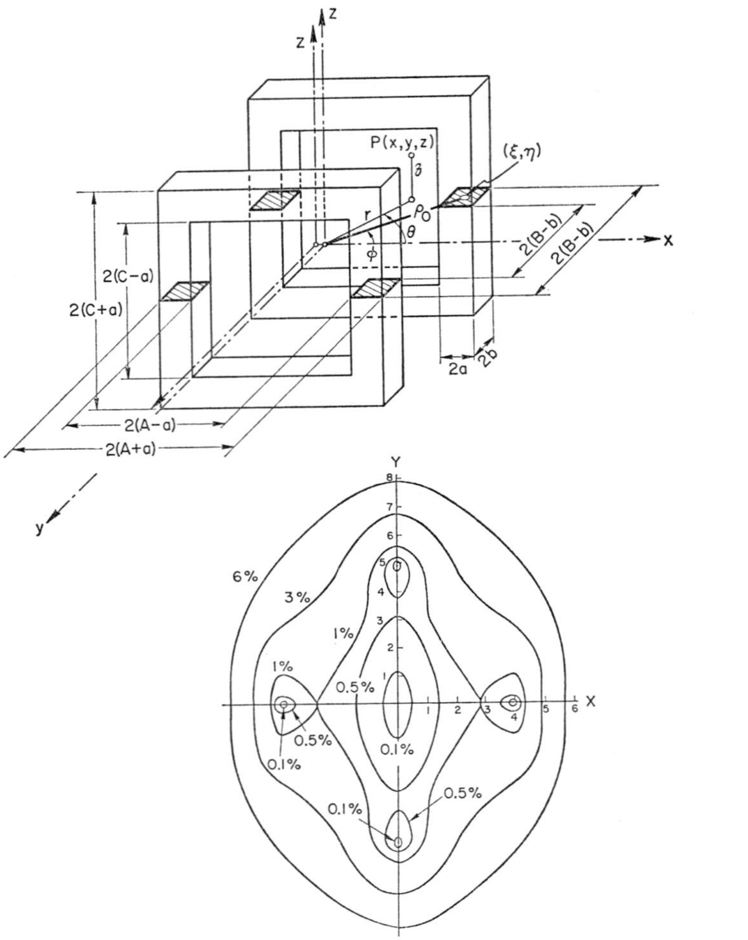

b

Fig. 2.1.29a. Finite size rectangular coil arrangement

Fig. 2.1.29b. Lines of constant field due to the coil arrangement of Fig. 2.1.29a

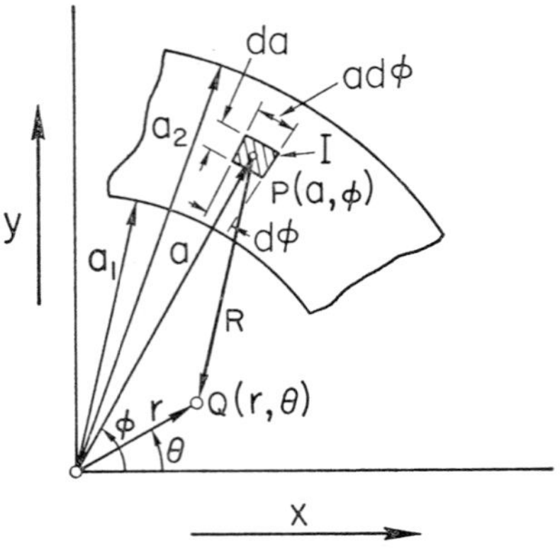

Fig. 2.1.30. Section of an axially symmetric coil

For $r < a$: $\quad R^2 = a^2 \left[1 - \left(\frac{r}{a}\right) e^{j(\varphi-\theta)}\right] \left[1 - \left(\frac{r}{a}\right) e^{-j(\varphi-\theta)}\right]$,

$$\ln R = \ln a + \tfrac{1}{2} \ln \left[1 - \left(\frac{r}{a}\right) d^{j(\varphi-\theta)}\right] + \tfrac{1}{2} \ln \left[1 - \left(\frac{r}{a}\right) e^{-j(\varphi-\theta)}\right]. \qquad (2.1.71)$$

But since

$$\ln(1 - z) = -\left[z + \frac{z^2}{2} + \frac{z^3}{3} + \frac{z^4}{4} + \dots\right] = -\sum_{n=1}^{\infty} \frac{z^n}{n}$$

for $z^2 < 1$, we obtain

$$\ln R = \ln a - \sum_{n=1}^{n} \frac{1}{n} \left(\frac{r}{a}\right)^n \cos\left[n(\varphi - \theta)\right].$$

The magnetic vector potential $\boldsymbol{A}$ for a filamentary conductor is written as:

$$\boldsymbol{A} = -\frac{\mu_0 I}{2\pi} (\ln R) \cdot \hat{z}_0, \quad \text{which yields:}$$

$$A_z(r, \theta) = \frac{\mu_0 I}{2\pi} \sum_{n} \frac{1}{n} \left(\frac{r}{a}\right)^n \left[\cos n(\varphi - \theta)\right]. \qquad (2.1.72)$$

The constant term ($\ln a$) is eliminated since it is the gradient of a scalar function.

The field at a point $Q(r, \theta)$, due to a symmetric p-pole magnet (Fig. 2.1.31) is obtained by subdividing the $(x - y)$ plane into p regions, each extending over an angle $\pi/2p$. The vector potential at the point $Q(r, \theta)$, due to conductors at the locations:

$P(a, \varphi)$, $P(a, \pi/p - \varphi)$, $P(a, \pi/p + \varphi)$, $P(a, 2(\pi/p) - \varphi)$, $P(a, 2(\pi/p) + \varphi) \dots P(a, p(\pi/p) - \varphi)$,

each carrying the currents $+I, -I, -I, +I, \dots +I$, is obtained from

$$A_z(r, \theta) = A_z(r, \varphi) + A_z\left(r, \frac{\pi}{p} - \varphi\right) + A_z\left(r, \frac{\pi}{p} + \varphi\right)$$
$$+ A_z\left(r, 2\left(\frac{\pi}{p}\right) - \varphi\right) + \cdots A_z\left(r, p\left(\frac{\pi}{p}\right) - \varphi\right). \qquad (2.1.73)$$

Substituting Eq. (2.1.73) into Eq. (2.1.72) yields:

$$A_z(r, \theta) = \frac{\mu_0 I}{2\pi} \sum_{n=1}^{n} \frac{1}{n} \left(\frac{r}{a}\right)^n \sum_{m=0}^{p} (-1)^m \left\{\cos\left(n\left[m\left(\frac{\pi}{p}\right) - \theta - \varphi\right]\right)\right.$$
$$\left. + \cos\left(n\left[m\left(\frac{\pi}{p}\right) - \theta + \varphi\right]\right)\right\}, \qquad (2.1.74)$$

$$A_z(r, \theta) = \frac{\mu_0 I}{\pi} \sum_{n=1}^{n} \frac{1}{n} \left(\frac{r}{a}\right)^n \cos(n\varphi) \sum_{m=1}^{p} (-1)^m \left[\cos\left(n\frac{m\pi}{p}\right) \cos(n\theta)\right.$$
$$\left. + \sin\left(n\frac{m\pi}{p}\right) \sin(n\theta)\right]. \qquad (2.1.75)$$

For $n = kp$, with k being an integer, Eq. (2.1.74) is reduced to:

$$\boxed{A_z(r,\theta) = \frac{\mu_0 I}{\pi} \sum_{n=1}^{n} \frac{1}{n} \left(\frac{r}{a}\right)^n \cdot p \cdot \cos(n\varphi)\cos(n\theta)\,.} \tag{2.1.76}$$

To get the vector potential due to a coil, we integrate $A_z(r, \theta)$ over φ for $0 \leqq \varphi \leqq \pi/p$, and over the radius a from a_1 to a_2.

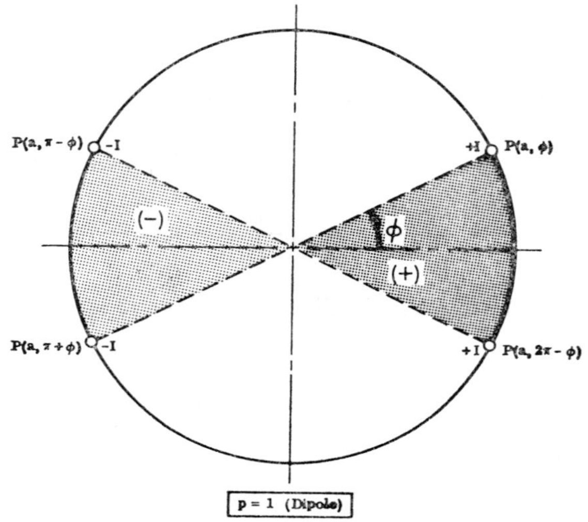

a

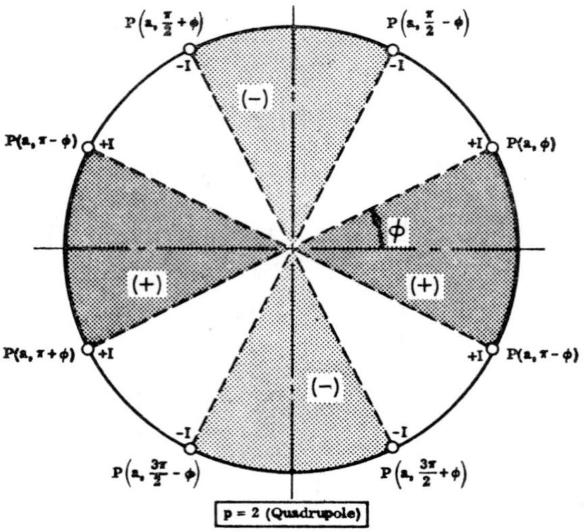

b

$$A_z(r,\theta) = \frac{\mu_0}{\pi} \sum_n \frac{p \cdot r^n \cos(n\theta)}{n} \int_{a_1}^{a_2} \int_0^{\pi/2p} \frac{\lambda J \cos(n\theta)}{a^n} \cdot a\, da\, d\varphi \qquad (2.1.76)$$

where we introduced the overall current density $\lambda J = I/(a\, da\, d\varphi)$.

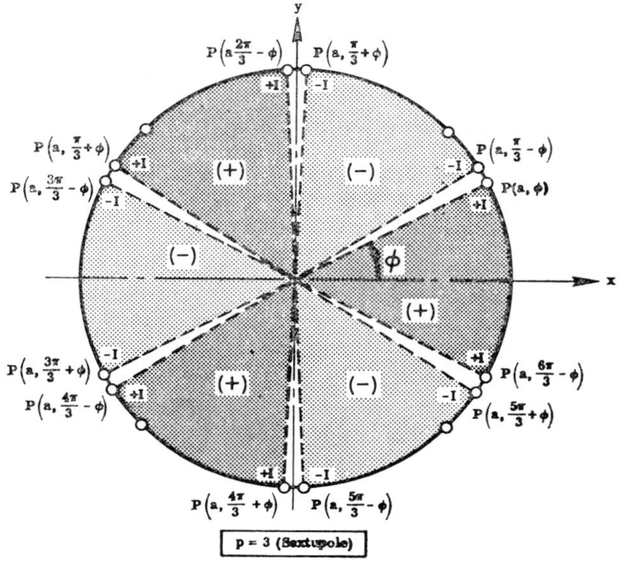

c

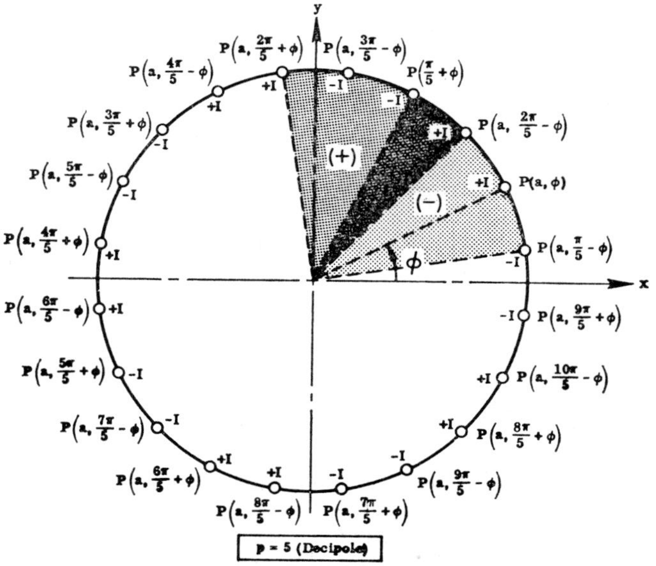

d

Fig. 2.1.31. Arrangement of current lines to generate multipoles

Case I: Current density over the coil cross-section is constant: $\lambda J = (\lambda J)_0$. From Eq. (2.1.76), we have:

$$A_z(r,\theta) = \frac{\mu_0}{\pi}(\lambda J)_0 \sum_n \frac{p \cdot r^n \cos(n\theta)}{n} \int_{a_1}^{a_2}\int_0^{\pi/2p} \frac{\cos(n\theta)}{a^{n-1}}\, d\theta\, da$$

$$= \frac{\mu_0}{\pi}(\lambda J)_0 \sum_n \frac{n \cdot r^n \cos(n\theta)}{n} \int_{a_1}^{a_2} \frac{da}{a^{n-1}}.$$

For $n = (2k+1)p$, with $k = 0, 1, 2, \ldots$

$$A_z(r,\theta) = \frac{\mu_0}{\pi}(\lambda J) \sum_{k=0}^{k} \frac{p \cdot r^{(2k+1)p} \cdot \cos[(2k+1)p \cdot \theta]}{(2k+1)p} \int_{a_1}^{a_2} \frac{da}{a^{(2k+1)p-1}}. \quad (2.1.77)$$

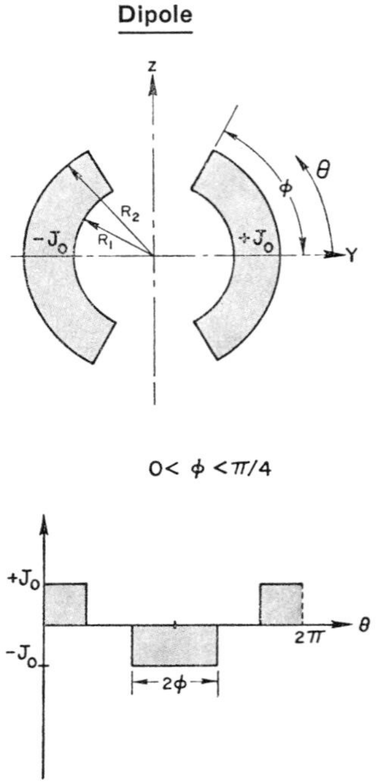

Fig. 2.1.32.
Dipole coil arrangement

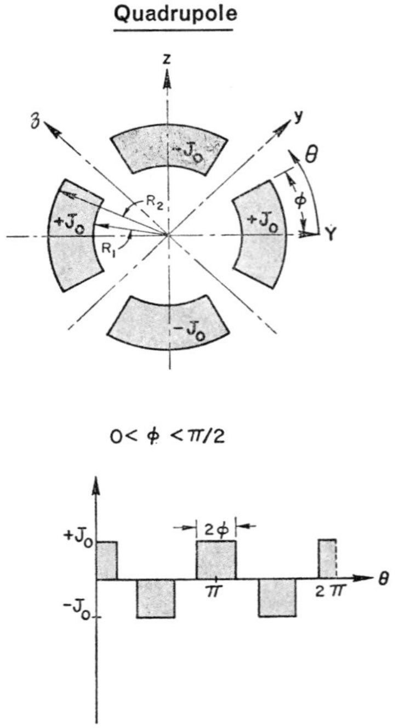

Fig. 2.1.33.
Quadrupole coil arrangement

For $p = 1$, corresponding to a dipole configuration (Fig. 2.1.32):

$$A_z(r,\theta) = \frac{\mu_0}{\pi}(\lambda J)_0 \cdot \frac{r^{(2k+1)}}{2k+1} \cos[(2k+1)\theta] \cdot \int_{a_1}^{a_2} \frac{da}{a^{(2k+1)-1}}.$$

The expression for the first term, $k = 0$, becomes:

$$A_z(r,\theta) = \frac{\mu_0}{\pi}(\lambda J)_0 \cdot r \cdot \cos(\theta) \int_{a_1}^{a_2} da = \frac{\mu_0}{\pi}(\lambda J)_0 \cdot r \cdot (a_2 - a_1)\cos(\theta) . \tag{2.1.78}$$

Since: $\boldsymbol{B} = \operatorname{curl} \boldsymbol{A}$, therefore: $B_r = \frac{1}{r}\frac{\partial A_z}{\partial \theta}$, $B_\theta = -\frac{\partial A_z}{\partial r}$,

we get the two field components:

$$B_r = -\frac{\mu_0}{\pi}(\lambda J)_0 (a_2 - a_1)\sin(\theta) , \quad B_\theta = -\frac{\mu_0}{\pi}(\lambda J)_0 (a_2 - a_1)\cos(\theta) . \tag{2.1.79}$$

For $p = 2$, corresponding to quadrupole configuration (Fig. 2.1.33):

$$A_z(r,\theta) = \frac{2\mu_0}{\pi}(\lambda J)_0 \frac{r^2}{2} \cdot (\cos 2\theta) \int_{a_1}^{a_2} \frac{da}{a} = \frac{2\mu_0}{\pi}(\lambda J)_0 \frac{r^2}{2} \cdot \cos(2\theta) \cdot \ln\left(\frac{a_2}{a_1}\right). \tag{2.1.80}$$

The field components are obtained as usual:

$$\begin{aligned} B_r &= \frac{2\mu_0}{\pi}(\lambda J)_0 \cdot r \cdot \ln\left(\frac{a_2}{a_1}\right) \cdot \sin(2\theta) , \\ B_\theta &= \frac{2\mu_0}{\pi}(\lambda J)_0 \cdot r \cdot \ln\left(\frac{a_2}{a_1}\right) \cdot \cos(2\theta) . \end{aligned} \tag{2.1.81}$$

If the coil cross-section does not extend over $\pi/2p$, but a fraction of it, say over the angle φ, the general equation for the vector potentials is written:

$$A_z = \frac{\mu_0}{\pi}(\lambda J)_0 \cdot \sum_n \frac{p \cdot r^n \cos(n\theta)}{n^2} \sin(n\varphi) \int_{a_1}^{a_2} \frac{da}{a^{n-1}} . \tag{2.1.82}$$

Substituting for $n = (2k + 1)p$ in Eq. (2.1.82), we get the equations for multipoles.

For a *dipole coil* configuration with constant current density distribution over each coil section, we get with $p = 1$ for $(r < a)$ and $(k = 0)$:

$$A_z(r,\theta) = \frac{\mu_0}{\pi}(\lambda J)_0 \cdot r \cdot \sin\theta \cdot (a_2 - a_1) \cdot \sin\varphi \tag{2.1.83}$$

with the two field components:

$$\left.\begin{aligned} B_r &= -\frac{\mu_0}{\pi}(\lambda J)_0 (a_2 - a_1)\sin\theta \sin\varphi \\ B_\theta &= -\frac{\mu_0}{\pi}(\lambda J)_0 (a_2 - a_1)\cos\theta \cos\varphi \end{aligned}\right\} . \tag{2.1.84}$$

If we choose $\varphi = 2\pi/3$, we get for $(r < a)$:

$$\left.\begin{aligned} B_r &= -\frac{\sqrt{3}}{\pi}\mu_0 (\lambda J)_0 (a_2 - a_1)\sin\theta \\ B_\theta &= -\frac{\sqrt{3}}{\pi}\mu_0 (\lambda J)_0 (a_2 - a_1)\cos\theta \end{aligned}\right\} . \tag{2.1.85}$$

For $r = x$ and $\theta = 0$ (median plane):

$$B_x = 0\,; \quad B_y = -\frac{\sqrt{3}}{\pi}\,\mu_0\,(\lambda J)_0\,(a_2 - a_1)\,.$$

The same calculation procedure gives the field components outside the coil area and within the coil conductor area. The results for a dipole coil are summarized:

$$a_1 < r < a_2 \left\{ \begin{aligned} B_r &= -\frac{\sqrt{3}}{\pi}\,\mu_0(\lambda J)_0 \left\{(a_2 - r) + \frac{r^3 - a_1^3}{3\,r^2}\right\} \sin\theta \\ B_\theta &= -\frac{\sqrt{3}}{\pi}\,\mu_0(\lambda J)_0 \left\{(a_2 - r) - \frac{r^3 - a_1^3}{3\,r^3}\right\} \cos\theta \end{aligned} \right\} \tag{2.1.86}$$

$$r > a_2 \left\{ \begin{aligned} B_r &= -\frac{\sqrt{3}}{\pi}\,\mu_0(\lambda J)_0 \left(\frac{a_2^3 - a_1^3}{r^2}\right) \sin\theta \\ B_\theta &= -\frac{\sqrt{3}}{\pi}\,\mu_0(\lambda J)_0 \left(\frac{a_2^3 - a_1^3}{r^2}\right) \cos\theta \end{aligned} \right\} \tag{2.1.87}$$

The required coil mmf per pole for a dipole coil is obtained from:

$$NI = (\lambda J)_0\,(a_2^2 - a_1^2)\cdot\varphi\,, \quad = (\lambda J)_0\,(a_2^2 - a_1^2)\,\frac{2\,\pi}{3}\,.$$

The outer coil radius is obtained from:

$$a_2 = a_1 + \frac{\pi}{\sqrt{3}\cdot\mu_0}\,\frac{B_y\,(0{,}0)}{(\lambda J)_0}\,. \tag{2.1.88}$$

The field components and gradient of a cylindrically-shaped *quadrupole* are calculated from Eq. (2.1.82) by using $p = 2$ and $\varphi = \pi/3$.

$$r < a_1 \left\{ \begin{aligned} B_r &= -\frac{\sqrt{3}}{\pi}\,\mu_0(\lambda J)_0 \ln\left(\frac{a_2}{a_1}\right)\cdot r\cdot\sin(2\,\theta) \\ B_\theta &= -\frac{\sqrt{3}}{\pi}\,\mu_0(\lambda J)_0 \ln\left(\frac{a_2}{a_1}\right)\cdot r\cdot\cos(2\,\theta) \end{aligned} \right\}. \tag{2.1.89}$$

The field gradient is, accordingly:

$$G\,(0) = \frac{\sqrt{3}}{\pi}\,\mu_0\cdot(\lambda J)_0 \ln\left(\frac{a_2}{a_1}\right). \tag{2.1.90}$$

For all other field points, we obtain:

$$a_1 < r < a_2 \left\{ \begin{aligned} B_r &= -\frac{\sqrt{3}}{\pi}\,\mu_0\cdot(\lambda J)_0 \ln\left[r\cdot\ln\left(\frac{a_2}{a_1}\right) + \frac{r^4 - a_1^4}{4\,r^3}\right] \sin(2\,\theta) \\ B_\theta &= -\frac{\sqrt{3}}{\pi}\,\mu_0\cdot(\lambda J)_0 \ln\left[r\cdot\ln\left(\frac{a_2}{a_1}\right) - \frac{r^4 - a_1^4}{4\,r^3}\right] \cos(2\,\theta) \end{aligned} \right\}, \tag{2.1.91}$$

$$r > a_2 \left\{ \begin{aligned} B_r &= \frac{\sqrt{3}}{2\,\pi}\,\mu_0\cdot(\lambda J)_0\cdot\frac{a_2^4 - a_1^4}{r^3}\sin(2\,\theta) \\ B_\theta &= \frac{\sqrt{3}}{2\,\pi}\,\mu_0\cdot(\lambda J)_0\cdot\frac{a_2^4 - a_1^4}{r^3}\cos(2\,\theta) \end{aligned} \right\}. \tag{2.1.92}$$

The outer coil radius is obtained from:

$$a_2 = a_1 \exp\left(\frac{-\pi G}{\sqrt{3}\,\mu_0 (\lambda J)_0}\right) \tag{2.1.93}$$

and the quadrupole ampere-turns per pole are calculated from:

$$(IN)_\varphi = \varphi\,(a_2^2 - a_1^2)\,(\lambda J)_0 = \frac{\pi}{3}\,(a_2^2 - a_1^2)\,(\lambda J)_0 . \tag{2.1.94}$$

Case II: The current density distribution is chosen to be according to $\lambda J = (\lambda J)_0 \cdot \cos(p\varphi)$.

Inserting the current density distribution in Eq. (2.1.76), we obtain:

$$A_z(r,\theta) = \frac{\mu_0}{\pi}(\lambda J)_0 \cdot \sum_n \frac{p \cdot r^n \cos(n\theta)}{n} \int_{a_1}^{a_2} \int_0^{\pi/2p} \frac{\cos(n\varphi)\cos(p\varphi)}{a^{n-1}}\, d\varphi\, da$$

$$= \frac{\mu_0}{\pi}(\lambda J)_0 \cdot \sum_n \frac{p \cdot r^n \cos(n\theta)}{n} \int_{a_1}^{a_2} \left[\frac{\sin[(n+p)\varphi]}{n+p} + \frac{\sin[(n-p)\varphi]}{n-p}\right]_0^{\pi/2p} . \tag{2.1.95}$$

For $n = (2k+1)\,p$ with $k = 0, 1, 2, \ldots$, all values of $A_z(r,\theta)$ are zero for $n \neq p$. For $n = p$, Eq. (2.1.76) becomes:

$$A_z(r,\theta) = \mu_0 (\lambda J)_0\, r^p \cos(p\theta) \int_{a_1}^{a_2} \cdot \frac{da}{a^{p-1}} .$$

For $p = 1$ (dipole):

$$A_z(r,\theta) = \frac{\mu_0}{4}(\lambda J)_0 \cdot r \cdot \cos(\theta)\,(a_2 - a_1) . \tag{2.1.96}$$

The two field components are, accordingly:

$$\left.\begin{aligned} B_r &= \frac{\mu_0}{4}(\lambda J)_0 \cdot (a_2 - a_1)\sin\theta \\ B_\theta &= \frac{\mu_0}{4}(\lambda J)_0 \cdot (a_2 - a_1)\cos\theta \end{aligned}\right\} . \tag{2.1.97}$$

For $p = 2$ (quadrupole):

$$A_z(r,\theta) = \frac{\mu_0}{8}(\lambda J)_0\, r^2 \cdot \ln(\alpha)\cos(2\theta) , \tag{2.1.98}$$

$$\left.\begin{aligned} B_r &= -\frac{\mu_0(\lambda J)_0}{4} \cdot r \cdot \ln(\alpha)\sin(2\theta) \\ B_\theta &= -\frac{\mu_0(\lambda J)_0}{4} \cdot r \cdot \ln(\alpha)\cos(2\theta) \end{aligned}\right\} . \tag{2.1.99}$$

The field gradient is written:

$$G(0) = \frac{\mu_0(\lambda J)_0}{4} \cdot \ln(\alpha) . \tag{2.1.100}$$

2.1.9 Multipole Coils with Rectangular Aperture

Rectangular coils, producing rectangular or square apertures, have been treated by Panofsky and Hand [19], Septier [20], and Dayton *et al.* [21]. We present in abbreviated form, some useful equations to calculate square-type multipoles:

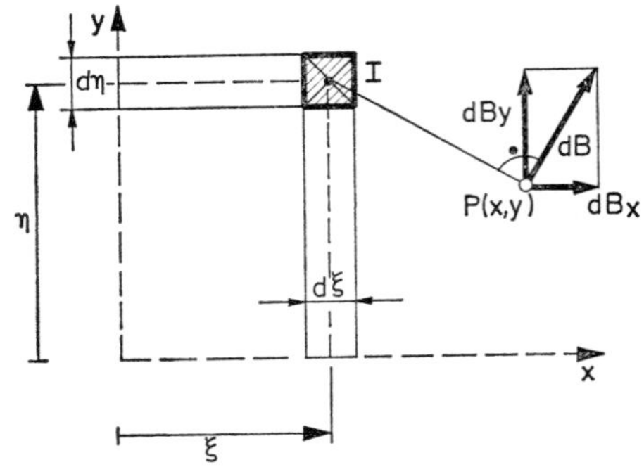

Fig. 2.1.34.
Field due to a current element

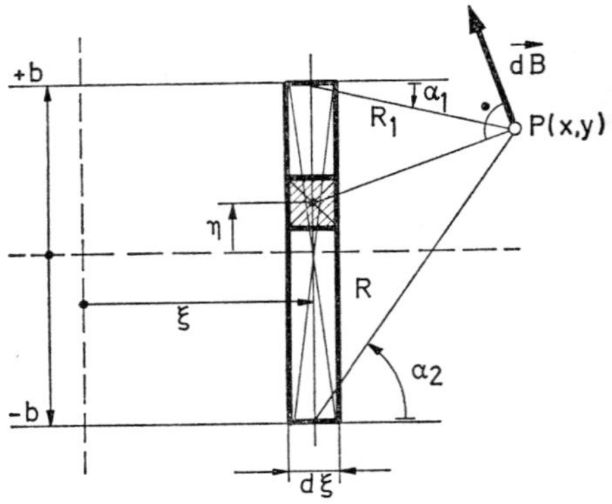

Fig. 2.1.35.
Field due to a current bar

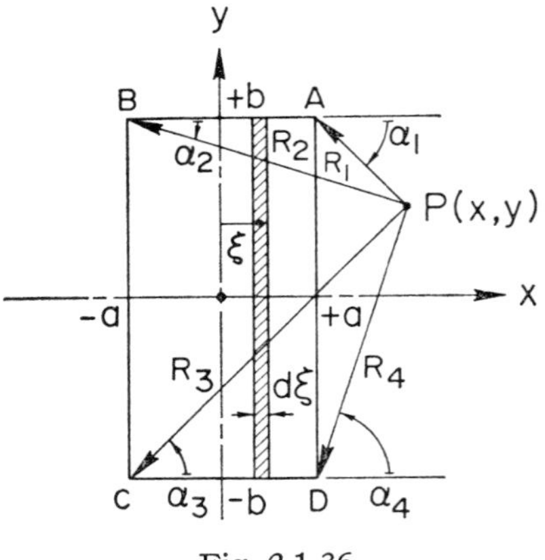

Fig. 2.1.36.
Field due to a current block

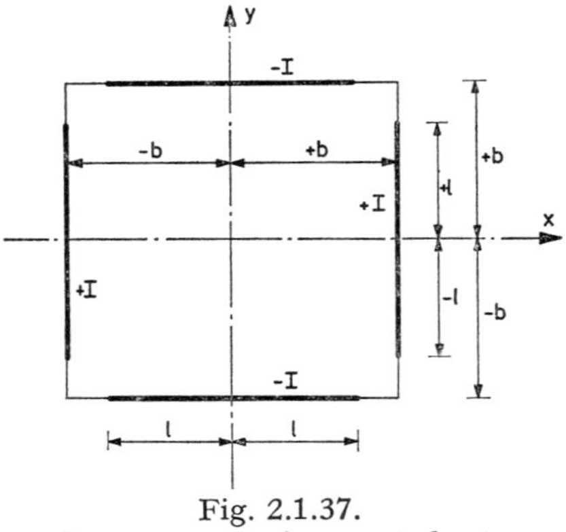

Fig. 2.1.37.
Arrangement of current sheets to generate a quadrupole field

Referring to Fig. 2.1.34, let us assume that a current element of cross-sectional area $dA = d\xi\, d\eta$ carries a uniform current density, λJ; with $I = \lambda J\, d\xi\, d\eta$, the field at a point $P\ (x, y)$ is given by: $dB = \frac{\mu_0}{2\pi} I \cdot \varrho_0$, which has the two components:

$$\left.\begin{aligned} dB_x &= \frac{\mu_0 \lambda J}{2\pi} \frac{y-\eta}{(x-\xi)^2 (y-\eta)^2}\, d\xi\, d\eta \\ dB_y &= +\frac{\mu_0 \lambda J}{2\pi} \frac{x-\xi}{(x-\xi)^2 + (y-\eta)^2}\, d\xi\, d\eta \end{aligned}\right\}. \qquad (2.1.101)$$

To obtain the magnetic field due to a bar of width d but length $2b$ at a point $P(x, y)$ (shown in Fig. 2.1.35), we integrate Eq. (2.1.101),

$$\left.\begin{aligned} \Delta B_x &= \frac{\mu_0 \lambda J}{2\pi} d\xi \ln \frac{(x-\xi)^2+(y-b)^2}{(x-\xi)^2+(y-b)^2} \\ \Delta B_y &= -\frac{\mu_0 \lambda J}{2\pi} d\xi \left[\cot^{-1}\left(\frac{x-\xi}{y+b}\right) - \cot^{-1}\left(\frac{x-\xi}{y-b}\right)\right] \end{aligned}\right\}. \quad (2.1.102)$$

Finally, the magnetic field due to a rectangular bar of infinite length is obtained when we integrate Eq. (2.1.102) over the width of the bar assumed to be $2a$. With the nomenclature of Fig. 2.1.36, we have, after integration:

$$\left.\begin{aligned} B_x &= \mu_0 \frac{\lambda J}{2\pi} \cdot \left\{ \frac{x+a}{2} \ln \frac{(x+a)^2+(y-b)^2}{(x+a)^2+(y+b)^2} \right. \\ &+ \frac{x-b}{2} \ln \frac{(x-a)^2+(y+b)^2}{(x-a)^2+(y-b)^2} + (y+b)\left[\tan^{-1}\left(\frac{y+b}{x+a}\right) \right. \\ &\left.\left. - \tan^{-1}\left(\frac{y+b}{x-a}\right)\right] + (y-b)\left[\tan^{-1}\left(\frac{y-b}{x-a}\right) - \tan^{-1}\left(\frac{y-b}{x+a}\right)\right]\right\} \\ B_y &= \mu_0 \frac{\lambda J}{2\pi} \cdot \left\{ \frac{y+b}{2} \ln \frac{(x+a)^2+(y+b)^2}{(x-a)^2+(y+b)^2} \right. \\ &+ \frac{y-b}{2} \ln \frac{(x-a)^2+(y-b)^2}{(x+a)^2+(y-b)^2} + (x+a)\left[\tan^{-1}\left(\frac{x+a}{y-b}\right) \right. \\ &\left.\left. - \tan^{-1}\left(\frac{x+a}{y+b}\right)\right] + (x-a)\left[\tan^{-1}\left(\frac{x-a}{y+b}\right) - \tan^{-1}\left(\frac{x-a}{y-b}\right)\right]\right\} \end{aligned}\right\}. \quad (2.1.103)$$

Combination of points, tapes, or bars results in dipoles, quadrupoles, or any desired multipoles.

Calculation of a Square-Type Quadrupole

We calculate the field due to four thin, straight, current sheets arranged as shown in Fig. 2.1.37. The length of each current sheet is $2l$, and the currents in the sheets are $\pm I$. The two field components at an arbitrary point P are given by:

$$\left.\begin{aligned} B_x &= -\frac{\mu_0 I}{4\pi l} \left\{ \frac{1}{2} \ln \frac{[(x+b)^2+(y+l)^2]\,[(x-b)^2+(y+l)^2]}{[(x+b)^2+(y-l)^2]\,[(x-b)^2+(y-l)^2]} \right. \\ &+ \left(\tan^{-1}\left(\frac{b-y}{l+x}\right) - \tan^{-1}\left(\frac{b+y}{l-x}\right)\right) \\ &\left. + \left(\tan^{-1}\left(\frac{b+y}{l+x}\right) - \tan^{-1}\left(\frac{b-y}{l-x}\right)\right)\right\} \\ B_y &= -\frac{\mu_0 I}{4\pi l} \left\{ \frac{1}{2} \ln \frac{[(x+l)^2+(y-b)^2]\,[(x+l)^2+(y+b)^2]}{[(x-l)^2+(y-b)^2]\,[(x-l)^2+(y+b)^2]} \right. \\ &+ \left(\tan^{-1}\left(\frac{l-y}{b-x}\right) - \tan^{-1}\left(\frac{l+y}{b+x}\right)\right) \\ &\left. + \left(\tan^{-1}\left(\frac{l-y}{b+x}\right) - \tan^{-1}\left(\frac{l+y}{b-x}\right)\right)\right\} \end{aligned}\right\}. \quad (2.1.104)$$

On the axis of symmetry (median plane $0\,x$), we obtain from Eq. (2.1.104):

$$B_x = 0\,; \quad B_y = -\frac{\mu_0 I}{4\,\pi l} \cdot \left\{ \ln \frac{(x+l)^2 + b^2}{(x-l)^2 + b^2} + 2\left(\tan^{-1}\left(\frac{l}{b-x}\right) - \tan^{-1}\left(\frac{l}{b+x}\right)\right)\right\}, \tag{2.1.105}$$

which, after some modification:

$$B_y = -\frac{\mu_0 I}{4\,\pi l} \cdot \left\{ \tanh^{-1}\left(\frac{2\,l x}{b^2 + l^2 + x^2}\right) + \tan^{-1}\left(\frac{2\,l x}{b^2 + l^2 - x^2}\right)\right\}. \tag{2.1.106}$$

If one uses the notation $\dfrac{l}{b} = u\,; \quad \dfrac{x}{b} = v\,,$

the Eq. (2.1.106) is modified into:

$$B_y = -\frac{\mu_0 I}{2\,\pi l} \cdot \left\{ \tanh^{-1}\left[\frac{2\,u}{1+u^2} \cdot v \cdot \left(1 + \frac{v^2}{1+u^2}\right)^{-1}\right] + \tan^{-1}\left[\frac{2\,u}{1+u^2} \cdot v \cdot \left(1 - \frac{v^2}{1+u^2}\right)^{-1}\right]\right\}. \tag{2.1.107}$$

The field gradient on the x-axis is obtained if we differentiate Eq. (2.1.107):

$$G\,(x) = \frac{\partial B_y}{\partial x} = \frac{2\,\mu_0 I}{\pi b^2 (1 + u^2)} \cdot \frac{1 + \dfrac{3 - u^2}{(1+u^2)^3} \cdot v^4}{1 + \dfrac{2\,v^4 (6\,u^2 - v^4 - 1) + v^8}{(1+u^2)^4}}. \tag{2.1.108}$$

Special case, $u = 1$:

$$B_y(y = 0) = -\frac{\mu_0 I}{2\,\pi b} \cdot \left\{ \tan h^{-1} \frac{2\,v}{2 + v^2} + \tan^{-1} \frac{2\,v}{2 - v^2}\right\}, \tag{2.1.109}$$

$$G\,(x) = -\frac{\mu_0 I}{\pi b^2} \cdot \left(1 + \frac{v^4}{4}\right)^{-1}. \tag{2.1.110}$$

As a further example, we present the quadrupole field due to four rectangular coils (Fig. 2.1.38). The quadrupole consists of four coils, each having a current density λJ and the dimensions $2\,a \cdot 2\,l$, while the aperture has a rectangular area of $(2\,b)^2$.

The field on the $0\,x$-axis is obtained similar to the above case:

$$\begin{aligned} B_y = \frac{\mu_0 (\lambda J)}{2\,\pi} \cdot \Bigg\{ & l \cdot \left[\ln \frac{(x+b+2\,a)^2 + l^2}{(x-b-2\,a)^2 + l^2} - \ln \frac{(x+b)^2 + l^2}{(x-b)^2 + l^2}\right] \\ & + b \ln \frac{(x+l)^2 + b^2}{(x-l)^2 + b^2} - (b + 2\,a) \ln \frac{(x+l)^2 + (b+2\,a)^2}{(x-l)^2 + (b+2\,a)^2} \\ & + 2\,b \left(\tan^{-1}\left(\frac{x+b}{l}\right) + \tan^{-1}\left(\frac{x-b}{l}\right)\right) \\ & - 2\,(b + 2a) \left(\tan^{-1}\left(\frac{x+b+2\,a}{l}\right) + \tan^{-1}\left(\frac{x-b-2\,a}{l}\right)\right) \\ & + 2\,l \Bigg[\left(\tan^{-1}\left(\frac{x+l}{2+2\,a}\right) + \tan^{-1}\left(\frac{x-l}{2+2\,a}\right)\right) \\ & - \left(\tan^{-1}\left(\frac{x+l}{b}\right) + \tan^{-1}\left(\frac{x-l}{b}\right)\right)\Bigg] \end{aligned} \tag{2.1.111}$$

$$+ 2x\left[\left(\tan^{-1}\left(\frac{x+b}{l}\right) - \tan^{-1}\left(\frac{x-b}{l}\right)\right)\right.$$
$$- \left(\tan^{-1}\left(\frac{x+b+a}{l}\right) - \tan^{-1}\left(\frac{x-b-a}{l}\right)\right)$$
$$+ \left(\tan^{-1}\left(\frac{x+l}{b+2a}\right) - \tan^{-1}\left(\frac{x-l}{b+2a}\right)\right)$$
$$\left.\left. - \left(\tan^{-1}\left(\frac{x+l}{b}\right) - \tan^{-1}\left(\frac{x-l}{b}\right)\right)\right]\right\}. \qquad (2.1.112)$$

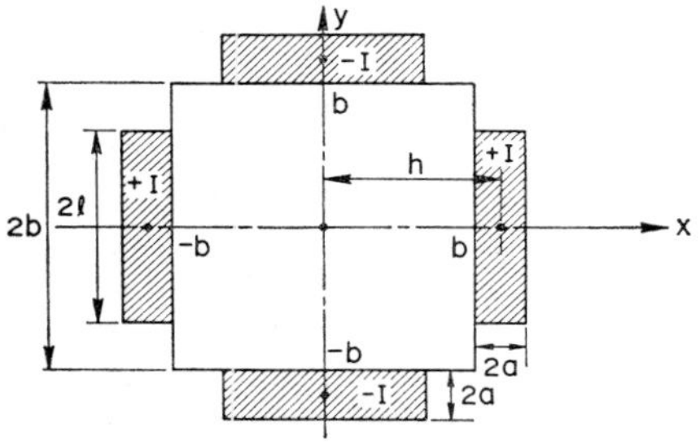

Fig. 2.1.38.
Coil blocks to form a rectangular quadrupole

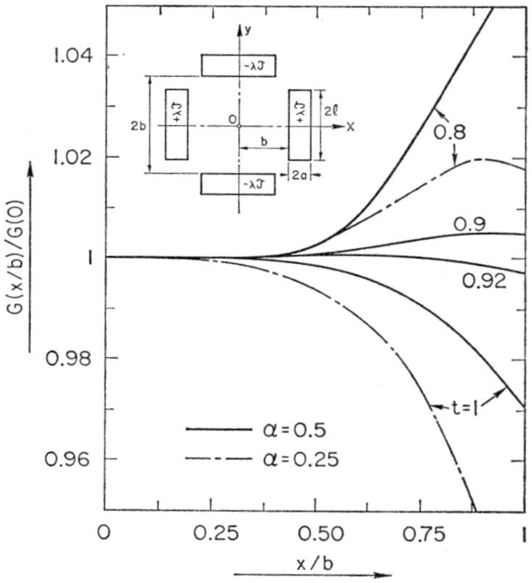

Fig. 2.1.39.
Field gradient within the aperture of a rectangular quadrupole

The expression for the field gradient is relatively simple. Putting:

$$\frac{l}{b} = t\,; \quad \frac{a}{b} = \alpha\,; \quad \frac{x}{b} = u\,,$$

we obtain:

$$G(0) = \frac{2\,\mu_0(\lambda J)}{\pi}\left[\tan^{-1}\left(\frac{1-t^2}{-2t}\right) - \tan^{-1}\frac{(1+2\alpha)^2 - t^2}{2(1+2\alpha)^2\cdot t}\right] \qquad (2.1.113)$$

and:

$$G(u) = \frac{\mu_0(\lambda J)}{\pi}\cdot\left[\left(\tan^{-1}\left(\frac{1-t^2+u^2}{2t}\right) + \tan^{-1}\left(\frac{1-t^2-u^2}{2t}\right)\right.\right.$$
$$\left.\left. - \tan^{-1}\left(\frac{(1+2\alpha)^2 - t^2 + u^2}{2(1+2\alpha)\,t}\right) + \tan^{-1}\left(\frac{(1+2\alpha)^2 - t^2 - u^2}{2(1+2\alpha)\,t}\right)\right)\right]. \qquad (2.1.114)$$

We have calculated curves of $G(u)/G(0)$ for two values of $\alpha = \frac{a}{b} =$ 0.25 and 0.5 in function of $u = \frac{x}{b}$ in Fig. 2.1.39. For $\alpha = 0.5$, we obtain for $t = 0.92$, a homogeneity of better than $3 \cdot 10^{-3}$.

Remark. It may be mentioned that for complicated coil geometries, computer programs are available, such as MAFCO [17], for two-dimensional cases.

2.2 Magnetic Fields due to Coils in Proximity to Ferromagnetic-Materials

2.2.1 Introduction

Utilization of ferromagnetic materials in combination with high-field coils has been considered by Bitter [22], Sampson [23], Zijlstra [24], and others. Placing ferromagnetic materials around coils serves several purposes:

Air-core magnets, whether superconducting or conventional, generate high external (fringing) fields, which, in practical cases, are troublesome; they affect measuring equipment and instruments around the magnet, influence data-taking, and reduce free mobility in laboratories. Placing iron around the coils enhances the aperture field and reduces the total field energy. If several superconducting magnets, such as synchrotron magnets or beam transport magnets, are placed near each other, and if there is no iron around each coil, the coils will be coupled electromagnetically. The quenching of one coil may affect adjacent magnets. Having the iron shield in proximity to each coil, the magnets can be magnetically decoupled. Fringing fields can be shielded or shunted by means of auxiliary current-carrying conductors or sheets placed around the original coil (which is combersome and not economical), or by using iron, iron alloys, or rare earth materials.

As the relative permeability of these ferromagnetic materials is, in general, much higher than air (even at fields exceeding 2.1 T), the mmf required by the coils to produce a certain field in the working region will be smaller if ferromagnetic materials are used. Iron is presently the most common and inexpensive material being widely used in magnets. It serves, in addition, as a support structure to withstand magneto-mechanical forces. Iron also influences the field distribution in space, and the field configuration in the aperture.

The calculation of iron magnetization for three-dimensional configurations has not yet been performed successfully. Two-dimensional coil-iron configurations have been calculated utilizing sophisticated computer programs, fast computers with large storage facilities. Even in simple geometries, iron is not uniformly magnetized. Thus, methods of field computation, assuming uniform magnetization or infinite permeability in iron, may be useful where the iron is not saturated or as a crude approximation, but is inadequate for final field computations.

The main advantage of superconducting coils is to generate high magnetic fields without exorbitant power consumption. At central field values in the order of 5 T (i.e., presently a field strength selected for beam transport magnets), the iron is saturated and the assumption of infinite permeability leads to errors.

The choice, shape, and position of the ferromagnetic material in proximity to the superconducting coils depends on factors discussed in chapter 6.

Basically, the field distribution in space is found from Maxwell's equation:

$$\operatorname{curl}\left(\frac{1}{\mu}\boldsymbol{B}\right) = +\boldsymbol{J} \tag{2.2.1}$$

which is modified to:

$$\frac{1}{\mu}\operatorname{curl}\boldsymbol{B} + \nabla\frac{1}{\mu}\cdot\boldsymbol{B} = +\boldsymbol{J}\,. \tag{2.2.2}$$

By using the relation:

$$\boldsymbol{B} = \operatorname{curl}\boldsymbol{A}\,, \tag{2.2.3}$$

we get:

$$\operatorname{curl}\left(\frac{1}{\mu}\operatorname{curl}\boldsymbol{A}\right) = +\boldsymbol{J}\,. \tag{2.2.4}$$

Equation (2.2.4) is a "nonlinear" Poisson equation, where $1/\mu$ is a function of the magnitude of $\boldsymbol{B}$. We confine the field calculation to a two-dimensional problem, which means that all field variables are independent of the third coordinate, z. When $\boldsymbol{A}$ is found from Eq. (2.2.4), the flux density $\boldsymbol{B}$ can be found from $\boldsymbol{A}$, using Eq. (2.2.3). $\boldsymbol{B}$ and the field $\boldsymbol{H}$ are related by the inverse of the permeability $\nu = 1/\mu$, according to $\boldsymbol{H} = \nu\boldsymbol{B}$.

Rectangular Coordinates

For the two-dimensional case, we assume that $\boldsymbol{A}$ and $\boldsymbol{J}$ have a z-component only. This means that $A = A_z(x,y)$ and $J = J_z(x,y)$. Thus, Eq. (2.2.4) reduces to:

$$\nabla\cdot\nu\nabla A_z(x,y) + J_z(x,y) = 0 \tag{2.2.5}$$

which can be written explicitly:

$$\frac{\partial}{\partial x}\left(\nu\frac{\partial A}{\partial x}\right) + \frac{\partial}{\partial y}\left(\nu\frac{\partial A}{\partial y}\right) + J = 0\,. \tag{2.2.6}$$

In iron, $\boldsymbol{A}$ obeys the equation:

$$\frac{\partial}{\partial x}\left(\nu\frac{\partial A}{\partial x}\right) + \frac{\partial}{\partial y}\left(\nu\frac{\partial A}{\partial y}\right) = 0\,. \tag{2.2.7}$$

From the relations:

$$|\boldsymbol{B}| = |\operatorname{curl}\boldsymbol{A}| = |\nabla A\,(x,y)| = \left[\left(\frac{\partial A}{\partial x}\right)^2 + \left(\frac{\partial A}{\partial y}\right)^2\right]^{\frac{1}{2}}, \tag{2.2.8}$$

we are able to calculate the field distribution in space. Since μ or ν depend only on the magnitude of field, we may write:

$$\frac{1}{\mu}|\boldsymbol{B}| = \nu\,|\boldsymbol{B}| = \nu\left[\left(\frac{\partial A}{\partial x}\right)^2 + \left(\frac{\partial A}{\partial y}\right)^2\right]^{\frac{1}{2}}.$$

Since the curl and gradient of any vector are orthogonal, we may use Eq. (2.2.8) to express the normal and tangential components of the fields (denoted by the subscripts n and t) as follows:

$$B_n = (\mathrm{curl}\,\boldsymbol{A})_n = (\nabla \boldsymbol{A})_t\,; \quad H_t = (\nu\,\mathrm{curl}\,\boldsymbol{A})_t = (\nu \nabla \boldsymbol{A})_n\,. \tag{2.2.9}$$

We consider an interface between two regions, 1 and 2, such as the iron-air interface. Since $\boldsymbol{A}_1$ must be continuous, then $\boldsymbol{A}_1 = \boldsymbol{A}_2$, where $\boldsymbol{A}_1$ and $\boldsymbol{A}_2$ are evaluated adjacent to the interface. This being the case, the derivatives of A on either side of the interface must be equal to:

$$\nabla \cdot (\hat{n} \times \boldsymbol{A}_1) = \nabla \cdot (\hat{n} \times \boldsymbol{A}_2) \tag{2.2.10}$$

which can be modified to:

$$\hat{n} \cdot (\mathrm{curl}\,\boldsymbol{A}_1) = \hat{n} \cdot (\nabla \times \boldsymbol{A}_2) \tag{2.2.11}$$

and is the same as $\hat{n} \cdot \boldsymbol{B}_1 = \hat{n} \cdot \boldsymbol{B}_2$.

The relation on the tangential components of $\boldsymbol{H}$ can be found in terms of $\boldsymbol{A}$ by integrating the normal components of the equation:

$$\nabla \cdot \nu \cdot \mathrm{curl}\,\boldsymbol{A} + \boldsymbol{J} = 0$$

over the plane surface of the interface region.
The equation

$$\int_S \hat{n} \cdot \mathrm{curl}\,(\nu\,\mathrm{curl}\,\boldsymbol{A})\,da + \int_S \hat{n}\,\boldsymbol{J}\,ds = 0$$

is modified into:

$$\oint_S \hat{n} \cdot (\nu\,\mathrm{curl}\,\boldsymbol{A})\,dl + \int_S \hat{n}\,\boldsymbol{J}\,ds = 0\,.$$

Assuming the region of the remaining surface integral forms a thin rectangle of length h (parallel to the interface) and the width d (perpendicular to the interface), then for $d \to 0$; the surface contribution vanishes and the line integral becomes the tangential component of $\nu\,\mathrm{curl}\,\boldsymbol{A}$ on each side of the interface. This means that:

$$\hat{n} \cdot (\nu_1\,\mathrm{curl}\,\boldsymbol{A}_1) - \hat{n} \cdot (\nu_2\,\mathrm{curl}\,\boldsymbol{A}_2) = 0 \tag{2.2.12}$$

or

$$\hat{n} \cdot (\boldsymbol{H}_1 - \boldsymbol{H}_2) = 0\,.$$

Cylindrical Coordinates

In cylindrical coordinates (axially symmetric cases), we assume that A and J have a θ component only. Then we get, from Eq. (2.2.4):

$$\mathrm{curl}\,(\nu\,\mathrm{curl}\,\boldsymbol{A}) = \left\{\frac{\partial}{\partial r}\left[\frac{\nu}{r}\frac{\partial}{\partial r}(rA_\theta)\right] + \frac{\partial}{\partial z}\left[\frac{\nu}{r}\frac{\partial}{\partial z}(rA_\theta)\right]\right\}. \tag{2.2.13}$$

Combining $\mathrm{curl}\,\nu\,\mathrm{curl}\,\boldsymbol{A} + \boldsymbol{J} = 0$ with Eq. (2.2.13), we obtain:

$$\frac{\partial}{\partial r}\left[\frac{\nu}{r}\frac{\partial}{\partial r}(rA_\theta)\right] + \frac{\partial}{\partial z}\left[\frac{\nu}{r}\frac{\partial}{\partial z}(rA_\theta)\right] + J_\theta = 0\,. \tag{2.2.14}$$

It is also possible to describe the magnetic field by means of a scalar potential $V\,(x, y)$. In order to do so, we introduce the magnetic moment $\boldsymbol{M}\,(x, y)$, which is related to the current density by:

$$\boldsymbol{J} = \mathrm{curl}\,\boldsymbol{M}\,. \tag{2.2.15}$$

From: $\operatorname{curl} \boldsymbol{H} = \boldsymbol{J}$,

it follows that: $\operatorname{curl} (\boldsymbol{H} - \boldsymbol{M}) = 0$. (2.2.16)

The field is now irrotational for the vector $(\boldsymbol{H} - \boldsymbol{M})$, which can be considered as the gradient of a scalar V^*.

$$\boldsymbol{H} - \boldsymbol{M} = \operatorname{grad} V^*$$

or
$$\boldsymbol{H} = \operatorname{grad} V^* + \boldsymbol{M} . \tag{2.2.17}$$

If the permeability μ is constant and uniform, then: $\operatorname{div} \boldsymbol{H} = 0$ and applying it to Eq. (2.2.17), we get:

$$\nabla^2 V^* = -\operatorname{div} \boldsymbol{M} . \tag{2.2.18}$$

Equation (2.2.15) is not sufficient to denote the vector $\boldsymbol{M}$ completely. Other conditions must be chosen to determine $\boldsymbol{M}$. For the two-dimensional case, we assume that $\boldsymbol{M}$ has only an x-component. This means:

$$M_y(x,y) = 0 ; \quad M_x(x,y) = M(x,y) .$$

From Eq. (2.2.15), we obtain: $-\dfrac{\partial M(x,y)}{\partial y} = J(x.y)$

which, integrated, yields: $M(x,y) = -J(x,y)\, dy + C$.

The boundary condition for V^* can be found from Eq. (2.2.17), through integration along the boundary:

$$V_s^* - V_{s0}^* = \int_{s_0}^{s} (\boldsymbol{H} - \boldsymbol{M})\, ds . \tag{2.2.19}$$

If the contour is an iron surface, the tangential component of the field H_s depends on the degree of saturation. At low field levels, H_s is very small and can be neglected. However, in high-field superconducting magnets, H_s has a large value!

The Eq. (2.2.4) is written explicitly for two-dimensional cases:

$$\frac{\partial^2 A}{\partial x^2} + \frac{\partial^2 A}{\partial y^2} - \frac{1}{\mu}\frac{\partial \mu}{\partial x}\cdot\frac{\partial A}{\partial x} - \frac{1}{\mu}\cdot\frac{\partial \mu}{\partial y}\cdot\frac{\partial A}{\partial y} + \frac{1}{\mu} J = 0 .$$

This is of the type:

$$a \cdot \frac{\partial^2 A}{\partial x^2} + b \frac{\partial^2 A}{\partial y^2} + c \frac{\partial A}{\partial x} + d \frac{\partial A}{\partial y} + e - u + f = 0 .$$

$a, b \neq 0$ and $a \ldots f$ may be functions of x, y and A. Only in special cases, where the boundary of the two-dimensional region is particularly simple, are analytical solutions available. In other cases, no analytical solution can be found; this has led to the development of finite difference methods [25], which, with the use of high-speed large storage capacity computers, have become practicable for routine work.

The essence of the method is to superimpose upon the region of interest a net or a grid, and replace the partial differential equation to be solved by an approximate difference equation at each nodal point of the grid (discretization method). The field distribution is obtained by solving a set of simultaneous difference equations.

The problem of finding an effective method of solving the resulting large number of equations has received intensive study by a number of various laboratories. A number of methods have been proposed.

The method based on using scalar potentials in the air region and the vector potential in the iron region is given by Christian [26]. The problem is solved by using scalar potential in air, and assuming infinite iron permeability. This solution in the air region is used to find the vector potential on the iron boundary. Having found this iron boundary value, the vector potential in iron is obtained. The solution in the iron region is then used to find the scalar potential in the air boundary. The cycle is repeated until the process converges to an answer.

An alternative solution is the method of magnetic vector potential throughout the problem.

The *Vector Potential Method* has the advantage of simplicity and flexibility in treating different magnet geometries.

The *Two-Potential Method* is superior to the vector potential method if a wide permeability range is considered. The convergence rate is faster, and a greater accuracy is obtained than in the case where the vector potential method alone is used. If the iron saturation can be considered only a small perturbation on the infinite permeability solution, then one can expect very accurate results from the two-potential method. The disadvantage of this method is the complexity in programming.

Applied Numberical Methods

There are several computer programs solving numerically two-dimensional "nonlinear Poisson" equations. In Table 2.2.1, some of the better known programs are presented.

There are several reasons for classifying these programs (given in Table 2.2.1). Each group has its specific need and resources. By the time a particular program is written and becomes operational, it appears that the program is not capable of solving all the new problems which have accumulated in the meantime. Hence, revision and modification to the original codes gradually appear. Under such a state of flux, a general comparison of the program becomes either difficult or misleading. Thus, instead of reviewing the programs, three available computer programs are compared by computing the same magnet.

The recent versions of TRIM, LINDA, and NUTCRACKER II were used to calculate the field generated by a magnet illustrated in Fig. 2.2.1. Information relating similar features of each program is presented in Table 2.2.2. The same magnetization curve was used in all three programs. For field points in air, the programs agree at least to the third significant figure.

The original seven-digit precision of the IBM computer used with NUTCRACKER II (single precision) limited the convergence of the problem to $c_2 = 4 \times 10^{-6}$, whereas the same calculation with LINDA and TRIM yielded values of $c_2 \cong 10^{-7}$ on a CDC computer. The NUTCRACKER code converges in fewer iterations, and the total computational time shows a marked decrease. This may also be due to the fact that the IBM 361/91 is from three to ten times faster than the CDC-6600,

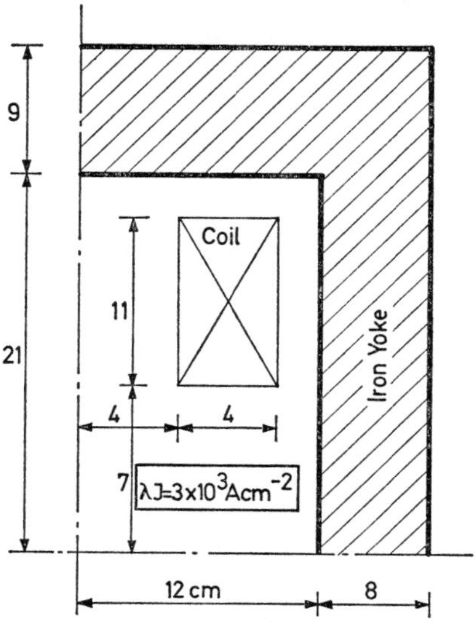

Fig. 2.2.1. Coil and iron yoke arrangement chosen for field computation

Table 2.2.1. *Computer programs available to solve nonlinear Poisson equations*

Research Establishment	Name of Program	Authors	Method
Electrical Engrg. Dept., University of Colorado	not given [27, 28, 29]	Erdelyi, Ahamed, Trutt	Vector Potential
Lawrence Radiation Laboratory (LRL), Berkeley	TRIM [30], POISSON [31], PISA [32], LINDA [33]	Winslow, Concus, Halbach, Colonias, Dorst	Vector Potential, Two-potential method
MURA (ANL)	SYBYL [34], FIELD [35], TINKER [36]	Christian, Dahl, Parzen	Two-potential method
CERN	MARE [37], MAGNET [38]	Perin, Van der Meer, Iselin	Two-potential method
SLAC	NUTCRACKER I [39], NUTCRACKER II [40]	Burfine, Anderson, Brechna	Vector Potential
NBL	GRACY [41]	Parzen, Dahl	Vector Potential
LRL, Berkeley	MAFCO [18]	Perkins, Brown	Scalar Potential
SLAC	MARC II [42]	Borglum, Anderson	Scalar Potential

depending upon the operation being performed. TRIM and LINDA each smooth their final finite difference solutions with a least square polynomial, whereas NUTCRACKER uses the finite difference solution as it stands.

The greatest disadvantage of NUTCRACKER is the fitting of the coil and iron geometry to the square mesh. Errors due to geometrical discrepancies can affect the computational results which have been found when compared to measurements.

Table 2.2.2. *Comparison of TRIM, LINDA, and NUTCRACKER*

Code	NUT-CRACKER	TRIM	LINDA
Type of potential	Vector	Vector	Scalar
Mesh type	square	triangular	square
Difference Operator	5-point	7-point	5-point
Initial Conditions			
Potentials	0.0	0.0	0.0
Permeabilities	1000	1000	1000
SOR factor ω	1.93	1.94	1.5
Optimization Data			
Recalculate ω	each 6 its	each 25 its	each 30
Recalculate $\nu = 1/\mu$	each 3 its	...	...
Magnetization data	40 points	31 points	40 points
Convergence criterion	4×10^{-6}	10^{-6}	rms of residuals 8×10^{-3}
Iterations to converge	203	323	181 in air coil regions, 278 in iron regions
Computer	IBM 360/91	CDC 6600	CDC 600
Maximum number of mesh points	2.25×10^{4}	4000	9450
Storage required for above	297 K	106 K	136 K
Precision	7 digits	14 digits	14 digits
Computation time for 2400-point problem (*Fig.* 2.2.1)			
Program initialization	1.5 sec	31.3 sec	1.5 sec
Iterative portion	38 sec	227 sec	165 sec
Edit (smoothing of solution)	—	120.5 sec	6.0 sec
Total CPU time	39.5 sec	378.8 sec	172.5 sec

2.2.2 Direct Current Magnetization Curves

In the course of solving a magnetostatic problem, the permeability of a point in iron must be recalculated many times during the iteration process. Any large-sized magnet requiring 10^4 nodes or more may need between 250 to 500 iterations for convergence. This means one must calculate the permeability $2.5 \times 10^6 \dots 5 \times 10^6$ times. The usual method is to find the values of permeability from tabulated experimental data and use extrapolation methods to obtain intermittent values.

Magnetization properties of iron in function of impurities are reported by Gerold [43] and Brechna [44]. Cold-work (cold rolling), grain orientation, and annealing procedures have a profound effect upon magnetization properties [45]. Thus, any computer code should use, as a subroutine, experimental data for the particular iron used for the magnet being considered. Modern steels use, in d.c. magnets, an impurity content

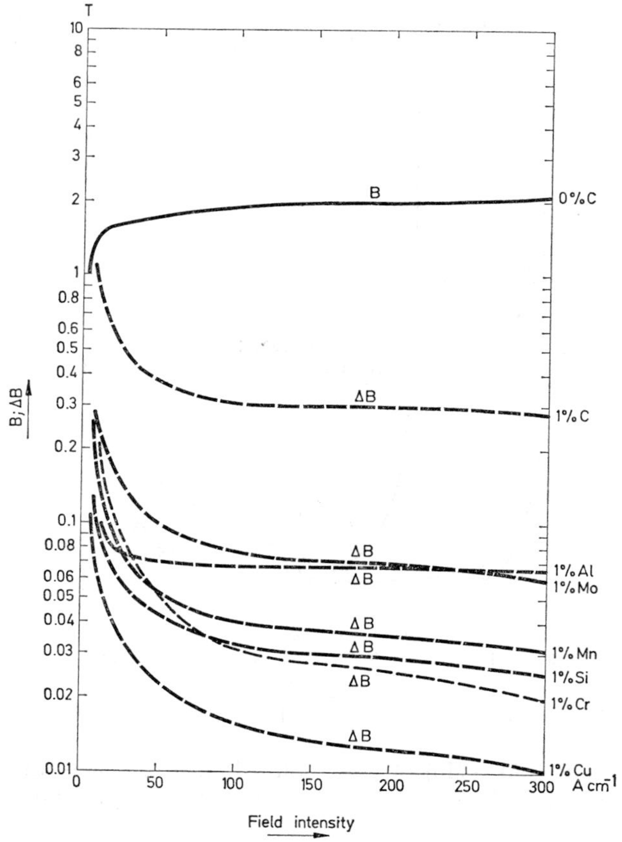

Fig. 2.2.2. The effect of impurities on the magnetization of steels: B is the magnetization curve of a pure iron. ΔB is the field reduction due to impurities. Carbon has the most adverse effect on the magnetization

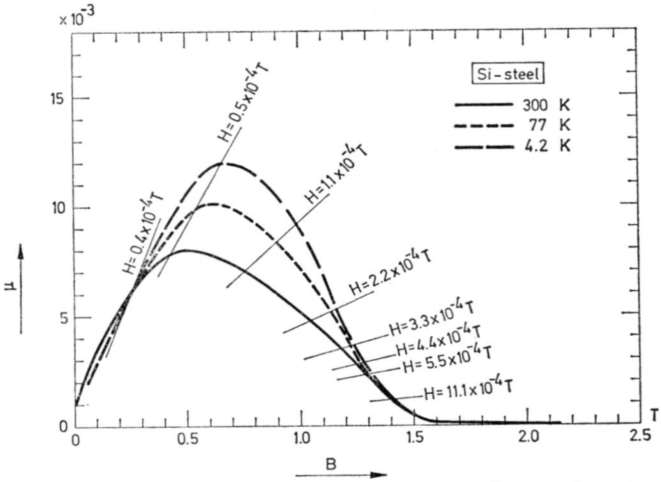

Fig. 2.2.3a. Permeability values vs. magnetic field at various temperatures for 1.8% Si-Steels

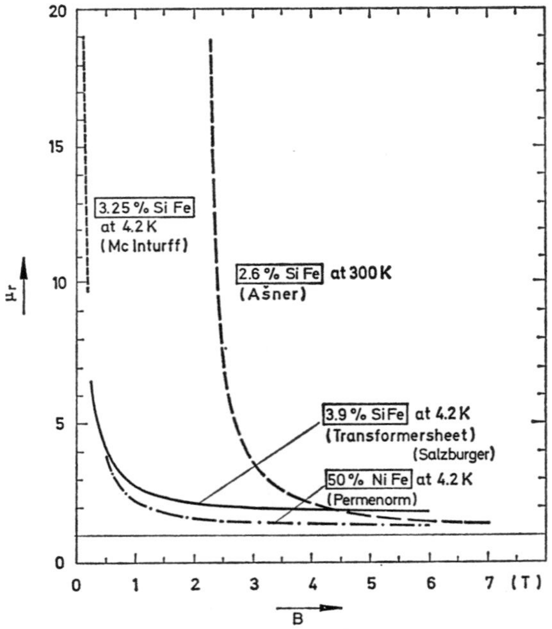

Fig. 2.2.3b. Relative permeability values vs. transverse magnetic field

of (N_2 + S + P + Al + Mo) of less than 0.1%, (Cr + Cn + Mn + Ni + Si) less than 0.7%. Carbon, the most dangerous impurity, is limited to less than 0.1%.

Figure 2.2.2 shows magnetization curves of pure iron and of iron with impurities. Figure 2.2.3a and 2.2.3b illustrate permeability curves at various temperatures [46, 47].

The experimental relation observed for specific magnetization at high field is given by

$$M = M_s\left(1 - \frac{a}{H} - \frac{b}{H^2}\right) + \varkappa_0 H \,, \tag{2.2.20}$$

a and b are material constants.

M_s is the saturation magnetization and $\varkappa_0$ the initial susceptibility of iron. The saturation magnetization is given by Danan *et al.* [48] as: $M_s = 2.1936$ T.

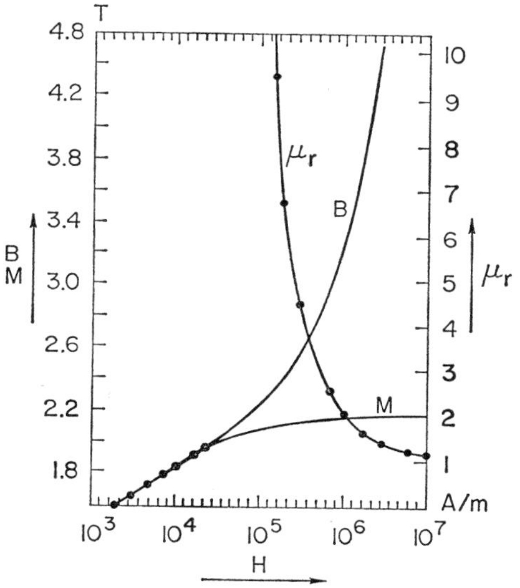

Fig. 2.2.4. Magnetization M, field B and relative permeability μ_r versus applied field

Typical magnetization curves, showing B, μ_r, and r as functions of H, are illustrated in Fig. 2.2.4. As it is difficult to determine analytically the B (H) relationship covering the whole field range, it has been found convenient to divide (for descriptive purposes) the $B - H$ curve into four regions. With some loss in accuracy, these regions are in the field intervals given below:

Region I: $0 \leqq H\ (A\ \mathrm{m}^{-1}) \leqq 1.6 \cdot 10^2$; $0 \leqq M\ (T) \leqq 9.148 \cdot 10^{-1}$
Region II: $1.989 \cdot 10^2 \leqq H\ (A\ \mathrm{m}^{-1}) \leqq 1.592 \cdot 10^4$;
$1.1 \leqq M\ (T) \leqq 1.910$
Region III: $2 \cdot 10^4 \leqq H\ (A\ \mathrm{m}^{-1}) \leqq 1.000 \cdot 10^8$; $1.95 \leqq M\ (T) \leqq 2.1936$
Region IV: $10^8 \leqq H\ (A\ \mathrm{m}^{-1}) \leqq 10^{10}$; $M\ (T) \cong 2.1936$

Each region may be fitted by second-order splines involving low-order polynomials. The position of "joints" can be optimized by a program developed by Smith [49]. The errors due to curve fitting are estimated to be less than 1%.

2.2.3 Difference Equations

Solution of Laplace's and Poisson's equations can be obtained by replacing the differential equations for the vector potential by a set of difference equations for each point of some chosen grid structure.

The difference equation is developed by expanding the potential at the point $P\ (x_0, y_0)$ in a Taylor series and deriving expressions for higher-order values such as $(\partial^2 A/\partial x^2)_P$ and $(\partial^2 A/\partial y^2)_P$, which are substituted in the Poisson's equation. If the potential at a point P_0 is A_0, the potential at any other point $P\ (x, y)$ close to $P_0(x_0, y_0)$ is given by:

$$\begin{aligned} A\ (x,y) &= A_0 + \frac{1}{1!}\left[(x - x_0)\frac{\partial}{\partial x} + (y - y_0)\frac{\partial}{\partial y}\right] A_0 \\ &+ \frac{1}{2!}\left[(x - x_0)^2\left(\frac{\partial}{\partial x}\right)^2 + (x - x_0)\ (y - y_0)\left(\frac{\partial}{\partial x}\right)\left(\frac{\partial}{\partial y}\right)\right. \\ &\left. + (y - y_0)^2\left(\frac{\partial}{\partial y}\right)^2\right] A_0 + \cdots \end{aligned} \tag{2.2.21}$$

To obtain the five unknowns:

$$\left(\frac{\partial A}{\partial x}\right)_0; \quad \left(\frac{\partial A}{\partial y}\right)_0; \quad \left(\frac{\partial^2 A}{\partial x^2}\right)_0; \quad \left(\frac{\partial^2 A}{\partial x\, \partial y}\right)_0; \quad \left(\frac{\partial^2 A}{\partial y^2}\right)_0,$$

one needs the values of A_1; $A_2 \ldots A_5$ for the vector potential $\boldsymbol{A}$ at five points in the neighborhood of $P_0\ (x_0, y_0)$. If symmetry conditions prevail, it can be shown that only two points besides $P_0(x_0, y_0)$ are necessary to obtain the first derivative of $\boldsymbol{A}$, three or four points to find the second derivative. Symmetry conditions reduce the number of unknowns by one.

2.2.4 The Grid System

The differential Eq. (2.2.4) can be solved numerically only by a difference method, or the vector potential $\boldsymbol{A}$ or the scalar potential V can be obtained numerically only for certain points separated in space. These points will be node points of a more-or-less regular grid system. The area to be considered must be covered by a sufficiently fine mesh to obtain the computational accuracy required.

There are several types of grids: grids with rectangular, square, and triangular mesh units. If sections of circles are used as basic mesh units, the system is called a polar grid arrangement.

In Fig. 2.2.5, we illustrate a small section of a rectangular grid system. The grid points and the vector potentials are numbered continuously with I and K. The position of each mesh point is indicated through

$$x_{I+1} = x_I + p_I\,; \quad y_{K+1} = y_K + q_K\,.$$

The distances p_I and q_K are such that:

(a) The boundary of the area to be considered or boundaries of sections are accurately contained in the grid system.

(b) The discretization factor R of the differential Eq. (2.2.4) is sufficiently small.

In Fig. 2.2.6, a section of the area to be considered is indicated by $S_{I,K}$ and is attached to the point $P_{I,K}$. The boundary between the two areas having different characteristics must go through the node points, if the boundary conditions of the differential equation which have to be met at the boundaries are to be fulfilled.

A polar grid system is illustrated in Fig. 2.2.7. In this system,

$$r_{K+1} = r_K + q_K \quad \text{and} \quad \theta_I = \theta_{I+1} + p_1 .$$

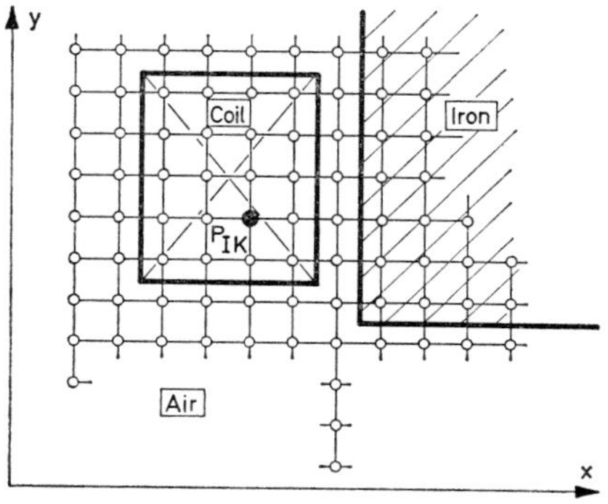

Fig. 2.2.5. Rectangular mesh arrangement for two dimensional field computation

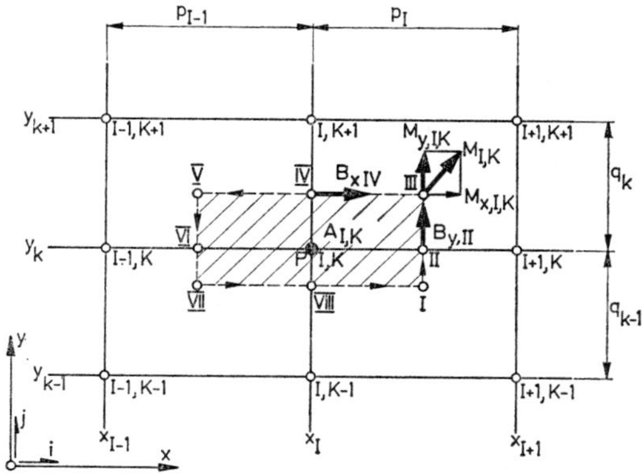

Fig. 2.2.6. Mesh arrangement over a magnet

The triangular grid system is the most flexible system, but also the most inaccurate one, if the system structure is not symmetrical (Fig. 2.2.8). The choice of the grid system and the mesh size determine $S_{I,K}$. We have, however, to make a few more assumptions: The section $S_{I,K}$ must be homogeneous. This means that within the boundaries, given by connecting lines between node points, the electrical and magnetic characteristics of the section are unchanged. Along the borderline, the vector potential $\boldsymbol{A}$ or the scalar potential V should change linearly. Thus the flux density $\boldsymbol{B}_{I,K}$ has, in a rectangular grid system, the following components:

$$B_{x_{I,K}} = \frac{\partial A}{\partial y} \approx \frac{A_{I,K+1} + A_{I+1,K+1} - A_{I,K} - A_{I+1,K}}{2\,q_K}, \qquad (2.2.22)$$

$$B_{y_{I,K}} = \frac{\partial A}{\partial x} \approx \frac{A_{I,K+1} + A_{I,K} - A_{I+1,K+1} - A_{I+1,K}}{2\,p_I}. \qquad (2.2.23)$$

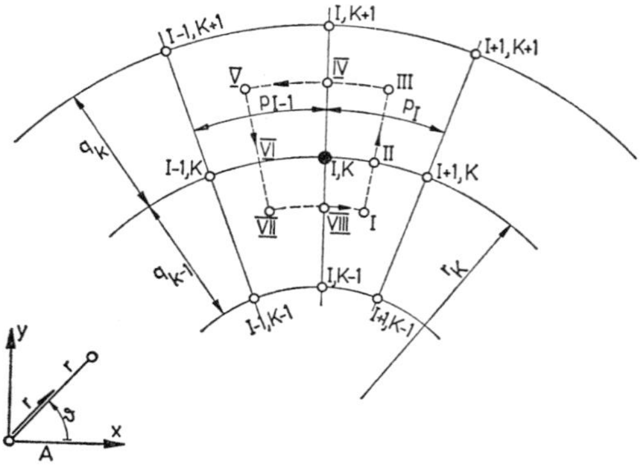

Fig. 2.2.7.
Mesh arrangement in polar coordinates

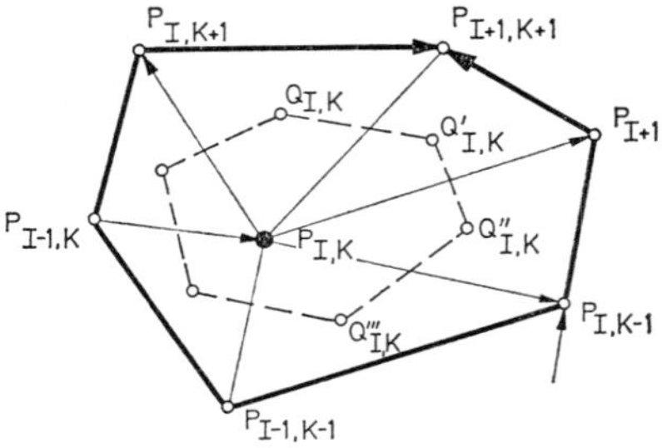

Fig. 2.2.8.
Triangular mesh arrangement

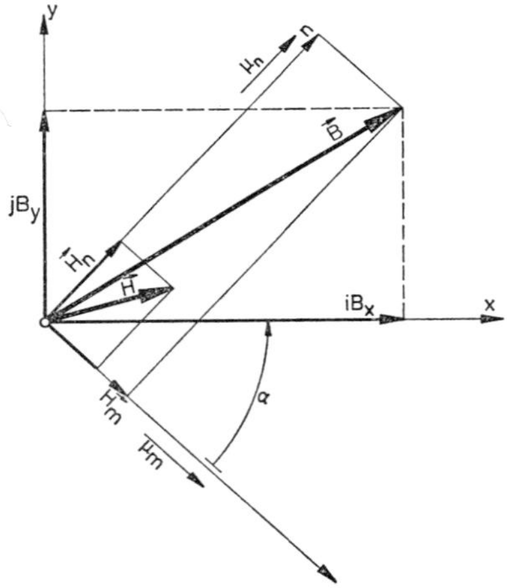

Fig. 2.2.9. Permeabilities at different crystal axis

The flux density in a polar grid system has the components:

$$B_{r_{I,K}} = \frac{\partial A}{r\,\partial\theta} \approx \frac{A_{I,K+1} + A_{I,K} - A_{I+1,K} - A_{I+1,K+1}}{2\,(y_K + 0.5\,q_K)\,p_I}, \qquad (2.2.24)$$

$$B_{\theta\,I,K} = -\frac{\partial A}{\partial r} \approx \frac{A_{I,K} + A_{I+1,K} - A_{I,K+1} - A_{I+1,K+1}}{2\,q_K}. \qquad (2.2.25)$$

For an axially-symmetric system, we may write:

$$B_{r\,I,K} = \frac{\partial A}{\partial z} = \frac{A_{I,K+1} + A_{I+1,K+1} - A_{I,K} - A_{I+1,K}}{2\,q_K}, \quad (2.2.26)$$

$$B_{z\,I,K} = \frac{-\partial A}{r\,\partial r} = \frac{(A_{I,K} + A_{I,K+1}) \cdot r_I - (A_{I+1,K} + A_{I+1,K+1}) \cdot r_{I+1}}{p_I\,(r_I + r_{I+1})}. \quad (2.2.27)$$

The permeability $\mu_{I,K}$ is a scalar in an isotropic iron. As we can neglect hysteresis effects, the value of $\mu_{I,K}$ depends only on the iron magnetization and thus on $\boldsymbol{B}_{I,K}$. In non-isotropic iron, such as grain-oriented silicon steels, widely used in pulsed magnets, the iron permeability has various values in the different crystal axes. In computing the differential Eq. (2.2.4), the various permeability values cannot be considered and, though results may suffer in accuracy, we assume ideal anisotropy which considers only two μ values being perpendicular to each other, as shown in Fig. 2.2.9.

Rectangular Grid

We place a rectangular grid over a region S_h to be considered; the nodes of the grid are points of S_h. In order to evaluate the potential at a particular point, generally a suitable set of difference equations are sought. We treat two cases:

(a) Equal Distance between Node Points (Constant Permeability)

Referring to Fig. 2.2.10, we apply Eq. (2.2.21) to the points, 1, 2, 3, 4, where $x - x_0$ and $y - y_0$ are set equal to $\pm\, h$, the mesh length or height. The points are now placed at the "nodes" of a rectangular network, and the derivatives of $\boldsymbol{A}$ are given by:

$$\left(\frac{\partial A}{\partial x}\right)_0 = \frac{1}{2\,h}\,(A_1 - A_3)\,; \qquad \left(\frac{\partial A}{\partial y}\right)_0 = \frac{1}{2\,h}\,(A_2 - A_4)\,;$$

$$\left(\frac{\partial^2 A}{\partial x^2}\right)_0 = \frac{1}{h^2}\,(A_1 + A_3 - 2\,A_0)\,; \qquad \left(\frac{\partial^2 A}{\partial y^2}\right)_0 = \frac{1}{h^2}\,(A_2 + A_4 - a\,A_0)\,.$$

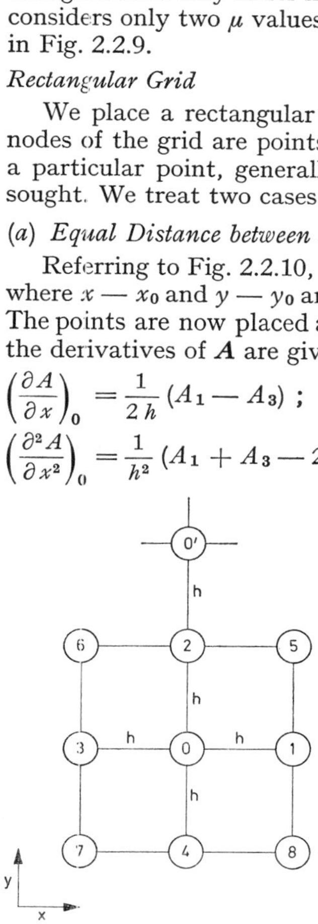

Fig. 2.2.10.
Equidistant node arrangement

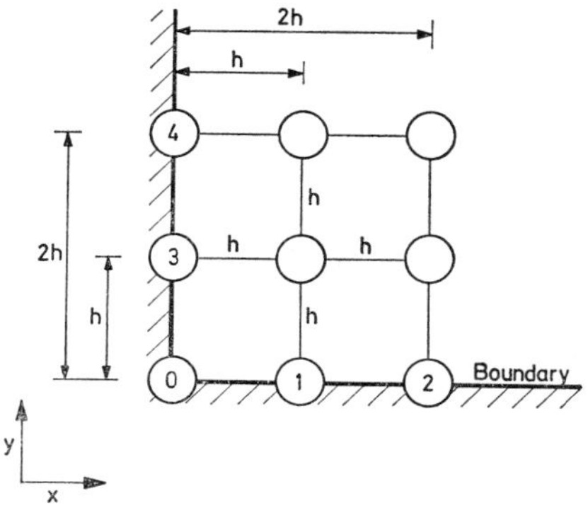

Fig. 2.2.11.
Node arrangement at a boundary

The Laplacian of $\boldsymbol{A}$ is written for a two-dimensional configuration:

$$\nabla^2 \boldsymbol{A} = \frac{\partial^2 A}{\partial x^2} + \frac{\partial^2 A}{\partial y^2} = \frac{1}{h^2} [A_1 + A_2 + A_3 + A_4 - 4 A_0] . \tag{2.2.28}$$

If $\nabla^2 \boldsymbol{A} = 0$, then: $A_1 + A_2 + A_3 + A_4 = 4 A_0$.

Applying Eq. (2.2.21) to diagonal points 5 ... 8, we obtain:

$$\left(\frac{\partial^2 A}{\partial x \, \partial y}\right)_0 = \frac{1}{4 h^2} [A_5 - A_6 - A_7 - A_8] . \tag{2.2.29}$$

Instead of expressing the derivatives in terms of a symmetric point P_0, we may also encounter the situation shown in Fig. 2.2.11, where the nodal point is a corner. In this case, we get from Eq. (2.2.21):

$$\left(\frac{\partial A}{\partial x}\right)_0 = \frac{1}{2 h} [-3 A_0 + 4 A_1 - A_2] , \tag{2.2.30}$$

$$\left(\frac{\partial A}{\partial y}\right)_0 = \frac{1}{2 h} [-3 A_0 + 4 A_3 - A_4] . \tag{2.2.31}$$

(*b*) *Unequal Distance between Node Points (Constant Permeability)*

We refer to Fig. 2.2.12, where the distance $\overline{01} = ph$ and $\overline{02} = qh$, p and q being multiplication factors. We substitute in Eq. (2.2.21) for x, the value

$$x = x_0 + ph$$

and obtain:

$$A_1 = A_0 + ph \left(\frac{\partial A}{\partial x}\right)_0 + \frac{1}{2!} (ph)^2 \left(\frac{\partial^2 A}{\partial x^2}\right)_0 + \frac{1}{3!} (ph)^3 \left(\frac{\partial^3 A}{\partial x^3}\right)_0 + \cdots$$

Inserting in Eq. (2.2.21) for x the value of $x = x_0 - h$, yields:

$$A_3 = A_0 - h \left(\frac{\partial A}{\partial x}\right)_0 + \frac{1}{2!} h^2 \left(\frac{\partial^2 A}{\partial x^2}\right)_0 - \frac{1}{3!} \left(\frac{\partial^3 A}{\partial x^3}\right)_0 + \cdots - \cdots ,$$

forming the summation:

$$A_1 + p A_3 = (1 + p) A_0 + \frac{1}{2!} \cdot p h^2 (1 + p) \left(\frac{\partial^2 A}{\partial x^2}\right)_0 - \frac{1}{3!} p h^3 (1 - p^2) \left(\frac{\partial^3 A}{\partial x^3}\right)_0 + \frac{1}{4!} p h^4 (1 + p^4) \left(\frac{\partial^4 A}{\partial x^4}\right)_0 + - \cdots \tag{2.2.32}$$

For small values of h, higher powers of h can be ignored. The second derivative of A is then:

$$h^2 \left(\frac{\partial^2 A}{\partial x^2}\right)_0 = \frac{2 A_1}{p (1 + p)} + \frac{2 A_3}{1 + p} - \frac{2 A_0}{p} . \tag{2.2.33}$$

For a regular mesh with $p = 1$, one obtains the equation derived in (a). If we substitute for $y = y_0 + qh$, we obtain A_2. Substituting $y = y_0 - h$, we obtain A_4. By using the same technique used above to obtain Eq. (2.2.33), we calculate $A_2 + q A_4$, and get:

$$h^2 \left(\frac{\partial^2 A}{\partial y^2}\right)_0 = \frac{2 A_2}{q (1 + q)} + \frac{2 A_4}{1 + q} - \frac{2 A_0}{q} . \tag{2.2.34}$$

Substituting Eq. (2.2.33) and (2.2.34) in Poisson's Equation for constant permeability in Cartesian coordinates:

$$\left(\frac{\partial^2 A}{\partial x^2}\right)_0 + \left(\frac{\partial^2 A}{\partial y^2}\right)_0 = -\mu J ,$$

we get:

$$\boxed{2\left[\frac{A_1}{p\,(1+p)} + \frac{A_2}{q\,(1+q)} + \frac{A_3}{1+p} + \frac{A_4}{1+q} - \left(\frac{1}{p} + \frac{1}{q}\right) A_0\right] + h^2 \mu J = 0 .} \tag{2.2.35}$$

For $p = q = 1$ and $J = 0$, we have the Laplace equation for a square mesh obtained in Eq. (2.2.28). For $J \neq 0$ and $p = q = 1$, we get Poisson's equation:

$$\boxed{A_1 + A_2 + A_3 + A_4 - 4\,A_0 + h^2 \mu J = 0 .} \tag{2.2.35a}$$

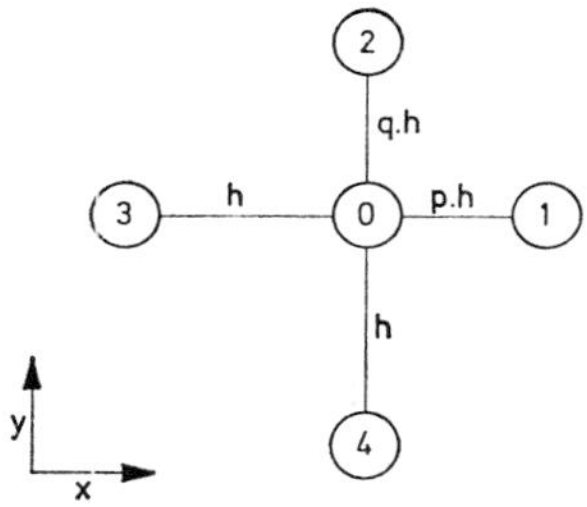

Fig. 2.2.12.
Unequal distances between nodes

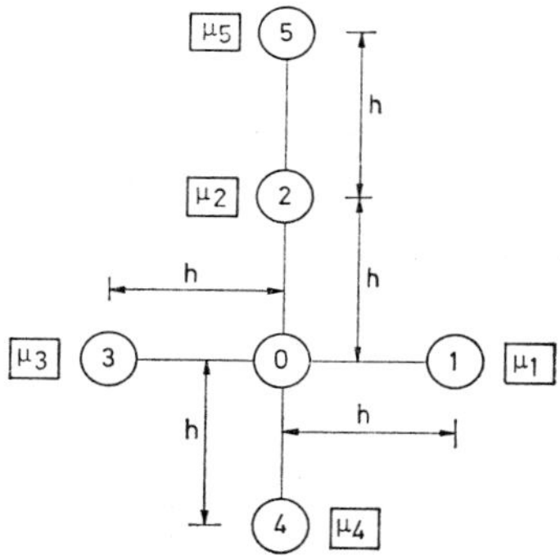

Fig. 2.2.13.
Equidistant nodes having different permeabilities

(c) Equal Distance between Node Points (Variable Permeability)

The permeability at each node point is different, according to the magnetization values of the ferromagnetic material. Referring to Fig. 2.2.13, the first partial derivatives, multiplied by the inverse permeability between nodes, are approximated by the relations:

$$\left.\begin{aligned}
\left(\frac{1}{\mu}\frac{\partial A}{\partial x}\right)_a &= \frac{1}{2\,h}\left(\frac{1}{\mu_0} + \frac{1}{\mu_1}\right)(A_1 - A_0) \\
\left(\frac{1}{\mu}\frac{\partial A}{\partial x}\right)_b &= \frac{1}{2\,h}\left(\frac{1}{\mu_0} + \frac{1}{\mu_3}\right)(A_0 - A_3) \\
\left(\frac{1}{\mu}\frac{\partial A}{\partial y}\right)_c &= \frac{1}{2\,h}\left(\frac{1}{\mu_0} + \frac{1}{\mu_2}\right)(A_2 - A_0) \\
\left(\frac{1}{\mu}\frac{\partial A}{\partial y}\right)_d &= \frac{1}{2\,h}\left(\frac{1}{\mu_0} + \frac{1}{\mu_4}\right)(A_0 - A_4)
\end{aligned}\right\} . \tag{2.2.36}$$

The second derivatives at each center node are approximated by:

$$\frac{\partial}{\partial x}\left(\frac{1}{\mu}\frac{\partial A}{\partial x}\right) = \frac{1}{2\,h^2}\left[\left(\frac{1}{\mu_0}+\frac{1}{\mu_1}\right)(A_1 - A_0) - \left(\frac{1}{\mu_0}+\frac{1}{\mu_3}\right)(A_0 - A_3)\right],$$

$$\frac{\partial}{\partial y}\left(\frac{1}{\mu}\frac{\partial A}{\partial y}\right) = \frac{1}{2\,h^2}\left[\left(\frac{1}{\mu_0}+\frac{1}{\mu_2}\right)(A_2 - A_0) - \left(\frac{1}{\mu_0}+\frac{1}{\mu_4}\right)(A_0 - A_4)\right].$$

Applying these two equations to the nonlinear Poisson equation, we get from:

$$\frac{\partial}{\partial x}\left(\frac{1}{\mu}\frac{\partial A}{\partial x}\right) + \frac{\partial}{\partial y}\left(\frac{1}{\mu}\frac{\partial A}{\partial y}\right) + J = 0\,,$$

the magnetic vector potential at 0:

$$\boxed{A_0 = \frac{\left(\frac{1}{\mu_0}+\frac{1}{\mu_1}\right)A_1 + \left(\frac{1}{\mu_0}+\frac{1}{\mu_2}\right)A_2 + \left(\frac{1}{\mu_0}+\frac{1}{\mu_3}\right)A_3 + \left(\frac{1}{\mu_0}+\frac{1}{\mu_4}\right)A_4 + 2h^2 J_0}{\frac{1}{\mu_1}+\frac{1}{\mu_2}+\frac{1}{\mu_3}+\frac{4}{\mu_0}}.} \tag{2.2.37}$$

2.2.5 Field Intensity in Rectangular Coordinates

The field intensity $\boldsymbol{H}$ is obtained from the finite difference solution by an approximation to the curl of the magnetic vector potential:

$$\boldsymbol{H} = \frac{1}{\mu}\operatorname{curl} A_z(x,y)\cdot \hat{k}\,,$$

with $\hat{k}$ the unit vector in z-direction.
The magnetic field intensity is expressed in the form:

$$\boldsymbol{H} = \frac{1}{\mu}\left[\frac{\partial A_z}{\partial y}\hat{\imath} - \frac{\partial A_z}{\partial x}\hat{\jmath}\right]$$

and thus in the center of the node, we may write for $\boldsymbol{H}$:

$$\boldsymbol{H} = \frac{1}{\mu_0}\left[\frac{A_2 - A_4}{h}\hat{\imath} - \frac{A_1 - A_3}{h}\hat{\jmath}\right] \tag{2.2.38}$$

and on the boundary:

$$\boxed{\boldsymbol{H} = \frac{1}{\mu_0}\left[\frac{-3\,A_0 + 4\,A_2 - A_5}{2\,h}\hat{\imath} - \frac{A_1 - A_3}{2\,h}\hat{\jmath}\right].} \tag{2.2.39}$$

In these equations, the truncation error is in the order (h^2), written as $0\,(h^2)$.

2.2.6 Finite Difference Equations in Cylindrical Coordinates

The method is illustrated in Fig. 2.2.14. With respect to an arbitrary point 0 (r,z), the correspondence between $\boldsymbol{A}$ and the approximated values of $\boldsymbol{A}$ at the various node points is given as:

$$\begin{aligned} A_\theta(r,z) &= A_0, & A_\theta(r;\,z+h) &= A_1,\\ A_\theta(r+h;\,z) &= A_2, & A_\theta(r;\,z-h) &= A_3,\\ A_\theta(r-h;\,z) &= A_4, & A_\theta(r;\,z+2\,h) &= A_5. \end{aligned}$$

The first partial derivatives, multiplied by the inverse value of the permeability and radial distance, are:

$$\left.\begin{aligned}
\left[\frac{1}{\mu r}\frac{\partial}{\partial z}(rA_\theta)\right]_a &= \frac{1}{2h}\left(\frac{1}{\mu_1 r_1}+\frac{1}{\mu_0 r_0}\right)(r_1A_1-r_0A_0)\\
\left[\frac{1}{\mu r}\frac{\partial}{\partial r}(rA_\theta)\right]_b &= \frac{1}{2h}\left(\frac{1}{\mu_2 r_2}+\frac{1}{\mu_0 r_0}\right)(r_2A_2-r_0A_0)\\
\left[\frac{1}{\mu r}\frac{\partial}{\partial z}(rA_\theta)\right]_c &= \frac{1}{2h}\left(\frac{1}{\mu_3 r_3}+\frac{1}{\mu_0 r_0}\right)(r_0A_0-r_3A_3)\\
\left[\frac{1}{\mu r}\frac{\partial}{\partial r}(rA_\theta)\right]_d &= \frac{1}{2h}\left(\frac{1}{\mu_4 r_4}+\frac{1}{\mu_0 r_0}\right)(r_0A_0-r_4A_4)
\end{aligned}\right\}. \qquad (2.2.40)$$

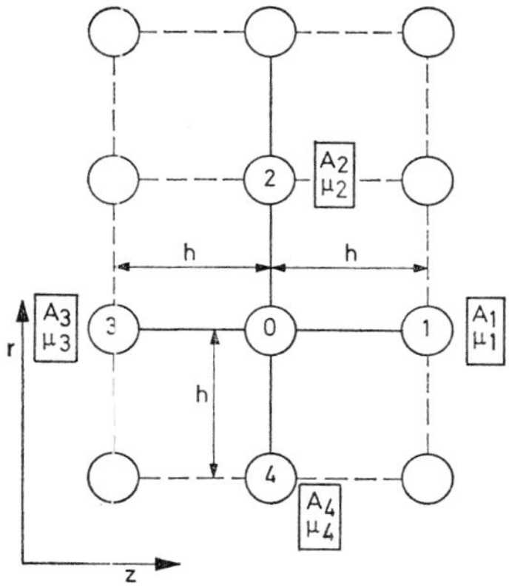

Fig. 2.2.14.

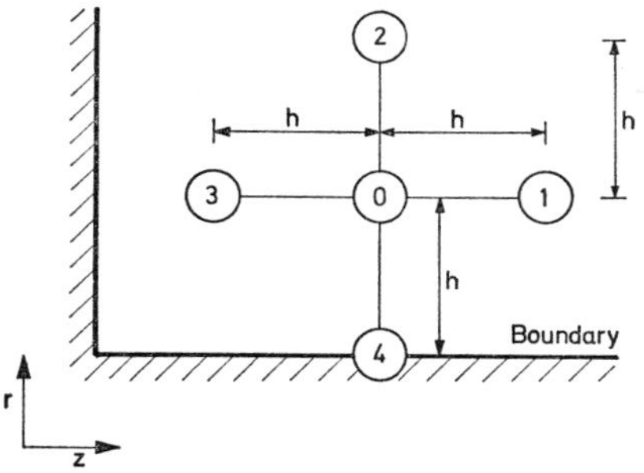

Fig. 2.2.15.

The second partial derivatives at the central node are, accordingly:

$$\left[\frac{\partial}{\partial r}\left(\frac{1}{\mu r}\frac{\partial}{\partial r}(rA_\theta)\right)\right]_0 = \frac{1}{2h^2}\left[\left(\frac{1}{\mu_2 r_2}+\frac{1}{\mu_0 r_0}\right)(r_2A_2-r_0A_0) - \left(\frac{1}{\mu_0 r_0}+\frac{1}{\mu_4 r_4}\right)\cdot(r_0A_0-r_4A_4)\right], \qquad (2.2.41)$$

$$\left[\frac{\partial}{\partial z}\left(\frac{1}{\mu r}\frac{\partial}{\partial r}(rA_\theta)\right)\right]_0 = \frac{1}{2h^2}\left[\left(\frac{1}{\mu_1 r_1}+\frac{1}{\mu_0 r_0}\right)(r_1A_1-r_0A_0) - \left(\frac{1}{\mu_0 r_0}+\frac{1}{\mu_3 r_3}\right)\cdot(r_0A_0-r_3A_3)\right]. \qquad (2.2.42)$$

The Poisson equation, with variable permeability in cylindrical coordinates, was given by:

$$\frac{\partial}{\partial r}\left[\frac{1}{\mu r}\frac{\partial}{\partial r}(rA_\theta)\right]+\frac{\partial}{\partial z}\left[\frac{1}{\mu r}\frac{\partial}{\partial z}(rA_\theta)\right]+J_\theta=0\,, \qquad (2.2.14)$$

where the solution of this equation, in terms of A_0, may be written:

$$A_0 = \frac{\left(\frac{1}{\mu_2 r_2} + \frac{1}{\mu_0 r_0}\right) r_2 A_2 + \left(\frac{1}{\mu_4 r_4} + \frac{1}{\mu_0 r_0}\right) r_4 A_4 + \left(\frac{1}{\mu_1 r_1} + \frac{1}{\mu_0 r_0}\right) r_1 A_1 + \left(\frac{1}{\mu_3 r_3} + \frac{1}{\mu_0 r_0}\right) r_3 A_3 + 2h^2 J_0}{r_0 \left(\frac{1}{r_1 \mu_1} + \frac{1}{r_2 \mu_2} + \frac{1}{r_3 \mu_3} + \frac{1}{r_4 \mu_4} + \frac{1}{r_0 \mu_0}\right)}. \tag{2.2.43}$$

On the symmetry axis, the magnetic vector potential is zero. In the notation of Fig. 2.2.15, we have:

$$r_1 + r_3 = r_0 = h\,, \quad r_2 = 2\,h\,, \quad r_4 = 0\,.$$

Thus:

$$A_0 = \frac{\frac{4}{3}\left(\frac{1}{\mu_2} + \frac{1}{\mu_0}\right) A_2 + \left(\frac{1}{\mu_1} + \frac{1}{\mu_0}\right) A_1 + \left(\frac{1}{\mu_3} + \frac{1}{\mu_0}\right) A_3 + 2h^2 J_0}{\frac{2}{3}\cdot\frac{1}{\mu_2} + \frac{2}{\mu_4} + \frac{1}{\mu_1} + \frac{1}{\mu_3} + \frac{14}{3\,\mu_0}}. \tag{2.2.44}$$

2.2.7 Field Intensity in Cylindrical Coordinates

The field intensity is expressed by

$$\boldsymbol{H} = \frac{1}{\mu} \operatorname{curl} A_\theta \cdot \hat{\varepsilon}_2\,,$$

or in explicit form:

$$\boldsymbol{H} = \frac{1}{\mu}\left[-\frac{\partial A_\theta}{\partial z}\,\hat{\varepsilon}_1 + \frac{1}{r}\frac{\partial}{\partial r}(r A_\theta)\,\hat{\varepsilon}_3\right], \tag{2.2.45}$$

where $\hat{\varepsilon}_1$, $\hat{\varepsilon}_2$, and $\hat{\varepsilon}_3$ are unit vectors in r, θ, and z directions.

The finite difference equation for $\boldsymbol{H}$ at the center node is thus:

$$H = \frac{1}{2\,\mu_0}\left[\left(\frac{A_1 - A_3}{h}\right)_1 \hat{\varepsilon}_1\right] + \frac{1}{2\,\mu_0 r_0}\left[\left(\frac{r_2 A_2 - r_4 A_4}{h}\right)\right] \cdot \hat{\varepsilon}_3\,. \tag{2.2.46}$$

On the symmetry axis $r = 0$, the field intensity is obtained by the limiting process of l'Hospital's rule:

$$\boldsymbol{H} = -\frac{1}{\mu}\frac{\partial A_\theta}{\partial z}\,\hat{\varepsilon}_1 + \frac{2}{\mu}\frac{\partial A_\theta}{2\,r}\,\hat{\varepsilon}_3\,,$$

which can be expressed in a finite difference form by:

$$\boldsymbol{H} = -\frac{1}{2\,\mu_0}\left(\frac{A_1 - A_3}{h}\right)\hat{\varepsilon}_1 + \frac{1}{\mu_0}\left(\frac{-3\,A_0 + 4\,A_2 - A_5}{h}\right)\hat{\varepsilon}_3\,. \tag{2.2.47}$$

2.2.8 Field Problem as a Set of Simultaneous Equations

The aim is to determine the field at any point of a square mesh in terms of other node points. To derive expressions for a larger area with many node points, we expand the vector potential at the point 0 in the x-direction in terms of a Taylor's series.

Node 1:

$$A_1 = A_0 + \frac{h}{1!}\frac{\partial A_0}{\partial x} + \frac{h^2}{2!}\frac{\partial^2 A_0}{\partial x^2} + \frac{h^3}{3!}\frac{\partial^3 A_0}{\partial x^3} + \cdots \quad (2.2.48)$$

Node 2:

$$A_2 = A_0 + \frac{2h}{1!}\frac{\partial A_0}{\partial x} + \frac{4h^2}{2!}\frac{\partial^2 A_0}{\partial x^2} + \frac{8h^3}{3!}\frac{\partial^3 A_0}{\partial x^3} + \cdots \quad (2.2.49)$$

To determine the first derivative in terms of the vector potentials $A_0, A_1, A_2, \ldots$ we multiply Eq. (2.2.48) by 4 and subtract Eq. (2.2.49) from it.

$$\frac{\partial}{\partial x} A_0 = \frac{1}{2h}(4A_1 - A_2 - 3A_0) + \frac{2}{3}h^2\frac{\partial^3}{\partial x^3}A_0 + \cdots \quad (2.2.50)$$

and in the same way:

$$\frac{\partial}{\partial y} A_0 = \frac{1}{2h}(4A_3 - A_4 - 3A_0) + \frac{2}{3}h^2\frac{\partial^3}{\partial y^3}A_0 + \cdots \quad (2.2.51)$$

If the mesh size, and consequently h, is chosen to be small, the higher order terms of Eqs. (2.2.42) and (2.2.43) can be neglected. We may regard these higher terms as error coefficients or "residuals" of the difference equations.

For example, in a 25-point star, the procedure is the same as explained. We obtain:

$$\begin{aligned} A_1 &= A_0 + h\frac{\partial}{\partial x}A_0 + \frac{h^2}{2!}\frac{\partial^2}{\partial x^2}A_0 + \frac{h^3}{3!}\frac{\partial^3}{\partial x^3}A_0 + \cdots, \\ A_2 &= A_0 + (2h)\cdot\frac{\partial}{\partial x}A_0 + \frac{(2h)^2}{2!}\frac{\partial^2}{\partial x^2}A_0 + \frac{(2h)^3}{3!}\frac{\partial^3}{\partial x^3}A_0 + \cdots, \end{aligned} \quad (2.2.52)$$

and:

$$A = A_0 = (3h)\cdot\frac{\partial}{\partial x}A_0 + \frac{(3h)^2}{2!}\frac{\partial^2}{\partial x^2}A_0 + \frac{(3h)^3}{3!}\frac{\partial^3}{\partial x^3}A_0 + \cdots \quad (2.2.53)$$

By the process of elimination, we obtain $\partial/\partial x\,(A_0)$ and $\partial/\partial y\,(A_0)$ in terms of A_1, A_2, A_3, and A_0.

The Poisson equation for a point 1 inside a 25-node grid (Fig. 2.2.16) is obtained:

$$A_6 + A_2 + A_{25} + A_{11} - 4A_1 + h^2\mu J = 0$$

and for a point 2 of the grid:

$$A_5 + A_3 + A_{24} + A_1 - 4A_2 + h^2\mu J = 0\,.$$

Similar equations can be obtained for the remaining points of the grid. The potentials are connected among themselves and to certain boundary conditions imposed upon the system by a set of linear difference equations, which have to be solved simultaneously.

2.2.9 Boundaries with Different Permeabilities

Two square-mesh grids, indicated by pure and primed numbers, are separated by the surface $A - 0 - B$, the media having the permeability μ and μ', respectively, as seen in Fig. 2.2.17. Referring to Eq. (2.2.39), which corresponds to the first derivatives at the point 0, the boundary conditions at 0 are:

$$\frac{1}{\mu}\left(\frac{\partial A}{\partial y}\right)_0 = \frac{1}{\mu'}\cdot\left(\frac{\partial A'}{\partial y}\right)_0 \tag{2.2.54}$$

and $A_0 = A'_0$.

Applying these conditions to Eq. (2.2.50), we get the relation:

$$3\left(1 + \frac{\mu}{\mu'}\right) A_0 = (4\,A_3 - A_4) + \frac{\mu}{\mu'}(4\,A'_3 - A'_4) \tag{2.2.55}$$

and: $\quad 4\,A_3 = A_1 + A_2 + A_0 + A_4\,; \quad 4\,A'_3 = A'_1 + A'_2 + A_0 + A'_4\,.$

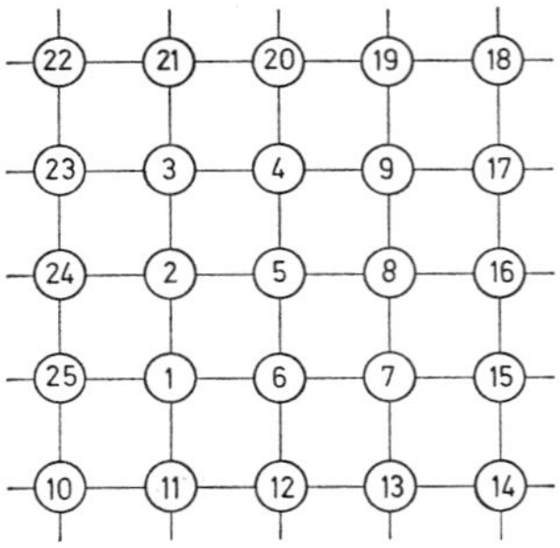

Fig. 2.2.16.
25 node mesh arrangement

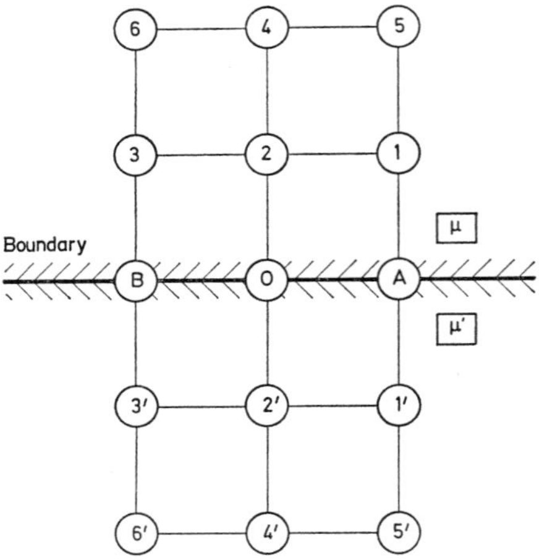

Fig. 2.2.17.
Node arrangement over a boundary having different permeabilities at eachside

The Eq. (2.2.55) can now be rearranged as:

$$2\left(1 + \frac{\mu}{\mu'}\right) A_0 = A_1 + A_2 + \frac{\mu}{\mu'}(A'_1 + A'_2)\,. \tag{2.2.56}$$

The relations for the corner points A and B are in analogy to Eq. (2.2.56):

$$\begin{aligned} 3\left(1 + \frac{\mu}{\mu'}\right) A_A &= (4\,A_1 - A_5) + \frac{\mu}{\mu'}(4\,A'_1 - A'_5) \\ 3\left(1 + \frac{\mu}{\mu'}\right) A_B &= (4\,A_2 - A_6) + \frac{\mu}{\mu'}(4\,A'_2 - A'_6) \end{aligned} \tag{2.2.57}$$

which leads to:

$$4\,A_3 = A_5 + A_6 + A_A + A_B\,, \quad 4\,A'_3 = A'_5 + A'_6 + A_A + A_B\,. \tag{2.2.58}$$

Adding the two Eqs. (2.2.57) and using the relations (2.2.58), we get the vector potential at 0:

$$4\left(1+\frac{\mu}{\mu'}\right)A_0 = 2\,A_3 + A_A + A_B + \frac{\mu}{\mu'}\left[2\,A_3' + A_A + A_B\right]. \tag{2.2.59}$$

2.2.10 Right-Angle Boundary with Different Permeabilities on Each Side

With reference to Fig. 2.2.18, the vector potential A_0 is given in the usual way:

$$4\,A_0 = A_A + A_B + A_2 + A_3\,. \tag{2.2.60}$$

However, in reality, the gradients of the vector potentials in the vicinity of point 0 are of extreme importance, and a coarse mesh, as considered in Eq. (2.2.60), may lead to considerable error. We apply the diagonal Eq. (2.2.30) to the point 0:

$$3\left(1+\frac{\mu}{\mu'}\right)A_0 = (4\,A_1 - A_6) + \frac{\mu}{\mu'}(4\,A_1' - A_6')\,. \tag{2.2.61}$$

Combining this equation with the diagonal equation at A and A', we obtain the rapidly converging relation:

$$2\left(1+\frac{\mu}{\mu'}\right)A_0 = (A_2 + A_3) + \frac{\mu}{\mu'}(A_A + A_B)\,, \tag{2.2.62}$$

Actually, Eq. (2.2.62) does not give very accurate numbers, but is used as a first approximation.

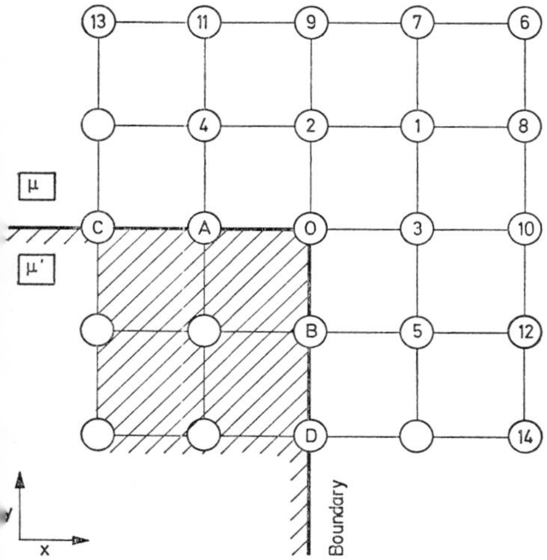

Fig. 2.2.18.
Rectangular shaped boundary

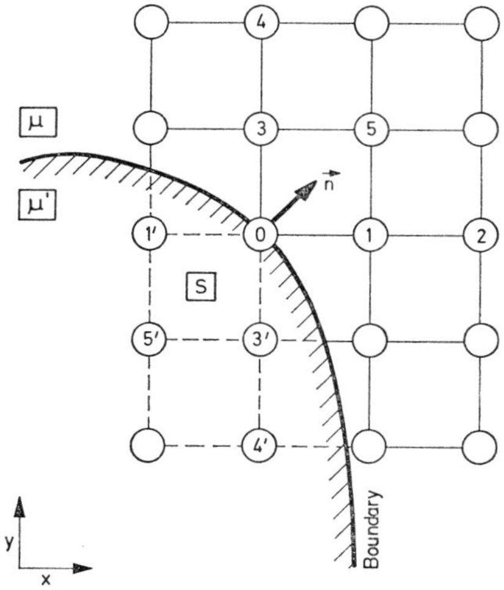

Fig. 2.2.19.
Curved shaped boundary

2.2.11 Curved Boundaries

Among a number of methods to obtain the vector potential at a point 0 of the boundary, the following treatment is simple and rapidly converging. Referring to Fig. 2.2.19, the normal unit vector $\boldsymbol{n}$, pointing outward from the surface S_h at a point 0, is used to solve the boundary problem. By using the components n_x and n_y of the normal unit vector and the boundary condition (Eq. (2.2.54)), we may write:

$$n_x \frac{\partial A}{\partial x} + n_y \frac{\partial A}{\partial y} = \frac{\mu}{\mu'} \left[n_x \frac{\partial A'}{\partial x} + n_y \cdot \frac{\partial A'}{\partial y} \right]. \tag{2.2.63}$$

In order to find the derivatives, we expand A_1 and A_2 by:

$$A_1 = A_0 + h \left(\frac{\partial A}{\partial x}\right)_0 + \frac{h^2}{2!} \left(\frac{\partial^2 A}{\partial x^2}\right)_0 + \frac{h^3}{3!} \left(\frac{\partial^3 A}{\partial x^3}\right)_0, \tag{2.2.64}$$

$$A_2 = A_0 + 2\,h \left(\frac{\partial A}{\partial x}\right)_0 + \frac{4\,h^2}{2!} \left(\frac{\partial^2 A}{\partial x^2}\right)_0 + \frac{8\,h^3}{3!} \left(\frac{\partial^3 A}{\partial x^3}\right)_0. \tag{2.2.65}$$

By multiplying Eq. (2.2.64) by 4 and subtracting Eq. (2.2.65), we obtain:

$$\left(\frac{\partial A}{\partial x}\right)_0 = \frac{1}{2\,h} [4\,A_1 - A_2 - 3\,A_0] + \frac{2}{3} h^3 \left(\frac{\partial^3 A}{\partial x^3}\right)_0, \tag{2.2.66}$$

$$\left(\frac{\partial A}{\partial y}\right)_0 = \frac{1}{2\,h} [4\,A_3 - A_4 - 3\,A_0] + \frac{2}{3} h^3 \left(\frac{\partial^3 A}{\partial x^3}\right)_0. \tag{2.2.67}$$

Assuming the terms of third order and higher are negligible, we may use the derivatives and, after some manipulation, using Eq. (2.2.63), we may write:

$$3\left(1 + \frac{\mu}{\mu'}\right)(n_x + n_y)\,A_0 = n_x \left[(4\,A_1 - A_2) + \frac{\mu}{\mu'}(4\,A_1' - A_2')\right] + n_y \left[(4\,A_3 - A_4) + \frac{\mu}{\mu'}(4\,A_3' - A_2')\right]. \tag{2.2.68}$$

Hence, we may find A_0 from the values of its eight neighbors and the direction of the normal. Eq. (2.2.68) assumes that μ' is uniform in the region near the boundary.

2.2.12 General Boundary Condition

In solving magnetostatic problems, one may simplify the problem or reduce the size of it by introducing the following conditions:

(*a*) *Dirichlet Condition*

The magnetic vector potential at the boundary is either a constant or a known function of the coordinate points, i.e., $\boldsymbol{A}$ = const or $\boldsymbol{A} = f(x, y)$. In the Dirichlet problem, one must give the value of the potential at the boundary points. As long as the boundary happens to coincide with a straight mesh line, the regular five-point Eq. (2.2.20) can be used. However, if the boundary is curved, such as in S in Fig. 2.2.19, there are several approximations to S_h. The first of these is called the interpolation of degree zero. One picks a value of $\boldsymbol{A}$ on S_h which corresponds to the closest points on S_h. Thus, the irregular points Q_1 and Q_2 of Fig. 2.2.20 are given by:

$$A_2 = A\ (Q_1)\ ; \quad A_1 = A\ (Q_2)\ .$$

The truncation error of this approximation is of the order $0\ (h)$ as $h \to 0$.

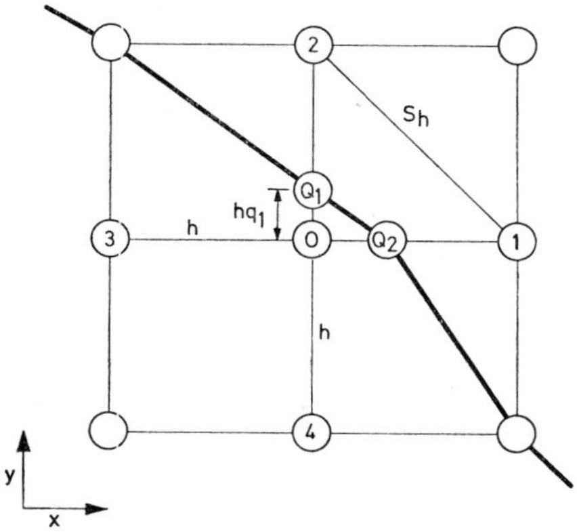

Fig. 2.2.20.

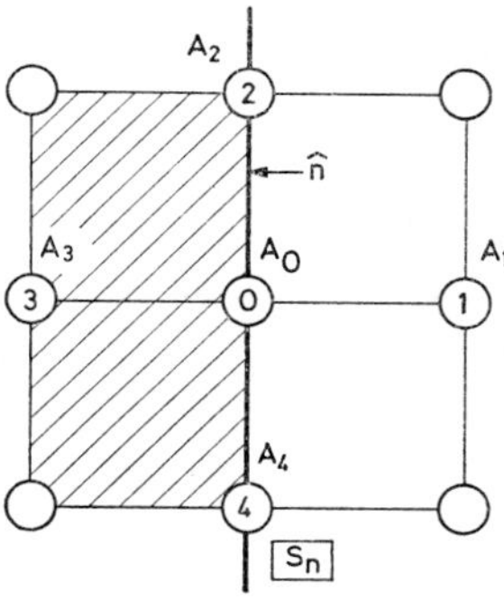

Fig. 2.2.21.

Another scheme, given by Collatz [49], consists of linear interpolation between the values A_S and A_Q. From Fig. 2.2.20, we get:

$$A_0 = \frac{h \cdot A\ (Q_1) + h q_1 \cdot A_S}{h + h q_1}\ . \tag{2.2.69}$$

The reason for finding A_0 instead of A_2 is that for the latter we have:

$$A_2 = \frac{h \cdot A\ (Q_1) + (h q_1 - h)\ A_0}{h q_1}\ . \tag{2.2.70}$$

The quantity $hq_1 - h$ may not be positive and may lead to difficulties in solving system equations. The second method gives truncation errors in the order $0\ (h^2)$ as long as $\boldsymbol{A}$ is sufficiently smooth near the boundary (interpolation of degree one).

(b) Neumann Condition

The gradient of the magnetic vector potential at the boundary surface is zero. The field lines penetrate the boundary surface at right angles. Thus the tangential field component is zero.

$$\partial \boldsymbol{A}/\partial \hat{n} = 0\ ; \quad B_t = 0\ .$$

Suppose the problem is symmetric about the line S_n, as shown in Fig. 2.2.21; then the field lines must pass perpendicularly to S_h. Hence, $B_t = 0$, and we require that $\partial \boldsymbol{A}/\partial \hat{n} = 0$. This condition is satisfied if we state:

$$\boldsymbol{A}\ (x, y) = \boldsymbol{A}\ (-\ x, y) \tag{2.2.71}$$

or equivalently: $A_1 = A_3$.

If the regions to the left of $\hat{n}$ in Fig. 2.2.21 were a material of infinite permeability, then the above equations would also be used.

(*c*) *Symmetry Condition*

The magnetic vector potential is unchanged when the signs of the coordinates are reversed, i.e.,

and alternately,

$$\left.\begin{aligned} \boldsymbol{A}(x,y) &= \boldsymbol{A}(-x,y) \\ \boldsymbol{A}(x,y) &= \boldsymbol{A}(x,-y) \end{aligned}\right\}. \tag{2.2.72}$$

(*d*) *Anti-Symmetry Condition*

The magnetic vector potential changes its sign when the sign of the coordinates is changed, i.e.,

and alternately,

$$\left.\begin{aligned} \boldsymbol{A}(x,y) &= -\boldsymbol{A}(-x,y) \\ \boldsymbol{A}(x,y) &= -\boldsymbol{A}(x,-y) \end{aligned}\right\}. \tag{2.2.73}$$

In all cases, (b), (c) and (d), we may use a five-point operator on such symmetry or antisymmetry axes, provided the sign of the potential is changed appropriately. For a square mesh, the equations are still of a positive type and have an associated error in the order of $0\,(h^2)$.

2.2.13 Solution of Difference Equations

As seen for any practical potential problem, the number of difference equations obtained is substantial. Each equation contains few terms. Reducing the number of equations to one or a few solvable is impractical. However, due to the few terms in each equation, numerical computation methods (generally called *iterative* methods) are used to solve a set of simultaneous equations. There are at least three types of iterative methods:

(a) Straight relaxation
(b) Over-relaxation
(c) Under-relaxation

The iterative method is based on a technique of successive approximations. In electromagnetic or magnetostatic problems, the method converges to an acceptable and correct solution, when the initial approximations or guesses are logical.

The relaxation method is based on continuous modification of the potential values until all the simultaneous equations are satisfied. General relaxation rules are difficult to formulate, but the relaxation methods are sufficiently simple, and some basic experience will lead to the solution of numerical problems.

2.2.14 Concept of Residuals

When the potentials $A_0 \ldots A_4$ (e.g., in a five-point star) are chosen to satisfy Poisson's equation, the right-hand side of Eq. (2.2.35a) is zero. When, however, the potentials do not satisfy Poisson's equation, the right-hand side may differ from zero by a quantity R_0, known as "Residual", and is expressed as:

$$\boxed{A_1 + A_2 + A_3 + A_4 - 4\,A_0 + h^2 \mu J = R_0 \,.} \tag{2.2.74}$$

The computational method is directed toward a systematic reduction of R_0 towards zero. It is generally satisfactory to set as an upper R_0 value some arbitrary acceptable small number. The choice of R_0 has an influence on the number of iterations to achieve the requested accuracy and the computation time.

The relaxation procedure consists of changing the starting values of the unknowns one or more at a time until the residuals become negligibly small. At the node under consideration, the value of A_K is changed by δA_K; then the residual is changed by $-\delta A_K$, while all other residuals change by $C_K \delta A_K$. Hence, to reduce R_0, we change A_K by $\delta A_K = R_0$. This means, in Eq. (2.2.74), that the residual R_0 is reduced to zero by the change of A_0 by $R_0/4$. This change in A_0 causes four changes in the residual, one at each adjacent node of the star. Applying Eq. (2.2.74) to each of these nodes in turn, it can be seen that the change they experience is again $R_0/4$.

This prodecure in reducing the residual is known as *point relaxation*, compared to *line relaxation*, where several points are relaxed simultaneously. The use of line relaxation methods is indicated over a group of nodes when the residuals have the same sign.

2.2.15 A Computational Method

The first step in computation is to establish boundary conditions, as discussed in Section 2.2.12. Thus, if the problem is symmetric upon one or several boundary surfaces, only a section of the problem must be computed. This is of particular importance if the available storage capacity of the computer would prevent the choice of small mesh sizes jeopardizing the accuracy of the result.

As stated, the nonlinear Poisson equation is written in Cartesian coonrdiates:

$$\frac{\partial}{\partial x}\left(\frac{1}{\mu}\frac{\partial A}{\partial x}\right) + \frac{\partial}{\partial y}\left(\frac{1}{\mu}\frac{\partial A}{\partial y}\right) + J = 0 \tag{2.2.75}$$

where the material permeability is expressed by $\mu = \mu\,(|\text{curl}\,A|)$. Investigating Eq. (2.2.75) over the area A and the contour C, we may write it in integral form:

$$\iint\limits_A \left[\frac{\partial}{\partial x}\left(\frac{1}{\mu}\frac{\partial A}{\partial x}\right) + \frac{\partial}{\partial y}\left(\frac{1}{\mu}\frac{\partial A}{\partial y}\right)\right] dx\,dy + \iint\limits_A J\,dx\,dy = 0\,. \tag{2.2.76}$$

Applying Green's Theorem to this equation, we obtain:

$$\int_C \left(\frac{1}{\mu}\frac{\partial \boldsymbol{A}}{\partial x} dy - \frac{1}{\mu}\frac{\partial \boldsymbol{A}}{\partial y} dx\right) + \iint_A J\, dx\, dy = 0 . \tag{2.2.77}$$

According to Eq. (2.2.36), we may write Eq. (2.2.77) in the form of difference equations:

$$\begin{aligned} &\frac{h_2 + h_4}{2}\left[\left(\frac{1}{\mu_1} + \frac{1}{\mu_0}\right)\frac{A_1 - A_0}{2\,h_1} - \left(\frac{1}{\mu_3} + \frac{1}{\mu_0}\right)\frac{A_0 - A_1}{2\,h_3}\right] \\ &+ \frac{h_1 + h_3}{2}\left[\left(\frac{1}{\mu_2} + \frac{1}{\mu_0}\right)\frac{A_2 - A_0}{2\,h_2} - \left(\frac{1}{\mu_4} + \frac{1}{\mu_0}\right)\frac{A_0 - A_4}{2\,h_4}\right] \\ &+ J\,\frac{(h_2 + h_4)}{2}\cdot\frac{(h_1 + h_3)}{2} = 0 . \end{aligned} \tag{2.2.78}$$

We assume, for simplicity, the current density to be constant. Introducing the abbreviations

$$\alpha_1 = \frac{1}{h_1\,(h_1 + h_3)}\cdot\left(\frac{1}{\mu_1} + \frac{1}{\mu_0}\right), \qquad \alpha_2 = \frac{1}{h_2\,(h_2 + h_4)}\cdot\left(\frac{1}{\mu_2} + \frac{1}{\mu_0}\right),$$

$$\alpha_3 = \frac{1}{h_3\,(h_1 + h_3)}\cdot\left(\frac{1}{\mu_3} + \frac{1}{\mu_0}\right), \qquad \alpha_4 = \frac{1}{h_4\,(h_2 + h_4)}\cdot\left(\frac{1}{\mu_4} + \frac{1}{\mu_0}\right),$$

we can simplify Eq. (2.2.78):

$$A_0 = \frac{1}{\alpha_1 + \alpha_2 + \alpha_3 + \alpha_4}\,[\alpha_1 A_1 + \alpha_2 A_2 + \alpha_3 A_3 + \alpha_4 A_4 - J] , \tag{2.2.79}$$

or:

$$\sum_{i=1}^{4} \alpha_i A_i - A_0 \sum_{i=1}^{4} \alpha_1 + J = 0 .$$

Using the notation:

$$\alpha_0 = -\sum_{i=1}^{4} \alpha_i ,$$

we may write:

$$\boxed{\sum_{i=0}^{4} \alpha_i A_i + J = 0 .} \tag{2.2.80}$$

The second derivatives of the potential are given by:

$$\left.\begin{aligned} \frac{\partial}{\partial x}\left(\frac{1}{\mu}\frac{\partial A}{\partial x}\right)\bigg|_{x,y} &= \frac{1}{h_1 + h_3}\left[\left(\frac{1}{\mu_1} + \frac{1}{\mu_0}\right)\frac{A_1 - A_0}{h_1} - \left(\frac{1}{\mu_3} + \frac{1}{\mu_0}\right)\frac{A_0 - A_3}{h_3}\right] \\ \frac{\partial}{\partial y}\left(\frac{1}{\mu}\frac{\partial A}{\partial y}\right)\bigg|_{x,y} &= \frac{1}{h_2 + h_4}\left[\left(\frac{1}{\mu_2} + \frac{1}{\mu_0}\right)\frac{A_2 - A_0}{h_2} - \left(\frac{1}{\mu_4} + \frac{1}{\mu_0}\right)\frac{A_0 - A_4}{h_4}\right] \end{aligned}\right\} . \tag{2.2.81}$$

Boundary conditions lead to the following simplifications:

Symmetry condition: $A_2 = A_4$ and $A_1 = A_3$,

Antisymmetry conditions: $A_2 = -A_4$ and $A_1 = -A_3$,

Neumann condition: $\partial \boldsymbol{A}/\partial \hat{n} = 0$, which yield:

$$\frac{A_2 - A_4}{h_2 + h_4} = B_x = \frac{\partial A}{\partial y} = 0 \qquad \therefore A_2 = A_4\,,$$

$$\frac{A_1 - A_3}{h_1 + h_3} = B_y = \frac{\partial A}{\partial x} = 0 \qquad \therefore A_1 = A_3\,.$$

The relation used quite commonly at each interior mesh point after $(n+1)$ iterations is given by: $A_0^{(n+1)} = A_0^{(n)} - \omega R^{(n)}$

with: (2.2.82)

$$R^{(n)} = \left[\frac{1}{\alpha_0}\alpha_1 A_1^{(n)} + \alpha_2 A_2^{(n)} + \alpha_3 A_n^{(n+1)} + \alpha_4 A_4^{(n+1)} + \alpha_0 A_0^{(n)} - J\right].$$

Referring to Figs. 2.2.22a, b and 2.2.23a, b, the computation is performed from left to right, from row to row, or in each row, moving upwards in the grid. n is the iteration number and ω is the over-relaxation num-

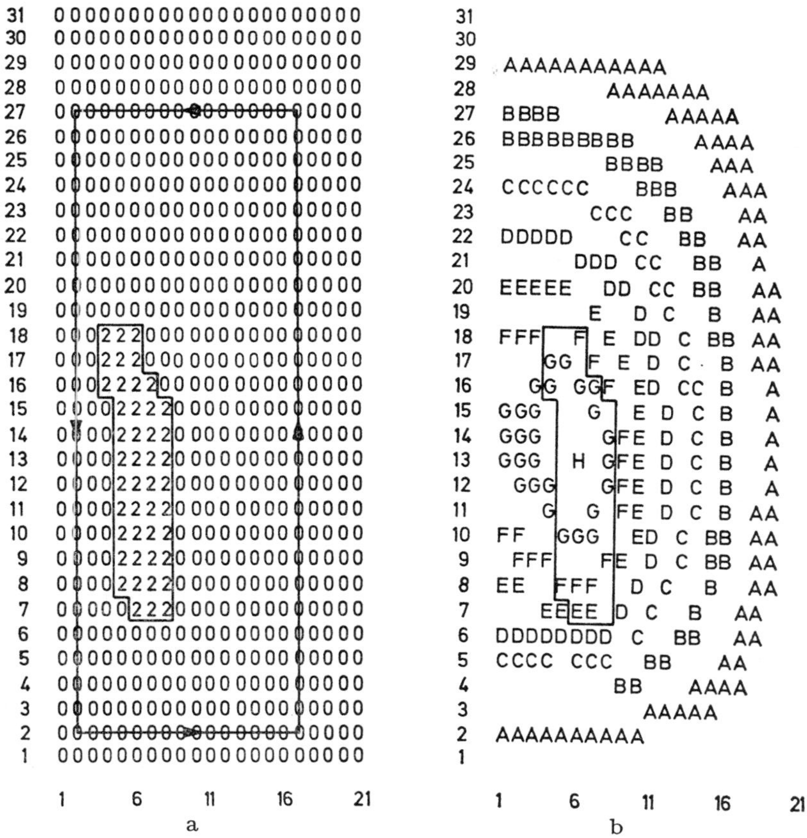

Fig. 2.2.22a. Computer data input for a coil *0* air, *2* coil

Fig. 2.2.22b. Lines of constant vectorpotential for a coil illustrated in Fig. 2.2.22a

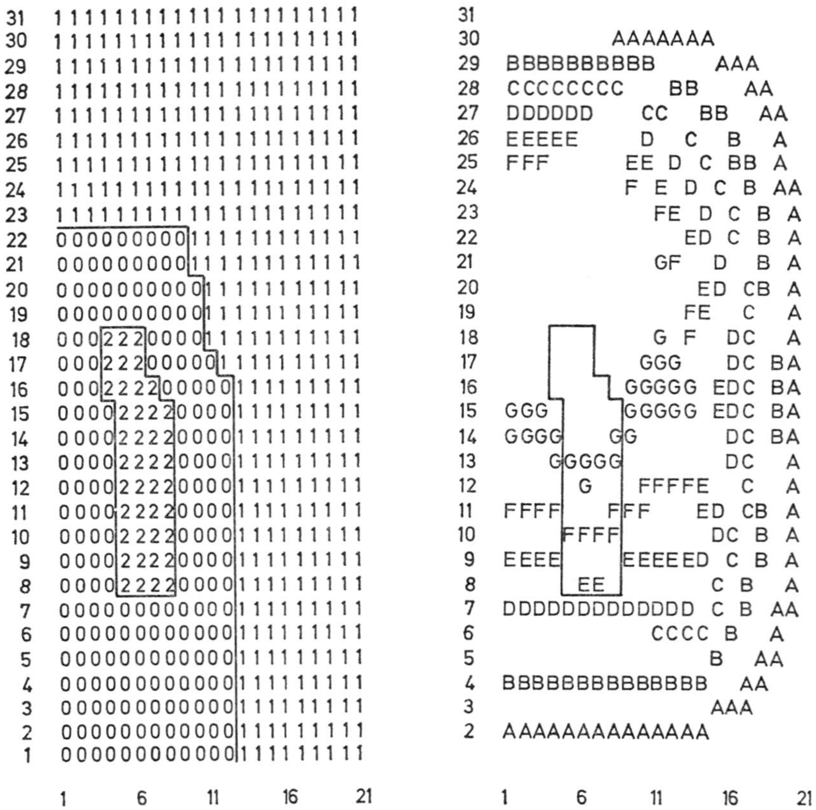

a b

Fig. 2.2.23 a. Two dimensional iron bound magnet geometry
0 air, *1* iron, *2* coil
Fig. 2.2.23 b. Lines of constant vectorpotential

ber. If the problem is linear, which, in magnet problems, means that the permeability is assumed to be independent of the magnetic field strength, then the convergence rate can be increased by the proper choice of ω.

For a rectangle of $(p+1)\cdot(q+1)$ points, a good approximation to the optimum ω have been found by Frankel and Young [51]

$$\omega_F = 2 - \sqrt{2}\,\pi\left[\frac{1}{p^2}+\frac{1}{q^2}\right]^{\frac{1}{2}}. \tag{2.2.83}$$

For a stable problem, ω_F is chosen less than ω_{opt}, which has values of $1 < \omega_{\text{opt}} < 2$.

To illustrate the influence of ω on the error as a function of the iteration numbers three values of ω were chosen: $\omega = 1.00$; $\omega = 0.8\,\omega_F$ and $\omega = \omega_F$. The relative error, ε, is plotted in Fig. 2.2.25 as a function of iteration number. We see the problem in stable for values of $\omega \leqq \omega_F$. In

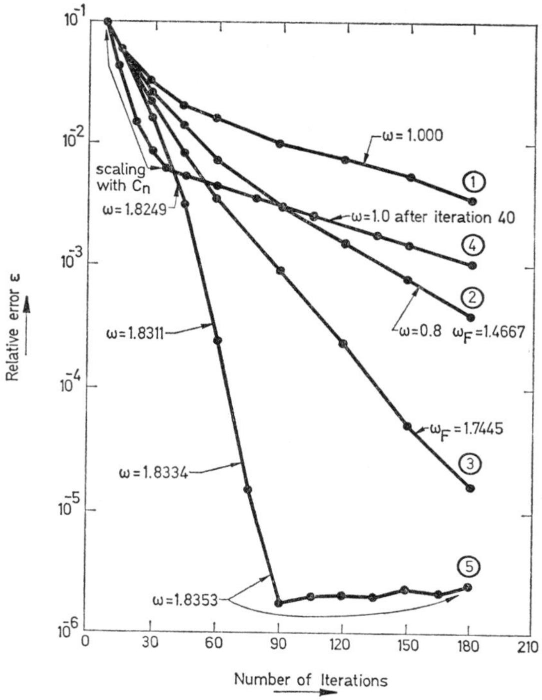

Fig. 2.2.24.

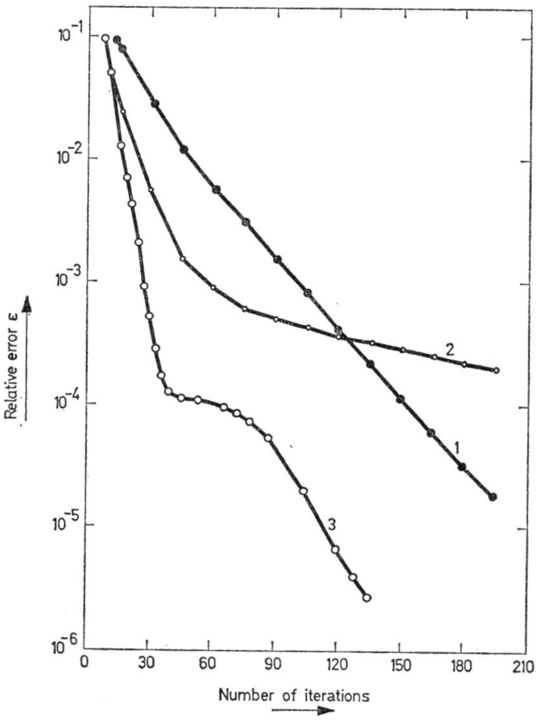

Fig. 2.2.25.

this case, ω_F must be smaller than ω_{opt}. Curve (5) has already converged after 90 iterations and oscillates around the value of 2×10^{-6} with further iterations.

Varga [51] extended the original successive over-relaxation theory (SOR) of Frankel and Young to P-cyclic matrices and gives a relationship between the modulus of the maximum eigen value of T_{SOR} the iteration matrix, denoted by λ, and the value of SOR parameter ω

$$(\lambda + \omega - 1)^P = \lambda^{P-1} \cdot \omega^P \gamma \,. \tag{2.2.84}$$

Here is γ the maximum eigen value of T_J. From this value, it may be shown that an optimum value for ω for the case $P = 2$ is:

$$\boxed{\omega_{opt} = \frac{2}{1 + (1 - \gamma^2)^{\frac{1}{2}}}.} \tag{2.2.85}$$

The next step to solve Poisson's equation is given by Ahamed and Erdelyi [28]. The ratio:

$$\boxed{C_n = \frac{\iint\limits_S \boldsymbol{J} d\boldsymbol{s}}{\oint \boldsymbol{H} d\boldsymbol{l}}.} \tag{2.2.86}$$

must be calculated during the first iterations. If $C_n = 1$, one obtains the modified Ampere's Law:

$$\oint \boldsymbol{H} d\boldsymbol{l} = \iint\limits_S \boldsymbol{J} d\boldsymbol{s} \,.$$

The number of initially used Ampere-turns in the system determines the current $I = \iint\limits_S \boldsymbol{J} d\boldsymbol{s}$. However, $\oint \boldsymbol{H} d\boldsymbol{l} = NI$ is calculated separately, using $\nabla \times \boldsymbol{A}$ and the numerical values of the vector potential.

If the calculated values of $\boldsymbol{H}$ are too large, C_n will be smaller than unity, which yields a smaller $\nabla \times \boldsymbol{A}$ and hence a reduction of $\oint \boldsymbol{H} d\boldsymbol{l}$. Ideally, C_n should be unity, but in practice, this is seldom achieved due to oscillations around the true value. With $0.96 < C_n < 1.04$, most problems are solved with no difficulty.

The method of Ahamed and Erdelyi [28] is applied to the problem of Fig. 2.2.24. The results are illustrated in curve (4). During the first 40 iterations, the value of C_n was calculated along the path shown in Fig. 2.2.22a. After 40 iterations, C_n was essentially stable and varied in the range of $0.995 < C_n < 1.005$.

A third step is under-relaxation of the permeability at each point in the grid for nonlinear problems, such as the magnet shown in Fig. 2.2.23a, b. Fig. 2.2.25 shows errors as a function of iteration numbers. In this figure, curve (1) is calculated according to the equation of Winslow. In the beginning, $\omega < \omega_{opt}$ $(0.8\times\omega_F)$ was chosen. After each iteration a new value of ω is found from λ_i, given by Eq. (2.2.84). The iron permeability is calculated after each iteration and under-relaxed by $\gamma = 0.85$.

Curve (2) is illustrated with $\omega = 1.0$ in iron. After 47 iterations, $\omega = 1$ is used throughout. The permeability calculated after each iteration is relaxed by $\gamma = 0.85$.

Curve (3) is a combination of the two previous curves.

2.2.16 The Iron-Air Interface

The boundary condition on the iron-air interface can be given by:

$$\left(\frac{\partial A}{\partial n}\right)_{\text{air}} = \frac{1}{\mu}\left(\frac{\partial A}{\partial n}\right)_{\text{iron}}, \tag{2.2.87}$$

where $\partial A/\partial n$ is the derivative of the vector potential along the normal-to-the-iron surface (see Neumann Condition). Continuity of A across the boundary guarantees that the condition of the normal component of B is satisfied.

The boundary condition (Eq. 2.2.87) is replaced by a difference equation. Using mesh points on the air side, we may write:

$$\left(\frac{\partial A}{\partial n}\right)_{\text{air}} = \sum_{i=0}^{2} \beta_i A_i \,. \tag{2.2.88}$$

The three mesh points used here include the boundary points being considered and its two neighbors on the air side that lie closest to the normal.

Referring to Fig. 2.2.26, the point 1 is the boundary point at which we want to find $\partial A/\partial n$. Point 0 is the associated interior point.

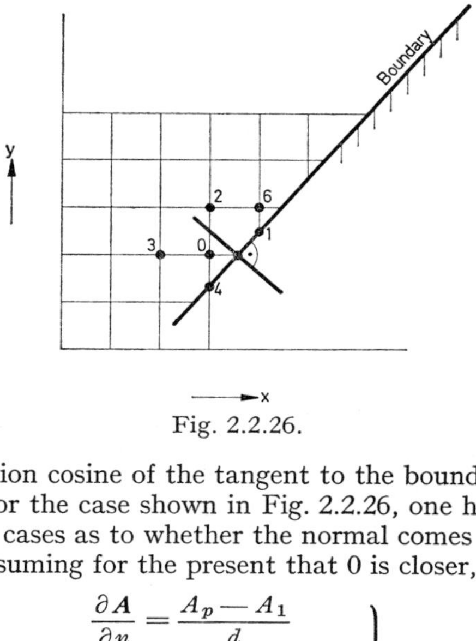

Fig. 2.2.26.

Let the direction cosine of the tangent to the boundary be indicated by C_x and C_y. For the case shown in Fig. 2.2.26, one has to distinguish between the two cases as to whether the normal comes closer to point 0 or to point 6. Assuming for the present that 0 is closer, then:

$$\left.\begin{aligned} \frac{\partial A}{\partial n} &= \frac{A_p - A_1}{d} \\ \text{with:}\qquad A_p &= a_2 A_2 + a_0 A_0 \\ a_2 &= (y_p - y_0)/r_{20} \\ a_0 &= 1 - a_2 \end{aligned}\right\} . \tag{2.2.89}$$

r_{20} is the positive distance between 2 and 0. x_p and y_p are the coordinates relative to point 1 of point p, where the normal intersects the x coordinate line between 0 and 2. x_0, y_0 are the coordinates of point 0 relative to point 1. We can eliminate y_p from Eq. (2.2.89) by using $x_p = x_0$ and the relation

$$x_p c_x + y_p c_y = 0 .$$

We find:

$$a_2 = -\frac{1}{r_{20}}(x_0 c_x + y_0 c_y) \cdot \frac{1}{c_y}; \quad d = |x_0/c_y|$$

$$\beta_2 = a_2/d; \quad \beta_0 = (1 - a_2)/d; \quad \beta_1 = -(\beta_2 + \beta_0) . \qquad (2.2.90)$$

To test whether one should use point 0 or point 6, one checks if $0 < a_2 < 1$; if true, one uses point 0, and if not true, one uses point 6.

In several computer programs, the outer iron boundary is also chosen as the boundary of the mesh. This method assumes that no magnetic flux escapes from the iron (no leakage). This means that $\boldsymbol{A}$ is constant along the outer boundary of the iron. For convenience at the outer border line we use, $\boldsymbol{A} = 0$. This method has the advantage of simplicity. In superconducting magnets with high interior magnetic field, some flux will always leak out past the iron, and the assumption of the outer boundary is not adequate. However, the iteration mesh cannot extend to infinity either, and some boundary condition is required that will allow the mesh to be extended at some reasonable point. Parzen [41] assumes that at some distance of the geometrical magnet center of a dipole magnet, the vector potential will of the form

$$\boldsymbol{A} \sim f(\theta)\, r ,$$

from which one obtains the relation at large r:

$$\boxed{\frac{\partial A}{\partial r} + \frac{1}{r} A = 0 .} \qquad (2.2.91)$$

This equation is used to obtain the iteration equation for the points on the mesh boundary. This equation can be generalized for quadrupole or multipole magnets.

Using a cylindrical mesh, Eq. (2.2.91) gives a different relation:

$$A_0\left(1 + \frac{h}{r}\right) - A_3 = 0 \qquad (2.2.92)$$

where h is the mesh interval in the r direction, A_0 is the vector potential at the mesh boundary point, and A_3 is the vector potential at the nearest neighboring mesh point along the r direction. This relationship satisfies the diagonal dominance requirement for the iteration matrix.

To compute Eq. (2.2.91) for a square mesh, we assume only that part of the mesh boundary on which x is constant. Then we can replace Eq. (2.2.91) by the difference equation:

$$A_0\left(\frac{c_x}{h} + \frac{c_y}{l} + \frac{1}{r}\right) - c_x A_3 - c_y A_4 = 0 \qquad (2.2.93)$$

with: A_3 the left horizontal neighbor with a lower x coordinate, and A_4 the lower vertical neighbor with a lower y coordinate. h and l are the horizontal and vertical mesh intervals. The choice of which neighbor to use in computing $\partial \boldsymbol{A}/\partial y$ is determined by the diagonal dominance requirement. A similar difference equation is found for the other sides of the rectangular mesh boundary.

Application of Eq. (2.2.91) requires that the mesh allow for some air region past the iron shield. How large an air region to allow can be determined by doing computer runs using different air region extents and noting the effect on the field results. The choice of adequate air boundaries is of particular importance in high-homogeneity magnets, where full accessibility in axial and radial directions is required.

As an example, Fig. 2.2.27 illustrates an eight-order coil without iron. The field of this axially-symmetric magnet was computed by the method

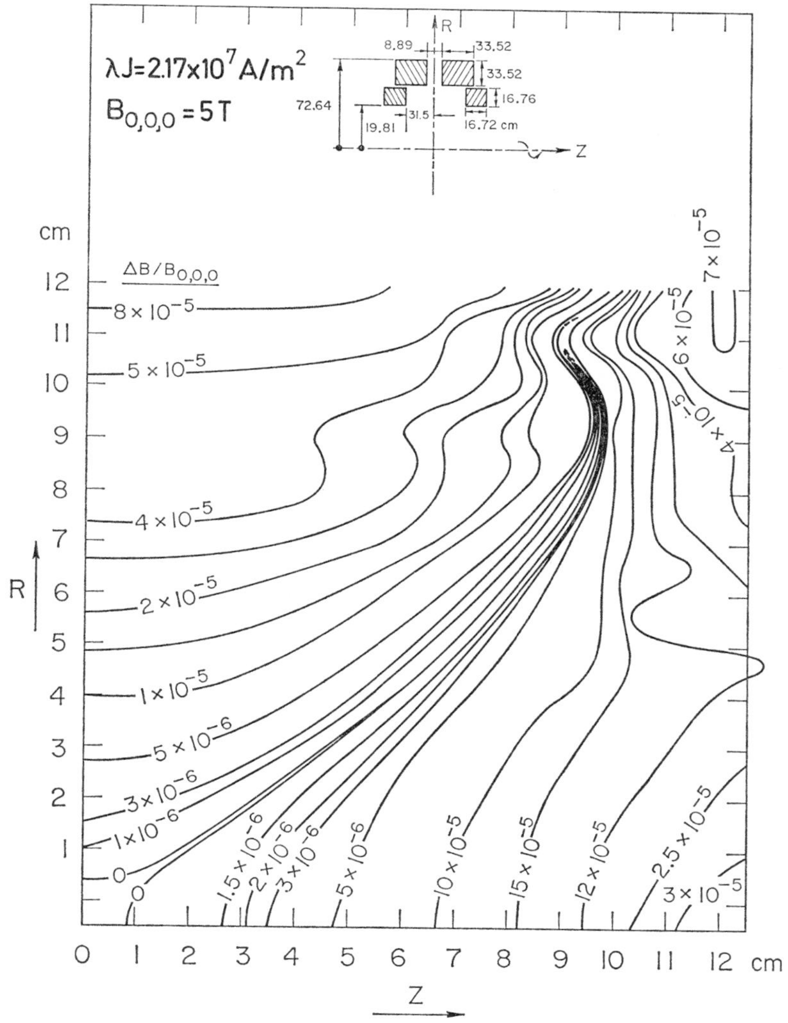

Fig. 2.2.27. Computed lines of constant field in an eight order coil

of scalar potentials (Mark II) and with the vector potential method, using overall $\mu_r = 1$ and constant current density. The requirement was to achieve, over an axial and radial distance of $\pm$ 5 cm, a homogeneity of 10^{-5}. The central field is 5 T.

Introducing an iron shield around the coil, as shown in Fig. 2.2.28 yielded a higher field of 5.45 T, but also lead to a field distortion (shown

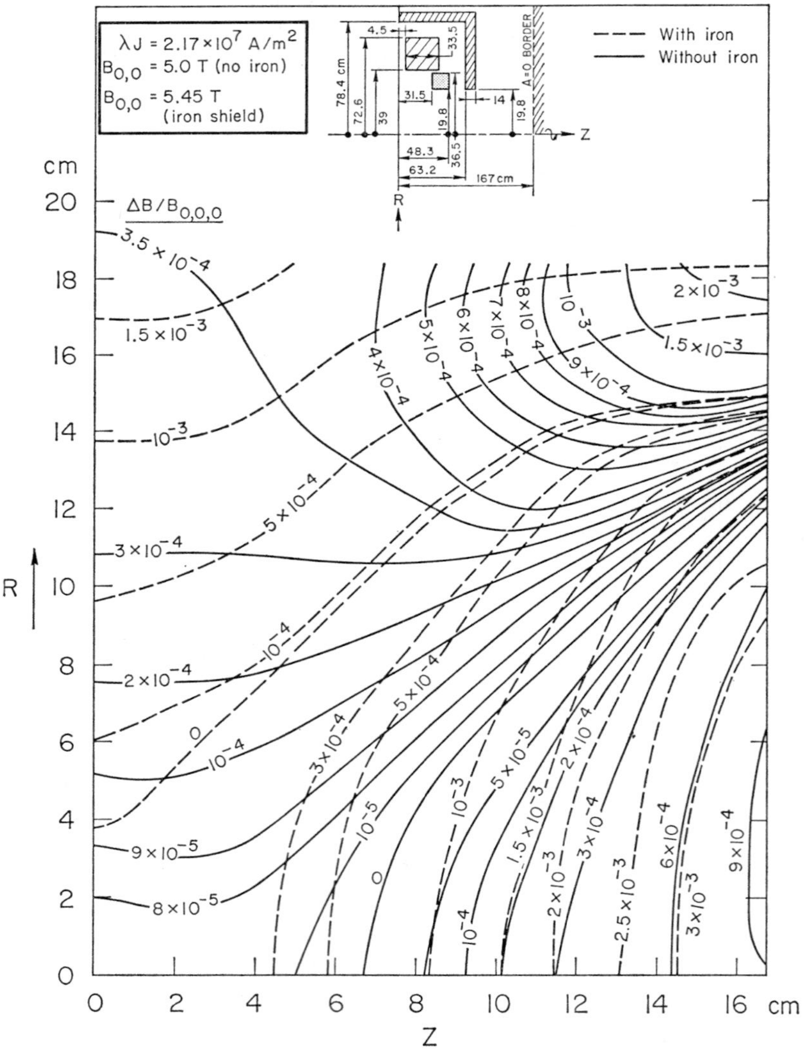

Fig. 2.2.28. Computed lines of constant field in an iron bound 8th order coil (Fig. 2.2.27). The line distortion is due to the inaccurate position of the boundaries and the computational precision

by plotting lines of constant field homogeneity). Reasons are summarized below:

1. The variable permeability computation must converge to an error of less than 10^{-7} if the data must be trustworthy. Single precision used in the NUTCRACKER [39] code yielded an error of $\sim 10^{-6}$ after 500 iterations. The error function was oscillating around this number and the problem did not converge further. Double precision improved the accuracy and an error of 10^{-7} was obtained.

2. With the SLAC computer (361/91), a magnetostatic problem with 150×150 nodes, or any combination not exceeding 2.25×10^4 nodes, can be solved. Requiring double precision reduces the area (due to the limitation of storage capacity) to 100×100 mesh points. If, in an open type magnet (with or without ferromagnetic shields), the potential boundary ($\boldsymbol{A} = 0$, or $\boldsymbol{A} =$ constant) is close to the magnet extension (as in Fig. 2.2.28), the computational errors are in the order of 10^{-2} or 5×10^{-3}. Contour plots for lines of constant $\Delta \boldsymbol{B}/\boldsymbol{B}$ are distorted. In order to minimize the influence of the boundaries, the extensions of the grid must be quite large, which is not possible due to storage limitations and computation time required. (The duration of each iteration was an average of about 3 seconds.)

An attempt to improve this deficiency is the following: Using large mesh units, which reduces calculation accuracy but also the size of the magnet within the grid, the distribution of $\boldsymbol{A}$ within the boundary of 100×100 mesh units is calculated, choosing as outer boundary the value of $\boldsymbol{A} = 0$. If the extension of the grid ($\boldsymbol{A} = 0$ lines to the axis) is about 10 times the size of the magnet (in this particular case, as the x axis is the axis of symmetry, the r axis a Neumann boundary; thus only ¼ of the problem must be considered), then only large mesh sizes must be chosen, in order to be able to compute the field within the magnet aperture. Thus the magnet size compared to the total problem is small.

By choosing smaller mesh units, the magnet size is enlarged. To fit the new problem within the computer storage capacity, the outer boundary contour will have $\boldsymbol{A}$ values calculated at the first run. By repeating this procedure several times, accuracte field values can be obtained.

One disadvantage of NUTCRACKER is that the coil and iron must be fitted within the mesh sizes. Scaling the mesh to minimize geometric errors proved the greatest problem. Programs with two potential methods eliminated the boundary problem.

An effective method to shield the fringing field from leaking out of the iron is by placing several iron shields separated by air gaps around the coil. Each time the interface relation (Eq. 2.2.87) must be applied. The axial and radial field components of an axially symmetric magnet with a central field of 7 T with an iron yoke is given in Fig. 2.2.29 and Fig. 2.2.30.

As seen, the field leaks specifically where the iron terminates at a certain distance from the axis of symmetry. Already at a distance of 1.3 m from the iron surface (at the right side of the magnet), a field of 0.1 T is calculated. The field leakage on top of the magnet is also impressive.

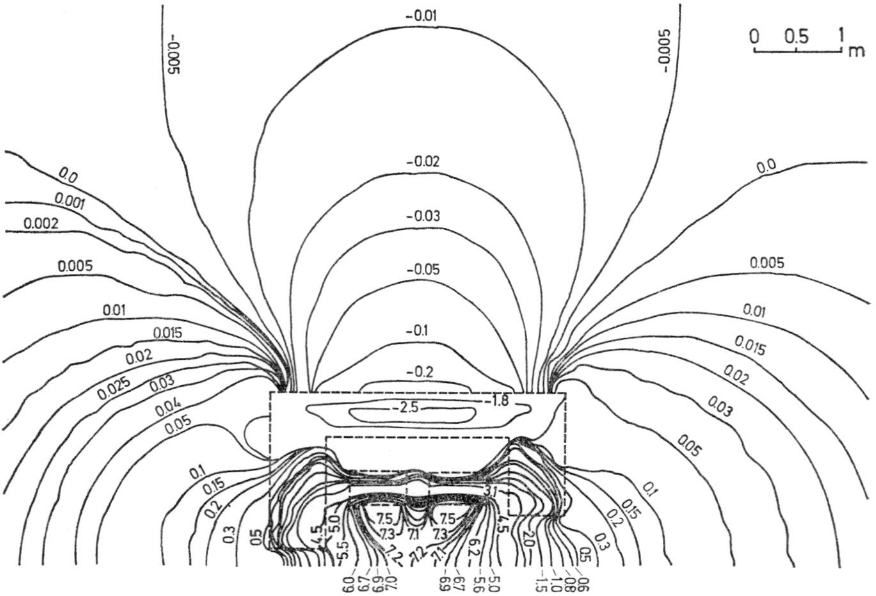

Fig. 2.2.29. The distribution of the axial field component in an iron bound axially symmetric magnet (Numbers are in T)

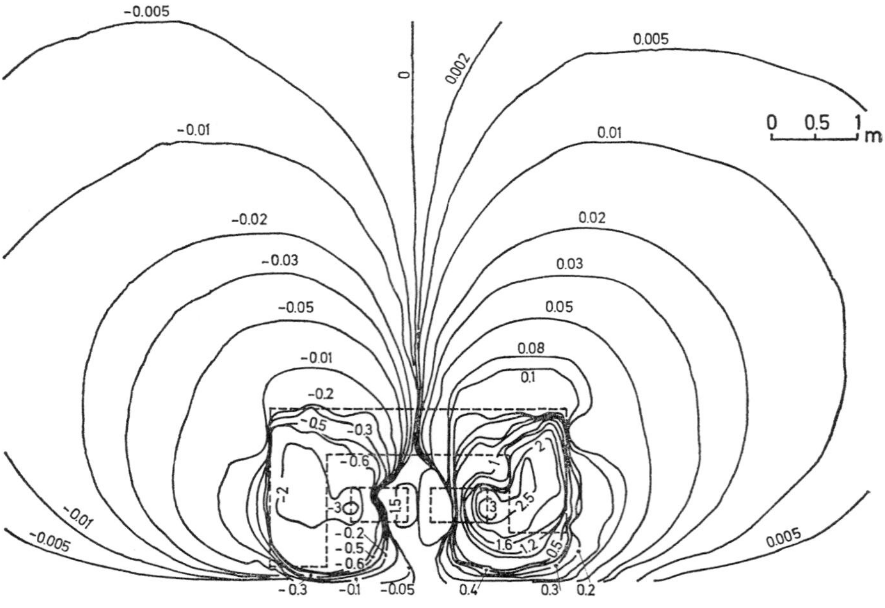

Fig. 2.2.30. The distribution of the radial field component in an iron bound axially symmetric magnet

The iron yoke having a thickness of 0.52 m does not shield all the field. The assumption of $\boldsymbol{A} = 0$ on the iron-air interface for ease of calculation does lead in high field magnets (part of the iron is not even saturated!) to calculation errors.

Thus the most accurate computational method will be to place the border line ($\boldsymbol{A} = 0$) at infinity. This would mean that the net of node points must be extended to infinity, which is impossible due to the finite capacity of the computer storage. Expanding the mesh sizes will, as we saw in the example of Fig. 2.2.28, lead to errors.

If we place a circular border line around the magnet with a radius r_0 such that outside the border line (Fig. 2.2.31a) the relative permeability is constant, then the transformation

$$r \cdot R = r_0^2 \quad \text{and} \quad \theta = \theta$$

will place the polar mesh distribution outside the border line within a finite polar mesh system having an outer circular boundary of radius r_0 [53]. The infinity point P_∞ will fall in the center of the transformed mesh system (Fig. 2.2.31b).

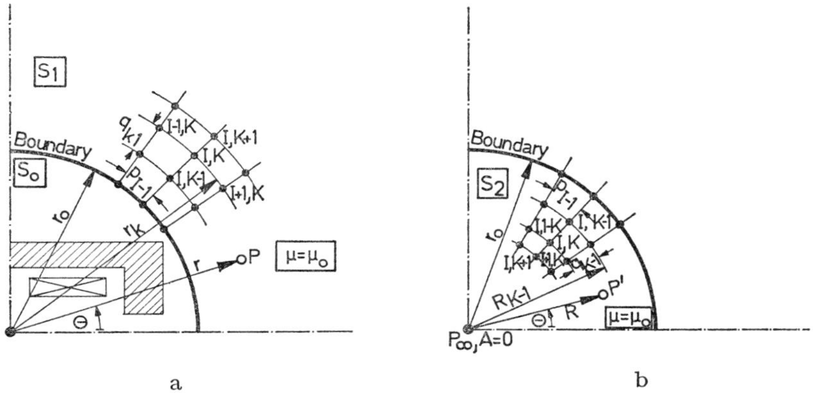

Fig. 2.2.31a and b. Conformal transformation of the border line to eliminate boundary effects on field distribution

At P_∞, the magnetic vector potential is zero. On the boundary the vector potential distribution is preserved. The computation can be limited to the areas S_0 and S_2 for the infinite problem after the transformation. This transformation is valid if μ is unchanged and the current density is zero, which is true outside the magnet iron yoke.

For $I = a$ and $\mu = \mu_0$, we get from the Poisson-Equation in polar coordinates:

$$\frac{\partial^2 A}{\partial r^2} + \frac{1}{r} \cdot \frac{\partial A}{\partial r} + \frac{1}{r^2} \frac{\partial^2 A}{\partial \theta^2} = 0.$$

Replacing r by r_0^2/R, we obtain the differential equation for S_2

$$\frac{\partial^2 A}{\partial R^2} + \frac{1}{R} \frac{\partial A}{\partial R} + \frac{1}{r} \frac{\partial^2 A}{\partial \theta^2} = 0$$

which is independent of r_0. Both equations are Laplace equations and the transformation conforms. The usual computational methods described above can be used for the area S_2.

2.2.17 Examples and Results of Numerical Computations

2.2.17.1 Superconducting Dipole Magnets [23]

Two designs studied for beam transport purposes at Brookhaven National Laboratory are presented as a first example. In Fig. 2.2.32a, the coil consists of two intersecting ellipses. The coil area is approximated and the overall current density is assumed to be constant. In Fig. 2.2.32b the current density is assumed to be of the cos θ type. The contour of the boundary between the coil and the iron shell (Case a) has been chosen to minimize the sextupole component of the field at high and low fields. The coil consists of blocks of conductor arranged around a circu-

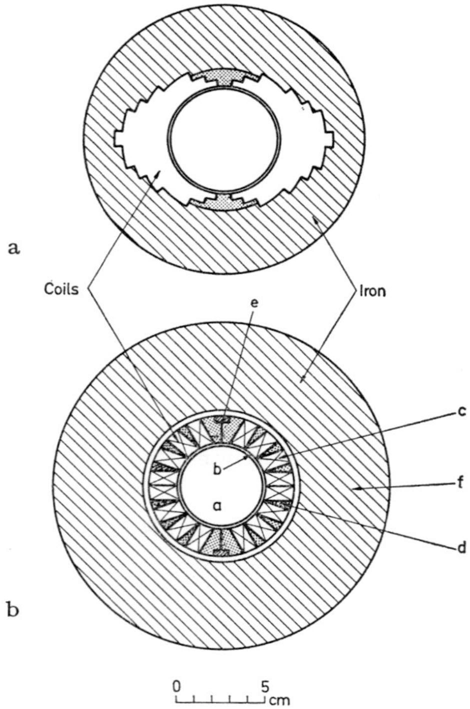

Fig. 2.2.32a. Dipole magnet. The coil cross-section is a modified version of two intersecting ellipses (constant current density)

Fig. 2.2.32b. Dipole magnet. The current density distribution in the coil is according to cos (θ) law.

a Aperture, *b* Coil former, *c* Coil block, *d* Spacer, *e* Invar bars placed between coil halves to prevent conductor motion when energized, *f* Iron shell lar aperture with suitable current density to produce the desired field in the aperture.

Calculation results of the computer program GRACY and the field measurements with magnets calculated according to this code are presented since they illustrate the effect of iron shields around dipole coils.

The field enhancement due to the iron shell is given in Fig. 2.2.33. With Δ being the radial thickness of the iron shell, the aperture field is successively enhanced with increasing Δ, at a constant current, while the radial gap between the shell and the outer coil radius is kept at a constant value of 0.127 cm. At a central field of 4 T the field enhancement due to the iron shell of $\Delta = 2.54$ cm, compared to the same coil with no iron shell at a nominal current of 10^3 A is about 29%. We also note, that for sufficiently thick iron shells (> 6 cm) the field-current-relation is practically linear up to a field of 4 T.

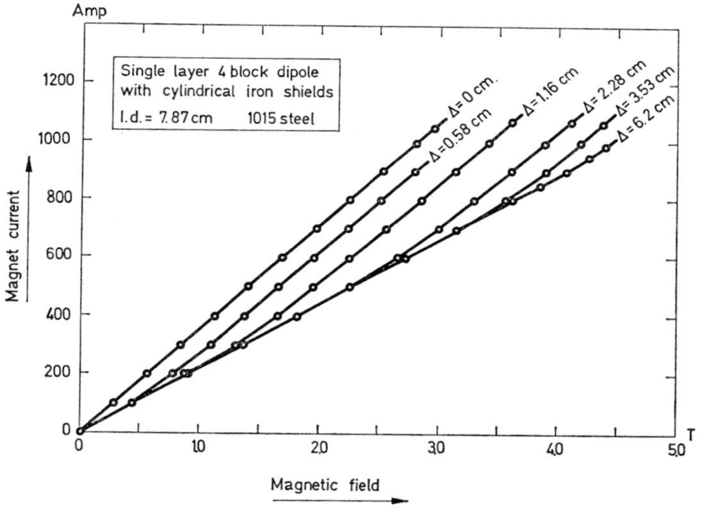

Fig. 2.2.33. The aperture field of a dipole coil for various iron thicknesses

The maximum possible field contribution from the iron shell in proximity to an optimized dipole with 5 cm aperture is obtained from Fig. 2.2.34, where the ratio of the total field to the field due to the coil mmf is plotted against the coil radial thickness. In the range of 4 to 6 T total central fields the contribution of the iron may be as high as 38 to 50%.

Also of importance is the magnitude of the external field near the outside surface of the iron shield. The field values obtained by measurements are an important indication of how far the field boundary ($A = 0$) must be placed around the iron. Figure 2.2.35 illustrates measurements. The field is measured 0.3 cm from the iron surface. The external field of a 2.54 cm thick shield is relatively small ($< 10^{-2}$ T) when the central field is 2 T. For thicker shields (with a radial thickness of 5.04 cm), the external field is again smaller than 10^{-2} T at a correspondingly high aperture field of 4 T.

The field distribution and, specifically, the saturation profile in the iron shield is shown in Fig. 2.2.36. The yoke material used is 3% Si-steel and has a saturation value of 1.99 T. In this particular case, the border line ($\boldsymbol{A} = 0$) is taken 5 cm from the outer surface. For the close type iron shield case, this distance seems to be satisfactory. The field perturbation due to this border is negligible.

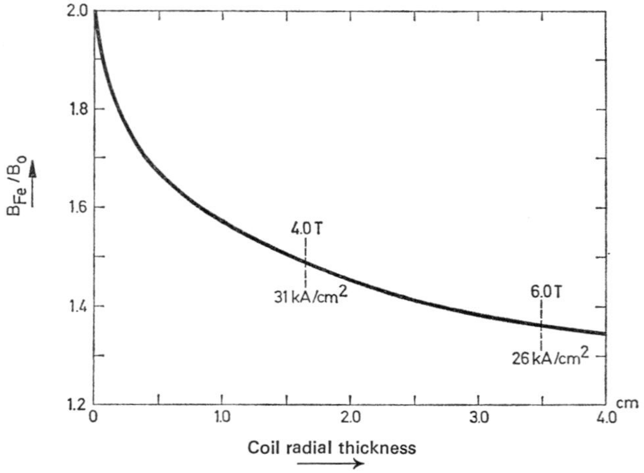

Fig. 2.2.34. Field enhancement due to the iron shell in a 5 cm aperture dipole magnet vs. radial coil thickness. The dashed lines indicate a single layer 4 T, and a double layer 6 T magnets

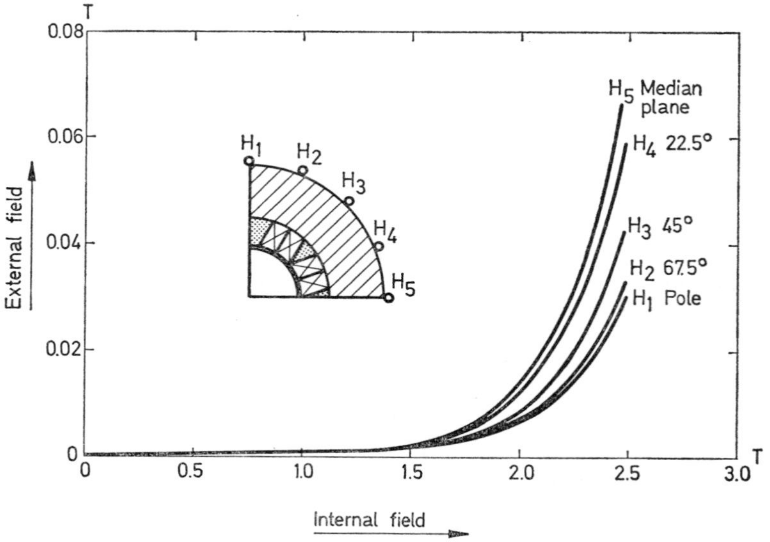

Fig. 2.2.35. Fringing field measured at a 0.3 cm distance from the outer boundary of a 2.5 cm thick iron shield

2.2.17.2 Superconducting Quadrupole Magnet

The quadrupole magnets studied with the computer program NUTCRACKER are illustrated in Fig. 2.2.37 (a—h). The coils are assumed symmetrical and obey the $^2/_3$ rule. The current density is constant with the volume of 1.7×10^8 A/m². The coil inner diameter is also kept constant at $2\,a_1 = 0.2$ m. The field gradient as a function of the radial position is given in Fig. 2.2.37, top. The iron thickness is changed, as well as its position. As NUTCRACKER uses vector potentials and square meshes, coil and iron had to approximated. However, the field enhancement is quite evident.

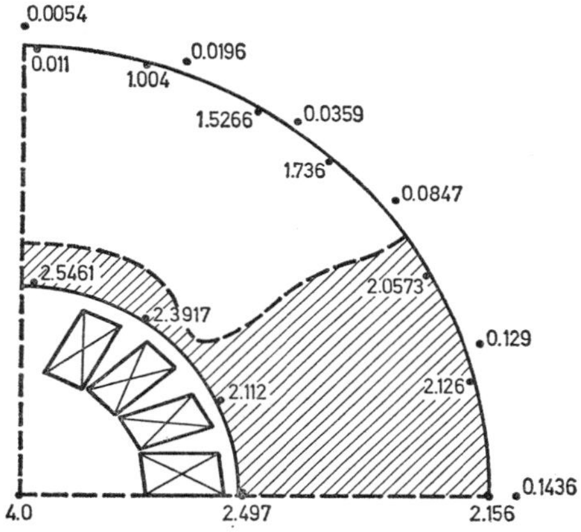

Fig. 2.2.36. Field distribution in the iron yoke of a 4 T dipole magnet. The numbers indicate flux density values in Tesla

2.2.17.3 Axially-Symmetric Magnets

The magnet considered is shown in Fig. 2.2.38 and has an open structure. Although the coil consists of superconducting composite conductor, the overall current density is kept at a low value of 2.5×10^7 A/m². Using the iron yoke as a flux return path, the central field is enhanced 16.5%, while the fringing field could be reduced to about 0.084 T at points B and C. Additional iron shields increased the central field from 5.96 T ro 7.1 T at the central point A, a gain of 19%! The influence of the iron on the field distribution is given in Fig. 2.2.39. The field homogeneity could be improved somewhat, and is, over the volume with $x = \pm\,0.25$ m, $R = \pm\,0.5$ m, about $\pm\,1.5\%$!

2.2.17.4 General

To summarize the influence of the iron return yoke on the magnetic field within the aperture, one may draw several conclusions:

Iron shields placed around dipole coils enhanced, as shown in Fig. 2.2.34, the aperture field by more than 40%!

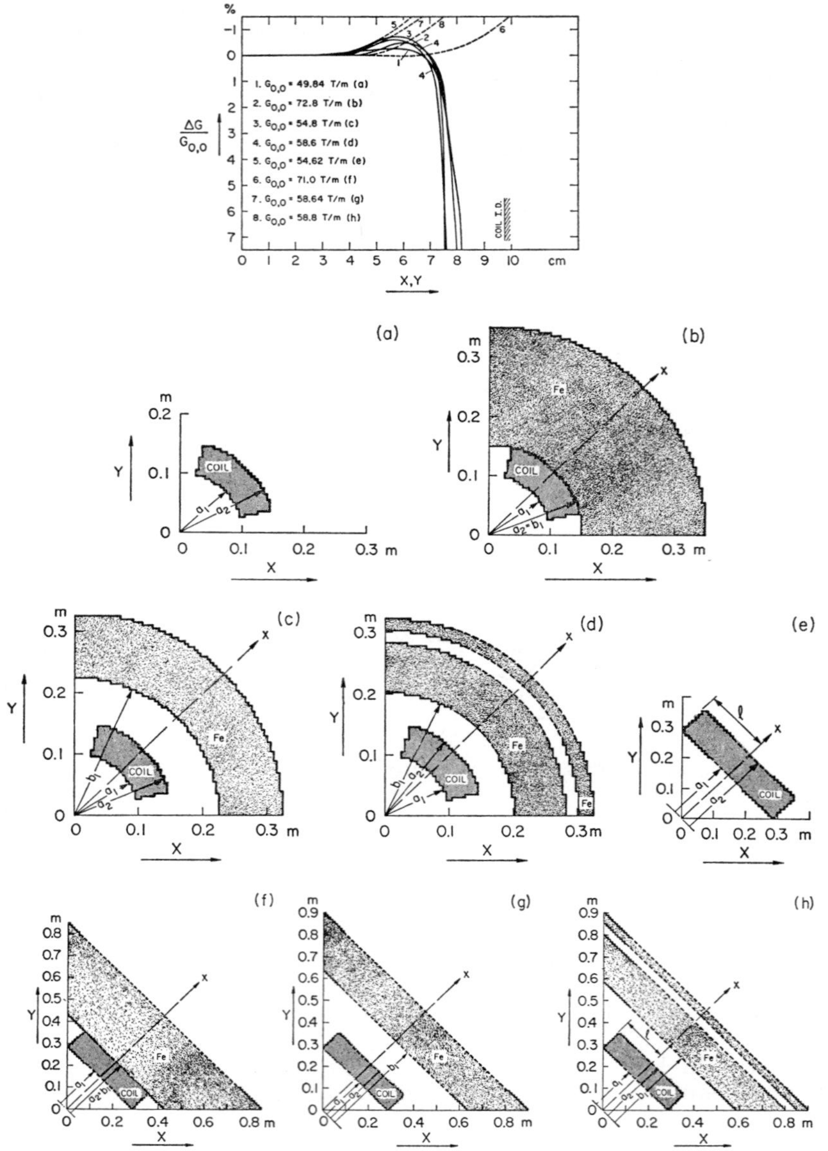

Fig. 2.2.37. Field gradient distribution over the median plane of a quadrupole with different yoke thicknesses and shapes

The stored energy in the magnet is reduced considerably, which (in case of pulsed synchrotron magnets) yields the greatest advantage of lower driving voltages. The field leakage outleak from the external surface of the iron yoke is also reduced.

With respect to the field distribution, we found two contradicting results so far.

(a) In the quadrupole magnet and in the axially symmetric magnet the gradient, resp., the field homogeneity within the useful aperture was improved.

(b) In dipole coils, the field was more distorted. We may write an expression for the field according to

$$B = B\,(0,0)\;[1 + C_2 \cdot a^2 + C_4 \cdot a^4 + \ldots C_{2n} \cdot a^{2n}]\,.$$

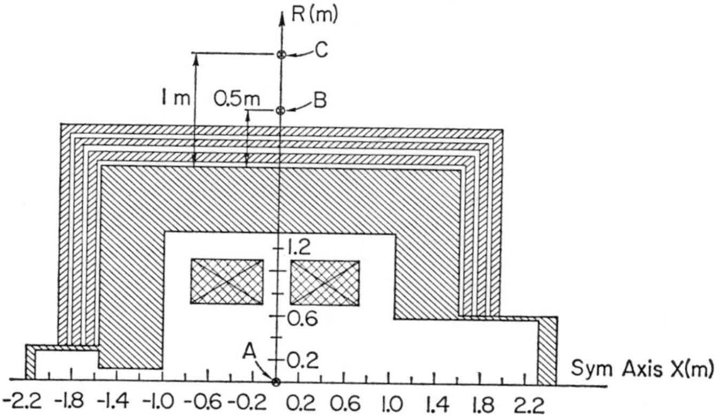

Fig. 2.2.38. Field enhancement due to iron shells of an axially symmetric magnet

Field at	A (T)	B (T)	C (T)
Coil: no Iron, no Shields	5.96	0.25	0.17
Coil + Iron Yoke:	6.96	0.084	0.046
Coil + Iron (Yoke and Shields):	7.1	0.018	0.010

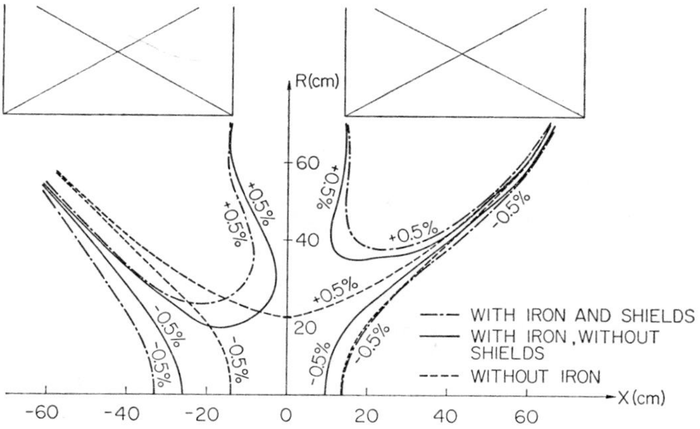

Fig. 2.2.39. Lines of constant flux density in the bore of an axially symmetric magnet

Specifically troublesome is the term C_2a^2, which is a sextupole component. This component may be small (Fig. 2.2.40), but still may introduce nonlinear effects in particle orbits. Parzen and Jellet [54] have shown that by changing the radial thickness of iron, a proper iron radial thickness can be found, which make the sextupole field disappear at a particular central field. The same conclusions are valid for iron bound solenoids and quadrupoles. The disturbing fact remains, however, that even if the sextupole components can be eliminated at a certain field level (higher terms are small and less troublesome), which is of advantage in d.c. magnets. The iron does not eliminate sextupole components at other field levels other the one which is selected for a particular field.

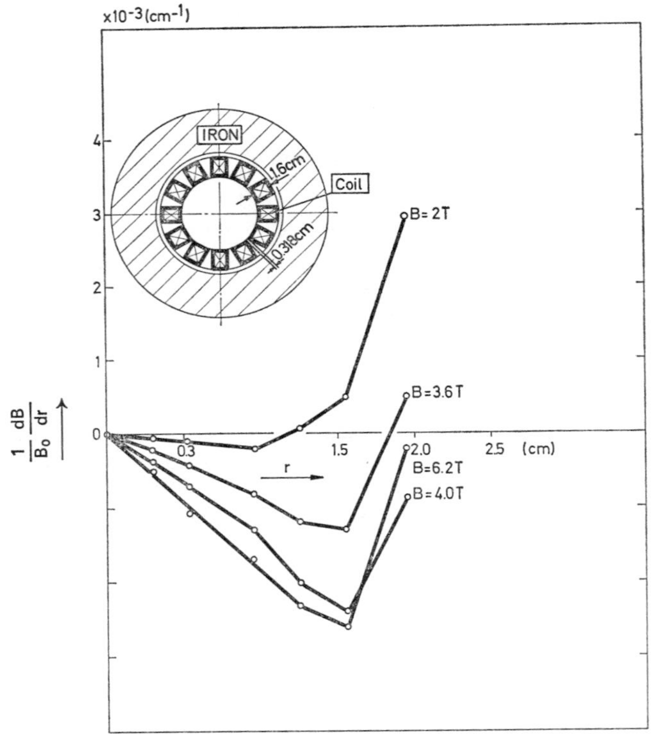

Fig. 2.2.40. Distribution of flux density at different excitation levels, computed over the median plane of dipole magnets

In pulsed synchrotron magnets, the field is raised continuously over the injection time from zero to a peak value, which is selected for superconducting synchrokons between 4 and 6 T. This will cause non-linearities and orbit distortions. The problem is serious enough to be of concern.

To illustrate the effect of iron on field distribution and stored energy, the Tables 2.2.3 to 2.2.5 present a few dipole magnet configurations. In Table 2.2.4, the sextupole field could be eliminated.

Table 2.2.3. *Variation of the C_2 and C_4 in the median plane of a dipole as a function of the iron radial thickness*

		B = 4 T			B = 6 T		
a_1 (cm)	a_3 (cm)	$\Delta_{r,Fe}$ (cm)	C_2 (cm^{-2})	C_4 (cm^{-4})	$\Delta_{r,Fe}$ (cm)	C_2 (cm^{-2})	C_4 (cm^{-4})
2.06	3.81	3.81	1.9×10^{-3}	1.44×10^{-4}	4.76	6.65×10^{-4}	7.2×10^{-5}
2.06	3.81	4.70	2.3×10^{-4}	9.35×10^{-5}	5.08	3.84×10^{-4}	7.76×10^{-5}
2.06	3.81	5.08	7.0×10^{-4}	7.4×10^{-5}	5.7	3.34×10^{-4}	6.4×10^{-5}
2.54	4.445	2.54	5.72×10^{-4}	6.8×10^{-5}	5.08	7.96×10^{-4}	3.53×10^{-5}
4.445	6.35	6.35	8.8×10^{-4}	1.7×10^{-5}	7.62	3.65×10^{-4}	7.45×10^{-6}

Table 2.2.4. *Critical radial iron thickness to eliminate sextupole components at 4 T and 6 T*

		B = 4 T		B = 6 T	
a_1 (cm)	a_2 (cm)	$\Delta_{r,Fe}$ (cm)	C_4 (cm^{-4})	$\Delta_{r,Fe}$ (cm)	C_4 (cm^{-4})
2.06	3.81	4.7	9.6×10^{-5}	5.461	6.7×10^{-5}
2.54	4.445	5.715	4.8×10^{-5}	6.35	3.36×10^{-5}
4.445	6.25	8.38	1.2×10^{-5}	9.42	7.2×10^{-6}

Table 2.2.5. *Stored energy in kJ/m for various dipole apertures for fields at 4 T and 6 T*

			Stored Energy	
a_1 (cm)	a_2 (cm)	$\Delta_{r,Fe}$ (cm)	4 T (kJ/m)	6 T (kJ/m)
2.08	3.81	5.08	18.4	45.4
2.54	4.445	2.54	29.9	
2.54	4.445	3.81	28.1	
2.54	4.445	5.08	26.3	
2.54	4.445	6.35	26.0	
4.445	6.35	5.08	73.51	182.3

2.3 Field Calculation of Iron-Bound Air-Core Magnets

The purpose of this section is to obtain approximated field values in the magnet aperture by assuming either uniform magnetization or infinite permeability in the iron shielding. The field distribution around air-core magnets shielded by iron is treated by Beth [55] and Blewett [56]. They apply the method of complex variables given in Section 2.1 for the two-dimensional geometries to obtain the field distribution.

2.3.1 Current Sheet

The potential inside a current sheet with the radius $r = a$, when surrounded by a ferromagnetic material of constant relative permeability μ in the region $r \geqq b$, producing a $2n$-pole field, is given by:

$$W_1 = C_n \cdot Z^n . \tag{2.3.1}$$

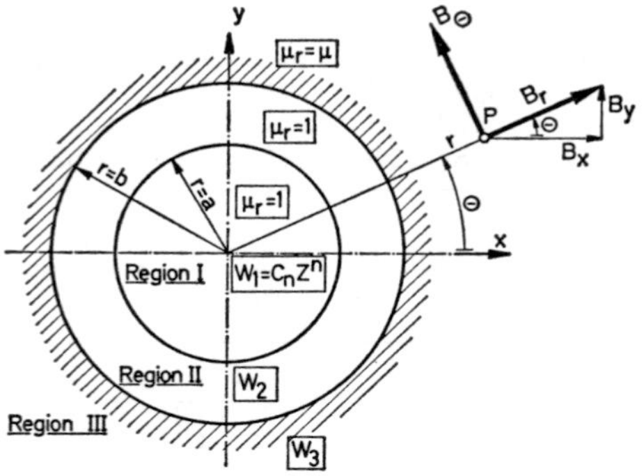

Fig. 2.3.1.

If a multipole is considered, the field in region III decreases with Z^{-n}. In region II, a mixture of Z^n and Z^{-n} is to be expected (Fig. 2.3.1). According to Beth [55], one may write:

$$W_2 = \alpha C_n \cdot Z^n + \beta C_n \cdot Z^{-n} , \tag{2.3.2}$$

$$W_3 = \gamma C_n \cdot Z^{-n} . \tag{2.3.3}$$

The magnetic field $B = B_y + jB_x$ is given by:

Region I:

$$B_1 = \mu_0 \frac{dW}{dZ} = n\mu_0 C_n \cdot Z^{n-1}$$

Region II:

$$B_2 = n \cdot \mu_0 \cdot C_n [\alpha \cdot Z^{n-1} - \beta \cdot Z^{-n-1}]$$

Region III:

$$B_3 = - n\mu\mu_0\gamma C_n \cdot Z^{-n-1}$$

with:

$$C_n = - \frac{I_0}{\pi a^n}\left[1 + \frac{\mu - 1}{\mu + 1}\left(\frac{b}{a}\right)^{-2n}\right].$$

The circular sheet carries a current perpendicular to the paper (x, y-plane), which has a distribution of $I = I_0 \cdot \cos(n\theta)$. The components are given by:

$$\alpha = \frac{1}{1 + \frac{\mu + 1}{\mu - 1}\left(\frac{a}{b}\right)^{-2n}}, \qquad \beta = \frac{1}{a^{-2n} + \left(\frac{\mu - 1}{\mu + 1}\right) b^{-2n}},$$

$$\gamma = \frac{2\mu}{(\mu + 1)\, a^{-2n} + (\mu - 1)\, b^{-2n}}.$$

For $\mu \to \infty$:
$$C_n = -\frac{I_0}{2\,a^n}\left[1 + \left(\frac{a}{b}\right)^{2n}\right];$$

$$\alpha = \frac{1}{1 + \left(\frac{a}{b}\right)^{-2n}}, \qquad \beta = \frac{1}{a^{-2n} + b^{-2n}} = \frac{a^{2n} \cdot b^{2n}}{a^{2n} + b^{2n}},$$

$$\gamma = \frac{2}{a^{-2n} + b^{-2n}} = \frac{2\,a^{2n}\,b^{2n}}{a^{2n} + b^{2n}}.$$

Region I

$$B_1 = -\frac{\mu_0 I_0}{\pi a^n} \cdot n\left[1 + \left(\frac{a}{b}\right)^{2n}\right] \cdot Z^{n-1}. \tag{2.3.4}$$

with: $Z = r \cdot e^{j\theta}$, and $B_1 = B_y + jB_x$

$$B_1 = -\left(\frac{n}{a}\right)\left(\frac{\mu_0 \cdot I_0}{\pi}\right) \cdot \left(\frac{r}{a}\right)^{n-1}\left[1 + \left(\frac{a}{b}\right)^{2n}\right] e^{jn\theta}. \tag{2.3.5}$$

Or:

$$B_y = -\left(\frac{n}{a}\right)\left(\frac{\mu_0 I_0}{\pi}\right) \cdot \left(\frac{r}{a}\right)^{n-1}\left[1 + \left(\frac{a}{b}\right)^{2n}\right] \cos\left[(n-1)\,\theta\right], \tag{2.3.6}$$

$$B_x = -\left(\frac{r}{a}\right)\left(\frac{\mu_0 I_0}{\pi}\right) \cdot \left(\frac{r}{a}\right)^{n-1}\left[1 + \left(\frac{a}{b}\right)^{2n}\right] \sin\left[(n-1)\,\theta\right]. \tag{2.3.7}$$

The radial and azimuthal field components are given by:

$$B_r = B_x \cos\theta + B_y \sin\theta\,,$$
$$B_\theta = -B_x \sin\theta + B_y \cos\theta\,.$$

Hence:

$$B_r = -\left(\frac{n}{a}\right) \cdot \frac{\mu_0 I_0}{\pi}\left(\frac{r}{a}\right)^{n-1}\left[1 + \left(\frac{a}{b}\right)^{2n}\right] \sin(n\,\theta)\,, \tag{2.3.8}$$

$$B_\theta = -\left(\frac{n}{a}\right) \cdot \frac{\mu_0 I_0}{\pi}\left(\frac{r}{a}\right)^{n-1}\left[1 + \left(\frac{a}{b}\right)^{2n}\right] \cos(n\,\theta)\,. \tag{2.3.9}$$

The term,

$$-\left(\frac{n}{a}\right)\frac{\mu_0 I_0}{\pi}\left(\frac{r}{a}\right)^{n-1} \cdot \left(\frac{a}{b}\right)^{2n} \sin(\theta)$$

is the contribution to the radial field by the presence of iron.

Region II

In the same manner as for Region I, we obtain:

$$B_y = -\left(\frac{n}{a}\right)\frac{\mu_0 I_0}{\pi} \cdot a^{n+1}\left\{\frac{r^{n+1}}{b^{2n}} \cos\left[(n-1)\,\theta\right] - \frac{1}{r^{n+1}} \cos\left[(-n-1)\,\theta\right]\right\}, \tag{2.3.10}$$

$$B_x = -\left(\frac{n}{a}\right)\frac{\mu_0 I_0}{\pi} \cdot a^{n+1}\left\{\frac{r^{n-1}}{b^{2n}} \sin\left[(n-1)\,\theta\right] + \frac{1}{r^{n+1}} \sin\left[(n+1)\,\theta\right]\right\}. \tag{2.3.11}$$

The radial and axial field components:

$$B_r = -\left(\frac{n}{a}\right) \frac{\mu_0 I_0}{\pi} a^{n+1} \left\{\frac{r^{n-1}}{b^{2n}} + \frac{1}{r^{n+1}}\right\} \sin(n\theta), \tag{2.3.12}$$

$$B_\theta = -\left(\frac{n}{a}\right) \frac{\mu_0 I_0}{\pi} a^{n+1} \left\{\frac{r^{n-1}}{b^{2n}} + \frac{1}{r^{n+1}}\right\} \cos(n\theta). \tag{2.3.13}$$

Region III

$$B_r = -2\left(\frac{n}{a}\right) \cdot \frac{\mu\mu_0 I_0}{\pi} \cdot \left(\frac{a}{r}\right)^{n+1} \sin(n\theta), \tag{2.3.14}$$

$$B_\theta = -2\left(\frac{n}{a}\right) \cdot \frac{\mu\mu_0 I_0}{\pi} \cdot \left(\frac{a}{r}\right)^{n+1} \cos(n\theta). \tag{2.3.15}$$

To illustrate the effect of placing a concentric circular iron shield close to a thin current sheet (cylindrical winding), using the above equations we plotted the flux in Fig. 2.3.2. Assuming $\mu_r = \infty$, the flux at the inner surface penetrates the iron at right angles. The flux density at the iron surface is a maximum opposite the poles. This value is indicated by $B = B_{\max,1}$. If the iron is placed too close to the coil, $B_{\max,1}$ will be very high. By increasing the inner iron radius, keeping the coil radius constant at some radial distance from the coil, $B_{\max,1}$ can be varied to a value such that it does comply to a present limiting number which, may be the iron saturation flux density.

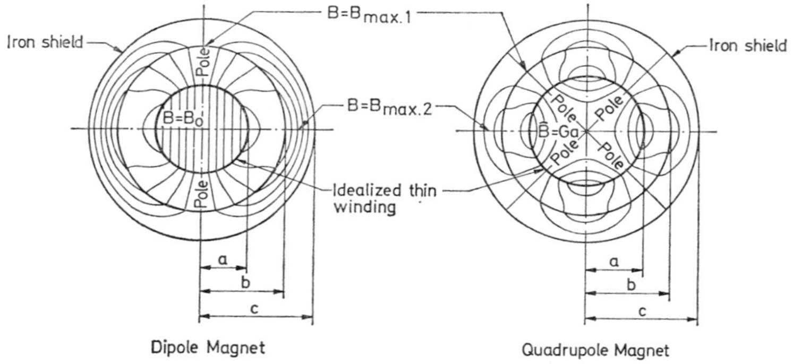

Fig. 2.3.2. Fluxlines in iron bound current sheets

The iron collects all the flux entering a half-pole and routes to the adjacant half pole of opposite sign. The total flux carried by the iron is maximum at a point midway between the two poles. The flux density at this area is indicated by $B = B_{\max,2}$. The iron must be thick enough such that the flux density $B_{\max,2}$ does not exceed the limiting value.

2.3.2 Coils of Finite Thickness

Replacing the current sheet by an overall current density $(\lambda J_0)\, a\, da$ in a coil with a finite radial thickness $(a_2 - a_1)$ where a_1 is the inner and a_2 the outer coil radii, and integrating the field values obtained for a thin sheet over a, we get:

For $r \leqq a_1$: (within the aperture)

$$B_r = \frac{\mu_0(\lambda J_0)}{\pi} \cdot r^{n-1} \cdot \left\{ \frac{1}{2-n} (a_2^{2-n} - a_1^{2-n}) + \frac{1}{(2+n)\, b_1^{2n}} \cdot (a_2^{2+n} - a_1^{2+n}) \right\} \cdot \sin(n\theta)\,,$$

$$B_\theta = \frac{\mu_0(\lambda J_0)}{\pi} \cdot r^{n-1} \cdot \left\{ \frac{1}{2-n} (a_2^{2-n} - a_1^{2-n}) + \frac{1}{(2+n)\, b_1^{2n}} \cdot (a_2^{2+n} - a_1^{2+n}) \right\} \cdot \cos(n\theta)\,. \tag{2.3.16}$$

If the coil of finite thickness is divided into sections (Fig. 2.23), Eq. (2.3.16) should be modified to:

$$B_r = \frac{\mu_0(\lambda J_0)}{\pi} \cdot (\cos\alpha_1 - \cos\alpha_2 + \cos\alpha_3 - + \cdots) \cdot r^{n-1} \left\{ \frac{1}{2-n} (a_2^{2-n} - a_1^{2-n}) + \frac{1}{(2+n)\, b_1^{2n}} (a_2^{2+n} - a_1^{2+n}) \right\} \sin(n\theta)\,,$$

$$B_\theta = \frac{\mu_0(\lambda J_0)}{\pi} \cdot (\cos\alpha_1 - \cos\alpha_2 + \cos\alpha_3 - + \cdots) \cdot r^{n-1} \left\{ \frac{1}{2-n} \cdot (a_2^{2-n} - a_1^{2-n}) + \frac{1}{(2+n)\, b_1^{2n}} (a_2^{2+n} - a_1^{2+n}) \right\} \cos(n)\theta\,. \tag{2.3.17}$$

For $b_1 \geqq r \geqq a_2$: (between coil and iron shield)

$$B_r = \mu_0 \cdot \frac{(\lambda J_0)}{\pi} \cdot r^{n-1} \left\{ \frac{1}{(2+n) \cdot r^{2n}} \cdot (a_2^{2+n} - a_1^{2+n}) + \frac{1}{(2+n) \cdot b_1^{2n}} (a_2^{2+n} - a_1^{2+n}) \right\} \cdot \sin(n\theta)\,,$$

$$B_\theta = \mu_0 \cdot \frac{(\lambda J_0)}{\pi} \cdot r^{n-1} \left\{ \frac{1}{(2+n) \cdot r^{2n}} \cdot (a_2^{2+n} - a_1^{2+n}) - \frac{1}{(2+n) \cdot b_1^{2n}} (a_2^{2+n} - a_1^{2+n}) \right\} \cdot \cos(n\theta)\,. \tag{2.3.18}$$

Within the winding:

$a_1 \leqq r \leqq a_2$:

$$B_r = \mu_0 \frac{(\lambda J_0)}{\pi} \cdot r^{n-1} \cdot \left\{ \frac{1}{2-n} (a_2^{2-n} - r^{2-n}) + \frac{1}{(2+n)\, r^{2n}} \cdot (r^{2+n} - a_1^{2+n}) + \frac{1}{(2+n)\, b_1^{2n}} (a_2^{2+n} - a_1^{2+n}) \right\} \sin(n\theta)\,,$$

$$B_\theta = \mu_0 \frac{(\lambda J_0)}{\pi} \cdot r^{n-1} \cdot \left\{ \frac{1}{2-n} (a_2^{2-n} - r^{2-n}) - \frac{r^{2+n} - a_1^{2+n}}{(2+n) \cdot r^{2n}} + \frac{1}{(2+n) \cdot b_1^{2n}} (a_2^{2+n} - a_1^{2+n}) \right\} \cdot \cos(n\theta)\,. \tag{2.3.19}$$

In all the above equations, b_1 is the inner radius of the iron shield.

2.3.3 Special Cases

We calculate the magnetic fields in dipole and quadrupole configurations and compare the obtained results with Section 2.1.

Dipole. The field pattern for $n = 1$ is given by:

$r \leqq a_1$:

$$B_r = \mu_0 \cdot \frac{(\lambda J_0)}{\pi} \cdot \left\{ (a_2 - a_1) + \frac{1}{3\, b_1^2} (a_2^3 - a_1^3) \right\} \sin(\theta)\,,$$
$$B_\theta = \mu_0 \cdot \frac{(\lambda J_0)}{\pi} \cdot \left\{ (a_2 - a_1) + \frac{1}{3\, b_1^2} (a_2^3 - a_1^3) \right\} \cos(\theta)\,, \tag{2.3.20}$$

$a_1 \leqq r \leqq a_2$:

$$B_r = \mu_0 \frac{(\lambda J_0)}{\pi} \cdot \left\{ (a_2 - r) + \frac{1}{3\, r^2} (r^3 - a_1^3) + \frac{1}{3\, b_1^2} (a_2^3 - a_1^3) \right\} \sin(\theta)\,, \tag{2.3.21}$$
$$B_\theta = \mu_0 \frac{(\lambda J_0)}{\pi} \cdot \left\{ a_2 - r - \frac{1}{3\, r^2} (r^3 - a_1^3) + \frac{1}{3\, b_1^2} (a_2^3 - a_1^3) \right\} \cos(\theta)\,,$$

$a_2 \leqq r \leqq b_1$:

$$B_r = \mu_0 \frac{(\lambda J_0)}{\pi} \cdot \left\{ \frac{1}{3\, r^2} (a_2^3 - a_1^3) + \frac{1}{3\, b_1^2} (a_2^3 - a_1^3) \right\} \sin(\theta)\,,$$
$$B_\theta = \mu_0 \frac{(\lambda J_0)}{\pi} \cdot \left\{ \frac{1}{3\, r^2} (a_2^3 - a_1^3) - \frac{1}{3\, b_1^2} (a_2^3 - a_1^3) \right\} \cos(\theta)\,. \tag{2.3.22}$$

The field pattern of a dipole with an aperture field of 6 T and an aperture diameter of $2\,a_1 = 10$ cm is given in Fig. 2.3.3. The current density of a coil with the radial thickness of $(a_2 - a_1) = 3$ cm is $(\lambda J_0) = 3 \times 10^4$ A/cm². The coil is surrounded by an iron shield of 10 cm thickness which has an inner radius of $b = 10$ cm.

Quadrupole

The field pattern for $n = 2$ cannot be obtained directly from Eq. (2.3.16) due to the undetermined character of the equations at $n = 2$. Blewett [56] introduced a term Δ, such as when n approaches 2, one sets $n = 2 - \Delta$. The first term of Eq. (2.3.16) becomes:

$$\frac{1}{2 - n} (a_2^{2-n} - a_1^{2-n}) = \frac{1}{\Delta} (a_2^{\Delta} - a_1^{\Delta}) = \frac{1}{\Delta} (e^{\Delta \ln a_2} - e^{\Delta \ln a_1})\,.$$

For $\Delta \to 0$, the term becomes: $\ln(a_2/a_1)$. Thus:

For $r \leqq a_1$:

$$B_r = \mu_0 \frac{(\lambda J_0)}{\pi} \cdot \left\{ \ln\left(\frac{a_2}{a_1}\right) + \frac{1}{4\, b_1^4} (a_2^4 - a_1^4) \right\} r \sin(2\,\theta)\,,$$
$$B_\theta = \mu_0 \frac{(\lambda J_0)}{\pi} \cdot \left\{ \ln\left(\frac{a_2}{a_1}\right) + \frac{1}{4\, b_1^4} (a_2^4 - a_1^4) \right\} \cdot r \cos(2\,\theta)\,. \tag{2.3.23}$$

For $b \geqq r \geqq a_2$:

$$B_r = \mu_0 \frac{(\lambda J_0)}{\pi} \left\{ \frac{1}{4\,r^4} (a_2^4 - a_1^4) + \frac{1}{4\,b_1^4} (a_2^4 - a_1^4) \right\} \cdot r \sin(2\,\theta)\,,$$
$$B_\theta = \mu_0 \frac{(\lambda J_0)}{\pi} \left\{ \frac{1}{4\,r^4} (a_2^4 - a_1^4) - \frac{1}{4\,b_1^4} (a_2^4 - a_1^4) \right\} r \cos(2\,\theta)\,, \tag{2.3.24}$$

For $a_1 \leqq r \leqq a_2$:

$$B_r = \mu_0 \frac{(\lambda J_0)}{\pi} \cdot \left\{ \ln\left(\frac{a_2}{r}\right) + \frac{1}{4\,r^4} (r^4 - a_1^4) + \frac{1}{4\,b_1^4} (a_2^4 - a_1^4) \right\} r \cdot \sin(2\,\theta)\,,$$
$$B_\theta = \mu_0 \frac{(\lambda J_0)}{\pi} \cdot \left\{ \ln\left(\frac{a_2}{r}\right) - \frac{1}{4\,r^4} (r^4 - a_1^4) + \frac{1}{4\,b_1^4} (a_2^4 - a_1^4) \right\} r \cos(2\,\theta)\,. \tag{2.3.25}$$

Figure 2.3.4 illustrates a quadrupole magnet with $2\,a_1 = 10$ cm aperture diameter, and a maximum field of $B_r = 6$ T at the coil inner radius $r = a_1$. The iron shield has a radius of $b = 10$ cm. The coil radial thickness is $(a_2 - a_1) = 2$ cm. The overall current density in the coils is $(\lambda J_0) = 3 \times 10^4$ A/cm². The current density in the superconductor is limited by the magnitude of the field in the wire. The field magnitude is obtained from:

$$|B| = [B_r^2 + B_\theta^2]^{\frac{1}{2}}\,.$$

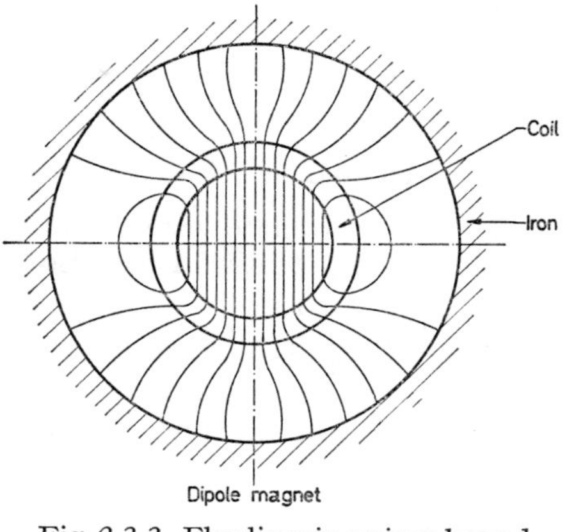

Fig. 2.3.3. Fluxlines in an iron bound dipole coil of finite cross-section

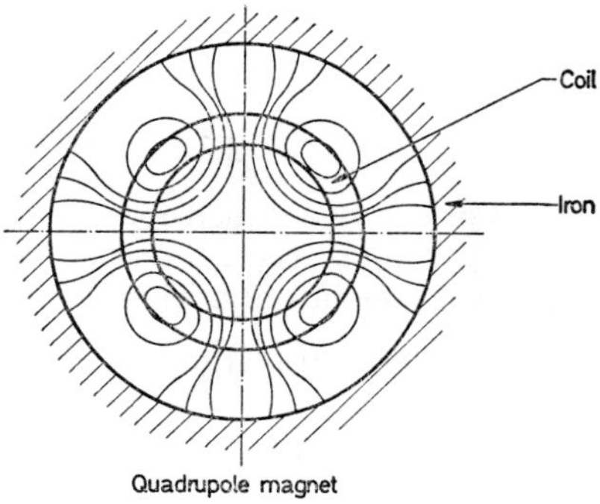

Fig. 2.3.4. Fluxlines of an iron bound quadrupole coil of finite cross-section

Using Eq. (2.3.19), we get:

$$|B| = \mu_0 \frac{(\lambda J_0)}{\pi} \cdot r^{n-1} \cdot \left\{ \left[\frac{a_2^{2+n} - a_1^{2+n}}{(2+n) \cdot b_1^{2n}} + \frac{1}{2-n} (a_2^{2-n} - r^{2-n}) \right.\right.$$
$$\left. + \frac{r^{2+n} - a_1^{2+n}}{(2+n)\, r^{2n}} \right]^2 - 4 \cos^2(n\theta) \left[\frac{1}{2-n} \cdot (a_2^{2-n} - r^{2-n}) \right.$$
$$\left.\left. + \frac{1}{(2+n)\, b_1^{2n}} (a_2^{2+n} - a_1^{2+n}) \right] \cdot \left[\frac{1}{(2+n)\, r^{2n}} (r^{2+n} - a_1^{2+n}) \right] \right\}. \tag{2.3.26}$$

For the dipole with $n = 1$:

$$|B| = \mu_0 \cdot \frac{(\lambda J_0)}{\pi} \cdot \left\{ \left[\frac{a_2^3 - a_1^3}{3\, b_1^2} + (a_2 - r) + \frac{r^3 - a_1^3}{3\, r^2} \right]^2 - \frac{4}{3} \cos^2 \theta \left[(a_2 - r) + \frac{1}{3\, b_1^2} (a_2^3 - a_1^3) \right] \left[\frac{r^3 - a_1^3}{r^2} \right] \right\}^{\frac{1}{2}}.$$

For a quadrupole coil with $n = 2$:

$$|B| = \mu_0 \frac{(\lambda J_0)}{\pi} \cdot \left\{ \left[\frac{a_2^4 - a_1^4}{4\, b_1^4} + \ln\left(\frac{a_2}{r}\right) + \frac{r^4 - a_1^4}{4\, r^4} \right]^2 - 4 \cos^2 (2\, \theta) \left[\frac{a_2^4 - a_1^4}{4\, b_1^4} + \ln\left(\frac{a_2}{r}\right) \right] \left[\frac{r^4 - a_1^4}{4\, r^4} \right] \right\}^{\frac{1}{2}}.$$

2.3.4 The Coil Ampere-Turns

As in most practical cases, we assume that the coils in a magnet have the same excitation. The current flows through each coil section such that the polarity of each pole changes cyclically. The ampere-turns per coil are given by:

$$(NI)_p = (\lambda J_0) \int_{a_1}^{a_2} \int_0^{\pi/2n} \cos (n \theta)\, d(\theta)\, a\, da\,.$$

The ampere-turns per coil are thus:

$$(NI)_p = \frac{(\lambda J_0)}{2\, n} (a_2^2 - a_1^2)\,. \tag{2.3.27}$$

2.3.5 The Magnetic Vector Potential

For the two-dimensional case of a cylindrical coordinate system with (λJ_0) being in the z-direction, the magnetic vector potential $\boldsymbol{A}$ has also a z-component only. Thus:

$$A_r = 0, \quad A_\theta = 0 \quad \text{and} \quad \boldsymbol{B} = \operatorname{curl} \boldsymbol{A} = \frac{1}{r} \frac{\partial A_z}{\partial \theta} \hat{r} - \frac{\partial A_z}{\partial r} \hat{\theta}$$

with $\hat{r}$ and $\hat{\theta}$ being the unit vectors in r and θ directions. Further, we have

$$B_r = \frac{1}{r} \frac{\partial A_z}{\partial \theta}; \quad B_\theta = -\frac{\partial A_z}{\partial r}$$

which, after integration, gives:

$$A_z = \int B_r r\, d\theta = -\int B_\theta\, dr\,.$$

We write the expressions for the vector potentials as we did for the fields.

For $r \leqq a_1$:

$$A_z = -\mu_0 \cdot \frac{(\lambda J_0)}{\pi n} r^n \left\{ \frac{1}{(2 - n)} (a_2^{2-n} - a_1^{2-n}) + \frac{1}{(2 + n)\, b_1^{2n}} (a_2^{2+n} - a_1^{2+n}) \right\} \cdot \cos (n \theta)\,. \tag{2.3.28}$$

For $a_1 \leqq r \leqq a_2$:

$$A_z = -\mu_0 \cdot \frac{(\lambda J_0)}{\pi n} r^n \left\{ \frac{1}{(2-n)} (a_2^{2-n} - r^{2-n}) + \frac{1}{(2+n)\, r^{2n}} \cdot (r^{2+n} - a_1^{2+n}) + \frac{1}{(2+n)\, b_1^{2n}} (a_2^{2+n} - a_1^{2+n}) \right\} \cos(n\theta) . \tag{2.3.29}$$

For $a_2 \leqq r \leqq b$:

$$A_z = -\mu_0 \cdot \frac{(\lambda J_0)}{\pi n} r^n \left[\frac{1}{(2+n)\, b_1^{2n}} (a_2^{2+n} - a_1^{2+n}) \right] \left[1 + \left(\frac{b_1}{r} \right)^{2n} \right] \cos(n\theta) . \tag{2.3.30}$$

Special case: Quadrupole magnet with $n = 2$.

For $r \leqq a_1$:

$$A_z = -\mu_0 \frac{(\lambda J_0)}{2\pi} \cdot r^2 \cdot \left[\frac{1}{4 b_1^4} (a_2^4 - a_1^4) + \ln\left(\frac{a_2}{a_1} \right) \right] \cos(2\theta) .$$

For $a_1 \leqq r \leqq a_2$:

$$A_z = -\mu_0 \frac{(\lambda J_0)}{2\pi} \cdot r^2 \left[\frac{1}{4 b_1^4} (a_2^4 - a_1^4) + \frac{1}{4 r^4} (r^4 - a_1^4) + \ln\left(\frac{a_2}{a_1} \right) \right] \cos(2\theta) .$$

For $a_2 \leqq r \leqq b_1$:

$$A_z = -\mu_0 \frac{(\lambda J_0)}{2\pi} \cdot r^2 \left[\frac{1}{4 b_1^4} (a_2^4 - a_1^4) \right] \left[1 + \left(\frac{b}{r} \right)^4 \right] \cos(2\theta) .$$

2.3.6 The Inner Radius of the Iron Shield

The maximum field entering or leaving the inner surface of the iron shield can be calculated from Eq. (2.3.18) by using $r = b_1$ and $\sin(n\theta) = \pm 1$.

$$B_r)_{r=b} = \frac{2 \mu_0 (\lambda J_0)}{\pi} \left[\frac{1}{(2+n)\, b_1^{1+n}} (a_2^{2+n} - a_1^{2+n}) \right] .$$

Since (λJ_0) is fixed by the choice of the superconductor, the magneto-mechanical forces, and the limiting strength of the composite conductor energized in a continuous or intermittent service, we may select $B_r)_{r=b}$ and calculate b_1. In our assumptions of $\mu_r = \infty$ and not saturated iron, we may as well use $B_r)_{r=b_1} = 2\,\mathrm{T}$. Although the calculation of the iron shield is based on large μ_r values, any desired $B_r)_{r=b_1}$ value may be selected. If μ_r approaches unity (saturated iron), then computer calculations are necessary. For $\mu_r \gg 1$, the inner iron radius is given by:

$$b_1 = \left\{ \frac{\mu_0 (\lambda J_0)}{B_r)_{r=b_1}} \cdot \frac{2}{\pi} \left[\frac{a_2^{2+n} - a_1^{2+n}}{2+n} \right] \right\}^{\frac{1}{1+n}} . \tag{2.3.31}$$

Eq. (2.3.31) requires the knowledge of a_2, the outer coil radius. The calculation of a_2 with iron shield for an arbitrary n is quite difficult. A special case of calculating a_2 is presented below:

From the magnitude of the aperture field

$$|B_a| = [B_r^2 + B_\theta^2]^{\frac{1}{2}}; \quad r \leqq a_2$$

which can be expressed explicitly as:

$$|B_a| = \mu_0 \cdot \frac{(\lambda J_0)}{\pi} \cdot r^{n-1} \left[\frac{1}{(2+n)\, b_1^{2n}} \left(a_2^{2+n} - a_1^{2+n}\right) + \frac{1}{2-n} \left(a_2^{2-n} - a_1^{2-n}\right) \right], \tag{2.3.32}$$

we can calculate a_2 for any desired B_a value.

If there is no iron shield around the coil, then $b \to \infty$, and:

$$|B_a| = \frac{\mu_0 (\lambda J_0)}{\pi} \cdot r^{n-1} \cdot \left[\frac{1}{2-n} \left(a_2^{2-n} - a_1^{2-n}\right) \right].$$

For $n = 1$ (dipole):

$$a_2 = a_1 + \frac{\pi |B_a|}{\mu_0 (\lambda J_0)},$$

For $n = 2$ (quadrupole):

$$a_2 = a_1 \exp \left[\frac{2 |B_a|/r}{\mu_0 (\lambda J_0)} \right].$$

If an iron shield is foreseen, then

For $n = 1$:

$$|B_a| = \mu_0 \frac{(\lambda J_0)}{\pi} \left[\frac{1}{3\, b_1^2} \left(a_2^3 - a_1^3\right) + (a_2 - a_1) \right]$$

and for $n = 2$:

$$|B_a| = \mu_0 \frac{(\lambda J_0)}{\pi} r \left[\frac{a_2^4 - a_1^4}{4\, b_1^4} + \ln \left(\frac{a_2}{a_1} \right) \right].$$

To determine b_1 and a_2, we need two equations for each n. If we assume that the maximum aperture field is equivalent to the field at the inner iron radius, then from Eq. (2.3.31), we obtain:

For $n = 1$:

$$a_2 = a_1 + \frac{2 |B_a| - |B_r|)_{r=b_1}}{\mu_0 (\lambda J_0)} \cdot \pi$$

and

$$b_1^2 = \frac{2 \mu_0 (\lambda J_0)/\pi}{3 |B_r|)_{r=b_1}} \cdot \left[a_2^3 - a_1^3\right].$$

For $n = 2$:

We obtain from Eq. (2.3.18) for $r = b$ and $\sin 2\theta = 1$:

$$|B_r|)_{r=b_1} = \frac{2 \mu_0}{\pi} \cdot (\lambda J_0) \cdot \frac{1}{4\, b_1^3} \left(a_2^4 - a_1^4\right).$$

The maximum field at $r = a_1$ was derived above. We get for $n = 2$:

$$a_2 = a_1 \exp \left[\frac{\pi}{\mu_0 (\lambda J_0)} \left(G - \frac{|B_r|)_{r=b_1}}{2\, b_1} \right) \right]$$

and:

$$b_1 = \left[\frac{2 \mu_0 (\lambda J_0)}{\pi |B_r|)_{r=b_1}} \cdot \frac{a_2^4 - a_1^4}{4} \right]^{\frac{1}{3}}.$$

In both cases, for $n = 1$ and $n = 2$, we have two equations and two unknowns. To solve for a_2 and b_1 explicitly, we have to use an iterative method until the solutions converge.

2.3.7 Iron Radial Thickness

Assuming the whole return flux is contained within the iron shield, and postulating that the flux enters (or leaves) the inner shield surface at $\theta = 0$ and $\theta = \pi/2n$, then an increment of radial flux (expressed in polar-coordinates) is given by:

$$d\varphi_s = lB_r r\, d\theta$$

with l the axial length of the shield in z direction. Since $B_r = \frac{1}{r}\frac{\partial A_z}{\partial \theta}$,

then:
$$\varphi_s = l \cdot \left[A_z\left(b_1, \frac{\pi}{2n}\right) - A_z(b_1, 0)\right]$$

which, after using Eq. (2.3.30) for $r = b_1$, yields

$$\varphi_s = \mu_0 \frac{2\,(\lambda J_0)\cdot l}{n\pi}\cdot\left[\frac{a_2^{2+n} - a_1^{2+n}}{(2+n)\, b_1^n}\right]. \tag{2.3.33}$$

The radial thickness of the shield which carries this flux, resulting in an average flux density $B_{r,s}$, is given by:

$$b_2 - b_1 = \frac{2\,\mu_0(\lambda J_0)}{\pi\cdot nB_{r,s}}\left[\frac{a_2^{2+n} - a_1^{2+n}}{(2+n)\, b_1^n}\right] \tag{2.3.34}$$

for all n values; b_2 is the outer shield radius. As Eq. (2.3.34) is based on a two-dimensional solution, the axial length of the iron shield must be at least equal to the effective magnet length. In order to retain all the flux within the magnet, the iron shield extends about two gap hights over the coil ends.

2.3.8 Stored Energy

In the case where $\mu_r \to \infty$, the iron shield cannot store any energy. This assumption, although not true in actual cases, greatly simplifies the calculation of the energy density given by:

$$\frac{dE}{dV} = \frac{B^2}{2\,\mu_0} \quad (\mathrm{J/m^3}).$$

The field volume within the iron shield is given by:

$$dV = lr^2 dr\, d\theta .$$

We can write for the coil arrangement discussed above the maximum aperture field derived in Eq. (2.3.32) in the form:

$$|B_a| = C_n \cdot r^{(n-1)}$$

with

$$C_n = \frac{\mu_0(\lambda J_0)}{\pi}\cdot\left[\frac{1}{(2+n)\, b_1^{2n}}\,(a_2^{2+n} - a_1^{2+n}) + \frac{1}{(2-n)}\cdot(a_2^{2-n} - a_1^{2-n})\right] \tag{2.3.35}$$

for $n \neq 2$.

$$C_n = \frac{\mu_0 (\lambda J_0)}{\pi} \cdot \left[\frac{a_2^4 - a_1^4}{4\, b_1^4} + \ln\left(\frac{a_2}{a_1}\right)\right]$$

for $n = 2$.

It is seen that C_n is independent of r, θ. Thus, the stored energy within the aperture per unit length is given by:

$$\frac{dE}{l} = \frac{C_n^2}{2\,\mu_0} r^{2n-1} \cdot dr\, d\theta\,,$$

which, integrated, gives:

$$E/l = \frac{\pi \cdot a_1^{2n}\, C_n^2}{2\, n\, \mu_0} \quad (\mathrm{J/m}) \qquad (2.3.36)$$

for all values of n.

Special Cases:

Dipole, $n = 1$:

$$E_a/l = \frac{\mu_0 (\lambda J_0)^2}{4} \left[\frac{a_2^3 - a_1^3}{3\, b_1^2} + (a_2 - a_1)\right]^2.$$

Quadrupole, $n = 2$:

$$E_a/l = \frac{\mu_0 (\lambda J_0)^2 a_1^4}{8} \left[\frac{a_2^4 - a_1^4}{4\, b_1^4} + \ln\left(\frac{a_2}{a_1}\right)\right]^2.$$

The stored energy per unit length in the coil region is given by:

For $n = 1$:

$$\frac{E}{l} = \frac{\mu_0 (\lambda J_0)^2}{4} \cdot \left\{(a_2^2 - a_1^2) \left[\left(\frac{a_2^3 - a_1^3}{3\, b_1^2} + a_2\right)^2 + \frac{a_2^4 + a_1^4}{9\, a_2^2} + \frac{5\, a_1^2}{9}\right] - \frac{4}{3}(a_2 - a_1)\left[\frac{a_2^3 - a_1^3}{3\, b^2} + a_2 + \frac{a_1}{3}\right]\right\}.$$

For $n = 2$:

$$\frac{E}{l} = \frac{\mu_0 (\lambda J_0)^2}{8} \left\{(a_2^4 - a_1^4) \left[\left(\frac{a_2^4 - a_1^4}{4\, b_1^4} + \ln\left(\frac{a_2}{a_1}\right) + \frac{1}{4}\right)^2 + \frac{1}{16}\left(\frac{a_1}{a_2}\right)^4 + \frac{1}{8}\right] - 2 \ln\left(\frac{a_2}{a_1}\right) a_2^4 \left[\left(\frac{a_2^4 - a_1^4}{4\, b_1^4} + \frac{1}{4}\right) + \frac{a_1^4}{4}\right] - a_2^4 \left(\ln\left(\frac{a_2}{a_1}\right)\right)^2\right\}.$$

The stored energy between the coil and the iron shield is given by:

For $n = 1$:

$$E/l = \frac{\mu_0 (\lambda J_0)^2}{4} \left[\frac{1}{3}(a_2^3 - a_1^3)\right]^2 \left[\frac{1}{a_2^2} - \frac{a_2^2}{b_1^4}\right]$$

For $n = 2$:

$$E/l = \frac{\mu_0 (\lambda J_0)^2}{8} \left[\frac{1}{4}(a_2^4 - a_1^4)\right]^2 \cdot \left[\frac{1}{a_2^4} - \frac{a_2^4}{b_1^8}\right].$$

The total stored energy of a dipole magnet per unit length can be expressed for $n = 1$ by:

$$E/l = \frac{\mu_0 (\lambda J_0)^2}{36} \cdot \left\{ (a_2^3 - a_1^3)^2 \left(\frac{1}{a_2^2} - \frac{2}{b_1^2} \right) + 2\, a_2^4 + g a_1^4 - 10\, a_1^3 a_2 - \frac{a_1^6}{a_2^2} \right\} \ (\mathrm{J/m}) .$$

The total stored energy for a quadrupole magnet ($n = 2$) per unit length is given by:

$$E/l = \frac{\mu_0 (\lambda J_0)^2}{128} \cdot \left\{ (a_2^4 - a_1^4) \left[\frac{a_1^4}{a_2^4} + \frac{2\, a_1^4}{b_1^4} \right] + 3\, a_2^4 - 16\, a_1^4 \ln \left(\frac{a_2}{a_1} \right) - \frac{a_1^8}{a_2^4} - 2\, a_1^4 \right\} \ (\mathrm{J/m}) .$$

2.3.9 Magnetic Fields due to Axially-Symmetric Iron Distribution

Consider a magnetic dipole $\boldsymbol{M}$ oriented at an angle θ with the z-axis. The basic equation for a dipole configuration shown in Fig. 2.3.5 is given by [22,57]:

$$\boxed{\boldsymbol{B}\,(x) = \frac{3\, \hat{n}\, (\boldsymbol{M} \cdot \hat{n}) - \boldsymbol{M}}{|x|^3},} \tag{2.3.37}$$

where $\hat{n}$ is the unit vector perpendicular to x.

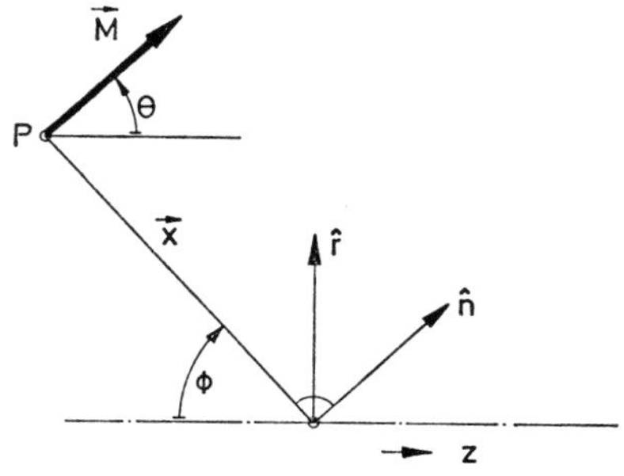

Fig. 2.3.5.

The two field components are given by:

$$B\,(z) = \boldsymbol{B}\,(x) \cdot \hat{z} = \frac{3\, (\hat{n} \cdot \hat{z})\, (\boldsymbol{M} \cdot \hat{n}) - \boldsymbol{M} \cdot \hat{z}}{|x|^3}, \tag{2.3.38}$$

$$B\,(r) = \boldsymbol{B}\,(x) \cdot \hat{r} = \frac{3\, (\hat{n} \cdot \hat{r})\, (\boldsymbol{M} \cdot \hat{n}) - \boldsymbol{M} \cdot \hat{r}}{|x|^3}. \tag{2.3.39}$$

The two field components, in terms of θ, ϕ and φ, are expressed as:

$$B_z = \frac{3 \cdot \cos \varphi\, [\boldsymbol{M} \cdot \cos (\phi + \theta)] - \boldsymbol{M} \cos \theta}{|x|^3} \tag{2.3.40}$$

and

$$B_r = \frac{-3\, \boldsymbol{M} \sin \varphi\, [\cos (\phi + \theta) - \boldsymbol{M} \sin \theta]}{|x|^3}. \tag{2.3.41}$$

These two equations can be reduced to:

$$B(z) = [(3\cos^2\phi - 1)\cdot\cos\theta - 3\sin\phi\cos\phi\sin\theta]\,\frac{M}{|x^3|}, \qquad (2.3.42)$$

$$B(r) = [(3\sin^2\phi - 1)\cdot\sin\theta - 3\sin\phi\cos\phi\cos\theta]\,\frac{M}{|x|^3}. \qquad (2.3.43)$$

Rotation of a dipole around a circle of radius a:

Referring to Fig. 2.3.6, we have:

$$l = [a^2 + r^2 - ar\cos\psi]^{\frac{1}{2}}$$

$$x = [z^2 + l^2]^{\frac{1}{2}}$$

$$\sin\phi = \frac{l}{x}; \quad \cos\varphi = \frac{z}{x}; \quad \cos\gamma = \frac{a\cos\psi - r}{l}.$$

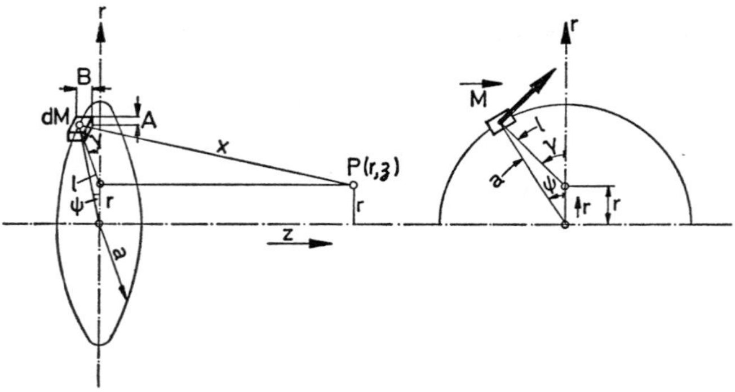

Fig. 2.3.6.

Using the notation given in Fig. 2.3.6, we can integrate an elemental dipole through 2π in order to produce a magnetic ring. The field of this ring can be obtained at any point (z, r). The cross-section of the ring is taken to be $A \cdot B$, which, when rotated through an angle $d\psi$, subtends a differential arc $a\,d\psi$. The differential dipole volume is given by $d\boldsymbol{M} = A \cdot B a\,d\psi \cdot \boldsymbol{M}'$.

The differential field, using the substitutions for $\sin\varphi$ and $\cos\varphi$, is then:

$$dB_z = \left[\left(3\frac{z^2}{x^2} - 1\right)\cos\theta - 3\frac{zl}{x^2}\sin\theta\right]\cdot\frac{aAB\boldsymbol{M}'d\psi}{x^3},$$

or:

$$B_z = \int_0^{2\pi}\left[\frac{(3z^2 - x^2)\cos\theta - 3zl\sin\theta}{x^5}\right]\cdot aAB\boldsymbol{M}'d\psi$$

with $x^2 = z^2 + l^2$.

$$B_z = \int_0^{2\pi}[(2z^2 - l^2)\cos\theta - 3zl\sin\theta]\,\frac{aAB\boldsymbol{M}'d\psi}{x^5}.$$

In the same manner:

$$B_r = \int_0^{2\pi} [(2\,l^2 - z^2)\sin\theta - 3\,zl\cos\theta]\,\frac{a\,A\,B\,\boldsymbol{M}'}{x^5}\cdot\frac{(a\cos\psi - r)}{l}\cdot d\psi\,.$$

On the axis, we can perform the integration with no difficulty:

$$B_z\,(r=0) = 2\,\pi a\cdot AB\left[\left(\frac{3\,z^2}{x^5} - \frac{1}{x^3}\right)M_z - \frac{3\,az}{x^5}\cdot M_r\right] \quad (2.3.44)$$

with:

$$M_z = \boldsymbol{M}\cdot\cos\theta\,,\quad M_r = \boldsymbol{M}\cdot\sin\theta\,.$$

2.4 Calculation of Forces

The design of superconducting magnets and magnet systems requires a careful optimization of a number of parameters which, in varying degrees, affect each other. It is not sufficient to consider only physical parameters without examination of mechanical arrangements and assembly possibilities. In addition to coil optimization by means of appropriate shaping of coil sections, adjustment of current density according to the field distribution, the mechanical forces on the conductor, joints, physical arrangements, and reinforcements must be known.

The forces on the coils or, specifically, the current-carrying conductor, may be divided into three categories:

1. Mechanical forces due to coil-winding, prestressing, and support.
2. Thermomechanical forces due to uneven material contraction during cool-down.
3. Magnetomechanical forces when the coil is energized.

In this section, we treat these forces and the mechanical stresses on the conductor briefly. While the first two stress categories are treated in literature and are well understood, the third category is treated by a number of authors, such as Lontai [58], Léon [59], Kilb [60], and others.

2.4.1 Forces due to Coil-Winding

To prevent a coil from unwinding or becoming loose while in operation and losing its original stiff structure, coils are wound, in general, with a certain experimentally predetermined constant stress. The prestressing on the conductor is accomplished by means of adequate torque motors. An elaborate arrangement of coil-winding is shown in Fig. 7.2.8. The picture illustrates the winding of a double pancake of 3.7 m diameter on a horizontal turntable. The superconductor, a stainless steel band for support, insulation tapes, and a coolant strip are sandwiched during the winding. Adjacent to the coil, the stress on the conductor is provided by a specially designed gadget. The conductor is cleaned during winding.

To prevent the coil from unwinding during power failures, adequately designed guiding arms, equipped with microswitches, are provided.

It is difficult to give the proper stress values on the conductor during winding. A certain amount of experience, as well as the superconductor characteristics, cold work of the substrate, and eventual damage to the insulation due to conductor pressure, does establish an upper limit to the pretension. If the superconductor is subjected to high mechanical tension, it shows a certain degradation. Figure 2.4.1 shows the critical current enhancement and subsequent degradation as a function of stresses [61] at 4.2 K. The prestress must be in accordance with the final stress on the coil due to magnetomechanical and thermal contraction forces.

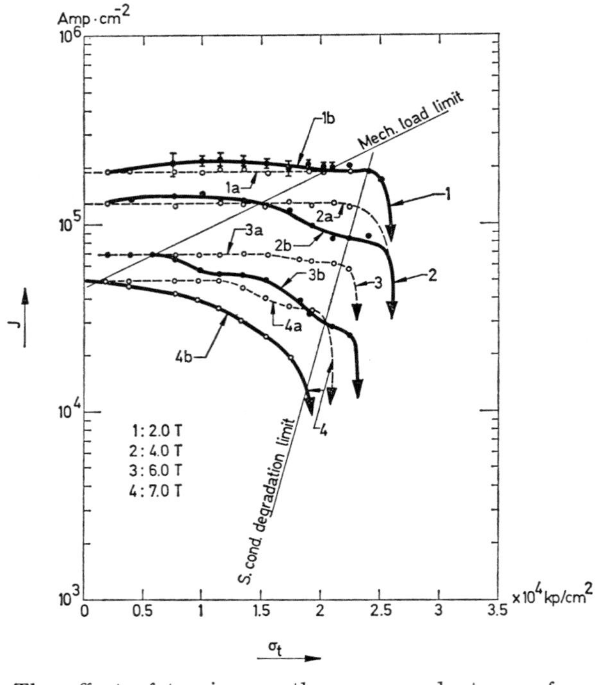

Fig. 2.4.1. The effect of tension on the superconductor performance. The critical current density degrades rapidly when a tension limit is reached. This limit is below the yield strength of the NbTi wire.

If a constant tension is provided on the conductor during winding (as long as the plastic limit of the composite conductor is not reached), we may use Hooke's Law to determine the stress on the conductor:

$$\sigma = E \cdot \varepsilon \tag{2.4.1}$$

with E the Young's modulus and ε the strain on the conductor. The direct proportionality between stress σ and strain ε is not fulfilled in the plastic region of the material. Large superconducting magnets are often strained to the limit of elasticity of materials; it will be adequate to study the behavior of some composite materials also in the plastic region.

The $\sigma - \varepsilon$ diagram of a few materials, such as copper and aluminium, is given in Fig. 2.4.2. According to Hooke's Law, the stress is proportional to the strain in the first portion of the curve. In a second portion, the stress is increased more slowly than the strain. Reaching the stress limit, σ_B, the stress decreases by further increases of ε until the material breaks.

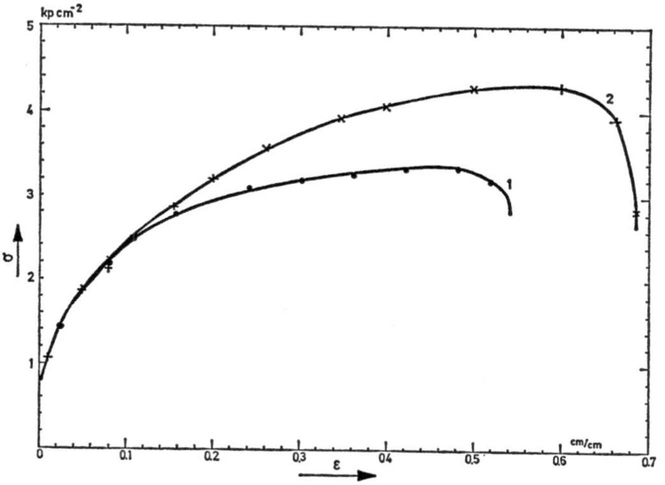

Fig. 2.4.2. Stress-strain diagram for aluminum *1* and copper *2*

If the strain is reduced before σ_B is reached (Fig. 2.4.2), a non-reversible situation occurs. The material shows a hysteretic behavior. It is prestrained when the stress again is increased. The material properties are altered. In magnet design, this situation must definitely be guarded carefully.

The $\sigma - \varepsilon$ curve of most metals can be approximated by either of the two relations (Fig. 2.4.3):

$$y = \tan h\,(x)$$

and

$$y = \frac{2}{\pi} \tan^{-1}\left(\frac{\pi x}{2}\right).$$

We can use the following relationship:

$$\sigma = \sigma_0 \tan^{-1}(A\,\varepsilon)\,. \tag{2.4.2}$$

To determine σ_0 and A, Parsch [62] has used the following assumptions:

For:

$$\sigma = \sigma_B \quad \text{it follows that} \quad \varepsilon \to \infty\,.$$

For:

$$A\,\varepsilon \ll 1 \to \sigma = \sigma_0 \cdot (A\,\varepsilon) = E_0 \cdot \varepsilon\,.$$

In the elastic region, $A\,\varepsilon \cong 1$, the Eq. (2.4.2) represents Hooke's Law. The stress-strain diagram is then represented analytically as:

$$\sigma = \sigma_B \cdot \frac{2}{\pi} \tan^{-1}\left(\frac{E_0}{\sigma_B} \cdot \frac{\pi}{2}\right). \tag{2.4.3}$$

The second assumption was with:

$$\sigma = \sigma_B \tan h\,(A \cdot \varepsilon)\,.$$

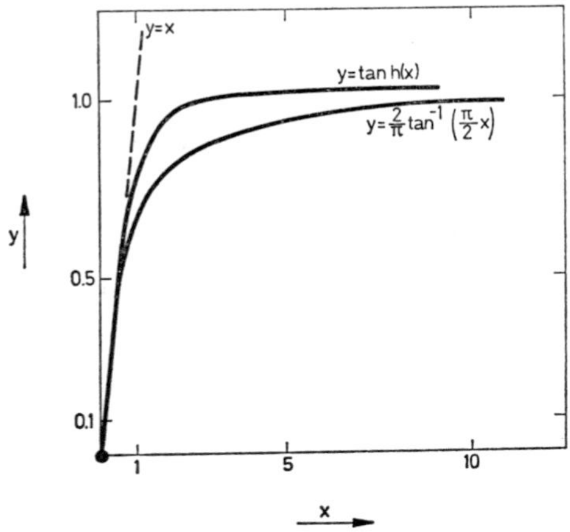

Fig. 2.4.3. Approximated stress-strain diagram

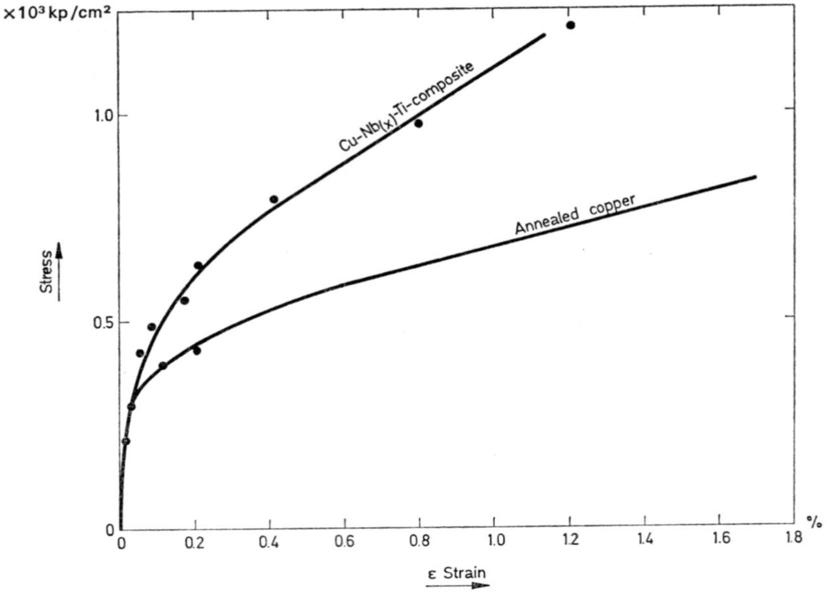

Fig. 2.4.4. Stress-strain curves for annealed copper ($E = 8.14 \times 10^5$ kpcm^{-2}) at 4.2 K, and copper-NbTi composite conductor ($A_{\mathrm{cu}}/A_{\mathrm{NbTi}} = 31/1$), at 4.2 K

With the same conditions as above, one obtains:

$$\sigma = \sigma_B \tan h \left(\frac{E_0}{\sigma_B} \varepsilon \right). \tag{2.4.4}$$

Of course, the two relations do not reproduce experimental data exactly. In the elastic region, the linear relationship between σ and ε prevails, while in the plastic region, Eqs. (2.4.3) and (2.4.4) may yield better results. The error due to the simplification does not, however, exceed 10%.

With reference to the above, the tension applied to the conductor should be around 300 kp/cm². In a composite conductor, this stress will produce different strain values in the superconductor and in the matrix material due to different values of E. In Fig. 2.4.4, stress-strain curves of Nb_xTi and Cu composite conductors are illustrated. Nb_xTi has a much higher yield strength than, for example, annealed copper. Tensile strength and resistivity of copper, copper alloys, aluminium, and Nb(50 at %) Ti are given in Table 2.4.1.

Table 2.4.1. *Tensile strength and resistivity of copper, copper alloys, aluminium, and Nb (50 at %) Ti*

Material	Ultimate tensile strength (kp/cm^{-2})			Resistivity (10^{-8} Ohm · m)		
	293 K	78 K	4.2 K	293 K	78 K	4.2 K
OFHC copper, hard	2070	3570	4320	1.66	0.218	0.015
OFHC copper, annealed	2000	3360	3790	1.61	0.20	0.012
ETP copper, hard	2285	3650	4650	1.72	0.26	0.019
ETP copper, annealed	2050	3400	4100	1.69	0.24	0.015
Cr-copper (0.7% Cr)	4200	6290	—*	2.1	0.618	—*
Zr-copper (0.15% Zr)	3860	5790	—	1.9	0.432	—
Be-copper (2% Be)	8560	10200	—	6.5	3.65	—
Be-copper (0.55% Be)	5720	7860	8720	3	1.29	—
Al, pure (99.99%) (commercial)	850	1640	2900	2.53	0.38	0.101
Al (99.6 + %), wrought	1355	2100	2550	2.8	—	—
Nb (50 at %) Ti (wire)	—	—	9700	75	33	—

* Data not measured.

The yield strength of annealed OFHC copper is 640 kp/cm² at 293 K, 790 kp/cm² at 78 K, and 1000 kp/cm² at 4.2 K. The modulus of elasticity of soft OFHC copper, annealed, is 1.28×10^6 kp/cm² at 293 K and 1.7×10^6 kp/cm² at 4.2 K, while the E modulus of Nb (60%) Ti is about 1.4×10^6 kp/cm² at 4.2 K and depends mainly on the Ti content.

Aluminium (99.99%) is, in comparison to copper, much softer: It has a (0.2%) yield strength of about 400 kp/cm² at 293 K. The yield strength increases only slightly to 640 kp/cm² at 4.2 K. The E modulus of aluminum is 0.71×10^6 kp/cm² at 293 K and 0.8×10^6 kp/cm² at 4.2 K.

2.4.2 Forces due to Thermal Contraction

When the superconductor is cooled, forces due to differential thermal contraction occur between the superconductor and the matrix material, between the composite conductor and the coil-winding or support structure, and, between layers or turns of the coil due to the different conductor length (as in axially-symmetric systems).

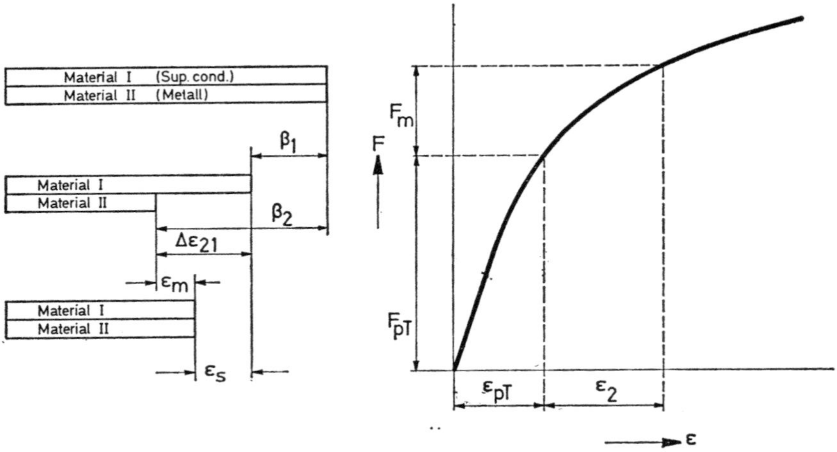

Fig. 2.4.5.

In a composite conductor, where the bond between the matrix and the superconductor must be of excellent quality, the contraction forces in the elastic region are according to:

$$F_{Ts} = \varepsilon_s \cdot E_s \cdot A_s \quad \text{(superconductor)}$$
$$F_{Tm} = \varepsilon_m \cdot E_m \cdot A_m \quad \text{(matrix)}$$

where one material is subjected to compression, the other to elongation. If one material is stressed beyond the plastic limit, then one gets, with Fig. 2.4.5:

$$F_{Ts} + F_{Tm} = 0 \tag{2.4.5}$$

$$F_{Ts} = \varepsilon_s \cdot E_s \cdot A_s \tag{2.4.6}$$

$$F_{Tm} = A_m \cdot \sigma_{B2} \cdot \frac{2}{\pi} \tan^{-1}\left(\frac{E_{0,m}}{\sigma_{B2}} \cdot \frac{\pi}{2}\, \varepsilon_m\right) \tag{2.4.7}$$

$$\Delta\varepsilon_{21} = \varepsilon_m - \varepsilon_s .$$

The forces F_{Ts} and F_{Tm} are calculated from:

$$\varepsilon_m = \Delta\varepsilon_{21} - \frac{A_m \cdot \sigma_{B2}}{A_s \cdot E_s} \cdot \frac{2}{\pi} \tan^{-1}\left(\frac{E_{0,m}}{\sigma_{B_2}} \cdot \frac{\pi}{2} \cdot \varepsilon_m\right) \tag{2.4.8}$$

$$\varepsilon_s = \varepsilon_m - \Delta\varepsilon_{21} .$$

In the case of a three-component conductor, we may use the same approach to determine the forces:

$$F_{T_1} + F_{T_2} + F_{T_3} = 0\,, \tag{2.4.9}$$

$$F_{T_1} = \varepsilon_1 \cdot A_1 E_1\,, \tag{2.4.10}$$

$$F_{T_2} = A_2 \sigma_{B_2} \cdot \frac{2}{\pi} \tan^{-1}\left(\frac{E_{0,2}}{\sigma_{B_2}} \cdot \frac{\pi}{2}\, \varepsilon_2\right). \tag{2.4.11}$$

$$F_{T_3} = \varepsilon_3 \cdot A_3 \cdot E_3\,, \tag{2.4.12}$$

$$\varepsilon_2 - \varepsilon_1 = \Delta\,\varepsilon_{21}\,; \quad \varepsilon_3 - \varepsilon_2 = \Delta\,\varepsilon_{32}\,.$$

The solution:

$$\varepsilon_2 = \frac{-\Delta\,\varepsilon_{21} \cdot A_1 E_1 + \Delta\,\varepsilon_{32} A_3 E_3}{-A_1 E_1 - A_3 E_3} + \frac{1}{-E_1 A_1 - E_3 A_3} \cdot A_2 \sigma_{B_2} \cdot \frac{2}{\pi} \tan^{-1}\left(\frac{E_{0,2}}{\sigma_{B_2}}\, \frac{\pi}{2}\, \varepsilon_2\right), \tag{2.4.13}$$

$$\varepsilon_1 = \varepsilon_2 - \Delta\,\varepsilon_{21}\,; \quad \varepsilon_3 = \varepsilon_2 + \Delta\,\varepsilon_{32}\,.$$

2.4.3 Magnetomechanical Forces F_m

We first consider a two-component conductor. Assuming that material 1 happens to be in the elastic region, and material 2 in the plastic region, we obtain the system of equations:

$$F_{m,1} + F_{m,2} = F_m\,, \tag{2.4.14}$$

$$F_{m,1} = \varepsilon_1 \cdot A_1 \cdot E_1\,, \tag{2.4.15}$$

$$F_{m,2} = A_2 \cdot \sigma_{B,2} \cdot \frac{2}{\pi} \tan^{-1}\left(\frac{E_{0,2}}{\sigma_{B,2}} \cdot \frac{\pi}{2} \cdot \varepsilon_2\right), \tag{2.4.16}$$

$$\varepsilon_1 = \varepsilon_2 = \varepsilon\,.$$

We give the equations for F_m in Section 2.4.5, but determine first $F_{m,2}$ from:

$$F_{m,2} = A_2 \cdot \sigma_{m,2} \cdot \tan^{-1}\left[\frac{E_{0,2}}{\sigma_{B,2}} \cdot \frac{\pi}{2} \cdot \frac{F_m - F_{m,2}}{E_1 \cdot A_1}\right] \tag{2.4.17}$$

$$F_{m,1} = F_m - F_{m,2}\,.$$

2.4.4 Magnetomechanical Forces due to Winding Pretension

This case may happen if the conductor is prestressed during winding, or due to thermal contraction, or both. Although all three types of stress may occur simultaneously, we cannot superpose them by mere addition if the material is stressed beyond the elastic region. We obtain with Fig. 2.4.5:

$$F = A_2 \sigma_{B,2} \cdot \frac{2}{\pi} \tan^{-1}\left[\frac{E_{0,2}}{\sigma_{B,2}} \cdot \frac{\pi}{2}\, (\varepsilon_{pT,2} + \varepsilon_2)\right] = F_{pT,2} + F_{m,2}$$

F_{pT} is the force due to pretension.

With: $F_{m,1} + F_{m,2} = F_m$, $F_{m,1} = \varepsilon_1 A_1 E_1$,

$$F_{m,2} = A_2 \cdot \sigma_{m,2} \cdot \frac{2}{\pi} \tan^{-1}\left[\frac{E_{0,2}}{\sigma_{m,2}} \cdot \frac{\pi}{2} (\varepsilon_{pT,2} + \varepsilon_2)\right] - F_{pT,2}. \quad (2.4.18)$$

The forces are obtained from:

$$F_{m,2} = A_2 \cdot \sigma_{m,2} \cdot \frac{2}{\pi} \tan^{-1}\left[\frac{E_{0,2}}{\sigma_{m,2}} \cdot \frac{\pi}{2} \left(\frac{F_m - F_{m,2}}{E_1 A_1} + \varepsilon_{pT,2}\right)\right] - F_{pT,2}. \quad (2.4.19)$$

If the conductor consists of material components, such as are usual in a.c. cables where copper, cupro-nickel, and superconductor are used, we obtain in the same manner:

$$F_{m,1} + F_{m,2} + F_{m,3} = F_m, \quad F_{m,1} = \varepsilon_1 \cdot A_1 \cdot E_1,$$

$$F_{m,2} = A_2 \cdot \sigma_{m,2} \cdot \frac{2}{\pi} \tan^{-1}\left[\frac{E_{0,2}}{\sigma_{m,2}} \cdot \frac{\pi}{2} (\varepsilon_2 + \varepsilon_{pT,2})\right] - F_{pT,2}, \quad (2.4.20)$$

$$F_{m,3} = \varepsilon_3 A_3 E_3, \quad \varepsilon_1 = \varepsilon_2 = \varepsilon_3 = \varepsilon.$$

The solution of this case is given by:

$$F_{m,2} = A_2 \cdot \sigma_{m,2} \cdot \frac{2}{\pi} \tan^{-1}\left[\frac{E_{0,2}}{\sigma_{m,2}} \cdot \frac{\pi}{2} \left(\frac{F_m - F_{m,2}}{E_1 A_1 + E_3 A_3} + \varepsilon_{pT,2}\right)\right] - F_{pT,2}, \quad (2.4.21)$$

$$\varepsilon = \frac{F_m - F_{m,2}}{E_1 A_1 + E_3 A_3},$$

$$F_{m,1} = \varepsilon_1 A_1 E_1 = \varepsilon A_1 E_1, \quad F_{m,3} = \varepsilon_3 A_3 E_3 = \varepsilon A_3 E_3.$$

In a conductor with n material components, one obtains the general equations:

Forces due to thermal contraction:

$$F_{T,i} = \frac{A_i \cdot E_i}{\sum_{k=1}^{n} A_k E_k} \cdot \sum_{k=1}^{n} A_k E_k \cdot \Delta_{i,k} \quad (2.4.22)$$

$$i \neq k \quad \text{and} \quad n \neq i$$

and to Lorentz forces:

$$F_{m,i} = F_m \cdot \frac{A_i E_i}{\sum_{k=1}^{n} A_k E_k}. \quad (2.4.23)$$

If all the components are in the elastic region, the forces are superposed by a mere additional process.

2.4.5 Magnetomechanical Forces in Cylindrical Geometrics

Considerable study has been devoted in recent years to the mechanical effects of the electromagnetic forces in magnet coils. The knowledge of the force distribution in superconducting coils is of particular interest since the Lorentz forces are higher by at least an order of magnitude in composite conductors than in conventional copper or aluminum conductors. Among a large number of investigators, we have selected a few who have analyzed stresses due to electromagnetic forces in coils: Kuznetsov [63] has considered "body forces" and suggested that the effect of insulation between turns should be neglected. Bitter [64] introduced the magnetic pressure concept and used the thick cylinder theory for analysis.

Giauque and Lyon [65] considered stresses in helical layers and the effect of friction in preventing unwinding. Leon [59], Kilb and Westendrop [60] derived expressions for radial and circumferential stresses. Lontai and Marston [58] discussed the use of mechanical and electrical regionalization in coils to obtain optimum electromechanical efficiency. This method can be combined also to optimize current density in superconducting coils.

Recently, a number of computer codes have been developed, among which the codes from BNL [66] and SLAC [67] have shown to be quite useful. The magnetic body forces in radial and axial directions are calculated by these codes:

$$\boxed{F_r = J \cdot B_z/9.81\,, \quad F_z = J \cdot B_r/9.81} \quad (\mathrm{kp/m^3})$$

where the conductor current density is expressed in $\mathrm{A/m^2}$ and the field in Teslas.

In designing a magnet, it may be desirable to perform a preliminary approximate calculation, which is particularly simple for solenoids if the thick-wall cylinder approximation is used.

For a coil of inner radius a_1, outer radius a_2, carrying a uniform current density J. Appleton, *et al.* [68] have derived a simple equation for circumferential stresses:

$$\boxed{\sigma_c = a_1 \cdot (\lambda J_0) \cdot \frac{B_{a1}(\alpha^2 + \alpha - 2) + B_{a2}(2\alpha^2 - \alpha - 1)}{b\left(\frac{a}{a_1}\right)\ln\alpha}\,.} \tag{2.4.24}$$

A more accurate calculation is made by solving the stress equations for a ring of radius r and thickness dr, shown in Fig. 2.4.6a.

$$\frac{d\sigma_r}{dr} + \frac{(\sigma_r - \sigma_c)}{r} + F_r(r,z) = 0\,, \tag{2.4.25}$$

$$\frac{d\sigma_z}{dz} + F_z(r,z) = 0\,. \tag{2.4.26}$$

The equations of strain displacement are obtained from Fig. 2.4.7b:

$$\varepsilon_c = \frac{2\pi(r+u) - 2\pi r}{2\pi r} = \frac{u}{r}\,, \qquad \varepsilon_r = \frac{r + du - r}{dr} = \frac{du}{dr} \tag{2.4.27}$$

and
$$\varepsilon_z = \frac{\sigma_z}{E} - \frac{m}{E}(\sigma_c + \sigma_r)\,. \tag{2.4.28}$$

ε_z is obtained from Hooke's Law.

Stress-strain relations

$$\sigma_u = \frac{E}{(1-2m)(1+m)} \cdot [\varepsilon_u + m(-\varepsilon_u + \varepsilon_v + \varepsilon_w)] \tag{2.4.29}$$

which, expanded into our case of axially-symmetric systems, yields:

$$\left.\begin{aligned}
\sigma_c &= \frac{E}{(1-2m)(1+m)} \cdot [\varepsilon_c + m(-\varepsilon_c + \varepsilon_r + \varepsilon_z)] \\
\sigma_r &= \frac{E}{(1-2m)(1+m)} \cdot [\varepsilon_r + m(-\varepsilon_r + \varepsilon_c + \varepsilon_z)] \\
\sigma_z &= \frac{E}{(1-2m)(1+m)} \cdot [\varepsilon_z + m(-\varepsilon_z + \varepsilon_c + \varepsilon_r)]
\end{aligned}\right\} \tag{2.4.30}$$

where we assume E is the same in all three directions in the coordinate system. We also used the same value of m for the Poisson's ratio. As the coil structure is not homogeneous, the correct values of E and m in axial and radial directions should be used.

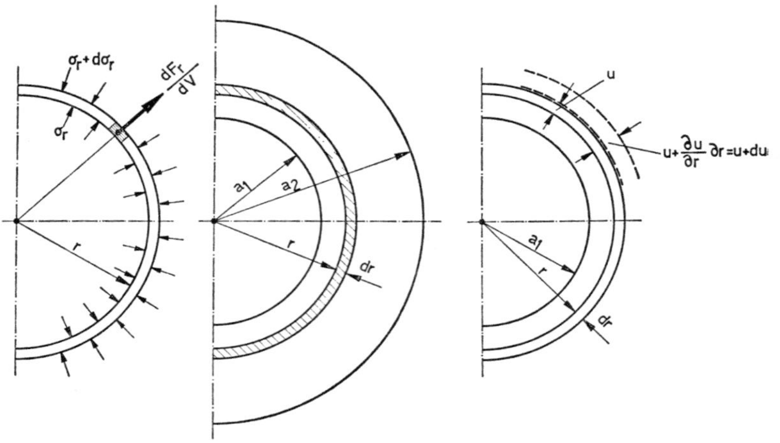

Fig. 2.4.6.

Stress-displacement relations

$$\left.\begin{aligned}
\sigma_c &= \frac{E}{(1-2m)(1+m)} \cdot \left[\frac{u}{r} + m\left(-\frac{u}{r} + \frac{du}{dr} + \varepsilon_z\right)\right] \\
\sigma_r &= \frac{E}{(1-2m)(1+m)} \cdot \left[\frac{du}{dr} + m\left(-\frac{du}{dr} + \frac{u}{r} + \varepsilon_z\right)\right] \\
\sigma_z &= \frac{E}{(1-2m)(1+m)} \cdot \left[\varepsilon_z + m\left(-\varepsilon_z + \frac{u}{r} + \frac{du}{dr}\right)\right]
\end{aligned}\right\}. \tag{2.4.31}$$

Combining Eq. (2.4.31) with Eq. (2.4.25), we obtain for the simplest case of $\varepsilon_z \cong$ constant,

$$\frac{(1-m)\,E}{(1-2\,m)\,(1+m)} \cdot \left[\frac{u}{r} - \frac{du}{dr} - r \cdot \frac{d^2u}{dr^2}\right] = rF_r(r,z)\,,$$

which, rearranged, is written:

$$\boxed{\frac{d^2u}{dr^2} + \frac{1}{r}\frac{du}{dr} - \frac{u}{r^2} = -\frac{(1-2\,m)\,(1+m)}{(1-m)\,E} \cdot F_r(r,z)\,.} \tag{2.4.32}$$

The general solution of this equation is:

$$u = C_1 \cdot r + \frac{C_2}{r} - \frac{(1-2\,m)\,(1+m)}{2\,E\,(1-m)}\, r \int F_r(r,z)\,dr + \frac{(1-2\,m)\,(1+m)}{2\,E\,(1-m)}\,\frac{1}{r}\int r^2 F_r(r,z)\,dr\,. \tag{2.4.33}$$

Calculating the values of u/r and du/dr, assuming $\varepsilon_z \cong 0$, we get for the radial stress:

$$\begin{aligned}\sigma_r &\cong \frac{E}{(1-2\,m)\,(1+m)} \cdot \left[(1-m)\frac{du}{dr} + m\frac{u}{r}\right] \\ &= \frac{E}{1+m}\,\frac{1}{(1-2\,m)}\left(C_1 - \frac{C_2}{r^2}\right) - \frac{1}{2\,(1-m)}\int F_r(r,z)\,dr \\ &\quad - \frac{1-2\,m}{2\,(1-m)} \cdot \frac{1}{r^2} \cdot \int r^2 F_r(r,z)\,dr\,.\end{aligned} \tag{2.4.34}$$

The two integration constants are obtained from boundary conditions:

$$\sigma_r\big|_{r=a_1} = \sigma_r\big|_{r=a_2} = 0\,.$$

The circumferential stress is obtained in the same manner:

$$\begin{aligned}\sigma_c &\cong \frac{E}{(1-2\,m)\,(1+m)} \cdot \left[(1-m)\frac{u}{r} + m\frac{du}{dr}\right] \\ &= \frac{E}{(1+m)} \cdot \left(\frac{1}{1-2\,m}C_1 + \frac{C_2}{r^2}\right) - \frac{1}{2\,(1-m)}\int F_r(r,z)\,dr \\ &\quad + \frac{1-2\,m}{2\,(1-m)} \cdot \frac{1}{r^2}\int r^2 F_r(r,z)\,dr\,.\end{aligned} \tag{2.4.35}$$

Middleton and Trowbridge [69] have obtained the expressions:

$$\boxed{\begin{aligned}\sigma_r &= \frac{\sigma_c(a_1) - m \cdot \sigma_z(a_1)}{2}\left(1 - \frac{a_1^2}{r^2}\right) - \frac{1+m}{2}\int_{a_1}^{r} F_r(r,z)\,dr \\ &\quad - \frac{1-m}{2\,r^2}\int_{a_1}^{r} r^2 F\,(r,z)\,dr + \frac{m}{r_2}\int_{a_1}^{r} r\,\sigma_z dr\,.\end{aligned}} \tag{2.4.36}$$

In this expression, $\varepsilon_z \neq 0$. The circumferential stress is given accordingly:

$$\sigma_c = \sigma_c(a_1) - m\,\sigma_z(a_1) - \sigma_r + m\,\sigma_z - (1 - m)\int_{a_1}^{r} F_r(r,z)\,dr\,. \qquad (2.4.37)$$

Computer programs to calculate σ_r and σ_c are written [67]. Some results are illustrated in Fig. 2.4.7.

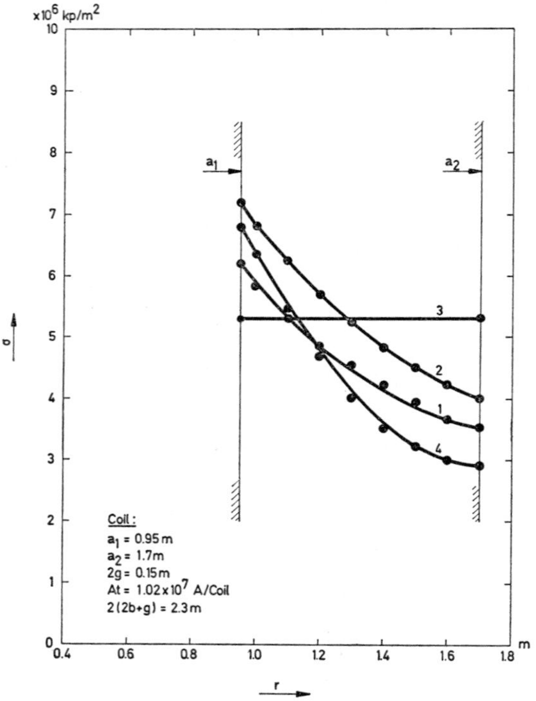

Fig. 2.4.7. Stressdistribution in an axially symmetric coil: *1* Calculation according Eq. (2.2.24), *2* Eq. $\sigma_\theta = J/4\,a \ln\alpha \cdot (B_1 + B_2)\cdot(\alpha^2 - 1)\,a_1^2$, *3* Eq. $\sigma_\theta = J/4\,(B_1 + B_2)\,(a_2 + a_1)$, 4 Eq. (2.4.37)

2.4.6 Stresses due to Thermal Contraction

When the coil undergoes the cooling process, the various radial turns shrink differently, exerting a radial inward force. Of special interest is the case where the conductor material and the coil support structure have different contraction coefficients. The calculation of the radial and tangential stresses is similar to the case given in Section 2.4.4. The axial stresses are non-existent in this case.*

* This calculation is also directly applicable for stress calculations due to winding pretension.

As the coil consists of a metallic conductor, reinforcements, and insulation materials, the equivalent modulus of elasticity of the coil E_q must be obtained. This is given by:

$$\boxed{\frac{1}{E_q} = \sum_{i=1}^{n}\left(\frac{1}{t_i}\right)\cdot\sum_{i=1}^{n}\frac{t_i}{E_i}.} \qquad (2.4.38)$$

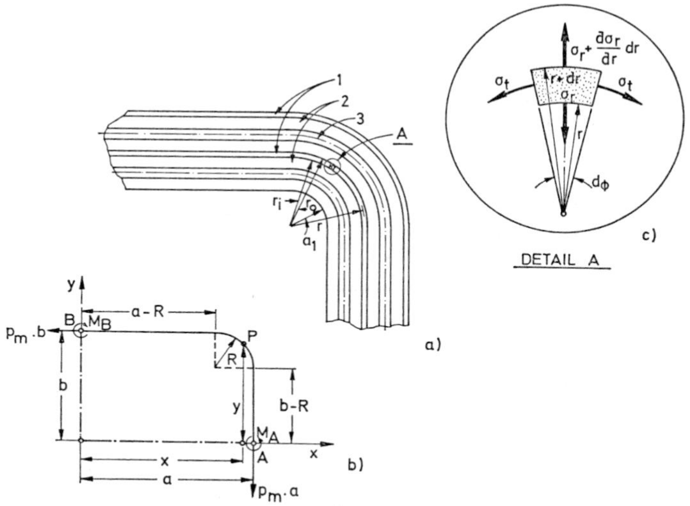

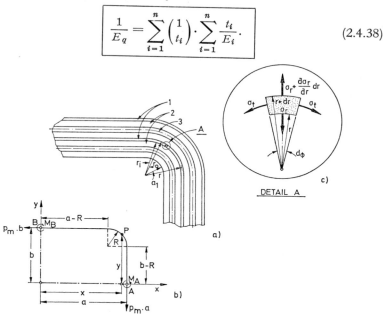

Fig. 2.4.8. Two dimensional coil geometry
a Cross-section through a corner 1. Insulation, 2. Conductor, 3. Coolant passage; *b* Coil mechanical neutral axis, *c* Stress distribution

In the particular case of an axially-symmetrical coil, the boundary stress values are known. The stress distribution in an element shown in Fig. 2.4.8 is given by:

$$\sigma_r \cdot r\,d\phi + \sigma_t \cdot r\,d\phi - \left(\sigma_r + \frac{d\sigma_r}{dr}\,dr\right)(r + dr)\,d\phi = 0\,. \qquad (2.4.39)$$

Neglecting all higher terms, by considering a small angle $d\phi$, we get:

$$\frac{d\sigma_r}{dr} = \frac{\sigma_r - \sigma_t}{r}.$$

With:

$$m_{rt} = m_{tr} \cdot E_r\,, \quad m_{rt} = m_{tr} = m$$

and

$$k = \left(\frac{E_t}{E_r}\right)^{\frac{1}{2}}$$

we obtain the stress-strain relation:

$$\varepsilon_r = \frac{\sigma_r}{E_r} - \frac{\sigma_t}{m \cdot E_t} = \frac{1}{E_r}\left(\sigma_r - \frac{\sigma_t}{m k^2}\right), \qquad (2.4.40)$$

$$\varepsilon_t = \frac{\sigma_t}{E_t} - \frac{\sigma_r}{m \cdot E_t} = \frac{1}{k^2 E_r}\left(\sigma_t - \frac{\sigma_r}{m}\right), \qquad (2.4.41)$$

where we used the simplification $m_{t,r} = m_{r,t} = m$ for the Poisson Ratios.

Introducing the displacements u at the radius r and $u + \frac{du}{dr}\,dr$ at $r + dr$, we get:

$$\varepsilon_r = \frac{du}{dr}; \qquad \varepsilon_t = \frac{u}{r}.$$

Thus the stresses:

$$\sigma_r = \frac{k^2 m^2}{k^2 m^2 - 1} \cdot E_r \left(\varepsilon_r + \frac{1}{m}\, \varepsilon_t \right) = \frac{k^2 m^2}{k^2 m^2 - 1} \cdot E_r \left(\frac{du}{dr} + \frac{1}{m} \cdot \frac{u}{r} \right). \tag{2.4.42}$$

in radial direction, and

$$\sigma_t = \frac{k^2 m^2}{k^2 m^2 - 1} \cdot E_r \left(\frac{1}{m}\, \varepsilon_r + k^2\, \varepsilon_t \right) = \frac{k^2 m^2}{k^2 m^2 - 1} \cdot E_r \left(\frac{1}{m}\, \frac{du}{dr} + k^2 \frac{u}{r} \right) \tag{2.4.43}$$

in tangential direction.

Substituting Eqs. (2.4.42) and (2.4.43) into Eq. (2.4.39), we obtain the differential equation for the coil displacement:

$$\boxed{\frac{d^2 u}{dr^2} + \frac{1}{r}\, \frac{du}{dr} - \frac{k^2}{r^2}\, u = 0,} \tag{2.4.44}$$

which is the same as Eq. (2.4.32) without the body force part and the assumption here is that $E_r \neq E_t$.

The general solution of this equation is:

$$u = C_1 \cdot r^k + C_2 \cdot r^{-k}. \tag{2.4.45}$$

Thus:

$$\sigma_r = \frac{k^2 m}{k^2 m^2 - 1}\, E_r \left\{ C_1 (km + 1)\, r^{k-1} - C_2 (km - 1)\, r^{-k-1} \right\},$$

$$\sigma_t = \frac{k^3 m}{k^2 m^2 - 1}\, E_r \left\{ C_1 (km + 1)\, r^{k-1} + C_2 (km - 1)\, r^{-k-1} \right\}. \tag{2.4.46}$$

With the boundary conditions: $(\sigma_r)_{r=a_1} = -p_i$, $(\sigma_r)_{r=a_2} = -p_0$

and using the relation $\alpha = a_2/a_1$, we obtain:

$$\sigma_r = \frac{a_1^2}{\alpha^{k-1} - \alpha^{-k-1}} \cdot \left\{ a_1^{-k-1} (p_i \alpha^{-k-1} - p_0)\, r^{k-1} - a_1^{k-1} (p_i \alpha^{-k-1} - p_0)\, r^{-k} \right\}, \tag{2.4.47}$$

$$\sigma_t = \frac{k\, a_1^2}{\alpha^{k-1} - \alpha^{-k-1}} \cdot \left\{ a_1^{-k-1} (p_i \alpha^{-k-1} - p_0)\, r^{k-1} - a_1^{k-1} (p_i \alpha^{k-1} - p_0)\, r^{-k-1} \right\}. \tag{2.4.48}$$

The shear stress is obtained from $\sigma = \sigma_r + \sigma_t/2$, which is a maximum at $r = a_1$

$$\sigma_{\max} = \frac{1}{2\,[\alpha^{k-1} - \alpha^{-k-1}]} \cdot \left\{ (p_i \alpha^{-k-1} - p_0)\,(k - 1) + (p_i \alpha^{k-1} - p_0)\,(k + 1) \right\}. \tag{2.4.49}$$

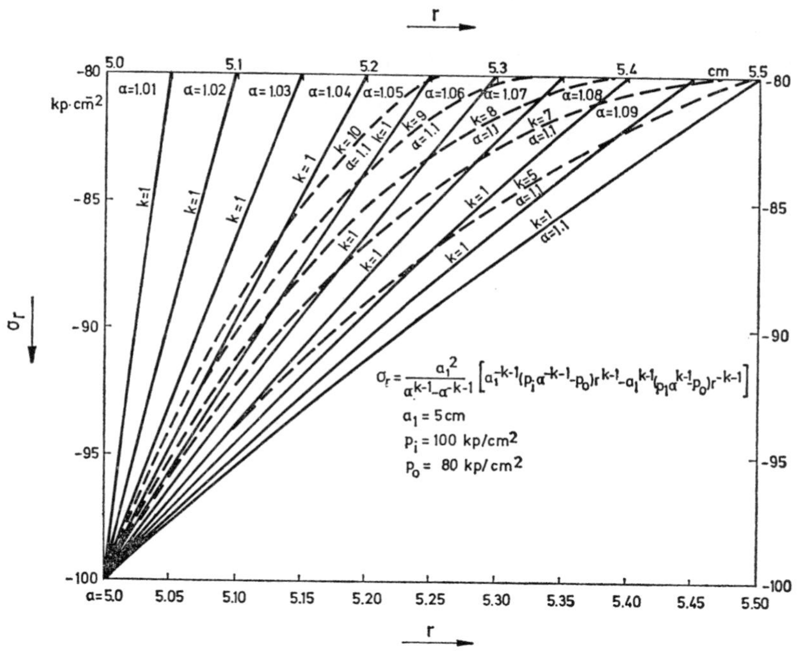

Fig. 2.4.9a.

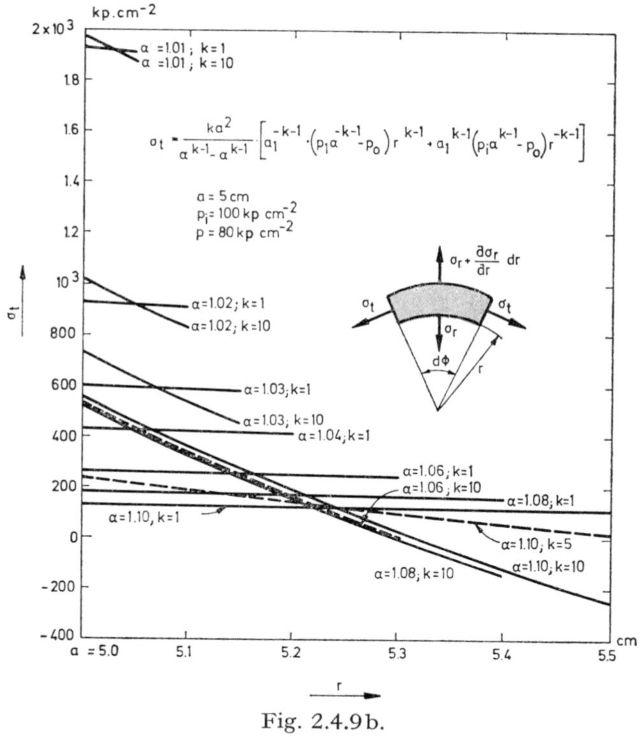

Fig. 2.4.9b.

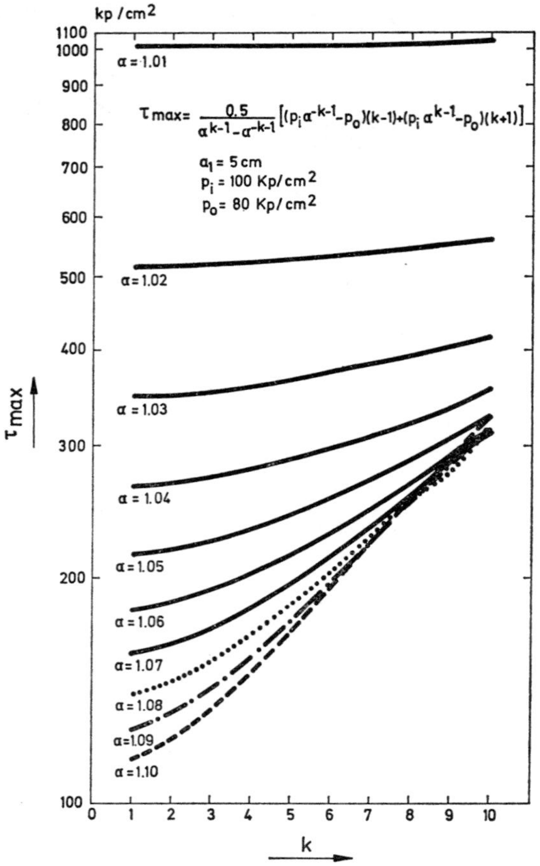

Fig. 2.4.9c.

The coil deformation is obtained from Eq. (2.4.44):

$$u = \frac{1}{k^2 m E_r [\alpha^{k-1} - \alpha^{-k-1}]} \cdot \{(km - 1)\, a_1^{-k+1} (p_i \alpha^{-k-1} - p_0) \cdot r^k$$
$$+ (km + 1)\, a_1^{k+1} (p_i \alpha^{k-1} - p_0) \cdot r^{-k}\}\,. \qquad (2.4.50)$$

Figures 2.4.9a—d illustrate, for one specific coil with $a_1 = 5$ cm, the stresses σ_r, σ_t, σ_{max}, and the deformation u.

To obtain the whole stress distribution over the coil, the values obtained in this chapter should be superimposed to the stress values calculated in Section 2.4.5.

2.4.7 Forces for a Dipole Coil Configuration

The dipole field generated by two intersecting current circles was given in Section 2.1.3. The field within the coil aperture of Fig. 2.5b was expressed by Eq. (2.1.27). We assumed that the coil current density was (λJ) and constant. If we ignore the coil end effects, then for a pure two-

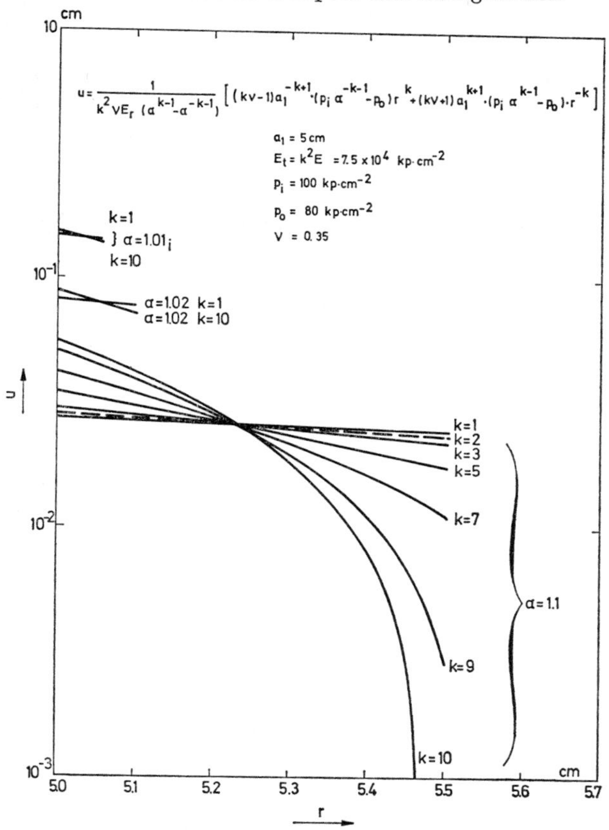

Fig. 2.4.9d.

dimensional coil configuration (shown in Fig. 2.4.10), the resulting force in horizontal direction is given by:*

$$F_x = \iint_A (\lambda J H_1 \cdot \cos\varphi - \lambda J H_2 \cdot \cos\psi)\, dA \tag{2.4.51}$$

$$= \frac{\mu_0}{2} (\lambda J)^2 \iint_A \left(r \cdot \cos\varphi - \frac{a^2}{\varrho} \cdot \cos\psi\right) dA\,. \tag{2.4.52}$$

This integral is solved by Atherton [70] and Beth [71], and given in Eq. (2.4.53):

$$F_x = \frac{\mu_0}{2} (\lambda J)^2 \cdot \left\{2\, a^3 \cdot \sin\beta - \frac{1}{4} c^3 \cdot \tan\beta - a^2 \left[\frac{c}{2} \cdot \sin 2\beta + \left(\frac{a^2}{c}\right) \cdot \left\{\sin^{-1}\left(\frac{c}{a}\right) \cdot \sin\beta + \frac{1}{2} \sin\left(2 \sin^{-1}\left(\frac{c}{a}\right)\right) \cdot \sin\beta\right\}\right]\right\}, \tag{2.4.53}$$

where $c = 2\, x_0 = 2\, a \cdot \sin\gamma$.

* Due to symmetry there is no force in the vertical or y direction.

Since

$$\sin^{-1}\left(\frac{c}{a}\right)\sin\beta = \sin^{-1}(2\sin\gamma\cdot\cos\gamma) = 2\gamma\,, \quad \text{and} \quad \beta = \left(\frac{\pi}{2}\right) - \gamma,$$

we may write:

$$\boxed{F_x = -\frac{\mu_0}{2}(\lambda J)^2\cdot\left\{2\cos\gamma - 2\sin^3\gamma\cot\gamma - \left[\sin\gamma\sin2\gamma + \frac{1}{2}\sin\gamma\left(2\gamma + \frac{1}{2}\sin4\gamma\right)\right]\right\}.} \tag{2.4.54}$$

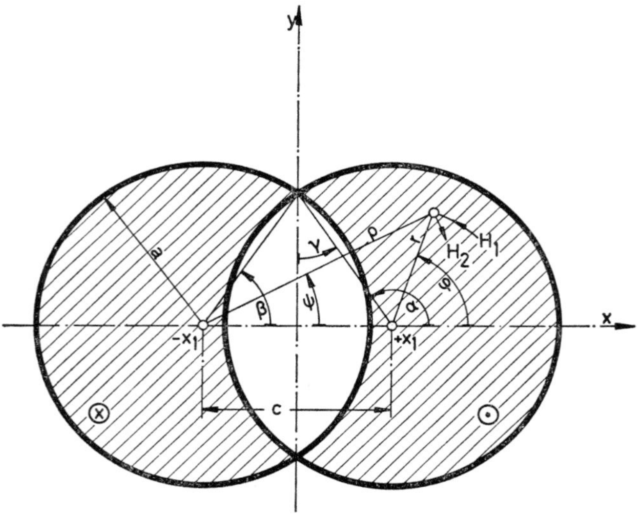

Fig. 2.4.10. To calculate forces in a dipole coil

2.4.8 Force Equations for Multipole Coils

Expressions for forces acting on the conductor of a magnet can be obtained from Maxwell's stress tensor. For our particular two-dimensional case, we express the force density by $\mu_0 \boldsymbol{J}\cdot\boldsymbol{H}$ acting on a conductor. The force per unit length $\boldsymbol{F} = F_x + jF_y$ can be given by:

$$F = \mu_0 \int JH\,dA\,. \tag{2.4.55}$$

In chapter 2.3, we had calculated the field for multipole arrangements assuming infinite iron permeability, or multipole arrangements without ferromagnetic return shields. If one assumes constant current density over the coil by inserting the value of $B = \mu_0 H$ in Eq. (2.4.55), one can calculate the radial and azymuthal force components for a multipole arrangement.

2.4.9 Forces in Spherical Coils

Forces in spherical coils are calculated generally for uniform internal or external pressure and are applicable at the equator ($\theta = \pi/2$) only. The exact evaluation of stresses and deformations in a sphere is complex in nature. Equations suitable for computer calculations are given below.

Referring to Fig. 2.4.11 the equations relating the stress components and strains for spherical geometries when shear is neglected are:

$$r \frac{\partial \sigma_r}{\partial r} + 2 \sigma_r - \sigma_\theta + \sigma_\varphi + r F_r = 0 \qquad (2.4.56)$$

$$\sigma_\theta - \sigma_\varphi + \frac{\partial \sigma_\theta}{\partial \theta} \tan(\theta) + r \tan(\theta) \cdot F_\theta = 0 \qquad (2.4.57)$$

$$\frac{\partial \sigma_\varphi}{\partial \varphi} = 0 \qquad (2.4.58)$$

$$\varepsilon_r = \frac{\partial w}{\partial r} = \frac{1}{E} [\sigma_r - m (\sigma_\varphi + \sigma_\theta)] \qquad (2.4.59)$$

$$\varepsilon_\theta = \frac{1}{r} \left(\frac{\partial v}{\partial \theta} + w \right) = \frac{1}{E} [\sigma_\theta - m (\sigma_r + \sigma_\varphi)] \qquad (2.4.60)$$

$$\varepsilon_\varphi = \frac{1}{r} (v \operatorname{cotan}(\theta) + w) = \frac{1}{E} [\sigma_\varphi - m (\sigma_\theta + \sigma_r)] . \qquad (2.4.61)$$

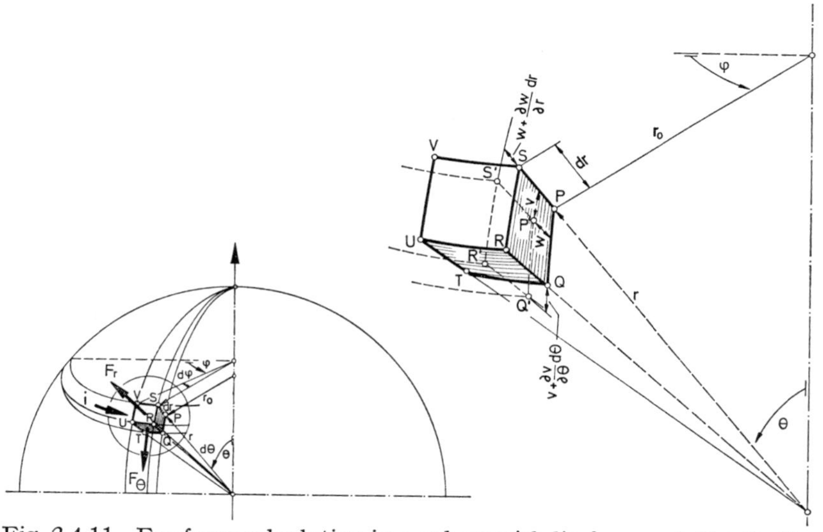

Fig. 2.4.11. For force calculation in a sphere with lin θ current distribution

In these equations v is the tangential and w the radial displacement of a volume element.

The body forces are expressed by:

$$F_r = \mu_0 (\lambda J_0)^2 \left[- \frac{a_1^4}{12 r^2} - \frac{2}{3} \alpha a_1 + \frac{3}{4} r \right] \cdot \sin^2(\theta) \qquad (2.4.62)$$

$$F_\theta = \mu_0 (\lambda J_0)^2 \left[- \frac{a_1^4}{6 r^2} + \frac{2}{3} \alpha a_1 - \frac{1}{2} r \right] \sin(\theta) \cdot \cos(\theta) . \qquad (2.4.63)$$

The displacement vector $\boldsymbol{u}$ combines both tangential and radial displacements:

$$\boldsymbol{u} = (u_r, u_\theta, 0) \equiv (w, v, 0) .$$

The displacements are related to the body forces $\boldsymbol{F} = (F_r, F_\theta, 0)$ by:

$$\boxed{\begin{aligned} 2\,(1-m)\,(\operatorname{grad}\cdot\operatorname{div}\boldsymbol{u}) - (1-2\,m)\,(\operatorname{curl}\cdot\operatorname{curl}\boldsymbol{u}) \\ = -2\,\boldsymbol{F}\,\frac{(1+m)\,(1-2\,m)}{E}\,. \end{aligned}} \qquad (2.4.64)$$

From these relations stresses and strains may be obtained.

2.4.10 Forces in Toroidal Coils

If one considers the non-uniform field distribution within the toroid the equations to obtain strains and displacements become equally complex as in the case of a sphere. The assumption of uniform field distribution within the bore gives a first order approximation to the stress calculations. Taking the maximum value of B at the inner coil radius we may write for the hoop and radial stresses:

$$\sigma_\theta = r_1 \cdot B_m \cdot (\lambda J_0) \cdot \frac{\beta}{\beta - \alpha} \cdot \frac{2-\beta}{2\,(1-\beta)} \qquad (2.4.65)$$

$$\sigma_r = r_1 \cdot B_m \cdot (\lambda J_0) \cdot \frac{\beta}{2\,(\beta - \alpha)}\,,$$

with R_0 the major radius, r_1 the inner coil radius and r_2 the outer coil radius, $\alpha = r_1/R_0$; $\beta = r_2/R_0$.

2.5 Calculation of Heating

Joule's heating is encountered in superconducting magnets for the following reasons:

1. If the superconductor is driven normal in a portion of the coil; this may be through a number of reasons, such as a flux jump, conductor movement, failures in the current source, etc., the current will flow partially through the substrate (low resistivity normal metal) and partially through the (now normal) superconductor.

2. In a.c. or pulsed magnets, the coil is heated due to a.c. losses in the superconductor, eddy current losses in the composite, and auxiliary losses.

In both cases, the heat must be removed by the coolant, being either in direct contact with the conductor (heat transfer), or by conduction (through substrate, insulations), and finally by convection (heat transfer).

The temperature distribution in a coil arrangement with internal heat sources is given by the Poisson's equation:

$$k_x \frac{\partial^2 T}{\partial x^2} + k_y \frac{\partial^2 T}{\partial y^2} + k_z \frac{\partial^2 T}{\partial z^2} + w_v = 0\,, \qquad (2.5.1)$$

where w_v denotes the losses per unit volume.

Even in simple coil geometries, the general solution of Eq. (2.5.1) will be quite a task.

The thermal conductivities in the three directions are different due to the conductor and insulation arrangements. The thermal conductivities are also functions of temperature.

The diffusion of heat along the conductor has been considered by Broom [72] and Brechna [73]. Wilson [74] considers the heat diffusivity along the conductor and perpendicular to it. Other simplified models assuming the thermal conductivity of the complex structure to be constant over certain temperature ranges give a crude estimate of the heat distribution. Thermal conductivity values for a variety of materials are available and are given in chapter V. Heat transfer values of helium is also given in the same section. Thus, choosing a suitable model the distribution of heat can be calculated with a sufficient degree of accuracy.

As seen from Fig. 2.5.1a–f, any coil arrangement can be subdivided into a number of rectangular portions, where the heat generated in the coil can be removed by the coolant. In such a rectangle (Fig. 2.5.2), we assume symmetrical conditions and neglect the heat conduction along the conducot. This assumption is justified in coils with axial symmetry, or if the length of the conductor is such that we may ignore heat conduction along the length of the wire. This assumption is not true in transposed multicore conductors. Thus, from Eq. (2.5.1)

$$k_z \frac{\partial^2 T}{\partial z^2} = 0 .$$

We have to solve a two-dimensional equation:

$$\boxed{k_x \frac{\partial^2 T}{\partial x^2} + k_y \frac{\partial^2 T}{\partial y^2} + w_v = 0 .} \tag{2.5.2}$$

We give the following boundary conditions:

$$T - T_{mc} = -\frac{k_x}{h_x} \cdot \frac{\partial T}{\partial x} \quad \text{for} \quad x = +\frac{a}{2},$$

$$T - T_{mc} = +\frac{k_x}{h_x} \cdot \frac{\partial T}{\partial x} \quad \text{for} \quad x = -\frac{a}{2},$$

$$T - T_{mc} = -\frac{k_y}{h_y} \cdot \frac{\partial T}{\partial y} \quad \text{for} \quad y = +\frac{b}{2},$$

$$T - T_{mc} = +\frac{k_y}{h_y} \cdot \frac{\partial T}{\partial y} \quad \text{for} \quad y = -\frac{b}{2}.$$

Using the substitutions

$$T - T_{mc} = \frac{w_v}{2\, k_x} \cdot \theta\, ; \quad y = \left(\frac{k_y}{k_x}\right)^{\frac{1}{2}} \cdot y_1$$

and inserting them into Eq. (2.5.2), we obtain:

$$\frac{\partial^2 \theta}{\partial x^2} + \frac{\partial^2 \theta}{\partial y_1^2} + 2 = 0 . \tag{2.5.3}$$

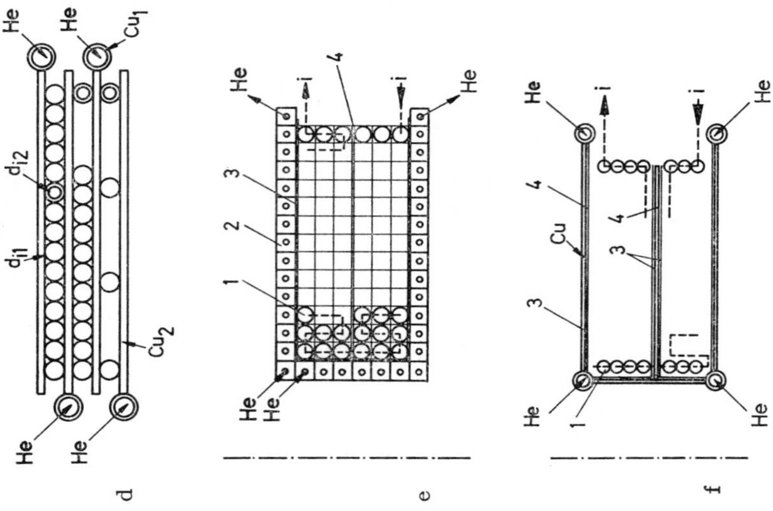

Fig. 2.5.1 d, e and f. Coil with heat drains

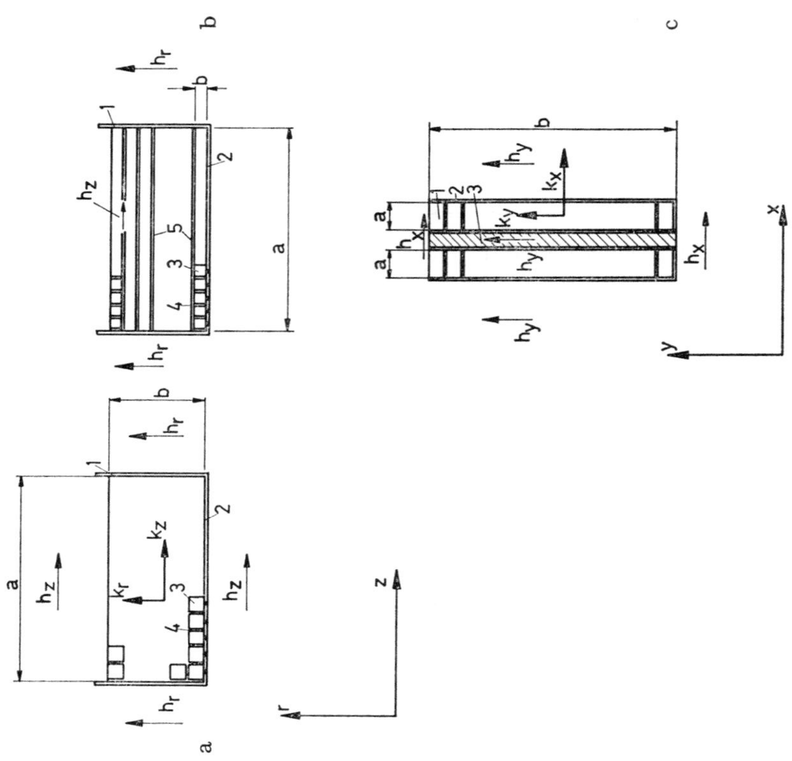

Fig. 2.5.1 a, b and c. Coil arrangements

The two boundary conditions are now:

$$\theta = \mp \frac{k_x}{h_x} \frac{\partial \theta}{\partial x} \qquad \text{for} \quad x = \mp \frac{a}{2}$$

$$\theta = \mp \frac{(k_x k_y)^{\frac{1}{2}}}{h_y} \cdot \frac{\partial \theta}{\partial y_1} \quad \text{for} \quad y = \mp \frac{b}{2}.$$

If we make the substitution:

$$\theta = \tau + \left[\left(\frac{a}{2}\right)^2 - x^2 + a \cdot \frac{k_x}{h_x}\right], \tag{2.5.4}$$

we obtain:

$$\frac{\partial^2 \tau}{\partial x^2} + \frac{\partial^2 \tau}{\partial y_1{}^2} = 0\,. \tag{2.5.5}$$

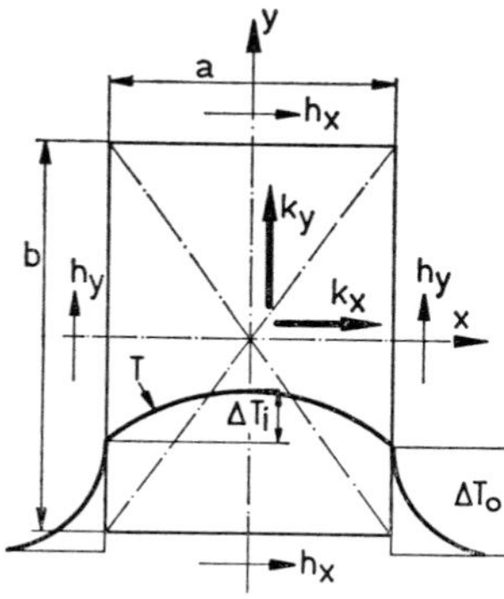

Fig. 2.5.2. Temperature distribution in a rectangular coil

The new boundary conditions corresponding to Eq. (2.5.5) are now

$$\tau = \frac{k_x}{h_x} \cdot \frac{\partial \tau}{\partial x} \quad \text{for} \quad x = +\frac{a}{2}, \tag{2.5.6}$$

$$\tau = \frac{k_x}{h_x} \cdot \frac{\partial \tau}{\partial x} \quad \text{for} \quad x = -\frac{a}{2}, \tag{2.5.7}$$

$$\tau = x^2 - \left(\frac{a}{2}\right)^2 - a\frac{k_x}{h_x} - \frac{(k_x \cdot k_y)^{\frac{1}{2}}}{h_y} \cdot \frac{\partial \tau}{\partial y_1} \quad \text{for} \quad y = +\frac{b}{2}, \tag{2.5.8}$$

$$\tau = x^2 - \left(\frac{a}{2}\right)^2 - a\frac{k_x}{h_x} + \frac{(k_x \cdot k_y)^{\frac{1}{2}}}{h_y} \cdot \frac{\partial \tau}{\partial y_1} \quad \text{for} \quad y = -\frac{b}{2}. \tag{2.5.9}$$

Assuming the temperature distribution is according to:

$$\tau = \sum_{n=1}^{n} A_n \cos(n x) \cosh(n y_1) \tag{2.5.10}$$

and inserting Eq. (2.5.10) into Eq. (2.5.6), we get:

$$\sum_n A_n \cdot \cos\left(n \frac{a}{2}\right) \cosh(n y_1) = +\frac{k_x}{h_x} \cdot n \sum_n \sin\left(n \frac{a}{2}\right) \cosh(n y_1)\,,$$

which yields:

$$\operatorname{cotan}\left(n \frac{a}{2}\right) = \frac{k_x}{h_x} \cdot n\,, \tag{2.5.11}$$

written more suitably:

$$\boxed{n\frac{a}{2} = \left(\frac{h_x}{k_x}\frac{a}{2}\right)\operatorname{cotan}\left(n\frac{a}{2}\right).} \tag{2.5.12}$$

The Fourier coefficient "n" is obtained graphically.

Inserting Eq. (2.5.10) in Eq. (2.5.8), we obtain:

$$\sum_{n=1}^{\infty} A_n\cdot\left\{\cosh\left[\left(\frac{k_x}{k_y}\right)^{\frac{1}{2}}\cdot n\cdot\frac{b}{2}\right] + \frac{(k_x k_y)^{\frac{1}{2}}\cdot n}{h_x}\cdot\sinh\left[\left(\frac{k_x}{k_y}\right)^{\frac{1}{2}}\cdot n\cdot\frac{b}{2}\right]\right\}\cos(n x) = x^2 - \frac{a^2}{2} - a\cdot\frac{k_x}{h_x}. \tag{2.5.13}$$

To calculate the coefficient A_n, we have to consider:

$$\int_{-a/2}^{+a/2} \cos(m x)\cdot\cos(n x)\,d x = 0\,, \quad \text{for} \quad m \neq n\,,$$

$$\int_{-a/2}^{+a/2} \cos^2(n x)\,d x = \frac{a}{2}\left[1 + \frac{\sin(n a)}{(n a)}\right],$$

$$\int_{-a/2}^{+a/2} \left[x^2 - \left(\frac{a}{2}\right)^2 - a\cdot\frac{k_x}{h_x}\right]\cos(n x)\,d x = -\frac{4\sin(n\,a/2)}{n^3}.$$

Thus:

$$A_n = a^2\cdot\frac{\sin\left(n\frac{a}{2}\right)}{\left(n\frac{a}{2}\right)^3}\cdot\frac{1}{1+\frac{\sin(na)}{na}}\cdot\frac{1}{\left[1+\frac{(k_x k_y)^{\frac{1}{2}}}{h_x}\cdot\tan h\left[\left(\frac{k_x}{k_y}\right)^{\frac{1}{2}} n\frac{b}{2}\right]\right]}. \tag{2.5.14}$$

If we chose the form:
$$A_n = \frac{P_n}{Q_n}a^2\,,$$

where:
$$\boxed{P_n = \frac{\sin\left(n\frac{a}{2}\right)}{\left(n\frac{a}{2}\right)^3}\,\frac{1}{1+\frac{\sin(na)}{na}},} \tag{2.5.15}$$

$$\boxed{Q_n = \left\{1+\frac{(k_x k_y)^{\frac{1}{2}}}{h_x}\cdot\tanh\left[\left(\frac{k_x}{k_y}\right)^{\frac{1}{2}}\cdot n\frac{b}{2}\right]\right\}\cos\left[\left(\frac{k_x}{k_y}\right)^{\frac{1}{2}}\cdot n\frac{b}{2}\right].} \tag{2.5.16}$$

Thus the temperature distribution is given from Eq. (2.5.4):

$$\theta = \left(\frac{a}{2}\right)^2 - x^2 + a\cdot\frac{k_x}{h_x} - \sum_n \frac{P_n}{Q_n}a^2\cos(n x)\cosh(n y_1)\,, \tag{2.5.17}$$

or from:

$$T - T_{mc} = \frac{w_v}{2\,k_x} \cdot \theta \quad \text{we get:}$$

$$T = T_{mc} + \frac{w_v}{2\,k_y}\left\{\left(\frac{a}{2}\right)^2 - x^2 + a \cdot \frac{k_x}{h_x} - a^2 \sum_n \frac{P_n}{Q_n} \cdot \cos(n\,x) \cdot \cosh\left[\left(\frac{k_x}{k_y}\right)^{\frac{1}{2}} n\,y\right]\right\}. \tag{2.5.18}$$

We are interested primarily in the following temperatures:

$$\Delta T_i = T(x = 0, y = 0) - T\left(x = \frac{a}{2},\quad y = 0\right),$$

or:

$$\Delta T_i = \frac{w_v}{2\,k_x}\left(\frac{a}{2}\right)^2\left\{1 - 4\sum_n\left(\frac{P_n}{Q_n}\right)\left(1 - \cos\left(n\,\frac{a}{2}\right)\right\}. \tag{2.5.19}$$

The temperature rise in the boundary layer is obtained from:

$$\Delta T_0 = T\left(x = \frac{a}{2};\quad y = 0\right) - T_{mc},$$

or:

$$\Delta T_0 = w_v \cdot \frac{a}{2\,h_x}\left\{1 - \frac{2\,h_x}{k_x}\left(\frac{a}{2}\right)\sum_n\left(\frac{P_n}{Q_n}\right)\left(\cos\left(n\,\frac{a}{2}\right)\right\}. \tag{2.5.20}$$

Thus, the "hot spot" temperature within the coil is found from

$$T_{hs} = \Delta T_i + \Delta T_0 + T_{mc}. \tag{2.5.21}$$

The total temperature drop over the coil cross-section is given by:

$$\boxed{\Delta T = \frac{w_v}{2\,k_x}\left(\frac{a}{2}\right)^2\left\{1 - 4\sum_{n=1}^{n}\frac{P_n}{Q_n}\right\} + w_v\,\frac{a}{2\,h_x}.} \tag{2.5.22}$$

The series P_n and Q_n are rapidly converging. Only two or three terms of the series will give an accurate solution within a few percent. If we assume $b \gg a$ or $k_x \gg k_y$, which means that we may ignore the cooling effect at the long side of the area, then $\sum P_n/Q_n$ will be zero, and we may write as a first approximation:

$$\boxed{\Delta T' = w_v \cdot \left[\frac{1}{2\,k_x}\left(\frac{a}{2}\right)^2 + \frac{a}{2\,h_x}\right],} \tag{2.5.23}$$

from which the coil width is obtained for a permissible temperature rise from the bulk helium temperature:

$$\frac{a}{2} = -\frac{k_x}{h_x} \pm \left[\left(\frac{k_x}{h_x}\right)^2 + 2\,k_x \cdot \frac{\Delta T'}{w_v}\right]^{\frac{1}{2}}. \tag{2.5.24}$$

To give an example, we assume that the coil is immersed in a pool of liquid helium. For vertical coolant passages, we may assume a value of

$h = 1.6 \text{ W/cm}^2 \text{ K}$. The coil is potted in a thermosetting, filled with an inorganic powder such as Al_2O_3. The thermosetting has an average thermal conductivity of 4×10^{-3}W/ cmK at 4.5 K. If the heat generated in the magnet is about 10^{-2} W/cm³ and we assume that the temperature rise in the coil does not exceed 0.2 K, then the coil width will be according to Eq. (2.5.24) $a/2 = 0.4$ cm, but according to Eq. (2.5.22), $a/2 =$ 0.41 cm.

In new composite conductors the superconducting filaments are twisted. If a multicore cable or braid is used as a conductor, the strands are transposed. This would mean that due to the transposition and twisting, each superconductor and each strand will be at the cable surface after a certain distance. In the above calculation, the longitudinal heat conduction along the conductor must be introduced. If the path length for the heat to be conducted to the surface of the coil to the conductor surface due to the transposition is short, then the actual transverse thermal conductivity of the coil will be much higher than anticipated above. One strand will always be periodically at the surface of the conductor, and in flat pancakes, also at the surface of the coil. The internal temperature drop ΔT_i will be very much smaller than calculated.

References Chapter 2

[1] Beth, R. A.: BNL Rep. AADD. **66** (1965).
[2] Yourd, R. B., Halbach, K.: UCRL preprint 18378 (1968).
[3] Halbach, K.: UCRL preprint 18947 (1969).
[4] Beth, R. A.: I.E.E.E. Trans. Nucl. Sci. NS **14,** 386 (1967).
[5] Beth, R. A.: BNL Rep. AADD. **102** (1966).
[6] Hart, P. J.: Universal Tables for Magnetic Fields. New York: American Elsevier 1967.
[7] Gauster, W. F., Garrett, M. W.: ORNL. 3652 (1964).
[8] Durand, E.: Electrostatique et Magnétostatique. Masson et Cie. Editeurs 1953.
[9] Garrett, M. W.: J. Appl. Phys. **22,** 1091 (1951).
[10] Girard, B., Sausade, M.: Nucl. Instr. and Methods **25,** 269 (1964).
[11] Brechna, H., Perot, J.: SLAC Publ. 739 (1970).
[12] Perot, J.: SLAC-TN-69-20 (1969).
[13] Green, M. A.: U.C.I.D. 3493 (1971).
[14] Harris, F. K.: NBS Paper RP 716. **13** (1934).
[15] Lyddane, R. H., Ruark, A. E.: Rev. Sci. Instr. **10,** 253 (1939).
[16] Grant, W. J. C., Strandberg, M. W. P.: Rev. Sci. Instr. **36,** 343 (1965).
[17] Perkins, W. A., Brown, J. C.: UCLRL 7744 Rev. II (1964).
[18] Lari, R. J.: Argonne Rep. R. JL-7 (1966).
[19] Hand, L. N., Panofsky, W. K. H.: Rev. Sci. Instr. **30,** 927 (1959).
[20] Septier, A.: CERN 60-6 (1960).
[21] Dayton, I. E., et al.: Rev. Sci. Instr. **25,** 485 (1954).
[22] Bitter, F.: Rev. Sci. Instr. **7,** 479 (1936); Rev. Sci. Instr. **8** (1937).
[23] Sampson, W. B., et al.: Particle Accel. **1,** 173 (1970).
[24] Zijelstra, H.: Experimental Methods im Magnetism. Vol. IX, North Holland Publishing Co, Amsterdam (1967).

[25] Forsythe, G. E., Wasow, W. R.: Finite-Difference Methods for Partial Differential Equations. New York: John Wiley 1960.
[26] Christion, R. S.: The LINDA Program and an earlier version SIBYL (No written description of program).
[27] Ahamed, S. V.: The Computor Journal **8**, 73 (1965).
[28] Ahamed, S. V., Erdelyi, E. A.: I.E.E.E. Trans. on Power Apparatus and Systems. PAS **85**, 61 (1966).
[29] Trutt, F. C., et al.: I.E.E.E. Trans. PAS **3**, 665 (1968).
[30] Winslow, A.: J. of Comp. Phys. **1**, 149 (1967).
[31] Dorst, J. H.: I.E.E.E. Trans. Nucl. Sci. NS **12**, 412 (1965).
[32] Halbach, K.: Proc. Second Int. Conf. on Magnet Techn. (1967).
[33] Colonias, J.: U.C.R.L. 17340 (1967).
[34] Dahl, P. F., et al.: I.E.E.E. Trans. Nucl. Sci. NS **12**, 408 (1965).
[35] Winslow, A. M.: Proc. Int. Symp. on Magnet Techn. Stanford 171 (1965).
[36] Dahl, P. F., Parzen, G.: Proc. Int. Conf. on High Energy Acc. Frascati 165 (1965).
[37] Perin, R., Van der Meer, S.: CERN Rep. 67-7 (1967).
[38] Iselin, C.: Proc. of the Third Internatl. Magnet Conf. Hamburg 83 (1970).
[39] Burfine, E. A., et al.: SLAC Rep. **56** (1966).
[40] Anderson, L. R.: Thesis Oregon State Univ. (1969).
[41] Parzen, G., Jellitt, K.: BNL int. Rep. BNL 15107 AADD. **165** (1970).
[42] Borglum, D., Anderson, L. R.: SLAC unpublished Rep. (1964).
[43] Gerold, E.: Stahl und Eisen **613** (1931).
[44] Brechna, H.: Proc. Second Int. Conf. on Magnet Techn. Oxford 305 (1967).
[45] Doke, T., et al.: University of Tokyo to be published (1971).
[46] McInturff, A. D., Claus, J.: BNL Rep. AADD. **162** (1970).
[47] Salzburger, H.: Siemens Repot FL SL 11-09436, (32301), (1972).
[48] Danan, H., et al.: J. of Appl. Phys. **39**, 669 (1968).
[49] Smith, L. B.: SLAC Rep. **95** (1969).
[50] Collatz, L.: Functional Analysis and Numerical Mathematics. New York: Academic Press 1966.
[51] Frankel, S. R.: Math. Tables and Other Aids to Comp. **4**, 65 (1950).
[52] Varga, R.: Matrix Iterative Analysis. Englewood Cliffs, N.J.: Prentice Hall 1962.
[53] Reichert, K.: Habilitationsschrift Univ. (TH) Stuttgart 1968.
[54] Parzen, G., Jellett, K.: BNL Rep. 15586, AADD. **173** (1971).
[55] Beth, R. A.: BNL Rep. AADD. **220** (1966).
[56] Blewett, J. P.: Proc. 1968 Summer Study on Supercond. Devices, Part III, p. 1042. Brookhaven 1968.
[57] Lotsch, H. K. V. et al: Techn. Inf. T 8–1825/501 N. Am. Rockwell, Annaheim, Calif. (1968) and Z. Angew. Math. Phys. **20**, N. 4 (1969).
[58] Lontai, L. M., Marston, P. G.: Proc. Int. Symp. on Magnet Techn. 723 (1965).
[59] Léon, B.: A.E.R.E. Trans. 1056 (1964).
[60] Kilb, R. W., Westendorp, W. F.: General Electric Rep. C **67**, 440 (1967).
[61] Brechna, H., Haldemann, W.: SLAC Publ. 337 (1967).
[62] Parsch, A., Bogner, G.: Siemens Erlangen, Akten-Notiz FL 52 (1967).
[63] Kuznetsov, A. A.: Soviet Phys. Techn. Phys. **5**, 555 (1960).
[64] Bitter, F.: High Magnetic Fields, p. 85. New York: John Wiley 1962.
[65] Giauque, W. F., Lyon, D. N.: Rev. Sci. Instr. **31**, 374 (1960).

[66] Culwick, B. B.: BNL Rep. HH 05-0, BBC (1965).
[67] Skarpaas, K.: SLAC Rep. 106, 85 (1969).
[68] Appelton, A. D., et al.: Proc. Second Int. Conf. Magnet Techn. Oxford 553 (1967).
[69] Middleton, A. J., Trowbridge, C. W.: Proc. Second Int. Conf. Magnet Techn. Oxford 140 (1967).
[70] Atherton, D.: J. Appl. Phys. **39,** 1411 (1968).
[71] Beth, R. A.: J. Appl. Phys. **40,** 2445 (1969).
[72] Broom, R. F., Rhoderick, R. H.: Brit. J. of Appl. Phys. **11,** 292 (1960).
[73] Brechna, H.: SLAC Publ. 182 (1966).
[74] Wilson, M. N.: Rutherford RPP.

3. Phenomena and Theory of Superconductivity

*G. D. Cody**

3.1 Theory

3.1.1 Introduction

Superconductivity was discovered by Kamerlingh Onnes in 1911, but the major theoretical developments have occurred since 1957, the year in which the microscopic theory of superconductivity was presented by Bardeen, Cooper and Schrieffer [1] (BCS). Soon after BCS, Gor'kov [2] extended the BCS theory to include spatial variation of its parameters, and placed the earlier work of Ginzburg and Landau [3] and Abrikosov [4] on a microscopic basis (GLAG). In recent years, there have been several excellent reviews of both theory and experiment in superconductivity, but the review of Shoenberg [5] is authoritative for research prior to 1957. Recent reviews that are of particular value for experimentalists are that of Lynton [6] for experimental work up to 1964, that of Kuper [7] for a particularly clear presentation of theory and the two volume compendium edited by Parks [8] that summarizes research up to 1967.

The present chapter is concerned with the critical field and currents of Type II superconductors. It is intended to summarize the present state of research in these areas, to provide a physical understanding of the phenomenon, and to point out the areas where our knowledge is incomplete, and where further research is required. Before we proceed to this task we outline in the following section, a broad overview of the theory of superconductivity. This section will provide the necessary background for the detailed consideration of the subsequent sections. It will also furnish a convenient table of formulae and relationships that will be useful for any consideration of superconducting applications. For motivation and further detail, the reader is referred to the aforementioned reviews, and the subsequent sections of the chapter.

3.1.2 Free Electron Theory

The starting point for the present theory of superconductivity is the free electron theory of metals [9]. This theory is essentially a fluid model of an electron gas in a potential well, however, many of its results can

* RCA Laboratories Princeton N.J.

be expressed in general forms that may be expected to hold for more realistic theories of metallic behaviour.

If we consider a metal with an electron concentration, n (cm^{-3}), we obtain from Fermi statistics for free electrons

$$n = \frac{k_F^3}{3\pi^2}, \tag{3.1.1}$$

where

$$k_F = \frac{P_F}{\hbar} = \frac{m v_F}{\hbar} \tag{3.1.2}$$

and k_F, P_F, and v_F are the Fermi wave vector, momentum and velocity respectively. The quantity m can be considered an effective mass. A basic parameter of the theory is the density of states for both spins at the Fermi level, $N(E_F)$, where

$$N(E_F) = \frac{4\, m k_F}{h^2} = \frac{S}{4\,\pi^2 \hbar v_F}. \tag{3.1.3}$$

In Eq. (3.1.3) the quantity S is the area of the Fermi surface. The Fermi level, E_F, in the free electron theory is given by

$$E_F = \frac{P_F^2}{2\,m}. \tag{3.1.4}$$

The density of states $N(E_F)$ can be experimentally determined, apart from many body effects [10], from either the electronic specific heat or paramagnetic spin susceptibility. It can be shown that the total specific heat c_p at low temperatures can be written in the normal state

$$c_p = \gamma T + b T^3. \tag{3.1.5}$$

The second term is the phonon contribution to the specific heat, which is given in the Debye theory by

$$b = \frac{1944}{\theta^3 V_m}, \tag{3.1.6}$$

where the units of b are J/cm^3 K^4, θ is the Debye temperature, and V_m the molar volume (cm^3).

The first term is the electronic contribution, and γ is given by

$$\gamma = \frac{\pi^2}{3} k^2 N(E_F), \tag{3.1.7}$$

where k is Boltzmann's constant. The paramagnetic spin susceptibility χ can be similarly related to $N(E_F)$ by

$$\chi = \frac{\beta^2}{\mu_0}\left(\frac{N}{E_F}\right), \tag{3.1.8}$$

where β is the Bohr-magneton. As noted, both Eqs. (3.1.7) and (3.1.8) are modified by many-body effects, and the correction is particularly large for superconductors with high T_c.

An additional expression relating the free electron parameters to experiment is given by the electrical conductivity. The well known formula for the conductivity

$$\sigma = \frac{n}{m} e^2 \tau \tag{3.1.9}$$

can be recast in a form related to the Fermi surface area S by expressing the relaxation time τ, as l/v_F, where l is the mean free path.

One obtains

$$\frac{\sigma}{l} = \frac{e^2 S}{12 \pi^3 \hbar} . \tag{3.1.10}$$

The quantity S can also be determined by anomalous skin-effect measurements [11].

Some of the more interesting superconductors are transition metals, and for these metals the accepted model is, that two bands are necessary to account for their behaviour. One band is a low density of states s- or p-band with an effective mass and Fermi velocity not very different from the free electron value. The other band is a high density of states d-band, with a large effective mass and low Fermi velocity. For specific heat and susceptibility the total density of states is a sum of the contributions from each band, and the d-band contribution dominates. For the case of the electrical conductivity, it is generally assumed that the s-electrons carry the current, but that the dominant mode of scattering is to final states in the d-band. Hence Eq. (3.1.10) would be written for transition metals as

$$\frac{\sigma}{l_{s-d}} = \frac{e^2 S_s}{12 \pi^3 \hbar} . \tag{3.1.11}$$

3.1.3 Zero Field Properties of a BCS Superconductor

The basic parameter of the zero field superconducting state is the energy gap $\Delta(T)$, which vanishes at T_c, and increases rapidly just below T_c according to the relation

$$\Delta(T) \approx 1.74 \, \Delta(0) \left[1 - \left(\frac{T}{T_c}\right)\right]^{\frac{1}{2}} . \tag{3.1.12}$$

At about $T/T_c \approx 1/2$, it is within a few percent of its $T = 0$ K value of $\Delta(0) = 1.76 \, kT_c$. The quantity $\Delta(T)$ represents a gap in the spectrum of excited states centered at the Fermi level, such that the energy of an electron excited from the ground state is given by

$$E^2 = \left\{ \left[\frac{\hbar k^2}{2m} - E_F\right]^2 + \Delta^2(T) \right\} . \tag{3.1.13}$$

The ground state of a superconductor, the "superconducting electrons" consists of electron pairs, and hence the creation of an excited electron — a quasi-particle or "normal" electron — requires an energy of $2 \Delta(T)$. At $T = 0$ K, there are no thermally excited electrons, and all the conduction electrons are paired in the superconducting ground state.

From the preceeding discussion, it is clear that the normalized energy gap can be considered as an order parameter of the superconducting transition with a value of zero at T_c, and rising to unity at $T = 0$ K. Indeed the temperature dependence of $\Delta(T)$ close to T_c is quite similar to the semi-empirical order parameter of the two fluid model of Gorter and Casimir [12]. In this theory the concentration of "superconducting electrons" (pairs of BCS) is given by a quantity x where

$$x = \left[1 - \left(\frac{T}{T_c}\right)^4\right]^{\frac{1}{2}}. \tag{3.1.14}$$

Close to T_c, x has a temperature variation similar to Eq. (3.1.12):

$$x \cong 2\left[1 - \left(\frac{T}{T_c}\right)\right]^{\frac{1}{2}}. \tag{3.1.15}$$

The energy gap at $T = 0$ K, and T_c are related to fundamental properties of the metal through the BCS relation

$$T_c = 1.14 < \omega > \exp\left(\frac{-2}{N(E_F)\,V}\right), \tag{3.1.16}$$

where $< \omega >$ is a characteristic frequency of the metal of the order of the Debye Temperature $k\theta/\hbar$, and V is the pairing potential arising from the electron-phonon interaction assumed responsible for superconductivity. The superconducting isotope shift, $T_c \sim m^{-\frac{1}{2}}$, where m is the ion mass, arises from the mass dependence of $< \omega >$, and the mass independence of $N(E_F)\,V$.

A more realistic expression than Eq. (3.1.16) is given by McMillan [10].

Physically the energy gap $\Delta(T)$ is responsible for the vanishing electrical resistance of the superconductor since it is centered at the Fermi level even when the Fermi sphere is displaced due to a current. It is also controlling in any process involving thermal activation of quasi particles (normal electrons) out of the ground state. At low temperatures the electronic specific heat, electronic thermal conductivity and ultrasonic attenuation can be related in many instances solely to the concentration of normal quasi particles since the paired electrons in the ground state do not carry entropy or scatter. In these cases, all these properties vary as $e(-\Delta/kt)$ at the lowest temperatures.

Direct measurements of the energy gap can be made either through tunneling or obtained from an analysis of the frequency dependence of high frequency electromagnetic absorption. In general the agreement with the BCS value $(2\,\Delta(0) = 3.52\,kT_c)$ is good, and shows that theory is not sensitive to details of band structure [12].

The rapid variation of the energy gap close to T_c expressed in Eq. (3.1.12) leads to an enhancement of the electronic specific heat at T_c above its value in the normal state. In the BCS theory

$$\left(\frac{c_s - c_n}{c_n}\right)_{T_c} = 1.43\,. \tag{3.1.17}$$

Similarly the two-fluid model of Gorter and Casimir can be shown from Eq. (3.1.15) to lead to

$$\left(\frac{c_s - c_n}{c_n}\right)_{T_c} = 2.00 \,. \tag{3.1.18}$$

Experiment tends to confirm Eq. (3.1.17), although as in the case of the energy gap there is some variation around the BCS value.

If we consider the normalized energy gap as an order parameter we can raise the question of how rapidly the order parameter can vary with distance at, for example, a normal-superconducting boundary. This question is crucial for Type II superconductors; it is also important for the magnetic state of ellipsoidal Type I superconductors (the intermediate state). The Ginzburg-Landau theory which is applicable to both Type I and Type II superconductors introduces a characteristic distance $\xi^*(T)$ over which the order parameter can vary. The quantity $\xi^*(T)$ was later shown to be proportional, for pure superconductors, to a characteristic distance that occurs in the electrodynamics of the BCS theory, ξ_0, the Pippard coherence distance. The quantity ξ_0 is given in the BCS theory by the relation

$$\xi_0 = \frac{\hbar v_F}{\pi \Delta(0)} \,. \tag{3.1.19}$$

The quantity ξ_0, for reasonable values of v_F and $\Delta(0)$ is of the order of several thousand Å. However, for transition metals with high T_c and low v_F, ξ_0 can be of the order of hundreds of Å or less. Moreover, for materials where the mean free path l is less than ξ_0, it can be shown that $\xi^*(T)$ is reduced considerably.

3.1.4 Superconductors in an Applied Field

Superconductors can be divided into two categories in terms of their magnetic behaviour. Type I superconductors exhibit perfect diamagnetism up to a critical field $H_c(T)$, whereupon they undergo a transition to the normal state. For a state of perfect diamagnetism the flux density $\boldsymbol{B}$ is equal to zero, and hence from the usual magneto-static theory for samples with no demagnetization fields ($\boldsymbol{B} = \mu_0 \boldsymbol{H}_0 + \boldsymbol{M}$)

$$\boldsymbol{M} = -\mu_0 \boldsymbol{H}_0 \,, \tag{3.1.20}$$

where $\boldsymbol{M}$ is the magnetic moment per unit volume, and $\boldsymbol{H}_0$ is the applied field. From thermodynamics

$$M = -\left(\frac{\partial g}{\partial H_0}\right)_T, \tag{3.1.21}$$

where $g(H_0,T)$ is the Gibbs free energy, we thus obtain

$$g_n = g_s(0,T) + \frac{1}{2\mu_0} H_c^2(T) \,, \tag{3.1.22}$$

where the subscripts n and s denote the normal and superconducting states respectively. Thus $H_c^2(T)/2\mu_0$ is the free energy difference between the normal and superconducting states respectively. The field $H_c(T)$ is thus denoted as the thermodynamic critical field.

From thermodynamics, the entropy s is given by $s = -(\partial g/\partial T)_{H_0}$ and hence

$$s_s - s_n = \frac{H_c}{\mu_0} \cdot \frac{dH_c}{dT} . \tag{3.1.23}$$

Furthermore, the specific heat $c = T \cdot (\partial S/\partial T)$ and

$$\left(\frac{c_s - c_n}{\gamma T_c}\right)_{T_c} = \frac{1}{\mu_0 \gamma}\left(\frac{dH_c}{dT}\right)^2_{T_c} \tag{3.1.24}$$

and from Eq. (3.1.17)
$$\left(\frac{dH_c}{dT}\right)^2_{T_c} = 1.43\, \mu_0 \gamma . \tag{3.1.25}$$

Thermodynamics thus establishes a connection between the field $H_c(T)$ and thermal quantities such as specific heat and entropy. The specific heat can be integrated to give the magnitude and temperature dependence of $H_c(T)$. Conversely the temperature dependence of $H_c(T)$ can be used to obtain the entropy and specific heat. In the BCS theory the field $H_c(T)$ is approximately parabolic, in the two fluid model it is exactly so. Thus from $H_c(T) \sim H_c(0)\,(1-(T/T_c)^2)$ we obtain from Eq. (3.1.23) the fact that the entropy of the superconducting state is less than the normal state, and hence is the more ordered state. An exact relation from BCS relates the thermodynamic critical field to the energy gap:

$$\frac{1}{2\mu_0} H_c^2(0) = \frac{1}{4} N(E_F)\, \Delta^2(0) . \tag{3.1.26}$$

A slight penetration of the magnetic field at the surface of the specimen modifies the perfect diamagnetism of Type I superconductors below $H_c(T)$. The distance in which the applied field falls to zero is denoted the penetration depth $\lambda(T)$, and from BCS, for pure superconductors as $T \to T_c$

$$\lambda = \frac{\lambda_L}{\sqrt{2}} \cdot \left[1 - \left(\frac{T}{T_c}\right)\right]^{-\frac{1}{2}} , \tag{3.1.27}$$

where λ_L, the London penetration depth is given by

$$\lambda_L = \left(\frac{m}{\mu_0 n e^2}\right)^{\frac{1}{2}} , \tag{3.1.28}$$

where m is the mass, and n the density of conduction electrons. In magnitude λ_L is the order of hundreds of Å for reasonable material parameters. From BCS, and the free electron theory we obtain the useful relation

$$H_c(0)\, \lambda_L \xi_0 = \left(\frac{3}{2\pi^2}\right)^{\frac{1}{2}} \cdot \frac{h}{2e} . \tag{3.1.29}$$

Prior to the BCS theory a new electrodynamic was introduced by H. and F. London to explain the Meissner effect. This theory introduced, in a natural manner, an additional equation to the usual Maxwell equations:

$$\operatorname{curl} \boldsymbol{J} = -\frac{\boldsymbol{B}}{\mu_0 \lambda_L^2} , \tag{3.1.30}$$

where J is the current density. When Eq. (3.1.30) is combined with the Maxwell equation

$$\operatorname{curl} \boldsymbol{H} = \boldsymbol{J}, \tag{3.1.31}$$

we obtain solutions which exhibit an exponential decrease in H over the distance λ_L at plane boundaries. Using $\boldsymbol{B} = \operatorname{curl} \boldsymbol{A}$ and $\boldsymbol{E} = -(\partial \boldsymbol{A}/\partial t)$ we can rewrite Eq. (3.1.30) as

$$\boldsymbol{J} = -\frac{\boldsymbol{A}}{\mu_0 \lambda_L^2} \tag{3.1.32}$$

and

$$\boldsymbol{J} = \frac{j n e^2 \boldsymbol{E}}{m \omega}, \tag{3.1.33}$$

where $e^{j\omega t}$ is the assumed time dependence of $\boldsymbol{A}$ and $\boldsymbol{E}$ (the vector potential and electric field respectively). Eq. (3.1.33) for supercurrent flow replaces the usual current field relation

$$\boldsymbol{J} = \frac{n e^2 \tau \boldsymbol{E}}{m}. \tag{3.1.34}$$

The apparent singularity of Eq. (3.1.33) at d.c. arises from the fact that the current arises from the applied field under these circumstances Eq. (3.1.32) and not by the time variation of the electric field.

An outstanding success of the theory of superconductivity prior to BCS was the insertion of the two fluid result for the temperature variation of the "superconducting electrons" into the formula for λ_L [Eq. (3.1.28)] leading to the predicted temperature dependence for $\lambda_L(T)$ as

$$\lambda_L(T) = \lambda_L \left[1 - \left(\frac{T}{T_c}\right)^4\right]^{-\frac{1}{2}}, \tag{3.1.35}$$

which is in good agreement with the BCS temperature dependence for λ, but is a factor of $\sqrt{2}$ smaller than Eq. (3.1.27) close to T_c. The two fluid model can also be used to construct a consistant electrodynamics at high frequencies where Eqs. (3.1.33) and (3.1.34) are combined to give

$$J = \left\{\frac{j n e^2}{m \omega}\left[1 - \left(\frac{T}{T_c}\right)^4\right]^{-\frac{1}{2}} + \frac{n e^2 \tau}{m}\left(\frac{T}{T_c}\right)^4\right\} E. \tag{3.1.36}$$

Eq. (3.1.36) can be usefully applied for an estimation of losses at high and low frequencies, however, for precise calculations the BCS results should be used [13].

The London's result Eq. (3.1.32) represents a local relation between field and current, and was shown by Pippard [14], prior to BCS to apply only when $\lambda \gg \xi_0$. Except at T_c, when $\lambda \to \infty$, so called "London" superconductors are the exception, and for the majority of elemental superconductors ($\lambda_L < \xi_0$) it is necessary to assume an integral relation between A and J. Such non-local superconductors are denoted "Pippard" superconductors.

However, to account for his experimental results for alloys Pippard introduced a new coherence distance ξ which depended on the mean free path as well as the intrinsic properties of the metal where

$$\frac{1}{\xi} = \frac{1}{\xi_0} + \frac{0.7}{l}. \tag{3.1.37}$$

Furthermore, under these circumstances Eq. (3.1.28) had to be modified to read for London superconductors ($\lambda \gg \xi$)

$$\lambda(0) = \lambda_L \left(\frac{\xi_0}{\xi}\right)^{\frac{1}{2}}. \qquad (3.1.38\,\text{a})$$

For Pippard superconductors ($\lambda \ll \xi$)

$$\lambda(0) \cong 0.7\,\lambda_L \left(\frac{\xi}{\lambda_L}\right)^{\frac{1}{3}}. \qquad (3.1.38\,\text{b})$$

In both cases the temperature dependence is close to $[1 - (T/T_c)^4]^{-\frac{1}{2}}$.

3.1.5 Type II Superconductors

A Type I superconductor has been defined by the empirical observation that it undergoes a magnetic transition from the Meissner state ($B = 0$) to the normal state ($B = \mu_0 H$) at a critical field $H_c(T)$. A more precise definition follows from a study of the relative contributions of field and order parameter penetration to the surface energy at a normal-superconducting interface. It can then be shown that if $\xi^*(T) > \lambda(T)$, the surface energy is positive and the material behaves as a Type I superconductor. However, if $\xi^*(T) < \lambda(T)$ the surface energy is negative and the material breaks up into a vortex structure with field penetration above a critical field $H_{c1}(T)$. This state is denoted the mixed state and persists up to a field $H_{c2}(T)$ whereupon the material goes normal. The theory of Abrikosov suggested that the mixed state was a unique thermodynamic state and this has been confirmed by later experiment on defect free specimens. Superconductors that exhibit such a magnetic transition are known as Type II superconductors.

From the GLAG theory, a relation can be found between $\xi^*(T)$ and material parameters. It can be shown that

$$\xi^*(T) = \frac{\lambda(T)}{\sqrt{2}\,\varkappa(T)}, \qquad (3.1.39)$$

where $\varkappa(T)$ is known as the Ginzburg-Landau parameter. In the limit of $T \to T_c$

$$\varkappa(T_c) = \lim_{T \to T_c} \frac{2\sqrt{2}\,\pi\lambda^2(T)}{\varphi_0} H_c(T), \qquad (3.1.40)$$

where $\varphi_0 \equiv (h/2e)$ is the unit of flux quantization ($\varphi_0 \approx 2 \times 10^{-15}$ Vs).

From the previous expressions, one can show that

$$\varkappa(T_c) = 0.96\,\frac{\lambda_L}{\xi_0} + 0.7\,\frac{\lambda_L}{l} \qquad (3.1.41\,\text{a})$$

and from BCS and the free electron theory

$$\varkappa(T_c) = \frac{3.02 \times 10^{-3}\,T_c}{\gamma^{\frac{1}{2}} \cdot v_F^2} + 0.75\,\varrho_0\gamma^{\frac{1}{2}}. \qquad (3.1.41\,\text{b})$$

In Eq. (3.1.41b) ϱ_0 is the residual (low temperature limit) resistivity in units of 10^{-6} Ohm · cm, γ is the coefficient of the electronic specific heat in units of 10^{-3} J/cm³ K² and v_F is the Fermi velocity in units of 10^8 cm per s. The first term in Eq. (3.1.41) represents an intrinsic property of the metal; the second term depends on residual resistivity and hence the impurity content. If $\varkappa(T_c) > 1/\sqrt{2}$, the material is thus a Type II superconductor, and as noted, this condition can arise from either the intrinsic properties of the metal, or through such extrinsic properties as resistivity. Examples of both sources of Type II behaviour exist, although the second variety, extrinsic Type II superconductors, is the most common.

The exact theories of the mixed state [15] show that the entrance field $H_{c1}(T)$ is given by $(\varkappa_3(T) \gg 1)$.

$$H_{c1}(T) = \frac{H_c(T) \cdot \ln \varkappa_3(T)}{\sqrt{2}\, \varkappa_3(T)} \tag{3.1.42}$$

and that the mixed state persists to a field $H_{c2}(T)$ where

$$H_{c2}(T) = \sqrt{2}\, \varkappa_1(T)\, H_c(T) \tag{3.1.43}$$

and that the magnetization M close to $H_{c2}(T)$ is given by

$$M = \frac{B - H_{c2}}{1.16\,[2\,\varkappa_2^2(T) - 1]}\,. \tag{3.1.44}$$

The quantities $\varkappa_{1,2,3}(T)$ approach $\varkappa(T_c) \equiv \varkappa$ as $T \to T_c$, but their detailed temperature dependence is considerably different than simple substitution of BCS or two-fluid expressions in Eq. (3.1.39). For example, in the case of the crucial quantity $\varkappa_1(T)$ we have for the exact theory $\varkappa_1(0)/\varkappa \approx 1.2$, whereas for BCS $\varkappa_1(0)/\varkappa \approx 1.1$, and for the two-field model $\varkappa_1(0)/\varkappa = 2$.

There are two other critical fields that play an important role in the magnetic behaviour of Type II superconductors. For $H_{c2} < H < H_{c3}$ a superconducting sheath of a thickness of about ξ^*, exists at the surface of the superconductor parallel to the applied field. In the limit $T \to T_c$, the surface superconducting field is given in terms of H_{c2} by the relation

$$H_{c3} = 1.695\, H_{c2} = 2.4\, \varkappa H_c\,. \tag{3.1.45}$$

The field H_{c3}, denoted the surface critical field, can also be observed in Type I superconductors for which $1/2.4 < \varkappa < 1/\sqrt{2}$. The surface sheath is observable through its small but finite current capacity, but can be eliminated through plating of surface or perpendicular orientation of the applied field.

For high $\varkappa$ materials with moderate T_c, the upper critical field H_{c2} is not determined entirely by the GLAG theory, but is strongly influenced by the paramagnetic spin susceptibility of the normal state and its absence in the superconducting state. Under these circumstances the GLAG critical field is denoted H_{c2}^* and H_{c2} can be considerably less than H_{c2}^*. A measure of the paramagnetic limitation is given by the field B_p where the spin paramagnetic energy of the normal state is equal to the superconducting condensation energy $(H_c^2(0)/2\,\mu_0)$. From BCS and the free electron theory

$$B_p = 1.84\, T_c(\mathrm{T})\,. \tag{3.1.46}$$

A measure of the paramagnetic effect is given by the quantity α where

$$\alpha = \frac{\sqrt{2}\, H_{c2}^{*}}{B_p}. \tag{3.1.47}$$

It is important to note that at T_c, the spin paramagnetism of the normal state is equal to that of the superconducting state due to the vanishing energy gap at T_c, and in this limit $H_{c2} \to H_{c2}^{*}$ and all the GLAG relationships apply.

Before we conclude this short survey of superconductivity we consider the specific heat and thermal conductivity of the mixed state. The specific heat is most conveniently found from the thermodynamic relation.

$$c_s = -T\left(\frac{\partial^2 g_s}{\partial T^2}\right)_V \tag{3.1.48}$$

where g_s is obtained from the thermodynamic relation

$$M = -\left(\frac{\partial g_s}{\partial B_0}\right)_{T,V} \tag{3.1.49}$$

where we assume that $B \gg H_{c1}$, and that M is given by Eq. (3.1.44). Thus

$$g_s(B,T) = g_n(T_c) - \frac{[B - H_{c2}(T)]^2}{2.32\,\mu_0[2\,\varkappa_2^2(T) - 1]}. \tag{3.1.50}$$

From Eq. (3.1.50) it can be shown that the entropy and specific heat in the mixed state, unlike the Meissner state, is field dependent. Indeed at low temperatures a fair approximation is

$$c_s \cong \gamma T \frac{B}{H_{c2}}. \tag{3.1.51}$$

and the electronic specific heat can not be ignored for a Type II at high fields. Similar conclusions apply to the thermal conductivity, but as is always the case for the thermal conductivity, the exact theory is quite complicated.

Before we proceed to a detailed discussion of the critical fields and currents of Type II superconductors we summarize the preceeding discussion [16].

3.1.6 Summary of Free-Electron, BCS and GLAG Formulae

1. Fermi wave vector, k_F, momentum P_F, and velocity v_F for electrons of mass m, and concentration n:

$$k_F = (3\,n\pi^2)^{\frac{1}{3}}$$

$$P_F = \hbar k_F = m v_F\,.$$

2. Total density of states, $N\,(E_F)$ in terms of electronic specific heat ($c = \gamma T$) and area of Fermi surface, S:

$$N\,(E_F) = 3 \cdot \frac{\gamma}{(\pi k)^2}, \qquad N\,(E_F) = \frac{4\,m k_F}{h^2}, \qquad N\,(E_F) = \frac{S}{4\,\pi^3 \hbar v_F}.$$

3. Pauli spin susceptibility χ:

$$\chi = N(E_F)\,\frac{\beta^2}{\mu_0}; \qquad \beta = \frac{\mu_0}{4\pi}\cdot\frac{e h}{m} = 1.16529 \times 10^{-29}\ \text{Vs}\cdot\text{m}\,.$$

4. Electrical conductivity σ in terms of scattering time τ, mean free path l and Fermi surface area:

$$\sigma = \frac{n e^2 \tau}{m}; \quad \frac{\sigma}{l} = \frac{n e^2}{m v_F} = \frac{e^2}{12\pi^3\hbar}\cdot S\,.$$

5. BCS energy gap $2\Delta(T)$:

$$2\Delta(0) = 3.52\,kT_c$$

$$(\Delta(T))_{T\to T_c} = 1.74\,\Delta(0)\,[1 - (T/T_c)]^{\frac{1}{2}}\,.$$

6. Pippard coherence distance ξ:

$$\frac{1}{\xi} = \frac{1}{\xi_0} + \frac{0.7}{l}; \qquad \xi_0 = \frac{\hbar\cdot v_F}{\pi\,\Delta(0)}\,.$$

7. Free energy difference between normal and superconducting states:

$$g_n(T) = g_s(0) + H_c^2(T)/2\,\mu_0$$

$$H_c(T) \approx H_c(0)\left[1 - \left(\frac{T}{T_c}\right)^2\right]; \quad \left(\frac{dH_c}{dT}\right)_{T_c} = -\frac{1.74\,H_c(0)}{T_c} \qquad \text{(BCS)}$$

$$H_c(T) = H_c(0)\left[1 - \left(\frac{T}{T_c}\right)^2\right] \qquad \text{(Gorter-Casimir)}$$

$$\frac{1}{2\,\mu_0}\,H_c^2(0) = \frac{1}{4}\,N(E_F)\cdot\Delta^2(0)\,. \qquad \text{(BCS)}$$

8. Specific heat jump at T_c:

$$\Delta c = \frac{1}{\mu_0}\cdot T_c\cdot\left[\frac{\partial H_c(T)}{\partial T}\right]^2_{T_c}$$

$$\frac{\Delta c}{c_n} = 1.43 \qquad \text{(BCS)}$$

$$\frac{\Delta c}{c_n} = 2.00 \qquad \text{(Gorter-Casimir)}$$

9. Penetration depth:

$$(\lambda(T))_{T\to T_c} = (\lambda_L\sqrt{2})\,[1 - (T/T_c)]^{-\frac{1}{2}} \qquad \text{(BCS)}$$

$$\lambda_L = \left(\frac{1}{\mu_0}\cdot\frac{m}{n e^2}\right)^{\frac{1}{2}}$$

$$\lambda(0) = \lambda_L\left(1 + \frac{0.7\,\xi_0}{l}\right)^{\frac{1}{2}} \qquad \text{(London limit)}$$

$$\lambda(0) = 0.7\,\lambda_L\cdot\left(\frac{\xi}{\lambda_L}\right)^{\frac{1}{3}} \qquad \text{(Pippard limit)}$$

10. General BCS relation

$$\xi_0\cdot\lambda_L\cdot H_c(0) = \frac{\sqrt{3}}{2\pi^2}; \qquad \varphi_0 = \frac{h}{2e} \approx 2\times 10^{-15}\ \text{Vs}\,.$$

11. GLAG parameters:

$$\xi^*(T) = \frac{\lambda(T)}{\sqrt{2}\,\varkappa(T)}$$

$$(\varkappa(T))_{T\to T_c} = \frac{2\sqrt{2}\,\pi}{\varphi_0}\cdot\lambda^2(T)\cdot H_c(T)\,,$$

$$\varkappa(T_c) = \varkappa = 0.96\,\frac{\lambda_L}{\xi_0} + 0.7\,\frac{\lambda_L}{l}$$

$$\varkappa = 3.02\times 10^{-3}\,\frac{T_c}{\gamma^{\frac{1}{2}} v_F^2} + 0.753\,\varrho_0\gamma^{\frac{1}{2}}$$

(γ [mJ/cm^3 K^2] ; ϱ_0 [10^{-6} Ohm · cm] ; v_F [10^8 cm/s]) .

12. GLAG mixed state:

$$H_{c1}(T) = H_c(T)\cdot\frac{\ln[\varkappa_3(T)]}{\sqrt{2}\,\varkappa_3(T)}$$

$$H_{c1}(T) = \varphi_0\cdot\frac{4\pi\lambda^2(T)}{\ln[\varkappa_3(T)]}$$

$$H_{c2}(T) = \sqrt{2}\,\varkappa_1(T)\cdot H_c(T)$$

$$H_{c2}(T) = \frac{\varphi_0}{\mu_0\,[\xi^*(T)]^2}$$

$$H_c(T) = \frac{\varphi_0}{\mu_0\lambda(T)\,\xi^*(T)}$$

$$M = \frac{B - H_{c2}}{1.16\,[2\varkappa_2^2(T) - 1]}$$

$$(\varkappa_{1,2,3})_{T\to T_c} \to \varkappa$$

$$\left(\frac{dH_{c2}}{dT}\right)_{T_c} = \sqrt{2}\,\varkappa\left(\frac{dH_c}{dt}\right)_{T_c}.$$

13. Surface critical field:

$$H_{c3} = 1.695\,H_{c2} = 2.4\cdot\varkappa\cdot H_c\,.$$

14. Paramagnetic limitations on H_{c2}:

$$B_p = 1.84\,T_c$$

$$\alpha = \sqrt{2}\,\frac{H_{c2}^*}{H_p}$$

$\alpha = 0.235\,\varrho_0\gamma$ (Extrinsic Type II; γ (μJ/cm^3 K^2); ϱ_0 (Ohm · cm)

$$\alpha = 0.533\left(\frac{dH_{c2}}{dT}\right)_{T_c}$$

$$\left(\frac{dH_{c2}^*}{dT}\right)_{T_c} = \left(\frac{dH_{c2}}{dT}\right)_{T_c} = \sqrt{2}\,\varkappa\left(\frac{dH_c}{dT}\right)_{T_c}.$$

15. Free energy in mixed state ($B \sim H_{c2}$):

$$g_s(H,T) = g_n(T) + \frac{[B - H_{c2}(T)]^2}{2.32\,\mu_0\,[2\varkappa_2^2(T) - 1]}\,.$$

3.2 Critical Fields of Type II Superconductors

3.2.1 Introduction

The purpose of this section is to discuss the critical fields of Type II superconductors. In order to develop a physical basis for the concept of the critical field, we start with a discussion of the magnetic behaviour of Type I superconductors, and particularly the intermediate state. The approach to Type II behaviour will be through the vortex model of De Gennes which follows naturally from a consideration of surface energies in a Type I superconductor. Finally, we will cover recent results from the Gor'kov microscopic Ginzburg-Landau equation which accurately predicts the effects of temperature, purity, normal electron spin paramagnetismus and variable coupling on the critical field behaviour of a large class of Type II superconductors. The exceptions to this theory, will be discussed, and possible sources of their unique behaviour. Finally, we will consider recent experimental observations of critical field behaviour in materials of practical interest.

3.2.2 Magnetostatics and Thermodynamics of Type I Superconductors

Superconductors exhibit zero resistance below the transition temperature, T_c, but when relatively defect free, exhibit a diamagnetism that is field dependent and reversible [6]. This last fact, although of little practical utility is the basis for the theoretical treatment of the superconducting state. For a uniformly magnetized ellipsoid three fields are introduced in addition to the microscopic local field $b\ (r)$ and the applied field B_0. These three fields are B, the flux density or average local field $\overline{b}(r)$, M, the dipole moment per unit volume or magnetization and $\boldsymbol{B}$ the internal field. From magnetic theory:

$$\boldsymbol{B} = \mu_0 \boldsymbol{H} + \boldsymbol{M} = \mu_0 \boldsymbol{H}_0 + (1 - N)\, \boldsymbol{M}\,. \tag{3.2.1}$$

In Eq. (3.2.1), N is the demagnetization coefficient of the ellipsoid. For a long cylinder, a sphere and a disk, $N = 0$, $^1/_3$ and 1, respectively, where the applied field, $\boldsymbol{H}_0$, is along the symmetry axis of the ellipsoid.

From considerations of the magnetic work, we are led to consider the specific Gibbs free energy g, which is a function of $\boldsymbol{H}_0$, and T and is a minimum under equilibrium conditions at constant external field and temperature. From the thermodynamic treatment,

$$dg = -S\,dT - M\,dH_0\,. \tag{3.2.2}$$

In Eq. (3.2.2) S is the specific entropy. The Gibbs free energies of two phases in equilibrium can be shown to be equal, and this criteria for phase equilibria, is critical for understanding the gross features of field effects in superconductors.

For Type I superconductors, when $N = 0$, it is observed experimentally that $M_s = (-\mu_0) \cdot H_0$ up to a critical field $H_c(T)$ [6]. Thus $B = 0$ in the superconducting state. Above this field the magnetization of the superconductor, $M_s(T)$, is the same as that of the normal state, $M_n(T) =$

$\chi_n \mu_0 H_0$ (the paramagnetic susceptibility, χ_n, is of the order of $(4\pi \cdot 20) \times 10^{-6}$ for transition metals, but is considerably less for nontransition metals). The discontinuous change of magnetization at $H_c(T)$, defines a Type I superconductor, as well as the thermodynamic critical field $H_c(T)$.

If we consider Eq. (3.2.2) we can arrive at another view of $H_c(T)$ which explains its nomenclature. From simple integration and use of the equilibrium criteria we see that

$$g_n(H_c,T) \approx g_n(0,T) = g_s(0,T) + \frac{1}{2\mu_0} H_c^2 = g_s(H_c,T) \qquad (3.2.3)$$

and thus $H_c/2\mu_0$ represents the free energy difference between the normal and superconducting state. Fig. 3.2.1 shows the free energy and the magnetization ($M = -(\partial g/\partial B_0)_T$) of a Type I superconductor and illustrates the content of Eq. (3.2.3). The effect of the field on the free energy of the superconductor is to destroy the energy gained by the initial condensation to the ordered superconducting state ($H_c^2/2\mu_0$). From Eqs. (3.2.3) and (3.2.2) one can obtain the entropy $S = -(\partial g/\partial T)_{B_0}$ and it can be easily seen that the superconducting state is the more ordered state ($S_s < S_n$ for $T < T_c$). Furthermore, in zero field $S_s(T_c) = S_n(T_c)$, which implies a second order transition. In a finite field the entropy remains at its zero field value, up to $H_c(T)$ but jumps discontinuously at T $(B_0 = H_c(T))$, to the normal entropy (a first order transition).

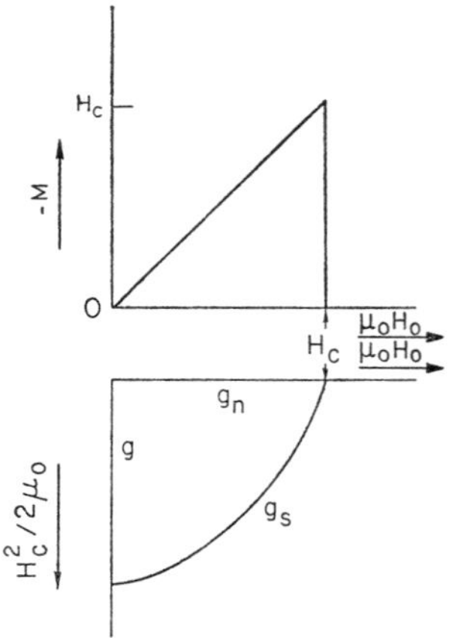

Fig. 3.2.1.
Magnetization and free energy (Type I)

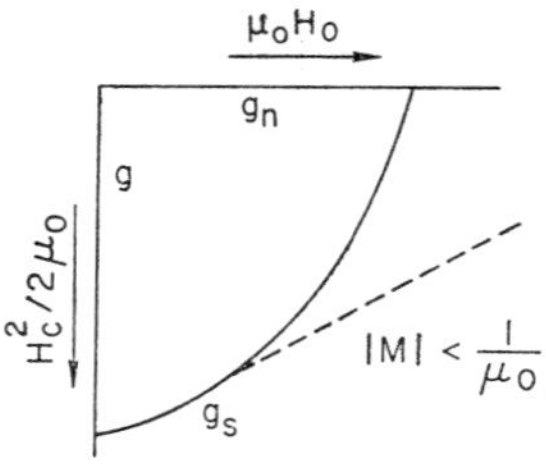

Fig. 3.2.2.
Effect of reduction of magnetization on free energy

The absence of a field dependence to the entropy is accounted for by the observation that since $B = 0$ up to $B_0 = H_c(T)$, the field plays no role in changing the order of the superconductor. Of course, this ignores the fact that there cannot be a discontinuity in the field at the surface of the superconductor. The field extends a distance into the superconductor, given by the penetration depth $\lambda(T)$. A penetration depth, λ_L, was first introduced by London who derived on the basis of a dissipationless fluid the following equation for the field penetration, which when combined with Maxwell's equations, are in semi-quantitative agreement with experiment

$$\lambda_L^2 \operatorname{curl} (\operatorname{curl} \boldsymbol{B}) = -\boldsymbol{B}\,, \tag{3.2.4}$$

where

$$\lambda_L^2 = \frac{1}{\mu_0} \frac{m}{n e^2} = \frac{3}{\mu_0} \cdot \frac{1}{N(E_F) \cdot v_F^2 e^2}\,. \tag{3.2.5}$$

In Eq. (3.2.5), m and n are the mass per particle and the density of the fluid respectively; in the second form of λ_L the terms $N(E_F)$ and v_F are the total density of states and Fermi velocity respectively. This second form is of greater generality than the fluid model. It is interesting to note that Eq. (3.2.5) is invarient under the transformation to pairs appropriate to the BCS theory of superconductivity ($m \to 2\,m$, $e \to 2\,e$, $n \to n/2$).

From Eqs. (3.2.4) and (3.2.5) we can associate $H_c(T)$ with the kinetic energy of the shielding currents that maintain the condition $B = 0$ in the bulk. It is simple to show that the kinetic energy density $E_{K,b}$ is given by

$$E_{K,b} = \frac{1}{2} n m v_d^2 = \frac{1}{2} \frac{m J^2}{n e^2} = \frac{\mu_0}{2} \lambda_L^2 \cdot J^2 = \frac{1}{2\,\mu_0} B^2\,, \tag{3.2.6}$$

where v_d is the drift velocity and we have used the solution for Eq. (3.2.4) a plane boundary: $J = (B/\mu_0 \lambda_L)$. At $B = H_c(T)$, the kinetic energy density of the shielding electrons equals the quantity $g_n - g_s(0,T)$ and the material goes normal.

It clearly costs the superconductor to expell a magnetic field. Consider Fig. 3.2.2 where again we show the free energy of a Type I superconductor. If there were a way to achieve a lower magnetization $M = -(\partial g/\partial B_0)_T$ as shown by the dashed curve, the material could achieve a lower free energy and would presumably remain superconducting to considerably higher fields. Experimentally such a situation has been known since the thirties for systems where one dimension approached a size of the order of λ ($\approx$ 300—600 Å). Under these conditions, field exclusion is not complete, $B = \bar{b} \pm 0$, and M is appreciably smaller than for bulk materials. Indeed critical fields considerably higher than $H_c(T)$ were observed and were correlated with the London expressions with qualitative and semi-quantitative agreement [6]. From these experiments, an important question emerges: namely, why a bulk specimen did not subdivide into domains of the order of λ, and hence remain superconducting to fields much higher than $H_c(T)$.

3.2.3 Intermediate State of Type I Superconductors

A model for such a domained structure was developed in the thirties to explain magnetization data on ellipsoidal specimens where $N \neq 0$.

While neither the experiments nor the model led to critical fields higher than $H_c(T)$, it led through the work of Pippard [17, 18], to what might be described as the physical basis of Type II superconductivity. Curve (1) in Fig. 3.2.3 shows the magnetization and free energy of a bulk specimen with $N \neq 0$.

The magnetization in the state for which $B = 0$ (the Meissner state) is of the form $M = -B_0/(1-N)$ and hence the free energy rises considerably faster than the curve (2) for $N = 0$. If this high moment state persists, the specimen would enter the normal state at a field considerably less than $H_c(T)$. In fact the material follows the curve (3), branching off from (1) at $B_0 = H_c(1-N)$ and finally enters the normal state close to $H_c(T)$. In this new state, the intermediate state, $M = (-1/N)(H_c - B_0)$ and

$$g(B,T) = g_s(0,T) + \frac{1}{2\mu_0} H_c^2 - \frac{1}{2\mu_0 N} \cdot (H_c - B)^2 . \qquad (3.2.7)$$

In the new state Peierls [19] suggested that the system is subdivided into normal domains where $b = H_c$, and superconducting domains where $b = 0$. Thus if x is the concentration of normal domains

$$x = 1 - \frac{1}{N}\left(1 - \frac{B_0}{H_c}\right) \qquad (3.2.8)$$

and x goes from 0 at $B_0 = H_c(1-N)$ to 1 at $B_0 = H_c$. It is easily shown that the entropy of the intermediate state is a continuous function of field and it, as well as M, go continuously to normal state values at $B = H_c$. In view of the presence of normal domains this is not a surprising effect.

Given the possibility of domains, it is not apparent why they can not reduce in size to such a degree that $M \to 0$, as for thin films which would permit the intermediate state to persist to fields above H_c. However, far

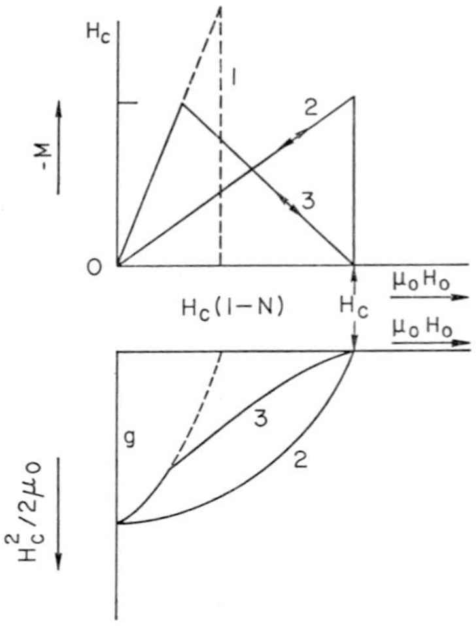

Fig. 3.2.3.
Magnetization and free energy of the intermediate state

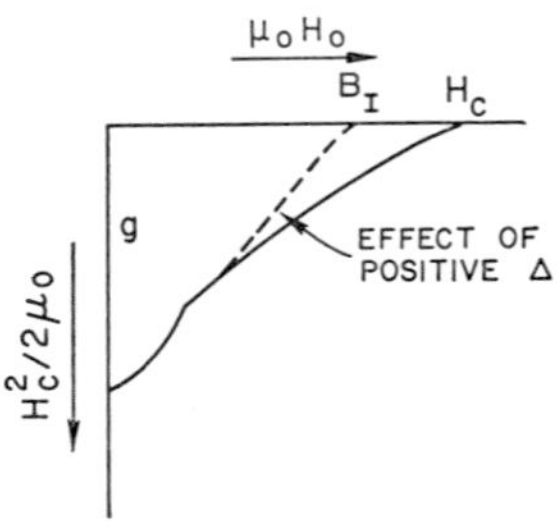

Fig. 3.2.4. Effect of positive surface energy on free energy of the intermediate state

from this being the case, a careful examination of the magnetic transition of the intermediate state shows that the critical field, $B_{\perp}$, is slightly less than $H_c(T)$.

In order to understand this phenomenon, an additional property has to be introduced, the surface energy. In the intermediate state with many normal-superconducting interfaces, it is clear that there may be surface tension or surface energy effects resisting the increase of the interface. If the surface energy is α, an additional free energy $\alpha f(x)$ will be lost to the superconductor and the critical field will be less than $H_c(T)$ as shown in Fig. 3.2.4.

Pippard [17] gave a physical picture of the source of this surface energy. Consider the superconducting state to be associated with an order parameter, n_s. The order parameter may be identified with the density of superconducting electrons, or with the energy gap. In either case it is a quantity that varies from its full value at $T = 0$ K, to zero at T_c. We consider a normal-superconducting boundary to be defined by $n_s = 0$, but we can not permit n_s to change discontinuously at the interface. As in the case of field penetration, we are led to consider a fundamental length ξ^*, over which the order parameter can vary.

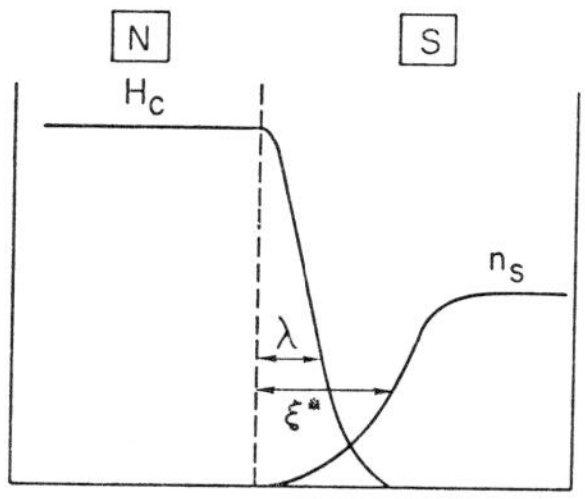

Fig. 3.2.5. Normal-superconducting boundary

In Fig. 3.2.5, we consider a normal-superconducting boundary where $B = H_c(T)$ in the bulk of the normal metal ($n_s = 0$) and $B = 0$ in the bulk of the superconductor ($n_s = n_s(T)$). It is easy to see that there is a gain of energy of order $\lambda H_c^2/2\,\mu_0$ per unit area due to flux penetration. Conversely there is a loss of condensation energy of the order, $\xi^* H_c^2/2\,\mu_0$, per unit area, where ξ^* is the distance over which the order parameter varies. Thus the surface energy is given by $\alpha = (\xi^* - \lambda)\, H_c^2/2\,\mu_0 = \Delta H_c^2/2\,\mu_0$. For a positive surface energy ($\xi^* > \lambda$) we can not gain energy by infinite subdivision. The minimum size of a domain is of the order of ξ^*, and the limit of $x \to 1$ for the intermediate state corresponds to the vanishing of the last domain, not its shrinking in size. For thick disks ($N \approx 1$) in a perpendicular field one can show that the critical field $B_{\perp}$ [20], is given by

$$B_{\perp}/H_c \approx (1 - (\Delta/d)^{\frac{1}{2}})\,. \tag{3.2.9}$$

And this relation, as well as the slope of $(\partial M/\partial B)_B = B_{\perp}$, can be used to determine Δ.

Pippard [14] further suggested a relationship for pure metals between ξ^* and the transition temperature and Fermi velocity. As later given by BCS [21], this is

$$\xi^* \approx \xi_0 = \frac{\hbar v_F}{\pi \Delta}, \tag{3.2.10a}$$

where 2Δ is the energy gap ($2\Delta = 3.56\, kT_c$). Calculations show that for most pure metals, $\xi_0 \approx 2$—10,000 Å, whereas $\lambda \approx 200$—600 Å. Hence $\Delta \approx 0$.

From these considerations we have been led to consider two fundamental lengths for a superconductor, ξ^* and λ. If we define their ratio as $\sqrt{2}\,\varkappa = \lambda/\xi^*$, we see that for $\sqrt{2}\,\varkappa > 1$, $\Delta < 0$, and one might expect a general depression of the free energy below even the Meissner curve for $N = 0$. Furthermore the experience with the intermediate state suggests that such a negative surface energy state might be a proper thermodynamic phase — homogeneous and reversible.

3.2.4 The Mixed State of Type II Superconductors

It is one of the triumphs of solid state physics that the negative surface energy state was predicted mathematically before there was any recognition of experimental justification. However, it remains a paradox, that the experimental evidence of such a state was largely ignored both prior to the theory and to its eventual experimental "verification". The theory of course, is the Ginzburg-Landau theory [22] published in 1950, "discovered" in 1961 [13], and which has received numerous verifications since that time, but some of whose predictions were observable in the experiments of Schubnikov [24] in 1937. The G-L theory, the theory of the microstructure of the negative surface energy state due to Abrikosov [4], and the microscopic theory of Gor'kov [2] (the GLAG theory) have all been the subject of numerous reviews, and only the results will be considered in detail in the present paper.

In its original form the theory was based on an expansion of the free energy close to T_c in terms of an order parameter, and was explicity designed to take into account spatial variations of the order parameter as occur in the intermediate state and thin films. It was shown by Gor'kov to be a natural consequence of the BCS microscopic theory when the energy gap was permitted spatial variations. As developed by Abrikosov [4], Gor'kov [2], De Gennes [25], Maki [26], Helfand and Werthamer [28—31], Eilenberger [32] and others, it is a sophisticated non-linear highly mathematical theory which is in very good agreement with experiment.

Despite the mathematical complexity of the theory, there is a physical approach to the magnetic behaviour of Type II superconductors that follows naturally from our previous discussions of surface energy. As developed by De Gennes [33], this approach considers a specific domained structure — a vortex of circulating superconducting electrons around a normal state core. In the following sections we will follow this approach, and will later make contact with the exact mathematical results of Maki [27], Helfand and Werthamer [28—31] and Eilenberger [32].

The lamellar domains [34] of the intermediate state are clearly not the only configuration appropriate to a domained structure. Again the task is to lower the magnetic energy while doing minimum damage to the condensation energy. Let us consider a normal core of radius ξ^*, with a field dropping off from its maximum value at the center as in Fig. 3.2.6. To proceed further we require a generalization of the London equations to take into account the normal core, and the circulation electrons about that core.

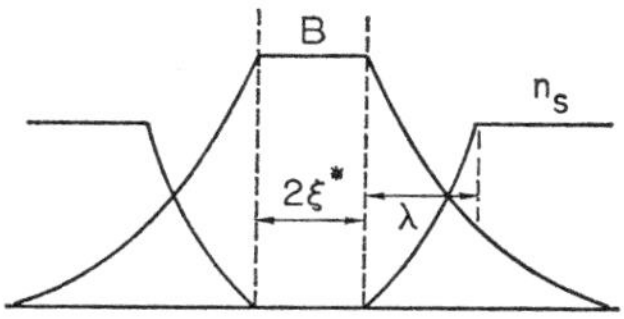

Fig. 3.2.6. Schematic of vortex structure

The usual form of London's Equation (3.2.4), when combined with the Maxwell Equation curl $\boldsymbol{H} = \boldsymbol{J}$, can be written as

$$\operatorname{curl} \boldsymbol{v} = -\frac{q}{m} \cdot \boldsymbol{B}\,, \tag{3.2.10b}$$

where $\boldsymbol{v}$, q, and m are the velocity, change and mass of the superconducting fluid ($J = nqv$). From $\boldsymbol{P} = m\boldsymbol{v} + \frac{q}{c}\boldsymbol{A}$, where $\boldsymbol{P}$ is the quantum mechanical momentum we can write Eq. (3.2.10b) as

$$\operatorname{curl} \boldsymbol{P} = 0\,. \tag{3.2.11}$$

However, from the presence of the normal core and the electron circulation we know that Eq. (3.2.11) can not hold everywhere. A suitable generalization of Eq. (3.2.11) is

$$\operatorname{curl} \boldsymbol{P} = q\Phi\delta(r)\,, \tag{3.2.12}$$

where $\delta(r)$ is a two dimensional δ function and Φ is the flux in the core.

Consideration of the circulation $\oint P \cdot dS$, about the core suggests Φ to be quantized in units of $h/2e$ ($q = 2e$) where $\varphi_0 = (h/2e) = (2 \times 10^{-15}\,\mathrm{Vs})$ is the unit of flux quantization. From Maxwell's equations and Eq. (3.2.2) we obtain as a generalization [35] of Eq. (3.2.4)

$$\lambda_L^2 \operatorname{curl}(\operatorname{curl} \boldsymbol{B}) + B = \varphi_0\delta(r)\,. \tag{3.2.13}$$

Using cylindrical symmetry, a solution of Eq. (3.2.13) for $\xi^* < r < \lambda_L$ is

$$B = \frac{\varphi_0}{2\pi\lambda^2}\log\frac{\lambda_L}{r}\,. \tag{3.2.14}$$

We next calculate the magnetic and kinetic energy of the vortex, per unit length of vortex. E_k (the core energy $(H_c^2/2\mu_0)\,\pi\xi^{*2}$ can be shown to be less than 12% of the magnetic and kinetic energy)

$$E_k = \int da\left(\frac{1}{2\mu_0}B^2 + \frac{1}{2}n\,m v^2\right) = \int da\left(\frac{B^2 \pm \lambda^2(\operatorname{curl} B)^2}{2\mu_0}\right). \tag{3.2.15}$$

Using the vector identity,

$$\nabla \cdot \big(B \cdot (\operatorname{curl} B)\big) = (\operatorname{curl} B)^2 - B \cdot \big(\operatorname{curl}\,(\operatorname{curl} B)\big) \tag{3.2.16}$$

and Eq. (3.2.13) we obtain

$$E_k = \frac{\lambda_L^2}{2\,\mu_0} \int\limits_{r=\xi^*} d s \cdot B \cdot (\operatorname{curl} B) = \frac{1}{4\,\pi\,\mu_0} \cdot \left(\frac{\varphi_0}{\lambda_L}\right)^2 \log\left(\frac{\lambda_L}{\xi^*}\right). \tag{3.2.17}$$

Noting that the flux density B is given by $n\varphi_0$, and neglecting the interaction between vortices we have for the total free energy

$$nF = \frac{1}{4\,\pi\,\mu_0} B \cdot \left(\frac{\varphi_0}{\lambda_L}\right)^2 \log\left(\frac{\lambda_L}{\xi^*}\right) \tag{3.2.18}$$

at equilibrium $B_0 = n\,\partial E_k/\partial B = (\varphi_0/4\,\pi\,\mu_0\,\lambda_L^2)\log\,(\lambda_L/\xi^*)$. Thus vortices will penetrate at $B_0 = H_{c1}$ where

$$\mu_0 H_{c1} = \frac{\varphi_0}{4\,\pi\,\lambda_L^2} \log\left(\frac{\lambda_L}{\xi^*}\right). \tag{3.2.19†}$$

We will find that the condition for $H_{c1} < H_c$ will be related to the previous negative surface energy requirement $\xi^* < \lambda_L$. We can further estimate the transition field where $M = 0$, (H_{c2}) as that field $(B \approx \mu_0 H)$ where the vortices overlap. Thus

$$H_{c2} \approx B \approx \frac{\varphi_0}{4\,\pi\,\xi^{*2}}. \tag{3.2.20}$$

The next step is to relate the fields H_{c1} and H_{c2} to H_c. In the G-L equations a parameter $\varkappa$ is introduced such that, as suggested previously $\xi^* = \lambda/\sqrt{2}\,\varkappa$. Furthermore

$$\varkappa = \lim_{T \to T_c} \frac{2\sqrt{2}\,\pi\,\lambda^2 H_c}{\varphi_0}. \tag{3.2.21}$$

If we ignore the restriction to temperatures close to T_c (and consider local superconductors where $\lambda \approx \lambda_L$) we can rewrite Eqs. (3.2.19) and (3.2.20) as

$$\boxed{H_{c1} = \frac{H_c}{\sqrt{2}\,\varkappa}\,(\ln \varkappa + 0.3)} \tag{3.2.22}$$

and

$$\boxed{H_{c2} = \sqrt{2}\,\varkappa H_c\,.} \tag{3.2.23}$$

Thus if $\varkappa > 1/\sqrt{2}$, $H_{c1} < H_c$ and $H_{c2} > H_c$, and we expect the superconductor to make a gradual magnetic transition from the state where $B = 0$ (Meissner state) at H_{c1} to the normal state at H_{c2} where a second order transition occurs. Materials for which $\varkappa > 1/\sqrt{2}$ are called Type II superconductors and the region between H_{c1} and H_{c2} where vortices penetrate is called the mixed state. Figs. 3.2.7 and 3.2.8 show the magnetization, free energy and phase diagram of this state.

† In this chapter, critical fields will be expressed whenever possible in Tesla, and the quantity μ_0 appearing in the log of Eq. (3.2.19) and similar equations will be often omitted.

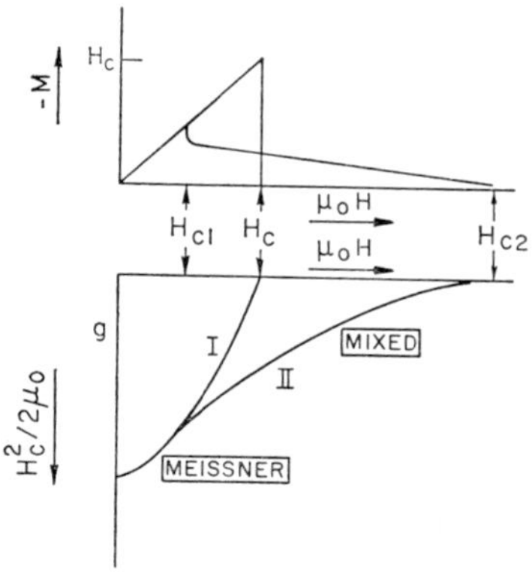

Fig. 3.2.7. Magnetization and free energy of Type I and Type II superconductors

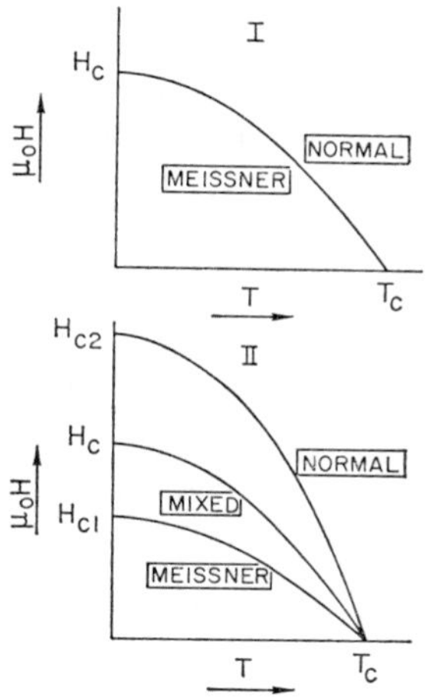

Fig. 3.2.8. Phase diagrams of Type I and Type II superconductors

The resemblances to the gross features of the intermediate state are obvious, however, it must be emphasized that there are fundamental differences. The intermediate state is only macroscopically second order; the last superconducting domain vanishes discontinuously at H_{c2}. However, the mixed state is *microscopically* second order at H_{c2}. In the intermediate state one expects supercooling i.e. a lowering of the field until the volume free energy can overcome the positive surface energy. In the mixed state there can be no supercooling due to the negative surface energy. Indeed in the original work of Ginzburg-Landau $H_{c2}(\sqrt{2}\,\varkappa < 1)$ is defined as the minimum field to which one can supercool a Type I superconductor. As we shall see this is only true for certain field orientations.

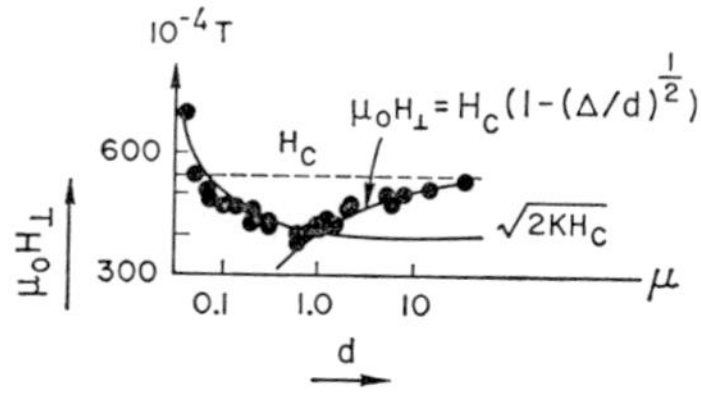

Fig. 3.2.9. Thickness dependence of perpendicular critical fields in thin Pb-films and foils

It is interesting to note that the vortex approach described here led Tinkham [36] to consider the question of what happened to thin disks of Type I superconductors in a perpendicular field. For example for low $\varkappa$, $\Delta \approx \xi^*$ and given Eq. (3.2.9), one expects $B_\perp$ to approach zero as $d \approx \xi^*$. Tinkham, from energy consideration, suggested that a vortex state at $B_\perp = \sqrt{2}\,\varkappa H_c$ might be energetically favorable and replace the intermediate state. The transition field B has been observed for lead [37], tin [38] and indium [38]. Fig. 3.2.9 shows experimental data for lead films. In this figure the rise in B for low thicknesses is related to a thickness dependence of $\varkappa$. However, the bulk value of $\varkappa$ ($< 1/\sqrt{2}$) can be obtained by extrapolation. This technique can be used instead of supercooling for the determination of $\varkappa$ in Type I materials.

3.2.5 Exact Theories of the Mixed State

The picture of the Type II superconductor is physically that of the reduction of magnetization due to the penetration of vortices with normal cores. The exact calculations of the Gor'kov equation leads to the following expressions for $H_{c1}(T)$, $H_{c2}(T)$ and the slope of the magnetization curve close to H_{c2} [27].

$$H_{c1}(T) = \frac{H_c(T)}{\sqrt{2}\,\varkappa_3(T)} \ln\left[\varkappa_3(T)\right], \tag{3.2.24}$$

$$H_{c2}(T) = \sqrt{2}\,\varkappa_1(T)\,H_c(T), \tag{3.2.25}$$

$$\left(\frac{\partial M}{\partial \mu_0 H}\right)_{Hc2} = \frac{-1}{(1.16)}\,\frac{1}{(2\varkappa_2^2 - 1)}. \tag{3.2.26}$$

where $\varkappa_1(T)$, $\varkappa_2(T)$ and $\varkappa_3(T)$ approach $\varkappa$ as $T \to T_c$ and [39, 40]

$$\varkappa \approx 0.96\,\frac{\lambda_L}{\xi_0} + 0.7\,\frac{\lambda_L}{l}, \tag{3.2.27}$$

$$= 0.96\,\frac{\lambda_L}{\xi_0} + 2.38\,\varrho_0\gamma^{\frac{1}{2}}. \tag{3.2.28}$$

In Eq. (3.2.27) l is the mean free path, and in Eq. (3.2.28) ϱ_0 is the residual resistivity (Ohm · cm) and γ (J/cm³ K²) is the coefficient of the linear term in the electronic specific heat. It should be emphasized that Eqs. (3.2.24) and (3.2.28) take band structure effects into account only in terms of a one band effective mass model. In this sence they are restricted in the same sense as BCS. However, the two terms for $\varkappa$ permit one to define an intrinsic Type II superconductor ($l \gg \xi_0$) and an extrinsic Type II superconductor ($l \ll \xi_0$). This classification coincides with the distinction between "clean" and "dirty" superconductors, and hence one expects poorer agreement with theory for clean materials where anisotropy, and multiple band effects have not been washed out by scattering.

Eqs. (3.2.27) and (3.2.28) permit one to give a more precise definition of ξ^*. Using the BCS results that

$$\lambda_{T \to T_c} \to \frac{\lambda_L}{\sqrt{2}}\,(1 - t)^{-\frac{1}{2}} \left(\frac{\xi_0}{\xi}\right)^{\frac{1}{2}}, \tag{3.2.29}$$

where

$$\frac{1}{\xi} = \frac{1}{\xi_0} + \frac{0.7}{l} \tag{3.2.30}$$

we obtain

$$\xi^* \approx \frac{\xi_0}{\sqrt{2}}(1-t)^{-\frac{1}{2}} \tag{3.2.31a}$$

$$\xi^* \approx \frac{(\xi_0 l)^{\frac{1}{2}}}{\sqrt{2}}(1-t)^{-\frac{1}{2}} \tag{3.2.31b}$$

in the clean and dirty limit respectively. From Eq. (3.2.31b) one notes that for impure metals $\xi^* \approx (\xi_0 l)^{\frac{1}{2}} \approx (v_F l)^{\frac{1}{2}}$ and is related to a diffusion length [25].

The original theory of G-L and Abrikosov was confined to temperatures close to T_c, where the order parameter was small and the free energy expansion might be expected to be valid. However, the order parameter is also small close to H_{c2}, and one might expect the theory to have a wider range of validity. Calculations of Gor'kov [2], Maki [26, 27] Werthamer [28—31], and Eilenberger [32], have shown that the temperature dependence of either $\varkappa_1(T)$, $\varkappa_2(T)$ or $\varkappa_3(T)$ is quite weak, and is of the order of 20% from 0 to T_c for extrinsic Type II superconductors. However, this result is correct only for low $\varkappa$ material. The general temperature and impurity dependence of $\varkappa_1(T)$ and $\varkappa_2(T)$ is quite complicated, and an additional physical mechanism, the magnetic nature of the normal state, has to be included before it can be discussed.

3.2.6 Paramagnetic and Impurity Effects on H_{c2}

Up to the present we have neglected the magnetic properties of the normal state. For transition metals such as Nb, or intermetallic compounds such as Nb_3Sn, V_3Si or V_3Ga the normal paramagnetic susceptibility, χ_N, is of the order of $4\pi \cdot 20 \times 10^{-6}$, with about equal orbital and spin contributions [41]. For a BCS superconductor the ground state consists of pairs of electrons with opposite spins, and hence the spin susceptibility is zero in the superconducting state. The unequal assignment of spin magnetic energy between the normal and superconducting states is the additional factor that has to be included to account in detail for the high field behaviour of Type II superconductors.

In the original suggestion of Clogston [42] and Chandrasekhar [43] the paramagnetic limiting field B_P was obtained by equating the magnetic energy of the normal state to the condensation energy i.e.

$$\frac{1}{2\mu_0}\chi_N B_P^2 = \frac{1}{2\mu_0} H_c^2 . \tag{3.2.32}$$

Using $\chi_N = (N/E_F)\, B_P^2/\mu_0$, where B_p is the Bohr magneton, as well as the BCS relations $H_c^2/2\mu_0 = (N/E_F)\,\Delta^2/4$; and $2\Delta = 3.5\, kT_c$ we obtain

$$B_P = 1.84\, T_c(T) . \tag{3.2.33}$$

In these papers the prediction was made that H_{c2} could not exceed B_P. A slightly more realistic calculation (although still incorrect) utilizes Fig. 3.2.10 to attain the effect of normal state spin paramagnetism.

If we ignore spin contributions in the superconducting state we have for the magnetization $M = \mu_0 \chi_s (B - H^*_{c2})$ where $\chi_s \approx 1/2\,\varkappa^2$ and therefore:

$$g_s(H) = g^0_n(H^*_{c2}) - \chi_s \cdot \frac{(B - H^*_{c2})^2}{2\,\mu_0}. \tag{3.2.34}$$

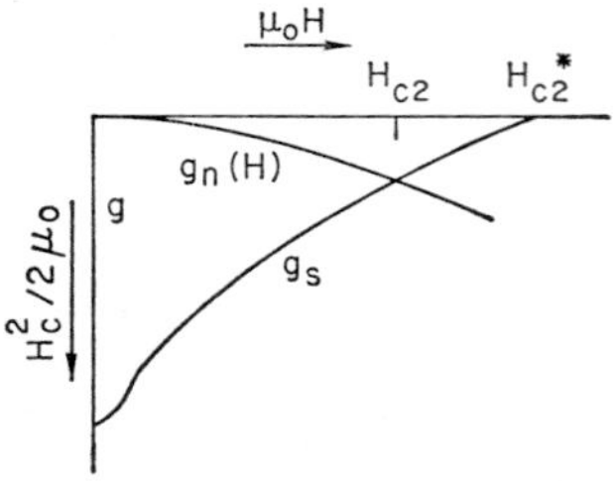

Fig. 3.2.10.
Free energy, including spin energy of normal state-first order transition

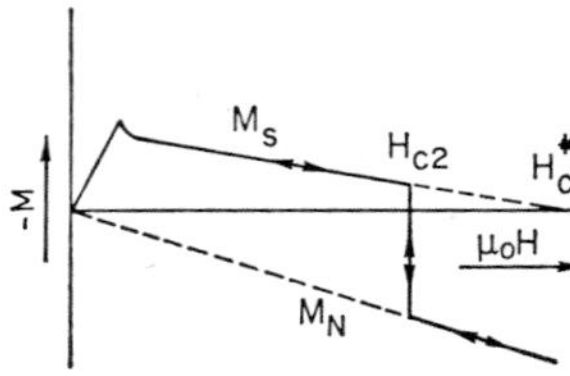

Fig. 3.2.11.
Magnetization corresponding to the free energy of Fig. 3.2.10. First order transition

In Eq. (3.2.34), $g^0_n(H^*_{c2})$ is the normal state free energy without the spin paramagnetism, $\chi_N B$. The quantity H^*_{c2} is the upper critical field in the absence of the paramagnetic limitation. If we include the spin paramagnetism in the normal state we have

$$g_n(B) = g^0_n(H^*_{c2}) - \frac{\chi_N B^2}{2\,\mu_0}. \tag{3.2.35}$$

Equating Eqs. (3.2.35) and (3.2.34) at the *transition field* H_{c2}, we obtain

$$\frac{H^*_{c2} - H_{c2}}{H_{c2}} = \left(\frac{\chi_N}{\chi_S}\right)^{\frac{1}{2}} = \frac{\sqrt{2}\,\varkappa}{B_P}\,H_c = \frac{H^*_{2c}}{B_P}, \tag{3.2.36}$$

where we have utilized Eq. (3.2.32). For a superconductor where $H^*_{c2} >$ 10 T, $T_c \approx 10$ K, we estimate $(H_{c2}/H^*_{c2}) \approx 0.6$. Fig. 3.2.11 shows the magnetization curve resulting from the above treatment and one notes the first order transition to the paramagnetic normal state at H_{c2}.

There are two objections to this approach. First it is not clear how the normal cores will modify the free energy of the superconducting state. In terms of the simple vortex model, the essentially normal material of the core will have a tendancy to lower the superconducting free energy, and lead to a magnetization that approaches that of the normal state at H_{c2}. Fig. 3.2.12 shows the expected free energy curve and magnetization. One notes that the transition is second order, and that M crosses the $\mu_0 H$ axis at a field where $g_s(B,T)$ has a maximum. Despite the change in the order of the transition, the magnitude of H_{c2} is not very different from the simple relation Eq. (3.2.36). A more significant effect arises from the fact that in highly disordered alloys there may be appreciable spin scattering in the superconducting state leading to a depairing of the spins in the superconducting state. This last effect, which is the dominant one, leads to a reduction of the paramagnetic effect on H^*_{c2} due to an equalization of the spin paramagnetism in the normal and superconducting states.

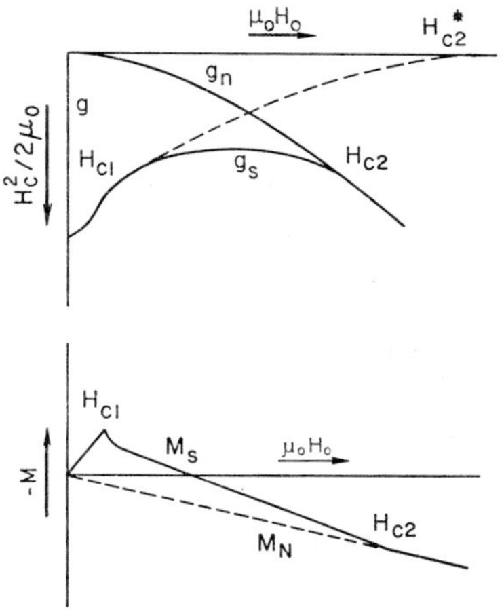

Fig. 3.2.12. Free energy and magnetization, including spin energy, of normal and superconducting states. Second order transition

The parameter to measure the effect of spin-orbit scattering can be estimated by comparing the energy uncertainty introduced by such scattering with the energy gap. If τ_{s0} is the spin orbit scattering time, from the uncertainty principle one expects depairing effects to be large when $\left(\hbar/\tau_{s0}(3.5\,kT_c)\right) \geqq 1$. The exact theories [27, 29] to be discussed introduce two parameters to describe scattering effects.

$$\lambda_{s0} \equiv \frac{\hbar}{3\,\pi k T_c \tau_{s0}}; \quad (3.2.37) \qquad \lambda \equiv \frac{\hbar}{2\,\pi k T_c \tau}, \quad (3.2.38)$$

where τ_{s0} is the spin-orbit scattering time, and τ is the spin independent scattering time. In the work of Maki [27] and Helfand, Werthamer and Hohenberg [28, 29], τ is identified with the transport scattering time. As suggested by Eq. (3.2.36) the basic parameter of the paramagnetic effect is

$$\alpha \equiv \frac{\sqrt{2}\,H_{c2}(0)}{B_P} \qquad (3.2.39)$$

and Maki [27] introduces a mixed parameter β_0 to characterize the paramagnetic effect, and its reduction by spin orbit scattering

$$\beta_0^2 \equiv \frac{\alpha^2}{1.78\,\lambda_{s0}}. \qquad (3.2.40)$$

To summarize the discussion to the present, we note that the results of the GLAG theory, Eqs. (3.2.24) and (3.2.28), are only exactly valid near T_c. Close to H_{c2}, at lower temperatures and when spin paramagnetism can be neglected, Eqs. (3.2.24) and (3.2.28), retain approximate validity, and departures from the exact theory will be measured by the parameter λ, Eq. (3.2.38). For high $\varkappa$ dirty materials the parameters α, Eq. (3.2.39), λ_{s0}, Eq. (3.2.37), and β_0, Eq. (3.2.40), determine the upper critical field and its temperature dependence.

Before we discuss the formal theories and their comparison with experiment it will be helpful to make three general observations about experimental results. First, it is important to note that in all the theories, paramagnetic effects vanish at T_c. Physically, this arises because close to T_c, quasi-particle excitations dominate the superconductor, and hence the mixed state is depaired. Thus if we use the notation:

$$H_{c2}(\alpha,\lambda,\lambda_{s0},T) \equiv H_{c2}(T)\,, \quad H_{c2}(0,\lambda,\lambda_{s0},T) \equiv H^*_{c2}(T)\,,$$

we obtain
$$\left(\frac{dH_{c2}}{dT}\right)_{T_c} = \left(\frac{dH^*_{c2}}{dT}\right)_{T_c} = \sqrt{2}\,\varkappa\left(\frac{dH_c}{dT}\right)_{T_c}, \tag{3.2.41}$$

where $\varkappa$ is given by Eqs. (3.2.27) and (3.2.28). Second, in the dirty limit, the parameter α does not have to be fit to the data but can be obtained independently. This observation follows from the theoretical temperature dependence of H^*_{c2}, which will be discussed presently, and Eqs. (3.2.41) and (3.2.28) plus some results from BCS. In the *dirty limit* one has

$$\alpha = 0.73\times10^{-3}\cdot(\varrho\gamma)\,, \tag{3.2.42a}$$

where ϱ is in Ohm · cm and γ in J/cm^3 K^2 is the coefficient of the linear term in the specific heat. Further algebra shows that for both intrinsic and extrinsic GLAG materials

$$\alpha = 5.33\times10^{-5}\left(\frac{-dH_{c2}}{dT}\right)_{T_c}. \tag{3.2.42b}$$

Hake [44] has presented an admirable review of the effect of spin paramagnetism on the mixed state, and this review has an appendix which lists a variety of formulae showing relationships between experimental and theoretical quantities.

The third observation is the most important and must be taken into account before any discussion of experiment. In general there are two ways of presenting experimental data. One can use the usual relation $H_{c2} = \sqrt{2}\,\varkappa_1 H_c$ and show $\varkappa_1$ as a function of temperature. For greater generality it is convenient to introduce a normalized quantity

$$\varkappa^*(t) \equiv \varkappa^*(t,\alpha,\lambda,\lambda_{s0}) \equiv \frac{\varkappa_1(T/T_c,\alpha,\lambda,\lambda_{s0})}{\varkappa}. \tag{3.2.43}$$

This mode of presentation yields two difficulties: first, it combines two quantities H_{c2} and H_c, and hence couples the mixed state with the zero field state, second; it requires the experimentalist to have data on $H_c(T)$ for the specimen under consideration, and this data is often not available. Werthamer, Helfand and Hohenberg [29] suggest the use of the normalized field

$$h^* = \frac{H_{c2}(T)}{-T_c\left(\dfrac{dH_{c2}}{dT}\right)_{Tc}} \tag{3.2.44}$$

for comparisons between theory and experiment, and they present data on impurity effects as well as spin paramagnetic effects in terms of h^* (to be distinguished from H^*_{c2}).

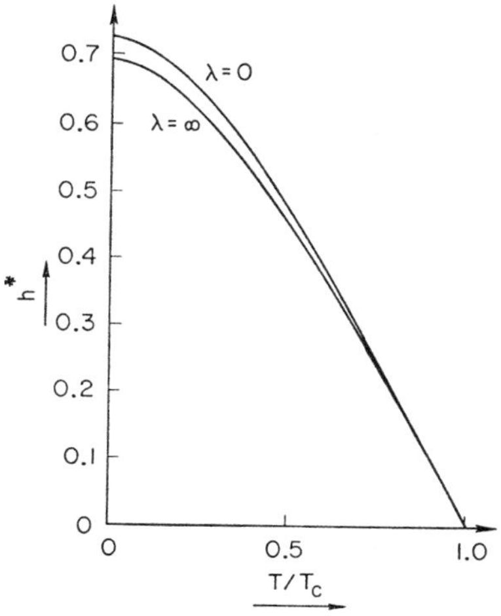

Fig. 3.2.13.
Reduced field h^* as a function of T/T_c (Ref. 218)

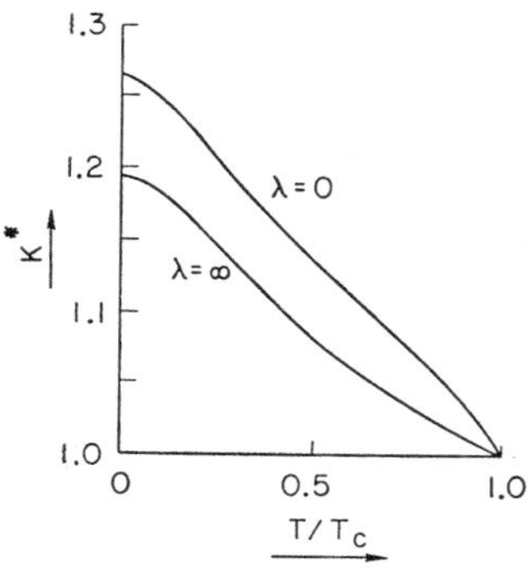

Fig. 3.2.14.
Reduced parameter $\varkappa^*$ as a function of T/T_c (Ref. 218)

Fig. 3.2.13 shows the results of the calculation of Helfand and Werthamer [18] for h^* as a function of T/T_c for $\alpha = 0$, and for various values of λ. Fig. 3.2.14 shows the same results expressed in terms of $\varkappa^*(t)$. One notes the weak impurity dependence of h^* and $\varkappa^*$. However, it must be emphasized that the calculations shown in Fig. 3.2.14 are based on BCS values for $H_c(T)$, and the small variation in $\varkappa^*(t)$ with temperature holds only for BCS like superconductors.

This observation can be seen more clearly if we note that

$$\varkappa^*(t) = -h^* \frac{T_c \left(\dfrac{dH_c}{dT}\right)_{T_c}}{H_c(T)} = -h^* \varphi(T) . \qquad (3.2.45)$$

Table 3.2.1. $\varkappa^*(0)$ *for BCS and non-BCS superconductors*

	$\varphi(0)$	$[\varkappa^*(0)]$ *pure*	$[\varkappa^*(0)]$ *dirty*
Sn	1.855	1.35	1.28
Pb	2.132	1.56	1.47
V	1.773	1.29	1.22
Nb_3Sn	2.000	1.46	1.38
Nb	2.000	1.46	1.38
BCS	1.737	1.27	1.20

For all superconductors, from the normalization of h^*, $\varkappa^* \to 1.0$ as $T \to T_c$. However, $\varkappa^*(0)$ will depend strongly on the value of $\varphi(0)$. Table 3.2.1 shows values of $\varphi(0)$ for Sn, Pb, V, Nb_3Sn and Nb. It also shows the value of $\varphi(0)$ predicted by BCS, and $\varkappa^*(0)$ for pure ($\lambda = 0$) and dirty ($\lambda = \infty$) materials.

From Table 3.2.1 we see that the variation in $\varkappa^*(0)$ between different materials is much larger than the variation induced by changing from completely pure to impure materials. Clearly the simple statement that $\varkappa^*(0) > 1.27$ does not in itself constitute a departure from theory. For completeness we note that

$$H_{c2}(T) = h^* \cdot \sqrt{2}\,\varkappa \left[-T_c \left(\frac{dH_c}{dT}\right)_{T_c}\right] \tag{3.2.45a}$$

and again any comparison between theory and experiment has to include a knowledge of $(dH_c/dT)_{T_c}$.

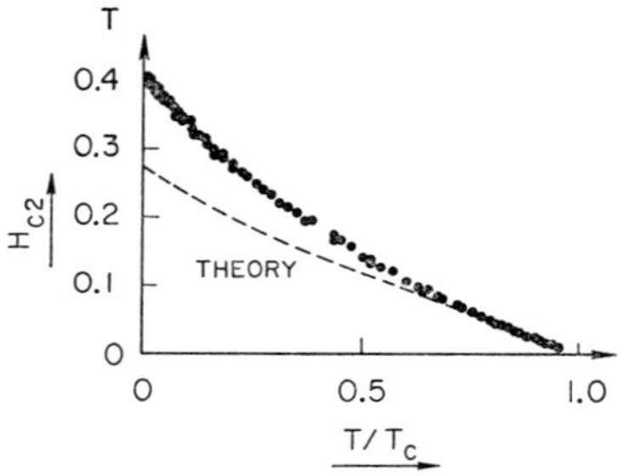

Fig. 3.2.15. Measured H_{c2} for Nb compared with theory of Hohenberg and Werthamer (Ref. 46)

The results shown in Figs. 3.2.13 and 3.2.14 are in excellent agreement with low $\varkappa$ materials where paramagnetic limitations are inoperative, and where impurity effects dominate. There is not good agreement for clean Type II superconductors such as V [45] and Nb [46]. Fig. 3.2.15 shows the departures from theory for Nb, and similar departures are seen for V. For Type I material such as Sn and In [37, 38], where measurements have been made of the transition fields of thin films in a perpendicular field, similar departures are noted. It is interesting to note that in all cases the departure from theory is expressed by the experimentalist as a preference for the two fluid temperature dependence of $\varkappa$ (Eq. (3.2.21)) i.e. $\varkappa = \varkappa_0 [1 + (T/T_c)^2]^{-1}$. For V, Nb and Sn use of the theory leads to $h^*(0) = 0.85$, 0.83, and 0.85 respectively whereas the maximum value of Fig. 3.2.13 is about 0.72. The discrepancy ($\approx 20\%$), although small, is well outside experimental and theoretical uncertainty. It cannot be accounted for on the basis of strong coupling effects, and according to Hohenberg and Werthamer [31] may arise from Fermi surface anisotropy effects. Before we leave the area of non-paramagnetically limited Type II superconductors, it is of interest to point out the extensive numerical calculations of Eilenberger [32] for $\varkappa_1$ and $\varkappa_2$ as a function of purity including both S-S and S-P scattering.

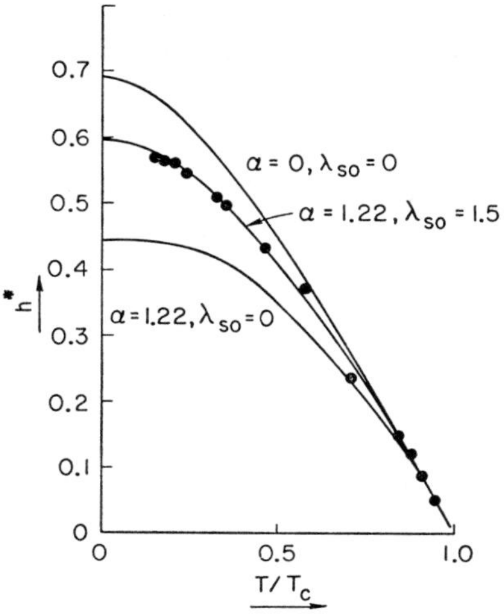

Fig. 3.2.16. Reduced field h^* as a function of T/T_c for Ti (0.56) Nb (0.44), (Ref. 217)

Although the small departures from theory for clean Type II superconductors are sufficiently large to bother theoreticians and experimentalists, the case of paramagnetically limited alloys (i.e. the dirty limit) appears to be in excellent shape. Fig. 3.2.16 shows the comparison between h^* for $Ti_{.56}Nb_{.44}$ and the predictions of the theory of Werthamer, Helfand, and Hohenberg [29]. In fitting the data, α is fixed by Eq. (3.2.42) and λ_{so} is chosen as a free parameter. The fit is excellent.

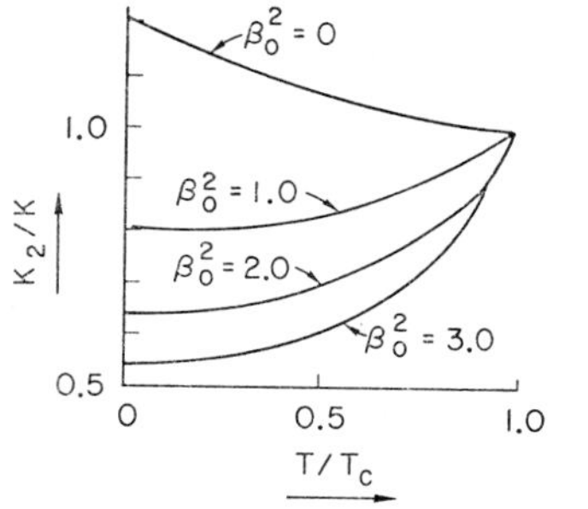

Fig. 3.2.17.
Reduced parameter $\varkappa_2/\varkappa$ for two alloys as a function of T/T_c compared to Maki theory (Ref. 44)

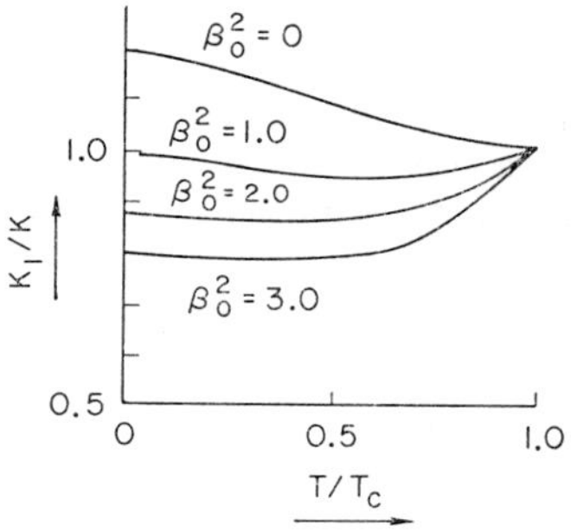

Fig. 3.2.18.
Reduced parameter $\varkappa_1/\varkappa$ as a function of T/T_c (Ref. 27)

Maki [27] has expressed his results in terms of $\varkappa_1/\varkappa$ and $\varkappa_2/\varkappa$ as a function of t and β_0 as shown in Figs. 3.2.17 and 3.2.18 and one notes the depression of $\varkappa_2(t)$ as well as $\varkappa_1(T)$ below $\varkappa$ where the paramagnetic effect is operative. In this connection it is important to note that $\varkappa_2(T)$ is defined by $-(M_s - M_n) = (H_{c2} - B)/[(2\varkappa_2^2(T) - 1)\ (1.16)]$. Fig. 3.2.19 shows a comparison between the Maki theory for $\varkappa_2(t)$ and experimental results of Hake [44]. The agreement appears satisfactory. Fig. 3.2.20 is an impressive example of the effect of spin paramagnetism from the work of Hake [44], where the free energy as well as the magnetization is plotted as a function of field for a Ti-16 Mo alloy with an H_{c2}^* of about 10 T, paramagnetically reduced to about 4 T.

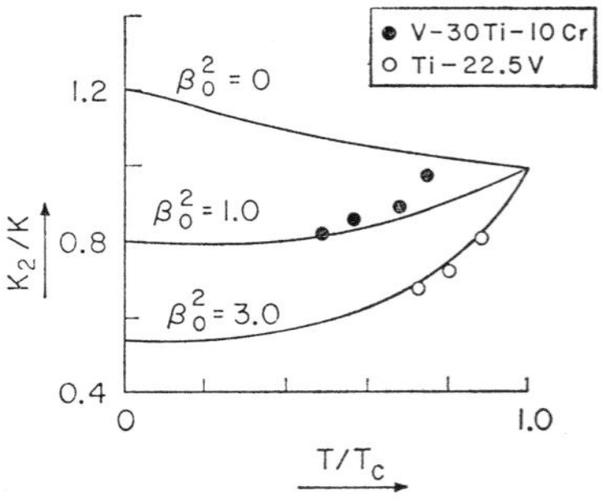

Fig. 3.2.19.
Reduced parameter $\varkappa_2/\varkappa$ for two alloys as a function of T/T_c compared to Maki theory (Ref. 44)

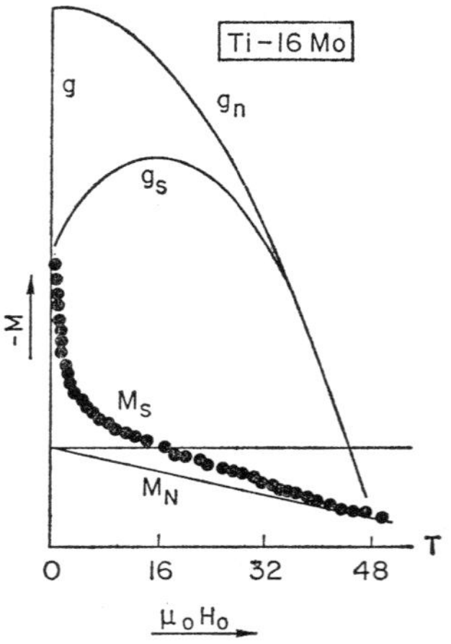

Fig. 3.2.20.
Magnetization and derived free energy for Ti (16%) Mo alloy (Ref. 44)

Fig. 3.2.21 from the paper of Hake shows a comparison of Werthamer, Helfand and Hohenberg [29] (WHH) computer solutions for $h^*(t = 0)$ as a function of α and λ_{s0}, compared to the analytic approximate form derived by Maki. The Maki [27] expression, corrected from the original paper, as quoted by Hake, is

$$h^*(t) = 1.39\, h_{c2}[1 + (1 + \beta_0^2 h_{c2})^{\frac{1}{2}}]^{-1}, \tag{3.2.46}$$

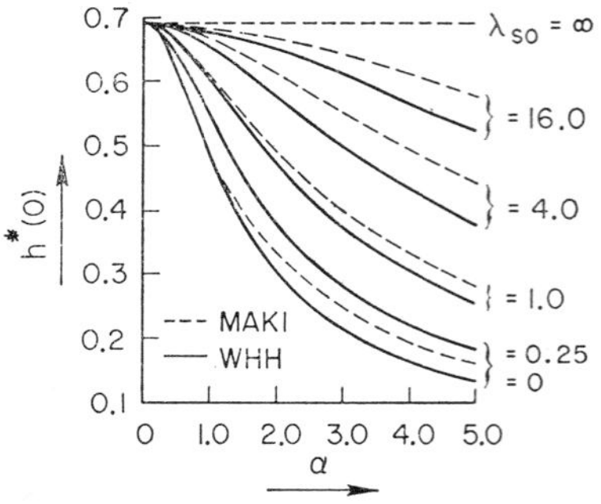

Fig. 3.2.21. The reduced field $h^*(0)$ as a function of α and λ_{s0} from theories of Maki and WHH (Ref. 44)

where β_0 is given by Eq. (3.2.40) and $h_{c2} = H^*_{c2}(t)/H^*_{c2}(0)$. A condition to be met for both theories is that the spin flip scattering time τ_{s0} be large compared to the transport scattering time τ; a condition which is physically realizable in high $\varkappa$ transition metal alloys. As shown in the paper of Hake one can get about equally good fits to either the Maki or WHH theory within the experimental uncertainties.

Finally it is important to note that for $\lambda_{s0} \to \infty$, there is no paramagnetic limitation of H_{c2}, and $H_{c2} \to H^*_{c2}$. Since $\lambda_{s0} \sim (Z^4)$, where Z is the average atomic number, one might expect high Z alloys to have less paramagnetic limitations and hence achieve the upper critical field predicted from Eq. (3.2.28).

3.2.7 Critical Fields of Intermetallic Compounds

From the previous discussion it is apparent that dirty Type II superconductors (extrinsic) are in excellent agreement with theory for both the paramagnetically limited and non-paramagnetically limited regime. Clean (intrinsic) Type II superconductors exhibit characteristic departures from theory at low temperatures. Intrinsic high $\varkappa$ material (if they exist) have yet to critically be examined in the paramagnetic regime, although the theory exists. The departure from theory for low $\varkappa$ materials have been ascribed to band structure and anisotropy effects, but no quantitative calculations have been made.

It is an interesting question whether similar departures from theory might be expected for the intermetallic compounds Nb_3Sn, V_3Si and V_3Ga. These materials show significant departures from "normal-metal" behaviour [41] above T_c, and there has been little success in correlating their superconducting behaviour with existing free electron theories.

One is inclined to believe that these materials exhibit band anisotropy and strong coupling, but they present experimental difficulties for both sample characterization and experiment. They are difficult to prepare in single crystal form, and the high critical fields and transition temperatures make it difficult to extract such parameters as the residual resistivity and electronic specific heat.

The Table 3.2.2 gives data for $(dH_{c2}/dT)_{T_c}$, $(dH_c/dT)_{T_c}$, γ, ϱ and $\varkappa$ for Nb_3Sn, V_3Si and V_3Ga, as well as a calculation of ($\varkappa_l = 0.7\,\lambda_L/l$). Unfortunately except in the case of Nb_3Sn, the H_{c2} data does not necessarily refer to the same specimen for which the ϱ is given. Moreover, resistivity data on sintered specimens is often overestimated due to problems of porosity. From the Table one notes that Nb_3Sn and V_3Si are intrinsic ($\varkappa > 2\,\varkappa_l$). However, the apparent intrinsitivity of $(dH_{c2}/dT)_{Tc}$ to sample preparation for Nb_3Sn, V_3Si and V_3Ga suggest that all are intrinsic and more so than is suggested by the calculation of $\varkappa_l$ from Eq. (3.2.28). For example, doubling the resistivity of Nb_3Sn has no measurable effect on the slope $(dH_{c2}/dT)_{T_c}$ [53].

Another example is the invariance of dH_{c2}/dT to large changes in stoichiometry for V_3Ga [51]. Furthermore, both B_3Si and Nb_3Sn (and perhaps V_3Ga) exhibit specific heat anomalies at low temperatures that casts some doubt one the quoted $\gamma's$. Since this table includes two ma-

Table 3.2.2. *Parameters of Type II intermetallic compounds*

	T_c (K)	ϱ (10^{-5} Ohm/cm)	γ (J/cm³ K²)	$(dH_{c2}/dT)_{Tc}$ (T/K)
Nb_3Sn	18	1.0 [31]	3.76 [37]	—1.80 [37]
V_3Si	17	0.4 [39]	7.58[c]	—1.90 [41]
V_3Ga	16	4.0 [38]	< 8.5[c]	—4.50 [41]

	$(dH_c/dT)_{T_c}$ (T/K)	$\varkappa^b$	$\varkappa_l^a$
Nb_3Sn	—0.0590 [37]	22 [37]	8.2
V_3Si	—0.0720 [40]	19	5.0
V_3Ga	—0.0870 [42]	37	< 49.0

[a] Eq. (3.2.28)
[b] Eq. (3.2.25) at T_c
[c] Best values derived by L. Vieland from Ref. (50) and Ref. (52)

terials of practical interest, and three of great theoretical interest it is of some importance to obtain critical field data on well characterized specimens.

Before leaving the intermetallics, it may be of value to consider two-band effects. Cohen has made some preliminary calculations of the dependence of $\varkappa$ on the material properties of a two band system with two energy gaps (e.g. a s gap and a d gap). Similar calculations have been made by Tilley [54] and Moskalenko [55]. Cohen expresses his results as a function of N_i, the density of states in the i-th band ($i = 1.2$); v_{Fi}, the Fermi velocity; T_c; and the energy gap Δ. A basic parameter is the quantity r where near T_c

$$\frac{\Delta_1^2}{\Delta_{\mathrm{BCS}}^2} = \frac{1 + \frac{N_2}{N_1} \cdot r^2}{1 + \frac{N_2}{N_1} \cdot r^4} \quad \text{and} \quad \Delta_2^2 = r^2 \Delta_1^2 \qquad (3.2.47/48)$$

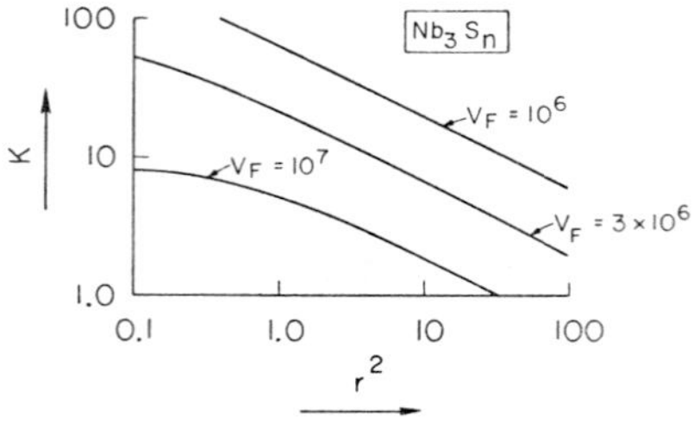

Fig. 3.2.22. Dependence of $\varkappa$, in two band model from Nb_3Sn on Fermi velocity and parameter r^2

and ΔBCS = ΔBCS(T), and ΔBDS(0) = 3.5 kT$_c$. The result for $\varkappa$ in the clean limit is

$$\varkappa = \varkappa_{01}^2 \frac{[1 + (v_{F1}^2/v_{F2}^2)\, r^2]}{[1 + (N_2 v_{F2}^2/N_1 v_{F1}^2)\, r^2]}, \tag{3.2.49}$$

where

$$\varkappa_{01}^2 = \left(\frac{0.96\, \lambda_{L1}}{\xi_{01}}\right). \tag{3.2.50}$$

Fig. 3.2.22 shows an approximately self consistant variation of $\varkappa$ with r for Nb_3Sn as a function of various d-band Fermi velocities. The curves are based upon paramagnetic susceptibility, Hall data and specific heat data for Nb_3Sn. Unfortunately, it is not possible to fit all the existing data, in particular λ and $\varkappa$ with this model. It is clear that more data is required to satisfactorily understand $\beta - W$ intermetallics.

3.2.8 Surface Superconductivity

Up to now, we have not touched upon a third critical field, H_{c3}. This field arises from a peculiar boundary conditions of the G-L equations for a field parallel to the surface. Under these conditions a superconducting sheath of thickness $\approx \xi^*$ can exist up to a field $H_{c3} = 1.695\, H_{c2}$ (Fig. 3.2.23). Furthermore, for Type I superconductors H_{c3} is the supercooling field. There is good experimental agreement with the ratio H_{c3}/H_{c2} when there is proper control of the surface. Calculations by Eilenberger and Ambegaokar [56] in the strong-coupling limit for pure superconductors close to T_c show that the ratio is maintained. Recent calculations of Lüders [57] for clean superconductors suggest that H_{c3}/H_{c2} at low temperatures can be as large as 2.2 due to the effect of electron reflection. Fischer [58] has interpreted microwave data for lead along these lines, and derived an $H_{c2}(T)$ for Pb which is in good agreement with the calculations of Werthamer and Helfand. However, in these measurements ($\varkappa < 1/\sqrt{2}$, $1.695\,\sqrt{2}\,\varkappa > 1$) H_{c2} is not directly measurable and there is some question of interpretation. Moreover given the disagreement of Nb and V with the calculations of Helfand and Werthamer [28] there is a question as to whether one should expect good agreement for Pb. Finnemore et al. [46] measured H_{c3}/H_{c2} for Nb, for samples with a residual ratio of 280 and 2,000 and found H_{c3}/H_{c2} extended to H_{c3}/H_{c2} of 1.8 and 2.0 respectively [63].

St. James [59] has derived the modification of H_{c3} when there are paramagnetic effects with the following expression at $T = 0$ K

$$\frac{H_{c3}(0,\alpha)}{H_{c2}(0,\alpha)} = \frac{1.695\,(1+\alpha^2)^{\frac{1}{2}}}{[1+(1.695\,\alpha)^2]^{\frac{1}{2}}}. \tag{3.2.51}$$

As before paramagnetic effects vanish close to T_c. When one includes spin orbit scattering one obtains

$$H_{c3}(T,\alpha,\lambda_{so}) = 1.695\, H_{c2}(T, 1.695\,\alpha, \lambda_{so}) \tag{3.2.52}$$

and as St. James points out simultaneous measurement of H_{c2} and H_{c3} should permit direct determination of α and λ_{so}. Eq. (3.2.52) does not appear to have received any experimental verification. Eq. (3.2.51) for $\alpha \approx 1$ suggest $H_{c3}(0,\alpha) \approx 1.2\, H_{c2}(0,\alpha)$, which is in fair agreement with measurements of Kim et al.

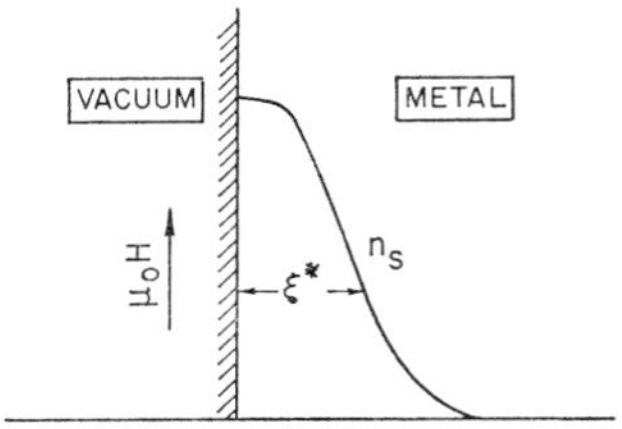

Fig. 3.2.23. Spatial variation of order parameter for surface superconductivity ($H_{c2} < H_c < H_{c3}$)

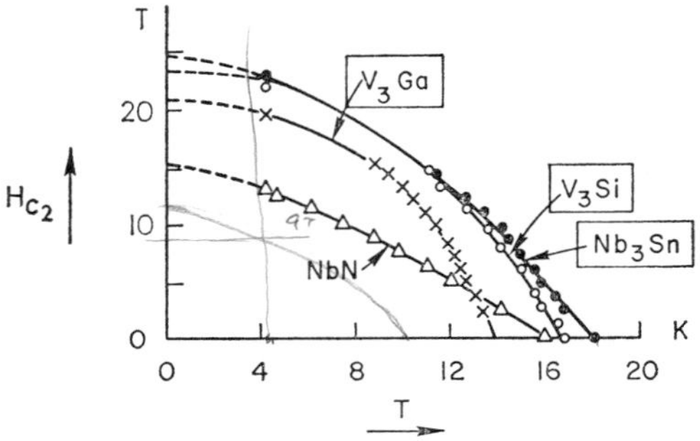

Fig. 3.2.24. Critical field curves of Nb_3Sn, V_3Si, V_3Ga and NbN (Ref. 61)

Finally, we would like to show some recently measured data on practical materials made by Sauer and Wizgall [61] (Fig. 3.2.24). Table 3.2.3 summarizes the results as well as giving some predictions of Hake [62] on possible values for H_{c2} if their properties could be maintained under massive alloying to raise ϱ. As noted, since we believe the intermetallics are inherently intrinsic, this part of the table should be considered with some discretion. In our own opinion the critical field of these materials is intimately related to their T_c, and poorly understood band structure and electronic interactions.

Table 3.2.3. *Upper critical fields of intermetallics*

Material	$H_{c2}(0)$ (T)	$H_{c2\,max}(T)$[a]
Nb_3Sn	24.5	88.0
V_3Si	23.5	85.0
V_3Ga	21.0	81.0
NbN	15.3	25.0

[a] After Hake [62]

It is quite possible that large enhancements in H_{c2} beyond that realized today will only follow the understanding that leads to increases in T_c above 20 K. Recent observations of critical fields (H_{c2}) above 20 T at 14 K and above 40 T at 4.2 K for $Nb_3Al_{0.8}Ge_{0.2}$ suggest that large increases in H_{c2} can accompany only modest, but significant, increases in $T_c (T_c = 21$ K) [64].

3.3 Critical Currents of Type II Superconductors

3.3.1 Introduction

In this chapter we will present a theory of the critical currents, which admittedly lacks microscopic rigor and is best described as a semi-quantitative model. Such models are common in solid state and soon are replaced or improved by further research and experimentation. Unfortunately, and perhaps due to the modest claims for the theory, there have been few attempts to relate its predictions to the microstructure of the superconductor, and in those cases where this has been done, there is often no indication of whether the results are general, or specific to a given material. Indeed the casual reader of any number of papers on the critical current can easily emerge with a sense of occasional and fortuitous triumphs, dispersed among a welter of ad-hoc calculations.

Despite this impression, we believe that the present model is well founded, and can give results in fair agreement with experiment in a variety of experimental situations, if reasonable assumptions are made of the relevant materials parameters. Indeed it appears that almost all published observations of the magnitude, field and temperature dependence of the critical current can be included within the present model. This is an impressive feat for any theory, and particularly so, when the complexity of the current carrying Type II superconductor is appreciated. One of the points of the present chapter, is that the surprising success of such a simple model implies that it can be used both to predict straight sample performance of practical material, from a knowledge of their microstructure of a given material. However, this conclusion hinges on further experimentation on well defined materials, as well as on further development of the theory.

In this chapter, we first establish the foundations of a model of the critical current of Type II superconductors as in Sections 3.3.2 and 3.3.3 and then illustrate its application to a variety of microstructures in Section 3.3.4. This section also serves as a review of literature pertinent to the model. The chief result of this section is a series of tables that display the parameters of the model, as well as examples of the magnitude and field dependence of the critical current to be expected at low temperatures.

In Section 3.3.5 we compare the predictions of the model with experiment, and find surprisingly good agreement for those cases where an effort has been made to characterise the material. This conclusion holds for materials as different as Pb-Bi alloys and Nb_3Sn. In Section 3.3.6 we examine an area of crucial importance for our evaluation of the model, its high temperature behaviour, and again find surprisingly good agreement with experiment.

Finally in Section 3.3.7 we briefly consider a.c. critical currents of Type II superconductors.

It is hoped that the present chapter will supply the background for the reader to understand the wide range of behaviour that the "straight sample" critical current of Type II superconductors can exhibit, that it will establish the fundamental basis of the present model, and finally suggest the value of further research on well defined materials.

3.3.2 Forces on Flux Lines

In the preceding Chapter 3.2 we have discussed the equilibrium properties of Type II superconductors in terms of a uniform lattice of flux vortices where the uniform flux density B is related to the vortex density n, by the equation $B = n\varphi_0 (\varphi_0 \equiv h/2e)$.

In the following sections we consider the non-uniform vortex lattice with a gradient in B (and n) arising from real currents. For the present we will not be concerned whether these currents arise from external sources or are induced in the specimen due to changes in the applied field.

The starting point in any discussion of the critical current of Type II superconductors is the notion of the force exerted on the vortex structure by a current density J. The most direct derivation of a model independent form for this force, is through thermodynamics [15] where we circumvent the essential non-equilibrium nature of the problem by considering small areas of the specimen assumed to be in local equilibrium. A basic parameter of this treatment, and indeed for any constant temperature process in a superconductor, is the Gibbs free energy density, g, whose differential can be written:

$$dg = -B\,dH\,. \tag{3.3.1}$$

Eq. (3.3.1) is easily derived from Maxwell's equations and a Legendre transformation of variables from B to H. It differs from the expression utilized in the preceding chapter by the inclusion of a term $\mu_0 H\,dH$ and the use of the internal field B rather than the applied field B_0.*

For the situation shown in Fig. 3.3.1, a two dimensional vortex structure, we are motivated to consider the vortices as a two dimensional gas, and to examine changes that can occur in the total free energy E_f per unit length of vortex. The quantity E_f is given by

$$E_f = fA = (g + BH)\,A\,, \tag{3.3.2}$$

where A is the area of the specimen occupied by the vortices and f is the free energy density. We note that f includes the energy required to create a vortex as well as vortex-vortex interactions. From Eqs. (3.3.1) and (3.3.2) we obtain for the change in E_f at constant temperature for a change in A and N (the total number of vortices in A)

$$dE_f = g\,dA + \varphi_0 H\,dN\,. \tag{3.3.3}$$

The form of Eq. (3.3.3) is analogous to what we would derive for the change in E_f for a two dimensional gas where A is again the area occupied by N particles. From Eq. (3.3.3) we can identify $\varphi_0 H$ as the energy required to add one vortex to the system and we can also identify $-g$ with the pressure P exerted by the gas of vortices at constant N. We

* In this chapter fields will be expressed whenever possible in Tesla. Thus, $\mu_0 H\,dH \rightarrow B\,dB/\mu_0$ whenever it is convenient.

thus obtain for the force F_y on the vortex lattice, per unit volume (Fig. 3.3.1)

$$F_z = \frac{-\partial P}{\partial y} = \frac{\partial}{\partial y}(f - BH) = -B\,\frac{\partial H\,(B)}{\partial y}\,. \tag{3.3.4}$$

In arriving at Eq. (3.3.4) we have used the equilibrium relation $(\partial f/\partial B) = H$, and hence defined $H\,(B)$ as the internal field in equilibrium with the local flux density B.

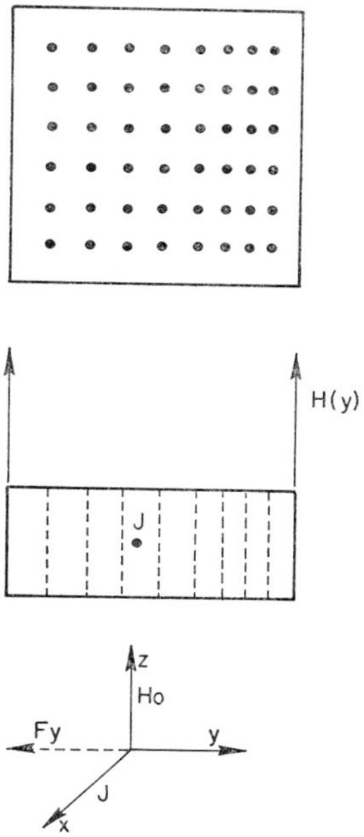

Fig. 3.3.1. Schematic of vortex lattice showing geometry of current density, field and force on vortex lattice

Before we consider the implications of Eq. (3.3.4) it will be useful to review the usual definitions of B and H, to point out the problems in interpretation that can occur in the case of a superconductor. As usual we start with the microscopic local field $h\,(r)$ and local current density $j\,(r)$ which obey the Maxwell equation [65]

$$\operatorname{curl} \bar{h}\,(r) = j\,(r)\,. \tag{3.3.5a}$$

Upon averaging we obtain

$$\operatorname{curl} B = \mu_0 \operatorname{curl} h\,(r) = \mu_0 \overline{j\,(r)}\,. \tag{3.3.5b}$$

For magnetic systems it is customary to write the right hand side of Eq. (3.3.5b) as

$$J = \overline{j\,(r)} = \frac{1}{\mu_0}(\operatorname{curl} M) + J_T = J_M + J_T\,. \tag{3.3.5c}$$

Table 3.3.1. *Defect distribution*

Dimension	Distribution	ϱ	ϱ_l	N	$1/l$
3	I [7]	$\nu^{\frac{1}{3}}/\lambda^2$	$\nu^{\frac{1}{3}}/n\lambda^2$	$n\lambda^2$	$\nu^{\frac{1}{3}}$
	II [16]	ν	ν/n	$n\nu^{-\frac{2}{3}}$	$\nu^{\frac{1}{3}}$
	III [9]	ν	ν/n	$n\nu^{-\frac{2}{3}}$	$\nu^{\frac{1}{3}}$
2	IV [16]	$n^{\frac{1}{2}}/D^2$	$1/n^{\frac{1}{2}}D^2$	$n^{\frac{1}{2}}D$	$1/D$
	V [13]	n/D	$1/D$	~ 1	$1/D$
	VI	$1/D^3$	$1/nd^3$	nD^2	$1/D$
1	VII [16], [17]	$Sn^{\frac{1}{2}}$	$S/n^{\frac{1}{2}}$	$n^{\frac{1}{2}}/S^{\frac{1}{2}}$	$1/S^{\frac{1}{2}}$
	VIII [18]	$\mathfrak{D}nS$	$\mathfrak{D}S$	$1/\mathfrak{D}S^{\frac{1}{2}}$	$1/S^{\frac{1}{2}}$
	IX [18]	$n\delta S$	δS	$1/\delta S^{\frac{1}{2}}$	$1/S^{\frac{1}{2}}$

ν = volume density of defects
n = vortex density
λ = penetration depth
D = grain size
$\mathfrak{D}$ = dislocation — vortex interaction distance
S = dislocation density
δ = vortex lattice deformation = $P_0 n^{\frac{1}{2}}/\overline{C}$
$\overline{C}$ = vortex lattice elastic constant (Eq. (3.3.22))
P_0 = pinning force

Table 3.3.2. *Magnitude of pinning energy and force*

Dimension	Defect	$U_0 = \varepsilon_i(J)$	$F_p = \varepsilon_i(N)$
3	normal cavity	$\varepsilon_0 = (H_c^2 b^3/6)\cdot 4\pi/\mu_0$	$\tau_0 = \varepsilon_0/b$
	normal cavity	$\varepsilon_1 = [(B - H_{c2})^2 b^3/12\varkappa^2]\cdot 4\pi/\mu_0$	$\tau_1 = \varepsilon_1/b$
	normal cavity	$\varepsilon_2 = [(\varphi_0 H_{c1}/4\pi \ln\varkappa)\, b \ln b/\xi^*]\cdot 4\pi/\mu_0$	$\tau_2 = \varepsilon_2/b$
	inhomogeneous cavity	$\varepsilon_3 = (2\,\Delta\lambda_L/\lambda_L)\,\varepsilon_2 + (\varepsilon_2/\ln b/\xi^*)\,(\Delta\lambda_L/\lambda_L - \Delta\varkappa/\varkappa)$	$\tau_3 = \varepsilon_3/b$
2	grain boundary[a]	$\varepsilon_4 = [\varphi_0 M D]\cdot(1/\mu_0)$	$\tau_4 = \varepsilon_4/\lambda$
	grain boundary	$\varepsilon_5 = \{\varepsilon_0, \varepsilon_1\}\, D^2/b$	$\tau_5 = \varepsilon_6/b$
	grain boundary thickness b	$\varepsilon_6 = \{\varepsilon_2, \varepsilon_3\}$	$\tau_6 = \varepsilon_5/b$
1	dislocation[b]	$\varepsilon_7 = \tau_7\,\mathfrak{D}^{\mathrm{b}}$	$\tau_7 = 10^{-13}\,(H_{c2}(T)/\varkappa H_c(0))^{\frac{3}{2}}\cdot(1 - B/H_{c2})$

[a] $M = [(H_{c2} - B)/8\pi\varkappa^2]\cdot 4\pi\mu_0$ [b] $\mathfrak{D}$ = dislocation interaction distance

This decomposition of J into two parts is designed to distinguish between magnetization currents, J_M and transport currents J_T. Magnetization currents are defined such that $\int_\Sigma J_M \cdot dS = 0$, where $\sum$ is any surface that passes completely through the specimen. That J_M satisfies this condition identically follows from the application of Stokes theorem and the observation that M vanishes outside the specimen. Furthermore, from the above definition of J_M, M can be shown to be given by the usual relation

$$\int_V M\, dV = (\mu_0/2)\,(r \cdot J_M)\, dV\,, \tag{3.3.5d}$$

where V is the volume of the specimen. Hence M is the magnetic moment per unit volume.

From the above equations we are led to the following equations

$$\operatorname{curl} B = \mu_0 J = \mu_0 (J_M + J_T)\,, \tag{3.3.6a}$$

$$\operatorname{curl} H = J_T\,, \tag{3.3.6b}$$

where $H = (B - M)/\mu_0$ is the internal field, which unlike B has its source only in the transport current. These equations plus $\nabla \cdot B = 0$ are then sufficient to completely specify the problem.

From the above derivation it should be clear that unique distinctions between transport and magnetization currents can only be made when the entire specimen is included or when the magnetization in the bulk is uniform. Neither condition applies to Eq. (3.3.4) since we are necessarily restricted to a local region, and the magnetization is not uniform where there is a gradient in B.

We can however write Eq. (3.3.4) without ambiguity as

$$F_y = -B \cdot \frac{dH}{dB} \cdot \frac{dB}{dy} = -B \cdot \frac{dH}{dB} \cdot \mu_0 \cdot J$$

or in vector notation

$$\boldsymbol{F} = \mu_0 \cdot (\boldsymbol{J} \times \boldsymbol{B}) \cdot \frac{dH}{dB}. \tag{3.3.7a}$$

In Eq. (3.3.7), H (B) is the local field external to the specimen in equilibrium with the local flux density B (y) and is to be calculated from the equilibrium theory at each position. The quantity J is then the total current density in the specimen. An alternative approach is to use Eq. (3.3.6b) and write

$$\boldsymbol{F} = \boldsymbol{J_T} \cdot \boldsymbol{B}\,, \tag{3.3.7b}$$

where J_T is the transport current. Eqs. (3.3.7a) and (3.3.7b) are equivalent forms for the Lorentz force.

We can finally rewrite Eqs. (3.3.7a) or (3.3.7b)

$$\boldsymbol{F_L} = \frac{\boldsymbol{F}}{n} = (\boldsymbol{J} \cdot \boldsymbol{\varphi_0}) \left(\frac{dH}{dB}\right) = \boldsymbol{J_T} \cdot \boldsymbol{\varphi_0} \tag{3.3.8}$$

where $\boldsymbol{F_L}$ is the force per unit length of vortex and $\boldsymbol{\varphi_0}$ is a vector in the direction of the vortex.

The different forms for Eqs. (3.3.7) and (3.3.8) arise from the difference between the total current J and the transport current J_T, which is usually considered to be the current inserted through leads. However, it is difficult to define J_T when the current is induced in the specimen through a change in the magnetic field. In these circumstances it is best to define J by Eq. (3.3.5b) and utilize Eq. (3.3.7a) since it then involves only the measured quantity B [66]. It should be noted that for $B > H_{c1}$ the quantity dH/dB is quite close to μ_0. In the present chapter, for simplicity we will use as the Lorentz force per unit volume F or per unit length F_L the relationships

$$\boldsymbol{F} = \boldsymbol{J} \cdot \boldsymbol{B}\,; \quad \boldsymbol{F}_L = \boldsymbol{J} \cdot \varphi_0\,, \tag{3.3.9}$$

where J is taken to be the local transport current, or total current depending on the experimental circumstances.

3.3.3 Flux Flow

The existence of the Lorentz force on a vortex line in the presence of a transport current suggests motion; and we will next consider the consequences of this motion. Under the action of a force, in the steady state, one expects a limiting velocity of the vortices v_L where $F_L = \eta \cdot v_L$ and η is a viscosity coefficient for the moving vortices [60]. Given the above force we can also derive an energy dissipation per unit volume $\mathfrak{E}$, given by $\mathfrak{E} = nF_Lv_L = JB \cdot v_L$. This result, derived simply from energetics suggests the existence of an electric field $\boldsymbol{E}$ of the form $E = B \cdot v_L$ where $|E| = n\varphi_0 v_L$. It is interesting to note the resemblance between the form for $\boldsymbol{E}$ and that derived from simple considerations of the effect of flux sweeping through the specimen. However careful analysis of the sources of the field $\boldsymbol{E}$ shows that the situation is considerably more complicated.

Whatever the source of $\boldsymbol{E}$, it is clear that in a direct measurement of the critical current, this field will manifest itself in a voltage whenever $B < H_{c1}$. Furthermore, if we consider a multiple connected ring. The electric field will lead to a decay of the trapped flux whenever the field exceeds H_{c1}.

For a Type II superconductor carrying a current in a field greater than H_{c1} we can thus define an effective resistance $\varrho_F = dE/dJ = (\varphi_0 B/\eta)$. The classic experiments of Kim et al. [60] on flux flow resistance shows further that $\varrho_F/\varrho_0 \approx B/H_{c2}$, where ϱ_0 is the normal resistance of the metal, and this relation has been qualitatively verified by microscopic theory (Fig. 3.3.2).

Physically it can be shown that the resistance, and the electric field, are derived from the presence of current in the normal cores of a moving vortex, and the result of Kim et al. [60] can be obtained from this assumption. If we consider a slab with a density n of cores or radius ξ^*, we obtain for the total power dissipation per unit volume $\mathfrak{E}$ in the slab with normal resistance ϱ_0

$$\mathfrak{E} = n\,\xi^{*2}J^2\varrho_0 = \frac{B}{\varphi_0}\,\xi^{*2}J^2\varrho_0\,. \tag{3.3.10a}$$

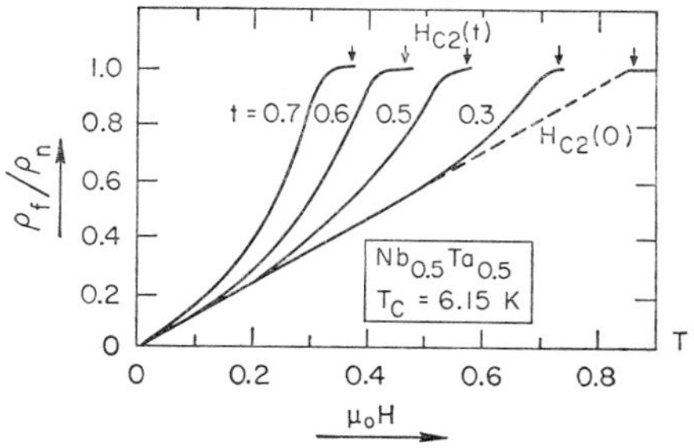

Fig. 3.3.2. Flux flow resistance in $Nb_{0.5}Ta_{0.5}$ after Kim, Hempstead and Strnad [60]

If we equate this to the previous expression we obtain

$$v_L = \frac{1}{\varphi_0} \cdot \varrho_0 \cdot J\, \xi^{*2}. \tag{3.3.10b}$$

and hence from $|E| = n\varphi_0 v_L$ we obtain

$$E \approx \frac{B}{H_{c2}} \varrho_0 J \tag{3.3.10c}$$

where we have used the relation $H_{c2} = \varphi_0/\xi^{*2}$. The above expression leads to $\varrho_F = \varrho_0(B/H_{c2})$ and suggests that the GLAG critical field and not the paramagnetically limited critical field determines the flux flow resistance [60] if we assume ξ^* is independent of paramagnetic effects. This assumption is probably valid for $B \ll H_{c2}$. Although the above argument emphasizes the physical source of the flux flow resistance, the exact theory is considerably more involved. In addition to the flux flow resistance the moving flux vortex exhibits a Hall voltage, and since it carries entropy, exhibits a variety of thermo-magnetic behaviour [67]. The theory of these effects, particularly in clean material (intrinsic Type II) is only in semi-quantitative agreement with experiment [68].

3.3.4 Thermally Activated Flux Creep

Given the theoretical conclusion and ample experimental observations that a moving vortex generates a resistance, the vortex structure has to be pinned against the Lorentz force for a Type II superconductor to support a transport current. The first, and most general approach to understanding the current carrying mixed state was that of Anderson and Kim [69] who introduced the four key concepts of the present theory: the Lorentz force, flux pinning defects, flux bundles and thermally activated flux creep.

The Anderson-Kim theory was the first attempt to obtain a microscopic model of the source of the critical current capacity of defect loaded Type II superconductors, as well as its field and temperature dependence. The model was successful semi-quantitatively explaining the initial experiments, and continues to be the structure on which later

Table 3.3.3. *Magnitude of pinning energies and forces*[a]

Dimension	Defect	$U_0(T=0)$ 10^{-12} ergs	$U_0(T=0)$ K	F_p (10^{-6} dynes)	Comments
3	$\varepsilon_0 =$	4	2.9×10^4	$\tau_0 = 4$	$b = 100$ Å
	$\varepsilon_1 =$	1	7.2×10^3	$\tau_1 = 1$	$b = 100$ Å
	$\varepsilon_2 =$	1.6	1.2×10^4	$\tau_2 = 1.6$	$b = 100$ Å
	$\varepsilon_3 =$	3.2	230	$\tau_3 = 3.2 \times 10^{-2}$	$\Delta\lambda_L/\lambda_L \sim 1\%$; $\Delta\varkappa/\varkappa = 0$
2	$\varepsilon_4 =$	3.9	2.8×10^4	$\tau_4 = 0.2$	$D = 500$ Å
	$\varepsilon_5 =$	$100 \to 25$	$(72 \to 18) \cdot 10^4$	$\tau_5 = 100 \to 25$	$b = 100$ Å
	$\varepsilon_6 =$	$8 \to 0.15$	$(6 \to 0.1) \cdot 10^4$	$\tau_6 = 8 \to 0.15$	$b = 100$ Å
1	$\varepsilon_7 =$	5×10^{-3}	36	$\tau_7 = 0.5 \times 10^{-2}$	$\mathfrak{D} = 100$ Å

[a] See text for source of parameters

Table 3.3.4. *Magnitude of* $\boldsymbol{\alpha} = JH$[a]

Dimension	Defect[b]	Distribution[c]	α (N/M^3)	$\lvert\mu\rvert$[a] (10^5 T · A/cm²)
3	ε_0	I	$\tau_0 \nu^{\frac{1}{3}}/\lambda^2$	10
$\nu = 10^{15}/\mathrm{cm}^3$		II	$\tau_0 \nu$	40
$b = 100$ Å	ε_3	I	$\tau_3 \nu^{\frac{1}{3}}/\lambda^2$	0.08
$\Delta\lambda_L/\lambda_L \sim 1\%$		II	$\tau_3 \nu$	0.32
2	ε_4	IV	$(B/\varphi_0)^{\frac{1}{2}} (\varphi_0 \lvert M\rvert/\lambda D)^{\mathrm{d}} (1/\mu_0)$	60
$D = 500$ Å	ε_5	VI	$\{\varepsilon_0, \varepsilon_1\}/b^3 D$	$320 \to 80$
	ε_6	IV	$\{\varepsilon_2, \varepsilon_3\} (B/\varphi_0)^{\frac{1}{2}} (1/b^2 D)$	$240 \to 45$
1	ε_7	VII	$\tau_7 S (B/\varphi_0)^{\frac{1}{2}}$	4
$S = 10^{11}/\mathrm{cm}^2$		VIII	$\tau_7 \mathfrak{D} (B/\varphi_0)^{\frac{1}{2}} S$	3
$\mathfrak{D} = 100$ Å		IX	$B \ll H_{c2} (B/\varphi_0)^{\frac{3}{2}} \tau_7^2 S \cdot [9\varkappa/B^3 H_{c2})^{\frac{1}{2}}) \cdot \mu_0$	0.5×10^{-3}
			$B \approx H_{c2}$ $0.8\,\mu_0/(1 - H/H_{c2})(BH_{c2})]$	

[a] See text for source of parameters
[b] See Table 3.3.2
[c] See Table 3.3.1
[d] $\lvert M\rvert = \{(H_{c2} - B)/8\pi\varkappa^2\} \cdot 4\pi\mu_0$

theoretical work is based. Unfortunately the theory is over endowed with constants and there have been only few experiments which relate the model to material properties. Hopefully this lack will be overcome.

The following sections consider the Anderson-Kim theory in detail, and show its relation to seemingly distinct formulations. Finally we will discuss the field and temperature dependence of the critical current predicted from the theory, and their agreement with experiment. As will be seen (Section 3.3.5), the agreement is sufficiently good as to suggest that the present model will be the starting point for any discussion of the critical current.

3.3.4.1 Thermally Activated Flux Creep

The Kim-Anderson model introduces vortex diffusion to account for the temperature, field and time dependence of the critical current. It assumes the existence of flux pinning sites which can be represented as potential wells. The basic process leading to flux motion; and hence observable voltages, is the thermal activation of bundles of vortices pinned

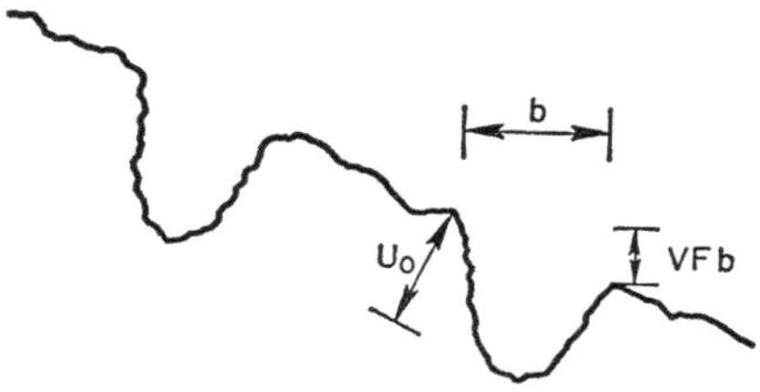

Fig. 3.3.3. Schematic of pinning defects in the presence of Lorentz force

in potential wells and acted upon by the Lorentz force due to the applied field and current (Fig. 3.3.3). The equation for the flux flow rate is a familiar one for any thermally activated diffusion process:

$$R = \omega \exp\left[\frac{1}{kT}(-U_0 + NF_L l b)\right]. \qquad (3.3.11\text{a})$$

In Eq. (3.3.11a), R is the vortex flow rate; ω is a characteristic vibrational frequency of a flux bundle; U_0 is a mean pinning energy; N is the number of vortices in a bundle; $F_L = \varphi_0 J$ (Eq. (3.3.9)); l is the average distance between pinning sites; and b is the mean width of the potential well. Eq. (3.3.11a) can be written as:

$$R = \omega \exp\left[\frac{1}{kT}(-U_0 + VFb)\right]. \qquad (3.3.11\text{b})$$

where $F = BJ$ (Eq. (3.3.9)) and $V = Nl/n$ is an effective volume for a pinned flux bundle (l/n is the volume of one pinned vortex between pinning sites). In both Eqs. (3.3.11a) and (3.3.11b) the second term in the exponential represents the decrease in the height of the potential well due to the Lorentz force acting over the distance, b.

From Eqs. (3.3.11), the Anderson-Kim model arrives at a critical current through the assumption of a critical creep rate $R_c (\ll \omega)$ defined by the voltage sensitivity of a particular experiment. Flux creep is present at all times for finite temperatures and it is only the exponential form of Eq. (3.3.11) that provides a definite value of the critical current. In this model the "critical state" of the material is one in which the current density has its limiting value (defined by R_c at every point where the field has penetrated. If the current density is exceeded at any point in the material due to a fluctuation, resistance occurs as flux flows, and the current density relaxes back to its critical value with a time dependence characteristic of Eq. (3.3.11). Such time dependent effects will be discussed in a later section. It should be emphasized that this model presents a definition of the critical state in terms of R_c that supports the magnetization theory of Bean and London [70].

In terms of the critical flow rate R_c, we obtain from Eqs. (3.3.11).

$$J_c \varphi_0 = \frac{U_0}{N\,l b}\left[1 - \frac{kT}{U_0} \ln\left(\frac{\omega}{R_c}\right)\right] \tag{3.3.12a}$$

and

$$J_c B = \frac{U_0}{V b}\left[1 - \frac{kT}{U_0} \ln\left(\frac{\omega}{R_c}\right)\right]. \tag{3.3.12b}$$

In the following discussion we will consider sufficiently low temperatures so that we can neglect the second term of Eq. (3.3.12). Later sections will cover the explicit temperature dependence of this term.

Eq. (3.3.12) can be written in a variety of ways, which add little to the model, but which emphasize aspects of either the impurity distribution or the nature of the pinning site. Thus we may write at low temperatures:

$$J_c \varphi_0 = \frac{U_0}{N\,l b} = \frac{P_0}{N l} = P_0 \varrho_l = \frac{P_0 \varrho}{n}. \tag{3.3.13a}$$

where $P_0 = U_0/b$ is a pinning force per defect
$\varrho_l = 1/Nl$ is a linear density of defects per pinned vortex
$\varrho = 1/V = n/Nl$ is a volume density of pinned vortex bundles.

Similarly Eq. (3.3.12b) can be written

$$J_c B = P_0 \varrho . \tag{3.3.13b}$$

The first task of the Kim-Anderson model was the relation of the right hand side of Eq. (3.3.13) to material parameters and to vortex-vortex interactions. Before we proceed, it may be illuminating to discuss a seemingly alternative model for flux pinning due to Silcox and Rollins [71]. These authors considered the force exerted upon a vortex when pinning prevents it from shifting with the rest of the flux lattice under the action of the Lorentz force. The force can be derived from the free energy of the equilibrium lattice, and has a particularly simple form in the limit of widely spaced vortices where only nearest neighbors are considered ($B \geqq H_{c1}$). After a fair amount of algebra, and several approximations, Silcox and Rollins [71] arrive at the relation:

$$\varphi_0 J_c = \frac{2}{3} \varrho_l P_0 , \tag{3.3.14}$$

which is almost numerically equal to Eq. (3.3.13). This agreement is hardly surprising since the forces between widely spaced vortices are again given by the Lorentz force. It does emphasize that the left hand side of Eq. (3.3.13) is determined by general arguments, and detailed calculations of forces are not fruitful. The field dependence of the critical current is not entirely determined by the force law, but by the interactions of the vortex bundle with the pinning center.

3.3.4.2 Calculation of the Effective Density of Vortex Pinning Sites

The original experiments of Kim et al. [72] showed that the critical current was fit best by an expression of the form $J_c \approx \alpha/(B + B_0)$, which reduced at high fields to the form $J_c B = \alpha$, where α is a materials constant. As can be seen from Eq. (3.3.13b), this result implies that the product $P_0 \varrho$ is independent of field at high fields. Silcox and Rollins arrived at the same result by assuming that the vortex lines intersect every pinning site, and hence $\varrho_l n = \nu$ is the volume density of defects. This assumption and the observation that $\nu = 1/l^3$ leads to a bundle size N that is field independent and given by $N = n l^2$, i.e. the flux bundle involved in the activation process contains the number of vortices between defects. Anderson and Kim [69] assume, however, that $N \approx n \lambda^2$, since only vortices within a penetration depth should be coupled together. Both theories imply $J_c B = \alpha$, but in the first case $\alpha \sim \nu$, and in the second $\alpha \sim \nu^{\frac{1}{3}}$. Although data on this point are limited, for the case of point defects introduced by neutron damage, the defect dependence is closer to linear. It is possible that there may be a transition from $\alpha \sim \nu^{\frac{1}{3}}$ for low defect concentration to $\alpha \sim \nu$ for high defect concentrations, but this has yet to be observed or even looked for.

Point defects, such as the amorphous "black spots" induced by neutron damage in Nb_3Sn [73] and Nb [74], or small cavities or normal precipitates, are three dimensional in character. Other kinds of defects — grain boundaries, the cell-like dislocation structure of heavily cold-worked materials, or large precipitates — are two dimensional in character since the pinning sites are distributed over a surface rather than localized and isolated in a volume. Indeed, the surface of the specimen is itself a two dimensional defect due to the role of image forces limiting the entrance and exit of vortices from the bulk of the specimen.

We next examine the defect distribution and bundle size for such two dimensional systems. From the previous discussion one might expect that $N \sim n D^2$ where D is the diameter of the grain cell or precipitate and furthermore set $l = D$. From Eq. (3.3.13)

$$J_c B = \frac{U_0}{D^3 b}. \tag{3.3.15}$$

However, since U_0 is confined to a surface one might further write $U_0 = u_0 D^2 b$ where u_0 is a pinning energy density. Thus

$$J_c B \approx \frac{U_0}{D}. \tag{3.3.16}$$

This is the simplest approach to obtain both an inverse D and an inverse B dependence for the critical current density. Gifkins et al. [13], on the other hand assume that N, the bundle size, exhibits no field dependence and that $\varrho_l \approx 1/D$ and derive from Eq. (3.3.13)

$$J_c \varphi_0 = \frac{U_0}{b\,D}. \tag{3.3.17}$$

There is only limited data available where both grain size and field dependence of J_c is determined. Neither Gifkins et al. [75] nor Love and Koch [76] determined the field dependence of the critical current, however, Gifkins et al. [75] observed an inverse grain size dependence of J_c while Love and Koch [76] found that J_c varied as the inverse two-thirds power of the grain size. Hanak and Enstrom [77] obtained fair agreement with an inverse grain size dependence but there was a suggestion of saturation of the critical current at small values of D.

Campbell et al. [78] arrive at the same grain size dependence as Eqs. (3.3.16) and (3.3.17), but the analysis is considerably different and they emerged with a unique field dependence. These authors consider as a model for a grain, a cube of dimension D. The pinning force per unit volume is the number of vortices per unit length of grain, $D n^{\frac{1}{2}}$, times the pinning force P_0 per defect (the grain boundary), divided by the volume of the grain, $1/D^3$. The pinning force is decomposed further into a pinning force per unit length of defect, P_L, times the length of the boundary D. Thus from Eq. (3.3.13).

$$J_c B \approx \left(\frac{B}{\varphi_0}\right)^{\frac{1}{2}} \cdot \frac{P_L}{D}. \tag{3.3.18}$$

An inverse grain size dependence of J_c is again obtained, but the field dependence apart from P_L is intermediate between that of Eqs. (3.3.16) and (3.3.17). This result is identical to that of Yasukochi et al. [79], and both imply a bundle size N given by $N \approx D n^{\frac{1}{2}}$. The grain size dependence of Eq. (3.3.18) is in excellent agreement with the experiments of Campbell et al. [78]. However, in order to account for the field dependence these authors invoked a specific field dependent P_L. This aspect of the problem will be discussed in the succeeding section.

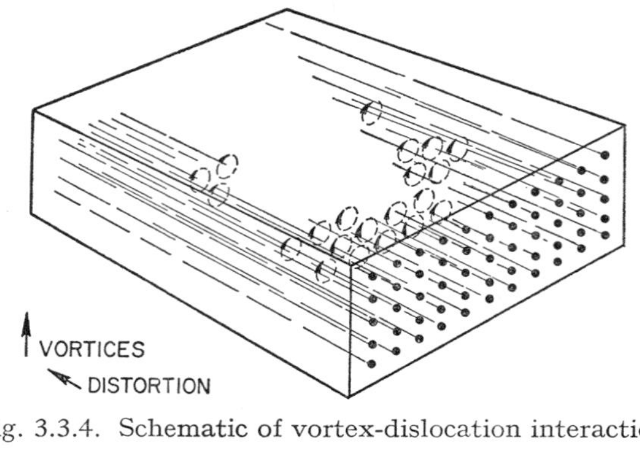

Fig. 3.3.4. Schematic of vortex-dislocation interaction

The final geometry to consider is a one dimensional one given by an array of dislocations at right angles to the vortex lattice (Fig. 3.3.4). A simple heuristic argument by Campbell et al. [78] gives both the field and dislocation density dependence where the dislocations are at right angles to the field. We consider a dislocation density S, where the dislocations have an effective width d. The length of line defect per unit volume is S and hence the number of vortex interactions per unit volume is S times the number of vortices per unit length, $n^{\frac{1}{2}}$. The total pinning force per unit volume is then given by $n^{\frac{1}{2}} S$ times the pinning force per defect $P_L d$. Hence from Eq. (3.3.13b)

$$J_c B = P_L \cdot d \left(\frac{B}{\varphi_0}\right)^{\frac{1}{2}} S \,. \tag{3.3.19}$$

This result implies a bundle size given by $N \approx n^{\frac{1}{2}}/S^{\frac{1}{2}}$. Eq. (3.3.19) has received no experimental justification. In the following paragraph we present a simple argument due to Fietz and Webb [80] which leads to a considerably different form then that of Eq. (3.3.19), but starts with the same geometrical structure of dislocations and vortices.

Fietz and Webb introduce a distance $\mathfrak{D}$ over which the interaction with a dislocation line can occur. They argue that the total number of vortex interactions per unit volume ϱ is given by the length of dislocation per unit volume S, times the distance $\mathfrak{D}$, times the vortex density n. If P_0 is the force per defect we have from Eq. (3.3.13)

$$J_c \varphi_0 = P_0 S \, \mathfrak{D} \tag{3.3.20}$$

and $N \approx 1/\mathfrak{D}\, S^{\frac{1}{2}}$. The results of Campbell et al. are only recovered if we permit $\mathfrak{D}$ to vary as $n^{-\frac{1}{2}}$, and hence permit the bundle size to increase with applied field. An alternative explanation of the difference between Eqs. (3.3.20) and (3.3.19) is that in the first case, the vortices are permitted to adjust their positions to the dislocation structure. In the Fietz and Webb treatment the vortex structure at this stage in their argument (it is later modified) is rigid and controls the geometry of the interaction. Clearly neither approach can be chosen in the absence of a detailed study of the interactions responsible for the bundle size N.

Up to now we have presented a variety of simple arguments for the field and defect density dependence of the critical current. The "theories" have a heuristic value, but in general have an arbitrary character that is unsatisfactory. Clearly, when the field and energy dependence of the interaction energy or force is considered, even more varied behaviour than $J_c \approx B^n, (n = 0, -1/2, -1)$ may be anticipated. Hence the general agreement with a particular field dependence of J_c found in many of the papers may be fortuitous and may not indicate a true test of our understanding of the flux pinning problem. From the different expressions for the flux bundle (N) found in each of the above models, it appears that a starting point for a microscopic theory has to be based on the elastic properties of the vortex lattice itself [81].

A considerably deeper analysis of the pinning problem has been made in the work of Fietz and Webb [80] following the work of Labusch [81]. They consider the case of dislocations and note that if the flux lattice were completely rigid there would be no pinning, for a random array of defects, since the gain in energy for a vortex entering a pinning site would be compensated by the loss of energy upon leaving. Indeed this observation has been implicitly assumed in all the above calculations since the flexibility of the flux lattice is chiefly responsible for the field dependence of N.

Webb and Fietz include lattice distortion by modifying Eq. (3.3.20) in the following manner. If we consider a vortex pinned by a force P_0, a measure of the distortion of the vortex lattice is given by the deformation δ ($\delta n^{\frac{1}{2}}$ is the strain), where $\delta = P_0 n^{\frac{1}{2}}/\overline{C}$ where $\overline{C}$ is a mean elastic constant of the two dimensional vortex lattice. A measure of the vortex lattice distortion is thus $\delta/\mathfrak{D}$, and Fietz and Webb [80] consider the first order contribution of lattice distortion to be Eq. (3.3.20) corrected by $\delta/\mathfrak{D}$. Thus

$$J_c \varphi_0 = \frac{P_0^2 n^{\frac{1}{2}} S}{\overline{C}} \tag{3.3.21}$$

and the number of vortices in a bundle is given by $N = l/\delta S^{\frac{1}{2}} = \overline{C}/P_0 n^{\frac{1}{2}} S^{\frac{1}{2}}$.

To proceed further, these authors utilize the calculations of Labusch [81] for the two dimensional vortex lattice where for $B \ll H_{c2}$

$$\frac{1}{\overline{C}} \approx \frac{5\,\pi^{\frac{1}{2}} \varkappa}{\left(\frac{H}{B}\right)^{\frac{1}{2}} \left(\frac{B}{H_{c2}}\right)^{\frac{3}{2}} \cdot H_{c2}^2} \cong \frac{1}{n^{\frac{3}{2}}}, \tag{3.3.22 a}$$

for $B \ll H_{c2}$

$$\frac{1}{\overline{C}} \approx \frac{0.76}{\left(\frac{B}{H_{c2}}\right)\left(1 - \frac{B}{H_{c2}}\right) H_{c2}^2} \cong \frac{1}{n\left(1 - \frac{B}{H_{c2}}\right)}. \tag{3.3.22 b}$$

Thus from Eq. (3.3.21), for $B \ll H_{c2}$

$$J_c \cong \frac{P_0^2}{B} S \tag{3.3.23 a}$$

and $N \cong \frac{n}{P_0 S^{\frac{1}{2}}}$. For $B \approx H_{c2}$

$$J_c \cong \frac{P_0^2 S}{B^{\frac{1}{2}}\left(1 - \frac{B}{H_{c2}}\right)} \tag{3.3.23 b}$$

and

$$N \cong n^{\frac{1}{2}} \cdot \frac{1 - \frac{B}{H_{c2}}}{P_0 S^{\frac{1}{2}}}.$$

The work of Fietz and Webb [80] is noteworthy for several reasons. First it is the first attempt to incorporate the elastic properties of the vortex lattice in the pinning problem, and although the treatment is only approximate, it is unique compared to previous work. Second, the result shows that pinning is the same for attractive as well as repulsive interactions. A similar conclusions was reached independently by Yamafugi and Irie [82] from an analysis of the a.c. behaviour of flux pinning, and is implicit in the work of Rosenblum and Gittleman [81]. Finally although it is necessary to explicitly invoke the field and temperature dependence of P_0 to account for the experimental data, the final agreement is good and suggests that the approach is correct.

From the present review of the theory of defect distributions it should be clear that it can be expected to apply only for defects of one kind. A complete theory has to include random and possibly overlapping defect distributions, interaction between defects, anisotropy, the elastic properties of the vortex lattice, and finally both the range and strength of interaction between defects and the vortex bundle. Such a task is indeed formidable, but is required if research on materials is to be directed toward controlled incorporation of defects to enhance the performance of superconducting materials in magnet applications.

Table 3.3.1 summarizes the results of the present section. The following section considers the nature of several proposed pinning sites.

3.3.4.3 Nature of Vortex Pinning Sites

In the preceding section we assumed the existence of vortex pinning sites, and found that published work on the effective distribution of these sites looked rigor. The present section examines these pinning sites in detail. The arguments of this section tend to be less definitive than the preceding one and are largely heuristic. Surprisingly as will be seen in Section 3.3.5 these estimates of pinning energies when combined with the distribution of Section 3.3.4.2 are in reasonable agreement with experiment.

The first estimate of the pinning energy U_0 (In what follows we will write $U_0 = \varepsilon_i$, where the ε_i are various sources of pinning energy. Furthermore we will write $P_0 = \tau_i$ for the equivalent pinning forces) was made by Anderson [69]. The reasonable assumption was made that this energy had to be some fraction of the maximum free energy difference between the normal and superconducting state:

$$\varepsilon_0 = \frac{H_c^2 V}{2\mu_0}, \tag{3.3.24}$$

where H_c is the thermodynamic critical field and V is the volume of the pinning defect. A better estimate would include the enhancement of the free energy of the mixed state due to flux penetration, which for high $\varkappa$, high T_c material (to avoid paramagnetic contributions to the normal state)

$$\varepsilon = \frac{1}{4\mu_0 \varkappa^2} \cdot V (B - H_{c2})^2, \tag{3.3.25}$$

which for low fields reduces to the previous expression. Reasonable values for V for three dimensional defects would be regions small compared to vortex separation, but large compared to the size of the core (10^{-18} — 10^{-15} cm^3).

Friedel et al. [84] calculated the free energy difference between a vortex in the mixed state and one centered on a normal defect. If we consider a spherical defect whose radius b is larger than the radius of the core (ξ^*), this free energy difference can be calculated in a manner similar to that of H_{c1}. For a spherical hole of radius b we obtain

$$\varepsilon_2 = \frac{\varphi_0}{\mu_0} \frac{H_{c1}}{\ln \varkappa} b \cdot \left\{ \ln \left(\frac{b}{\xi^*} \right) \right\}. \qquad (3.3.26)$$

For a superconducting region of radius b where $(\lambda_L)_b = \lambda_L + \Delta \lambda_L$ and $(\varkappa)_b = \varkappa + \Delta \varkappa$

$$\varepsilon_3 = \frac{2 \Delta \lambda_L}{\lambda_L} \varepsilon_2 + \frac{\varphi_0}{\mu_0} \frac{H_{c1}}{\ln \varkappa} \left(\frac{\Delta \lambda_L}{\lambda_L} - \frac{\Delta \varkappa}{\varkappa} \right). \qquad (3.3.27)$$

The expressions for ε_0, ε_1, ε_2 and ε_3 are such that the pinning force is easily obtained by simple dividing out the radius of the defect b.

For extended two dimensional defects, Campbell et al. [78] calculate the pinning force by equating it with the image force at a normal or superconducting boundary. The original calculations of the image force only considered isolated vortices and were clearly valid only near H_{c1}. Campbell et al. [78] estimate the pinning force per unit length of pinned vortex from thermodynamic considerations and obtain (see Eq. (3.3.18))

$$P_L = \frac{\varphi_0 M}{\mu_0 \lambda} = \frac{\tau_4}{D}, \qquad (3.3.28)$$

where M is the reversible magnetization ($|M| \approx -(B - H_{c2})/2\varkappa^2$) and λ is the penetration depth. The pinning energy is thus given by

$$\varepsilon_4 = \frac{\varphi_0 M P}{\mu_0}, \qquad (3.3.29)$$

since the force varies over a distance λ. Campbell et al. [78] also note in the derivation of τ_4 and ε_3, that a rigid lattice would not be pinned, but account for the existence of the above force in terms of a distortion of the flux lattice within a penetration depth of the defect. Unfortunately they neglect to include such distortion explicitly in the arguments that lead to Eq. (3.3.18). An alternative calculation of the pinning energy for two dimensional defects can be made if it is assumed that a grain boundary is a continuous region of perturbed material of thickness b and diameter D. The quantities ε_0, ε_1, ε_2 and ε_3 suitably scaled, can then be used as estimates for U_0 (see Eq. (3.3.16)).

Fietz and Webb utilize an expression of the pinning force for dislocation interactions derived from a second-order elastic strain perturbation of the condensation energy in the vortex core [85]. The pinning force is given in dynes by

$$\tau_7 \approx 10^{-8} \left(\frac{H_{c2}(T)}{\varkappa H_c(0)} \right)^{\frac{3}{2}} \cdot \left(1 - \frac{B}{H_{c2}} \right) \qquad (3.3.30)$$

and for the pinning energy

$$\varepsilon_7 \approx \mathfrak{D} \tau_7, \qquad (3.3.31)$$

where $\mathfrak{D}$ is the distance over which interactions with a dislocation line can occur.

The calculation of the vortex-dislocation interaction is derived from the small change in the elastic constants that occurs at the superconducting transition. This change in elastic constant then introduces a small change in the elastic energy of the dislocation. This calculation may only be accurate within an order of magnitude, but the field and temperature dependence may still be a good approximation.

Table 3.3.2 summarizes the present discussion of pinning energies and forces and Table 3.3.3 presents the magnitude of the energies and forces for the representative high field superconductor Nb_3Sn. In this table we have used [47, 86—88], $\varkappa(T_c) \approx 20$, $\varkappa(0) \approx 34$, $H_{c2}(0) \approx 24$ T, $H_{c1}(0) \approx 0.1$ T, $H_c(0) \approx 0.5$ T, $\lambda = 2{,}000$ Å, $\xi^* \approx 70$ Å and the calculations are at T $= 0$ K and for $B/H_{c2} \approx 1/2$. In this table we also give the temperature equivalent of each pinning energy U_0. It is apparent that with the possible exception of grain boundaries and dislocations U_0 can be considerably larger than kT, suggesting that the low temperature limit of Eq. (3.3.12) may be an excellent approximation over a wide temperature range. We will return to this point in a later section.

In Table 3.3.4 we combine the results summarized in Tables 3.3.1 through Table 3.3.3 to give an estimate for the product JB $(= \alpha)$ in the limit of low temperatures and again for Nb_3Sn. As a calibration point for the examination of this table we note: that single crystal specimens of Nb_3Sn can have values of α as low as 0.1×10^5 T. A/cm² [87]; that pure polycrystalline specimens of Nb_3Sn have $\alpha \approx 1 \times 10^5$ T. A/cm² [89]; that Nb_3Sn tape doped for solenoid fabrication has $\alpha \approx 100 \times 10^5$ T. A/cm² [77]; and that neutron damage in Nb_3Sn leading to amorphous "spots" of diameter 100 Å with a concentration of 10^{15} cm^{-3}, produces an increase in α from 1×10^5 T. A/cm² to 10×10^5 T. A/cm² [73].

A comparison of the numerical values of α in Table 3.3.4 with the above representative experimental values leads to the surprising, and perhaps disconcerting, result that these crude estimates for the magnitude of the pinning energies are relatively good. Indeed, with the possible exception of dislocations, almost all of the simple heuristic models can be used to reasonably account for the broad range of α found in Nb_3Sn. Similar conclusions would emerge from considerations of Nb-Zr or Nb-Ti. The Kim-Anderson model can easily account for the magnitude of α; a more crucial test is the explicit field and defect dependence of α. The following section examines the experiments that support and were the motivation for the models summarized in Tables 3.3.1 to 3.3.4.

3.3.5 Low Temperature Experimental Results on the Field and Defect Dependence of the Critical Current Density

The following section is not meant to be a complete summary of the literature, but is designed, through selected examples, to illustrate the preceding discussion. With few exceptions, the examples will not be chosen from commercial "magnet" wire or ribbon, since little attempts has been made to define these materials and they often exhibit characteristics that suggests multiple defect structures. Furthermore, the presence of instabilities often obscure the features of interest. However, in some cases, practical materials, notably Nb_3Sn exhibit critical current behaviour that is in excellent agreement with the simple Kim-Anderson model.

3.3.5.1 High Field Materials

The early work of Kim et al. on Nb-Zr and sintered Nb_3Sn [72], showed that low α specimens of these materials were in good agreement with the simple Lorentz law $J_c B = \alpha \approx$ const. These early measurements were accomplished through tube magnetization studies, and their interpretation was unnecessarily complicated. Later direct measurements of Cody and Cullen [89] showed that for low α specimens of chemically transported Nb_3Sn, the agreement with this simple expression was extremely good. Fig. 3.3.5 shows the reciprocal of the critical current as a function of magnetic field over the range 0 to 10.5 T $(\approx 1/2\,(H_{c2}))$ at 4.2 K. The current density is in excellent agreement with the expression:

$$J_c = \left(\frac{1{,}4 \times 10^5}{B + 0.06}\right) \qquad \left(\frac{\text{A}}{\text{cm}^2}\right). \tag{3.3.32}$$

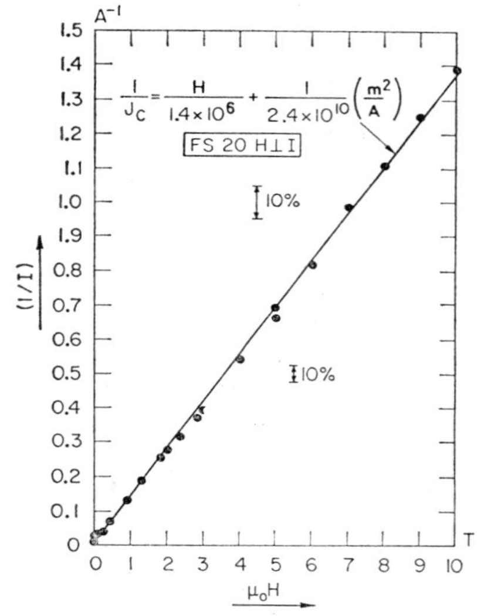

Fig. 3.3.5. Reciprocal values of critical current of Nb_3Sn at 4.2 K as a function of field [89]

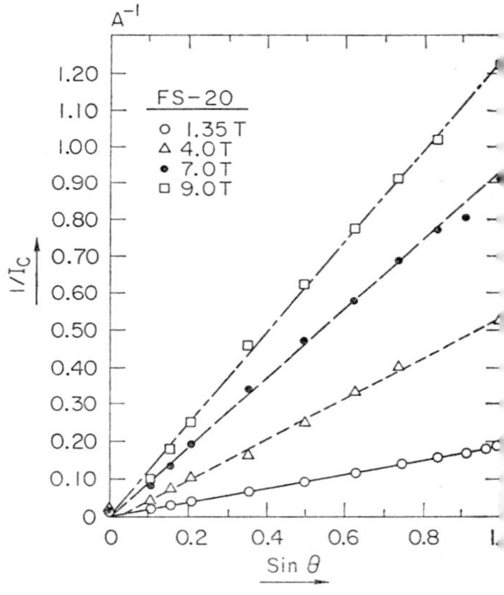

Fig. 3.3.6. Reciprocal values of critical current of Nb_3Sn at 4.2 K as function of field and angle [89]

Fig. 3.3.6 shows that one can generalize the above expression to the form

$$J_c = \left(\frac{1.4 \times 10^5}{B \sin\theta + 0.06}\right), \qquad \left(\frac{\text{A}}{\text{cm}^2}\right), \tag{3.3.33}$$

where θ is the angle between the current axis and the field direction. Again this is just what would be expected on the basis of the simple Lorentz force law $J \cdot B = \alpha \approx$ const, and this data represents the least ambiguous confirmation of Eq. (3.3.13e) with $P_0 \varrho$ independent of field at high fields ($B \gg H_{c1}$). It is interesting to note that the constant in the denominator of Eqs. (3.3.32) and (3.3.33) (denoted by B_0 in the early work of Kim) is very close to H_{c1} (4.2 K), and this observation holds for all law α specimens that exhibit the $1/H$ dependence at high fields. This observation has yet to receive any theoretical justification.

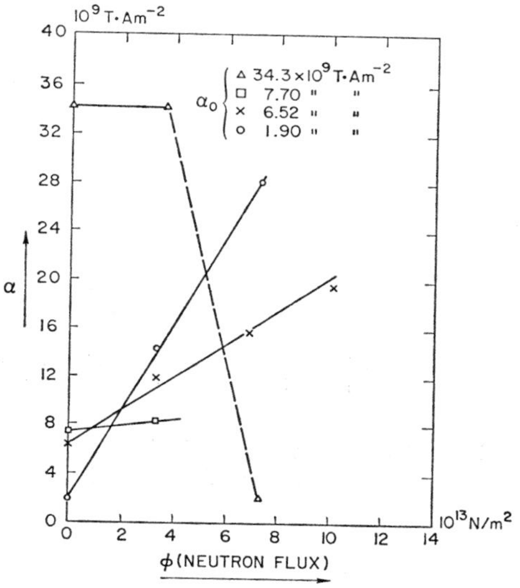

Fig. 3.3.7. The quantity ($\alpha = JB$) as a function of fast neutron irradiation [73]

The critical current density of low α specimens of Nb_3Sn can be increased considerably by fast neutron [73] or proton irradiation [90]. Electron microscopy of the specimens shows that the damage consists of amorphous spots of about 100 Å diameter and at a concentration of about 10^{15}/cm^{-3} for 10^{18} neutrons per cm^2. As noted previously, the increase in α ($\approx 10\times10^5$ T. A/cm^2) is in fair agreement with Table 3.3.4 (ε_0, I, II). Fig. 3.3.7 shows that the initial increase in JB at any field is close to linear with flux, but as shown by Figs. 3.3.8 and 3.3.9 where $1/I$ is plotted against $\mu_0 H$, a characteristic break in the linear variation of $1/I$ with $\mu_0 H$ occurs for higher dosages for both neutron and proton damage. In general α increases above its low field value. For neutron damage there is some evidence to suggest that for high α specimens, neutron irradiation actually causes a decrease in α (Fig. 3.3.7). For proton irradiation, the data shows a maximum in the critical current as a function of flux. Both these effects presumably follow from an interaction between residual and induced defects, but no theoretical explanation has been given. The neutron and proton damage studies involved room temperature irradiation. The research of Coffey et al. [91] with low temperature irradiation shows that there are a variety of annealing stages for deuteron damage between low temperatures and room temperature [91]. The effect of radiation damage on the critical current appears to be a potentially useful tool for the metallurgist as well as the physicist.

In the commercial production of Nb_3Sn wire or tape, by chemical transport impurities are added which can be shown to decrease the grain size. Hanak and Enstrom [77] were able to correlate the critical current at 7 T with the inverse of the grain size, determined by *x*-ray analysis.

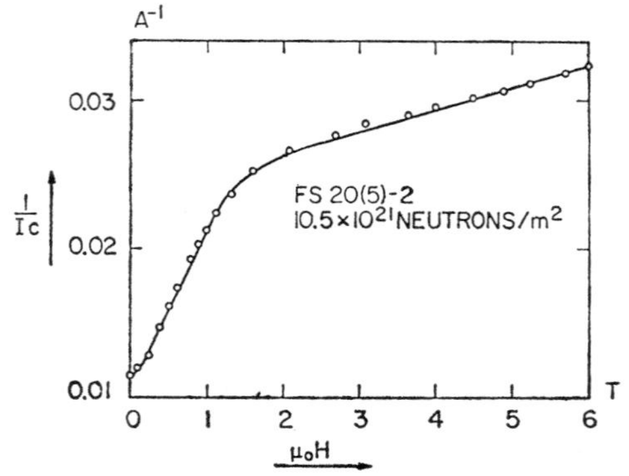

Fig. 3.3.8. The reciprocal values of the critical current as a function of field for Nb_3Sn with high concentration of neutron induced defects [73]

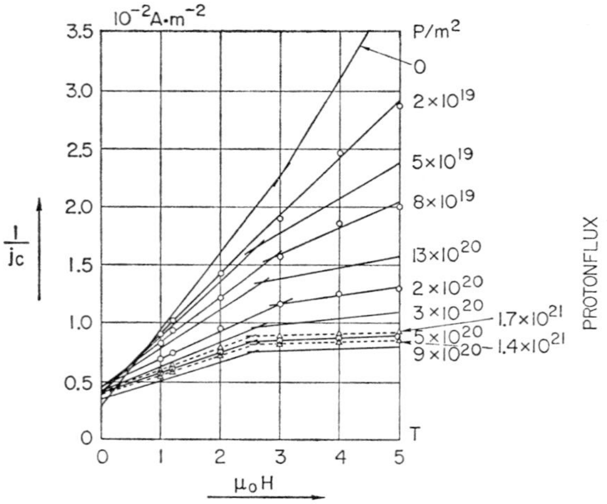

Fig. 3.3.9. Reciprocal of critical current density as a function of field for Nb_3Sn specimens for a variety of proton irradiation levels [90]

Fig. 3.3.10 shows a plot of α ($= JB$) against $1/D$, and the fit to a straight line is good, with evidence for saturation around $D \approx 300$ Å. The equation of the line is given by

$$JB \approx \frac{150}{D} \approx \frac{15}{P} \qquad \left(\frac{\mathrm{T \cdot A}}{\mathrm{cm}^2}\right), \tag{3.3.34}$$

which for $D \sim 500$ Å leads to $JB \sim 30 \times 10^5$ (T · A/cm²), within a factor of two of values estimated in Table 3.3.4 (e.g. ε_4, IV, where $JB \sim 60 \times 10^5$

(T · A/cm²). In this connection it is interesting to note that the formula of Campbell et al. [78] for grain boundary vortex pinning (Eqs. (3.3.18) and (3.3.28)) would predict for high $\varkappa$ materials, a maximum in the product JB at $H_{c2}/3$. There is experimental evidence for a wide variety of chemically transported Nb_3Sn tapes that such a maximum occurs between 7 and 10 T [77].

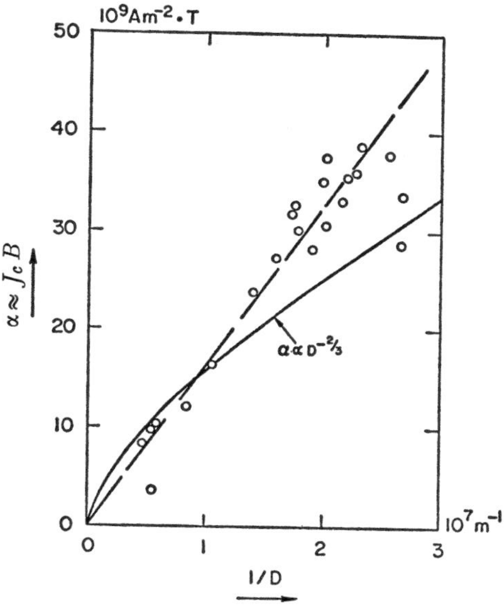

Fig. 3.3.10. The quantity ($\alpha = JB$) as a function of grain size for Nb_3Sn at 4.2 K [77]

Other grain boundary studies have been made on high field materials. Particularly noteworthy is the recent work of Love and Koch [76] which reported a dependence of critical current on grain size in Nb-Zr alloys that goes as $D^{-\frac{2}{3}}$ rather than D^{-1}. If we force their data to fit a D^{-1} dependence we obtain:

$$JB \approx \frac{0.5 \times 10^2}{D}, \qquad \left(\frac{\mathrm{T \cdot A}}{\mathrm{cm}^2}\right). \tag{3.3.35}$$

Substitution of appropriate values for Nb-25% Zr ($\varkappa \approx 28$, $H_c(0) \sim 0.25$ T, $\lambda \approx 2{,}000$ Å, $B = 2$ T) in Eqs. (3.3.18) and (3.3.28) leads to $JB \approx 0.1 \times 10^2/D$, (T · A/cm²) which is only a factor of 5 less than Eq. (3.3.35). However, in connection with Nb-Zr it is worth noting that Fietz et al. [92] in investigations of the critical current of commercial wire were best able to fit their data on the critical current density by an expression of the form

$$J = \left(\frac{a}{b}\right) e^{-\frac{B}{b}} + c\,, \tag{3.3.36}$$

which is not directly derivable from any of the preceding discussion. Presumably the peculiar form of Eq. (3.3.36) results from multiple defects.

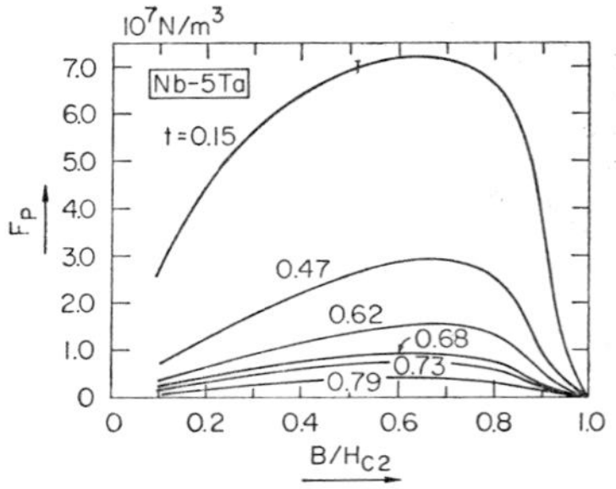

Fig. 3.3.11.
Pinning force for Nb-Ta as a function of field and reduced temperature [80]

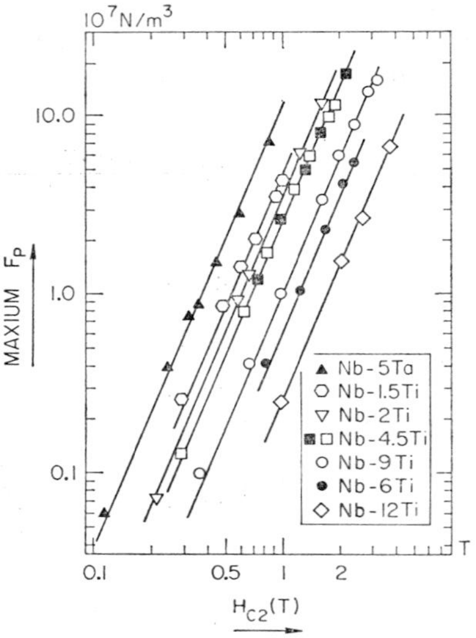

Fig. 3.3.12.
Pinning force for Nb-Ta and Nb-Ti alloys as a function of the critical field H_{c2} [80]

A very definitive study of the effect of dislocations on high field material was made by Fietz and Webb [80] on heavily cold worked Nb-Ta and Nb-Ti alloys. Particularly noteworthy is their result that the quantity JB was of the form

$$JB = g(\varkappa)\, H_{c2}^{\frac{5}{2}} \cdot f\left(\frac{B}{H_{c2}}\right), \tag{3.3.37}$$

where $\varkappa$ is the Ginzburg-Landau parameter and $g(\varkappa) \approx 1/\varkappa^2$. Fig. 3.3.11 shows the scaling with respect to B/H_{c2} for one alloy at a variety of temperatures and Fig. 3.3.12 shows the scaling with respect to $H_{c2}^{5/2}$ on a log-log plot of the maximum JB against $H_{c2}(T)$. Substitution of the relevant expressions from Tables 3.3.1 to 3.3.4 supports the form of Eq. (3.3.37) i.e. at small B

$$JB \approx [H_{c2}(T)]^{\frac{5}{2}} \cdot \left(\frac{B}{H_{c2}}\right)^{\frac{1}{2}} \cdot \frac{1}{\varkappa^2}, \tag{3.3.38}$$

for $B \approx H_{c2}$

$$JB \approx [H_{c2}(T)]^{\frac{5}{2}} \left(\frac{B}{H_{c2}}\right)^{\frac{1}{2}} \left(1 - \frac{B}{H_{c2}}\right) \cdot \frac{1}{\varkappa^3}. \tag{3.3.39}$$

It should be noted that the field dependence of Eq. (3.3.39) is equivalent to that suggested by Campbell et al. for grain boundary pinning, but the low field expression (Eq. (3.3.38)) differs considerably. The field dependence of Eqs. (3.3.38) and (3.3.39) is in fair agreement with that shown in Fig. 3.3.11 and the agreement with the temperature dependence is remarkable. Substitution of the pinning force and values for $\varkappa$ and H_{c2} for the Nb-5% Ta alloy at $t = 0.15$ leads, at low fields, to

$$JB \approx 0.6 \times 10^2 \left(\frac{B}{H_{c2}}\right)^{\frac{1}{2}}, \quad \left(\frac{\mathrm{T \cdot A}}{\mathrm{cm}^2}\right). \tag{3.3.40}$$

The curve shown in Fig. 3.3.11 varies approximately as

$$JH \approx 80 \times 10^2 \left(\frac{B}{H_{c2}}\right)^{\frac{1}{2}}, \quad \left(\frac{\mathrm{T \cdot A}}{\mathrm{cm}^2}\right). \tag{3.3.41}$$

There is thus a discrepancy of about two orders in the magnitude of JB. However, the expression for JB is quite sensitive to the pinning force (Eq. (3.3.23)) and the discrepancy may not be surprising in view of the complexity of the vortex-dislocation interaction.

3.3.5.2 Low Field Materials

There appears to be a scarcity of definitive critical current data on low field alloys. This may arise from the difficulty of incorporating defects in these usually soft materials, or possibly because of the limited field range available between H_{c1} and H_{c2}. Indeed materials such as Nb_3Sn and V_3Si are ideally suited for comparison with theory since they can be obtained in single crystal form and because of their high H_{c2} and T_c. However the low temperature structural transformation of some specimens of Nb_3Sn and V_3Si presents a complication, and has been shown for V_3Si, to effect the critical current [93]. As noted, Gifkin et al. [75] examined the critical current of recrystallized Pb-Tl alloys and inferred the grain size dependence of the critical current density. Although it is impossible from their published data to infer the magnitude of JB, they did establish that the remanant flux in a magnetization measurement (and hence JB) varied inversely as the grain size.

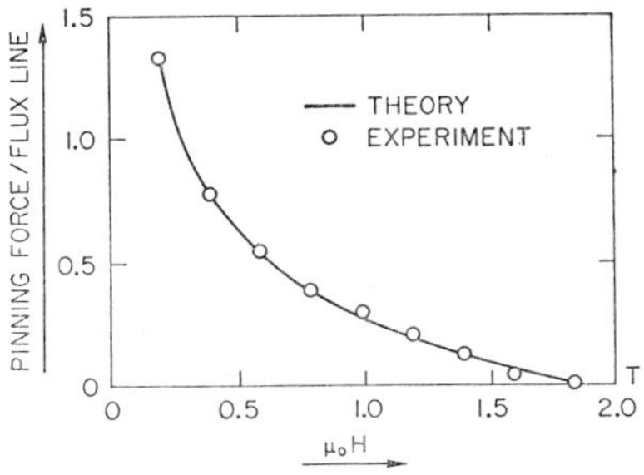

Fig. 3.3.13. Pinning force as a function of field for Pb-Bi alloys from magnetization measurements [78]

The most definitive study of the relation of grain size to pinning was the research of Campbell et al. [78]. These authors examined Pb-Bi alloys and utilized for the internal grain boundaries, the normal Bi precipitates of the epsilon phase. They measured both magnetization and critical current and found excellent agreement with the form of Eq. (3.3.18) after the proper substitution for $P_0(\tau_4$ of Table 3.3.2). The theoretical magnitude of J_c is about an order of magnitude larger than the empirical expression

$$J_c = 33\left(\frac{M}{4\pi}\right)\left(\frac{1}{B^{1/4}D}\right), \qquad \left(\frac{\mathrm{A}}{\mathrm{cm}^2}\right) \tag{3.3.42}$$

found to fit the experimental data. Fig. 3.3.13 shows the critical current obtained from magnetization as a function of B compared with the quantity $M/B^{1/4}$ (Table 3.3.3 (ε_4, IV)), and one notes the excellent agreement. Similar agreement is found for the directly measured critical current compared with the same function in Fig. 3.3.14 except for fields close to H_{c2} where the agreement is less satisfactory. This aspect of the data emphasizes the importance of direct measurement of J_c. Magnetization measurements are convenient because they permit utilization of bulk samples, however, they do represent averages over the sample and can obscure fine details. The fact that the critical current appears to go para-

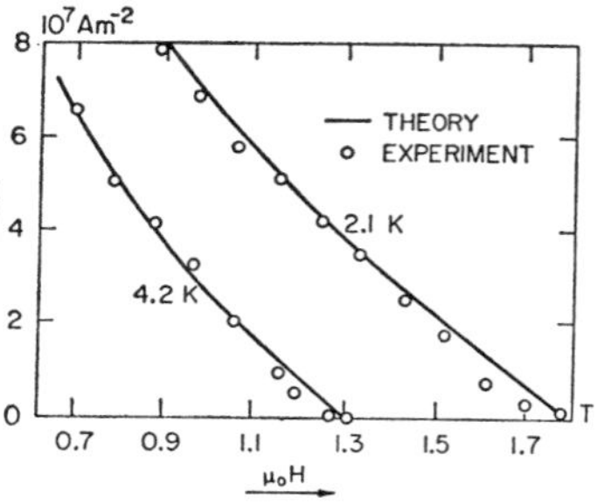

Fig. 3.3.14. Pinning force as a function of field for Pb-Bi alloys from direct critical current measurements [78]

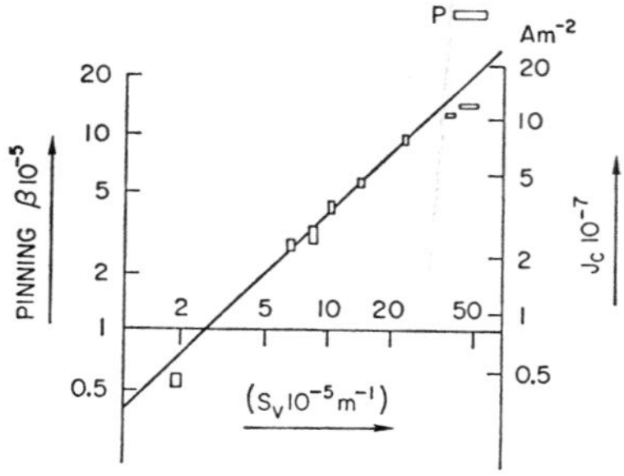

Fig. 3.3.15. Pinning force as a function of inverse grain size for Pb-Bi [78]

bolically to zero in Fig. 3.3.14 is a common feature of many measurements close to H_{c2} and may have a fundamental basis. This point will be discussed in a later section. Finally Fig. 3.3.15 shows the excellent agreement between the critical current and the inverse grain size. It is interesting that the same defect dependence is found for such diverse materials as Pb-Bi, Nb-30% Zr and Nb_3Sn.

3.3.5.3 Other Low Temperature Effects

Before we leave this consideration of the critical current at low temperatures it might be appropriate to consider a variety of phenomena that are common to experimentation at low temperatures, but not necessarily restricted to this temperature range. The first of these is the peak effect, a pronounced peak in the critical current observed close to H_{c2}.

Although this effect was first observed by LeBlanc and Little [94] as early as 1961, it has yet to receive a definitive explanation. Fig. 3.3.16 is an example of the peak effect in NbN in a recent paper by Maxwell et al. [95]. These authors attempt a qualitative explanation of the effect by associating the peak with the repulsion between pinned vortices which raise the potential barrier for a given vortex bundle. The rapid increase

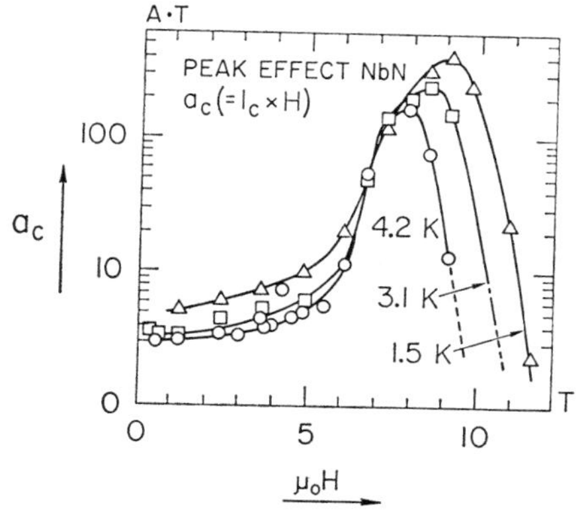

Fig. 3.3.16. Peak effect in NbN [95]

in the repulsive interaction at high fields is thus the basic mechanism for the peak in JB. Although no calculations are presented, this explanation does account for the fact that the peak only occurs for materials with intermediate pinning densities and pinning strengths (it is never observed for Nb_3Sn where the pinning is presumably always large). For low densities of defects the repulsion does not become operative at an individual pinning site, but only keeps the vortices uniformly spaced. For high densities and strong pinning the repulsive interactions presumably make little difference in the net pinning strength. If this observation is correct one might expect the peak effect to show up where dislocations and small variations in parameters are the source of pinning (Table 3.3.3, ε_3 and ε_4). Furthermore as Maxwell et al. [95] remark this explanation "would remove the possibility of improving the current carrying capacity of technically interesting materials by means of the peak effect".

An observation that does not fit into the main pattern of the Lorentz force model is the behaviour of the critical current close to H_{c2}. For a wide variety of materials J_c appears to approach zero not linearly but as

$$J_c \sim (B - H_{c2})^2 \,. \tag{3.3.43}$$

This observation is often obscured by the common display of the data on log-log plots, but is supported by a variety of experiments when the data is displayed in a non-logarithmic plot close to H_{c2} [96, 99]. A particularly noteworthy example is the work of Montgomery and Sampson on Nb_3Sn [88], where above a certain field, all the samples approach zero parabolically (Fig. 3.3.17).

An explanation of this phenomena, which is of practical importance in extrapolating toward H_{c2}, has yet to be made. However a simple model can supply the dominant feature of Eq. (3.3.43), namely J_c approaches zero at H_{c2}, with a vanishing slope. This model is based on the idea that the persistence of pinning toward high fields does not in itself

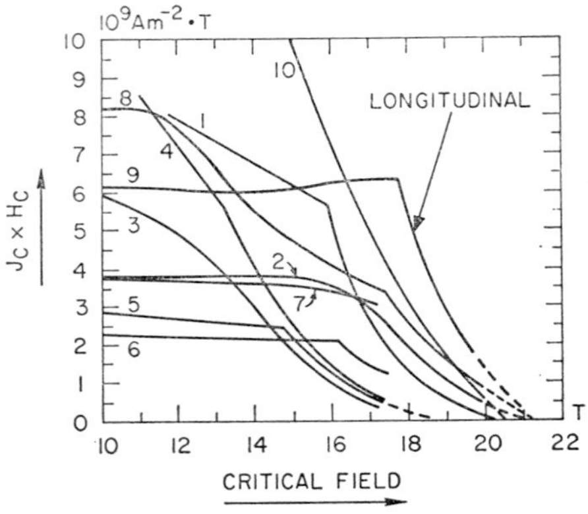

Fig. 3.3.17. The quantity ($\alpha = JB$) for Nb_3Sn at fields close to H_{c2} [88]

guarantee that the superconductor can support the current density. As is well known for Type I superconductors, there is a maximum current density, for which depairing occurs, and above which the material goes normal. For Type I material the current density can be shown to be given by

$$J_c = \frac{H_c(T)}{\mu_0 \lambda_L}. \tag{3.3.44}$$

An argument similar to that used in deriving Eq. (3.3.44) can be used to derive the equivalent expression for Type II materials.

From the GLAG theory, close to H_{c2}, the free energy density difference between normal and superconducting regions Δf is given by

$$\Delta f \approx (H_c^2/2\,\mu_0)\,(1 - B/H_{c2})^2. \tag{3.3.45}$$

The kinetic energy density E can always be written in terms of the current density J by the following relation

$$E = \tfrac{1}{2}\, n m v^2 = \left(\frac{\mu_0}{2}\right) \cdot \left(\frac{\mu_0}{n}\right) \cdot (J^2 \lambda_L^2)\,. \tag{3.3.46}$$

In Eq. (3.3.46) n is the concentration, m is the mass and v is the velocity of the superconducting electrons at the field B (n_0 is the concentration at zero field), and $\lambda_L = (m/\mu_0 n e^2)^{\frac{1}{2}}$ is the London parameter. From the GLAG theory $n = n_0(1 - B/H_{c2})$ and thus if we equate E to Δf we obtain

$$J_c \approx \frac{H_c(T)}{\mu_0 \lambda_L}\left(1 - \frac{B}{H_{c2}}\right)^{\frac{3}{2}}. \tag{3.3.47 a}$$

Eq. (3.3.47a), while not identical to Eq. (3.3.43) has similar behaviour close to H_{c2}, since J_c approaches zero at H_{c2} with zero slope. I would require careful measurements to distinguish between the square and three halves power of the exponent.

Clearly if pinning were such that the current density exceeded the value given by Eq. (3.3.7), the material would go normal and follow Eq. (3.3.47a) rather than Eq. (3.3.13). The data of Fig. 3.3.17 strongly suggest that such a mechanism is operative in Nb_3Sn, although it might be argued that despite the "break" in the curves, the parabolic drop in J_c is related to some pinning energy falling to zero at H_{c2}. However, for longitudinal orientation of the specimen, where there is no Lorentz force on the current density, the above mechanism would also serve as a cut-off on the current close to H_{c2}, and this is what is observed (Fig. 3.3.17).

Finally, it is worth remarking that Eq. (3.3.47) is almost identical to an expression derived for the current density of the superconducting sheath between H_{c2} and H_{c3} [98]. Such surface currents may also be significant in low $\varkappa$ Type II materials well below H_{c2}.

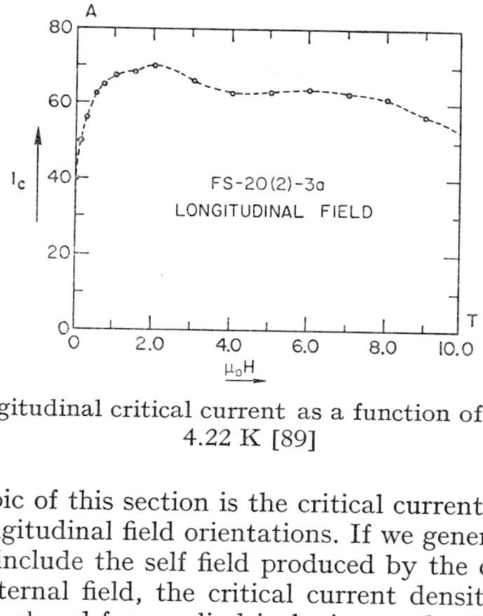

Fig. 3.3.18. Longitudinal critical current as a function of field for Nb_3Sn at 4.22 K [89]

The final topic of this section is the critical current for zero external field and for longitudinal field orientations. If we generalize the Lorentz force model to include the self field produced by the current it is clear that in zero external field, the critical current density should be proportional to the $\varkappa^{\frac{1}{2}}$ and for a cylindrical wire, to the three-halves power of the radius. This result, first observed in Nb_3Sn in 1964 [89] is amply confirmed in the recent work of Campbell et al. [78]. Furthermore one expects that the critical current in a finite longitudinal field should be field independent if the current can achieve a force-free configuration. Fig. 3.3.18 shows that this field independence is maintained in Nb_3Sn up to fields as high as $H_{c2}/2$. The maximum in the current as B increases from zero is undoubtedly associated with the redistribution of the current to permit Lorentz force-free flow. It is worth remarking on the sensitivity of the current to small misalignment. As shown in Fig. 3.3.19, a shift of 2° can drop the current to one half of its maximum value. The enhancement of the critical current for longitudinal orientations has yet to receive any practical application.

The past two sections have described the enormous generality of the Kim-Anderson model in the limit $T \to 0$. In general the agreement with field dependence and magnitude is surprisingly good. However, the large

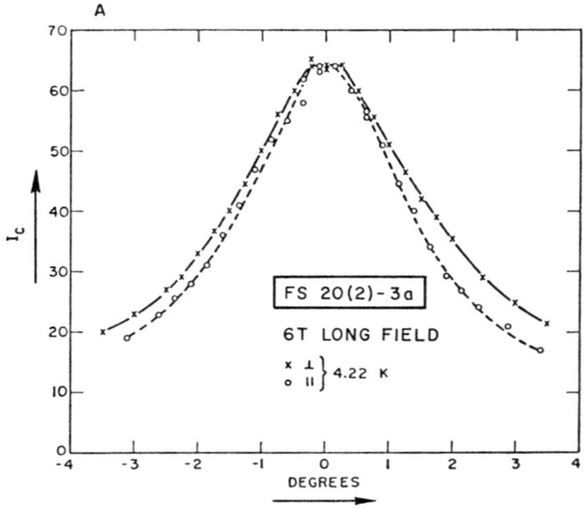

Fig. 3.3.19. Critical current for Nb_3Sn at 6 T as a function of angular deviation from longitudinal orientation [90]

number of undetermined parameters, as well as the lack of detailed measurements on well defined materials, has prevented a definitive test of the model. As will be seen in the next section, it is in the time and temperature dependence of the critical current that the model has received its chief justification.

3.3.6 Kim Anderson Theory at Finite Temperatures

Up to now we have only considered the theory of thermal activated flux creep at low temperatures, where the thermal activation represented by the last term in Eq. (3.3.12) can be neglected. The following discussion examines the consequences for the current density when this term is appreciable.

From Table 3.3.2 it is easy to see that the temperature dependence of the various ε_i which are field independent (ε_0, ε_2, ε_3, ε_5, ε_6) varies as $(1 - t^2)_n$ where $t = T/T_c$ and $n = 2$ or 1. For field dependent ε_i (ε_1, ε_6, ε_6) at constant B/H_{c2}, the temperature dependence is as $(1 - t^2)_n$ where $n = 1$, 2 or 1.5. For simplicity we will consider $n = 1$, and consider only the field independent ε_i. Under these circumstances

$$J B = \alpha(t) = C_1 (1 - t^2) - C_2(t) , \qquad (3.3.47\text{b})$$

where $C_1 = \alpha(0)$ and $C_2 = \dfrac{k T_c \alpha(0)}{\varepsilon_i(0)} \ln\left(\dfrac{\omega}{RC}\right)$. At low temperatures $\alpha(t)$

approaches α (0) linearly with t, a variation considerably different from the temperature dependence of the equilibrium thermodynamic quantities. At low temperatures

$$\frac{d\alpha(t)}{dt} = T_c \frac{d\alpha(t)}{dT} = -\alpha(0)\frac{kT_c}{\varepsilon_i(0)}\ln\left(\frac{\omega}{RC}\right), \quad (3.3.48)$$

but $\alpha(0) = \left(\varepsilon_i(0)/b\right)\cdot\varrho$ and hence

$$\frac{d\alpha(t)}{dt} = -\frac{k\cdot T_c P}{b}\ln\left(\frac{\omega}{RC}\right). \quad (3.3.49)$$

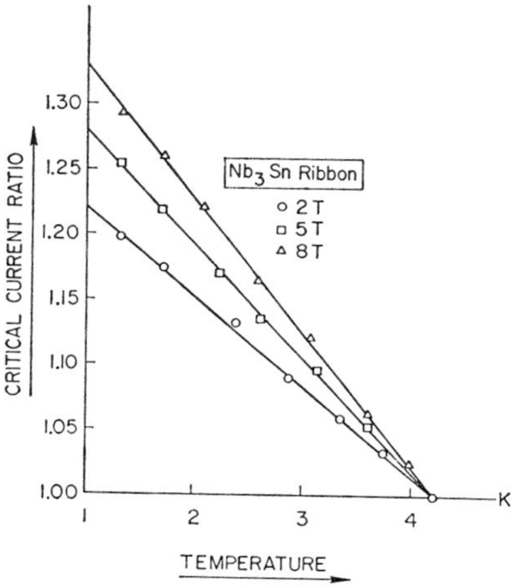

Fig. 3.3.20. Critical current as a function of temperature for Nb_3Sn ribbon [99]

The low temperature slope of α is thus independent of the size of the potential wells, and only depends on the density of defects and the quantity b. For a variety of Nb_3Sn specimens $T_c \frac{d\alpha}{dT} \approx -(1 \rightarrow 2)\,\alpha(0)$ (e.g. Fig. 3.2.20), which implies that for all these materials the potential well is the same. This result combined with the estimate $R_c \approx 10^{-3}$ (Eq. 3.3.10 for $E \approx 1\ \mu V/cm$, $b \approx 100$ Å and $\omega \approx 10^{12} \rightarrow 10^8$) implies that $kT_c/\varepsilon_i(0) \approx 4\times10^{-2}$ or $\varepsilon_i(0)/k \approx 500$ K.

Figs. 3.3.21 and 3.3.22 show $\alpha(T)$ as a function of temperature for samples [100, 101] with $\alpha \approx 10\times10^5$ $(T\cdot A/cm^2)$ and $\alpha \approx 0.1\times10^5$ $(T\cdot A/cm^2)$. In Fig. 3.3.22 the experimental points are fitted to Eq. (3.3.47b) and the agreement is quite good up to about 10 K. Above this temperature $\alpha(t)$ approaches zero at T_c (18 K) with diminishing slope. Presumably above 10 K, a transition is made to an alternative mode of pinning. The small value of $\varepsilon_i(0)$ derived from this data [101] for chemically deposited Nb_3Sn would narrow the pinning potential from Table 3.3.3 to quantities of the magnitude of ε_3, and the previously described grain size dependence would require two dimensional defects (distribution 3.3.4). However, while this result, shows the value of the temperature dependence of α in limiting the choices among pinning potentials it is hardly convincing. It does emphasize, however, that if the Kim-Anderson model is correct, a high T_c is no guarantee that high values of α will be maintained at high temperatures.

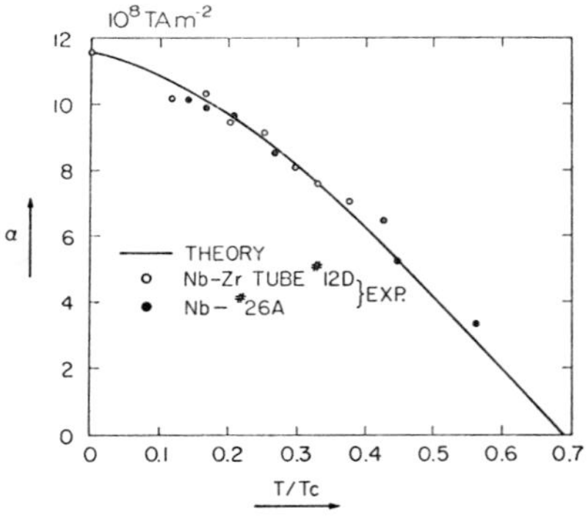

Fig. 3.3.21

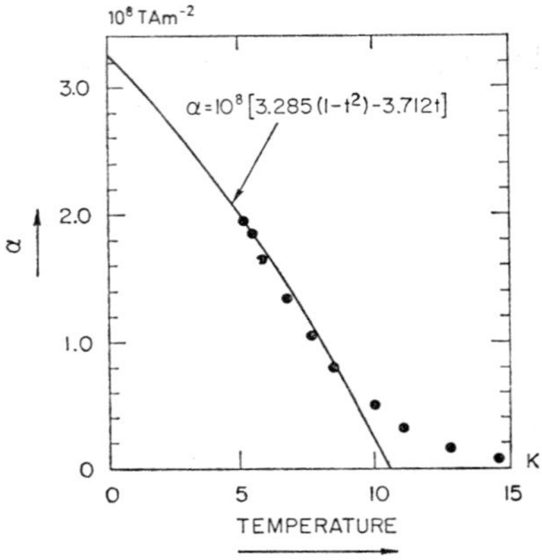

Fig. 3.3.22

Fig. 3.3.21. The quantity ($\alpha = JB$) derived from tube magnetization for Nb and Nb-Zr as a function of reduced temperature [69]

Fig. 3.3.22. The quantity ($\alpha = JB$) for Nb_3Sn crystalline samples as a function of temperature [101]

The most graphic verification of the thermally activated flux creep model is in the measurements on flux creep by Kim et al. [102], and recently by Beasily, Labusch and Webb [103, 104]. In these experiments we consider a perturbation of the equilibrium critical state and ask how the system, which undergoes an enhanced creep rate, returns to equilibrium. This approach explores another aspect of the Kim-Anderson theory, and when carefully done on well defined materials supplies excellent data on the parameters of the theory.

For simplicity we consider a slab with the current density independent of field. The expression for the creep rate is

$$R = \omega \exp\left(-\frac{U}{kT}\right) = \omega \cdot \exp\left[-\frac{1}{kT}\left(U_0 - J \cdot \varphi \cdot a \cdot b\right)\right]. \quad (3.3.50)$$

From the geometry of Fig. 3.3.23, if w is the distance between pinned vortices, L is a length in the x direction, and T is the width of the slab in the y direction

$$\frac{\partial N}{\partial t} = -\frac{L\,dy}{w}\frac{\partial R}{\partial y}, \quad (3.3.51)$$

where N is the number of vortices between Ly and $L(y + dy)$. Thus

$$\frac{\partial B}{\partial t} = -\frac{\varphi_0}{w} \cdot \frac{\partial R}{\partial y} \quad (3.3.52)$$

and from Maxwell's equations (ignoring the y dependence of w)

$$\frac{\partial J}{\partial t} = -\frac{\varphi_0}{\mu_0 w} \cdot \frac{\partial^2 R}{\partial y^2}, \tag{3.3.53}$$

or

$$\frac{\partial J}{\partial t} = -K_0 \frac{\partial^2 R}{\partial y^2} e^{K_1 J}, \tag{3.3.54}$$

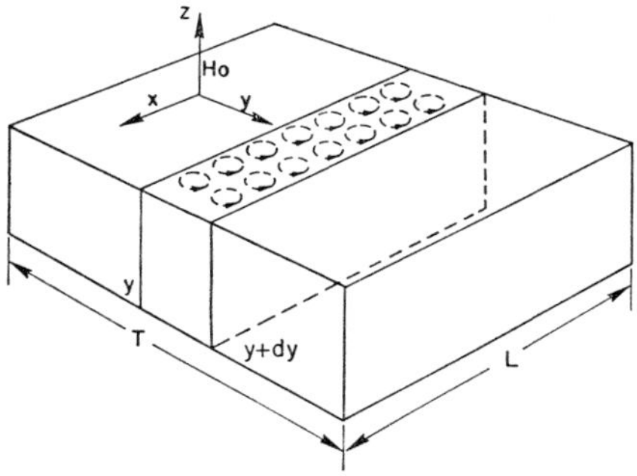

Fig. 3.3.23. Schematic of geometry for derivation of equations for flux creep

where $K_1 = (\varphi_0 b/kT)$ and $K_0 = (\omega e^{-U_0/kT})(\varphi_0/\mu_0 w)$. We can integrate Eq. (3.3.54) if we let $e^{K_1 J} = g(t) \cdot f(y)$ and obtain

$$K_1 J(y,t) = \ln \frac{C_2 y^2 + C_1 y + C_0}{2 K_1 K_0 C_2 t}, \tag{3.3.55}$$

where C_2, C_1 and C_0 are numerical constants. Furthermore if $\Phi(t)$ is the flux in the slab at time t

$$\frac{\partial \Phi}{\partial t} = \frac{\varphi_0 \cdot L}{w} (R(0,t) - R(T,t)) \tag{3.3.56}$$

but

$$R(y,t) = \omega e^{-U_0/kt} \cdot e^{K_1 J(y,t)}. \tag{3.3.57}$$

Thus

$$\frac{\partial \Phi}{\partial t} = \frac{C_3 kt}{\varphi_0 b} \frac{1}{t} \tag{3.3.58}$$

and

$$\frac{\partial \Phi}{\partial \ln t} = C_3 \frac{kT}{U_0} J_c, \tag{3.3.59}$$

where C_3 is a constant related to the thickness (T) of the slab. From Eq. (3.3.59) we see that the flux in the slab, if once perturbed relaxes back to its equilibrium value with a characteristic logarithmic time dependence. Fig. 3.3.24 shows the time dependence of the flux in the experiments of Beaseley et al. [103], and the agreement with the logarithmic time dependence is striking. Since Eq. (3.3.59) is first order in (kT/U_0) it is clear that this and the earlier experiments are direct confirmation of the $1/kT$ temperature dependence of Eq. (3.3.12).

The exact solution of Beaseley et al. [103], where the field dependence of J_c and U_0 is accounted for, as well as magnetic hysteresis, is for cylindrical geometry (ϱ = radius)

$$\frac{\partial \Phi}{\partial \ln t} = \frac{\pi}{3} \varrho^3 \mu_0 \cdot \frac{kT}{U_0} J_c (1 \pm \delta) , \qquad (3.3.60)$$

$$\delta = \frac{1}{2} \cdot \mu_0 \cdot \varrho \left(\frac{1}{2} J_c \frac{\partial \ln U_0}{\partial B} \frac{-\partial J_c}{\partial B} \right) . \qquad (3.3.62\,a)$$

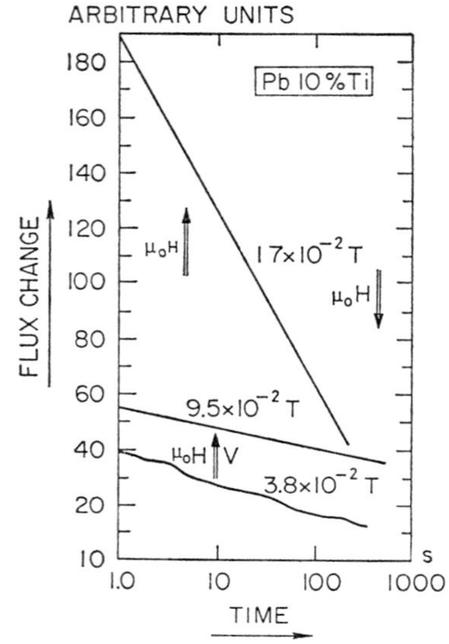

Fig. 3.3.24. Flux creep as a function of the logarithm of the time [103]

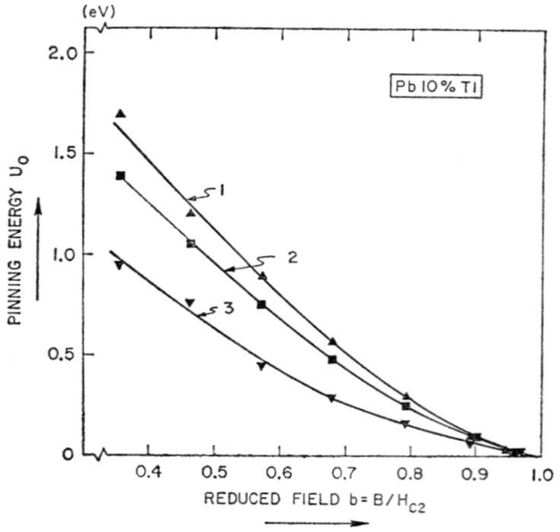

Fig. 3.3.25.
Pinning energy for Pb-10% Tl with a variety of cold work as a function of applied field [103]

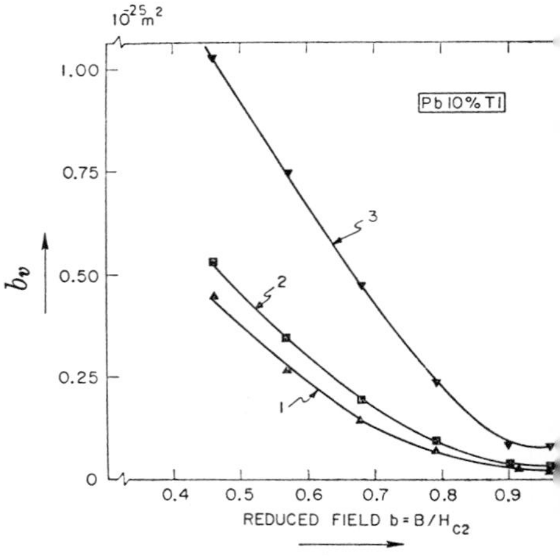

Fig. 3.3.26.
Pinning volume as a function of fiel
for samples of Pb-10% Tl for a variet
of cold working [103]

Beaseley et al. [103], through simultaneous measurements of J_c and creep rate could extract the magnitude, and field dependence of U_0, as well as that of N ($J_c = U_0/Nlb$) and bV ($J_c = U_0/bVn$). Indeed this work probably furnishes the most direct and precise determinations of these quantities in the literature. The specimens were a series of Pb-Tl alloys extruded at nitrogen temperatures.

Fig. 3.3.25 shows the field dependence of U_0 and Fig. 3.3.26 shows the field dependence of bV as a function of increasing cold work for a Pb-10% Tl. The magnitude of U_0, about an electron volt, is well within the magnitude of pinning energies given in Table 3.3.3 (suitably sealed for a material with $\varkappa \sim 2$, $H_c \sim 0.08$ T) with the exception of the pinning energy due to dislocations, where the discrepancy is more than two orders of magnitude. Unfortunately dislocations are the expected mode of pinning for these specimens.

One source of the discrepancy is that many dislocations are overcome simultaneously in depinning, another route is through a cellular structure of dislocations (i.e. grain boundaries). While both mechanisms can account for the value of U_0, it is unfortunate that microscopic examination of the specimens did not permit distinguishing between them. It is clear that while flux creep measurements are difficult, they are perhaps the only technique that permits the determination of the multiplicity of constants that enter into the Kim-Anderson model.

3.3.7 A.C. Effects

A. C. losses in superconductors are separated naturally into two regions $B < H_{c1}$ and $B > H_{c1}$. For $B < H_{c1}$ the losses can be calculated on the basis of the London theory [104] (or Mattis and Bardeen [105]) and the losses are only a function of the normal state conductivity, the fre-

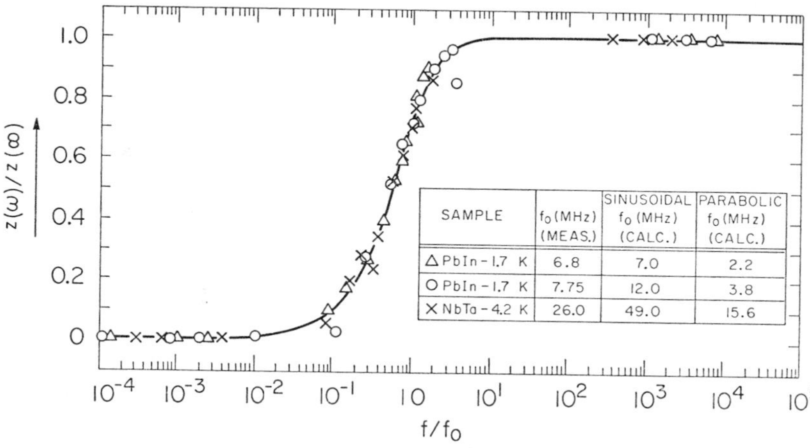

Fig. 3.3.27. Intrinsic impedance of Type II superconductors as a function of normalized frequency [83]

quency and the reduced temperature. For fields above H_{c1} losses can be related to hysteresis loops in the magnetization curves, and a variety of models exist for this case. For losses in the presence of d.c. bias field, the losses can be related to the size of a minor hysteresis loop. In general these losses (with d.c. bias field) can be reduced to quite small values if the a.c. field is made sufficiently small.

However, at sufficiently high frequencies (microwave range) measurements with small microwave fields in the presence of a d.c. field, give a surface resistance that has the full flux flow value ($\varrho_F = \varrho_N B/H_{c2}(0)$). This surprising result was first obtained by Gittleman and Rosenblum [83], and turns out to be an effective way of measuring flux flow without applying contacts or utilizing d.c. currents close to J_c.

The physical explanation of the microwave dissipation is the fact that at sufficiently high frequencies, the equation of motion of the vortices is dominated by the flux flow viscosity, and the pinning potential can be neglected. The theory defines a characteristic frequency

$$\omega_0 = \frac{(100)\,(2\,\pi J B \varrho)}{B^{\frac{1}{2}} \varphi_0^{\frac{1}{2}} H_{c2}}, \tag{3.3.62 b}$$

above which flux flow losses occur (ϱ in Ohm · cm, J in A/cm²). Fig. 3.3.27 shows experimental data on the surface impedance as a function of frequency at $B \approx H_{c2}/2$ for a variety of specimens. The agreement is good.

3.3.8 Conclusions

In this chapter we have introduced a model which, despite its simplicity, can account for the wide variety exhibited by the critical current of Type II superconductors as a function of field and temperature. The model appears capable of accounting for the magnitude of the critical current to better than an order of magnitude when reasonable estimates are made of relevant parameters. Given the complexity of the current carrying state, this is a remarkable achievement. However, if the theory is to ever be used for the optimization of the current capacity in practical materials further experimental and theoretical research is required. It is hoped that this chapter has not only supplied an understanding of the critical current, but has indicated the areas where further research is required. When it is realized that present materials are optimized if at all, only for 4.2 K operation in conventional solenoids the need for such research should be obvious.

References Chapter 3

[1] Bardeen, J., Cooper, L. N., Schrieffer, J. R.: Phys. Rev. **108,** 1175 (1957).
[2] Gorkov, L. P.: Soviet Phys. JETP **9,** 1364 (1959); **10,** 593 (1960); **10,** 998 (1960).
[3] Ginzburg, V. L., Landau, L. D.: J. Expt. Theor. Phys. (USSR) **20,** 1064 (1950).
[4] Abrikosov, A. A.: J. Phys. Chem. of Solids **2,** 199 (1957).

[5] Schoenberg, D.: Superconductivity. Cambridge: Cambridge University Press 1952.
[6] Lynton, E. A.: Superconductivity. London: Methuen 1964.
[7] Kuper, C.: An Introduction to the Theory of Superconductivity. Oxford: Clarendon Press 1968.
[8] Superconductivity, I and II ed. by R. D. Parks. New York: Marcel Dekker 1969.
[9] Olsen, J. L.: Electron Transport in Metals. New York: Interscience 1962.
[10] McMillan, W. L.: Phys. Rev. **167,** 331 (1968).
[11] Ziman, J. M.: Electrons and Phonons, p. 478. Oxford: Oxford University Press 1960.
[12] Douglass, D. H., Falicov, L. M.: Prog. in Low Temperature Physics, IV, p. 97, ed. by C. J. Gorter. Amsterdam: North Holland 1964.
[13] Bardeen, J., Schrieffer, J. R.: Prog. in Low Temperature Physics **3,** p. 170, ed. by C. J. Gorter. Amsterdam: North Holland 1961.
[14] Pippard, A. B.: Proc. Prog. Soc., A **216,** 547 (1953); Faber, E. E., Pippard, A. B.: Proc. Roy. Soc. A **231,** 336 (1955).
[15] De Gennes, R. G.: Superconductivity of Metals and Alloys. New York: Benjamin 1966.
[16] For an excellent summary of BCS and GLAC formulas see: Habe, R.: Phys. Rev. **159,** 356 (1967).
[17] Pippard, A. B.: Proc. Camb. Phil. Soc. **47,** 617 (1951); Phil. Mag. **43,** 273 (1952); Phil. Trans. Roy. Soc. A **248,** 97 (1955).
[18] Doidge, P. R.: Phil. Trans. Roy. Soc. A **248,** 553 (1956).
[19] Peierls, R.: Proc. Roy. Soc. A **155,** 613 (1936).
[20] Guyon, E., Caroli, C., Martinet, A.: J. Phys. Radium **25,** 683 (1964).
[21] Bardeen, J., Schrieffer, J. R.: Prog. in Low Temp. Phys. 3, p. 170, ed. by C. J. Gorter. Amsterdam: North Holland 1961.
[22] Ginzburg, V. L., Landau, L. D.: J. Expt. Theor. Phys. (USSR) **20,** 1064 (1950). An excellent review of the theory is Goodman, B. B.: Rep. on Prog. in Phys. II **29,** 445 (1966).
[23] Goodman, B. B.: IBM J. Research and Development **6,** 631 (1962).
[24] Shubnikov, L. W., Khotkevich, W. I., Shepdev, J. D., Riabinin, J. N.: J. Expt. Theor. Phys. (USSR) **7,** 221 (1937).
[25] De Gennes, P. G.: Rev. Mod. Phys. **36,** 225 (1964); Phys. Kondens. Materie **4,** 79 (1964).
[26] Maki, K.: Physics **1,** 127 (1964); Physics **1,** 21 (1964).
[27] Maki, K.: Phys. Rev. **148,** 362 (1966).
[28] Helfand, E., Werthamer, N. R.: Phys. Rev. **147,** 288 (1966).
[29] Werthamer, N. R., Helfand, E., Hohenberg, P. C.: Phys. Rev. **147,** 195 (1966).
[30] Werthamer, N. R., McMillan, W. L.: Phys. Rev. **158,** 415 (1967).
[31] Hohenberg, P. C., Werthamer, N. R.: Phys. Rev. **152,** 493 (1967).
[32] Eilenberger, G.: Phys. Rev. **153,** 584 (1967).
[33] De Gennes, P. G.: Unpublished Lecture Notes, Herzengnovi Summer School; De Gennes, P. G., Matricon, J.: Rev. Mod. Phys. **36,** 45 (1964).
[34] Faber, T. E.: Proc. Roy. Soc. A **248,** 460 (1958).
[35] Goodman, B. B.: Rep. on Progn. in Phys. **29,** 445 (1966).
[36] Tinkham, M.: Phys. Rev. **129,** 2413 (1963); Rev. Mod. Phys. **36,** 268 (1964).
[37] Cody, G. D., Miller, R. W.: Phys. Rev. Letters **16,** 697 (1966); Phys. Rev. Sept. 1968 to be published.

[38] Chang, G. K., Serin, B.: Phys. Rev. **145,** 274 (1966).
[39] Goodman, B. B.: Rev. Mod. Phys. **36,** 12 (1964).
[40] Berlincourt, T. G.: Rev. Mod. Phys. **36,** 19 (1964).
[41] Cohen, R. W., Cody, G. D., Halloran, J.: Phys. Rev. Letters **19,** 840 (1967).
[42] Clogston, A. M.: Phys. Rev. Letters **9,** 266 (1962).
[43] Chandrasekhar, B. S.: Appl. Phys. Letters **1,** 7 (1962).
[44] Hake, R. R.: Phys. Rev. **158,** 356 (1967).
[45] Radebaugh, R., Keesom, P. H.: Phys. Rev. **149,** 209 (1966); Phys. Rev. **149,** 217 (1966).
[46] Finnemore, D. K., Stromberg, T. F., Swenson, C. A.: Phys. Rev. **149,** 231 (1966).
[47] Vieland, L. J., Wicklund, A. K.: Phys. Rev. **166,** 424 (1968).
[48] Montgomery, L. J., Wizgall, H.: Phys. Letters **23,** 48 (1966).
[49] Hauser, J. J.: Phys. Rev. Letters **13,** 470 (1964).
[50] Kinzler, J. E., Maita, J. P., Levinstein, H. J., Ryder, E. J.: Phys. Rev. **143,** 390 (1966).
[51] Wernick, J. H., Morin, F. J., Hsu, F. S. L., Dorsi, D., Maita, J. P., Kunzler, J. E.: Proc. Conf. High Magnetic Fields, ed. by Kolm, LBX, Bitter, Mills, p. 609. New York: John Wiley 1962.
[52] Morin, F. J., Maita, J. P., Williams, H. J., Sherwood, R. C., Wernick, J. H., Kunzler, J. E.: Phys. Rev. Letters **8,** 275 (1964).
[53] Cooper, J. L.: RCA Review **25,** 405 (1964).
[54] Tilley, D. R.: Proc. Phys. Soc. **84,** 573 (1964).
[55] Moskalenkov, V. A.: Sov. Phys. JETP **24,** 780 (1967).
[56] Eilenberger, G., Amgebaokar, Vinay: Phys. Rev. **158,** 332 (1967).
[57] Lüders, G.: Z. für Physik **202,** 8 (1967).
[58] Fischer, G.: Phys. Rev. Letters **20,** 268 (1968).
[59] Saint-James, D.: Phys. Letters **23,** 177 (1966).
[60] Kim, Y. B., Hempstead, C. F., Strnad, A. R.: Phys. Rev. A **139,** 1163 (1965).
[61] Sauer, E., Wizgall, H.: Proc. Int. Conf. on High Magnetic Fields, p. 223. Grenoble 1966.
[62] Hake, R. R.: Appl. Phys. Letters **10,** 189 (1967).
[63] For further work on Nb see: Webb, G.: Sol. State. Comm. **6,** 33 (1968).
[64] Foner et al.: Phys. Letters A **31,** 349 (1970).
[65] Landau, L. D., Lifschitz, E. M.: Electrodynamics of Continuous Media, p. 113. Oxford: Pergamon 1966.
[66] Dew-Hughes, D.: Materials Science and Engineering **1,** 2 (1966); Evetts, J. E., Campbell, A. M., Dew-Hughes, D.: J. Phys. C **1,** 715 (1968).
[67] Stephen, M. J.: Phys. Rev. Letters **16,** 801 (1966).
[68] Fiory, A. T., Serin, B.: Proc. of Stanford Conf. on Superconductivity. Stanford 1969.
[69] Anderson, P. W., Kim, Y. B.: Rev. Mod. Phys. **36,** 39 (1964); Anderson, P. W.: Phys. Rev. Letters **1,** 309 (1962).
[70] Bean, C. P.: Phys. Rev. Letters **8,** 250 (1962); Rev. Mod. Phys. **47,** 31 (1964); London, H.: Phys. Letters **6,** 162 (1963).
[71] Silcox, J., Rollins, R. W.: Appl. Phys. Letters **2,** 231 (1963).
[72] Kim, Y. B., Hempstead, C. F., Strnad, A. R.: Phys. Rev. **129,** 528 (1963).
[73] Cullen, G. W., Novak, R. W.: J. Appl. Phys. **37,** 3348 (1966); Cullen, G. W.: Proc. 1968 Summer Study on Supercond. Devices, p. 437. Brookhaven 1968.

[74] Tucker, R. P., Ohr, S. M.: Phil. Mag. **17,** 643 (1967); Kernohan, R. H., Sekula, S. T.: J. Appl. Phys. **38,** 4904 (1967).
[75] Gifkins, K. J., Malseed, C., Rochinger, W. A.: Scripta Metallurgia **2,** 141 (1968).
[76] Love, G. R., Koch, C. C.: Appl. Phys. Letters **14,** 250 (1969).
[77] Hanak, J. J., Enstrom, R. E.: Proc. LT 10, II B, p. 10. Moscow 1966.
[78] Campbell, A. M., Evetts, J. E., Dew-Hughes, D.: Phil. Mag. **18,** 313 (1968).
[79] Yasukochi, K., Ogasawara, T., Usui, N., Ushio, S.: J. Phys. Soc. Japan **19,** 1649 (1964).
[80] Fietz, W. A., Webb, W. W.: To be published 1968.
[81] Labusch, R.: Phys. Stat. Sol. **19,** 715 (1967).
[82] Yamafuji, K., Irie, F.: Phys. Letters A **25,** 387 (1967).
[83] Gittleman, J. I., Rosenblum, B.: Phys. Rev. Letters **16,** 734 (1966); J. Appl. Phys. **39,** 2617 (1968).
[84] Friedel, J., De Gennes, P. G., Matrison, J.: Appl. Phys. Letters **2,** 119 (1963).
[85] Webb, W. W.: Phys. Rev. Letters **11,** 191 (1963).
[86] Cody, G. D.: RCA Review **25,** 414 (1964).
[87] Hanak, J. J., Halloran, J. J., Cody, G. D.: Proc. LT 10, II A, p. 373. Moscow 1966.
[88] Montgomery, D. B., Sampson, W.: Appl. Phys. Letters **7,** 108 (1965).
[89] Cody, G. D., Cullen, G. W., McEvoy, J. P.: Rev. Mod. Phys. **36,** 95 (1964); Phys. Rev. **132,** 577 (1963); Cody, G. D., Cullen, G. W.: RCA Review **25,** 466 (1964) and to be published.
[90] Wohlleben, K.: Z. f. angew. Phys. **27,** 92 (1968).
[91] Coffey, H. T., Keller, E. L., Patterson, A., Autler, S. H.: Phys. Rev. **155,** 355 (1967).
[92] Fietz, W. A., Beaseley, M. R., Silcox, J., Webb, W. W.: Phys. Rev. **136,** A 335 (1964).
[93] Brand, R., Webb, W. W.: To be published 1968.
[94] Le Blanc, M. A. R., Little, W. A.: Proc. LT 7, p. 362. Toronto: Univ. of Toronto Press 1961.
[95] Maxwell, E., Schwartz, B. B., Wizgall, H., Hechler, K.: J. Appl. Phys. **39,** 2568 (1968).
[96] Hiaton, J. W., Rose Innes, A. C.: Phys. Letters **9,** 112 (1964).
[97] Neugebauer, C. A., Ekvall, R. A.: J. Appl. Phys. **35,** 547 (1964).
[98] Abrikosov, A. A.: Sov. Phys. JETP **20,** 420 (1965).
[99] Sampson, W. W., Strongin, M., Thomson, G. M.: Appl. Phys. Letters **8,** 191 (1966).
[100] Aaron, P. R., Aklgren, G. W.: Advances in Cryogenic Engineering, Vol. 13, p. 21, and by Timmer Haus, K. D. New York: Plenum Press 1968.
[101] Halloran, J. J., Cody, G. D., Hanak, J. J.: To be published.
[102] Kim, Y. B., Hempstead, C. F., Strnad, A. R.: Phys. Rev. **131,** 2480 (1963).
[103] Beaseley, M. R., Labusch, R., Webb, W. W.: Phys. Rev. **181,** 682 (1969).
[104] Wipf, S. L.: Proc. 1968 Summer Study on Supercond. Devices, p. 511. Brookhaven 1968.
[105] Mattis, D. C., Bardeen, J.: Phys. Rev. **111,** 412 (1958); Miller, P. B.: Phys. Rev. **118,** 928 (1960).

4. Superconducting Alternating Current Magnets

4.1 Alternating Current Losses

4.1.1 Introduction

Speaking of alternating current high field devices for technical applications at the present time may be somewhat overoptimistic, if we consider only technical frequencies such as are usual in the generation and distribution of electrical energy (50 Hz). Due to fundamental problems, discussed later in this chapter, it is presently not feasible to sweep the magnetic field in a Type II superconductor with technical frequencies without producing excessive losses and conductor degradation.

Under technical application, one may consider the generation and conversion of electrical energy (generators and motors), high field pulsed magnets to be used in synchrotrons, inductive energy storage units, and energy transmission.

Superconducting a.c. energy transmission lines (underground cables) imploy either pure niobium tubes or multifilamentary composite conductors of NbTi alloys and copper and cupronickel matrices, or concentric layers of Nb_3Sn interspersed with normal metals such as Nb. As will be shown, self-field losses of these alloys due to transport currents are only weakly dependent on the size of the filaments, and increase slightly with the twist pitch. Thus, Nb_3ZrTi multifilament conductors may be useful in energy transmission cable manufacturing. Nb_3Sn layers are being tested also in underground cables and have exhibited promising results.

In generators, only the excitation coils are being currently wound with NbTi-Cu composites. Flux motion due to load fluctuation are low frequency phenomena and the transient flux linkage between the main and excitation coil is shielded by means of a high conductivity normal metal cylinder around the rotor.

Inductive energy storage coils are being developed for the storage of electrical energy and in fusion reactor type coils of toroidal or spherical shape. The major problem here is the slow and fast effective switching of large quantities of energy from the storage to the load without energy loss.

Pulsed magnets for use in high energy applications, such as synchrotron magnets, are being built by a number of laboratories in the U.S. and Europe. Remarkable progress has been made to reduce losses and make the use of composite conductors feasible in synchrotron and particle storage rings. The progress in the design and testing of these magnets is also directly applicable in electrical machines.

In solenoids, field variations in the order of 7 T/s have been achieved. In beam transport and accelerator magnets which have more complicated coil geometries, field changes of about 6 T/s have been obtained without generating excessive dissipative losses. Depending on the field amplitude, the size of the coil, and the stored field energy, superconducting magnets are pulsed with frequencies of $\leqq 1$ Hz. The main reasons for these low duty cycle operations are the a.c. losses generated in the superconductor and the normal metal matrix, and the degradation which prevents the superconductor to reach d.c. critical current densities.

There is, of course, considerable interest in the use of Type II superconductors, not only for a.c. magnets (which may be used in high energy accelerators), but also for the use of alternating current motors and generators, for the use in energy transmission underground lines, for the generation of auxiliary power (where conventional methods are not practicable), and for storage, transformation, and distribution of energy. Type I superconductors play an insignificant role in technical applications. The critical fields and currents of these materials are too low (only Pb with a critical field of 5.5×10^{-2} T at 4.2 K has been used in cavities for linear accelerators and particle separators). The principal task in using Type II superconductors for a.c. applications is to develop conductors (or materials with low a.c. losses) in which the instability (referred to as fluxjumping) is either absent, or the energy released by a fluxjump is reduced to such a low level that it does not lead to degradation.

A conductor with these properties will reach full d.c. critical current values without theoretically requiring means to stabilize it. In addition, the a.c. losses generated in a conductor, which is pulsed, should be sufficiently small, in order that the size of a refrigerator and the operating cost of the magnet system should be economically feasible.

We distinguish between two kinds of Type II superconductors:

Ideal Type II superconductors which are defect free.

In the mixed state, $H_{c1} \leqq B \leqq H_{c2}$, the bulk superconductor admits magnetic flux. This flux is bundled into fluxoids by means of a corresponding current pattern. The field maxima are found at the "normal" core of the fluxoids. The flux filaments threading the superconductor are free to move about in the superconductor, subject only to their mutual repulsive interaction and to a viscous drag [1]. If a transport current is impressed on the superconductor, it has to flow in addition to these circular currents. The fluxoids, having no preferred places in the ideal superconductor, move under the influence of Lorentz forces, causing dissipation. These superconductors carry only a small non-dissipative current as illustrated in Fig. (4.1.1d) for pure annealed niobium.

Imperfect Type II superconductors (hard superconductors) are able to carry very high currents as seen in Fig. (4.1.1c). The superconducting parameters ($T_c, \lambda, \xi \ldots$) have locally different values.

Diverging from the ideal Type II superconductor, the inhomogenities in a nonideal superconductor (dislocations, precipitations, grain boundaries) (Fig. 4.1.2) lead to a spatial variation of the free energy of a vortex

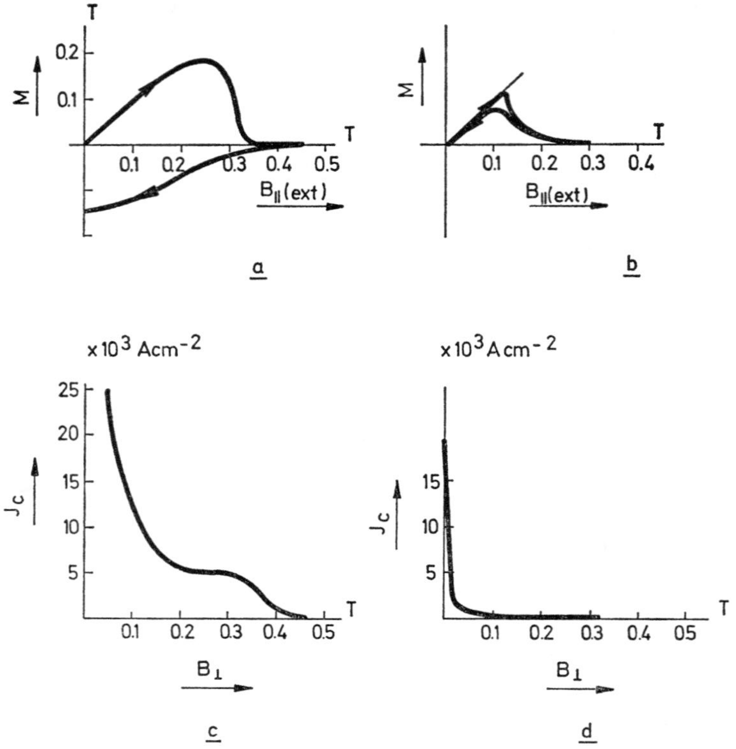

Fig. 4.1.1. Magnetization and critical currents in real (with pinning) and ideal (without pinning) Type II superconductors. *a* and *c*: Magnetization and critical currents of real Type II superconductors. *b* and *d*: Magnetization and critical currents of ideal Type II superconductors

$\mathfrak{F}(\boldsymbol{r})$ (Fig. 4.1.3). In $\mathfrak{F}(\boldsymbol{r})$, the interaction with the neighbouring fluxoids as well as the line tension of a curved fluxline is to be included. A Lorentz force, $\boldsymbol{f}_L$, can be regarded as an additional contribution to the free energy

$$\mathfrak{F}'(\boldsymbol{r}) = \mathfrak{F}(\boldsymbol{r}) + \boldsymbol{f}_L \cdot \boldsymbol{r}, \tag{4.1.1}$$

or:

$$\mathfrak{F}'(\boldsymbol{r}) = \mathfrak{F}(\boldsymbol{r}) + (\boldsymbol{J}_T \times \boldsymbol{\phi}_0) \cdot \boldsymbol{r}, \tag{4.1.2}$$

with $\boldsymbol{\phi}_0$ the fluxquantum in the direction of the vortex. The fluxoids are trapped in the potential minima of $\mathfrak{F}'$ and remain stationary until $\boldsymbol{f}_L$ exceeds the maximum pinning force $\boldsymbol{f}_p = \max\left(\nabla \mathfrak{F}(\boldsymbol{r})\right)$ and a minimum in $\mathfrak{F}'$ does not exist further. Dissipative fluxflow occurs at the "critical current" $\boldsymbol{J}_c$. The derivation of $\boldsymbol{f}_p$ and thus $\boldsymbol{J}_c$ from the individual fluxoid-defect interaction is extremely complicated by the fact that $\mathfrak{F}(\boldsymbol{r})$ is not rigid but dependent on the interfluxoid distance. Thus, interaction with the neighbouring fluxoids has to be taken into account. A theoretical approach to get an effective average pinning force $\langle f_p \rangle$ per unit length of fluxoids is given by Labusch [2].

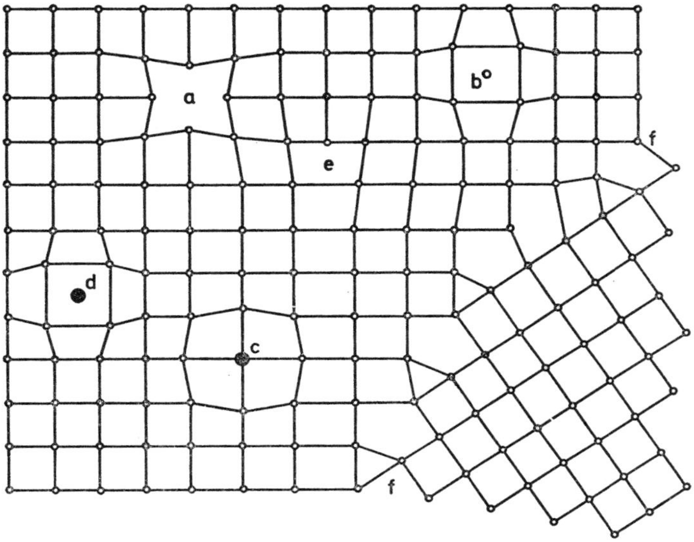

Fig. 4.1.2. Schematic representation of lattice defects. *a* Vacancy; *b* Interlattice atom; *c* Impurity in substitution position; *d* Impurity deposited within the lattice; *e* Step displacement; *f—f* Grain boundary

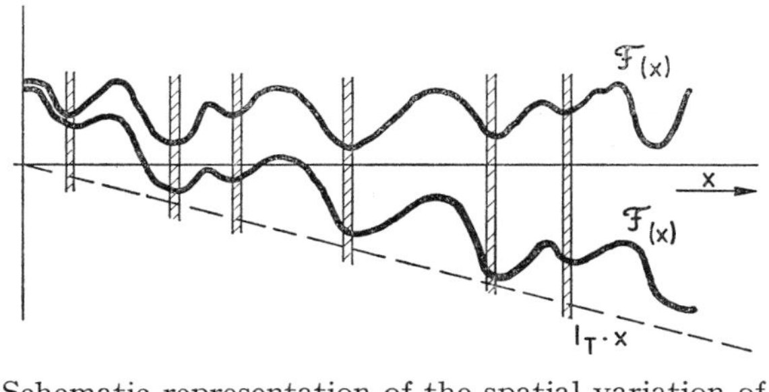

Fig. 4.1.3. Schematic representation of the spatial variation of free energy of a fluxline or a bundle of fluxlines in a real Type II superconductor without and with impressed transport currents

The value of J_c with $\langle f_p \rangle = J_c \cdot \phi_0$ can be exactly determined, as due to thermal activation, a number of fluxoids can move out of the potential minima at smaller currents than J_c. This leads to a (strongly current and temperature dependent) "flux creep"-voltage [3], expressed by

$$U \sim \exp (I_c/kT) . \tag{4.1.3}$$

Usually, J_c is related to a fixed fluxflow resistivity e.i. $\varrho_F = 10^{-12}$ Ohm · cm.

Up to now, the external force per vortex unit length was expressed by $f_L = J_T \cdot \phi_0$. This assumption leads to a total force on a conductor of volume V,

$$F_L = \int_V [B(r) \cdot J_T(r)] \, d^3r \,, \tag{4.1.4}$$

where we used $B = n\phi_0$ and n = fluxline density. Obviously, this is not identical with the classical magnetic force on the conductor located in a uniform field $\mu_0 H_{\text{ext}}$:

$$F_M = \mu_0 H_{\text{ext}} \cdot \int_V [J_T(r)] \, d^3r = \mu_0 H_{\text{ext}} \cdot I_T \cdot l \,, \tag{4.1.5}$$

with H_{ext}, the external field*.

The superconductor consists generally of a bulk region in which the distribution of the magnetic flux is controlled by pinning forces and surrounded by a thin vortex-free surface layer which can support a surface screening current. The screening current causes the effective applied field $B(r)$ at the boundary of the pinning force dominated region to be different from the externally applied field $\mu_0 H_{\text{ext}}$.

The superconducting surface sheath (due to the existance of screening currents) also exists in the mixed (Schubnikov) state. We can induce a persistent critical current in the surface sheath whose effect is to change the internal magnetic field $B(r)$ from the applied field.

When the external field is increased, the average internal magnetic field will be smaller than the external field; when H_{ext} is decreased, the average internal magnetic field will be larger than the equilibrium field ($I = 0$). Once a critical current I_c is established in the surface sheath by changing the applied magnetic field, one has to reverse the change in H in order to decrease the magnitude of the total current I, or to reverse its direction.

When the applied magnetic field is decreased below H_{c1}, the entrance into the Meissner state is delayed due to the enhanced internal magnetic field. A similar argument is applicable near H_{c2} for increasing or decreasing fields. A hysteresis appears, which is an intrinsic property of an ideal superconductor due to its surface properties, when the applied magnetic field is changed.

As $B(r)$ in the conductor differs from H_{ext}, the resulting difference must be attributed to the following additional effects:

The density gradient of flux lines leads to a net force on a fluxoid as the interfluxoid forces do not cancel.

For specimen with a demagnetization factor $N \neq 0$, the persistent supercurrents lead to a distortion of the magnetic field inside and outside the superconductor. The curved shape of fluxlines inside the superconductor gives an additional line tension force acting on the pinning centers.

---

* The field intensity indicated by H is expressed in (Am^{-1}) or (Acm^{-1}), the field B in (T).

From the free enthalpy $G\ (T, H)$, Friedel, de Gennes, and Matricon [4] have derived an expression for the correction to the force on a flux-line

$$f_L = \frac{\partial H\ (B)}{\partial B}\ [\boldsymbol{J}_T \cdot \phi_0]\,, \tag{4.1.6}$$

where $H\ (B)$ denotes the external field necessary to generate an internal equilibrium induction B in the ideal (unpinned) superconductor.

Eq. (4.1.6) must fail at least for low fields, $B \to H_{c1}$, where $\partial H\ (B)/\partial B \to 0$. In principle, the electromagnetic force must result from the Maxwell tensor which is not known, as it presumes detailed knowledge of the internal field.

For the total volume force acting on the fluxline lattice, we suggest setting the relation:

$$\boldsymbol{F}_L = \mu_0 \cdot \boldsymbol{J} \cdot \boldsymbol{H}_{\text{ext}}\,. \tag{4.1.7}$$

The effective pinning force density $\boldsymbol{F}_p$ counteracting $\boldsymbol{F}_L$ is thus

$$n\ \langle f_p \rangle = \boldsymbol{F}_p = \mu_0\ [\boldsymbol{J}_c \cdot \boldsymbol{H}_{\text{ext}}]\,. \tag{4.1.8}$$

Without the need for a detailed analysis of interfluxoid forces and of the line tension, these effects should automatically be included. The only assumption made here, for simplicity, is that the magnetic forces act on the fluxoids alone.

For superconductor geometries regarded here (thin filaments), the condition of $B \approx \mu_0 H_{\text{ext}}$ is always fulfilled.

Using Maxwells law curl $\boldsymbol{H} = \boldsymbol{J}$, the Eq. (4.1.7) can be written as

$$\boldsymbol{F}_L = \mu_0\ [\text{curl}\, \boldsymbol{H} \cdot \boldsymbol{H}_{\text{ext}}]\,. \tag{4.1.9}$$

Here, $\boldsymbol{H}$ is the local macroscopic field in the superconductor, where the microscopic field variations due to the vortex currents are averaged out.

Assuming the field (and thus the vortex axes) is in the z-direction and J_c is in the y-direction, Maxwells equation reduces to $J_y = dH_z/dx$. The expression for $\boldsymbol{F}_p(B, T)$ as a material constant of the individual superconductor defines a "critical field gradient"

$$\left.\frac{d\boldsymbol{H}}{dx}\right|_{\text{crit}} = \boldsymbol{J}_c\,. \tag{4.1.10}$$

If $\boldsymbol{J}_c$ and thus $\left.\frac{d\boldsymbol{H}}{dx}\right|_{\text{crit}}$ is exceeded, fluxlines or fluxline bundles overcome the pinning barriers and move in the opposite direction to the fluxoid density gradient dn/dx, with:

$$\left.\frac{dn}{dx}\right|_{\text{crit}} = \frac{\mu_0}{\phi_0} \cdot \left.\frac{dH_z}{dx}\right|_{\text{crit}},$$

until a new stable fluxoid arrangement is obtained.

Due to the same electrodynamic arguments which lead to the skin effect in a normal conductor, any change in the currents and thus the associated magnetic field always commences from the surface into the bulk superconductor.

In order to maintain the critical field gradient (assuming a quasi-stationary case of slow dH_{ext}/dt), the rising external field H_{ext} pushes the whole fluxfront into the superconductor as illustrated for a slab placed parallel to the field in Fig. 4.1.4a for the simple (but unrealistic) case of field independent J_c. The interior of the superconductor is screened by supercurrents flowing in a layer of thickness $x_p = H_{ext}/J_c$.

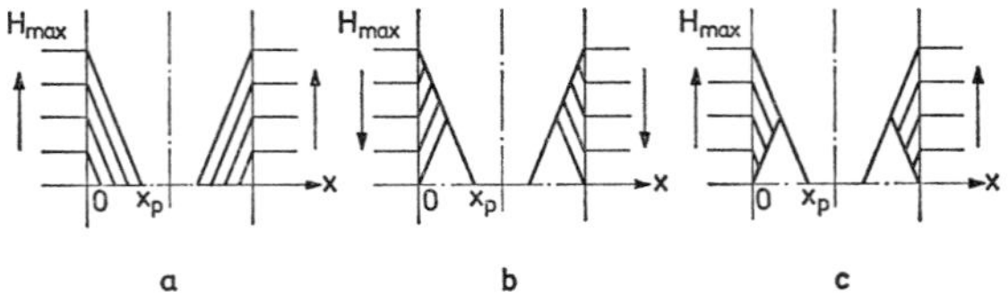

Fig. 4.1.4a—c. Field penetration into a Type II superconductor with pinning

Reducing H_{ext} below the peak value causes the fluxlines to move out of the superconductor, leading to an inverse critical field gradient and associated critical current density in the opposite direction. This condition is shown in Fig. 4.1.4b for subsequent values of H_{ext} after a maximum value H_{max} with a penetration depth x_p, given by

$$H_{max} = \frac{dH_z}{dx} \cdot x_p = J_c x_p \, ,$$

has been reached. A subsequent field rise yields a new pattern (Fig. 4.1.4c), which is different from the initial behaviour shown in Fig. 4.1.4a. This model for the magnetic behaviour of imperfect Type II superconductors was first formulated by BEAN [5] as the critical state model. It postulated that depending on the history of the external field, the current density can only be zero in regions where no flux has penetrated, or it attains the critical value $\pm J_c$.

Different to a normal conductor, regions of opposite flowing currents can coexist stationary in a superconductor. For H_{ext} cycled between $+H_{max}$ and $-H_{max}$, the corresponding field patterns are given in Fig. 4.1.5. If a transport current is impressed on the superconductor, the fields at both sides of the conductor are essentially different and the current pattern is displaced such that a net current I_T remains, as illustrated in Fig. 4.1.6.

In the case of full field penetration, i.e. if $J_c \cdot d/2 \geqq H_{max}$ is greater than the thickness d of the conductor, no field and current free regions exist.

4.1.2 Flux Profiles

For a real Type II superconductor, the critical current density J_c is a function if H and T. The flux profile $H(x)$, in the dimensional case, is expressed as the solution of the differential equation

$$\frac{dH(x)}{dx} = \pm J_c\,(H(x)) \, , \tag{4.1.11}$$

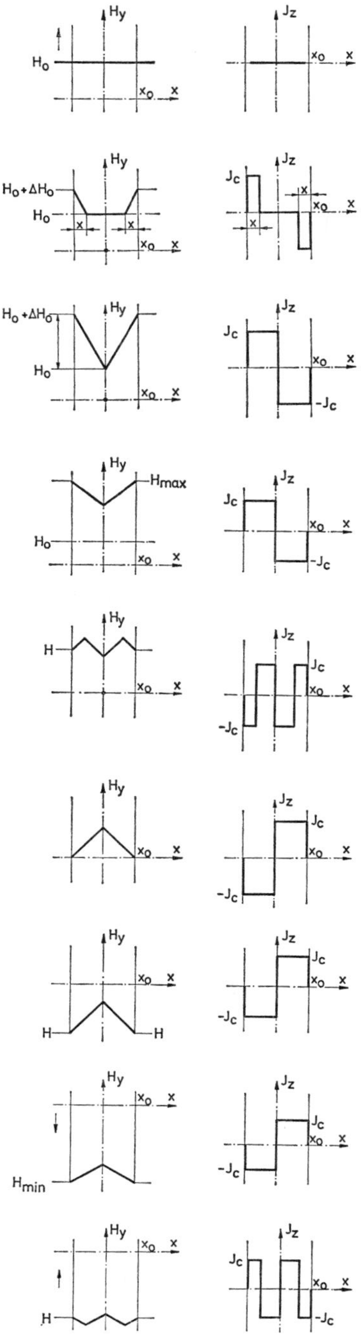

Fig. 4.1.5.
Field and current distribution in a Type II superconducting slab with the external field parallel to the surface of the slab, according to Bean's critical state model

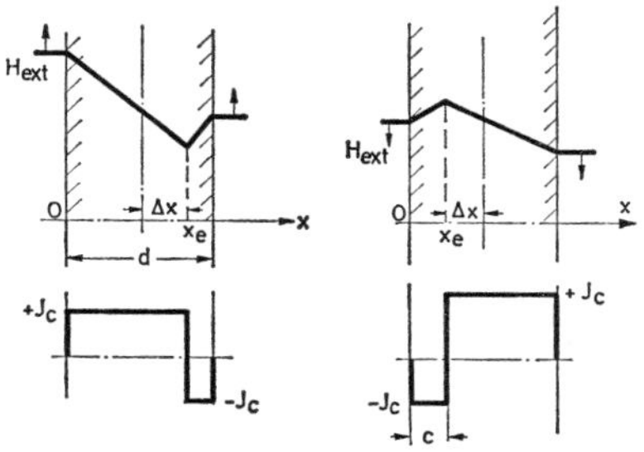

Fig. 4.1.6.
Field and current profiles in an infinite sheet due to the application of transport currents and external fields parallel to the sheet surface

were the (±) sign depends on the direction of J_c. For constant J_c, the linear field profile given in Fig. 4.1.5 (BEAN-model) was obtained.

A better approximation of the distribution of the current density in a real Type II superconductor is obtained by using the KIM [6] model, which relates the critical current density in the superconductor to the local magnetic field by:

$$\boxed{J_c = \frac{J_0 H_0}{H + H_0},} \tag{4.1.12}$$

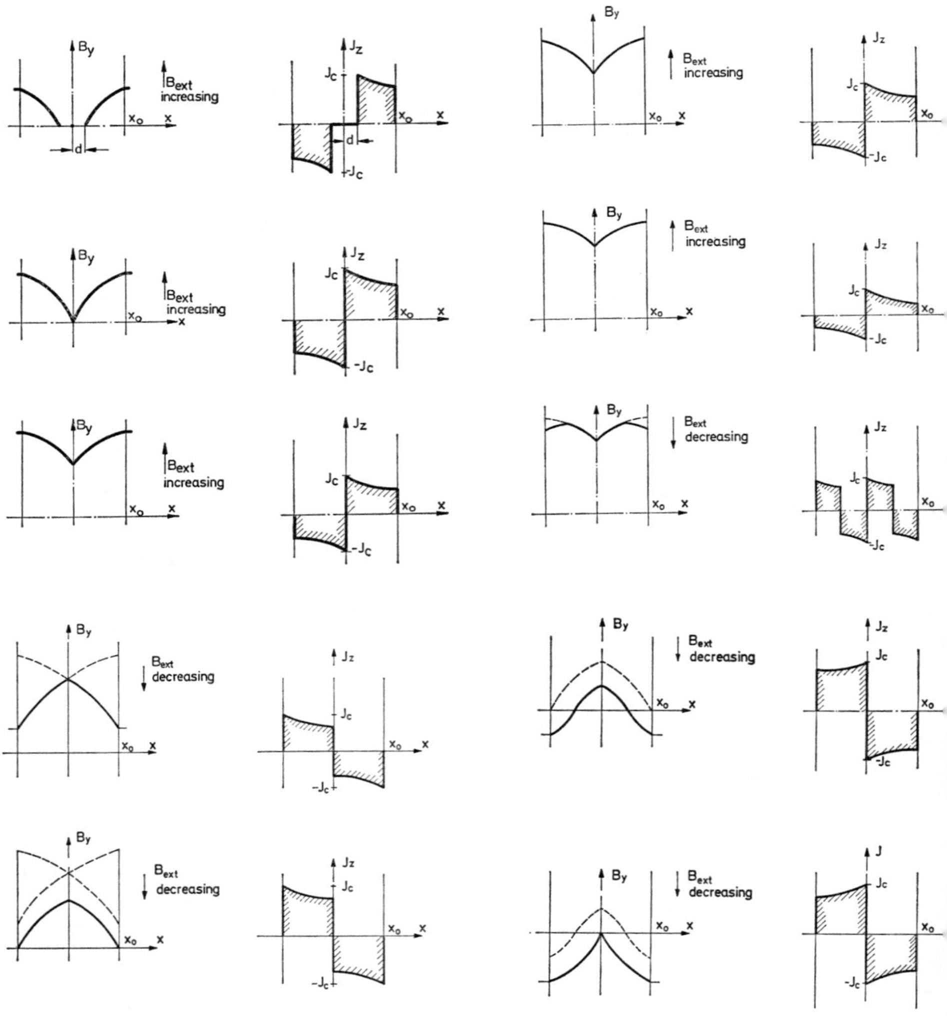

Fig. 4.1.7. Field profiles in an infinite slab due to increasing and decreasing external fields (Kim-model)

where J_0 and H_0 are material constants. J_0 is the current density at a zero transverse field. H_0 corresponds to the field at $J_0/2$.

Combining Eqs. (4.1.11) and (4.1.12), we obtain:

$$\frac{dH}{dx} = \pm \frac{J_0 H_0}{H + H_0}. \tag{4.1.13}$$

Integrating Eq. (4.1.13) and noting that at $x = 0$, $H(0) = H_{ext}$, we get the equation for the field profile in an imperfect Type II superconductor

$$H(x) = H_0 \left\{ \left[\left(1 + \frac{H_{ext}}{H_0}\right)^2 - \frac{2\,\mu_0 J_0 x}{H_0} \right]^{\frac{1}{2}} - 1 \right\}. \tag{4.1.14}$$

This distance from the surface at which H has dropped to zero is given by:

$$x_p = \frac{H_{ext}}{J_0 H_0} \left[\frac{1}{2} H_{ext} + H_0 \right]. \tag{4.1.15}$$

Fig. 4.1.7 illustrates the field profile in a slab for increasing and decreasing external magnetic fields parallel to the surface of the slab.

4.1.3 Thin Superconducting Tapes and Filaments

For superconducting specimen with a small thickness d, the external field penetrates into the whole conductor and is screened only partially in the middle of the conductor by an amount

$$\Delta H = \frac{d}{2} J_c .$$

Thus, $H(0) = H_{ext} \pm \Delta H$.

If the condition $\quad \Delta H \ll H_{ext}$

holds, i.e. the conductor thickness is such that

$$d \ll \frac{2\,H_{ext}}{J_c},$$

then $J_c(H)$ can be replaced by $J_c(H_{ext})$ and the approximation to a linear field profile in the superconductor is valid.

This is always satisfied for multifilamentary wires, where fine superconducting filaments (5 ... 50 μm) are embedded in a normal metal matrix. With typical values of $d = 10\,\mu m$ and $J_c = 2 \times 10^5$ A/cm², we get $\Delta H = 100$ A/cm, whereas the external field in a superconducting magnet is in the range of several 10^4 A/cm.

4.1.4 Finite Size Slabs and Cylindrical Conductors Located in a Transverse External Field

In the one dimensional model treated above, curl $\boldsymbol{H}$ was expressed by dH_z/dx, and all the other contributions were assumed to be zero. This is only true for an infinitely extended slab with a surface parallel to the (uniform) external field. In all other conductor configurations, the magnetization distorts the field and current configuration in a complicated manner and cannot be treated by elementary methods (Fig. 4.1.8). The case of a thin strip conductor with its face perpendicular to the field is evaluated by Morgan [7].

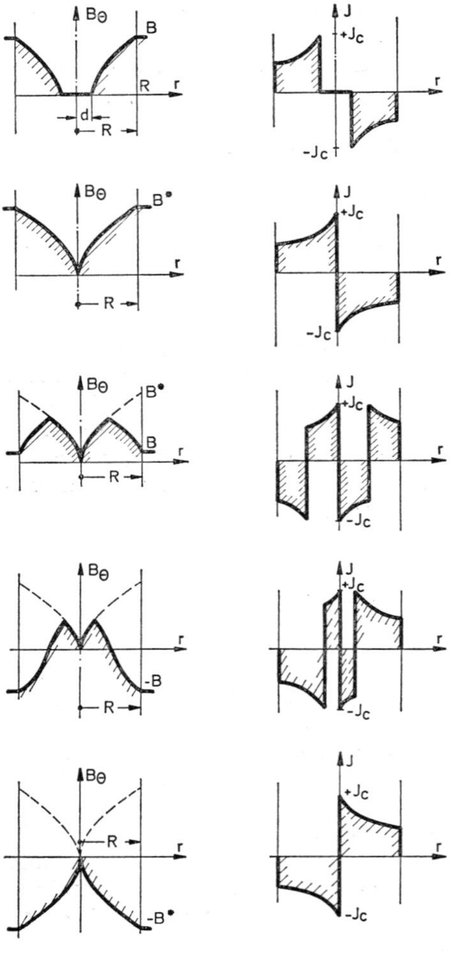

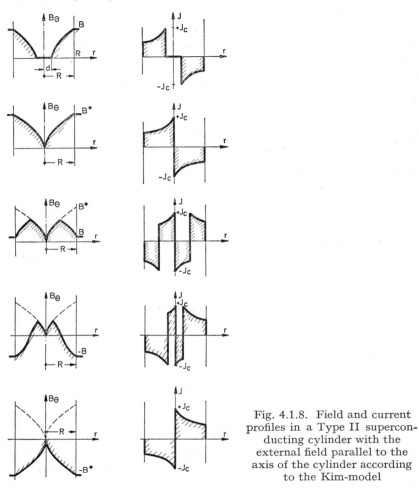

Fig. 4.1.8. Field and current profiles in a Type II superconducting cylinder with the external field parallel to the axis of the cylinder according to the Kim-model

Clearly, the one dimensional model, as described remains only a good approximation as long as the curvature of fieldlines inside and outside of the conductor can be neglected. This is true if the screening field fulfills the condition

$$\Delta H = J_c \frac{d}{2} \ll H_{\text{ext}} .$$

This condition always holds for multifilamentary conductors used for a.c. magnet applications.

4.1.5 Methods of Calculating Hysteretic Losses due to Alternating Fields

For an imperfect Type II superconductor located in a varying external field, the critical state model has the consequence of a varying magnetic flux in the material (Fig. 4.1.9). Essentially, the changing flux must penetrate the conductor surface due to the motion of fluxlines and generates "hysteretic losses". The dissipative mechanism of a moving fluxline is not well understood, but as the energy loss results entirely from the magnetic field, loss computation is equivalent to evaluation of the electromagnetic energy dissipated during cycling of the field. Different approaches to solve this problem must essentially yield a unique result, as they describe the same physical process:

(a) Integration of the Poynting vector $\boldsymbol{S} = \boldsymbol{E}_s \times \boldsymbol{H}$ over the conductor surface and one cycle. The electrical field $\boldsymbol{E}_s$ is given by the induction law curl $\boldsymbol{E} = -\dot{\boldsymbol{B}}$.

(b) Integration of Joule's energy $\boldsymbol{J}_c \cdot \boldsymbol{E}$ over the conductor volume, where $\boldsymbol{E}$ has the same origin as in (a).

(c) Evaluation of the area under the magnetization curve for a full cycle $\oint M(H) \cdot dH\, dV$, where M is the magnetization due to currents $\pm I_c$ in the sample.

(d) Volume integration of the mechanical work performed to the moving fluxline lattice by the Lorentz force. The power loss per volume is expressed by $p_v = (\boldsymbol{J}_c \times \boldsymbol{B}) \cdot \boldsymbol{v}$, where $\boldsymbol{v}$ is the velocity of the moving fluxoids and can be obtained by using the conservation law for the fluxline density div $(|\boldsymbol{B}| \cdot \boldsymbol{v}) = \dot{\boldsymbol{B}}$.

4.1.6 Hysteretic Losses in Slabs

For the simplest case, where the critical current density is independent of the magnetic field within the conductor, BEAN calculated hysteretic losses in slabs parallel to the external field. The transport current is not considered in his computation.

If the field penetrates from both sides into the slab of thickness d, full field penetration is obtained at

$$H_s = J_c \cdot \frac{d}{2}. \tag{4.1.16}$$

For a probe exposed to a variable external field H_{ext} and cycled between the two values of $-H_{max}$ and $+H_{max}$, we may distinguish between two cases:

(a) *Incomplete Flux Penetration*: $H_{max} < H_s$

Integrating the magnetization $\oint M\, dH\, dV$ over one cycle, the hysteretic losses can be expressed by:

$$W_{h1} = \frac{2}{3} \mu_0 V \frac{H_{max}^3}{H_s}, \tag{4.1.17}$$

$$\boxed{W_{h1}/A = \frac{4}{3} \mu_0 \frac{H_{max}^3}{J_c},} \tag{4.1.18}$$

with V the volume and A the area of the slab. In each half section of the slab, the losses $W_{hl}/2$ are dissipated.

Both equations do not consider the presence of screening fields at the surface of the superconductor, and the fact that in the Meissner phase, $H \leqq H_{c1}$, no losses are produced. If the screening field at the surface is ΔH, we can write for the hysteretic losses:

$$W_{hl}/A = \frac{4}{3}\mu_0 (H_m - \Delta H - H_{c1})^2 (H_m + 2\Delta H + 0.5 H_{c1}) \cdot \frac{1}{J_c} . \tag{4.1.19}$$

This equation shows that the hysteretic losses are reduced if ΔH is increased, which is possible if superconducting surfaces with high pinning strength are produced. For Nb_3Sn surfaces, values of $\mu_0 \Delta H = 0.05$ to 0.1 T have been achieved, which in comparison to $\mu_0 H_m$, are negligable. It is, however, possible to increase $\mu_0 \Delta H$ to 0.2 ... 0.5 T. The construction of multilayer conductors, where cylindrical layers of Type II superconductors are interspersed with layers of either a normal conductor or layers of Nb, are being considered for this reason in underground power transmission cables for a.c. application.

(b) *Complete Flux Penetration*: $H_{max} \geqq H_s$

In this case, no current-free region exists in the slab. Integration of the magnetization yields:

$$\boxed{W_{hl} = 2\mu_0 H_s H_{max} V \left[1 - \frac{2}{3}\frac{H_s}{H_{max}}\right].} \tag{4.1.20}$$

For field values: $H_{max} \gg H_s$, this equation reduces to:

$$W_{hl} = \mu_0 d J_c H_{max} V . \tag{4.1.21}$$

If, however, the external field is cycled between the values 0 and $+ H_{max}$, we obtain:

$$W_{hl} = \frac{\mu_0 d}{2} \cdot J_c \cdot H_{max} V . \tag{4.1.22}$$

For a cycle in which the field is swept from $- H_{max}$ to $+ H_{max}$ and back to $- H_{max}$, the energy loss can be expressed from Eq. (4.1.21) by:

$$W_{hl} = \mu_0 \cdot d^2 \cdot b \cdot l \cdot J_c \cdot H_{max} , \tag{4.1.23}$$

$$W_{hl} = \mu_0 \cdot d \cdot l \cdot I_c \cdot H_{max} , \tag{4.1.24}$$

$$W_{hl} = 2\mu_0 \cdot d \cdot b \cdot l \cdot H_s \cdot H_{max} . \tag{4.1.25}$$

In these equations, we used the volume of the slab V to be $b \cdot d \cdot l$. The Eq. (4.1.21) is also derived by Hancox [8].

For real superconductors, the case of J_c independent of field is unrealistic. The more realistic model assumes the current density to be field dependent according to Kim [6] as given by Eq. (4.1.12).

For this particular case, the field profile within the conductor was given by the Eq. (4.1.14) and Fig. 4.1.7.

If the external field is swept in the range between H_{max} and H_{min}, the corresponding field values within the superconductor can be obtained from the equation:

$$H^2 + 2\,H_0 H - H_{ext}^2 - 2\,H_0 \cdot H_{ext} \pm 2\,J_0 H_0 x = 0\,.$$

The field penetration depth in the slab, as shown in Fig. 4.1.9, is obtained from

$$x = \pm \frac{1}{2\,J_0 H_0}\,[(H + H_0)^2 - (H_{ext} + H_0)^2]\,. \qquad (4.1.26)$$

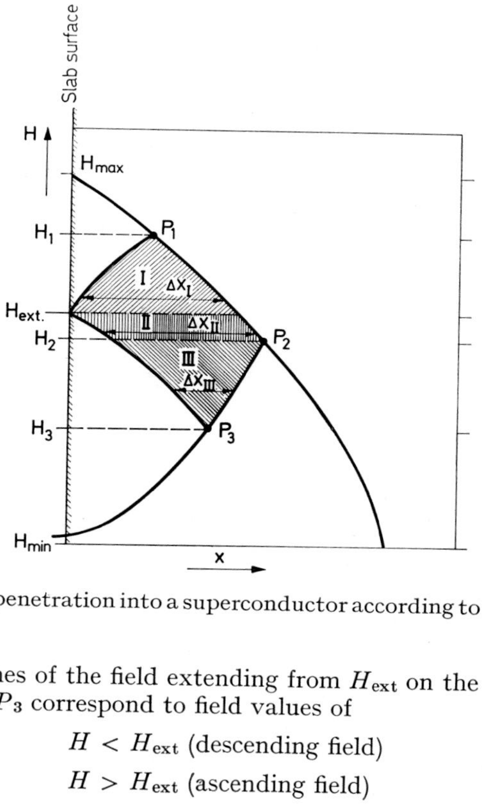

Fig. 4.1.9. Field penetration into a superconductor according to the Kim-model

The two branches of the field extending from H_{ext} on the surface of the slab to P_1 and P_3 correspond to field values of

$$H < H_{ext} \text{ (descending field)}$$
$$H > H_{ext} \text{ (ascending field)}$$

The field penetration in the three regions I—III are given by:

$$\Delta x_{I} = \frac{1}{2\,J_0 H_0}\,[2\,(H + H_0)^2 - (H_{ext} + J_0)^2 - (H_{max} + H_0)^2]\,,$$

$$\Delta x_{II} = \frac{1}{2\,J_0 H_0}\,[(H_{max} + H_0)^2 - (H_{ext} + H_0)^2]\,,$$

$$\Delta x_{III} = \frac{1}{2\,J_0 H_0}\,[2\,(H + H_0)^2 - (H_{ext} + H_0)^2 - H_0^2]\,.$$

Introducing the abbreviations:

$$h = H + H_0\,; \quad h_{\text{ext}} = H_{\text{ext}} + H_0\,; \quad \text{etc. and}$$

$$h_1 = \frac{1}{\sqrt{2}}\,(h_{\text{ext}}{}^2 + h_{\max}{}^2)^{\frac{1}{2}}\,; \quad h_2 = \frac{1}{\sqrt{2}}\,(h_{\max}{}^2 + h_{\min}{}^2)^{\frac{1}{2}}\,;$$

$$h_3 = \frac{1}{\sqrt{2}}\,(h_{\text{ext}}{}^2 + h_{\min}{}^2)^{\frac{1}{2}}\,,$$

we may write for the magnetization:

$$\Delta M = \frac{1}{2\,J_0 H_0}\int\limits_{h_1}^{h_{\text{ext}}} (2\,h^2 - h_{\text{ext}}^2 - h_{\max}^2)\,dh + \frac{1}{2\,J_0 H_0}\int\limits_{h_{\text{ext}}}^{h_2} (h_{\max}^2 - h_{\text{ext}}^2)\,dh$$

$$+ \frac{1}{2\,J_0 H_0}\int\limits_{h_2}^{h_3} (2\,h^2 - h_{\text{ext}}^2 - h_{\min}^2)\,dh\,,$$

which, after integration and rearrangement, is written in the form:

$$\Delta M = \frac{1}{2\,J_0 H_0}\left[\frac{2}{3}\,h_{\text{ext}}^3 + \frac{\sqrt{2}}{3}\,(h_{\text{ext}}^2 + h_{\max}^2)^{\frac{3}{2}} - \frac{\sqrt{2}}{3}\,(h_{\text{ext}}^2 + h_{\min}^2)^{\frac{3}{2}} + \frac{\sqrt{2}}{3}\,(h_{\max}^2 + h_{\min}^2)^{\frac{3}{2}} - 2\,h_{\text{ext}}\cdot h_{\max}^2\right]. \tag{4.1.27}$$

As usual, we calculate the hysteretic losses from

$$W_{\text{hl}} = \mu_0 \int\limits_{H_{\min}}^{H_{\max}} M\,dH_{\text{ext}} = \mu_0 \int\limits_{h_{\min}}^{h_{\max}} M\,dh_{\text{ext}}$$

and obtain the hysteretic losses per unit surface for a semiinfinite slab.

$$W_{\text{hl}} = \frac{\mu_0}{2\,J_0 H_0}\,(H_{\min} + H_0)^4\left[f^4\left(\frac{1}{4} + \frac{1}{4\sqrt{2}}\ln\frac{1 + \sqrt{1 + f^2}}{(1 + \sqrt{2})\,f}\right) - \left(\frac{5}{12} - \frac{1}{4\sqrt{2}}\ln\frac{f + \sqrt{1 + f^2}}{1 + \sqrt{2}}\right) - f^2 - \frac{\sqrt{1 + f^2}}{6\sqrt{2}}\left(3\,f^2 - \frac{13}{2}\,f^2 + \frac{3}{2}\,f - 5\right)\right] \tag{4.1.28}$$

with
$$f = \frac{H_{\max} + H_0}{H_{\min} + H_0}\,.$$

It may be pointed out that Eq. (4.1.22) is true only if $H_{\max}$ and $H_{\min}$ have the same sign, as the analytical expressions for $J_c(H)$ are valid only for $H \geq 0$.

Eq. (4.1.22) is particulary useful to calculate losses in superconducting magnetic shields (incomplete penetration).

4.1.7 Application to Multifilamentary Conductors

Superconductors used for a.c. applications have diameters in the μm range. For these superconductors, the case of complete fluxpenetration is applicable.

As seen in section 4.1.2, the assumption of $\Delta H \ll H_{ext}$ implies that linear field penetration is valid, and the equations derived from Bean's model are correct for slabs of finite size as well as for cylindrical conductors as shown in sections 4.1.2 and 4.1.3.

We regard a slab having a thickness d, a height b, and located parallel to the external field H_{ext}. A net transport current I_T is now admitted to pass through the slab. The electric center of the slab*, which was identical to the geometric center of the slab for $I_T = 0$, is displaced due to the transport current by a distance:

$$\Delta x = \frac{d}{2} \frac{I_T}{I_c} . \tag{4.1.29}$$

The field and current displacement was illustrated in Fig. 4.1.6 for increasing and decreasing fields. At points of field reversal, the simple field pattern is modified. These field perturbations alter the hysteretic losses, as written in Eq. (4.1.18). The additional loss contribution in this equation can be neglected only if $H \ll H_{max}$.

Obviously, for $\Delta x = d/2$, the critical current of the probe is attained, beyond which steady flux flow occurs across the superconductor. The power density dissipated in the superconductor is calculated for the case, that the external field changes with H_{ext} from

$$\boldsymbol{P} = \boldsymbol{E} \cdot \boldsymbol{J} .$$

For our particular geometry, $\boldsymbol{H} = H_z$; $\boldsymbol{J}_c = J_y$; $\boldsymbol{F}_L = F_{Lx}$, Maxwell's law for the electrical field curl $\boldsymbol{E} = \dot{\boldsymbol{B}}$ yields,

$$\frac{dE_y}{dx} = -\mu_0 \cdot \dot{H}_z(x) ,$$

where the internal field change $\dot{H}_z(x)$ can be set approximately as $\dot{H}_{ext}$ for thin probes. As at the electrical center x_e, the fluxline velocity is zero and the axial electric field $E_y(x_e)$ must vanish. E_y is obtained by simple integration with one boundary being x_e.

$$E_y(x) = -\mu_0 \dot{H}_{ext}(x - x_e) . \tag{4.1.30}$$

The sign of J_y is given by the position and sense of the field change

$$J_y = \begin{cases} +J_c & \text{for} \quad x < x_e ; \quad \dot{H} > 0 \quad \text{or} \quad x > x_e ; \quad \dot{H} < 0 \\ -J_c & \text{for} \quad x < x_e ; \quad \dot{H} < 0 \quad \text{or} \quad x > x_e ; \quad \dot{H} > 0 . \end{cases}$$

Thus, the energy dissipation rate per conductor length is given by:

$$\frac{p_{hl}}{l} = \mu_0 b J_c |\dot{H}_{ext}| \int_0^d |x - x_e| \, dx . \tag{4.1.31}$$

By introducing $x_e = \dfrac{d}{2} + \dfrac{I_T}{2 J_c b}$ and the slab volume $V = b \cdot l \cdot d$, the integral yields the conductor losses:

$$\boxed{p_{hl} = \mu_0 V J_c \cdot \frac{d}{3} \cdot |\dot{H}_{ext}| \cdot \left[1 + \left(\frac{I_T}{I_c}\right)^2\right]} \quad \text{(W)} , \tag{4.1.32}$$

* The electric center is defined as a plane within the conductor separating the two regions of opposite flowing I_c.

where $I_c = b \cdot d \cdot J_c$ is the critical short sample current at the corresponding field.

Hysteretic losses are usually related to one cycle if H_{ext} is changed periodically between the field values of $H_{\min}$ and $H_{\max}$:

$$H_{\min} \to H_{\max} \to H_{\min} \,.$$

The losses per cycle are independent of frequency and wave form, the only material property entering the loss equation is the $J_c(H)$-dependence.

$$\Delta W_{\text{hl}} = \frac{\mu_0}{2} V d \int_{H_{\min}}^{H_{\max}} J_c(H) \left[1 + \left(\frac{I_T(H)}{I_c(H)}\right)^2\right] dH \,. \tag{4.1.33}$$

The term I_T/I_c depends on the particular magnet and contributes only at the peak current to the losses in the high field region of a coil. Usually the term is small and can be neglected. Kim's model, relating the critical current and field, is a good approximation for Type II multifilamentary conductors. Entering J_c from Eq. (4.1.12) in Eq. (4.1.27), we obtain the well known formula for hysteretic losses in superconductors with small diameters:

$$\boxed{W_{\text{hl}} = \frac{\mu_0 J_0 H_0 V}{2} d \ln \left(\frac{H_{\max} + H_0}{H_{\min} + H_0}\right); \quad \text{(Ws/cycle)}\,.} \tag{4.1.34}$$

V is the total volume if J_c is referred to as the overall current density of the conductor, or it is the volume of the superconductor if J_c is the current density in the superconductor. If field reversal occurs during the field sweep (for instance $H_{\min}$ is negative), the losses have to be split into two parts: Losses in the region $0 \to H_{\max} \to 0$ and losses in the region $0 \to H_{\min} \to 0$.

Thus, the total losses are given by:

$$W_{\text{hl}} = \frac{\mu_0 J_0 H_0 V}{2} d \cdot \ln \left[\frac{(H_{\max} + H_0)\,(H_{\min} + H_0)}{H_0^2}\right]. \tag{4.1.35}$$

4.1.8 Hysteretic Losses in Cylindrical Shaped Superconductors

Loss calculations for slab geometries are simple and useful, but not correct when applied to magnets using composite conductors with superconducting tapes embedded in normal metal strips. As the field over the length of the tape is not always parallel to the surface of the superconductor, this type of conductor is not utilized in pulsed magnets due to enhanced dissipative losses.

The loss calculations are specifically applicable to cylindrical shaped filamentary conductors.

Coextruded superconductor and copper wires generally result in superconducting filament shapes having more or less circular cross sections. If the superconducting filament is distorted during extrusion, the use of the hydraulic diameter ($d = 4\,A/P$) in the calculation is a good approximation to obtain losses in superconductors, where A denotes the cross section and P the perimeter of the filament. With the restriction to thin filaments with a diameter of $d \ll 2\,H_{ext}/J_c$ and the transport current $I_T \ll I_c$ (valid for nearly the whole field sweep), the integration of $(J \cdot E)$ over a cylinder is per formed by Morgan [7] in cylindrical coordinates. Starting with curl $\boldsymbol{E} = -\mu_0 \dfrac{d\boldsymbol{H}_{ext}}{dt}$ in cylindrical coordinates (r, θ), we have

$$\frac{1}{r}\frac{\partial E_z}{\partial \theta} = -\mu_0 \cdot \frac{\partial H_r}{\partial t} \tag{4.1.36}$$

and:

$$\frac{\partial E_z}{\partial r} = \mu_0 \cdot \frac{\partial H_\theta}{\partial t}. \tag{4.1.37}$$

H_r and H_θ are substituted by H_{ext} in the case of a symmetric flux profile with $\Delta x = 0$.

Since

$$\frac{\partial H_r}{\partial t} = \dot{H}_{ext} \cdot \sin(\theta);$$

and

$$\frac{\partial H_\theta}{\partial t} = \dot{H}_{ext} \cdot \cos(\theta),$$

one obtains:

$$\frac{1}{r}\frac{\partial E_z}{\partial \theta} = \mu_0 \dot{H}_{ext} \cdot \sin(\theta) \tag{4.1.38}$$

and

$$\frac{\partial E_z}{\partial r} = \mu_0 \dot{H}_{ext} \cdot \cos(\theta). \tag{4.1.39}$$

The solution of Eq. (4.1.39),

$$E_z = \mu_0 \cdot \dot{H}_{ext} \cdot r \cdot \cos(\theta) \tag{4.1.40}$$

gives the field distribution within the cylinder. The power loss is given by:

$$P_{hl} = \boldsymbol{E} \cdot \boldsymbol{J} \cdot dV = E_z \cdot J_c \cdot dV, \tag{4.1.41}$$

where E_z is always directed along $J_z = \pm J_c$.

Substituting E_z from Eq. (4.1.40) in Eq. (4.1.41) we obtain for a conductor of length l the losses:

$$P_{hl} = 2\,\mu_0 J_c \left|\dot{H}_{ext}\right| \cdot l \cdot \int_0^{d/2} \int_{-\pi/2}^{+\pi/2} r^2 \cos(\theta)\, dr\, d\theta$$

$$\boxed{P_{hl} = \frac{8}{3\pi}\mu_0 V J_c \cdot \frac{d}{4} \cdot \dot{H}_{ext}} \quad \text{(W)} \tag{4.1.42}$$

with d the diameter of the superconductor and V its volume. The energy loss is given by the time integral of Eq. (4.1.42). Comparison with the corresponding Eq. (4.1.33) for a slab shaped conductor shows that hysteretic losses in a cylindrical conductor are slightly less, i.e. by a factor $8/3\,\pi \approx 0.85$, than in a slab of the same volume, i.e. the same current carrying capacity and thickness.

The factor $8/3\,\pi$ in Eq. (4.1.42) occurs in all loss equation for cylindrical geometries.

At a low frequency field sweep (≤ 1 Hz), hysteretic losses are the main contributor among the losses encountered in the conductor.

To evaluate hysteretic losses per cycle for the entire coil, Eq. (4.1.25) must be integrated over the coil volume. As H_{max} and H_{min} occur in logarithmic form, the losses per cycle can be obtained with good accuracy (about 20%), if H_{max} and H_{min} are just replaced by the mean field value over the entire coil.

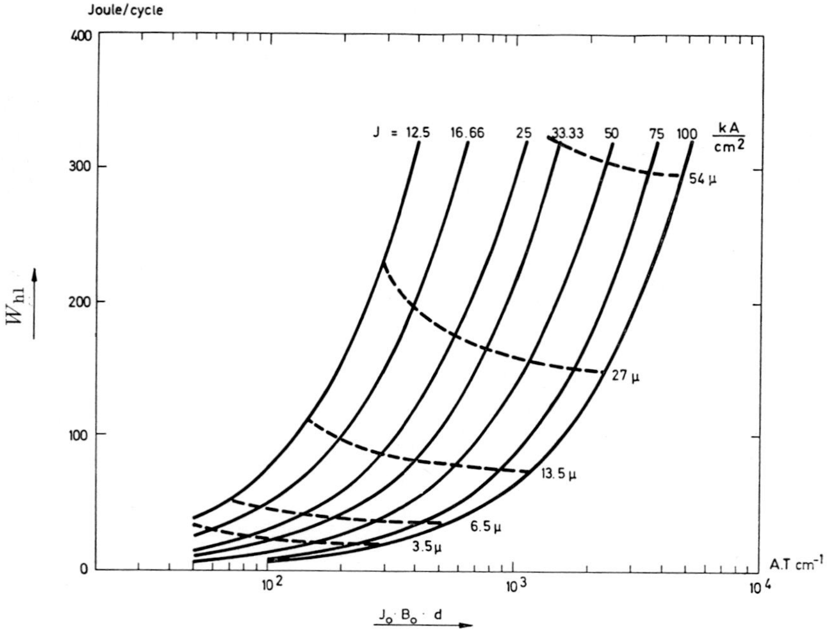

Fig. 4.1.10. Hysteretic losses vs. the characteristic conductor parameters $J_0 B_0 d$ for a 1 m long dipole magnet

Fig. 4.1.10 illustrates calculated hysteretic losses for a pulsed superconducting dipole magnet. The aperture field of the magnet is cycled from zero to 5 T. The coil cross-section is an approximation of intersecting ellipses; the magnet has an aperture of 8×11 cm² and a length of 100 cm. In the abscissa, the term $J_0 H_0 d$ contains all relevant conductor parameters*. For $\mu_0 H_0$, a typical value of 1 T is used in the calculation. The losses for different peak current densities (at 5 T) are represented by the solid lines. As the conductor volume decreases with increasing current carrying capacity $J_0 H_0$, we see that the total losses proportional to $J_0 H_0$ are not a strong function of the parameter $J_0 H_0$ for a given magnet.

This loss dependence is shown by the broken line with the filament diameter d as a parameter, where we have assumed that the current density at 5 T attains just the critical value. From the presented analy-

* Twisted multifilamentary conductor is assumed.

sis and the curves of Fig. 4.1.10, it is seen that the most promising attempt to reduce a.c. losses is in using thin filaments in the conductor. NbTi filaments having individual diameters of $\leq 5\ \mu m$ are commercially available. Several thousands of such filaments are drawn in a metallic matrix and are twisted. A number of such conductors (called composite conductors, or strands) are transposed to form a cable or braid for a.c. applications.

4.1.9 Hysteretic Losses in Coils Using Hollow Superconducting Filaments

It was possible to produce multifilament composites in long lengths (> 1 km) only by using NbTi. Nb_3Sn and V_3Ga have been used in multifilament composites (~ 350 filaments in Cu) of short length (100 to 300 m). The β-tungsten series is brittle and production of composites has been restricted to short lengths due to manufacturing difficulties.

An alternate solution is sputtering of thin layers on glass filament carriers or some other suitable material. The losses of such a turbular conductor, having an inner diameter d_1 and an outer diameter d_2, are obtained from Eq. (4.1.43):

$$W_{\mathrm{hl}} = \frac{(\pi)^2}{8} J_0 H_0 l_{sc} (d_2^3 - d_1^3) \cdot \ln \frac{H_m + H_0}{H_1 + H_0}. \tag{4.1.43}$$

Dividing the losses by the volume of the superconductor

$$V_{sc} = \frac{\pi}{4} (d_2^2 - d_1^2) \cdot l_{sc}$$

and referring to the same current carrying capacity in the conductor, we find that $W_{\mathrm{hl}}/I_{T\,\max}$ depend only the ratio d_1/d_2:

$$\frac{W_{\mathrm{hl}}}{I_T} \sim \frac{d_2^3 - d_1^3}{d_2^2 - d_1^2} = d_2 \frac{1 - (d_1/d_2)^3}{1 - (d_1/d_2)^2}. \tag{4.1.44}$$

Compared to a filament with $d_1 = 0$, losses are increased by the factor $1 - \left(\frac{d_1}{d_2}\right)^3 \Big/ 1 - \left(\frac{d_1}{d_2}\right)^2$ for the same critical current density and an identical total current.

4.1.10 Losses in Composites

Losses in superconducting coils, when exposed to time varying magnetic fields, have several origins:

— Eddy current losses in the conductor matrix.
— Self field losses.
— Hysteretic losses.
— Additional losses due to non-uniform magnetic fields.

In Sections 4.1.1.6 and 4.1.7, we had treated only hysteretic losses. In the following, we discuss the origin and the magnitude of the other losses occurring in a.c. coils. We consider, however, only twisted multifilament conductors.

4.1.11 Eddy Current Losses in the Conductor Matrix

In any metallic conductor, eddy currents are induced if $\dot{B} \neq 0$. In a composite conductor, these induced currents are modified by the superconducting filaments, which do not admit (first order observation) a resistive longitudinal component of the electrical field.

To calculate eddy current losses, the matrix material, the current distribution in the matrix depending on the conductor geometry, and the rate of field rise must be known. We assume uniform distribution of filaments throughout the composite and an external field perpendicular to the conductor axis, as is usually the case in coils:

(a) We regard, for the time being, only the outer layer of filaments embedded in a cylindrical composite. This layer has a radius $D/2$, and the filaments are twisted with a pitch of length l_p.

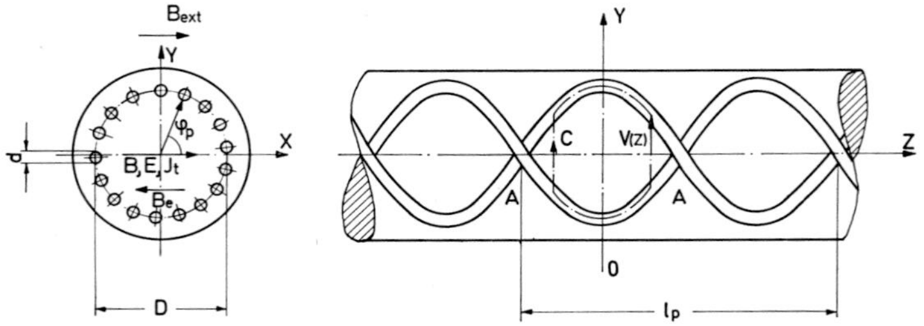

Fig. 4.1.11. Schematic representation of a composite conductor with twisted filaments

The magnetic field B, acting in the cylinder, is assumed to be homogeneous and perpendicular to the axis of the composite*. The field is composed of the external field B_{ext} and an additional field B_i produced by the induced currents in the filaments.

$$B = B_{ext} + B_i .$$

The voltage induced in the conductor is obtained from the contour integral (over C) taken along the electric centers of two oppositely placed filaments as shown in Fig. 4.1.11.

$$\oint_c \boldsymbol{E} d\boldsymbol{s} = -\dot{\Phi} = -\int \dot{B} \, dA = -2 D \sin\left(\frac{2\pi}{l_p} z\right) \cdot \dot{B} \frac{l_p}{2\pi}. \tag{4.1.45}$$

This voltage must drop entirely through the matrix in the two transverse passages $\overline{1-2}$ and $\overline{3-4}$. Relating the electrical potential $V(z)$ of the filaments to the axis of the composite, it is seen that at the crossoverpoint A at $z = l_p/4$, the potential $V(z)$ must be a maximum.

The voltage at $z = 0$ must be zero because of the antisymmetric condition at $z = 0$.

* The assumption of a uniform magnetic field is justified according to Eq. (4.1.62).

At an arbitrary point, the electrical potential, expressed by

$$V(z) = \dot{B}\,\frac{l_p}{2\pi}\,\frac{D}{2}\,\sin\left(\frac{2\pi}{l_p}\,z\right) \tag{4.1.46}$$

is the potential between the conductor axis and one filament, thus a quarter of the contourintegral given by z can be expressed by the twist angle ϕ_p between the filament axis and the plane perpendicular to B; i.e. $\phi_p = \pm\, 2\pi z/l_p$, where the sign corresponds to the sense of the twist. Thus, we get for each single filament:

$$V(\phi) = \pm\,\dot{B}\,\frac{l_p}{2\pi}\cdot\frac{D}{2}\,\sin(\phi_p) = \pm\,\dot{B}\,\frac{l_p}{2\pi}\,x\,, \tag{4.1.47}$$

if the sense of $\dot{B}$ is assumed to be in the x direction.

Eq. (4.1.47) is valid for each single filament in the outer layer. If we approximate the layer of filaments by an infinitesimal thin cylinder, then the boundary conditions for Eq. (4.1.47) are met by a homogeneous electric field in the cylinder in $\dot{B}$ direction.

$$E = E_x = \mp\,\dot{B}\left(\frac{l_p}{2\pi}\right). \tag{4.1.48}$$

The transverse current density in the composite is thus:

$$J_t = \mp\,\frac{1}{\bar{\varrho}_c}\left(\frac{l_p}{2\pi}\right)\cdot\dot{B}\,. \tag{4.1.49}$$

From these equations, we obtain the eddy current losses per volume occurring within the filament bundle of diameter D:

$$p_e = \frac{1}{\bar{\varrho}_c}\,E^2 = \frac{1}{\bar{\varrho}_c}\left(\frac{\dot{B}\,l_p}{2\pi}\right)^2, \tag{4.1.50}$$

where $\bar{\varrho}_c$ is the effective transverse resistivity of the composite.

(b) Only the outer layer of filaments in the cylindrical composite conductor was considered above. It is seen, however, from Eq. (4.1.47) that the diameter of the filament layer could be eliminated, indicating that the electric field E_x is independent of D. Thus, we may fill the entire composite cross-section with concentric layers of filaments and present a realistic model of a twisted multifilament conductor. As l_p is unchanged for all layers, the field configuration in the interior of the composite is not altered. However, the transverse effective resistivity $\bar{\varrho}_c$ must be averaged properly over the conductor, which is written for a single component matrix:

$$\bar{\varrho}_c = \varrho_{\text{matrix}}\cdot\left(\frac{w}{w-d}\right), \tag{4.1.51}$$

where d is the filament diameter and w is the distance between the centers of adjacent filaments.

It is assumed here, however, that the eddy currents flow through the matrix material and do not cross superconducting filaments due to the relatively high resistance of the interface layer between superconductor and the matrix. ϱ_{matrix} is the resistivity of the matrix material, including size effect, mechanical strain effects, and longitudinal magnetoresistance.

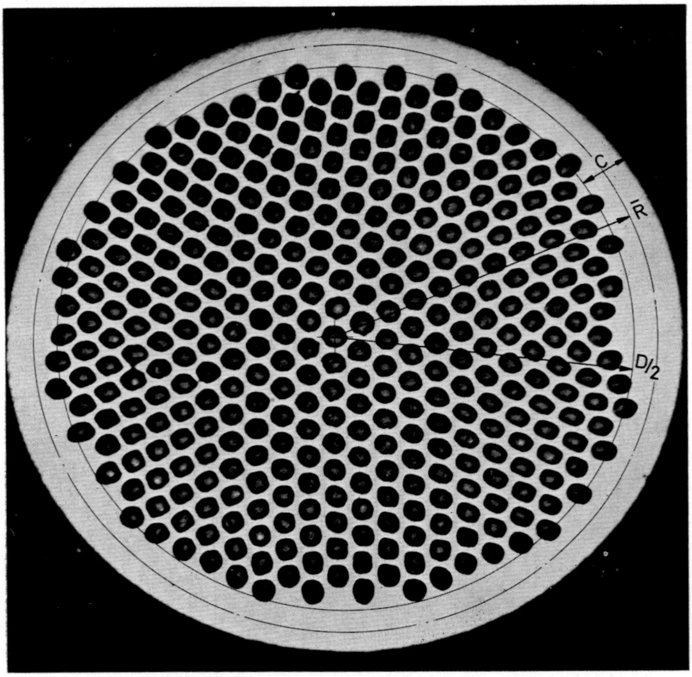

Fig. 4.1.12. Composite multifilamentary conductor

In presently available composite conductors, the filament bundle is still surrounded by a layer of "pure" matrix material of thickness C (see Fig. 4.1.12). The potential $V(\phi_p)$ in Eq. (4.1.47) causes additional currents in this part of the conductor such that the associate losses, p'_e/l per conductor length, are produced:

$$\frac{p'_e}{l} = \int_0^{2\pi} \overline{R}\, d\,\phi_p \cdot \left(\frac{d\,V(\phi_p)}{\overline{R}\, d\,\phi_p}\right)^2 \cdot \frac{C}{\varrho_{\text{matrix}}} = \frac{\pi}{4} D^2 \left(\frac{\dot{B}\, l_p}{2\,\pi}\right)^2 \cdot \frac{C}{\overline{R}} \cdot \frac{1}{\varrho_{\text{matrix}}}. \tag{4.1.52}$$

Eddy current losses per unit length for a cylindrical composite are thus given by

$$\boxed{\frac{p_e}{l} = \frac{\pi}{4} D^2 \left(\frac{\dot{B}\, l_p}{2\,\pi}\right)^2 \left(\frac{w-d}{w} + \frac{C}{\overline{R}}\right) \cdot \frac{1}{\varrho_{\text{matrix}}}.} \tag{4.1.53}$$

The surrounding layer (mean radius $\overline{R}$) can thus be taken into account simply by a modified resistivity.

The eddy current losses are inverse proportional to the effective resistivity. To calculate $\bar{\varrho}_c$, Fig. 4.1.13a illustrates a composite conductor with twisted filaments. A magnetization current flows across the matrix from a filament F_1 to a filament F_2 due to a variable transverse magnetic field. The current path is indicated by the dashed line PQ. From Maxwell's equation, we may write the relation between the current density and the changing magnetic field $\dot{B}$ for a surface A bound by a contour C:

$$\oint_C \varrho \boldsymbol{J} \, d\boldsymbol{l} = -\int_A \dot{\boldsymbol{B}} \, d\boldsymbol{A} \,. \tag{4.1.54}$$

Using the path $OP\,Q\,P'O'O$ for the contour C, we obtain:

$$\oint_C \varrho \boldsymbol{J} \, d\boldsymbol{l} = J_m \varrho_m \cdot (w - d) \, \frac{r \Delta\phi}{w} \,. \tag{4.1.55}$$

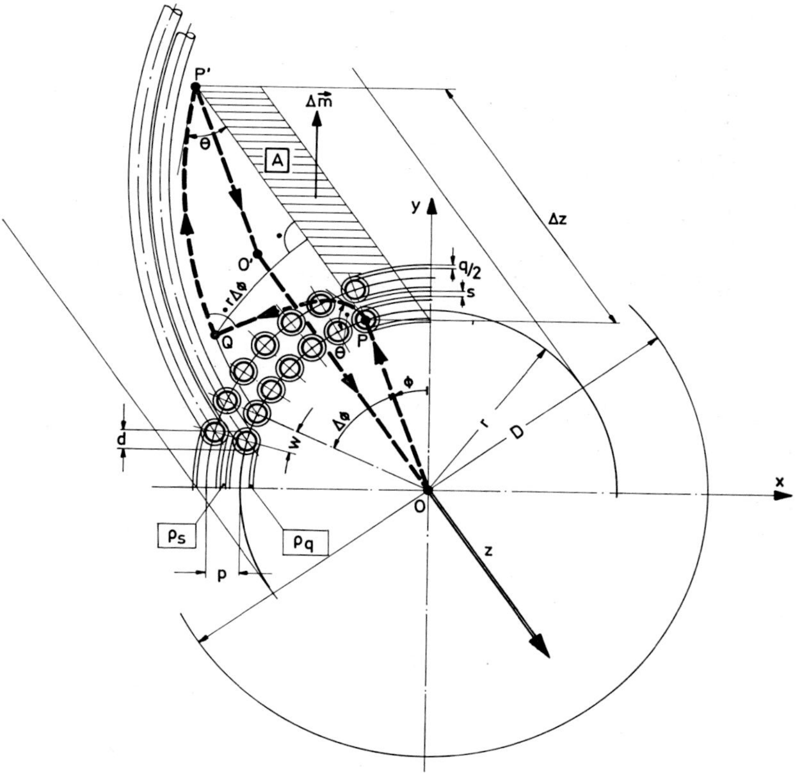

Fig. 4.1.13a

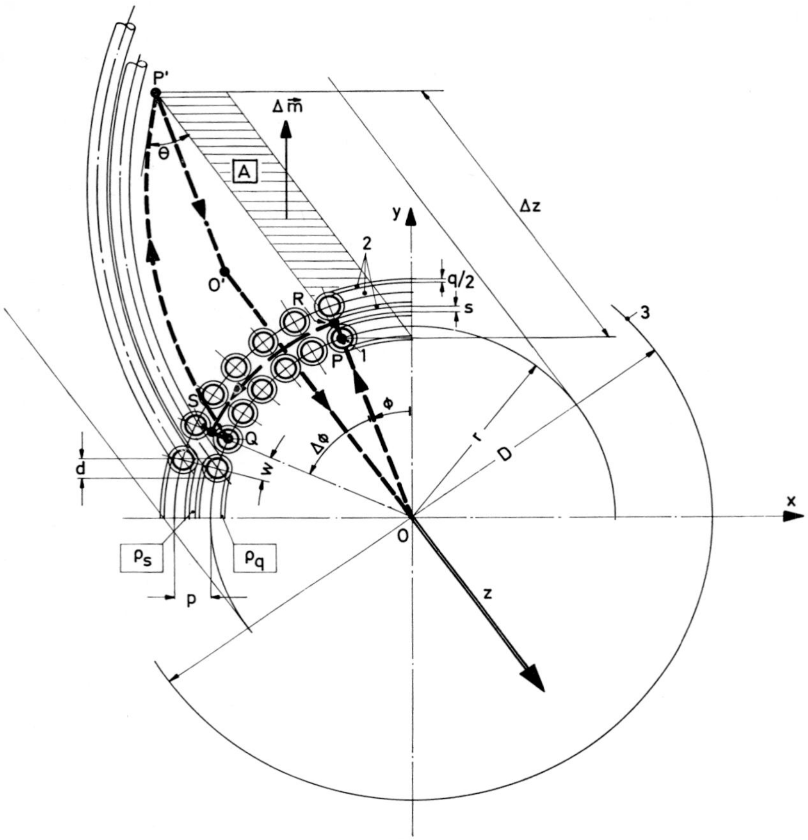

Fig. 4.1.13b

Fig. 4.1.13. Schematic representation of a composite conductor with twisted filaments to calculate $\bar{\varrho}_c$. *a* Integration path over $OPQP'O'O$, *b* Integration path over $OPRSQP'O'O$.
1 filament; 2 matrix material; 3 composite

ϱ_m is the resistivity of the matrix material. $r\Delta\phi/w$ gives the number of barriers of thickness $(w - d)$. Since

$$\int_A \dot{B}\, d\boldsymbol{A} \cong \dot{B}\cdot r \sin\phi\, dz$$

and

$$r\Delta\phi/\Delta z = \sin\theta\,,$$

we obtain:

$$\dot{B} r \sin\phi\,\Delta z = \frac{J_m\cdot\varrho_m\cdot(w-d)\cdot r\Delta\phi}{w}$$

and

$$J_m = \frac{\dot{B}\cdot r\sin\phi\,\Delta z\cdot w}{\varrho_m\cdot(w-d)\cdot r\Delta\phi} = \frac{\dot{B}\cdot r\sin\phi\cdot w}{\varrho_m\cdot(w-d)\sin\theta}\,. \tag{4.1.56}$$

The same consideration is applicable to the contour C' for an alternative path $O\,P\,R\,S\,Q\,P'O'O$, where the current passes also through a material of resistivity ϱ_q:

$$\int\limits_A \dot{B}\,dA \cong \dot{B} r \sin\phi\,\Delta z = \int\limits_R^S \varrho \boldsymbol{J}\,dl + \int\limits_P^R \varrho \boldsymbol{J}\,dl + \int\limits_S^Q \varrho \boldsymbol{J}\,dl$$

$$= J_s \varrho_s r \Delta\phi + \frac{\Delta J_s(\phi)\, s\cdot(p-d-s)\cdot\varrho_q}{2\,r\,\Delta\phi} - \frac{\Delta J_s(\phi+\Delta\phi)\, s\cdot(p-d-s)\,(p-d)\,\varrho_q}{2\,r\,\Delta\phi}.$$

Since:

$$\frac{1}{\Delta\phi}\left[\frac{\Delta J_s(\phi+\Delta\phi)-J_s(\phi)}{\Delta\phi}\right] \cong \frac{\partial^2 J_r(\phi)}{\partial\phi^2},$$

we have:

$$\dot{B} r \sin(\phi)\cdot\frac{\Delta z}{\Delta\phi} = J_s\cdot\varrho_s\cdot r - \frac{s\,(p-d-s)}{2\,r}\cdot\frac{\partial^2 J_s(\phi)}{\partial\phi^2},$$

or modified

$$\frac{\partial^2 J_s}{\partial\phi^2} - \frac{2\,J_s\cdot\varrho_s\cdot r^2}{s\,(p-d-s)\,\varrho_q} = -\frac{2\,\dot{B} r^3 \sin(\phi)}{s\,(p-d-s)\cdot\varrho_q\cdot\sin\theta}.$$

Applying the boundary condition $J_s(-\phi) = -J_s(\phi)$, we obtain the solution:

$$J_s(\phi) = \frac{2\,\dot{B} r^3}{s\,(p-d-s)\cdot\varrho_q + \varrho_s\cdot r^2}\cdot\frac{\sin\phi}{\sin\theta}, \tag{4.1.57}$$

where ϱ_q denotes the matrix resistivity in the radial direction adjacent to the filaments and ϱ_s the normal resistivity of the second matrix material. So far, we calculated the transverse and radial current densities J_m and J_s. These current densities are related to the longitudinal current density by

$$J(\phi) = \frac{J_m(\phi)\cdot d + J_s(\phi)\cdot s}{\sin\theta}, \tag{4.1.58}$$

where we assume the filaments to have a square shape. Referring to Fig. 4.1.13b, the magnetic moment of the current between ϕ and $\phi+\Delta\phi$ is given by:

$$\Delta m(\phi) = \mu_0 \int J\,dA = J\cdot r\Delta\phi\cdot A.$$

Since:

$$A = r\sin\phi\cdot\Delta z$$

$$\Delta m(\phi) = \mu_0\cdot\Delta z\cdot\Delta\phi\cdot r^2 J(\phi)\cdot\sin\phi.$$

Substituting for $J(\phi)$, the transverse and radial current density values and write the expressions for these current densities from Eqs. (4.1.56) and (4.1.57) in the above equation, one gets:

$$\Delta m(\phi) = \mu_0\cdot r^2 \Delta z\cdot\frac{\dot{B}}{\sin^2\theta}\cdot\sin^2\phi\cdot\Delta\phi\left[\frac{r\cdot d\cdot w}{(w-d)\,\varrho_m} + \frac{r^2}{(p-d-s)\cdot\frac{\varrho_q}{2\,r} + \frac{r}{s}\cdot\varrho_s}\right].$$

Integrating this expression over the circumference of the composite, we obtain the magnetic moment of the anullus with radius r:

$$m(r) = \mu_0 \cdot \frac{4 \cdot \dot{B} l^2}{\pi} \cdot \left[\frac{r \cdot d \cdot w}{(w - d) \cdot \varrho_m} + \frac{r^2}{\frac{(p - d - s)}{2r} \cdot \varrho_q + \frac{r}{s} \cdot \varrho_s} \right]. \tag{4.1.59}$$

Since the "magnetization" is defined as the magnetic moment per unit volume, i.e.

$$\boldsymbol{M} = \frac{\Sigma \boldsymbol{m}}{\Delta V},$$

where in our particular case: $\Delta V = \pi \frac{D^2}{4} \cdot \Delta z$ (D, the composite diameter), and the total magnetic moment is given by:

$$\sum_r m(r) = \frac{4 \mu_0 \cdot \dot{B} l^2}{\pi} \sum_r f(r)$$

with
$$f(r) = \frac{r \cdot d \cdot w}{(w - d) \varrho_m} + \frac{r^2}{\frac{(p - d - s)}{2r} \varrho_q + \frac{r}{s} \varrho_s}$$

the summation is carried out over the annulus radii.

The magnetization of the multifilamentary conductor can be calculated from the above:

$$\begin{aligned} M &= M_0 \left[1 + \frac{\dot{B} l^2}{\lambda J_c d} \frac{24}{\pi D^2} \sum_r f(r) \right] \\ &= M_0 \left| 1 + \frac{\dot{B} l^2}{\lambda J_c d} \cdot \frac{1}{\bar{\varrho}_c} \right|, \end{aligned} \tag{4.1.60}$$

where
$$M_0 \cong 2 \mu_0 \lambda \frac{d \cdot J_c}{3 \pi}$$

and

$$\boxed{\frac{1}{\bar{\varrho}_c} = \frac{24}{\pi D^2} \sum_r \left[\frac{r \cdot d \cdot w}{(w - d) \cdot \varrho_m} + \frac{r^2}{\frac{(p - d - s)}{2r} \varrho_q + \frac{r}{s} \cdot \varrho_s} \right].} \tag{4.1.61}$$

From this general equation, we can calculate the effective resistivity of a single metal matrix. In this case: $\varrho_m = \varrho_s$; $p - d - q = 0$; $s = w - d$. After some manipulation, one gets for the effective resistivity, $\bar{\varrho}_c = \frac{\pi}{3} \cdot \varrho_m \cdot \frac{w}{w - d} \cong \varrho_m \cdot \frac{w}{w - d}$, which was written above.
Let us now assume that no transport current I_T flows through the conductor. As the current density J_T within the filament bundle in the composite is homogeneous, there can be no net induced currents along the inner filaments.

All transverse currents must be collected by the filaments of the outer layer only. If there are N filaments in the outer layer of diameter D, the induced current I_F through each filament is:

$$\frac{d I_F}{d z} = \pm \frac{2\pi}{l_p} \cdot \frac{d I_F}{d \phi_p} = \frac{2\pi}{N} \cdot J_t \cdot \frac{D}{2} \sin(\phi_p)$$

or:

$$I_F(\phi_p) = \pm \int_0^{\phi_p} \frac{d I_F}{d \phi_p} \cdot d\phi_p = \pm \frac{\pi D}{N} J_t \cdot \frac{l_p}{2\pi} \cdot \cos(\phi_p) . \qquad (4.1.62)$$

The cos (ϕ_p) distribution of the induced currents generates a homogeneous self field B_i in the direction of the field B_{ext}, justifying the assumption of homogeneous B over the conductor.

$$B_i = \frac{1}{2} \mu_0 \cdot \frac{2\pi}{N} \cdot I_F(0) \cdot \frac{N}{\pi D},$$

$$B_i = \pm \mu_0 \cdot \frac{J_t}{2} \cdot \frac{l_p}{2\pi} = -\frac{\mu_0}{2} \left(\frac{l_p}{2\pi}\right)^2 \cdot \frac{1}{\bar{\varrho}_c} (\dot{B}_{\mathrm{ext}} + \dot{B}_i) . \qquad (4.1.63)$$

The azimuthal component of $I_F(\phi_p)$ produces a further component of B in the direction of the conductor axis. For $l_p \gg D$, this component can

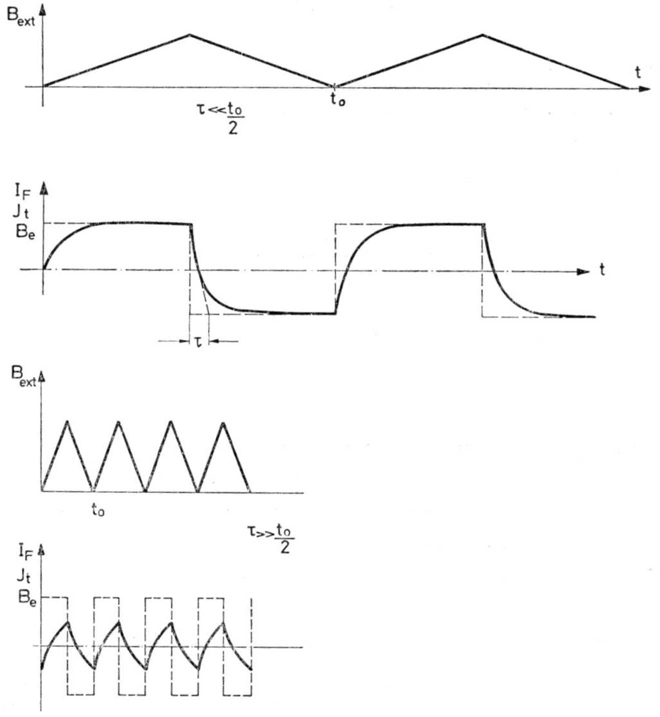

Fig. 4.1.14. Pulsed fields of triangular shape and induced currents into the superconducting filament

be neglected. The negative sign in Eq. (4.1.63) reflects the fact that the currents counteract the change of the external field.

If $\dot{B}$ is expressed by the change of transverse current density $\dot{J}_t$, one obtains the equation:

$$\frac{d J_t}{d t} + \frac{2}{\mu_0}\left(\frac{2\pi}{l_p}\right)^2 \cdot \bar{\varrho}_c \cdot J_t = \pm\, 2\, \dot{B}_{\text{ext}} \cdot \left(\frac{2\pi}{\mu_0 l_p}\right). \tag{4.1.64}$$

The homogeneous solution of Eq. (4.1.64) yields, after a sudden change in B_{ext} occurs, an exponential decay for J_t:

$$J_t(t) = J_t(0) \cdot e^{-t/\tau}, \tag{4.1.65}$$

where the time constant τ, is given by:

$$\boxed{\tau = \frac{\mu_0}{2} \cdot \left(\frac{l_p}{2\pi}\right)^2 \cdot \frac{1}{\bar{\varrho}_c},} \tag{4.1.66}$$

τ can be understood as the ratio of the inductance of the filament loops and the matrix resistance. For a triangular pulsed external field, Fig. 4.1.14 illustrates the transverse current (J_t) as well as the induced current (I_F) into the filaments, and the field B_e for two limiting cases of $\tau \ll t_0/2$ and $\tau \gg t_0/2$, with t_0 denoting the pulse length. With increasing ratio of τ/t_0, the variation of the external field in the conductor is shielded more and more by the induced currents in the filaments. For $\tau \ll t_0/2$, the losses in the matrix are modified only slightly by the transient phenomena. The current carrying capacity of the filament is limited by the critical current I_c. Beyond this value, flux crosses the outer layer of filaments. Energy is dissipated, leading to a longitudinal voltage drop. The next inner layer takes over the expressive part of the shielding currents and leads to a partial coupling between filaments.

By equating the induced screening current $I_F(\phi_p = 0)$ from Eq. (4.1.62), with the critical current I_c for a single filament, assuming no transport current is flowing, we get a relation between field rise and the "critical twist length" l_c.

$$I_{F,\max} = I_c = \frac{1}{\bar{\varrho}_c} \cdot \left(\frac{l_c}{2\pi}\right)^2 \cdot \frac{\pi D}{N} \cdot \dot{B}\,. \tag{4.1.67}$$

The critical twist length, l_c, is expresses for $\dot{B} \approx \dot{B}_{\text{ext}}$, i.e. for $\tau \ll t_0$, by:

$$\boxed{l_c^2 = (2\pi)^2 \cdot \bar{\varrho}_c \cdot I_c \cdot \frac{N}{\pi D} \cdot \frac{1}{|\dot{B}_{\text{ext}}|}.} \tag{4.1.68}$$

This equation is different by the numerical factor $(\pi^3/32)^{\frac{1}{2}}$ from the expression that Morgan [10] has obtained for a two filament conductor model (TV-cable).

For a sinusoidally varying field, expressed by $B_{\text{ext}} = \bar{B}_{\text{ext}} + \tilde{B}_{\text{ext}} \cdot \exp(j\omega t)$, we consider only the variable part, and get from Eq. (4.1.47)

$$J_t = \frac{2\pi}{\mu_0 l_p} \cdot \tilde{B}_{\text{ext}} \cdot \left(1 + \frac{1}{j\omega\tau}\right)^{-1} \tag{4.1.69}$$

and obtain the losses per unit volume from:

$$\boxed{p_e = \frac{|J_t|^2}{2}\,\bar{\varrho}_c = \frac{(\tilde{B}_{\text{ext}})^2}{4\,\mu_0\,\tau}\left(1 + \frac{1}{\omega^2\tau^2}\right)^{-1}} \quad \text{(W)} . \tag{4.1.70}$$

For $\omega \ll 1/\tau$ (low frequencies), p_e is proportional to ω^2. This is different from the frequency dependence of hysteretic losses, which are proportional to ω!

As p_e is also proportional to τ, high resistivity matrix materials and short twist pitch are recommended in contrary to the dynamic stability criterion, which requires low electrical resistivity materials.

With increasing frequency, eddy current losses approach the value of $p_e = (\tilde{B}_{\text{ext}})^2/4\,\mu_0\tau$. The field variation within the composite decreases with increasing frequency due to shielding currents.

From $B_{\text{ext}} = \bar{B}_{\text{ext}} + \tilde{B}_{\text{ext}} \exp(j\omega t)$, and $\tilde{B}_i = \tilde{B} - \tilde{B}_{\text{ext}} = -\tau \cdot \dot{\tilde{B}}$, we get for the alternating part of the field seen by the filaments

$$\tilde{B} = \tilde{B}_{\text{ext}} \cdot e^{j\omega t}\,(1 + j\,\omega\tau)^{-1} . \tag{4.1.71}$$

This part induces the common hysteretic losses due to flux movements in the superconductor. These losses depend only on the maximum and minimum field values.

The inner filaments of a bundle, which are not forced to carry screening currents, see only the shielded field, and the dissipated energy per cycle decreases with $1/(1 + j\,\omega\tau)$. This means that the average hysteretic loss rate approaches a constant value with increasing frequencies. This conclusion may be of interest for industrial (i.e. 50 Hz) applications.

Qualitative confirmation of this effect was obtained by McInturff [11], for the similar case of transposed braids with a metallic* insulation. At a field change of $\dot{B} \gtrsim 4$ T/sec, the dissipation rate approached a constant value.

The above model is applicable to metallic insulated cables where multifilament conductors are simply twisted. Transposed cables and braids will show a more or less complicated internal eddy current distribution pattern depending on the manufacturing procedure. For this case the model developed here will no longer be in quantitative agreement with the experimental data.

Other measurements by Critchlow et al. [19] on the effect of twist on the magnetization of the multistrand superconding NbTi composites show that the magnetization peak and the external field at which the peak occurs gradually decrease. This effect was observed at approximately one twist per cm. Further twisting produced little or no change in the magnetization curve. The magnetization of samples with one twist of less than 1 per cm, was field sweep dependent. The effect is in qualitative agreement with Eq. (4.1.68). Losses measured on the same samples show a marked reduction with increasing twist rate up to one twist per cm. Increasing the twist rate above this value did not lead to a further loss

* Metallic insulation refers to a solder of high electrical resistivity such as Sn-Ag, Sn-In, In(Tl) alloy, or others.

reduction. The samples tested had a wire diameter of 0.075 cm, a filament diameter of 7 μm. The matrix material used was OFHC copper. 68 strands formed a cable of 1.8 m length. Matrix to superconductor ratio was 3 : 1.

Fluxjumps were observed only at a field sweep rate of 0.2 T/s. At slower sweep rates no fluxjumps were triggert. In untwisted multistrand conductors the fluxjump stability is size dependent, as was expected. In conductors with a cupronickel matrix the twisting did not eliminate fluxjumping. Thus while a cupronickel matrix is advantageous in reducing transverse currents, it is detrimental to stability.

4.1.12 Self Field Losses

It may be pointed out, that in the above treatment of losses the effect of field inhomogeneities on the conductor was not considered. Specifically the self field effect due to transport currents in the wire has the same origin as the skin effect in the normal conductor, and tends to exclude current density changes in the interior of the wire. This effect is not compensated by simple twisting, but by transposition of filaments.

Fluxchanging induced by the self fields in a superconductor lead to movement of fluxoids within the superconductor and thus to dissipative losses. These losses are of electromagnetic nature and may be obtained by integrating the Poynting vector $\boldsymbol{S} = (\boldsymbol{E} \cdot \boldsymbol{H})$ over the conductor surface. The irreversible part of this energy, integrated over a cycle, is dissipated as heat.

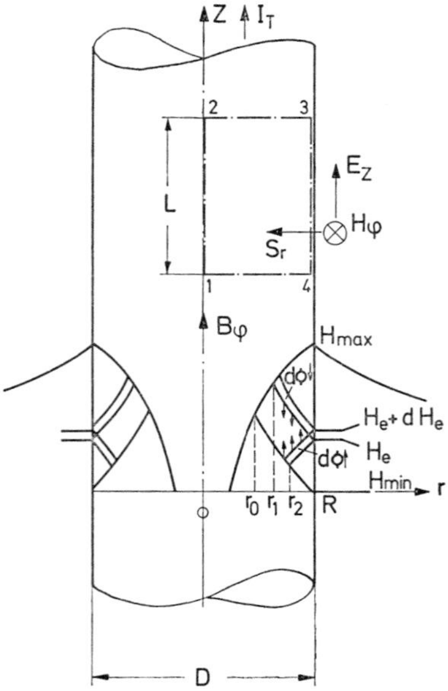

Fig. 4.1.15. Composite multifilamentary conductor

The component S_r of the Poynting vector, perpendicular to the conductor surface, is obtained from the azimuthal component of the self field $B_\theta(R) = \mu_0 I/2\pi R$ and the axial component of the electric field E_z at the conductor surface, when the flux penetrates the surface.

From Fig. 4.1.15, we obtain the voltage for a closed loop C:

$$V = \int \boldsymbol{E}\, d\boldsymbol{s} = -\dot{\Phi} = -\int \dot{B}_\varphi(r)\, dr \cdot L = E_z \cdot L\,. \qquad (4.1.72)$$

The electric field is zero on the conductor axis $\overline{1-2}$ and has no contribution to $\overline{2-3}$ and $\overline{4-1}$. The electric field, E_z, is directed such to exclude current changes from the interior of the conductor. Only if the critical current density is exceeded in the superconductor, an associated longitudinal voltage drop is generated.

The self field due to the transport current can penetrate further into the composite conductor.

Evidently, the critical state model, developed for a superconductor located in an external field, is also valid for the self field penetration produced by a transport current. The model is still valid for a multifilament conductor, if the integration paths $\overline{1-2}$ and $\overline{3-4}$ of Fig. 4.1.15 are chosen along the filaments. An external field can enter the matrix without flux flow (if $l_p \ll l_c$), whereas for the self field, the same conductor behaves as a compact filament.*

In magnets, the external field is much higher than the self field of the conductor. Thus, J_c is the same across the conductor and the Bean-model, modified for a cylindrical geometry, can be applied. At a peak current $I_{\max}$, the critical state current density $J = J_c$ penetrates the conductor up to a radius $r = r_0$. This current is given by $I_{\max} = \bar{J}_c \cdot \pi \cdot (R^2 - r_0^2)$, where R denotes the radius of the composite conductor. The associated self field $H_\theta(r)$ is given by curl $\boldsymbol{H} = \boldsymbol{J}$, which is expressed in the cylindrical geometry by

$$\frac{dH_\theta}{dr} + \frac{H_\theta}{r} = J_z \qquad (4.1.73)$$

and has the solution:

$$H_\phi(r) = \pm \frac{rJ_c}{2} \pm \frac{R}{r}\left(H_e + \frac{\bar{J}_c R}{2}\right). \qquad (4.1.74)$$

The $(\pm)$ sign depends on the sign of $J_z = \pm J_c$. The field H_e denotes the self field $H_\phi(R)$ at the conductor surface. During current cycling in the range between $I_{\min}$ and $I_{\max}$, self field patterns as shown in Fig. 4.1.16 are obtained. The nonlinear field $H_\phi(r)$ is the field generated in concentric tubes, each having a uniform current density $\pm \bar{J}_c$.

The change of the field $B_\phi(r)$ in the conductor is linked to the variation of the self field $H_e = I/2\pi R$ by:

$$\frac{dB_\phi(r)}{dt} = \mu_0 \frac{dH(r)}{dH_e} \cdot \frac{dH_e}{dt} = \mu_0 \frac{R}{r} \cdot \frac{dH_e}{dt}\,. \qquad (4.1.75)$$

The dissipative part of $\boldsymbol{S}$ is obtained from the difference of $\dot{\phi}$, if the transport current I_T is raised (field energy is entering the probe) and

* For simplification, the filamentary structure of the conductor is replaced by a single conductor having a mean current density $\bar{J}_c$.

then reduced (part of the field energy leaves the probe) as illustrated in Fig. 4.1.15.

Thus, from $E_z = V/L$, we obtain:

$$E_z^{\uparrow} - E_z^{\downarrow} = -\mu_0 \left[\int_{r_2}^{R} \frac{R}{r} dr - \int_{r_1}^{R} \frac{R}{r} dr \right] \left[\frac{dH_e}{dt} \right] = \mu_0 \left[\frac{dH_e}{dt} \right] R \cdot \ln \left(\frac{r_2}{r_1} \right). \tag{4.1.76}$$

The values of r_1 and r_2 are obtained from the relations:

$$H(r_1) = \frac{r_1}{2} \bar{J}_c + \frac{R}{r_1} \left(H_{\max} - \frac{\bar{J}_c R}{2} \right) = -\frac{r_1}{2} \bar{J}_c + \frac{R}{r_1} \left(H_e + \frac{\bar{J}_c R}{2} \right),$$

from which we calculate:

$$r_1 = R \left[1 - \frac{H_{\max} - H_e}{\bar{J}_c R} \right]^{\frac{1}{2}} \tag{4.1.77}$$

and similarly:

$$r_2 = R \left[1 - \frac{H_e - H_{\min}}{\bar{J}_c R} \right]^{\frac{1}{2}}. \tag{4.1.78}$$

Self field losses per cycle can be calculated for a conductor with a surface area of $2\pi RL$ by integrating $(E^{\uparrow} - E^{\downarrow}) \cdot H_e$ over half a cycle.

To simplify calculations, we have assumed that J_c is independent of the field:

$$W_{sf} = 2\pi R^2 L \int_0^{t_0/2} H_e (E_z^{\uparrow} - E_z^{\downarrow})\, dt$$

$$= \pi R^2 L \mu_0 \int_{H_{\min}}^{H_{\max}} H_e \left[\ln \left(1 - \frac{H_{\max} - H_e}{\bar{J}_c R} \right) - \ln \left(1 - \frac{H_e - H_{\min}}{\bar{J}_c R} \right) \right] dH_e, \tag{4.1.79}$$

which after integration is rearranged to the form:

$$W_{sf} = \pi \mu_0 \bar{J}_c^2 R^4 L \cdot \left\{ \left(1 - \frac{H_{\max} - H_{\min}}{\bar{J}_c R} \right) \cdot \left[\ln \left(1 - \frac{H_{\max} - H_{\min}}{\bar{J}_c R} \right) \right. \right.$$

$$\left. \left. + \frac{H_{\max} - H_{\min}}{\bar{J}_c R} \right] + \frac{1}{2} \left(\frac{H_{\max} - H_{\min}}{\bar{J}_c R} \right)^2 \right\} \text{ (Ws/cycle)}. \tag{4.1.80}$$

Modifying this equation for transport currents instead, one uses $I_c = \pi R^2 \bar{J}_c$, the critical short sample current, and introduces the transport current difference ΔI. This is the change in current from the lowest to the peak value expressed by:

$$\frac{H_{\max} - H_{\min}}{\bar{J}_c R} = \frac{\Delta I}{2\pi \bar{J}_c R^2} = \frac{\Delta I}{2 \bar{I}_c}.$$

For a transport current cycled between the values $+I_0$ and $-I_0$ (Fig. 4.1.16), Eq. (4.1.80) is identical to the expression derived by Hancox [8] for a cylindrical superconductor without external field.

The logarithmic term in Eq. (4.1.80) is expanded in a power series, and one obtains:

$$W_{sf} = \frac{\mu_0}{\pi} I_c^2 L \left\{ \frac{1}{2 \cdot 3} \left(\frac{\Delta I}{2 I_c} \right)^3 + \frac{1}{3 \cdot 4} \left(\frac{\Delta I}{2 I_c} \right)^4 + \frac{1}{4 \cdot 5} \left(\frac{\Delta I}{2 I_c} \right)^5 + \cdots \right\}, \tag{4.1.81}$$

or:

$$\boxed{W_{sf} = \frac{\mu_0}{4 \pi} \cdot (\Delta I)^2 L \left\{ \frac{1}{2 \cdot 3} \left(\frac{\Delta I}{2 I_c} \right) + \frac{1}{3 \cdot 4} \left(\frac{\Delta I}{2 I_c} \right)^2 + \cdots \right\}.} \tag{4.1.82}$$

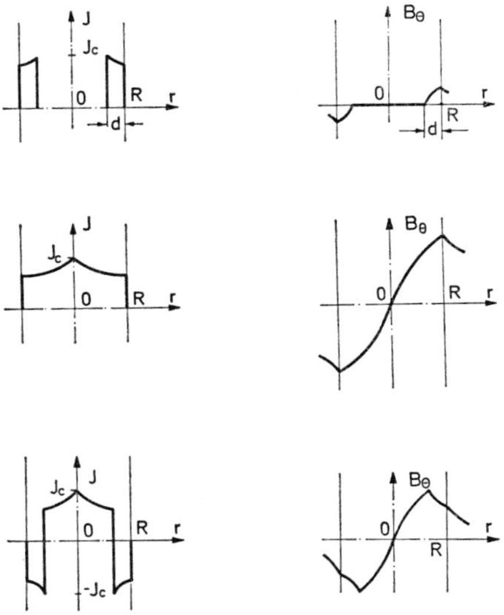

Fig. 4.1.16. Transport current and field profile in a superconducting cylinder

As $(\Delta I) \cdot L$ (total current change multiplied by the length of a strand) is nearly constant for a coil of a given geometry, the self field losses are proportional to ΔI and thus to the square of the diameter D of the composite conductor.

For $D \to \infty$ (as for a slab), the only remaining term in Eq. (4.1.80) is:

$$\boxed{W_{sf} = \frac{\mu_0}{6} \frac{\pi R L}{J_c} (H_{max} - H_{min})^3} \quad \text{(Ws/cycle)}, \tag{4.1.83}$$

which is essentially the expression Bean obtained for hysteretic losses in a slab with a surface area of $2 \pi R L$ located in an external field parallel to the surface of the slab and cycled between the field values H_{min} and H_{max}. Self field losses are treated by Brechna and Ries [12, 13].

4.1.13 Contribution of External Fields

In the above calculation we had omitted the influence of a superposed external field in addition to the self field in the Poynting vector at the conductor surface, which will now be rectified.

The axial field component $(B_{ext})_z$ is parallel to E_z and does not contribute to W_{sf}. The transverse field component can be expressed in the form of multipoles:

$$\begin{pmatrix} B_{ext,\phi} \\ B_{ext,r} \end{pmatrix} = \sum_{n=1}^{\infty} B_{ext,n} r^{n-1} \begin{pmatrix} \cos(n\phi) \\ \sin(n\phi) \end{pmatrix}. \qquad (4.1.84)$$

All terms of the series with $n > 1$ describe higher field multipoles. As mentioned, hysteretic- and eddy current losses can be expressed by a Poynting vector at the surface of the conductor due to the electric field $E_{z,ext}(\phi)$ for which a similar harmonic expansion as for $B_{ext,\phi}$ is valid with of $n \geqq 1$.

Combined with the self fields $E_{0,z} H_{0,\phi}$ (which are independent of ϕ in the cylindrical case), the Poynting vector can be expressed by:

$$S_{tot} = (E_{ext,z} + E_{0,z})(H_{ext,\phi} + H_{0,\phi}) .$$

S_{tot}, integrated over the conductor surface, gives only the contribution of the external field and of the self field as evaluated above, while the mixed terms yield no contribution. This means that the loss sources can be treated independently, even in the case of a non-uniform external field.

4.1.14 Discussion

The equations for self field losses were derived, assuming $\bar{J}_c$ is constant. If $\bar{J}_c$ changes with the applied field (e.g. due to the self field of a coil, where $B_{ext} \sim I$), Eq. (4.1.82) must be corrected by using $\bar{J}_c(B_{ext}(I))$. The two integration boundaries r_1 and r_2, must also be modified. Analytical integration over one cycle is no longer possible. Self field losses in a coil can be approximated if one uses $\bar{J}_c$ value at an average field over the entire coil and a complete cycle which is about a quarter of the aperture peak field. For a zero external field, losses were measured for a changing transport current in the range of $-I_0 \to 0 \to +I_0$ for 5 samples (Fig. 4.1.17). Losses for $I_0 = I_c$ are denoted by triangles. The broken line is calculated from Eq. (4.1.82); for $\Delta I = 2\,I_c$, which yields (per unit, conductor length), the relation:

$$\boxed{W_{sf,max} = \frac{\mu_0}{2\pi} I_c^2 .} \qquad (4.1.85)$$

Comparing the loss data in Fig. 4.1.17, measured for niobium filaments with self field values measured in a number of drawn Type II superconducting wires, all 0.025 cm in diameter (Fig. 4.1.18), the following observation is possible: The data for the two superconductors Pb and Nb are for as received materials. Polishing the surfaces of these materials yield-

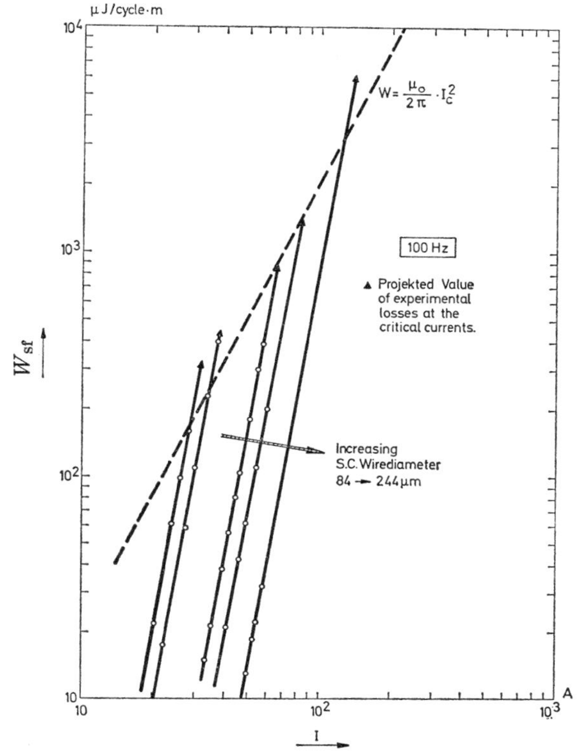

Fig. 4.1.17. Energy dissipation per unit length of specimen vs. peak transport current. The loss at the critical current for each specimen is indicated by a triangle [14]. The dashed line is according to Hancox's expression

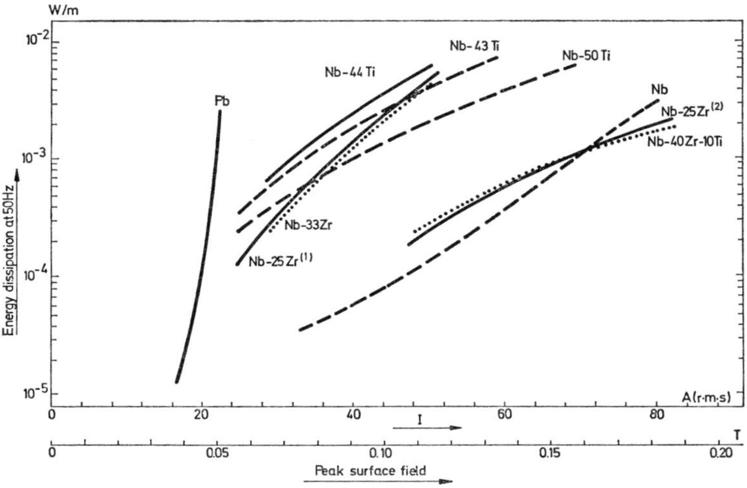

Fig. 4.1.18. 50 Hz self field losses for lead and Type II superconducting wires (wire diameter 0.025 cm)

ed a significant reduction in a.c. losses, whereas roughening the surface has the opposite effect [15]. The losses show a frequency dependence of f^n with $0.65 \leqq n \leqq 0.9$ and imply viscous motion of the penetration of flux structure during part of the a.c. cycle. The other Type II superconductors were obtained from several manufactorers. The standard NbTi and NbZr wires exhibit similar a.c. losses over the range of current investigated. It is interesting to note that the Nb-40Zr-10Ti wire exhibited the lowest loss rates compared to the other superconductors. The sample Nb-25Zr (2) was heat treated in vacuum for two hours at 650 °C.

The losses of NbTi and NbZrTi wires are proportional to frequency implying a hysteretic loss mechanism and show little dependence on surface condition. The lowest a.c. losses measured were in the field ranges of Pb (0–4×10^{-2} T), Nb (4×10^{-2} — 1.6×10^{-1} T) and Nb-40Zr-10Ti ($> 1.6\times10^{-1}$ T).

The effect of filament size and twist pitch on self field losses is measured by Eastham and Rhodes [16]. The tests indicate that the a.c. losses are practically independent of filament size in a composite conductor. Losses decrease somewhat with increasing twist pitch.*

4.1.15 Comparison between Self-Field and Hysteretic Losses

It was shown that the hysteretic losses are proportional to d, the diameter of the individual filament and the self field losses are proportional to the square of the diameter D of the strand or composite conductor. For practically all superconducting magnets, the ratio $\Delta I/2\,I_c \ll 1$, thus we may neglect all higher terms in Eq. (4.1.82). Assuming a constant critical current density as in the case of W_{sf}, we may write for the hysteretic losses:

$$W_{hl} = \mu_0 J_c \frac{d}{2} V \Delta H_{ext} \tag{4.1.86}$$

with V the conductor volume. ΔH_{ext} is the maximum range of the changing external field in one cycle.

The self field losses are given approximately by:

$$W_{sf} \cong \mu_0 \pi \frac{D^4 \bar{J}_c^2 L}{4 \cdot 112} \left(\frac{\Delta I}{I_c}\right)^3 = \mu_0 \frac{D^2 \bar{J}_c^2 V}{192} \left(\frac{\Delta I}{I_c}\right)^3 \tag{4.1.87}$$

Thus:

$$\boxed{\frac{W_{sf}}{W_{hl}} = \frac{D^2}{d} \frac{\bar{J}_c}{96} \frac{(\Delta I/I_c)^3}{\Delta H_{ext}}} \tag{4.1.88}$$

For a magnet with $\Delta H_{ext} = 20\times10^3$ A cm^{-1}, $J_c = 10^5$ A cm^{-2}, where ($J_c(H)$) is taken at one quarter of the peak field value) and $\Delta I/I_c = 0.5$ one obtains

$$\frac{W_{sf}}{W_{hl}} \sim 6.5\times10^{-3} \frac{D^2}{d}$$

where D and d are expressed in (cm).

* Losses decreased by a factor of about 2, when the twist pitch of a 61 filament wire was increased from 0 to 5.5 twist per cm.

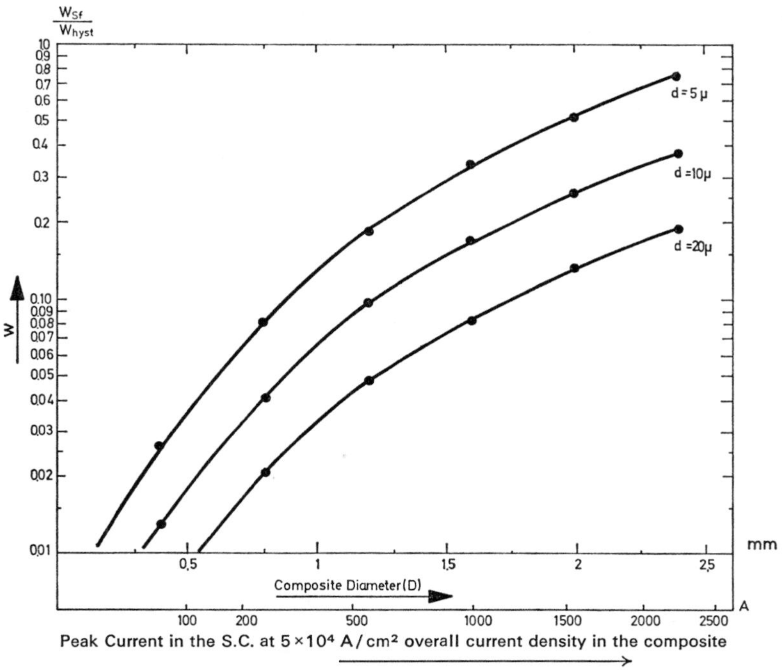

Fig. 4.1.19. Comparison between self field and hysteretic losses for a pulsed magnet with 5 T peak field and 5×10^4 A/cm^2 overall current density and vs. composite diameter

Fig. 4.1.19 illustrates this relation for a pulsed magnet with a 5 T peak field at 5×10^4 A cm^{-2} overall current density.

4.1.16 Modification of the Hysteretic Losses, if the Transport Current is not Zero

In Section 4.1.6, it was shown that the hysteretic losses in a superconducting slab, carrying a transport current I_T, is increased by the factor, $[1 + (I_T/I_c)^2]$.

Nearly the same factor is expected for cylindrical conductors.

As shown in the preceding chapters, according to Bean's critical state model, only the outer filaments in the penetration region carry the transport current. This is particularly true in the low field region of the coil, due to the high current carrying capacity of type II materials at low fields.

In the penetration regions ($r_0 - R$, $r_0 - r_1$, and $r_0 - r_2$, resp.), the filament is not in its critical state condition except near the current maximum (see Fig. 4.1.15). The reason is that the individual current per filament, given by its critical value at the field maximum, is conserved at low field levels, whereas I_c increases such that the correction factor $[1 + (I_T/I_c)^2]$ is zero or very small for a large portion of filaments during most of the time of the field cycle. The overall correction, which would be difficult to evaluate is expected to be small and can be neglected in comparison with other error sources such as $J_c(H)$, filament diameter, matrix resistivity, average field over the coil, etc.

4.2 Additional Effects in Twisted Multifilamentary Conductors

4.2.1 Axial Diffusion of the Self Field

As shown, changes of the transport current can be accepted by the inner filament circles only, if the flux density B_e produced by the self field changes in the space between adjacent filaments.

This requires that

$$\oint E\, dS = -\dot{\phi} \neq 0\,, \tag{4.2.1}$$

as shown a voltage appears at the outer filament circle only if the current exceeds the critical J_c value and the vortices start to move.

Even if the critical current is not exceeded in the outer filament circles, a voltage can appear across matrix material as a result of currents flowing between filaments.

In the following the derivation of the diffusion equation of the self field $B(r)$ inside a composite multifilament conductor is given. Here the twist induced current density in the noncritical inner part of the conductor is neglected temporarily but may be added to the result as a first approximation. The conclusion of this calculation may give one possible explanation of the degradation phenomens, observed in long conductors compared to short sample values: The self field

$$B_\phi(r) = \mu_0 \frac{I(r)}{2\pi r} \tag{4.2.2}$$

describes also the acceptance of current by the inner filament circles, where $I(r)$ is the portion of the transport current, which (in case of cylindrical geometry) flows within a cylinder or radius r.

We consider an area $dF = dz \cdot \Delta r$ between two filament circles (Fig. 4.2.1), of radii r_1 and $r_1 + \Delta r = r_2$. The paths $\overline{1-2}$ and $\overline{3-4}$ are located

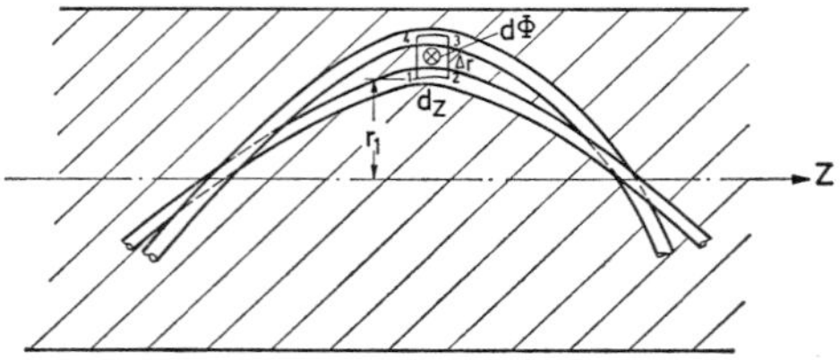

Fig. 4.2.1. Composite conductor with twisted filaments

in the electrical center of the filaments, to avoid potential differences over these passages. The flux $d\phi$ through dF depends on the current $I\ (r_1,z)$ flowing through filaments located within the cylinder with a radius r_1:

$$d\phi = \mu_0 dz\, \Delta r \cdot \frac{I\ (r,z)}{2\,\pi r}. \tag{4.2.3}$$

Eq. (4.2.3) is true only if $\Delta r \ll r_1, r_2$.

The induced voltages are given by:

$$U_{14} - U_{23} = \frac{\partial U}{\partial z}\, dz = -\,d\dot{\phi} = -\,\mu_0 \cdot \frac{dz}{2\,\pi r}\,\Delta r \cdot \frac{dI\ (r,z)}{dt}. \tag{4.2.4}$$

The voltage $U\ (z)$ between filament circles generates radial cross-currents flowing through the matrix, i.e. the transport current must be redistributed in the z coordinate between the outer ($r \geqq r_1$) filaments.

An external field (assumed homogeneous) produces a flux $d\phi$, but the additional induced eddy currents are only small if the twist pitch is short enough and yields a sinewave modulation of the current along the filament z (Section 4.1.10).

If the electrical conductivity of the matrix materials is γ, the radial current density in the matrix can be given by:

$$J_r(r,z) = \frac{U\ (z) \cdot \gamma}{\overline{\Delta r}}, \tag{4.2.5}$$

with $\overline{\Delta r}$ the averaged distance between filaments.

The current redistribution within a radius r is given by:

$$\frac{\partial I\ (r,z)}{\partial z} = 2\,\pi r\ U\ (z) \cdot \frac{\gamma}{\overline{\Delta r}}. \tag{4.2.6}$$

Combining Eq. (4.2.4) and Eq. (4.2.6), we obtain the differential equation for the current $I\ (r)$:

$$\frac{\partial^2 I\ (r,z)}{\partial z^2} = 2\,\pi r \frac{\gamma}{\Delta r} \cdot \frac{\partial U}{\partial z} = -\,\mu_0 \frac{\Delta r}{\overline{\Delta r}}\,\gamma\, \frac{dI\ (r,z)}{dt}. \tag{4.2.7}$$

The self field penetration in the filament bundle along with the current distribution (assuming there is no longitudinal voltage drop along the superconductor, which is true for $J < J_c$), is accomplished axially according to a diffusion equation with the diffusion constant:

$$D_m = \frac{\overline{\Delta r}}{\mu_0 \gamma \Delta r} \sim \frac{1}{\mu_0 \gamma}. \tag{4.2.8}$$

4.2.2 Solution of $I\ (r,\ z,\ t)$

1. For a conductor infinitely extended in the z-direction and carrying a transport current, there is no z-dependent solution of the current $I\ (r,z,t)$:

$$\frac{\partial I\ (r,z,t)}{\partial z} = 0\,.$$

Self field flux and transport currents can only penetrate radially the conductor through the flux flow produce.

2. The multifilament conductor is connected at one end ($z = 0$) to a normal metal, e.q. the superconducting coil end to the current lead.

The small change ΔI of the total transport current $I_{tot} = I(R,z,t)$, independent of already flowing equilibrium currents $I_0(r)$, shall be a step function at $t = 0$:

$$I(R,0,t) = I_0(r) + \Delta I\, \theta(t) \tag{4.2.9}$$

$$\theta = \begin{cases} 0 & \text{for} \quad t < 0\,. \\ 1 & \text{for} \quad t > 0 \end{cases}$$

If $I < I_c$, the filaments at the outer circle, will carry the entire current ΔI at $t = 0$. The currents in the inner filaments are forced radially outwards through the matrix by means of induced anticurrents at $z \geqq 0$ (discussed in Section 4.1.12).

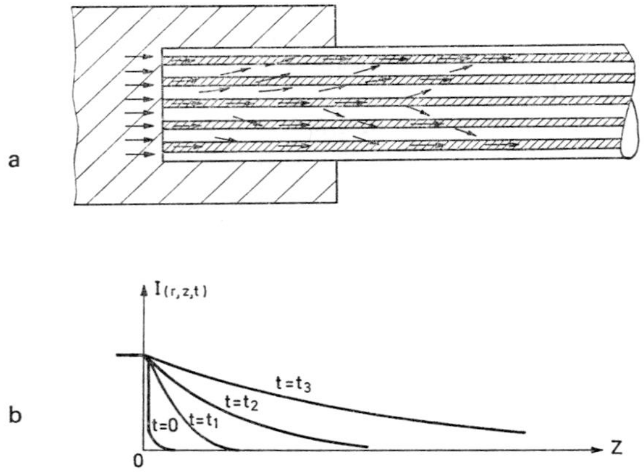

Fig. 4.2.2. Axial diffusion of a transport current through a normal joint into a composite conductor. *a* Schematic representation of the current paths *b* Current distribution vs. length of the composite

As illustrated in Fig. 4.2.2, the initial situation is developing according to the solution of Eq. (4.2.5), with the described boundary conditions at $t = 0$:

$$I(r,z,t) - I_0(r) = (I(r,0,0) - I_0(r))\left[1 - \frac{2}{\sqrt{\pi}} \int_0^{\xi} e^{-u^2}\, du\right] \tag{4.2.10}$$

where

$$\xi = \frac{z}{2\sqrt{D_m t}} = \frac{z}{2}\sqrt{\frac{\mu_0 \gamma}{t}}. \tag{4.2.11}$$

In the case of a nonideal joint (normal metal-superconductor), the current will flow even in the normal metal, as illustrated in Fig. 4.2.2, for $z < 0$.

In reality, the transient zone is not defined as well as derived for the idealized case, but the general behaviour is essentially the same.

From the ends of the composite, a zone of nonstationary current distribution penetrates the conductor which is no longer determined by the critical state model for the self field, but tends only for $t \to \infty$ to equilibrium current sharing between filaments as given by the transposition resistance between the current lead (normal conductor) and each single filament.

In a copper matrix with $R_{300}/R_{4.2} = 100$, the magnetic diffusion constant has a value of

$$D_m = \frac{1}{0.4\,\pi \cdot 10^{-8} \cdot 56 \cdot 10^4 \cdot 100} = 1.42\ \text{cm}^2\ \text{sec}^{-1}.$$

In Nb-Ti composite conductors, a high resistivity Cu-Ti interlayer is present between filaments and the copper matrix which decreases the effective matrix conductivity and thus increases D_m considerably.

3. A similar current profile is to be expected as for (2), if a disturbance occurs over a short section of a long conductor. If we assume that a flux jump occurs in the outer filament circles and with it, a longitudinal voltage is generated. A part of the transport current will be transferred to the next inner filament circle through the matrix. The corresponding self-field penetrates through the resistive part into the space between filaments, from where it is propagated axially to both sides according to the diffusion equation, even if the overloaded filaments carry stationary currents.

As the superconducting properties are not completely homogeneous along a composite conductor, it must be assumed that changes in transport current and the related self field distribution across the outer filament circles by flux flow or flux jumps usually occur in those "hot spots", and propagate from here on along the conductor.

With respect to instabilities, this situation is worse than the previous assumption of an uniformly penetrating front of current and self field, as the heating in the overcritical filaments and the current carrying matrix is concentrated around these "hot spots" in this dynamic model. The total self-field flux and thus the dissipated heat which must penetrate for a certain ΔI_T to be overtaken by an inner filament circle, depends on the velocity. The penetrating self-field gives the amount of current transferred to the inner part of the conductor and increases until the disturbed filaments recover again. If the heat is removed insufficiently, the heated spot expands and a quench occurs due to "self-field instability", even when the short sample current of the conductor is not reached. Low magnetic diffusivity D_m, a high thermal conductivity of the matrix, and a good bond between matrix and filaments reduce energy dissipation and favor a recovery without quench. In short samples, self-field can penetrate through the ends towards the middle of the wire in a reasonable time and a nearly uniform current density distribution across the conductor is obtained after a few diffusion time constants, related to the half length.

This effect may give a possible explanation for the often observed degradation of the critical current in coils compared to I_c-values measured in short samples. We apply a sudden current change ΔI_T at $t = 0$ to a conductor of length $2\,l$. The corresponding current value $\Delta I\,(r, l)$ in the middle of the conductor attains 50% of its final value at a diffusion time for which

$$1 - \frac{2}{\sqrt{\pi}} \int\limits_0^{\xi} e^{-u^2}\,du = 1 - \phi\left(\frac{l}{2\sqrt{D_m T_{ss}}}\right) = 0.25 \qquad (4.2.12)$$

is valid (penetration from both ends).

The appropriate diffusion time for a short sample with $2\,l = 10$ cm, and magnetic diffusivity of $D_m = 1.42\ \mathrm{cm^2\,s^{-1}}$ is:

$$T_{ss} = 6.62\ \mathrm{sec},$$

a time which may be considered small compared to the current rise time during short sample measurements, giving the full current carrying property of the composite.

On the other hand, if 100 m of this wire are wound into a coil, the diffusion time, proportional to the length square, yields

$$T_{lS} = 6.62 \cdot 10^6\ \mathrm{sec}.$$

In this case, current penetration imposes self field flux flow and quenching may occur even if the single filament is still stable in the external field.

4.2.3 Extension of the Self Field Model in Twisted Multi-Filament Conductors

In cylindrical multifilament conductors with twisted filaments, the transport current flowing through a filament has a screw type path on a cylindrical surface of radius r and forms a pitch angle ϕ_p with the conductor axis. For the pitch angle we may write:

$$\tan(\phi_p) = 2\,\pi r / l_p, \qquad (4.2.13)$$

with l_p the length of the twist pitch and r the distance to the conductor axis.

The transport current density has two components: an axial current density component J_T, and due to the twist, an azimuthal component J_ϕ. These two components are related to each other by:

$$J_\phi = J_T \tan(\phi_p). \qquad (4.2.14)$$

J_ϕ produces an additional solenoidal self field within the conductor in axial direction.

This field couples the twisted filaments like a mutual inductance and thus produces a modification of the transport current distribution.

The self field components are indicated by $B_\phi(r)$ and $B_z(r)$ in the conductor at a radius r. If $I\,(r)$ is the axial component of the transport current flowing within a cylinder of radius $r \leq R$; then B_z is generated only from the current $(I_T - I\,(r))$ outside r. If I_T is the total transport current, then

$$B_z(r) = \mu_0 \frac{I_T - I\,(r)}{l_p}. \qquad (4.2.15)$$

B_ϕ is produced only from $I\,(r)$ through the relation:

$$B_\phi(r) = \mu_0 \frac{I\,(r)}{2\,\pi r}. \tag{4.2.16}$$

$I\,(r)$ is determined from the boundary condition that no flux-changes, due to the self field, can occur through a plane passing through any two arbitrary filament axis.

This means that in a simplified model in which the heterogeneous current pattern is smoothed out by averaged values, the resulting self field must be parallel to the filament and thus forms an angle ϕ_p, with the conductor axis

$$\frac{\dot{B}_\phi}{\dot{B}_z} = \frac{\dot{I}\,(r)}{2\,\pi r} \cdot \frac{l_p}{\dot{I}_T - \dot{I}\,(r)} = \tan\,(\phi_p) = \frac{2\,\pi r}{l_p}. \tag{4.2.17}$$

From which we calculate $\dot{I}\,(r)$:

$$\dot{I}\,(r) = \frac{\dot{I}_T}{1 + \left(\frac{l_p}{2\,\pi r}\right)^2} = \frac{\dot{I}_T}{1 + \cot\text{an}^2\,(\phi_p)}. \tag{4.2.18}$$

Integrating $\dot{I}\,(r)$ with respect to time gives the value of $I\,(r,t)$:

$$I\,(r,t) = \frac{I_T(t) - I_T(0)}{1 + \cot\text{an}^2\,(\phi_p)} + I\,(r,0)\,. \tag{4.2.19}$$

The magnetic coupling, due to currents flowing through twisted filaments, forces the filaments at the inner conductor parts closer to the axis to overtake a part of transport current without the condition that accompanying self field has passed across the outer filaments by the flux flow mechanism [17].

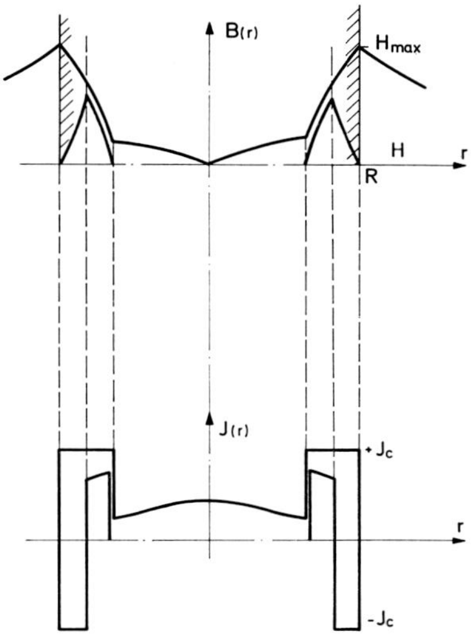

Fig. 4.2.3. Self field and current density patterns in a cylindrical Type II superconductor

Clearly, in the calculated self field formula, the inserted transport current ΔI is reduced by this fraction, which penetrates the conductor without energy dissipation.

Eq. (4.2.19) has been derived by using the condition that the electrical field along the filaments is negligible during the change of the transport current. This is true only in the conductor region, where the critical current in the filaments is not exceeded, i.e. in the interior cylindrical portion of the conductor; while in the outer region of self field penetration, the filaments carry the critical current.

The average current density is thus given by:

$$\bar{J}(r) = \frac{1}{2\pi r} \cdot \frac{dI(r)}{dr} = I_T \cdot \frac{4\pi l_p^2}{(4\pi^2 r^2 + l_p^2)^2} \qquad (4.2.20)$$

(with $I(r,0) = I_T(0) = 0$).

The self field $B(r)$ and the current density $J(r)$ are illustrated for maximum and minimum values of $I_T = I_{\max}$ and $I_T = I_{\min} = 0$ in Fig. 4.2.3.

4.3 Eddy Current Losses in Metallic Parts

4.3.1 Iron Losses in the Flux Return Path

For completion, one has to consider losses in the metallic coil reinforcements, in the metallic helium container, and in other metallic parts surrounding the coil. In superconducting beam transport magnets, the iron shell is placed within the helium container and must be cooled by the helium. Eddy current losses in all metallic parts can be calculated in analogy to Section 4.1.11. Iron losses (without end effects) are given by:

$$\boxed{p_{\text{core}} = G_{Fe}(\sigma_e f^2 B_m^2 + \sigma_h f B_m^h)}\,, \qquad (4.3.1)$$

σ_h and σ_e, are material constants due to hysteretic and eddy current effects in iron. The exponent h varies in the range of 1.6 to 2, G_{Fe} is the weight of the laminated core, and f the frequency of the pulsed field.

Core end losses can be reduced either by shaping the core end section (Rogowski shapes) or by extending the iron yoke about 2 times the aperture diameter over the coil ends.

4.3.2 Eddy Current Losses in the Metallic Cryostat

The cryostat design is considerably simplified, if the iron shell is placed around the coil and is cooled down to liquid helium temperature. The cryostat contains both the coil and the iron shell for the flux return path. The bore tube (if metallic) is linked to the flux as is the outer cryostat shell at liquid helium temperature. For a cylindrical tube of thickness Δr, inner radius r and resistivity ϱ the eddy current losses per unit length are given by:

$$\boxed{p_{\text{tube}} = \dot{B}^2 \cdot \pi \cdot r^3 \frac{\Delta r}{\varrho}} \quad (\text{W/m})\,. \qquad (4.3.2)$$

$\dot{B}$ is the rate of field change. The resistivity of stainless steel changes by about a factor of 1.8, compared to room temperature values. For a tube of $r = 5\times10^{-3}$ m, $\varrho = 5\times10^{-7}$ Ohm · m, $\dot{B} = 3$ T/s and $\Delta r = 10^{-3}$ m the eddy current losses will be 7.1 W/m, which is appreciable. Corrugated tubes permit the use of thinner walls. The bore tube may also be made of non-metallic materials, such as glass filament epoxy structures with H-film liners.

4.4 Multifilamentary Conductors

Superconducting wires used in pulsed magnets must be free from flux-jumping instability (see 4.7.1) and must exhibit low losses (4.1). Attempts to comply with both requirements were:

(a) Cables consisting of twisted monofilamentary copper coated superconductors interspersed with copper strands.

(b) Composite conductors with superconducting wires embedded in a copper matrix.

(c) Composite conductors with many fine filaments, transposed into cables.

The first two types of conductor are not suitable for pulsed operation; the coil does revert to a resistive state at currents much lower than predicted from short sample measurements of the same wire. This effect, called "degradation", results from a phenomenon occurring in the wire as fluxjumping, which is a sudden rearrangement of the pattern of the current flow, releasing enough energy to heat up portions of the superconductor above the transition temperature, driving it normal.

Fluxjumping and instability effect was eliminated by the development of filamentary conductors. Actually, if fluxjump instability is eliminated, additional stabilizing normal metal or internal coil cooling would not be necessary; this is achieved by subdividing the superconductor into sufficiently fine filaments ($\leq 5\times10^{-3}$ cm diameter), However, some normal metal is still necessary to provide mechanical support for the fine filaments. In new composite conductors, nearly 50% of the cross section can be filled by the superconductor, resulting in high current densities. At a field of 5 T, coil overall current densities in the order of 2×10^4 A/cm^2 in pulsed dipole magnets are common.

The first filamentary superconducting composite conductor was produced in 1968 and contained about 50 filaments of NbTi. Several manufacturers are now producing composite conductors containing 1000 filaments and some new conductors containing more than 10000 filaments, each having diameters ranging down to 5 μm.

These composites are incorporated into a cable or braid utilized in pulsed magnets. A 5 μm filament carries a maximum current of about 20 mA at a field of 5 T. Typical magnet currents are in the range of 2000 to 5000 A, so that 10^5 filaments must be connected in parallel. With 10^3 filaments per strand, one has to produce a cable with 100 strands. Cables with many strands have a poor packing factor and compacting the cable to required dimensions result in damage of individual composite wires which have diameters of only about 0.3 mm and are rather fragile. Reducing the number of strands in a cable can be achieved if the number of filaments in a composite is increased, or if the filament size is enlarged. However, as seen in Section 4.1, filament diameters larger than 10 μm are not recommended due to occurrence of high hysteretic losses.

The filaments of a strand are electrically connected by an electrically conducting matrix and are thus coupled with each other electrically in a changing magnetic field. They do not behave as individual filaments. The hysteresis of the composite is greater than the sum for the separate filaments, losses are increased and so is the fluxjump sensitivity. Stability is reduced.

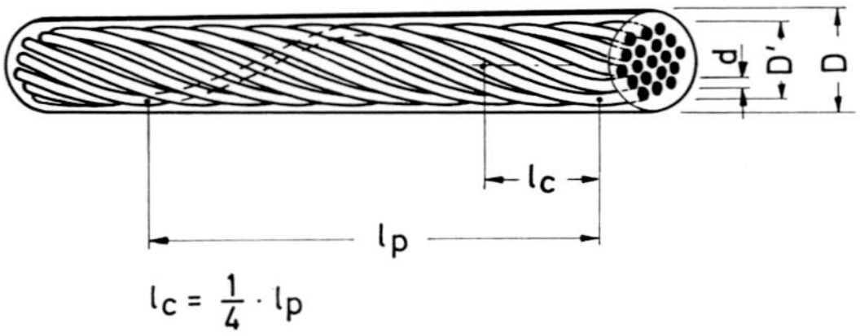

Fig. 4.4.1a. Schematic representation of a composite conductor with twisted filaments. l_c is the critical length of the twist pitch. D' the diameter of the outer filament circle

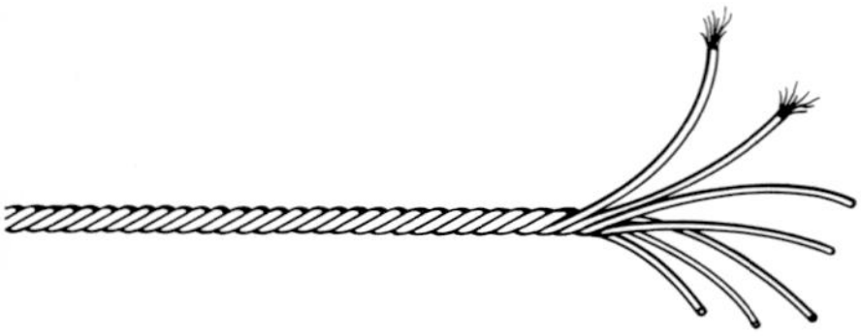

Fig. 4.4.1b. Flat cable produced from transposed composite strands

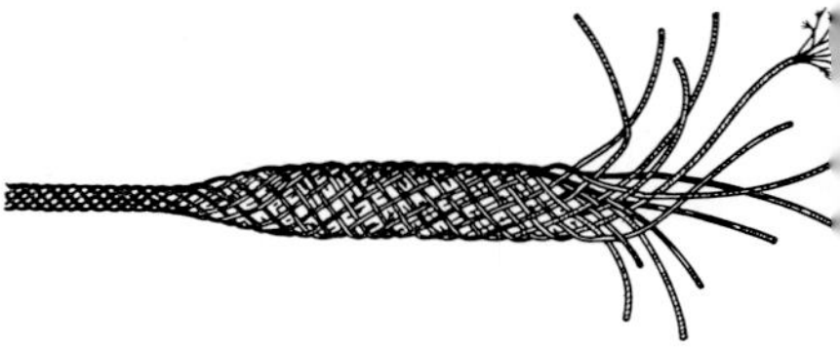

Fig. 4.4.1c. Braid

The coupling between filaments increases with the number of filaments. To ensure that filaments behave essentially independent of one another electrically, despite being connected together by a metallic matrix (Fig. 4.4.1a), the composite must be given an axial twist. The critical twist length of filaments was given in Eq. (4.1.68). With increased number of filaments, it is no longer possible to twist the composite tightly enough; therefore, the self field effect will be present. If the twist pitch is comparable or longer than the critical twist pitch, we modify Eq. (4.1.34) in the form:

$$W_{\mathrm{hl}} = \frac{\mu_0}{2} \cdot J_0 H_0 \cdot V \cdot d \cdot \ln\left(\frac{H_{\max} + H_0}{H_{\min} + H_0}\right)\left[1 + \frac{8}{3\sqrt{\lambda}}\left(\frac{l}{l_c}\right)^2\right]. \qquad (4.4.1)$$

Although self field effects are small in composites having a diameter of ∼ 0.1 cm with increasing number of filaments in a strand, the coupling effect will impose an upper limit to the number of strands. The critical coupling magnetic field for a single component matrix is obtained from Eq. (4.1.68):

$$\boxed{\dot{B}_c = 2\,\varrho_m \cdot \frac{J_c \cdot d}{l_c^2} \cdot \sqrt{\lambda} \cdot \frac{w - d}{w}}\,. \qquad (4.4.2)$$

The critical $\dot{B}_c$ is reached when the coupling currents flowing through the normal metal into a superconducting filament approch the current carrying capacity of the filaments.

It was noted that the magnetization of a twisted multifilamentary conductor exposed to a changing external magnetic field can be expressed by:

$$M \cong M_0 \left[1 + \frac{8}{3\sqrt{\lambda}} \frac{\dot{B}}{\dot{B}_c}\right] \qquad (4.4.3)$$

with M_0 the magnetization of a filament.

To determine the effects of Eq. (4.4.3) on the number of filaments N in a composite, one must consider that the minimum practical twist pitch is approximately five times the diameter of the composite. Reducing the twist pitch further has two ill effects:

(a) The wire (strand) may break or be damaged during twisting. At least, filaments may break.

(b) The effective current density in the filaments of the outer layer $(2\,r \leqq D)$ is lowered as these filaments are stretched due to twisting and their cross-sectional area is reduced. The ratio of the actual to the effective current density for the filaments, evently distributed within the composite, is given by:

$$\boxed{\frac{J_{c,\text{act}}}{J_{c,\text{eff}}} = \frac{2}{1 + \left(\frac{\pi^2 D^2}{l^2} + 1\right)^{\frac{1}{2}}}}\,. \qquad (4.4.4)$$

For:

$$l_p/D = 5: \frac{J_{c,\text{act}}}{J_{c,\text{eff}}} \cong 0.917\,,$$

with D the strand diameter.

This means that the actual current carrying capacity of the composite, due to the twist effect, is 91% of the effective short sample value.

It is usual to demand that the rate dependent contribution to magnetization in Eq. (4.2.3) is less than the magnetization of the filament M_0; otherwise, there would be no gain in subdividing the superconductor into filaments. If the ratio of the two magnetization terms are given by α_m, then: $M = M_0(1 + \alpha_m)$.

The maximum number of filaments is calculated, assuming $4\,l_c = l_p = 5\,D$, from Eq. (4.2.2) and Eq. (4.2.3).

$$N = \lambda \cdot \frac{D^2}{d^2} \cong \left(\frac{4\,\lambda}{5}\right)^2 \cdot \alpha_m \cdot \frac{\bar{J}_c \cdot \bar{\varrho}_c}{d \cdot \dot{B}}\,. \qquad (4.4.5)$$

To give an example, we assume for a NbTi composite conductor with a copper matrix, the following data:

$$\alpha_m = 0.5\ ;\quad \lambda = 0.5\ ;\quad d = 10\times10^{-6}\ \mathrm{m}\ ;\quad \dot{B} = 2\ \mathrm{T\ s^{-1}}\ ;$$
$$\bar{J}_c = 1.5\times10^{9}\ \mathrm{A\ m^{-2}}\ .$$

Thus: $N = 6\times10^{12}\,\bar{\varrho}_c$. For a copper matrix with $\bar{\varrho}_m \simeq 1.7\times10^{-10}$ Ohm · m, and the effective resistivity of $\bar{\varrho}_c \simeq 4\times10^{-11}$ Ohm · m, we obtain $N \simeq 240$ filaments.

Only the use of a high resistivity matrix such as cupro-nickel with a $\bar{\varrho}_c \simeq 7\times10^{-8}$ Ohm · m leads to greater numbers of filaments in a composite. In the above case, $N = 5.6\times10^5$! If the number of filaments is to be increased, a matrix material with high resistivity is needed. It will be shown, however (4.7), that the use of high resistivity matrices lead to an unstable conductor performance, even if the filament diameter is well below the values which adiabatic or dynamic stability criterion requires. Protection of the magnet is possible if copper is used, which indicates that high thermal diffusivity and low magnetic diffusivity (damping) are quite important. Thus, a two component matrix was proposed by RHEL

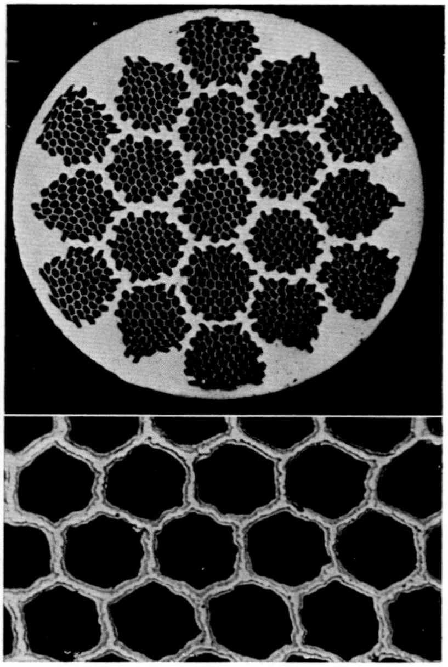

Fig. 4.4.2

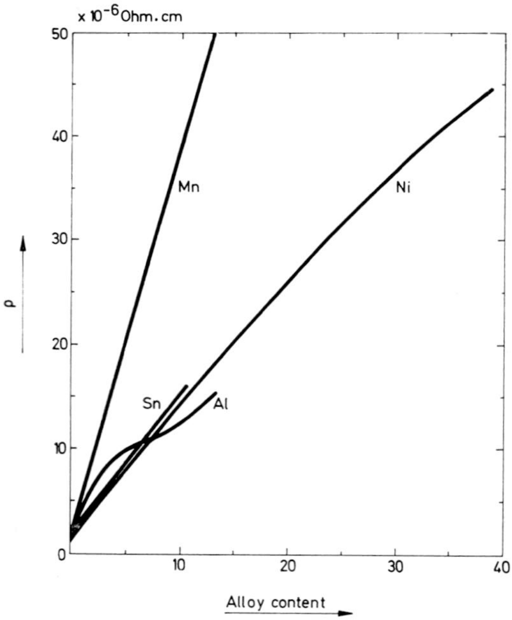

Fig. 4.4.3

Fig. 4.4.2. Cross section through a composite conductor. The superconducting filaments are separated by copper and cupronickel substrates (IMI)

Fig. 4.4.3. Resistivity of copper alloys at 300 K used as a matrix material in composite conductors (RHEL)

[18], where a cupro-nickel layer is provided around single filaments or clusters of filaments and the basic matrix material between filament clusters is copper, shown in Fig. 4.4.2. For cupro-nickel, resistivity values in the order of 5×10^{-8} Ohm · m have been measured. The resistivity of a few copper alloys vs. the alloy content is given in Fig. 4.4.3 measured at room temperature. At 4.2 K, the resistivity values are somewhat lower than at 300 K ($<$ factor 2).

Even in two component matrices, the effective resistivity measured in twisted superconducting composites is higher than expected in theory. Chritchlow [19] measured the effective resistivity of the copper matrix to vary in the range of $(3\text{—}10) \times 10^{-10}$ Ohm · m, rather than about 10^{-10} Ohm · m for copper. The reason for this discrepancy is seen in the high resistivity layer around superconductors, due to the reaction layer between copper and NbTi. Spurway et al. [20] measured the resistance between two superconducting filaments in a composite wire and found the effective resistivity of the matrix to be $(6\text{—}34) \times 10^{-10}$ Ohm · m when the distance between the filaments ranged from 17 μm to 3 μm.

If the spacing between filaments is greater than 40 μm (which is very unlikely in composites for pulsed magnets), the more expected resistivity value of $\sim 3 \times 10^{-10}$ Ohm · m gives satisfactory results. For three component matrices, Wilson [9] has calculated approximate values of the effective matrix resistivity as a function of the matrix material, the filament spacings, and the strand geometry.

4.4.1 Cables and Braids

Conductors used in pulsed superconducting magnets are multistrand cables (ropes), or braids, with each strand consisting of twisted superconducting filaments embedded in a two or three component matrix. The strand are twisted or transposed and shaped by compacting to desired configurations and to obtain higher packing factors and with it, high overall current densities. Figs. 4.4.1b and 4.4.1c illustrate schematically a cable and braid.

The size of the individual strands in a conductor needed for coil winding is determined by the type of conductor chosen.

A cable is made by twisting a number of strands in layers around a central wire. Each successive layer is twisted over the previous concentric layer, with the sense of twist reversed from layer to layer.

The central wire is generally a copper conductor. This cable has the disadvantage that the various layers are located in the space within the conductor in different places and are exposed to a non-uniform magnetic field; thus, it may carry different currents. In a simple cable consisting of 7 composite insulated conductors of identical geometry, number and size of strands, Smith [21] reported that the central strand (composite) carried a negative current when the field was pulsed with the rise time of a few seconds. All other 6 strands around the central strand carried positive currents. Simply twisted insulated cables have higher coupling losses and exhibit degradation.

Only fully transposed cables, in which each strand occupies all positions within the cable and has the same mutual inductance with respect to the assembly of all others, eliminate the above disadvantage, but they have a low packing factor ($< 50\%$) and a large number of crossover points. Strand breakage is observed when a large number of strands is used and the cable is compacted for higher current densities.

To overcome this handicap, two methods have been employed successfully:

(a) The number of strands is reduced to a possible minimum such that they fit a single layer. The strands are twisted axially over a central wire consisting of a soft insulated copper wire. The diameter of the central wire may be equal to the diameter of the strands (6 strands + 1 copper wire)* or it can be larger, in which case, more strands can be twisted around the central core. This implies that for higher currents, the size of individual strands is large. The cable is then compacted into a square or rectangular shape to desired dimensions. A cable having overall dimensions of $2.1 \times 2.6\ mm^2$ consisting of 12 composite strands with 0.54 mm diameter (1000 filaments at 12 μm per strand) twisted around a central copper wire carries a current of > 2000 A at 5 T. A dipole coil [22] with a central field of 4.5 T using such a cable is shown in Fig. 4.4.4.

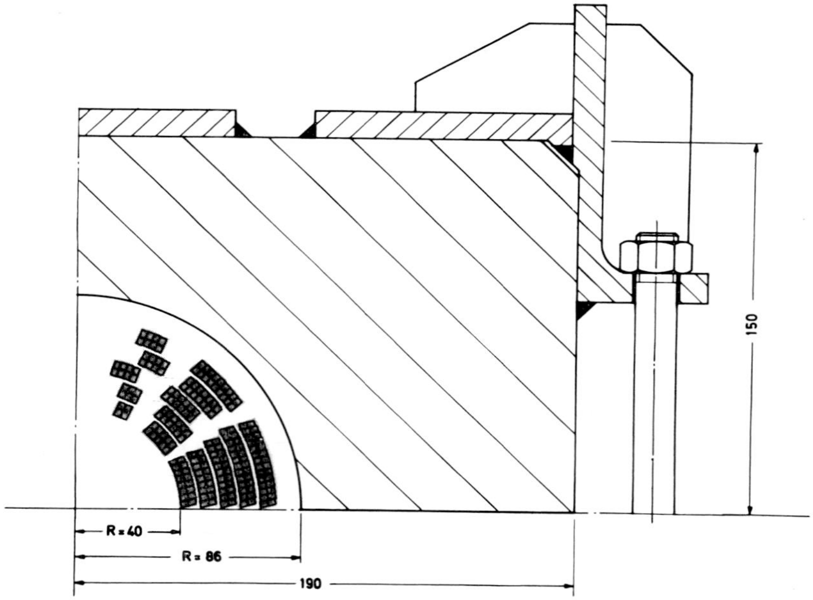

Fig. 4.4.4. Dipole magnet using square shaped cables (IEKP)

* If high strength cable is required, the central wire may be non-magnetic stainless steel or beryllium copper.

(b) A finite number of strands in a layer (with no central core) is twisted axially and then compacted to form a flat cable. In this case, the cable is also fully transposed. As examples, we quote two cables: A cable consisting of 6 strands with overall dimensions of 4×2.5 mm (strand diameter 1.1 mm) carries a current of 3000 A at a transverse field of 5 T. A flat cable having overall dimensions of 5×1.5 mm² consists of 15 strands with 0.625 mm diameter having a transposition pitch of about 4 cm. Each strand consists of 2035 filaments with 8.7 μm diameter. The filaments are twisted with a pitch of 4 mm. The metal packing factor of this flat cable is 70%; it carries 2550 A at 5 T. A dipole coil with a central field of 4.5 T using this flat cable is shown in Fig. 4.4.5. The dipole end section is illustrated in Fig. 4.4.6.

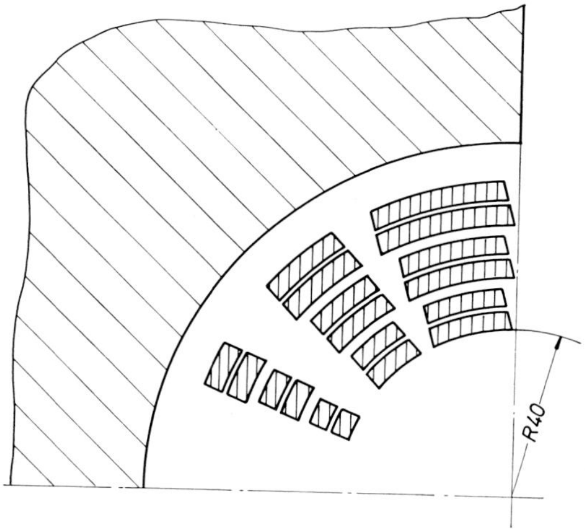

Fig. 4.4.5. Dipole magnet using flat cables (IEKP)

A braid is produced by interlacing a number of strands to form a hollow cylindrical structure as shown in Fig. 4.4.1c. The strands in a braid are fully transposed. However, the large crossover angles between strands result in a reduced mechanical rigidity of the conductor and a low packing factor as compared to cables. In addition, stress concentration at the crossover points can lead to strand damage during braid compaction.

A high degree of field uniformity is required in the useful coil aperture; thus, the maintenance of the conductor shape during and after coil winding is imperative. In general, braids are less rigid than cables and their cross-section is distored when exposed to winding stress. To maintain a field uniformity of 10^{-3} within the useful aperture, the required tolerance on the conductor shape is quite stringent. In a 2000 A cable with the cross-sectional area of 2.1×2.6 mm², one requires edge tolerance of $+0$, -0.1 mm! Winding semi-rigid braids or cables consisting of individually insulated strands into coil require special precautions. In flat braids, the shape is maintained during winding, only if they are sub-

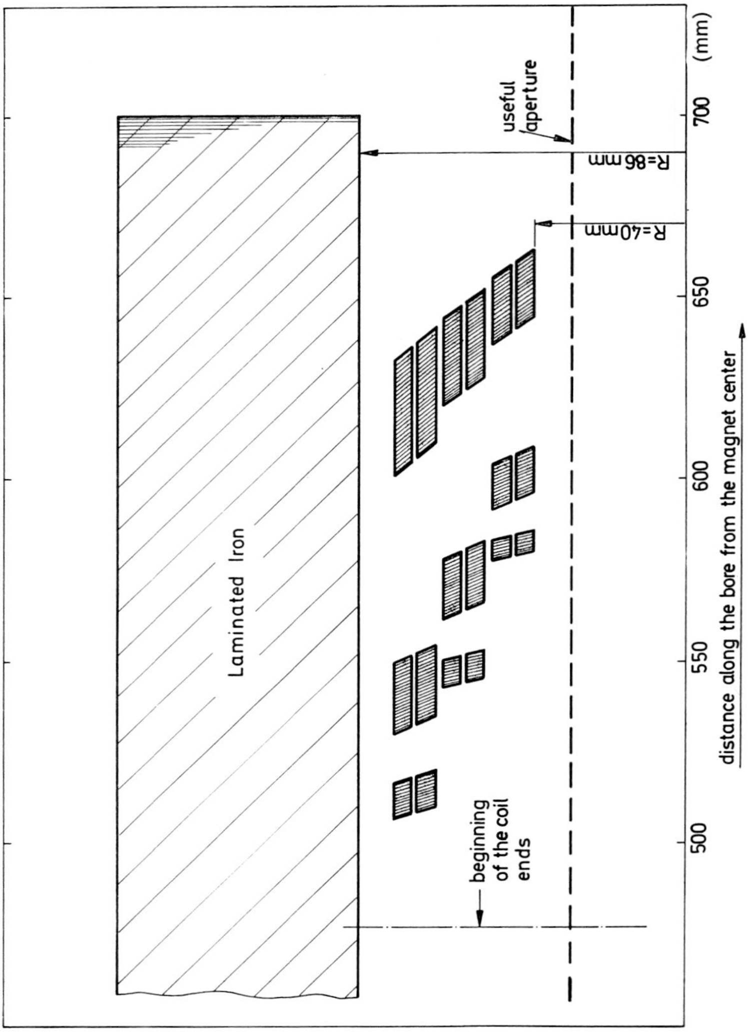

Fig. 4.4.6. Dipole end sections with the coil wound with flat cables. The coil end section prevents field enhancement

jected to uniform tension. A braid with the dimensions of 1.8×4.5 mm² carries about 2200 A at 5 T. The 42 strands have individual radial insulation of 20 μm. The braid retains its shape within $\simeq 0.1$ mm during winding, only if the tension of 300 kp/cm² could be kept during coilwinding. In addition, when the magnet is energized, individual strands in a non-impregnated cable or braid move under the influence of time variable Lorentz forces. This motion generates additional losses and a local temperature rise which may lead to degradation and conductor training.

The conduction of heat away from the superconductor to the bath may also limit the coil current carrying capacity as seen from Fig. 4.4.7, which illustrates the critical current in the conductor of a solenoid after successive quenches. To prevent wire motion under the influence of mag-

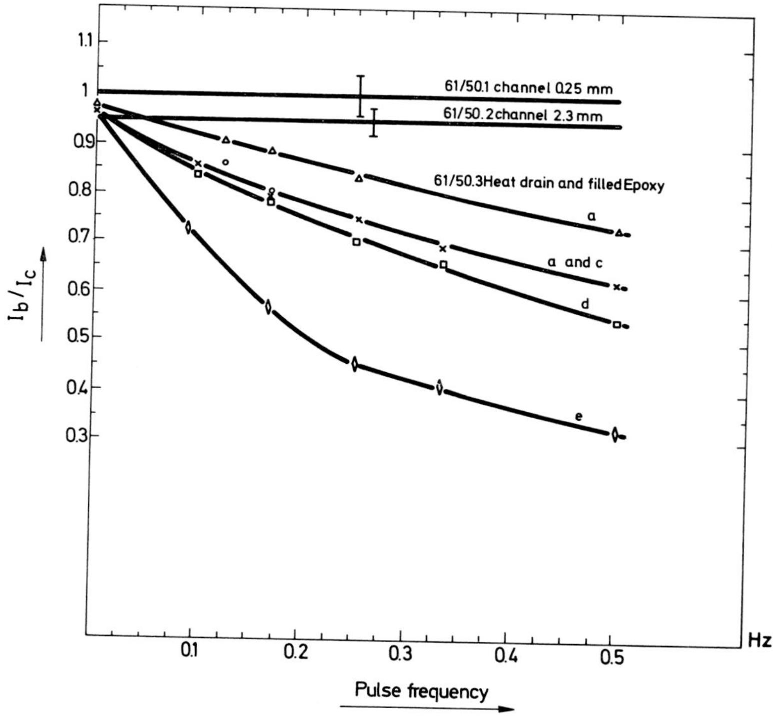

Fig. 4.4.7. Current degradation in coils due to heat drains (Saclay)

a △ Heat Drain 15 cm extending on each side of coil pacage into helium
b × Heat Drain, one side impregnated, other side 1.5 cm extending into He
c ○ Heat Drain, one side impregnated, other side 0.8 cm extending into He
d □ Heat Drain, one side impregnated, other side flash with coil side
e ♦ Coil fully impregnated

netomechanical forces, coils are either rigidly held in place (which requires an elaborate reinforcement structure) or they are potted in suitable thermosettings. Fig. 4.4.8 illustrates the energy dissipation in the same unpotted and after the test potted solenoid. If wire motion is eliminated, the energy dissipation is approximately proportional to the peak transverse magnetic field. If the conductor is allowed to move, a quadratic field dependency is observed [23]. There may be several methods to prevent wire motion; of these, we name three:

(i) The cable is preimpregnated with a "metallic" insulation. This is accomplished by pretinning (95% Sn to 5% Ag) individual, not insulated, strands and transposing them into a cable. The strands are heat treated in an inert atmosphere for longer than 100 hours at 325 °C. The non-compacted cable is passed through an indium thallium bath (10% Tl by weight) and compacted in the still warm condition to the desired shape. If the cable is being used for relatively higher pulse frequencies ($\geqq$ 1 Hz), the transverse cable resistance between strands can be increased by heat treatment of the filled cable at 300 °C during 10 to 100 hours.

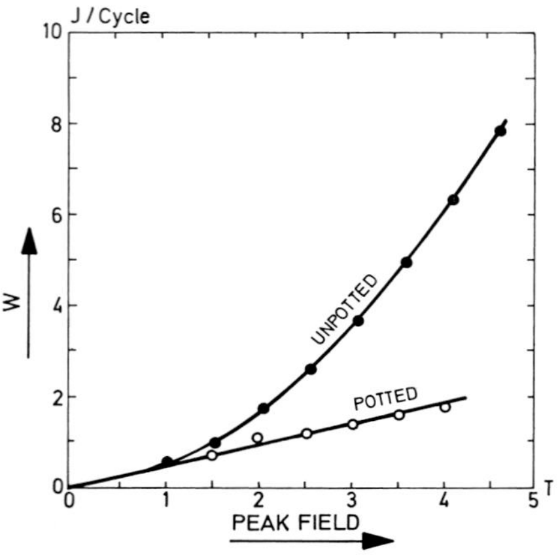

Fig. 4.4.8. Energy dissipation vs. peak field measured in small solenoids with ID = 2.5 cm; OD = 7.6 cm; length = 4.75 cm. The effect of conductor motion (unpotted coil) has a B^2 loss dependence, while the hysteretic losses are proportional to B in potted coils (BNL)

A reaction layer of Cu-Sn-Ag-In-Tl is formed around each individual strand. The thickness of this layer depends on the duration of heat treatment. The cable must be shaped again after the heat treatment. The reaction layer is, however, quite brittle and it cracks when the cable is wound around small radii ($\leqq$ 10 conductor diameter).

A possible way to avoid this is by winding the cable into the coil and heat treating the coil at 300 °C over a period longer than 24 hours. Metallic insulation increases adiabatic and dynamic stability (2.3) and the heat transfer properties. InTl has a thermal conductivity of 70×10^{-3} W cm^{-1} K^{-1} and a moderate electrical resistivity of 1.6×10^{-8} Ohm · m at 4.2 K. The presence of reaction layers, reduce adiabatic and dynamic stability.

The major difficulty arising from the use of metallic insulated cables is the coupling between strands. This is the same phenomenon which also causes eddy current losses in the individually twisted filaments. Additional eddy current losses occur in metallic insulated cables (braids), depending on its twist rate. Induced eddy currents have an additional contribution to the irreversible magnetization which increases with $\dot{B}$, and the twist length.

The coupling factor defined as the ratio of the magnetization of M of an untwisted composite, which is equivalent to a solid core, to the magnetization of the conductor being tested, which may be twisted, indicates the additional losses in the conductor and is plotted as a function of the conductor length in Fig. 4.4.9 at an applied transverse field of 3.1 T. During the test, the field was changed at a constant rate of 5×10^{-2} Ts^{-1} at which M was a maximum. Further increase in $\dot{B}$ did not affect M. The decoupling factor is a function of the field rise and the twist length. Fully insulated strands (organic insulation with $R \geqq 10^{15}$ Ohm) are totally decoupled. If the strands of a cable are coupled to any degree (e.g. because of shorts in crossover points, connections, or metallic insulations, etc.), say by higher than about 20% of the totally coupled case, then the cross currents flowing through these shorts or through the high resistivity metallic layers generate losses far beyond the individual strand contribution and dominate the coil losses. Fig. 4.5.1 gives the losses per cycle versus $\dot{B}$ for solenoids using metallic insulated cables. It may be noted that losses per cycle are increasing with the frequency in the low $\dot{B}$ range and decreasing with frequency at high $\dot{B}$ from the interior of the conductor. Recent loss data obtained by McInturff [23] using metallic insulated cables are shown in Fig. 4.5.3. Compared to calculated loss values which are applicable for organic insulated cables, losses due to a field sweep of 2.2 T s^{-1} are by a factor of 5 higher!

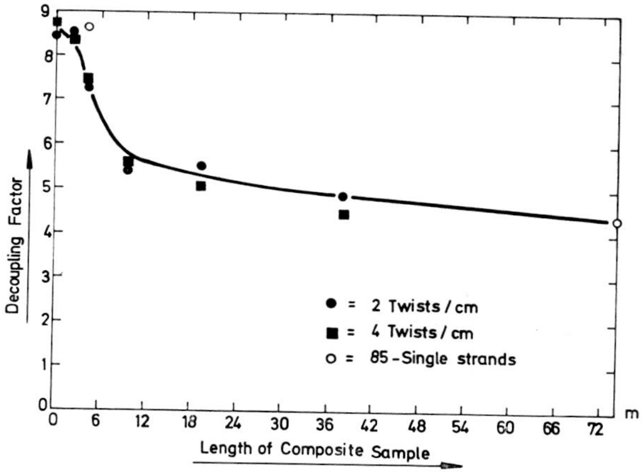

Fig. 4.4.9. Decoupling factor vs. length of a composite conductor at an applied transverse field of 3.01 T (BNL)

(ii) The cable or braid is preimpregnated with B-staged thermosetting or a thermoplastic material. The individual strands are fully insulated with an organic coating such as a thermosetting compatible to the B-stage material and to the final potting material used to impregnate the coil. During winding, the cable is heated slightly just to overcome the conductor stiffness. The cable is rigid enough to withstand a dimensional distortion during winding and is easy to wind. During the final potting, the B-staged thermosetting liquefies and mixes with the potting mixture to form a uniform compound.

The major disadvantage of this method is the additional application of heat during winding; this plasisizes the bonding agent and thus reduces the mechanical rigidity of the cable at the time when it is needed most. Maintaining cross-sectional tolerances is then difficult without prestraining.

(iii) By using a three component composite, the number of filaments in the wire can be increased to 5×10^5. Clearly, it is necessary that some kind of resistivity barriers (usually cupro-nickel) must be erected around each filament to prohibit transverse currents to flow from one filament to another and prevent coupling. Pure copper is always included in the composite to aid stable operation and, as a protection measure, to provide a low resistance current path should the superconductor revert to a normal state, e.g. because of current overload, refrigeration or power supply failure, etc. The copper in the matrix must be finely divided and

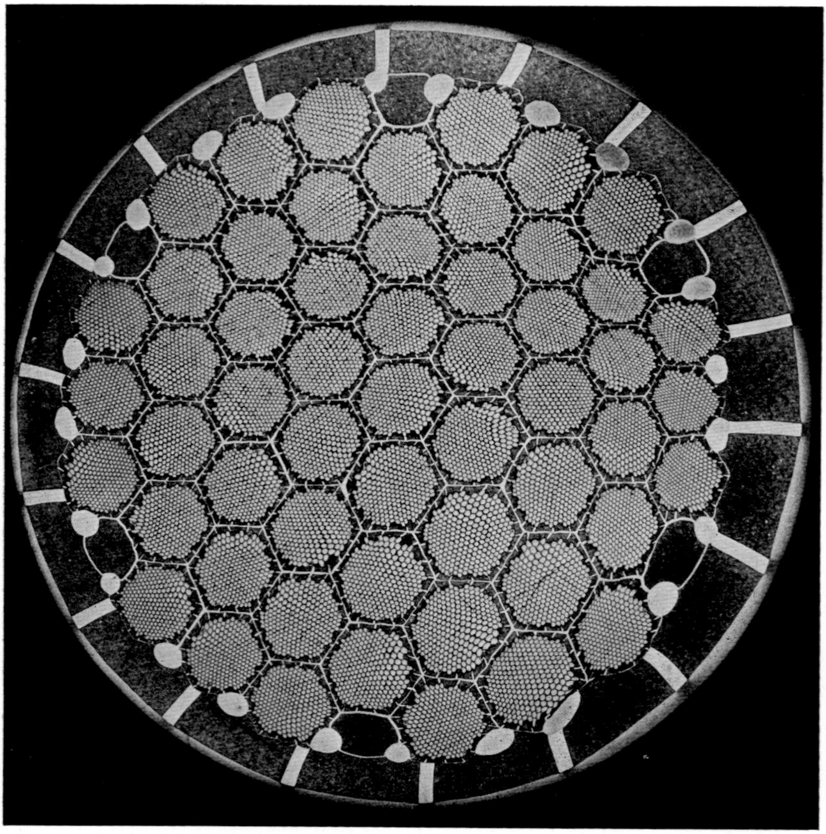

Fig. 4.4.10. Composite conductor with 13,255 NbTi filaments à 5 μm diameter. The conductor has an overall diameter of 0.1 cm. The matrix consists of copper (gray) and cupronickel (white) potions. To prevent coupling currents from flowing around filament clusters and at the outer sections cupronickel barriers are foreseen (IMI)

distributed around the filaments to minimize eddy currents. A conductor embodying these features of low losses, stability, and low coupling is shown in Fig. 4.4.10. The conductor has 13,255 Nb-Ti filaments, each of 5 μm diameter. It carries 470 A at 5 T and has a diameter of 0.1 cm. The conductor development is progressing rapidly and it is expected that by 1974, conductors with 10^5 filaments carrying 5000 A at 5 T will be available. This conductor type is comparable to the cable having a metallic insulation. If a conductor of this type becomes available, the dimensional stability is no more of concern.

Losses for both cases (i) and (iii) can be calculated from Eq. (4.1.70) or from Morgan's relation:

$$p/l_p = 4\left[B_0(w-d)\cdot l_p\cdot\frac{\omega}{4\pi}\right]^2 / \bar{\varrho}_c \quad (\mathrm{W/m})\,, \qquad (4.4.6)$$

with ω the pulse angular frequency, B_0 the peak field, and $\bar{\varrho}_c$ the effective resistivity of the matrix. w is the spacing between filaments (case (iii)) or strands (case (i)), while the matrix resistivity ϱ_m is either the resistivity of the solder including reaction layers (i) or the resistivity of copper and cupro-nickel (iii). Equations to determine $\bar{\varrho}_c$ are given in 4.1.11 and by Wilson [9]. The most simple form for a single component matrix and uniformly distributed filaments is given by

$$\bar{\varrho}_c = \varrho_m\,\frac{w-d}{d}\,.$$

Eq. (4.2.6) is applicable for low frequency pulsed fields only.

The average losses per unit length for a multicore (multifilamentary) conductor (type (i) or type (iii)) are given by:

$$\boxed{p_{\mathrm{av}} = \frac{D^2\cdot l_p\cdot\dot{B}^2}{16\,\pi\bar{\varrho}_c}\cdot\left(1-\frac{8}{3\,\pi}\cdot\frac{d}{D}\cdot\frac{\dot{B}^2}{\dot{B}_c^2}\right)} \quad (\mathrm{W/m})\,. \qquad (4.4.7)$$

In this equation, D may indicate either the cable diameter (i) or the strand diameter (iii); d is the strand diameter (i) or the filament diameter (iii).

As mentioned in (4.1), there is an upper limit in the composite size and the number of filaments in the composite due to self field stability and self field losses. This limit may be in about 10^5 filaments of 5 μm for the above 3 component conductor.

The choice of the filament size down to 5 μm is not only to reduces a.c. losses, but also to limit the remanent field in the magnet aperture to small acceptable values.

The residual fields are generated by circulating currents in the superconductor and are proportional to the filament diameter [24]. Residual fields measured in the aperture of a dipole with the coil inner diameter of 10 cm, after it had been energized to the maximum field of 3.9 T, and the field reduced afterwards to zero had the following components [22]:

$$\text{Dipole } B_1 = 4\times10^{-4}\ \mathrm{T};\quad \text{Sextupole } B_3 = 5.9\times10^{-4}\ \mathrm{T};$$
$$\text{Decapole } B_5 = 0.5\times10^{-4}\ \mathrm{T}.$$

These values were measured at an aperture radius of 5 cm. The filament diameter was 10 μm.

Metallic impregnated cables (type (i)) have operated better under pulsed conditions than have the fully insulated cables (type (ii)). The reason may be found in the better heat transfer properties of the solder and the rigidity of the cable against motion under the influence of Lorentz forces (Fig. 4.4.11). Fully insulated strands and cables, if potted (type (ii)), exhibit training after the first cooldown. After several quenches, short sample values are approached. Fig. 4.4.12 illustrates the behaviour of insulated and potted coils. After each cooldown, training effects have been observed in some cases. Metallic insulated cables have shown no training.

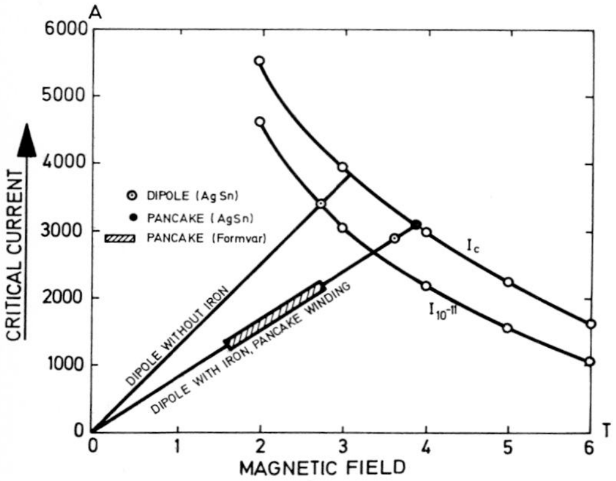

Fig. 4.4.11. Performance of small superconducting magnets (BNL)

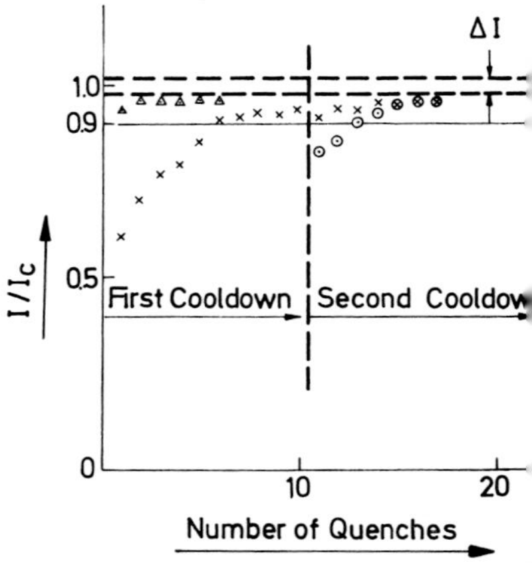

Fig. 4.4.12. Coil training (RHEL)

4.5 Comparison of Loss-Calculation with Experiments

Comparison of results obtained from theoretical considerations and experimental investigation requires the exact knowledge of the location and distribution of superconducting filaments within a single strand, in the cable or the braid, the physical properties of the matrix material, and the nature of impregnant used. It is not possible to introduce a simple universal theory which includes effects of various components mentioned. Difficulties in manufacturing a.c. cables have lead to discrepancies between theory and experiments (broken filaments, interturn, shorts, etc.). Published data on loss measurements [23, 26] agree reasonably well with theoretical calculations only in special cases. Generally due to lack of detailed informations, experimental data on a.c. magnets do not fit theoretical predictions. Good agreement between theory and experiment is reported, if all coil and conductor data are available.

Fig. 4.5.1 illustrates measured losses for solenoids using 3000 m of 0.02 cm diameter composite wires. The matrix consists of 210 filaments of NbTi, each having 8 μm diameter. The filaments are twisted with a pitch of 4 twists per cm. 33 strands are transposed in a braid of 0.6 cm width (5.2 cm transposition length) and impregnated in a silver-tin or tin-bismuth alloy. In the first case, the coupling losses dominate specific a.c. losses at low pulse frequencies; thus, the comparison to theory shows a great discrepancy. The addition of bismuth already reduces the coupling losses by a factor of 2.5 at a field rise time of 4.5 s (1 Ts^{-1}). Measured and calculated energy dissipation for a field rise time of 12 s is given

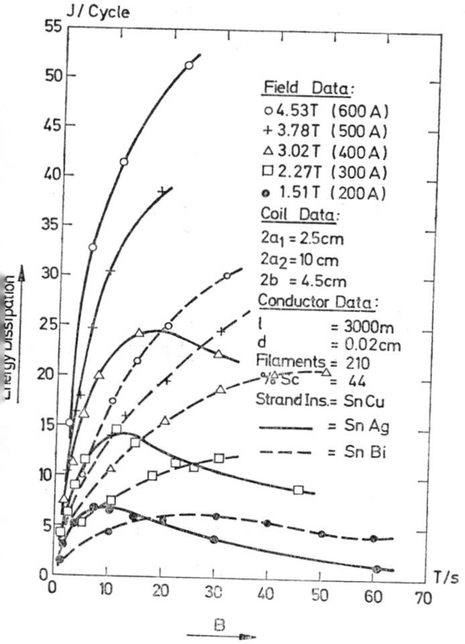

Fig. 4.5.1

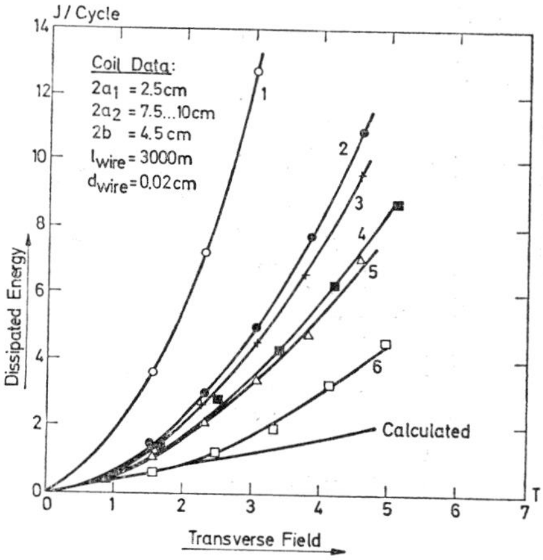

Fig. 4.5.2

Fig. 4.5.1. Hysteretic losses in solenoids. The losses are in SnAg soldered cables due to coupling between strands very high. The losses are smaller if SnBi is used as a solder (BNL)

Fig. 4.5.2. Hysteretic losses in solenoids (BNL).

1 Braid soldered in Sn(Bi). (Transpos. length 21 cm). Strand coating Cu(Sn).
2 Braid soldered in In(Tl). (Transpos. length 21 cm). Strand coating Cu(Sn).
3 Braid soldered in In(Tl). (Transpos. length 7 cm). Strand coating Cu(Sn).
4 Braid soldered in Sn(Bi). (Transpos. length 7 cm). Strand coating Cu(Sn). Cable heat treated.
5 Braid soldered in In(Tl) and heat treated. Transposition length 7 cm. Strand coating Cu(Sn).
6 Braid soldered in In(Tl) and heat treated. Transposition length 5 cm. Strand coating Cu(Sn).

in Fig. 4.5.2. The addition of a reaction layer due to a final heat treatment has reduced coupling losses by a factor of 7, as indicated by the direct comparison between curves 1 and 6 at a field of 3.8 T. The strands (curve 1) have a tin-copper coating and the braid is soldered in tin-bismuth;

Table 4.5.1. *Comparison between calculated and measured a.c. losses in solenoids* [27]

	I				II					III			
Coil Geometry	$2\,a_1 = 6$ cm				$2\,a_1 = 2.4$ cm					$2\,a_1 = 2.4$ cm			
	$2\,a_2 = 9$ cm				$2\,a_2 = 8.4$ cm					$2\,a_2 = 6.5$ cm			
	$2\,b = 5.6$ cm				$2\,b = 5.3$ cm					$2\,b = 5.3$ cm			
	$N = 2990$				$N = 5650$					$N = 6125$			
Conductor	IMI:				AIRCO:					VAC:			
	61 fil. à 42 μm				361 fil. à 13 μm					61 fil. à 34 μm			
	$I_0 H_0 = 650$ AT				$I_0 H_0 = 285$ AT					$I_0 H_0 = 650$ AT			
	$H_0 = 1$ T				$H_0 = 1$ T					$H_0 = 1$ T			
	$l = 705$ m				$l = 958$ m					$l = 856$ m			
	$V_{cond} = 112$ cm³				$V_{cond} = 152$ cm³					$V_{cond} = 82.4$ cm³			
	$D = 0.045$ cm				$D = 0.045$ cm					$D = 0.033$ cm			
Matrix	Cu:				Cu:					Cu:			
	$r = 180$				$r = 70$					$r = 100$			
	$\varrho_c = 4.7 \times 10^{-8}$ Ohm · cm				$\varrho_c = 7 \times 10^{-8}$ Ohm · cm					$\varrho_c = 10 \times 10^{-8}$ Ohm · cm			
	$l_{pitch} = 2.5$ cm				$l_{pitch} = 0.4$ cm					$l_{pitch} = 0.6$ cm			
	$\tau = 21 \times 10^{-3}$ s				$\tau = 1.45 \times 10^{-3}$ s					$\tau = 0.57 \times 10^{-3}$ s			
Hysteretic losses	$B_m = 4.2$ T				$B_m = 3.6$ T					$B_m = 5.2$ T			
Eq. (4.1.34)	$W_{hl} = 9.24$ Ws/cycle				$W_{hl} = 1.6$ Ws/cycle					$W_{hl} = 10.3$ Ws/cycle			
Self field losses	$\Delta I = 74$ A				$\Delta I = 36$ A					$\Delta I = 95$ A			
Eq. (4.1.82)	$I_c = 325$ A (at 1 T)				$I_c = 150$ A (at 0.9 T)					$I_c = 370$ A (at 1.25 T)			
	$W_{sf} = 7.3 \times 10^{-3}$ Ws/cycle				$W_{sf} = 2.5 \times 10^{-3}$ Ws/cycle					$W_{sf} = 16.5 \times 10^{-3}$ Ws/cycle			
Field rise time $t_0/2$ (s)	5.6	8.77	15	25	2.5	4	5	6.5	10	1.8	3.6	6.25	10
Eddy current losses Eq. (4.1.70) (Ws/cycle)	8.44	5.06	3.0	1.78	1.2	0.76	0.61	0.5	0.3	0.752	0.38	0.22	0.135
Total calculated losses	17.7	14.31	12.25	11.03	2.8	2.36	2.21	2.1	1.9	11.5	10.7	10.52	10.44
Total measured losses (Ws/cycle)	21	15.8	10.8	10.0	3.75	2.16	1.5	1.25	1.2	10.5	10.1	10	10

Table 4.5.2. *Comparison between calculated and measured a.c. losses in solenoids* [28]

Coil geometry	$2a_1 = 5$ cm $2a_2 = 9$ cm $2b = 7.5$ cm $N = 4000$											
Conductor	VAC: 61 fil. à 34 µm $I_0H_0 = 840$ AT $H_0 = 1$ T $l = 780$ m $V_{cond} = 67$ cm³ $D = 0.033$ cm											
Matrix	Cu: $r = 100$ $\varrho_c = 10 \times 10^{-8}$ Ohm · cm $l_{pitch} = 0.2$ cm $\tau = 0.07 \times 10^{-3}$ s				Cu: $r = 100$ $\varrho_c = 10 \times 10^{-8}$ Ohm · cm $l_{pitch} = 0.3$ cm $\tau = 0.14 \times 10^{-3}$ s				Cu: $r = 100$ $\varrho_c = 10 \times 10^{-8}$ Ohm · cm $l_{pitch} = 0.6$ cm $\tau = 0.57 \times 10^{-3}$ s			
Hysteretic losses Eq. (4.1.34)	$B_m = 4$ T $W_{hl} = 9.5$ Ws/cycle				$B_m = 4$ T $W_{hl} = 9.5$ Ws/cycle				$B_m = 4$ T $W_{hl} = 9.5$ Ws/cycle			
Self-field losses Eq. (4.1.82)	$< 10^{-2}$ (Ws/cycle)				$< 10^{-2}$ (Ws/cycle)				$< 10^{-2}$ (Ws/cycle)			
Field rise time $t_0/2$ (s)	1	2	5	10	0.5	1	2	5	0.5	1	2	5
Eddy current losses Eq. (4.1.70) (Ws/cycle)	70×10^{-3}	35×10^{-3}	14×10^{-3}	7×10^{-3}	0.32	0.16	0.08	0.032	1.28	0.64	0.32	0.13
Total calculated losses (Ws/cycle)	9.57	9.53	9.51	9.5	9.82	9.66	9.58	9.53	10.78	10.14	9.82	9.63
Total measured losses (Ws/cycle)	9.8	9.8	9.8	9.8	13.2	11.8	10.9	10.1	13.8	12.2	10.7	9.0

Table 4.5.3. *Comparison between calculated and measured a.c. losses in solenoids* [29]

	I				II			III			
Coil Geometry	$2a_1 = 2.16$ cm $2a_2 = 4.76$ cm $2b = 8.64$ cm $N = 5177$				$2a_1 = 2$ cm $2a_2 = 4.25$ cm $2b = 5$ cm $N = 2376$			$2a_1 = 2.12$ cm $2a_2 = 8.99$ cm $2b = 8.64$ cm $N = 1698$			
Conductor	IMI: 61 fil. à 28 μm $I_0H_0 = 300$ AT $H_0 = 1$ T $l = 563$ m $V_{cond} = 39.5$ cm³ $D = 0.03$ cm				IMI: 61 fil. à 28 μm $I_0H_0 = 300$ AT $H_0 = 1$ T $l = 223$ m $V_{cond} = 15.6$ cm³ $D = 0.03$ cm			Superan: 16 · 400 fil. à 7.5 μm $I_0H_0 = 1400$ AT $H_0 = 0.7$ T $l = 296$ m $V_{cond} = 150$ cm³ 16 strand transposed cable			
Matrix	Cu: $r = 180$ $\varrho_c = 4.7 \times 10^{-8}$ Ohm · cm $l_{pitch} = 0.25$ cm $\tau = 0.2 \times 10^{-3}$ s				Cu: $r = 180$ $\varrho_c = 4.7 \times 10^{-8}$ Ohm · cm $l_{pitch} = 0.25$ cm $\tau = 0.2 \times 10^{-3}$ s			Cu: $r = 100$ $\varrho_c = 7 \times 10^{-8}$ Ohm · cm $l_{pitch} = 0.2$ cm $\tau = 0.2 \times 10^{-3}$ s			
B_m (T)	1	2	3	4	1	2	3	1	2	3	4
Hysteretic losses Eq. (4.1.34) (Ws/cycle)	0.81	1.38	1.83	2.2	0.32	0.5	0.7	0.69	1.15	1.5	1.76
Field rise time $t_0/2$ (s)			2				2			2	
Eddy current losses Eq. (4.1.82) (Ws/cycle)	No frequency dependence obs. $B_m = 4$ T $W_e = 0.07$ Ws/cycle				No frequency dependence obs. $B_m = 3$ T $W_e = 0.016$ Ws/cycle			No frequency dependence obs. 0.007	0.029	0.064	0.114
Self field losses Eq. (4.1.70)	$\Delta I = 57$ A $I_c = 150$ A (at 1 T) $B_m = 4$ T $W_{sf} = 6 \times 10^{-3}$ Ws/cycle				$\Delta I = 59$ A $I_c = 170$ A (at 0.75 T) $B_m = 4$ T $W_{sf} = 2.2 \times 10^{-3}$ Ws/cycle			$\Delta I = 12$ A/strand $I_c = 51$ A (at 1 T) $B_m = 4$ T $W_{sf} = 1.35 \times 10^{-3}$ Ws/cycle			
Total calculated losses (Ws/cycle)	0.886	1.456	1.906	2.27	0.338	0.568	0.75	0.7	1.18	1.56	1.88
Total measured losses (Ws/cycle)	0.58	1.1	1.5	1.85	0.27	0.5	0.7	0.73	1.26	—	—

the strands in the braid (curve 6) are coated in copper-tin and the braid is impregnated in indium-thallium and heat treated to form a reaction layer around individual strands. The transposition pitch in these braids is 7 cm.

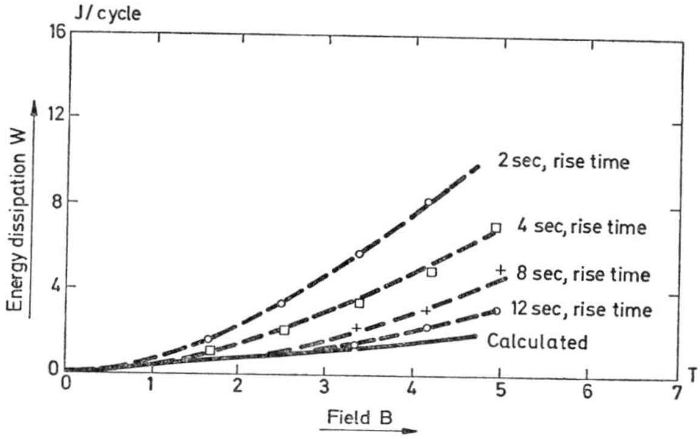

Fig. 4.5.3. Hysteretic losses in a solenoid according to Fig. 4.5.1. The braid is soldered in In(Tl) and heat treated to form a reaction layer, around strands coated individually in Cu(Sn)

Finally, Fig. 4.5.3 shows the energy dissipation of the solenoid using an indium-thallium soldered braid at different pulse times. At a maximum field of 5 T, the coupling losses are still a factor of 3.5 higher than for a braid with fully insulated strands.

A comparison between calculated and measured losses for the above solenoids are given in Fig. 4.5.4. The difference in both cases with fully insulated and metallic insulated strands is about a factor of two.

Better agreement between calculations and measurements were found in solenoids measured by CEN, IEKP and Siemens as shown in Tables 4.5.1, 4.5.2, and 4.5.3. However, it must be noted that in most cases the solenoids were wound with a single strand.

Discrepancies between measurements and calculations in dipoles are much larger than for solenoids due to their complicated winding shape.

The main reasons for this discrepancy is seen first from the assumed data of the effective resistivity in two and three component composites. As shown above, the eddy current and coupling losses are directly proportional to effective resistivity. The exact measurement of $\bar{\varrho}_c$ is difficult and pertinent data are still missing. The other source of errors in the calculation is the assumption of a model based on merely twisted strands and twisted filaments with uniform cross-section. Neither assumption is justified, as the strands are transposed and even in fully insulated strands, the presence of shorts in crossover points is difficult to avoid. The twisted filaments within the strand have non-uniform cross-sections. Conductors move under the influence of magnetomechanical forces. If the coils are not impregnated in a thermosetting or any fluid which may freeze at liquid helium temperature, the losses due to wire motion may be considerable.

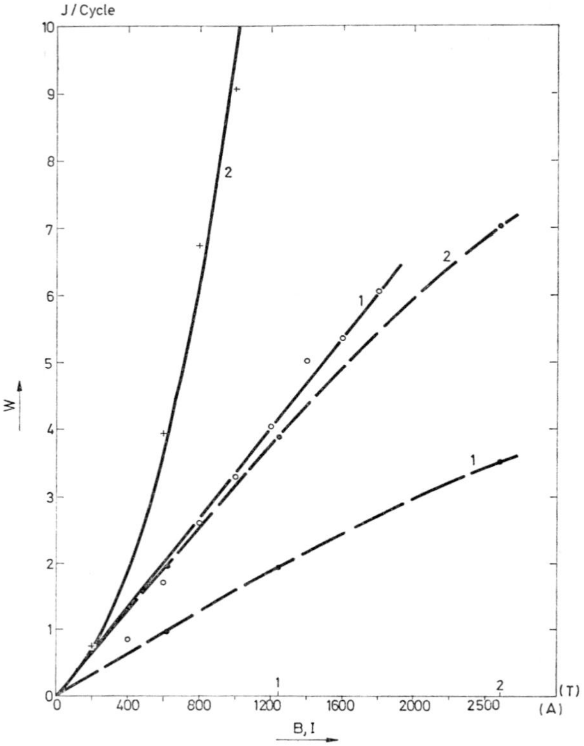

Fig. 4.5.4. Measured and calculated losses in solenoids according to Fig. 4.5.1.
1 full line: Organic insulated cable (no coupling between strands) measured; *dashed line:* Organic insulated cable, calculated.
2 full line: Ag(Sn) soldered cable (highly coupled) measured; *dashed line:* Ag(Sn) soldered cable, calculated.

4.6 Methods of Loss Measurement

Methods to measure a.c. losses in superconductors can be implied to test samples and specimen and to measure losses in coils. The most popular method to measure the additional helium boil-off due to a.c. losses is the calorimetric method. Direct loss measuring methods are by means of electric circuits.

4.6.1 Calorimetric Method

Fig. 4.6.1 illustrates the method of loss measurement of specimen schematically. The specimen is wound bifilar and placed in a calorimeter. The superconductor is energized from a pulse-generator, or a variable current, low frequency ($\leq$ 50 Hz) source. The helium boil-off due to a.c.

losses is measured by means of a gas flowmeter. The calorimeter is placed in a main liquid helium dewar, such that current leads are located inside the main cryostat prior to entering the calorimeter.

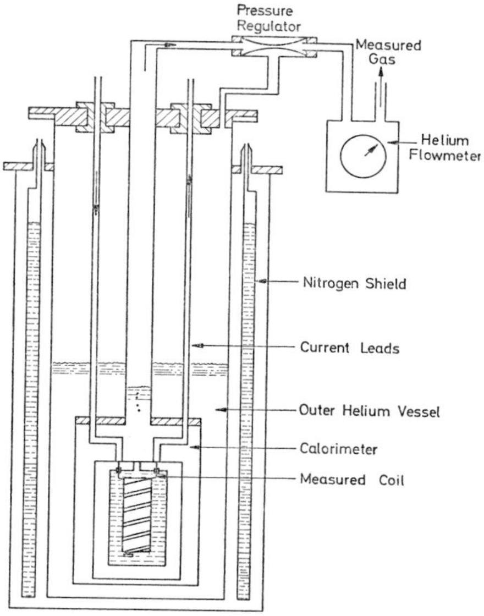

Fig. 4.6.1. Schematic arrangement of the calorimetric He boil-off method to measure a.c. losses

The sensitivity of the method (~1 mW) is moderate, if low frequency pulsing is applied. The measurements accuracy is improved by raising pulse frequency. It is established that a.c. losses in superconductors increase linearly with frequency and that the losses per cycle is frequency independant (not counting eddy current effects). The advantage of this method is its independence on magnetic energy stored in the coil. Its main disadvantage is the long time required until equilibrium conditions are established to measure losses. Only specimen and magnets of moderate size can be tested with this method.

4.6.2 Electric Methods

Several methods are used sucessfully and in the following descriptions of these methods are given. The basic principle of these methods is to measure a voltage which is proportional to the power delivered to the coil and integrating it over one or several cycles.

(a) The circuit shown in Fig. 4.6.2 is described by Gilbert et al. [30] and is used to measure a.c. losses in superconducting specimen and coils. The circuit consists of two major parts: A *multiplying circuit* to obtain instantaneous power flow between specimen (coil), and the *power supply* and an *integrator* to keep track of the energy balance into the superconducting magnet. The *multiplying circuit* operates as follows: It utilizes a Hall probe which generates an output voltage proportional to the product of input current to the Hall probe and the magnetic field around the probe. The test magnet is connected to the Hall probe through a resistor by a pair of potential leads inserted into the cryostat. Thus, the input current to the Hall probe is proportional to the magnet voltage.

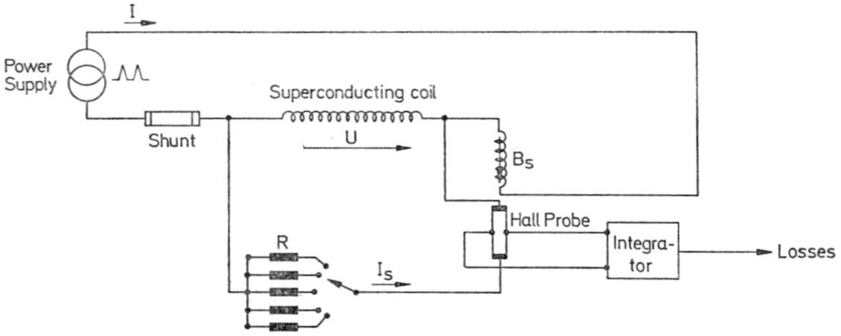

Fig. 4.6.2. Schematic arrangements of a.c. loss measurement electrical method using a Hall multiplier (LBL)

The Hall device is mounted in the gap of a coil which generates a magnetic field proportional to the transport current of the coil ($B_s = $ K.I.). As this coil is connected in series to the superconducting magnet to be tested, the Hall device is exposed to a magnetic field proportional to the current flowing through the superconducting magnet. The input to the Hall device is thus proportional to the voltage across the superconducting magnet and to the magnet current. The output voltage of the Hall device is proportional to the instantaneous value of the power into the superconducting magnet, and changes sign with the direction of the energy flow.

The *integrator* uses a solid state chopper operational amplifier with a feed back capacitor C and a series resistor R, such that its output voltage is

$$U_{\text{out}} = \frac{1}{RC} \int_0^T U_{\text{in}}(t)\, dt \,.$$

The drift in the integrator circuit appears as an equivalent power loss and must be kept small (1 mV-drift in 10^3 sec, with a dynamic range of 10 V).

The Hall voltage is integrated by means of an integrator. The output voltage is given by:

$$U_0 = \frac{s \cdot K}{\tau \cdot R} \cdot W \leqq \frac{s \cdot B_{s,\max} \cdot I_s}{\tau \cdot I_{\max} \cdot U_{\max}} \cdot W \,, \qquad (4.6.1)$$

where W are the losses to be measured, s is the sensitivity, and I_s the maximum current through the Hall probe; $B_{s,\max}$ is the maximum flux density determined by the non-linearity of the Hall probe. τ is the time constant of the feed-back loop of the integrator; and U_{max}, I_{max} the maximum voltage and maximum current, resp., applied to the superconducting coil during the cycle.

It is seen from the above relation, that the sensitivity of the method is limited by the characteristics of the Hall probe (s, I_s, $B_{s,max}$) and by the magnitude of the applied voltage and current. The circuit has a moderate sensitivity of about $\cong 5\%$ at $Q \leqq 100$.

The sensitivity of the system is also limited by the drift of the integrator, its non-linearity, and by the thermoelectric voltage of the Hall probe.

(b) To reduce the influence of the inductive term U_{max}, Hlasnik [31] and coworkers have proposed its compensation by means of a linear mutual inductance, whose primary side is in series to the superconducting coil (Fig. 4.6.3). The voltage U_d is essentially proportional to losses and can be amplified and applied to the Hallprobe.

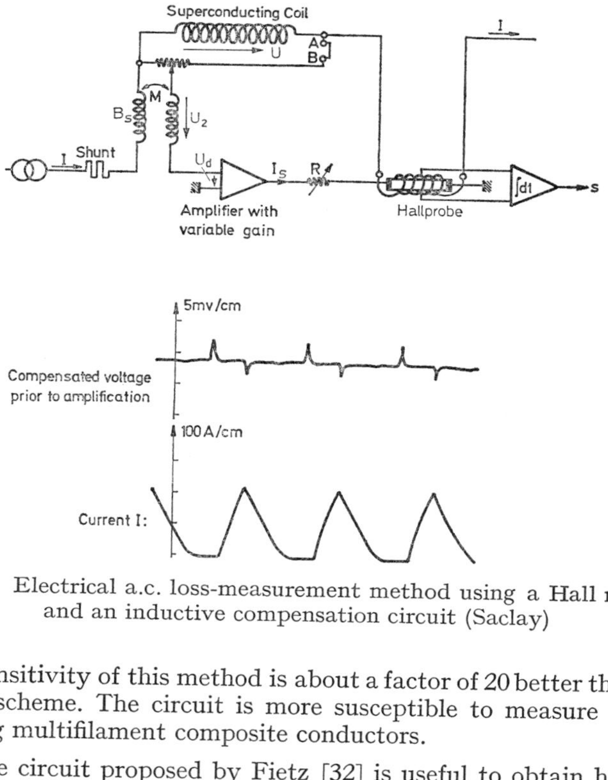

Fig. 4.6.3. Electrical a.c. loss-measurement method using a Hall multiplier and an inductive compensation circuit (Saclay)

The sensitivity of this method is about a factor of 20 better than in the previous scheme. The circuit is more susceptible to measure losses in coils using multifilament composite conductors.

(c) The circuit proposed by Fietz [32] is useful to obtain hysteretic losses in specimen. The sample is inserted into the gap of one of the two coils of identical cross sectional area and identical number of turns placed side by side in the bore of a solenoid with their axis parallel to the axis of the solenoid generating the external field (Fig. 4.6.4).

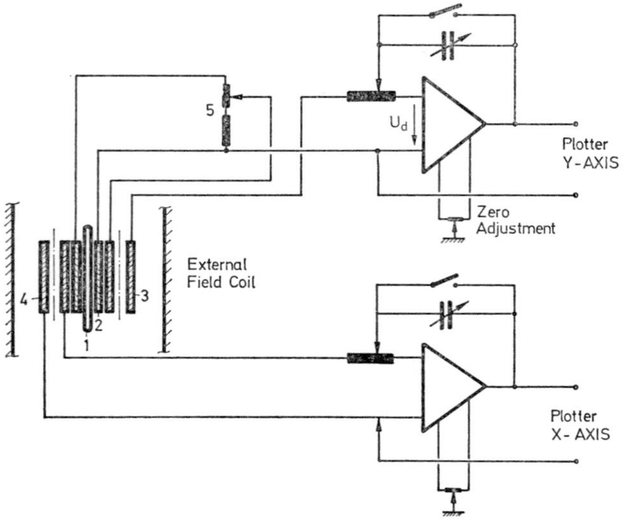

Fig. 4.6.4. Schematic diagram of a.c. loss measuring system (Fietz)

As long as the sample is not superconducting, the output induced voltage from the coils 2 and 3 are the same and cancel each other in the circuit. A coarse and fine potentiometer corrects small voltage deviations due to measuring coil errors.

The measuring coils 2 and 3 are designed to fit the specimen closely such that a good coupling between sample and coil is obtained. The output voltage from the coils (empty coil and coil with specimen) is amplified by an operational amplifier. The difference voltage U_d is proportional to the magnetization of the specimen. Measuring the external applied fields by means of a pickup coil 4, and tracing the amplified voltage U_d over the amplified pickup voltage, the magnetization curve shown in Fig. 4.6.5 is obtained. The area of the magnetization curve is proportional to the hysteretic losses for one cycle.

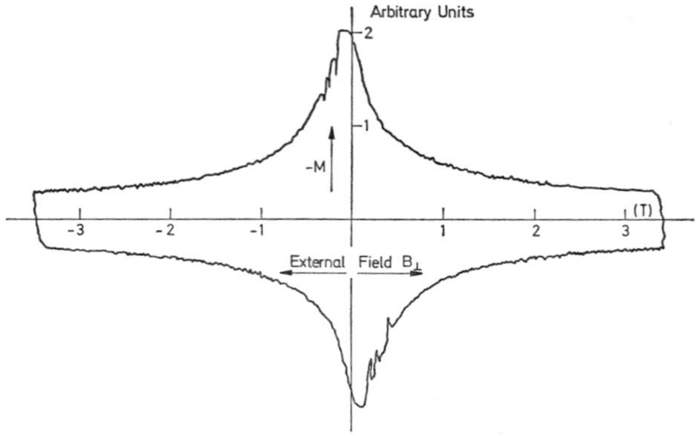

Fig. 4.6.5. Magnetization loop traced for Nb(50%)Ti

Error sources are due to deviations in the area or number of turns of the measuring coils, the positioning of these coils in the external field, unmatched flux densities (different Ampereturns in the coils), and mainly errors due to the drift of the operational amplifier.

The circuit also has been used by McInturff [33] to measure magnetization losses of NbTi wires and by Brechna [34] in modified form to measure magnetization of niobium specimen.

(d) An apparatus to measure low frequency losses is described by Sekula [35] and is shown in Fig. 4.6.6. An oscillator and a power amplifier produce a sinusoidal current in the primary circuit of the bridge. The primary part of the bridge includes the primary of a variable mutual inductance, an end-corrected coil, which generates a sinusoidal magnetic field in the region of the sample, a dummy coil, and a 1 Ohm resistor across which a voltage is developed to furnish a reference signal for the lock-in-amplifier and also to provide a resistive voltage component in the secondary part when desired. The rms primary current is measured by means of a voltmeter across a 0.1 Ohm standard resistor.

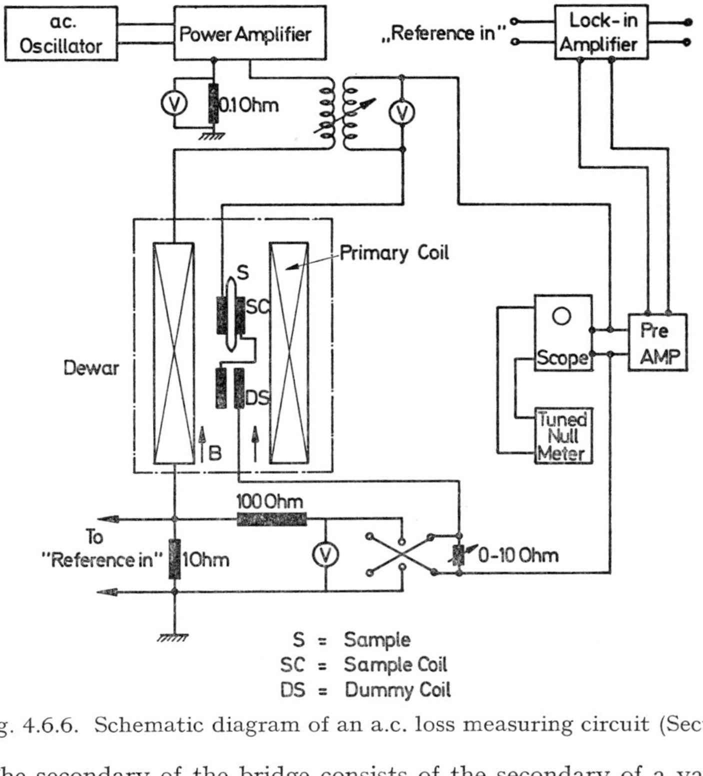

Fig. 4.6.6. Schematic diagram of an a.c. loss measuring circuit (Secula)

The secondary of the bridge consists of the secondary of a variable mutual inductance, the superconducting sample and dummy coils, and a variable resistance of 0 to 10 Ohms.

The circuit operates as follows:

A fraction of the resistive voltage developed across the nominal 1 Ohm resistance in the primary can be introduced into the secondary and permits 180° phase reversal. A high impedance voltmeter measures the components of the voltage across the resistive network and the variable mutual impedance secondary for calibration purposes. The primary and secondary coils are arranged in a helium dewar such that the secondary coils are in the uniform region of the end-corrected superconducting coil generating the external field. The sample is mounted such that it can be inserted into or taken out of the test coil during the experiment. The long axis of the coil is parallel to the direction of the d.c. and a.c. fields. The output of the bridge is connected to a PAR model CR-4 pre-amplifier and an oscilloscope. Losses are measured by means of the PAR model JB-5 lock-in-amplifier. The oscilloscope is used to visually observe the bridge output. If a superconducting sample of radius R is placed in a pickup coil of N turns, the induced voltage in the coil due to variable external magnetic field is $U(t) = -N \cdot d\phi/dt$. If the external field has a steady state part B and a sinusoidally varying part $b = b_0 \cos(\omega t)$, the losses in a cylinder of unit length is given by $W = \int b\, d\phi$, where $d\phi$ represents an increment of magnetic flux in the superconductor. The energy dissipated per cycle per unit area of the cylinder surface is given by:

$$W_s = -\frac{1}{2\pi NR} \int_0^{2\pi/\omega} b_0 \cdot \cos(\omega t)\, dt\,. \tag{4.6.2}$$

A Fourier expansion of $U(t)$ may be performed. It can be noted that because of the orthogonality of the trigonometric function, the only non-zero contribution of Eq. (4.6.2) arises from the term $U_\nu \cos(\omega t)$. U_ν is the corresponding Fourier coefficient that depends generally on the sample size and the mechanism of flux penetration. The surface losses per cycle is calculated to be:

$$W_s = \frac{-b_0}{2\omega} \cdot \frac{U_\nu}{NR}\,. \tag{4.6.3}$$

To measure losses in the sample, one may use the apparatus described and select the phase of the lock-in-amplifier to detect that component of the fundamental discussed.

The response of the sample in the Meissner region, i.e. $B + b_0 \leqq H_{c1}$ is the field of the first flux penetration. This is measured as follows: The bridge is initially nulled with both secondary coils empty. By inserting the specimen into the upper coil, a sinusoidal signal is observed at the bridge output. This signal is taken to correspond to a purely diamagnetic and hence lossless response within the sensitivity of the apparatus. The phase of the lock-in-amplifier is now adjusted to reject this signal so that only the component of the fundamental in quadrate is detected. By readjusting the calibrated bridge settings for null to prevent amplifier saturation, the sensitivity of the measurement can be increased. Under these new null conditions, a signal at the bridge output, observed on the oscilloscope, is proportional to the complex permeability of the specimen.

As the d.c. magnetic field is monotically swept above H_{c1} into the mixed state at a slow rate ($\sim 5 \times 10^{-4}$ T/s), a d.c. signal U_L is observed at the output of the lock-in-amplifier.

This voltage can be related after recalibration to U_ν in Eq. (4.6.3) by $U_L = \frac{1}{\sqrt{2}} U_\nu$.

(e) Another electric method to measure losses is discussed by Pech [36].

4.7 Magnetic and Thermal Instabilities

4.7.1 Introduction

Nonideal type II superconductors become resistive when the macroscopic critical current density (induced, or impressed by transport currents), limited by the fluxpinning strength, is exceeded. Current densities beyond J_c cause the flux to move, causing energy dissipation proportional to the flux flow resistivity; $\varrho_f = \varrho_n (B/H_{c2})$*, where ϱ_n is the normal state resistivity. The transport current density must exceed the depinning threshold value J_c before fluxlines can move. The resulting electric field is then approximately:

$$E = \varrho_f (J - J_c) . \tag{4.7.1}$$

The electric field E gives rise to power dissipation. However, the problem of instabilities which are of major concern in a.c. applications, deals with perturbances occurring in the voltage-current characteristic of the superconductor. This makes the superconductor revert to the resistive state before the true critical current (d.c. short sample value) is reached. One may speak in this case of a "degraded" critical current.

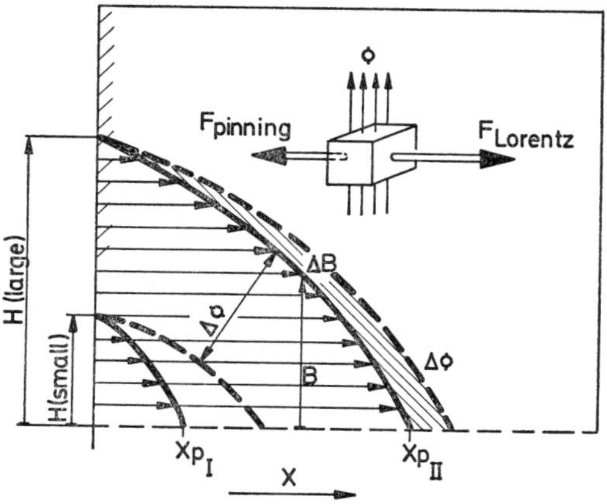

Fig. 4.7.1. Field penetration in a semi-infinite slab. Representation of flux distribution at an equilibrium condition pinning force equal to Lorentz force

* H_{c1}, H_{c2} are expressed in (T); H in A m^{-1}, or A cm^{-1}.

Table 4.7.1. *Properties of materials at 4.2 K*

Material	Density (δ) (g/cm^3)	c_p (Ws/g K)	$c_p \delta$ (Ws/cm^3 K)	k (W/cm K)	$D_{magn.}$ (cm^2/s)	$D_{therm.}$ (cm^2/s)
Al (0.997)	2.7	2.8×10^{-4}	7.56×10^{-4}	0.57	1.25	0.76×10^3
Cu (OFHC)	8.95	10^{-4}	8.95×10^{-4}	2.5	0.95	2.8×10^3
He	0.16 à 15 $kpcm^{-2}$	$c_v = 2.1$ à 15 $kpcm^{-2}$	0.336	0.263×10^{-3}	—	0.78×10^{-3}
Pb	11.25	8×10^{-4}	9×10^{-3}	20	0.32[a]	2.22×10^3
Na	1.009	1.7×10^{-3}	1.715×10^{-3}	48	0.135	2.8×10^4
In	7.28	1.2×10^{-3}	9.35×10^{-3}	8.5	2.47	9.1×10^3
Nb_3Sn	3.6	3.4×10^{-4}	1.22×10^{-3}	0.4×10^{-3}	240[b]	0.33
Nb (25%) Zr	5.8	1.8×10^{-4}	1.044×10^{-3}	1.7×10^{-3}	240	1.63
Nb (60%) Ti	5.6	2×10^{-4}	1.12×10^{-3}	1.5×10^{-3}	240	1.2

[a] $B > H_c$

[b] a.c. resistivity calculated for $\dot{B} = 2$ T/sec

A disturbance changes both Lorentz- and pinning forces. Fig. 4.7.1 schematically shows that a disturbance reduces the Lorentz force, since for a point in the superconductor with the same value of B, the field gradient is now smaller.

The pinning force is also changed, since the energy dissipation generates heat. Increasing temperature reduces the pinning strength of the material. If the reduction of the Lorentz force is larger than the reduction of the pinning force, the equilibrium is called "stable".

The disturbance* does not spread over larger areas of the specimen. If the decrease of the Lorentz force is smaller, than of the pinning force, the equilibrium is called "unstable". It has been shown by Wipf [37] that the magnetic flux can diffuse through the useful high field superconductor much faster than can the heat, produced by the dissipative process (Table 2.1.1, illustrates thermal and magnetic diffusivities of pure metals and alloys).

The superconductor can heat up appreciably, if no countermeasures are taken. If we designate by ΔB a sudden change in fluxdensity, then from $J_c(T) = \frac{1}{\mu_0} \cdot \frac{\partial B}{\partial X}$, we may obtain, according to Bean's model with $J_c = \text{const.}$, the penetration depth: $d = \frac{\Delta B}{\mu_0 J_c(T)}$.

An increase in temperature leads to a reduction in $J_c(T)$ in type II superconductors.

The field penetrates at a constant rate of B_0 further into the bulk superconductor, which leads to a greater temperature rise etc. ΔJ_c is negative and the fluxdensity-gradient in the shielding layer (d) becomes smaller. In order to keep the fluxdensity at the surface at the same level, more flux ϕ must be admitted in the superconductor. The fluxmovement dissipates energy, which further increases the temperature gradient. Thus a feedback cycle (Hart) [38] is obtained:

$$\Delta T \to \Delta J_c \to \Delta \phi \to \Delta Q \to \Delta T'.$$

If $\Delta T' < \Delta T$, then the temperature rise will asymptomatically reach a final value after many cycles, and the situation will be stable.

If $\Delta T' > \Delta T$, than the temperature rise will increase and a thermal runaway or a catastrophic fluxjump will occur. The quantitative evaluation of this heating process is complicated, as some heat escapes from the superconductor to the ambient through the thermal diffusion effect. The heating depends on the speed of the thermal diffusion by which heat is removed (thermal diffusivity $k/c_p\delta$), and the magnetic diffusion (magnetic diffusivity, ϱ_F/μ_0) by which $\Delta \phi$ is admitted.

4.7.2 Diffusion Equations

The one dimensional diffusion equation for any variable F is given by:

$$\frac{\partial^2 F}{\partial x^2} = \frac{1}{D} \cdot \frac{\partial F}{\partial t}, \tag{4.7.2}$$

* A disturbance can be a fluxjump or a sudden change in the transport current. Fluxjumping is generally referred to as induced current changes in the transport current, which lead to superconductor degradation.

with D being the diffusivity. In superconducting specimen, we have to concern ourselves with finite regions, say within

$$0 \leqq x \leqq 2w \tag{4.7.3}$$

then, if F has fixed values at each boundary, Eq. (4.7.2) has the solution

$$F = F_0 + \sum_{n=1}^{\infty} F_n e^{-n^2 t/\tau} \sin(n x/2w) , \tag{4.7.4}$$

in which τ is the time constant given by:

$$\tau = \frac{4}{\pi^2} \cdot \frac{w^2}{D} . \tag{4.7.5}$$

The coefficients F_n are determined by the value of F at $t = 0$.

$$F = \sum_{n=1}^{\infty} F_n \sin(n \pi x/2w) . \tag{4.7.6}$$

4.7.2.1 Magnetic Diffusivity

The variable F is the fluxdensity B (Fig. 4.7.2). As the magnetic diffusivity is defined by: $D_m = \varrho_F/\mu_0$, the field penetration time constant is given by:

$$\tau = \frac{4\mu_0}{\pi^2} \cdot \frac{w^2}{\varrho_F} , \tag{4.7.7}$$

where w is expressed in (m).

4.7.2.2 Thermal Diffusivity

The variable F is the temperature T.

With the thermal diffusivity being defined as: $D_{th} = k/c_p\delta$.

Where k is the thermal conductivity in (W/mK) and $(c_p\delta)$ is the heat capacity in (Ws/m³ K), the thermal time constant is thus:

$$\tau_{th} = \frac{4}{\pi^2} \frac{w^2}{k} \cdot (c_p\delta) . \tag{4.7.8}$$

Referring to Table 4.7.1, we see that for pure metals, the thermal diffusivity is in the order of 10^3 to 10^4 ($cm^2 \cdot s^{-1}$) where the magnetic diffusivity is about 1 ($cm^2 \cdot s^{-1}$).

In alloys and superconductors, the conditions are practically reversed. The thermal diffusivity is about 1 ($cm^2 \cdot s^{-1}$). The flux penetrates in alloys much faster than the heat can be removed. The specimen is heated up more readily and is fluxjump sensitive. The first idea was to combine pure metals with alloys, which lead to the development of composite conductors consisting of superconductors and a copper or aluminium matrix. The w^2-dependance of the time constant τ ensure that, in this case, heat can escape in such a composite conductor faster than flux can diffuse into the composite.

The mechanism by which the magnetic field penetrates a type II superconductor was given in chapter 4.1. We saw that currents are induced in the superconductor to oppose any change in the applied field.

These currents were referred to as shielding currents or magnetization currents. Since the current density of a superconductor is limited to a value of J_c, the field which can be shielded is also limited. Referring to Fig. 4.1.4, we see that complete shielding is possible only at low fields. After B has reached B_s, the shielding currents remain constant. This is true only if J_c is constant, otherwise we have to consider the field dependant current densities in the superconductor.

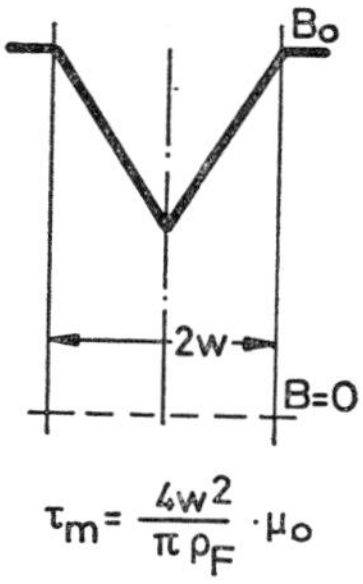

Fig. 4.7.2. Penetration of the magnetic field into a superconducting slab of with $2\,w$

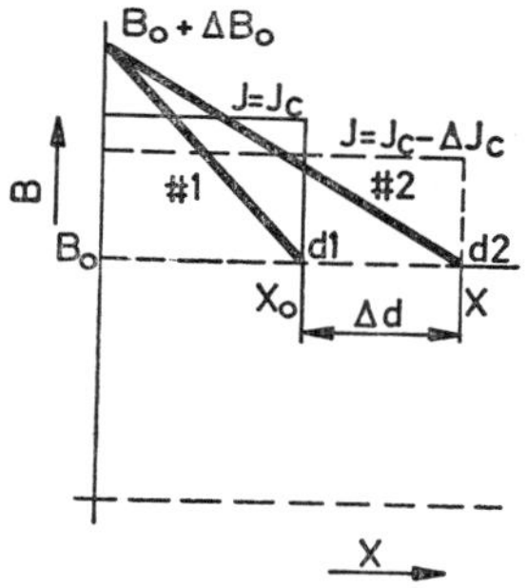

Fig. 4.7.3. Adiabatic stability: Field distribution prior (Profile 1) and after (Profile 2) a (small) temperature perturbation in a superconducting slab

The magnetization currents are generally unstable when the shielded field exceeds about 0.2 T. They decay rapidly (complete or partial) and release their energy as heat. This instability (fluxjump) occurs quite rapidly with τ_m and we can estimate the temperature rise from the losses per unit volume (Eq. (4.1.34) and (4.1.70)) and the heat capacity of the specimen:

$$\Delta T = \Delta p_v / c_p \delta \,.$$

4.7.3 Stability

We increase the applied field isothermally by an amount B_0, as shown in Fig. 4.7.3, and calculate the response of the system to a small disturbance (Δp_v) in the region of the disturbance as a function of the excluded field ΔB_0.

If ΔB_0 remains constant, then, due to the heating effect and consequently the decrease in J_c by ΔJ_c, the flux penetrates further from d_1 to d_2. First, we calculate the additional heat generated when the flux penetrates from profile 1 to profile 2.

Using the Poynting vector $\boldsymbol{S} = \boldsymbol{E} \cdot \boldsymbol{H}$ and subtracting from it that portion of this energy which is used for the increase of the magnetic field energy, then

$$W_{F,12} = \frac{1}{2\,\mu_0} \int_0^\infty [B_2^2(x) - B_1^2(x)]\, d\,x \,.$$

The resulting energy is the amount of heat generated in the specimen.

The power is obtained from the Poynting vector through integration:

$$\Delta P_{12} = \int_{t_1}^{t_2} S\,dt = \frac{1}{\mu_0} \int_{t_1}^{t_2} \dot{\phi}\,(x = 0) \cdot (B + \Delta B_0)\, dt$$

$$= \frac{1}{\mu_0} \cdot (B_0 + \Delta B_0)\, \Delta \phi_{12}(x = 0)\,. \qquad (4.7.9)$$

With $\Delta \phi_{12}(x)$ as the total flux crossing the surface from profile 1 to profile 2.

With $\Delta \phi_{12}(x = 0) = \frac{1}{2} \Delta B_0 (d_2 - d_1)$, as seen in Fig. 4.7.3, the total input energy is:

$$W_{12} = \frac{1}{2\,\mu_0} (B + \Delta B_0) \cdot \Delta B_0 \cdot (d_2 - d_1)\,, \qquad (4.7.10)$$

with:

$$d_2 = \frac{\Delta B_0}{\mu_0 J_c(T_2)}; \quad d_1 = \frac{\Delta B_0}{\mu_0 J_c(T_1)}\,. \qquad (4.7.11)$$

The total change in energy is thus:

$$\Delta W_{12} = \frac{1}{2\,\mu_0} \left(B_0 + \frac{\Delta B_0}{3}\right) \Delta B_0 (d_2 - d_1)\,. \qquad (4.7.12)$$

The total additional heat input per unit surface area is:

$$\Delta Q_{12} = \Delta P_{12} - \Delta W_{12}\,,$$

and the total heat input per unit volume is:

$$\Delta Q = \Delta Q_{12} - \Delta Q_0 = \Delta Q_0 + \frac{2}{3}\left(1 - \frac{d_1}{d_2}\right) \frac{\Delta B_0^2}{2\,\mu_0}\,. \qquad (4.7.13)$$

Thus the temperature rises:

$$\Delta T = \frac{\Delta Q}{c_p \delta} = \Delta T_0 + \frac{2}{3} \Delta B_0^2 \frac{1}{2\,\mu_0 c_p \delta} \cdot \left[1 - \frac{d\,(T_1)}{d\,(T_2)}\right]$$

$$= \Delta T_0 + \frac{2}{3} \frac{\Delta B_0^2}{2\,\mu_0} \cdot \frac{1}{c_p \delta} \left[1 - \frac{J_c(T_2)}{J_c(T_1)}\right], \qquad (4.7.14)$$

or:

$$\frac{\Delta T}{\Delta T_0} = 1 - \frac{2}{3} \cdot \frac{\Delta B_0^2}{2\,\mu_0} \cdot \frac{1}{c_p \delta} \left[-\frac{1}{J_c(T_1)} \cdot \frac{\partial J_c(T)}{\partial T}\right]. \qquad (4.7.15)$$

For any value of

$$\boxed{(\Delta B_0)^2 \leqq \frac{3}{2} \cdot 2\,\mu_0 \cdot c_p \delta \left[-\frac{1}{J_c} \cdot \frac{\partial J_c}{\partial T}\right]^{-1}} \qquad (4.7.16)$$

the temperature rise is the stable ΔT_0 value.

At the critical field, ΔB_0, the field penetration layer has reached a thickness of $\Delta B_0 / \mu_0 J_c$.

For a specimen thinner than twice this thickness, the condition of complete stability is always fulfilled. ΔB_0 is also called the screening field.

The decrease of J_c in a field implies that full stability condition is also fulfilled, once the field is high enough that $\mu_0 J_c R \leqq \Delta B_0$, where $2R$ is the diameter of the specimen. In the case of a cylinder with a transport current density J_T, and, in addition to an external field, the condition for full stability becomes: $(B + \mu_0 J_T R) \leqq \Delta B_0$, or for high fields:

$$\mu_0 R \,(J_c + J_T) \leqq \Delta B_0 \,. \tag{4.7.17}$$

The above simple calculation gives the necessary (but not sufficient) condition for fluxjumping.

The question is, will the specimen in which a fluxjump actually happens stay superconducting, or is it driven normal? The difference in magnetic field energy prior to and after the fluxjump is converted into the increase of enthalphy of the superconductor during the fluxjump. In the simplest case:

$$\frac{(B_{fjc})^2}{2\,\mu_0} = \int\limits_{T_1}^{T_2\,(B_{fjc})} (c_p \delta)\, dT \,. \tag{4.7.18}$$

Where B_{fjc} determines the smallest fluxjumping field, which drives the specimen normal. T_1 is the initial, T_2 the endtemperature. If a transport current is also present in the superconductor, the endtemperature is such that $J_c(T_2, B_{fjc}) = J_T$.

If a fluxjump happens at a field value smaller than B_{fjc}, it is called incomplete (the specimen may stay superconductive), otherwise it is complete or catastrophic (Fig. 4.7.4).

4.7.3.1 Temperature Rise from Fluxjump

The exact calculation of the temperature rise due to a disturbance is complicated, but to give an estimate of the order of magnitude, the equations derived above are adequate. To equate ΔT, we have to know ΔQ, or Δp_v, which we derived for a cylindrical wire, and the heat capacity of the specimen. Taking a NbTi wire with a diameter of $2R = 0.1$ cm and a current density of $J_c \cong 3 \times 10^5$ A cm^{-2}, then we obtain for linear field penetration and for $B = B_s$ i.e. full fluxpenetration, the relations:

$$p_v = \frac{5}{12\,\mu_0} \cdot B_s^2 \,, \tag{4.7.19}$$

or:

$$p_v = \frac{5\,\mu_0}{12} \cdot J_c^2 R^2 \,. \tag{4.7.20}$$

With the above values we get: $p_v = \frac{3\,\pi}{8} = 1.18$ J/cm^3.

If this energy is suddenly released in the superconductor, its temperature will rise to about 20 K. But if the same material is subdivided into filaments with individual diameters of $d_f = 2 \times 10^{-3}$ cm, then $p_v = 1.88 \times 10^{-3}$ J/cm^3 and the temperature rise will be $\Delta T = 0.5$ K.

If we use the Kim model, we obtain for full field penetration, where B is reduced to zero at $x = x_s$,

$$p_v = \frac{5}{9\,\mu_0} B_s^2 \tag{4.7.21}$$

or:

$$p_v = \frac{5}{9\,\mu_0}\,[B_0 \pm (B_0^2 - 2\,\mu_0 J_0 B_0 \cdot x_s)^{\frac{1}{2}}]^2\,. \tag{4.7.22}$$

The temperature rise will be slightly higher than in the previous case of linear fluxpenetration.

The subdivision of superconductor into filaments was first proposed by Chester [39].

A combination of both ideas, subdividing the superconductor into a large number of filaments and inbedding them in a matrix material of high thermal diffusivity, was proposed by Wilson et al. [40].

4.7.3.2 Adiabatic Stability

In the previous section, we had treated basically the stability criterion. We calculated also the critical shielding field, above which magnetization currents become unstable. For most type II superconductors, these fields are below 0.2 T.

If a perturbation occurs, the following chain of events are observed:

(a) ΔT_1 is increased, causing a reduction in J_c by say ΔJ_c.

(b) The decrease in J_c leads to an increase in Δd, a change in the penetration depth.

(c) Fluxmotion sets in, causing some amount of heat to be released.

(d) If the magnetic diffusivity is much larger than the thermal diffusivity, very little heat will escape from the superconductor (adiabatic condition) and the temperature will increase to ΔT_2.

If $\Delta T_2 > \Delta T_1$, the process of flux motion will continue until J_c is reduced to zero. The superconductor is driven normal. If $\Delta T_2 < \Delta T_1$, the system is stable against small perturbations. We can represent the above chain of events quantitatively for a linear fluxpenetration as follows:

$$\text{(a)}\quad \Delta J_c = \frac{dJ_c}{dT}\,\Delta T_1\,, \tag{4.7.23}$$

$$\text{(b)}\quad d = \frac{1}{\mu_0}\cdot\frac{B}{J_c}\,, \tag{4.7.24}$$

$$\Delta d = -\frac{1}{\mu_0}\cdot\frac{B}{J_c}\,\frac{\Delta J_c}{J_c}\,. \tag{4.7.25}$$

To obtain the losses per unit volume, we calculate the flux passing through a small area between d_1 and d_2 of Fig. 4.7.3

$$\phi(x) = \int_x^{x_0} \mu_0\cdot\frac{x}{x_0}\,\Delta d\,J_c\,dx = \int U\,dt = \mu_0 J_c\,\frac{\Delta d}{2\,x_0}\,(x_0^2 - x^2)\,. \tag{4.7.26}$$

The energy dissipation per unit surface area is given by:

$$p_s = \int_0^{x_0} J_c\cdot U\,dt\,,$$

which, for constant, J_c yields:

$$p_s = \mu_0 J_c^2 \cdot \frac{\Delta d}{3} \cdot x_0^2 \,. \tag{4.7.27}$$

The mean losses per unit volume are thus:

$$p_v = \mu_0 J_0^2 \frac{\Delta d}{3} x_0$$

or:

$$p_v = J_c B \frac{\Delta d}{3} . \tag{4.7.28}$$

The temperature rise due to these losses are:

$$\Delta T_2 = p_v / c_p \delta \,.$$

If ΔT_2, corresponding to profile No. 2, is smaller than ΔT_1, or if $\Delta T_2 / \Delta T_1 < 1$; which is equivalent to the criterion $\partial \Delta T / \Delta T < 0$, means that ΔT must decay with time. For this condition, the combination of the above equations including (a), and (b), which reflects the change in the critical current density and the change in the penetration layer, one obtains:

$$B < \left[3 \mu_0 (c_p \delta) \cdot \left(-J_c \Big/ \frac{\partial J_c}{\partial T} \right) \right]^{\frac{1}{2}} . \tag{4.7.29}$$

This relation was obtained also in Eq. (4.7.16).

The limiting penetration thickness can be calculated from

$$B = \mu_0 J_c \cdot x$$

or:

$$x < \left[\frac{3}{\mu_0} \cdot (c_p \delta) \cdot \left(-J_c \Big/ \frac{\partial J_c}{\partial T} \right) \right]^{\frac{1}{2}} \cdot \frac{1}{J_c} . \tag{4.7.30}$$

This equation is also valid for a conductor with a diameter D:

$$\boxed{J_c^2 D^2 < \frac{3}{\mu_0} (c_p \delta) \left(-J_c \Big/ \frac{\partial J_c}{\partial T} \right) .} \tag{4.7.31}$$

Both Eqs. (4.7.29) and (4.7.30) are criteria for adiabatic stability. For a specific layer thickness x, and for a given current density J_c, no flux-jump should occur.

To give a numerical example, we calculate the wire diameter for NbTi, carrying a current density of $J_c = 3 \times 10^5$ A/cm². Referring to Table 4.7.1, we gave for NbTi the heat capacity: $c_p \delta = 1.12 \times 10^{-3}$ Ws/cm³ K. The value of $\left(-J_c \Big/ \frac{dJ}{dT} \right) \simeq 4$ to 5 K, which gives a critical field of $B = 0.13$ T, and a superconducting wire diameter of $D = 3.5 \times 10^{-2}$ cm.

If the wire diameter is larger than the above value, fluxjump may occur earlier.

Eq. (4.7.29) suggests the following possibilities to prevent fluxjump:

(i) Increasing heat capacity $(c_p \delta)$ of the conductor: This implies the use of materials with high thermal capacity at temperatures below the critical temperature of the superconductor. Hancox [41] proposed lead as a substrate, Brechna [42] dense helium gas.

(ii) Increasing $\left(-J_c / \frac{dJ}{dT}\right)$: This is only possible if dJ_c/dT is positive.

This means that the pinning strength is increasing with temperature. Special treatment of NbTi has shown that the slope of the $J(T)$ curve near T_c will become positive. However, the current density of NbTi near T_c is very small and not of practical use. The alloy system Pb-In with fine dispersed tin percipitates has also shown values of $\partial J/\partial T > 0$. With rising temperature, specifically at 3.8 K, the tin becomes normal and the particles act as pinning centers.

(iii) Decreasing J_c: This asks for a reduction of the critical current density at the operating point. This is not attractive, as more superconducting material must be used to generate a certain field in a specified volume.

Considering the heat capacity versus temperature characteristics of a superconductor, alloys, or pure metals, we see that $c_p\delta$ rises quite rapidly with temperature (Fig. 5.4). Even at transition temperatures of the superconductor but higher than the liquid helium temperature, the heat capacity is increased substantially.

Fluxjump in the conductor may stop at this new condition. This effect of self-healing is called partial or incomplete fluxjump and indicates that not just any fluxjump may lead necessarily to a quench.

4.7.3.3 Dynamic Stability

In the above section, we saw that adiabatic stability was achieved by the fact that the energy dissipation due to fluxmotion or fluxjump can be reduced to a level at which it can be absorbed by the heat capacity of the conductor material itself.

In this case, we may assume that the original field profile and the current density distribution determines the initial fluxmotion and the heating rate for some time after the first temperature rise.

The dynamic stabilization criteria postulates that a system is dynamically stable, if the temperature rise due to a perturbation decays immediately towards an equilibrium temperature.

Upon a sudden increase in temperature, a momentary decrease in the critical current density and with it, a decrease in fluxjump is encountered, which also indicates that dynamic stability can prevent fluxjumps. This however, does not mean that the total current density in the superconductor is decreased. The flux begins to move through the superconductor, generating heat. Referring to Fig. 4.7.5 and assuming that ϱ_F is unaffected by temperature, we find the initial heating rate:

$$\dot{p}_v(x) = J \cdot U(x) = J_c \varrho_F (J_{c1} - J_{c2}) = J_{c1}^2 \varrho_F \left(-\frac{1}{J_c} \frac{\partial J_c}{\partial T}\right) \Delta T(x) . \tag{4.7.32}$$

The total heating rate is proportional to the local temperature rise. The dynamic stability of the system is defined by the relative magnitude of the heat generated to the heat removed.

Dynamic stability requires that high thermal conductivity metals are utilized in combination with the superconductor. In such a composite conductor which consists of superconducting sheets or filaments and a normal metal matrix, we can essentially derive the dynamic stability criterion in the same manner as we calculated the adiabatic stability measure:

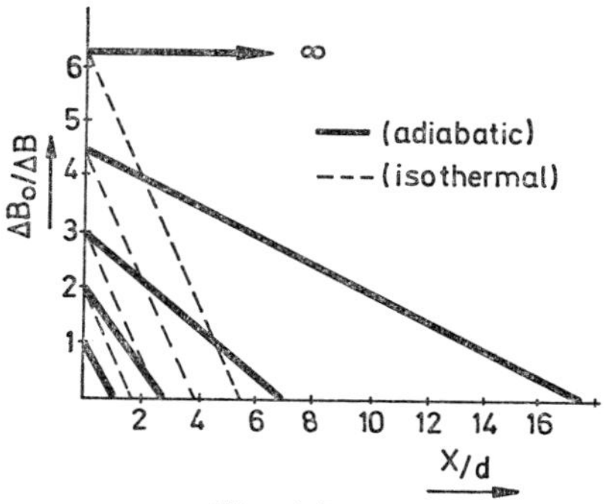

Fig. 4.7.4

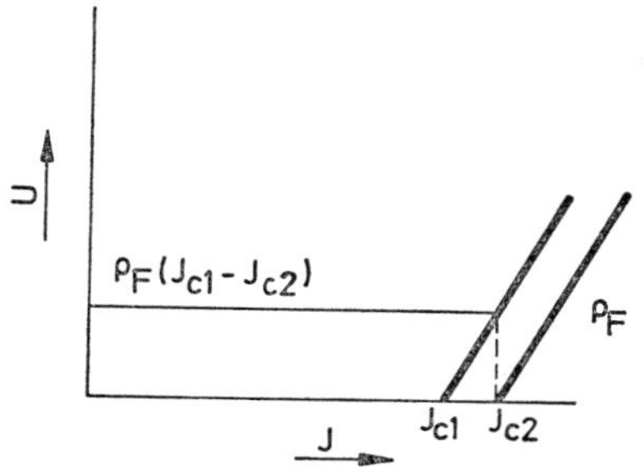

Fig. 4.7.5

Fig. 4.7.4. Isothermal and quasi-adiabatic field profiles after a small perturbation for a semi infinite slab

Fig. 4.7.5. Nonlinear relation between the electric field and the current density. ϱ_F represents the fluxflow resistivity for the superconductor. In composites, ϱ_F may also represent the effect of the matrix. U—J relation may be considered as one formulation of the critical state model

We commence with a flux perturbation causing a temperature rise ΔT_1, then the following steps are to be considered:

(i) The change in current density due to the temperature rise,

$$\Delta J_c = \left(\frac{dJ_c}{dT}\right) \Delta T_1 \,. \tag{4.7.33}$$

(ii) The variation in the fluxpenetration layer due to fluxmotion,

$$\Delta d = -\frac{1}{\mu_0} \frac{B}{\lambda_s J_c} \cdot \left(\frac{\Delta J_c}{J_c}\right). \tag{4.7.34}$$

(iii) The rate of heat released per unit volume in a filament,

$$\dot{p}_v \cong \frac{2\,\Delta p_v}{\tau_{\text{mag}}} \quad (\text{Ws} \cdot \text{s}^{-1}\,\text{cm}^{-3}) \,.$$

Whereas, according to the previous section:

$$\Delta p_v = J_c \cdot B \,\frac{\Delta d}{3} \,.$$

The magnetic time constant τ_{mag} is given by [Eq. (4.7.7)]

$$\tau_{\text{mag}} \cong \frac{4\,\mu_0}{\pi^2} \frac{x^2}{\varrho_n} (1 - \lambda_s) \,, \tag{4.7.35}$$

with λ_s, the space factor of the superconductor within the composite and $\varrho_n/(1 - \lambda_s)$ the effective resistivity of the matrix material.

(iv) The heat removal is assumed to be determined by the thermal conductivity k of the composite conductor; the heat transfer to helium is taken to be infinite. We assume complete flux penetration, i.e. $B > B_s$ and a field independant J_c.

The equation to be solved is the thermal diffusion Eq. (4.7.2) for the temperature distribution within the superconductor.

For a semi infinitive slab, the temperature distribution is according to:

$$k_s \frac{\partial^2 (\Delta T)}{\partial x^2} - (c_p \delta)_s \frac{\partial (\Delta T)}{\partial t} + p_v = 0 . \tag{4.7.36}$$

With the assumption*

$$p_v = a_s \Delta T = J_c^2 \varrho_F \left(-\frac{1}{J_c}\frac{dJ_c}{dT}\right) \Delta T \tag{4.7.37}$$

and the boundary conditions that at the surface of the superconductor ($x = \pm d/2$), the temperature at the superconductor is the same as in the substrate and that the temperature rise in the substrate is negligibly higher than in bulk helium, the general solution of Eq. (4.7.36) for $\tau_{\text{mag}} \gg \tau_{\text{therm.}}$ is:

$$\Delta T(x,t) = \sum_{n=0}^{\infty} A_n \exp \cdot \cos\left[\frac{(2n+1)\pi x}{d}\right] \left(\frac{a_s}{(c_p\delta)_s} - \frac{(2n+1)^2 \pi^2 k_s}{d^2 (c_p \delta)_s}\right) t . \tag{4.7.38}$$

The criterion for dynamic stability is that all terms of the series must decrease, i.e.:

$$\frac{a_s}{(c_p\delta)_s} - \frac{(2n+1)^2\pi^2 k_s}{d^2 (c_p\delta)_s} \leqq 0 \tag{4.7.39}$$

or:

$$\frac{d^2 \cdot a_s}{\pi^2 k_s} \leqq (2n+1) .$$

---

* If one assumes that a fraction of the current does flow through the superconductor in the normal state, then the voltage across this section of the composite is given by:

$$U = J_c \left(-\frac{1}{J_c}\frac{dJ_c}{dT}\right) \cdot d \cdot \left(\frac{d}{\varrho_F} + \frac{D}{\varrho_n}\right)^{-1} \Delta T$$

and the generated loss per volume is:

$$p_v = J_c^2 \left(-\frac{1}{J_c}\frac{dJ_c}{dT}\right) \cdot d \cdot \left(\frac{d}{\varrho_F} + \frac{D}{\varrho_n}\right)^{-1} \Delta T .$$

Generally: $d/\varrho_F \ll D/\varrho_n$, and can be neglected. Thus:

$$p_v \cong J_c^2 \left(-\frac{1}{J_c}\frac{dJ_c}{dT}\right) \cdot d \cdot \frac{\varrho_n}{D} \cdot \Delta T$$

or:

$$p_v \cong J_c^2 \left(-\frac{1}{J_c}\frac{dJ_c}{dT}\right) \frac{\lambda_s}{1-\lambda_s} \varrho_n \Delta T$$

which is equivalent to Eq. (4.7.40), except for a numerical constant.

For the particular case of $n = 0$, which is the most stringent requirement, we have:

$$\frac{d^2 \cdot a_s}{\pi^2 k_s} \leqq 1 .$$

From the above equation, we obtain:

$$\dot{p}_v = \frac{2\,\Delta p_v}{\tau_{\text{mag}}} = \frac{\pi^2}{6} \cdot \frac{\lambda_s}{1-\lambda_s}\,\varrho_n \cdot J_c^2 \left(-\frac{1}{J_c}\frac{dJ_c}{dT}\right) \Delta T_1 = a_s \Delta T_2 . \qquad (4.7.40)$$

From the dynamic stability criterion (4.7.39) for $n = 0$ and $\Delta T_2 = \dot{p}_v / a_s$, we obtain:

$$\frac{\Delta T_2}{\Delta T_1} = \frac{1}{6}\frac{d^2}{k_s} \cdot \frac{\lambda_s}{1-\lambda_s}\,\varrho_n \cdot J_c^2 \left(-\frac{1}{J_c}\frac{dJ_c}{dT}\right) < 1 ,$$

which gives the maximum permissible slab thickness for which a perturbation does not lead to a quench:

$$\boxed{J_c^2 d^2 \leqq 6\,k_s \frac{1-\lambda_s}{\lambda_s}\,\frac{1}{\varrho_n} \left(-\frac{1}{J_c}\frac{dJ_c}{dT}\right)^{-1} .} \qquad (4.7.41)$$

If in the slab-geometry, $\lambda_s = d/D$, denotes the ratio of the superconductor to the matrix material thickness*, we may modify this equation and calculate the thickness of the normal metal substrate for a given thickness of the superconductor.

$$\boxed{D \leqq \frac{\mu_0}{6}\,\frac{d^3 J_c^2}{(c_p \delta)_s} \cdot \left(-\frac{1}{J_c}\frac{dJ_c}{dT}\right) \cdot \frac{(D_m)_n}{(D_{\text{th}})_s} + d .} \qquad (4.7.42)$$

$(D_m)_n$ is the magnetic diffusivity of the substrate and $(D_{\text{th}})_s$ is the thermal diffusivity of the superconductor.

Inserting the value of J_c from $B = \mu_0 \cdot \lambda_s \cdot J_c \cdot x$ in Eq. (4.7.41), we find the screening field:

$$B_s^2 = 6\,\mu_0 \cdot \frac{(D_{\text{th}})_s}{(D_m)_n}\,\lambda_s (1-\lambda_s)\,(c_p \delta)_s \left(\frac{x}{d}\right)^2 \left(-\frac{1}{J_c}\frac{dJ_c}{dT}\right)^{-1} . \qquad (4.7.43)$$

To give a numerical example, we calculate the thickness of a NbTi-copper slab from the following known data:

$k_s = 10^{-3}$ W/cm K; $\varrho_n = 3 \times 10^{-8}$ Ohm · cm; $J_c = 2 \times 10^5$ A/cm², (at 5 T) and $\lambda_s = 0.5$. From Eq. (4.7.41), we obtain $d = 5 \times 10^{-3}$ cm.

The thickness of the copper matrix is obtained from $\frac{1-\lambda_s}{\lambda_s} = \frac{D-d}{d}$; which gives for $\lambda_s = 0.5$, the value of $D = 2\,d$.

Referring to Eq. (4.7.42) and Table 4.7.1, we obtain

with $(D_m)_n = 2.39$ cm² s⁻¹ and $(D_{\text{th}})_s = 1$ cm² s⁻¹,

the width of the copper matrix $D = 10^{-2}$ cm.

The maximum screening flux density is attained, if we set $x = d$, in Eq. (4.7.43), and obtain for the above example the value of $B_s \leqq 0.063$ T.

---

* The width of the slab is assumed to be infinity and, therefore, no edge cooling is considered.

Dynamical stability criterion for an infinitely long *cylindrical conductor* is obtained from the differential equation:

$$\frac{\partial^2 (\Delta T)}{\partial r^2} + \frac{1}{r}\frac{\partial (\Delta T)}{\partial r} + \frac{1}{r^2}\frac{\partial^2 (\Delta T)}{\partial \phi^2} - \frac{(c_p\delta)_s}{k_s}\cdot\frac{\partial (\Delta T)}{\partial t} + \frac{a_s}{k_s}\Delta T = 0\,, \tag{4.7.44}$$

where ΔT is the finite temperature rise due to the disturbance. Eq. (4.7.44) has the general solution:

$$\Delta T\,(t,r,\phi) = e^{\frac{a_s}{(c_p\delta)_s}t} \sum_{\lambda} e^{-\nu^2 \frac{k_s}{(c_p\delta)_s}t} \sum_{n=-\infty}^{+\infty} c_{\nu,n}\, e^{jn\phi}\cdot J_n(\nu r)$$

$$n = 0;\ 1;\ 2;\ \ldots \tag{4.7.45}$$

Using the boundary conditions, that at $t = 0$; $\Delta T = \Delta T_1$ and for $r = \pm\, d/2$; $\Delta T = 0$, $(h = \infty, k_n = \infty)$, the roots of the Bessel function J_n $(\nu d/2)$ gives the criterion for dynamic stability:

$$\frac{a_s}{k_s} < \nu^2\,. \tag{4.7.46}$$

The first root $(n = 0)$ is also the most stringent one:

$$\nu\cdot\frac{d}{2} = 2.4\,.$$

Inserting this value of ν in Eq. (4.7.46), we get:

$$a_s \leqq (2.4)^2\, k_s \left(\frac{d}{2}\right)^2. \tag{4.7.47}$$

As in the case of a semi-infinite slab, the ratio between final and initial temperature rise is given by:

$$\frac{\Delta T_2}{\Delta T_1} = \frac{\pi^2}{6}\cdot\frac{d^2}{23\,k_s}\cdot\frac{\lambda_s}{1-\lambda_s}\cdot\varrho_n\cdot J_c^2\left(-\frac{1}{J_c}\cdot\frac{dJ_c}{dT}\right) < 1\,, \tag{4.7.48}$$

from which we obtain the diameter of the superconducting cylinder:

$$\boxed{d^2 \leqq \frac{138}{\pi^2}\cdot\frac{k_s}{\varrho_n}\cdot\frac{1-\lambda_s}{\lambda_s}\cdot\frac{1}{J_c^2}\left(-\frac{1}{J_c}\cdot\frac{dJ_c}{dT}\right)^{-1}.} \tag{4.7.49}$$

With exception of a numerical constant, this equation is identical to Eq. (4.7.41) derived for a slab. If we assume that the superconductor is surrounded by a normal metal matrix of cylindrical shape of diameter D, this diameter is calculated for a given superconducting wire diameter d in the usual way:

$$D \leqq \mu_0\frac{\pi^2}{138}\, d^3\, \frac{(D_m)_n}{(D_{\text{th}})_s}\cdot\frac{1}{(c_p\delta)_s}\, J_c^2\left(-\frac{1}{J_c}\cdot\frac{dJ_c}{dT}\right) + d\,. \tag{4.7.50}$$

As an example, we calculate the diameter of the superconducting wire for $\lambda_s = 0.5$ and the data given for the slab:

$$d = 7.65\times 10^{-3}\ \text{cm}\,.$$

The diameter of the copper matrix is according to Eq. (4.7.50):

$$D = 15.3 \times 10^{-3} \text{ cm}.$$

For a cylindrical conductor, the equation for heat balance following a disturbance ΔT_1, assuming the heat transfer (h) from the conductor surface to the coolant is finite, is given by:

$$(c_p \delta) \cdot d \cdot \frac{\partial (\Delta T)}{\partial t} + a_s d \Delta T - h \Delta T = 0 , \tag{4.7.51}$$

where h is the coefficient of heat transfer from the surface of the conductor to the bath.

The condition for stability, $\partial (\Delta T)/\partial t < 0$, gives the dynamic stability criterion $a_s \cdot d < h$, or:

$$\frac{\pi^2}{6} \cdot \frac{\lambda_s}{1 - \lambda_s} \varrho_n \cdot J_c^2 \left(-\frac{1}{J_c} \frac{dJ_c}{dT} \right) \cdot d < h , \tag{4.7.52}$$

which yields the diameter of the superconductor:

$$\boxed{d \leqq \frac{6\,h}{\pi^2} \frac{1 - \lambda_s}{\lambda_s} \cdot \frac{1}{\varrho_n J_c^2} \left(-\frac{1}{J_c} \cdot \frac{dJ_c}{dT} \right)^{-1} .} \tag{4.7.53}$$

The superconducting filament diameter is proportional to h. For $h = 1$ W/cm² K, we obtain with the values given above, $d = 2.5 \times 10^{-3}$ cm.

Dynamic stability is also treated by Chester [39], Hancox [43] and Hart [38].

4.7.3.4 Steady State Stability

We consider a superconductor embedded in a low resistivity matrix material and postulate the early formation of a region of normally conducting material somewhere along the superconductor. The existence of a normal region can have many reasons such as a current pulse (power

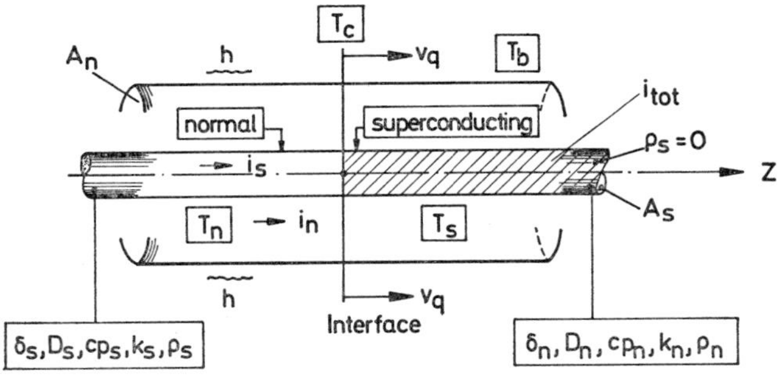

Fig. 4.7.6a. Composite conductor. To calculate steady state stability, it is assumed that each superconducting filament is embedded in a matrix of area A_n. T_n is the temperature of the normal region, T_s of the superconducting region, T_b is the bath temperature or of the surrounding substrate. T_c is the temperature of the interface. The composite conductor is assumed to be composed of a number of elements consisting of a superconductor and a normal metal

supply failure), local heating, a pulsed magnetic field, inhomogeneities in the metallurgical composition, or by a mechanical failure.

To calculate the propagation of the normally conducting region along the composite conductor, we assume (for simplicity's sake) that the significant variables occur along the wire such as the temperature, and analyze the heat balance equation for a one dimensional case. The one dimensional approach to solve the heat balance equation neglects the impedance to the radial heat flow from the conductor to the insulation, and ignores radial interface and boundary problems between superconductor and normal metal substrate.

Referring to Fig. 4.7.6a, we designate quantities and properties of the superconductor with a subscript (s) and of the substrate with (n), and may write the heat balance equation for an elementary length of conductor at z and t from first principles; term by term [44]:

$$\frac{d^2(\Delta T)}{dz^2} + \frac{1}{D_s} \cdot \left(\frac{1 + \frac{A_n}{A_s} \cdot \frac{D_s}{D_n} \cdot \frac{k_n}{k_s}}{1 + \frac{A_n}{A_s} \cdot \frac{k_n}{k_s}} \right) \cdot v_q \cdot \frac{d(\Delta T)}{dz} - \frac{h \cdot f \cdot \left(1 + \frac{A_n}{A_s}\right)^{\frac{1}{2}}}{k_s A_s^{\frac{1}{2}} \cdot \left(1 + \frac{A_n}{A_s}\right)} \cdot \Delta T$$

$$+ \frac{(i_s + i_n)^2 \cdot \varrho_s}{k_s A_s^2 \left(1 + \frac{A_n}{A_s} \cdot \frac{\varrho_s}{\varrho_n}\right)^2} \cdot \frac{1 + \frac{A_n \cdot \varrho_s}{A_s \cdot \varrho_n}}{1 + \frac{A_n \cdot k_n}{A_s \cdot k_s}} = 0 \,. \tag{4.7.54}$$

In Eq. (4.7.54) the quantities D, k, and h are used for diffusivity, thermal conductivity, and heat transfer to the bath; ϱ is the resistivity and v_q the propagation speed of the interface between region of normality and superconductivity. A is the cross sectional area of the superconductor, resp., normal metal substrate. $\Delta T = T - T_b$ is the temperature difference between superconductor and bulk helium. Since D, k, ϱ and h are temperature dependent, Eq. (4.7.54) can be solved numerically. However, in the range of temperatures encountered in the coil, we may assume that these quantities are constant. It is also feasible to divide the complete temperature range into several intervals where the quantities are nearly constant. The assumption of constant values of D, k, ϱ and h leads to quite useful informations on the heat propagation phenomena.

The solution of Eq. (4.7.54) with constant coefficients is given by:

$$\Delta T = \frac{(i_n + i_s)^2 \varrho_s}{h \cdot f \cdot A_s^{\frac{3}{2}}} \cdot \frac{1}{\left(1 + \frac{A_n}{A_s} \cdot \frac{\varrho_s}{\varrho_n}\right)\left(1 + \frac{A_n}{A_s}\right)^{\frac{1}{2}}}$$

$$+ \left\{C_1 e^{(m^2+n)^{\frac{1}{2}} z} + C_2 e^{-(m^2+n)^{\frac{1}{2}} z}\right\} e^{-mz} \,, \tag{4.7.55}$$

where

$$m = \frac{1}{2 D_s} \cdot \frac{1 + \frac{A_n k_n}{A_s k_s} \cdot \frac{D_s}{D_n}}{1 + \frac{A_n}{A_s} \cdot \frac{k_n}{k_s}} \,, \qquad n = \frac{h \cdot f}{k_s \cdot A_s^{\frac{1}{2}}} \cdot \frac{\left(1 + \frac{A_n}{A_s}\right)^{\frac{1}{2}}}{\left(1 + \frac{A_n}{A_s} \cdot \frac{k_n}{k_s}\right)} \,.$$

The integration constants C_1 and C_2 are calculated from the boundary values:

For $z < 0$ and specifically for $z \to \infty$, $\Delta T = \Delta T_n$ is finite and thus $C_2 = 0$. For $z = 0$; $\Delta T = \Delta T_c = \Delta T_n$, the critical temperature of the superconductor. Thus:

$$C_1 = \Delta T_c - \frac{(i_n + i_s)^2 \varrho_s}{h \cdot f \cdot A_s^{\frac{3}{2}}} \cdot \frac{1}{\left(1 + \frac{A_n}{A_s} \cdot \frac{\varrho_s}{\varrho_s}\right)\left(1 + \frac{A_n}{A_s}\right)^{\frac{1}{2}}}.$$

Inserting C_1 and C_2 in Eq. (4.7.55), we get:

$$(\Delta T)_n = \frac{(i_n + i_s)^2 \varrho_s}{h \cdot f \cdot A_s^{\frac{3}{2}}} \cdot \frac{1}{\left(1 + \frac{A_n}{A_s} \cdot \frac{\varrho_s}{\varrho_s}\right)\left(1 + \frac{A_n}{A_s}\right)^{\frac{1}{2}}} \left[1 - e^{[(m^2+n)^{\frac{1}{2}} - m]\, z}\right] + (\Delta T)_c \cdot e^{[(m^2+n)^{\frac{1}{2}} - m]\, z}. \qquad (4.7.56)$$

For $z > 0$, the wire is superconductive and $\varrho_s = 0$. Thus:

$$(\Delta T)_s = (\Delta T)_c \cdot e^{[-m - (m^2+n)^{\frac{1}{2}}]\, z}. \qquad (4.7.57)$$

Eqs. (4.7.56) and (4.7.57) contain the unknown parameter v_q. This is determined by utilizing the condition that the heat flow across the moving interface at $z = 0$ must be continuous. In the absence of a latent term, we may postulate:

$$\left.\frac{d(\Delta T)_n}{dz}\right|_{z=0} = \left.\frac{d(\Delta T)_s}{dz}\right|_{z=0}.$$

Applying this condition to Eqs. (4.7.56) and (4.7.57), we obtain the velocity of the interface

$$v_q = \frac{\sqrt{n}}{2} \frac{\frac{m'}{\Delta T_c} - 2}{\sqrt{\frac{m'}{\Delta T_c} - 1}} \cdot m'' \qquad (4.7.58)$$

with:

$$m' = \frac{(i_n + i_s)^2 \varrho_s}{h \cdot f \cdot A_s^{\frac{3}{2}}} \cdot \frac{1}{\left(1 + \frac{A_n}{A_s} \cdot \frac{\varrho_s}{\varrho_n}\right)\left(1 + \frac{A_n}{A_s}\right)},$$

$$m'' = 2\, D_s \frac{1 + \frac{A_n}{A_s} \cdot \frac{k_n}{k_s}}{1 + \frac{A_n}{A_s} \cdot \frac{k_n}{k_s} \cdot \frac{D_s}{D_n}}.$$

Eq. (4.7.58) enables us to define three limiting cases:

Case 1:

$$v_q = 0\,; \quad \frac{m'}{\Delta T_c} - 2 = 0\,.$$

This corresponds to a condition of static equilibrium. If $i_1 = i_s + i_n$, we may write:

$$i_1^2 = \frac{2\,h}{\varrho_s} \cdot f\,(T_c - T_b) \cdot A_s^{\frac{3}{2}} \left(1 - \frac{A_n}{A_s} \frac{\varrho_s}{\varrho_n}\right)\left(1 + \frac{A_n}{A_s}\right)^{\frac{1}{2}}. \qquad (4.7.59)$$

Modifying Eq. (4.7.59), noting that $\varrho_s \gg \varrho_n$, and using the relation combining cross section A to the cooled surface area of the conductor $S_n, f =$

$S_n/(A_n + A_s)^{\frac{1}{2}}$, we get the limiting current value for the steady state condition:

$$\boxed{i_1^2 \cdot \varrho_n = 2\,h\,(T_c - T_b) \cdot A_n \cdot S_n\,.} \tag{4.7.60a}$$

Case 2:

$$v_q = -\infty : \frac{m'}{\Delta T_c} = 1\,.$$

In this case, no quenching will occur. If i_2 is the current corresponding to this case, then:

$$i_2 = i_1/\sqrt{2}\,.$$

or:

$$i^2 \cdot S_n = h\,(T_c - T_b)\,A_n \cdot S_n\,. \tag{4.7.60b}$$

Case 3:

$$v_q > 0 : \frac{m'}{\Delta T_c} - 2 > 0$$

is the unstable case and the interface propagates along the superconductor.

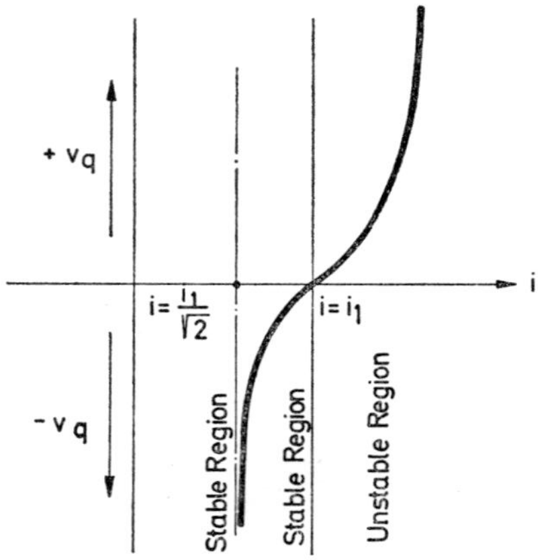

Fig. 4.7.6b. Propagation speed of normal superconductive interface vs. current

Fig. 4.7.6b illustrates the propagation of the normal region in the superconductor. $i_1 = i_c$ denotes the critical current in a steady state condition. The current region, $0 < i < i_2$, is the value for absolute (full) stabilization. No coil quenching will occur.

Steckly and Coworkers [43] have also developed the theory of steady state stability of composite superconductors. For a constant heat transfer coefficient h, they postulate:

$$\boxed{i^2 \cdot \varrho_n = \alpha \cdot h \cdot \Delta T \cdot A_n \cdot S_n}\,. \tag{4.7.61}$$

In this equation, α is a stability parameter, which was calculated in Eq. (4.7.60) to be 2, or 1, depending with region is chosen.

If a fluxjump occurs in the superconductor, the worst that can happen locally is for the current to be displaced into the matrix. If Eq. (4.7.61) is satisfied for $\alpha \leqq 1$, heat transfer to liquid helium will restore the temperature of the transient normal zone to some value below T_c and the current will begin to return to the superconductor. The time constant for this will be determined by the thermal diffusivity of the superconductor.

The stability parameter α is introduced, which characterizes the slope of voltage-current curve. (V—I) characteristic at the onset of the resistance of $\alpha \leqq 1$ indicates stable, $\alpha > 1$ unstable performance. The steady state equilibrium is determined by the maximum nucleate boiling heat flux q_{max}.

Tests conducted by Coffey and Gauster [46] show that the above simple model is not sufficient for a quantitative prediction of the V—I characteristic and that the q_{max} condition in general, is not a criterion for the termination of the steady state.

More recent assumptions use a variable heat transfer coefficient instead of a constant value.

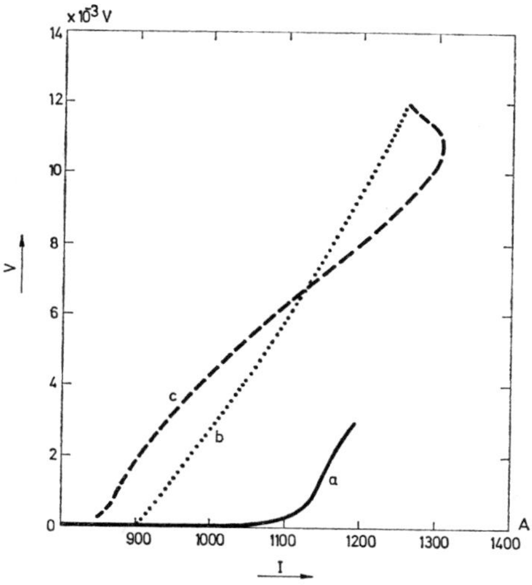

Fig. 4.7.7. Voltage current characteristics of a composite conductor. *a* Measured V—I characteristic of a NbTi-Cu composite with 252 filaments in copper at 5.5 T external field, *b* Calculated V—I characteristic for NbTi-Cu composite, assuming h = const, *c* Calculated V—I characteristic for NbTi-Cu composite, assuming the heat transfer coefficient $h = f(T)$

In Fig. 4.7.7, measured voltage-current characteristics of a short sample multistrand conductor are illustrated (curve *a*). In curve *b*, the (V—I) characteristic is calculated for an average heat transfer coefficient, where $\alpha < 1$. The difference between measured and calculated curves is appreciable.

The observed termination of the steady state ("take-off") corresponds to $q = 0.25\ \mathrm{W/cm^2}$, in contrast to the value of q_{max}, measured to be $0.985\ \mathrm{W/cm^2}$. In this case, dynamic stability criteria should be taken into account.

Curve c in Fig. 4.7.7 is calculated for variable heat transfer coefficients. The calculation indicates an initial decreasing current (unstable performance) and thus, even poorer agreement with the experimental data.

It was shown by Gauster [47], however, that the current flowing through the superconductor in the fluxflow state is not only dependent on T and B, but also on the voltage along the superconductor. In a composite conductor, the V-dependence can be neglected only if the stabilizer resistance R_m is much smaller than the flux flow resistance. For currents larger than I_c, this is the case. But experiments show at the onset of resistance that if $R_F \ll R_m$, V-dependence of I_c cannot be neglected, and thus the use of a model based on V, B, and T is indicated.

Cryostatic stability is the special form of the steady state stability criterion. The superconductor is formed into a composite with a considerable amount of copper or another suitable normal metal of high thermal conductivity. It is not unusual to have a ratio of copper to superconductor areas in the range of 10 to 100. The composite is immersed in a coolant and is cooled by boiling heat transfer.

The stabilization principle is simple: If a disturbance occur in the superconductor, such that it cannot carry a transport current, the current will flow entirely in the matrix. The heat which is then generated in the copper is dissipated to the liquid helium. If the temperature of the composite is less than the critical temperature of the superconductor in the prevailing condition of field strength, the transport current will return to the superconductor, allowing the superconducting operation to be resumed. The stability condition of Eq. (4.7.61) is valid.

The complication in using this equation is the discontinuity in the h values *vs* temperature. If the heat flux is increased beyond the peak nucleate boiling value, a temperature jump will occur and film boiling commences. The upward jump in the temperature difference between conductor surface and bulk liquid occurs at a "burn out" heat flux. If the heat flux is reduced below a "recovery value", typically about 0.3 W per cm² nucleate boiling is restored with a reduction in temperature difference. The choice of a critical h value is thus of significant importance. Generally, two conditions have to be considered:

(a) $I_c^2 \cdot \varrho_n < A_n S_n h_{cr} \cdot \Delta T$ the conductor is called fully stabilized.

(b) $I_c^2 \cdot \varrho_n < A_n S_n h_{cb} \cdot \Delta T$ the conductor is partially stabilized.

In case (a), the recovery current is smaller than the critical current and h_{cr} is not exceeded; superconducting operation will always be restored after any disturbance.

In case (b), the recovery current may be larger than the critical current, and $h_{cb} \Delta T$ is the critical burn out heat flux. Complete recovery to superconducting operation is not resumed unless the current is further decreased below I_c.

We can now discuss the problem of choosing the correct size of the superconducting filaments within the copper matrix. The filament diameter depends on two parameters: The heat transfer in the boundary layer between superconductor and matrix (h_b), and the conductivity of the superconductor k_s, which is quite poor. In the first case, the filament diameter may be calculated from the relation:

$$J_c^2 d_s < 4 \cdot \frac{A_n}{A_s} \cdot \frac{h_b}{\varrho_n} \cdot \left(- \frac{1}{J_c} \cdot \frac{d J_c}{dT} \right)^{-1}, \tag{4.7.62}$$

with A_n the cross-sectional area of the normal metal and A_s the area of the superconductor.

In the second case, the filament diameter is given by the previously calculated relation for dynamic stability:

$$J_c^2 d_s < \frac{138}{\pi^2} \cdot \frac{k_s}{\varrho_n} \cdot \frac{A_n}{A_s} \left(- \frac{1}{J_c} \cdot \frac{d J_c}{dT} \right)^{-1}. \tag{4.7.63}$$

We give typical values of parameters and calculate the filament diameter:

$$\begin{aligned} J_c &= 10^5 \text{ A cm}^{-2}, \\ h_b &= 0.03 \text{ W cm}^{-2} \text{ K}^{-1}, \\ A_n/A_s &= 10, \\ \varrho_n &= 3 \times 10^{-8} \text{ Ohm} \cdot \text{cm for copper.} \end{aligned}$$

From Eq. (4.7.62), we get $d_s < 2 \times 10^{-2}$ cm and from Eq. (4.7.63), $d_s < 4.8 \times 10^{-2}$ cm.

The heat transfer value given for the interface is certainly the lower limit for a codrawn copper-superconductor composite.

Thus, filament size is clearly limited by the heat transfer value in the interface and the bond between the superconductor and copper is of utmost importance.

From Eq. (4.7.61), we also see that compared to the dynamic stability criterion, the copper to superconductor ratio is very large and thus the overall current density in a cryostatically stabilized coil is small. Cryostatically stable conductors are only used in large d.c. magnets, where electromagnetic forces limit the overall magnet-current density, rather than the stability problems.

In d.c. magnets, where the field or current charging rate is not of prime concern, one may assume that twisting filaments within the composite may not be important. It could be shown in (4.7.1), however, that even at a very slow charging rate, shielding currents will flow across filaments through the matrix.

These shielding currents lead to a temperature rise in the matrix, as they would flow exactly as if the whole composite would be superconducting. To remedy the situation, the filaments in the composite are twisted or transposed with a twist pitch length l_c given in Eq. (4.7.68).

4.8 A.C. Magnet Fabrication Techniques

Mainly, we discuss engineering problems connected to superconducting a.c. magnet system and gives practical hints how to manufacture, wind coils, and operate a superconducting magnet. In various chapters, coil shapes (2) for d.c. and a.c. applications, methods of cooling and economy (4, 6), methods of loss reduction and loss measurements (4), and methods of coil protection (7) are discussed in detail.

The development of new multifilament conductors to be used in a.c. and d.c. magnets greatly influenced the design and manufacturing of superconducting magnets. In the past, overall current densities in superconducting d.c. coils were chosen according to the cryostatic stability criteria. These current densities for larger experimental magnets generating central fields higher than 5 T, seldom exceeded values of $5\ 10^3$ A/cm^{-2}. Composite conductors and the theory of dynamic stability made it possible to build magnets with overall current densities in the range of $(2—5)\times 10^4$ A/cm^{-2}, depending on the coil size and the forces on the conductor. In d.c. magnets for large experimental facilities, the problem of forces become predominant and conductor reinforcement methods had to be studied in great detail (6). In a.c. magnets, the problem of reducing a.c. losses to acceptable values in order to reduce refrigeration requirements is still a major task.

For both types of magnets, "traditional" methods of constructing and cooling superconducting magnets are abandoned and new winding methods more suited to the new conductor could be developed (2, 6). In this chapter we discuss, engineering problems pertinent to the new technology.

4.8.1 Coil Fabrication

Superconducting composites are expensive, owing to the complicated manufacturing process for their production. It is thus desirable to place superconductors either as close as possible near the bore of a magnet, or to use them as a backing field for saving d.c. power and achieve, in both cases, the highest possible current density. The shape of the coil reflects the requirement on the distribution of the magnetic field within the useful aperture of the magnet, shown in chapter 2. Coil cross-sections simulating intersecting circles or ellipses, variable current density distribution over the coil cross-section, optimization of current densities according to the field distribution are common for beam transport, accelerator type magnets, and experimental magnets. As the conductor has a finite cross-section (composite bars or tapes for d.c. applications, cables or braids for a.c. applications), the theoretical ideal shape is seldom approached. The ideal cross-section is approximated as good as possible, such that the sum of all higher term harmonics is less or equal to tolerable values dictated by the requirements of particular applications.

In a.c. magnets, the heat generated within the coil body by energy dissipation must be removed efficiently to prevent degradation and premature quenching. Coolant passages incorporated between layers or sections, and so-called "heat drains" are frequently used.

Coolant passages lower the overall current density and may pose some winding difficulties. Heat drains made of copper strips, copper meshes, or aluminum meshes must be thoroughly insulated from the active conductor and grounded to prevent capacitive charging and possible arcing due to quenches.

In d.c. magnets, coolant passages around the periphery of the coil are adequate to remove heat generated by sporadic fluxjumps, wire movements, or infrequent external disturbances.

Supporting the conductors against electromagnetic forces to reduce conductor movement (wire motion in the order of 10 μm under the influence of electromagnetic forces can cause a temperature rise of about 10 K) to a possible minimum and the improvement of the thermal conduction of pulsed coils is possible by "casting" the winding in suitable materials. Potting materials used to date have been resins, thermoplastics and thermosettings, waxes and compounds, and even liquid nitrogen in frozen form.

Training and degradation behaviour of coils reflect the inadequacy of the different potting materials. So far, epoxies have shown best performance only if they have been loaded with inorganic fillers. Conductor training after each thermal cycle has been reported by Smith [19] for unfilled epoxies, and is contributed to the strain energy of the potting material which is released when the frozen impregnant cracks. The strain energy is highest in pure epoxies. It is reduced by adding mineral fillers to the mixture. Flexibilizers and deluents are not suitable. The potting material must possess the following properties:

(a) The impregnant must have low viscosity in order to penetrate into all areas not occupied by conductors reinforcing and insulating tapes, etc.

It must have a long "pot life", which means it must be usable over the impregnation time, which may extend over several hours.

(b) The impregnant must be compatible to the conductor material. This implies, specifically, that it must thoroughly wet all the surfaces exposed to the potting mixture (the use of surface finishes with reactive or epoxy functional materials and compounds is recommended), and the thermal contraction coefficients of the cured casting and the conductor must match if possible; this may be achieved if the impregnant is loaded with inorganic fillers (see Fig. 6.2.8).

(c) The mechanical properties of the potting material must be such that it can withstand static and cyclic magnetomechanical and thermomechanical stresses during cooldown and operation.

(d) The potting material should not be subject to changes in phase or structure when cooled down to the operational temperature of the magnet, and must not fatigue over the lifetime of the magnet.

(e) The thermal conductivity of the casting material must be high, allowing the heat to be removed from the conductor. It must also provide sufficient electrical insulation, so that breakdown between layers or conductors are prevented.

Loaded or filled epoxies are best suited to fulfill these requirements. From a number of recipes, polyimids, epoxies with anhydride, and polyamine hardners have been most suitable.

The potting material in itself is not enough to withstand magnetomechanical stresses. High field magnets are, in addition, reinforced by means of nonmagnetic or nonmetallic support rings to ensure coil rigidity.

As an example, a dipole coil with a central field of 4.5 T and an overall current density of 2×10^4 A/cm^{-2} having an aperture radius $a_1 = 4$ cm; outer radius $a_2 = 7$ cm should illustrate the problems:

The Lorentz force, acting radially outward on the largest section of a composite coil, is 6×10^4 kp/m length of magnet. The coil is a complex mixture of several materials, as pointed above, and one may calculate the average value of 10^6 kp/cm^2 for the Young's module of the coil (about 9% of the coil volume is filled with pure epoxy). The epoxy is loaded to about 60% of the total weight with inorganic fillers. In addition, each coil layer is wrapped in glass tapes having a suitable surface finish to make them compatible to the epoxy. In such a coil, the strain energy dissipated was not enough to generate internal microscopic cracks during cooldown. After the first cooldown, the coil exhibits some training; subsequent cooldowns did not lead to a repetition of the training phenomenon. It may be pointed out that the tensile strength of glassfiber reinforced epoxies ($\sim$ 7000 kp · cm^{-2} at 4.2 K) is reduced by a factor of 2.5 when the inorganic filler was added. The viscosity of the mixture at impregnation temperature (80 °C) was two orders of magnitude higher than the pure epoxy mixture. (Three component mixtures of epoxy, hardner, and accelerator have initial viscosities of 20 to 50 cP at 80 °C.)

The tensile strength reduction of the epoxy mixture is partially compensated for, if the filler is heat-cleaned and chemically treated (surface finishes, also used for glassfiber tapes), and if epoxy functional materials are added to the epoxy mixtures. The maximum stress in a coil section (with 8 cm spacing between reinforcement rings) is 2.3×10^3 kp/cm^2. For a composite conductor with copper to superconductor ratio of 1.25 : 1, this corresponds to a strain of 0.4% at 4.5 K! When pulsed, the conductor is displaced about 3 μm.

The copper (annealed OFHC copper used as conductor matrix) is slightly work hardened and its residual resistivity ratio measured is about 150. When the magnet is thermomechanically cycled, the tensile strength of copper is reduced about 30%, and the residual resistivity ratio of 135 is obtained.

After 10^6 magnetomechanical cycles, the epoxy mixture exhibits a reduction of tensile strength by a factor of 2.3 from its initial value. The magnet dissipates 11 Watts of power per meter coil length, when the field is changed in 10 second cycles. The major loss contribution is due to hysteretic and eddy current effects. The dipole coil has an effective volume of 66.6×10^2 cm^3 and thus the average power density of 1.65×10^{-3} W/cm^3, corresponding to a heat flux of 1.4×10^{-3} W/cm^2, is obtained.

The heat generated in a coil section, assuming symmetric boundary conditions, is carried away by thermal conduction and heat transfer to the helium. The maximum temperature rise in the coil section over the bulk helium temperature is given in Fig. 5.7.10 and the approximation Eq. (2.5.23).

$$\Delta T = p_v \left[\frac{1}{2\,k} \left(\frac{a}{2} \right)^2 + \frac{a}{2\,h} \right],$$

p_v is the instantaneous loss value per unit volume, given by: (Eq. 4.1.22)

$$p_v = \frac{d}{2} \mu_0 J_c(B)\, \dot{H}_{\text{ext}}\,,$$

k is the thermal conductivity through the coil matrix* and h the heat transfer coefficient. The width of the coil section perpendicular to the coolant passage is denoted by a. With the values of $k = 4.5 \times 10^{-3}$ W/cm^{-1} K^{-1} at 4.2 K for filled epoxies, and using two phase forced helium cooling ($v = 0.2$ ms^{-1}), a heat flux of 0.6 W/cm^2 can be removed at a temperature rise of 0.2 K. If $a = 1.2$ cm, the average temperature rise will be $\Delta T = 0.06$ K.

Although it appears that the heat generated during field sweeping can be removed away from the conductor to the bath, exact evaluation of instantaneous local heating effects is not possible. The coil is wound on a glass filament epoxy coil former. Few magnets using composite conductors of large cross-section ($\geqq 1$ cm^2), or cables impregnated in thermosettings, are self-supporting; no coil former is necessary.

The ends of race track shaped coils are bent (in two dimensional flat coils) at some appropriate angle, or shaped on a cylindrical coil former in such a way that the field enhancement at the ends compared to field at the coil central section is small. It is now common that an iron shell (laminated for a.c. magnets) is placed around the coil. In high-field magnets, the iron is saturated if the shell is in closely positioned around the coil. The iron shell, located within the helium container, is cooled to liquid helium temperature. The eddy current and hysteretic losses of iron must also be removed by the coolant. The iron extends axially over the coil by about one or two times the aperture diameter or is shaped at the ends, in order to eliminate field distortion due to the iron shell end effects. The iron shell also prevents coil deflection (sagging) due to its own weight and due to any forces generated by excentricities between coil and external shell. It may be possible to design a magnet which ideally suits all the requirements imposed on the coils, but practically, one has to cope with manufacturing errors, conductor, tolerances, winding dislocations, approximations to the exact coil shape, the finite conductor size, and the manufacturing and winding tolerances. Thus, one has to accept some field perturbations (multipoles) within the useful magnet aperture.

---

* The thermal conduction through a cable is practically the same as for copper due to the transposition of individual strands which brings them periodically along the conductor length to the surface of the cable.

The field errors can be summarized under two categories:

(*a*) *Asymmetric Coil Aberrations*

1. Coil displacement errors:

Field errors due to shifting of one coil-section with respect to others, conductor displacements, dimensional changes of conductor cross-section, etc.

2. Coil and iron shell displacement errors.

3. Errors which occur when the magnetic and geometric centers do not coincide.

4. Field perturbations due to current leads, conductor cross-overs, electrical connections between sections, etc.

5. Field errors due to short circuits within the coil.

(*b*) *Symmetric Coil Aberrations*

1. Finite size effect of the conductor
2. Elastic and inelastic conductor motion
3. Residual field effects [24]
4. Iron shell saturation effects.

All errors may be compensated for by using correction coils or correcting magnet elements (such as quadrupoles in accelerators or in beam transport sections), provided the sum of all errors in the magnet aperture does not exceed $\Delta B/B \leqq 10^{-3}$.

For beam transport magnets, the following mechanical tolerances have been achieved:

Displacement errors between coil sections, particularly between coil halves (dipoles) and coil quarters (quadrupoles): $\pm$ 0.1 mm.

Mechanical tolerances in pancakes, doublepancakes or coil packages ($\leqq$ 10 mm thick): $\pm$ 0.04 mm.

Conductor tolerances (multifilament-composites)	$\pm$ 0.01 mm.
Conductor tolerances (cables and braids)*	$\pm$ 0.05 mm.

4.8.2 Electrical-Design

The electrical problem facing the magnet designer may be divided into three major areas:

(a) Transport of current to the magnet, during energizing.
(b) Internal coil insulation and coil composition.
(c) Energy dissipation or energy discharge and magnet protection.

The transport of current from an energy source, such as a power supply or energy storage, to the superconducting coil occurs over normal conductors, "current leads", which are joined to the superconductor by means of soldering, brazing, or swaging. The joint resistant must be in the order of $\leqq 10^{-8}$ Ohms to prevent heat from diffusing from the normal joint to the superconductor.

The design of current leads requires some care if the magnet is kept cool by means of bulk liquid helium. For such an operation, a number

* Mechanical tolerances in cables or braids are tested when the cable is subjected to a specified tensile force.

of leads have been developed by making use of the enthalpy of the gas passing through the lead with a small helium boiloff. Optimum loss values at 4.2 K of 1 mW/Amp/lead have been achieved. If the magnet is operated in a close refrigerator circuit, the design of current leads has to be coupled with the refrigeration circuit, since the enthalpy of the cold gas is better used in heat exchangers than in leads.

For example, a refrigeration of 25 W is required for a pair of 2000 A leads. This is due to the fact that helium gas must pass through each lead to cool them down. The warm gas does not pass the heat exchangers and the refrigerator, and as a result, the refrigerator sees this as an equivalent liquefaction. Only an additional small fraction of refrigeration power is required to remove the heat leaking down the leads and to remove Joules heating due to ohmic losses.

4.8.2.1 Current Leads

With increasing current into the superconducting magnets, the requirements for current leads entering a helium cryostat become more stringent. Current leads for rated currents of 10^4 Amp have been designed and tested (Fig. 4.8.1). It is of great importance that leads must have minimum helium evaporation rates. Current leads have been designed by a number of investigators [48, 49, 50]. The treatment of the problem is based on a number of simplified assumptions such as constant cross-section over the entire length of the lead, averaged temperature values of thermal conductivity k, electrical resistivity ϱ, and specific heat c_p. Even then the solution of the equation for steady heat flow is not transparent and more assumptions are necessary. We summarize the treatment of the problem briefly and present curves which have shown good agreement to experimental data: Under the condition of steady heat flow, all the heat created or entering on element dx of the current lead must be transferred to the gas. The temperature of the gas flowing form x to $x + dx$ is increased by dT. The Joule heat generated in the element by the electric current is $I^2 \varrho \cdot dx/A(x)$, where $A(x)$ is the cross-sectional area of the element. If m is the heat evaporation rate in (g/s), the equation of heat flow becomes:

$$\frac{dQ}{dx} + \frac{\varrho I^2}{A(x)} = m c_p \frac{dT}{dx}. \tag{4.8.1}$$

The heat in flux into the refrigerant is given by:

$$Q = k \cdot A(x) \frac{dT}{dx}, \tag{4.8.2}$$

which yield:

$$\frac{dQ}{dT} = m c_p - k \cdot \varrho \frac{I^2}{Q}. \tag{4.8.3}$$

In this equation, c_p, k, and ϱ are temperature dependent. At temperatures much below or much higher than the Debye temperature, we may use Wiedemann-Franz law and introduce the Lorentz-Number, L. The variation of $L(T)$ is much less than that of k or ϱ over the temperature

Fig. 4.8.1. Counterflow current leads for 10^4 A rated current in a test set up (SLAC)

range of interest, and we may assume L to be constant. From Eq. (4.8.3), we obtain:

$$\frac{dQ}{dT} = m c_p - I^2 \frac{L(T)\,T}{Q}. \qquad (4.8.4)$$

At the place where the lead enters the helium bath, Q has a value $Q_0 = m c_L$, with c_L the latent heat of the cryogen; thus:

$$\frac{dQ}{dT} = \frac{c_p}{c_L} Q_0 - I^2 \frac{L(T)\,T}{Q}. \qquad (4.8.5)$$

One can normalize the above equation by writing: $Q = Iq$; $Q_0 = Iq_0$, and we obtain:

$$\frac{dq}{dT} = \frac{c_p}{c_L} q_0 - \frac{L(T)\,T}{q}. \tag{4.8.6}$$

If the upper end of the lead (exposed to ambient temperature) is maintained to some fixed value T_1, there is a minimum value q (min) at which the evaporation rate of helium is minimum at a given current. From Eq. (4.8.2), we obtain for $A(x) = A = \text{const}$:

$$I \cdot \int_0^l \frac{dx}{A(x)} = \int_{T_B}^{T_1} \frac{k\,dT}{q},$$

or:

$$I \frac{l}{A} = \int_{T_B}^{T_1} \frac{k \cdot dT}{q}. \tag{4.8.7}$$

Referring to 5.3, it is seen that for helium gas, the ratio c_p/c_L is decreasing with increasing temperature; however, q is rapidly decreasing and at a temperature $T_{\max}$ at the upper end of the lead, we may assume this ratio is close to zero. For this particular temperature, Eq. (4.8.6) yields:

$$\frac{dq}{dT} = -\frac{LT}{q}$$

with the solution:

$$q = \sqrt{L}\,(T_m^2 - T^2)^{\frac{1}{2}}.$$

Thus, from Eq. (4.8.7):

$$I \frac{l}{A} = \frac{1}{\sqrt{L}} \int_{T_B}^{T_m} \frac{k\,dT}{(T_m^2 - T^2)^{\frac{1}{2}}}. \tag{4.8.8}$$

It can be shown that for most metals the thermal conductivity obeys the rule

$$\frac{1}{k} = \frac{\varrho\,(T=0)}{LT} + 3.35 \times 10^{-5}\,T^2. \tag{4.8.9}$$

In this relation, we used Matthiessen's rule and Wiedemann-Franz law. Thus:

$$\boxed{I \cdot \frac{l}{A} = \frac{1}{\sqrt{L}} \int_{T_B}^{T_m} \frac{dT}{\left(\frac{\varrho\,(T=0)}{L \cdot T} + 3.35 \times 10^{-5}\,T^2\right)(T_m^2 - T^2)^{\frac{1}{2}}}.} \tag{4.8.10}$$

The integral is of the type:

$$\boxed{I \cdot \frac{l}{A} = \frac{1}{3.35 \times 10^{-5}\,T_m^2 \sqrt{L}} \int_{t_B}^{1} \frac{t\,dt}{(t^3 + a)(1 - t^2)^{\frac{1}{2}}}}$$

with $a = \dfrac{\varrho\,(T=0)}{3.35 \times 10^{-5}\,L \cdot T_m^3}$ and t the reduced temperature $t = T/T_m$.

The analytical solution of this equation yields a complicated and not transparent result.

Selecting suitable values of q_0 at various temperatures Eq. (4.8.6) can be integrated numerically. The results of calculated optimized leads are illustrated in Fig. 4.8.2.

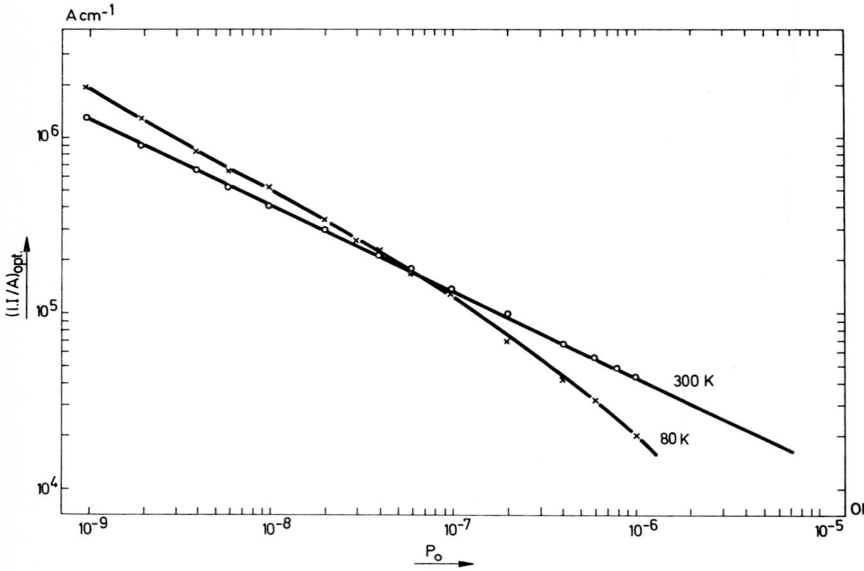

Fig. 4.8.2. Optimized values of Il/A for current leads as a function of ϱ_0 ($B = 0$)

Several methods to design optimized leads have been employed by different investigators. For high current leads (> 1000 A), the simple lead design is in the use of plane parallel copper plates separated by spacers to permit helium gas passage. The copper sheets have a thickness of about 1 mm each; their separation can be 0.5 mm. The area and number of sheets are obtained from Eq. (4.8.10) as the length of the lead is determined by the size of the magnet and the dewars. The copper sheets are attached at both ends to copper bars, or copper blocks and soldered at one side to the composite conductor and connected at the warm end to room temperature cables or bus bars.

4.8.2.2 Superconductor to Lead Joints

The electric current is passed from the normal metal lead to the superconductor by means of solder joints or pressure contacts. The electrical and thermal quality of the interface between superconductor and normal metal is important for the behaviour of a stabilized coil.

At the boundary between superconductor and normal metal (such as copper), heat is generated by ohmic losses. Generally, the copper to superconductor joint must be such that the resistance measured across the joint is less than 10^{-8} Ohms. The same or smaller resistance values must be obtained in joints within the coil or in series connections.

In stabilized conductors where the superconductor is embedded in a copper matrix, again the bond quality is important for the diffusion of heat from the superconductor to the matrix material. The bond is thus a measure for the dynamic stability of the superconductor. To determine the bond quality, test methods shown in Fig. 4.8.3a and 4.8.3b are com-

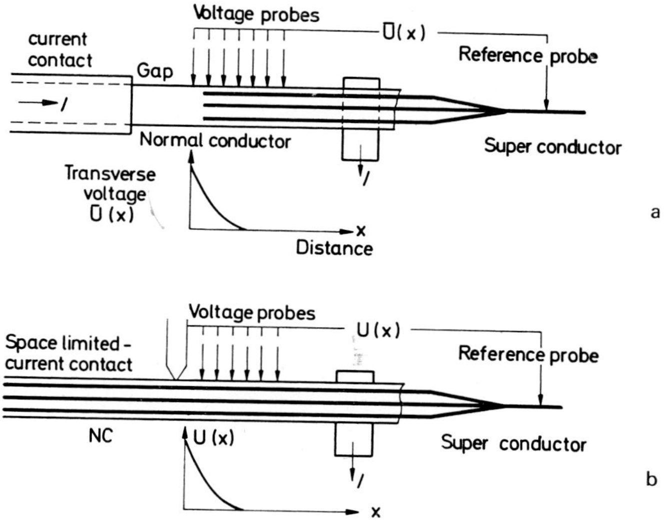

Fig. 4.8.3. Test methods to determine joint-quality between normal metal and superconductor. *a* Gap-method, *b* Current-contact method

mon [51]. The method shown in Fig. 4.8.3a is a gap method, which is applied to sandwich type conductors. The second method (Fig. 4.8.3b) is used to test bond resistance of multifilament conductors. A constant current is passed through the normal metal (e.g. current lead), whose width is small compared to the transfer length of the current. The local distribution of the transverse voltage $U(X)$ along the sample is given by the relation:

$$U(X) = R_T \frac{di}{dx}; \quad \text{with} \quad i(x) = I_0 \exp\left[-(R_L/R_T)^{\frac{1}{2}} x\right] \quad (4.8.11)$$

where R_T (Ohm · cm) is the specific transverse resistance and R_L (Ohm per cm) is the resistance of the normal metal per unit length. R_L and R_T can be calculated from the current voltage distribution. For a given R_L, the current has transferred within a few mm into the superconductor [52]. Joints between leads and composite conductors having a length of minimum 10 cm and joint resistances in the order of $R_T \leqq 10^{-8}$ Ohm · cm can be obtained readily (Fig. 4.8.4).

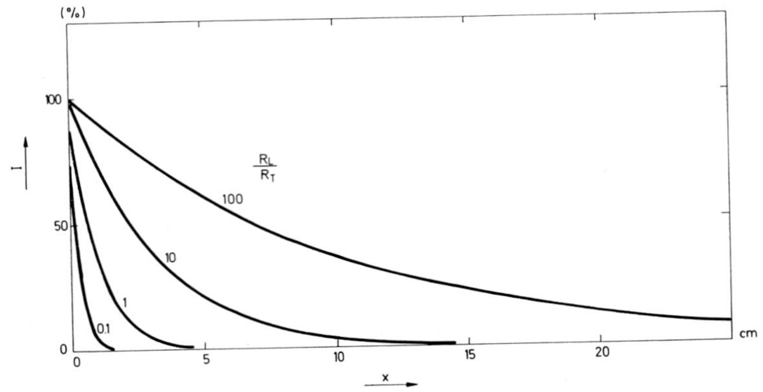

Fig. 4.8.4. Current decay vs. distance in normal metal to superconducting joints. Parameter is the ratio between longitudinal resistance along substrate (R_L) to transverse resistance (R_T) across the joint

4.8.3.3 Transient Voltages in Coils due to Quenches

The design of the coil structure is primarily aimed for generation of a field of specified amplitude and distribution in a given volume. As the magnet lifetime and with it the life of a system in which the magnet is a major part depends on the life of the conductor and its insulation, the knowledge of forces and electric fields within the coil must be known so that the winding distribution and the insulation is selected accordingly. If we ignore the ground and interturn capacitances of the magnet, the current, voltage, and temperature variation, after a quench has occurred, is illustrated in Fig. 4.8.5. The magnet considered here has a stored energy of about 100 kJ. The temperature rise in the conductor is about 70 K and the voltage rise small enough as not to be of concern. However, the magnet also has a ground capacitance and turn to turn or series capacitance. Due to these capacitances, when the magnet quenches, transient voltages appear over the coil which are higher than the terminal voltage. The insulation must withstand these transients, specifically if the transient voltage is above the corona threshold of the conductor insulation.

The employment of magnet safety circuits when a quench occurs has several reasons. It has to protect coil windings when the conductor temperature is increased, it has to limit the boiloff of helium at tolerable limits, and if several magnets located in proximity of each other are electrically connected, it should prevent a chain reaction, if one magnet quenches.

Although the voltage across the magnet terminals which drives the a.c. current through the conductor is low and in pulsed magnets seldom exceed a few hundred volts, it is known that due to a quench, internal voltage surges may appear which puncture the insulation locally and may lead to partial or total coil destruction.

To calculate the internal voltage distribution in a coil, the winding diagram given in Fig. 4.8.6 is used. The coil consists of ground capacit-

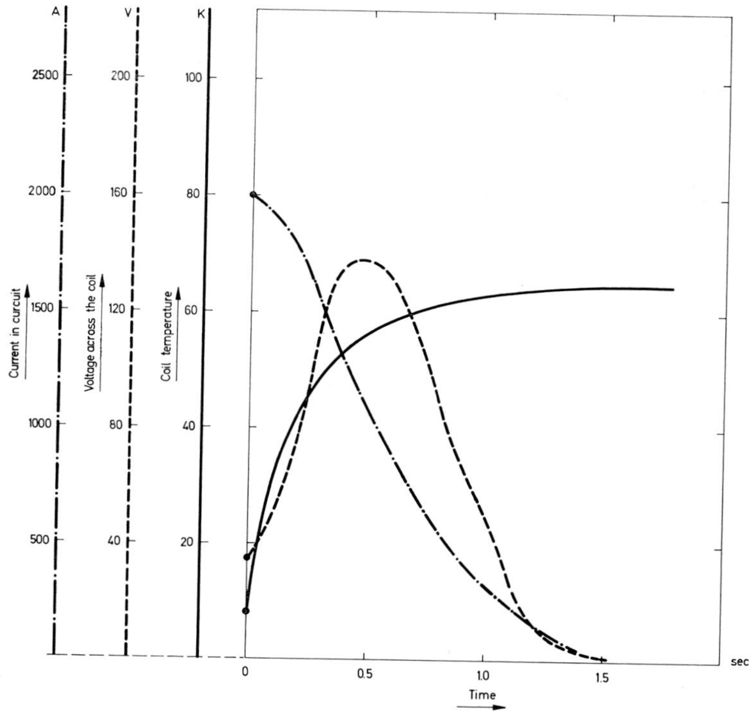

Fig. 4.8.5 Voltage, current and temperature variation after the occurence of a quench (coil capacitances neglected) over a dipole coil, 1 m long, 8 cm aperture diameter, with a central field of 4.5 T

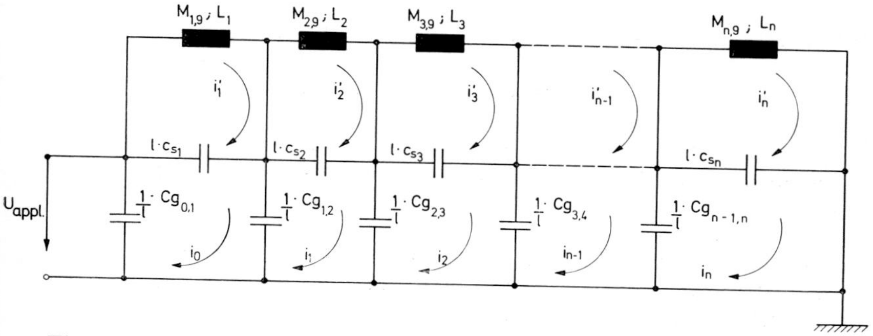

Fig. 4.8.6. Schematic representation of a magnet coil. L self-inductance, M mutual inductance, C_s series capacitance and C_g ground capacitance of a coil with the conductor length l

ances C_g, series or winding capacitances C_s, self inductivities L_i, and mutual inductivities M. If the coil length is measured axially from the grounded end towards the terminal voltage end, the voltage distribution is given by [53]:

$$\frac{\partial^4 u}{\partial z^4} - \frac{L}{l^2}\left(C_s \cdot \frac{\partial^4 u}{\partial z^2 \partial t^2} - \frac{C_g}{l^2} \frac{\partial^2 u}{\partial t^2}\right) = 0 . \tag{4.8.12}$$

Assuming the voltage change occurs in an infinitisimal short time, then the terminal voltage has a rectangular shape and therefore: $\partial^2 u/\partial t^2 = \infty$ at $t = 0$. Thus, we have for $u = u(z)$ at $t = 0$;

$$\frac{d^2 u_i}{d z^2} = \frac{C_g}{C_s} \cdot \frac{1}{l^2} u_i . \tag{4.8.13}$$

This equation can also be obtained by assuming that no current will flow through the coil inductance L at the time $t = 0$, and thus L can be assumed infinite. The solution of Eq. (4.8.13) is given by

$$\boxed{u_i = U_{\text{appl}} \cdot \frac{\sinh(\alpha z/l)}{\sinh(\alpha)}} \tag{4.8.14}$$

with:

$$\alpha = (C_g/C_s)^{\frac{1}{2}} \tag{4.8.15}$$

U_{appl} is the applied terminal voltage.

The initial voltage gradient is

$$\frac{d u_i}{d z} = \frac{\alpha}{l} \frac{\cosh(\alpha z/l)}{\sinh(\alpha)} U_{\text{appl}} . \tag{4.8.16}$$

The maximum voltage gradient occurs at the line end of the winding ($z = l$)

$$\left.\frac{d u_i}{d z}\right|_{\max} = U_{\text{appl}} \cdot \frac{\alpha}{l} \cdot \coth(\alpha) . \tag{4.8.17}$$

If the coil is not grounded, then the boundary conditions become: $U_i = U_{\text{appl}}$ at $z = l$ and $du/dz = 0$ at $z = 0$, the equation of initial voltage distribution is therefore:

$$U_i = U_{\text{appl}} \cdot \frac{\cosh(\alpha z/l)}{\cosh(\alpha)} , \tag{4.8.18}$$

and the maximum voltage gradient occurs at $z = l$:

$$\left.\frac{d u_i}{d z}\right|_{\max} = U_{\text{appl}} \cdot \frac{\alpha}{l} \cdot \tanh(\alpha) . \tag{4.8.19}$$

The above simplified calculation ($L = \infty$) illustrates the fundamental significance of α. The smaller α, the more the curved shape of the initial voltage distribution approximates a linear voltage distribution over the coil.

This implies that the electric stresses are equally distributed over the length of the winding and the high voltage stress at the coil line end is removed (Fig. 4.8.7).

Inversely, a high value of α produces a concentration of high stresses along a small section of the winding. For example, for $\alpha = 10$, about 65% of the applied voltage is distributed along the first 10% of the coil.

The final voltage distribution appears along the coil, when the voltage changes of the wave tail are small. In this case, no current will flow through the capacitances and the inductive and resistive elements only control the current. It follows that the final voltage distribution is linear and for the neutral coil end grounded, we have:

$$u_f = \frac{z}{l} \cdot U_{\text{appl}} . \tag{4.8.20}$$

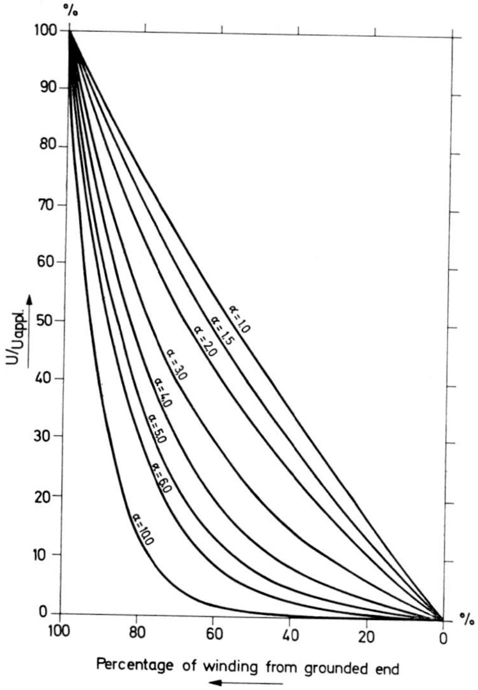

Fig. 4.8.7. Initial voltage distribution along the coil winding. The parameter $\alpha = (C_g/C_s)^{\frac{1}{2}}$ determines the stress over the winding

The following equations can be written for this case:

$$\frac{du}{dz} = -\frac{L}{l} \cdot \frac{di}{dt} . \tag{4.8.21}$$

By substituting u_f into this Eq. (4.8.21), we obtain:

$$i = \frac{U_{\text{appl}}}{L} \cdot t . \tag{4.8.22}$$

The transition from the initial to the final voltage distribution takes place in form of transient oscillations. To find the general solution for the voltage distribution, we proceed once more from Eq. (4.8.12). Assuming rectangular voltage change and grounded neutral end, we may write:

$$U(z,t) = U_{\text{appl}} \cdot \left[\frac{z}{l} + \sum_{n=1}^{\infty} (-1)^n \frac{2\alpha^2}{(n\pi)^2 + \alpha^2} \cdot \frac{\sin(n\pi z/l)}{n\pi} \cdot \cos(\omega_n t)\right]. \tag{4.8.23}$$

The envelope of all maximum voltages to ground can be obtained from Eq. (4.8.23).

$$U(z)_{\max} = U_{\text{appl}} \cdot \left[\frac{z}{l} + 2\alpha^2 \sum_{n=1}^{\infty} \frac{1}{(n\pi)^2 + \alpha^2} \cdot \frac{\sin(n\pi z/l)}{n\pi}\right]. \tag{4.8.24}$$

We see immediately that the maximum voltage gradient appears at $t = 0$, i.e. during the initial period where a voltage disturbance occurs for $z = l$. However, high voltage gradients can occur during the transition period in other parts of the winding.

The envelope of the maximum voltage gradients is derived from Eq. (4.8.23):

$$\left.\frac{du}{dz}\right|_{\max} = U_{\text{appl}} \cdot \left[\frac{1}{l} + \frac{2\alpha^2}{l} \sum_{n=1}^{\infty} \frac{1}{(n\pi)^2 + \alpha^2} \cdot \cos(n\pi z/l)\right]. \tag{4.8.25}$$

The maximum value of Eq. (4.8.25) occurs at $z = l$, corresponding to $t = 0$, and coincides to values obtained in Eqs. (4.8.16) and (4.8.17). The voltage amplitudes and voltage gradient decrease rapidly as n increases. Calculation of only a few terms of the above series is sufficient for most practical purposes [54] (Fig. 4.8.8). The Eqs. (4.8.23) to (4.8.25) refer to the case of rectangular voltage shapes, but can be extended by applying Duhamel's theorem [55].

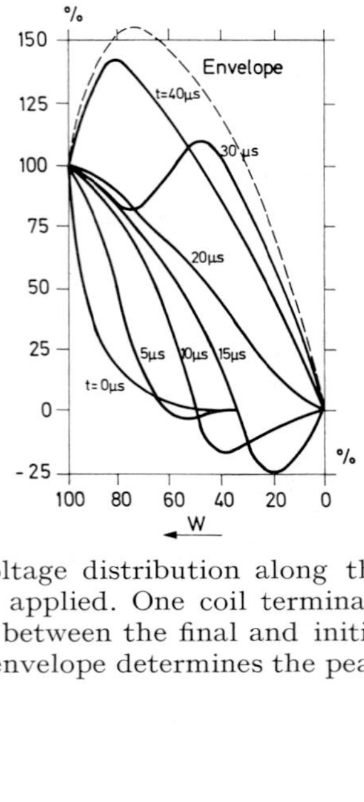

Fig. 4.8.8. Calculated voltage distribution along the coil winding after a square shaped voltage is applied. One coil terminal is grounded. Voltage oscillations are measured between the final and initial voltage distribution. The voltage envelope determines the peak voltages

The general conclusion reached above is about the influence of α. In coils with small currents but a large number of turns, the ground capacitance is large compared to the series capacitance, specifically if the grounded iron shell is placed within the helium container and is in close proximity of the coil. Methods of reducing α are to use less turns and to chose high currents, which may solve the problem of transient voltages, but a better method is to increase the series capacitance of the coil by interleaving the turns in pancakes.

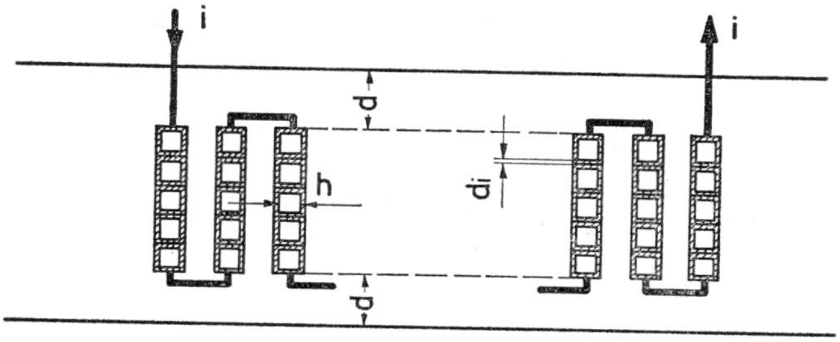

a

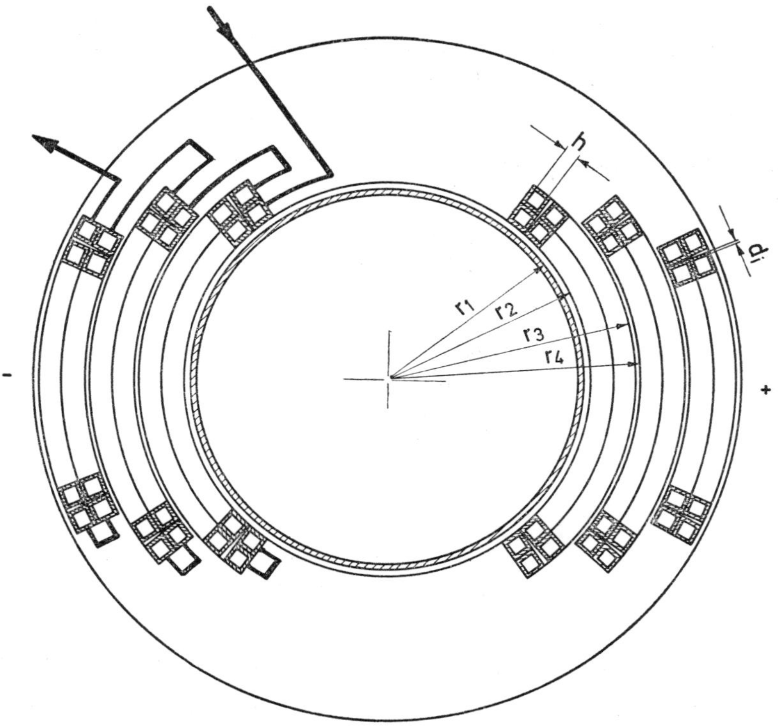

b

Fig. 4.8.9. Schematic representation of different coil configurations

For a dipole coil configuration given in Fig. 4.8.9, the ground capacitance is given by:

$$C_g = \frac{8.85\,A}{\sum_i \frac{1}{\varepsilon_i} \ln\left(\frac{r_{i+1}}{r_i}\right)} \quad \text{(pF)} \tag{4.8.26}$$

with A the surface area of the coil in (m^2) facing the ground electrode, ε_i the dielectric constant of the insulation or helium, r_i the radius or width of the insulating cylinder.

For a noninterleaved coil, according to Fig. 4.8.9, the series capacitance is given by:

$$C_{S1} = 4 \cdot \frac{N-1}{N^2} \cdot \frac{1}{n} \cdot \frac{8.85\,\pi}{d_i} \cdot d_m \cdot h \cdot \varepsilon_r \quad \text{(pF)} \tag{4.8.27}$$

with n the number of turns per double pancake, N the number of sections in a coil, d the insulation thickness between turns, m the main coil diameter, h the axial conductor height, and ε_r the relative dielectric constant of the conductor insulation.

The series capacitance between pancakes is expressed by

$$C_{S2} = \frac{4}{3\,N^2} \left[C_s'(N_1 - 1) + C_s''(N_2 - 1) + C_s'''(N_3 - 1) + \cdots\right], \tag{4.8.28}$$

where the partial capacitances of the coil end faces are given by:

$$C_s' = 8.85 \cdot \left[\frac{A_1}{d_1/\varepsilon_1} + \frac{A_{tot} - A_1}{\sum_i d_i/\varepsilon_{i,1}}\right],$$

$$C_s'' = 8.85 \cdot \left[\frac{A_2}{d_2/\varepsilon_2} + \frac{A_{tot} - A_2}{\sum_i d_i/\varepsilon_{i,2}}\right]$$

with $A_1, A_2, A_3 \ldots$ the faces of spacers between pancakes, d_i the axial thickness of spacers, A_{tot} the face of a pancake.

The total series capacitance is $C_s = C_{s1} + C_{s2}$.

If ΔU is the voltage across an interleaved element (Fig. 4.8.10a), then the voltage between neighbouring turns is approximately $\Delta U/2$. The effective series capacity C_s can be calculated from:

$$\frac{1}{2} C_s (\Delta U)^2 = \frac{1}{2} m\,(n - r - 1)\,C_t \left(\frac{\Delta U}{2}\right)^2$$

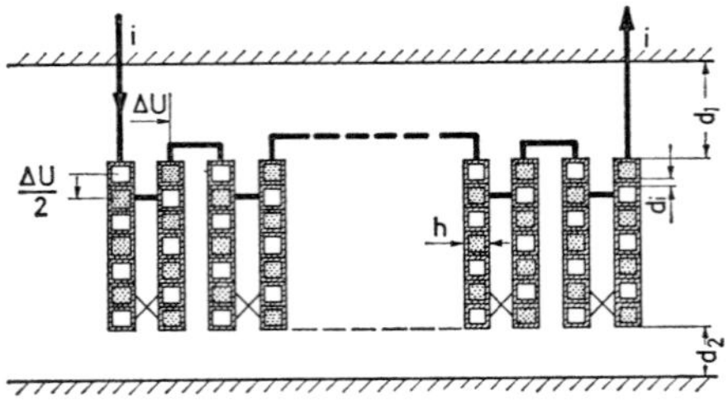

Fig, 4.8.10a

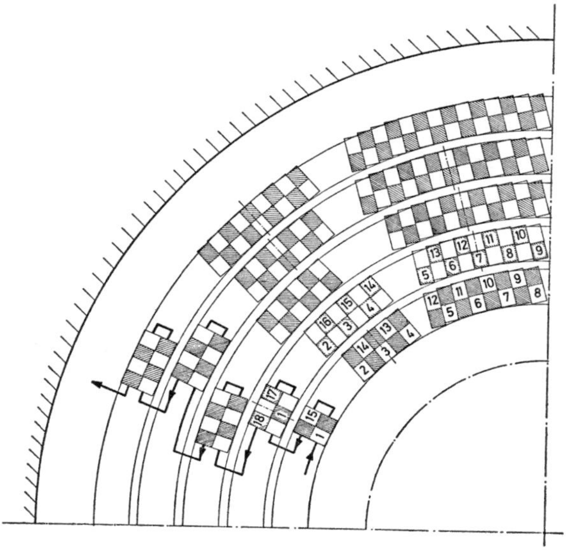

Fig. 4.8.10b

Fig. 4.8.10a and b. Interleaved magnet coil: Two conductors are wound in parallel and connected electrically in series

or:

$$C_s = \frac{m}{3}(n - r - 1)\, C_t$$

with:

$$C_t = 8.85 \cdot \varepsilon_r \pi \cdot d_m \cdot \frac{h}{d} \quad (\mathrm{pF})\,. \tag{4.8.29}$$

Where $m \geqq 2$ is the number of sections in an interleaved element, n the number of turns per section, r the number of gaps in a section, and C_t the actual capacitance between neighbouring turns.

It is easily seen that the series capacitance of interleaved coils is much larger than for conventional noninterleaved coils. Between turns, the voltage $\Delta U/2$ will persist. The disadvantage of interleaved windings is that two or more conductors in a pancake must be wound in parallel, necessitating at least the double number of joints between pancakes, or the double number of cross overs.

4.9 Irradiation Effects in Superconducting Magnets

4.9.1 Introduction

The superconducting magnets being placed in the accelerator ring are, in general, not exposed to high irradiation doses.

To determine the dose rates in synchrotron magnets, a 10^3 GeV proton synchrotron with 10^{13} ppp, pulse duration of 10 sec and a pause between pulses of 10 sec are taken as a model (GESSS) [21]. The calculation of the dose rates at first collision at the magnet where maximum energy absorbed is based on experimental data [56], and on Monte Carlo calcuation performed over the magnet volume [57].

Appreciable irradiation does may be expected in magnets in cases of
(i) mis-stearing,
(ii) beam scraping or beam ejection,
(iii) magnets placed close to internal or external targets, beam absorbers and septums.

The magnet coils are not protected by means of steel plates in the upstream and downstream faces. The dewar walls are thinner than one radiation length and we can expect that the scattered primary particles colliding with the cryostat wall will produce a primary collision directly with the superconducting coil.

We have assumed in our calculations that 10% of the beam particles are absorbed by the internal target. According to measurements by Citron et al. [54], one may expect an exponential decay of the absorbed irradiation energy in the radial direction (from the inner to the outer diameter) and nearly constant energy deposition in the magnet axial length.

The estimate of the production of fast neutrons, which are the most dangerous secondaries, is based on the data that for proton energies of $E > 30$ GeV, the number of secondaries will be: $n \cong 1.7 \cdot E^{\frac{1}{4}}$.

A Monte Carlo calculation provides the production of neutrons over the magnet volume. As the magnets have an axial length of more than one meter, we can assume that practically all the scattered beam energy is absorbed in the magnet. The calculated expected dose rates (at 10% beam loss due to the internal target) for proton energies in the range of 1000 GeV is about 2.5×10^{11} rads per year, averaged over the entire magnet, which has an inner bore radius of 0.05 m and an outer coil radius of 0.2 m. The maximum dose rate is expected to reach $\sim 2.3 \times 10^{11}$ rads per year at the magnet upstream face, adjacent to the bore. In the most endangered coil section, at heat flux due to mere proton irradiation of ~ 0.5 W/cm^2 into the helium may be expected!

Other effects of irradiation on magnet components depend on the type of particles, the energy of the particle, and production of secondaries. Irradiation effects with low energy deuterons on superconducting type II materials are reported by Coffey et al. [58], and electron irradiation effects by Brechna [59]. Dose rates up to 10^9 rads increase the critical current at 4.2 K only for NbZr and Nb_3Sn compositions. NbTi alloys have not shown an improvement in critical current density enhancement.

At low energy deposition ($\sim$ keV) in superconductors, vortices are moved and point defects either produced or moved within the lattice. At MeV energy levels, point defects, defect clusters, or lattice deformation are observed.

At energy levels in the GeV range, provided the superconductor is not destroyed through heating, defect clusters are either produced, or moved and changed.

The irradiation effects have two distinct influences which, in general, work together. At low dose rates per hour where the heating of the sample is trivial, the mere ionization effects tend to soften grain boundaries, as observed in the metal tempering process; thus, defects can move more readily within the lattice, producing a different composition of pinning cites, which for Nb_3Sn has been more favorable and has resulted in enhancing J_c. The increase of J_c, however, also yields a more unstable performance and the tendency towards fluxjumps.

Irradiation effects on matrix materials are also of interest. High purity aluminum and copper used for stabilization as a matrix material are work-hardened by irradiation: The resistivity of aluminum ($r = 2000$) is increased by a factor of ~ 120 at 4.2 K, the resistivity of copper ($r = 1000$) by about 60.

The irradiation tests were performed on thin specimen [60, 61], irradiated with 1,4 to 2,8 MeV electrons. A high resistivity matrix material such as cupronickel, which is also frequently used in combination with copper, did not change at dose rates up to $\sim 10^{12}$ rads when exposed to electron irradiation.

The effect of electron irradiation, gammas, and fast neutrons on a few insulation materials were reported by Brechna [62]. The particle energies used to irradiate samples were about 3 MeV. The effect of irradiation on insulation materials is, in general, polymerization, and thus increases in yield strength at low dose rates. The materials, however, become brittle and crush sensitive, and being under constant or intermittent stress, will not be able to support magnetomechanical strain. The most promising material tested so far is glass fiber-reinforced epoxies, where the ratio of organic to inorganic materials will be about 40/60% by weight.

Few of the glass filament-thermosetting materials tested at cryogenic temperatures withstood irradiation damage ($\leqq 25\%$ change) at doses of $\leqq 10^{12}$ rads.

The ionization effects on irradiated metals and superconductors are annealed to a large extend if warmed up to liquid nitrogen temperature or higher. Copper recovers its pre-irradiated properties more than 90% of its original value if warmed up to room temperature. At 77.3 K, the increase in resistivity is about a factor 6 compared to pre-irradiated values when irradiated with 4.7×10^{11} rads.

NbTi alloys also recover pre-irradiated current carrying capacity, if warmed up appreciably above their critical temperature.

The thermal effects on irradiated samples or magnets are of concern. As mentioned, at a yearly dose rate of $\sim 2.3 \times 10^{11}$ rads, a heat flux of ~ 0.5 W/cm^2 can be generated at the most endangered magnet areas. This heat must be removed entirely by the coolant through a refrigeration system. This may prove to be a limiting factor in the magnet performance, specifically if the superconducting coils are impregnated with thermosetting material; thus, the heat generated must removed by convection and heat transfer.

In this paper, we present methods of irradiation dose estimates in magnets where we confine the attention to protons* and briefly discuss

* Estimated electron dose rates in magnets are reported in Literature (59).

endangered magnet areas. We also discuss production of defects and the properties of superconducting magnets when exposed to high irradiation dose rates.

4.9.2 Energy Loss by Collisions

The incident beam passing through an internal target or at extraction areas and through collimators or slits will produce secondaries. A fraction of the primary particles (say about 10% of the proton beam) of the incident beam may collide with the magnet surfaces. The distribution of the deposited irradiation dose is shown for a dipole magnet with circular aperture in Fig. 4.9.1. The data given are extrapolated values measured at 20 GeV and adjusted for 1000 GeV protons. The secondary particles generated by the primaries within the magnet will propagate by subsequent interactions. The cascade develops within the magnet in lon-

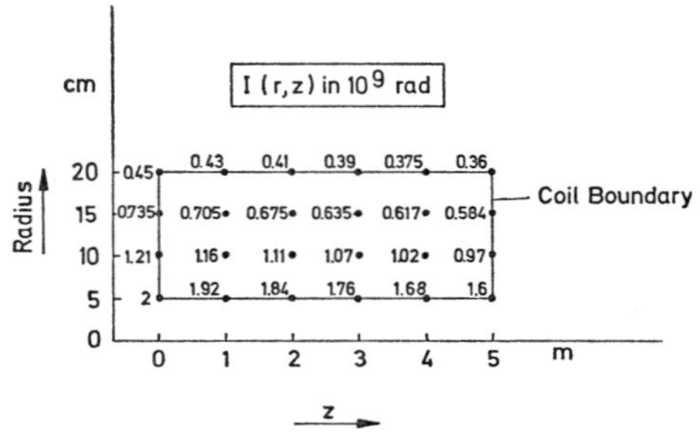

Fig. 4.9.1. Irradiation dose rate distribution over an axially symmetric coil

gitudinal and radial directions. In Fig. 4.9.1, the distribution of intensity over the magnet is illustrated. The incident beam distribution curve, indicated by (0 cm), is assumed to have an exponential radial decay. The intensity distribution is given by:

$$J(r) = 2\times 10^9 \text{ rad} \cdot e^{-\left(\frac{r - r_i}{r_0}\right)}. \qquad (4.9.1)$$

In lateral or axial magnet direction, the beam intensity is lost by less than 20%. The incident intensity over the magnet having an axial length of several meters is completely absorbed by the active and inactive magnet material. Secondary particles produced are estimated 15% thermal neutrons, 60% fast neutrons, 15% high energy particles (such as pions), and 10% γ-rays and ionization.

Among these secondaries, the fast neutrons are the most dangerous. The average number of secondaries, n, produced in a collision is approximated by the relation [63]:

$$\left.\begin{aligned} n &\cong 1.7 \cdot E^{\frac{1}{4}} && \text{for} \quad E > 30 \text{ GeV} \\ n &\cong 0.85 \cdot E^{\frac{1}{2}} && \text{for} \quad 3 < E < 30 \text{ GeV} \\ n &= \tfrac{5}{9} E - 0.17 && \text{for} \quad E < 3 \text{ GeV} \end{aligned}\right\} \qquad (4.9.2)$$

The data have been obtained empirically.

About one third of the energy of the primary particle is lost to the secondaries. The average secondary particle energy is thus: $E/3\,n$.

Using this approximation, the beam intensity in the magnet as a function of radial and axial distance is calculated. The average dose rate per year over the magnet due to primary particles (assumed 10% of the particles are lost in the internal target) is 2.4×10^{10} rads/year. This value is obtained for a beam energy of 1000 GeV and an intensity of 10^{13} ppp. The pulse duration is assumed 10 seconds with a 10 second pause between pulses.

The maximum dose rate due to primary incident particles alone is 2.5×10^{11} rads/year, assuming an 8000 hour machine operational schedule.

It is seen that the cascade build up occurs after one to three collision lengths and the distance increases logarithmically with the primary energy (Fig. 4.9.1). The maximum build up occurs in our particular case radiation lengths from the upstream surface of the magnet.

The effective energy loss per ionizing particle was estimated to be 3 MeV cm^2/g in the 1000 GeV accelerator.

The peak absorbed irradiation dose deposits in one cm^3 a heating power of 0.4 W (average magnet density $\sim$ 5 g/cm^3). Assuming only a cooling surface of 0.8 cm^2 per cm^3 is exposed to liquid helium and the heat is transferred through this surface, then the heat flux to be removed is 0.5 W/cm^2! The average heat flux over the magnet produced by irradiation is about 0.06 W/cm^2, and thus not of importance.

To give an example we calculate the temperature rise in the irradiated coil section of 10^{-2} m width terminated by coolant passages on two opposing surfaces. The casting has a thickness of 10^{-3} m on each side of the cable. Since in a fully transposed cable each strand will be at the surface every $5 \cdot 10^{-2}$ m to $10 \cdot 10^{-2}$ m along the conductor, we can represent the cable thermally by a solid metal conductor. The overall density of a coil section is about 5.10^3 kg/m^{-3}. Thus the energy deposites per unit volume is calculated from $w_v = 2{,}3 \cdot 10^{11} \cdot 10^{-2} \cdot 5 \cdot 10^3 / 8 \cdot 10^3 \cdot 3{,}6 \cdot 10^3 \cong 4 \cdot 10^5$ W/m^3.

With typical data of $k_{Cu} = 400$ Wm^{-1} K^{-1}; $k_{ins} = 2.10^{-1}$ Wm^{-1} K^{-1}; and $h = 3 \cdot 10^3$ Wm^{-2} K^{-1} we get from

$$\Delta T \cong w_v \cdot \left[\frac{1}{2\,k_{Cu}} \left(\frac{a}{2}\right)^2 + \frac{1}{2\,k_{ins}} (a_{ins})^2 + \frac{a + 2\,a_i}{4\,h}\right]$$

a temperature rise of 1,6 K! In the above equation $a = 5 \cdot 10^{-3}$ mm half the width of a section, and $a_i = 10^{-3}$ m for the outer casting layer. The above temperature rise is certainly not acceptable. Subdivision of the coil into narrower slices may help somewhat. Loading the thermosetting with mineral fillers will result in a temperature rise of 1 K, since $k_{ins} = 4.10^{-1}$ Wm^{-1} K^{-1} at 4, 4 K. But also this temperature rise will result in a drastic reduction of the critical current density of the superconductor as seen from Fig. 1.6a,b. In most endangered areas shielding of coils by means of thick iron plates is more appropriate.

Particles are lost in the accelerator and in the beam transport area by a variety of ways:

(a) Injection
(b) Beam Extraction
(c) Beam Collimation (Slits)
(d) Septum Magnets
(e) Thick and Thin Targets.

The areas where most particles are lost are in the septum magnets, beam extraction, which has an efficiency of maybe 50%. If the length of the collimator and slit is adequate, the lost fraction of the beam will be absorbed fully in the collimator and only muons may penetrate the surfaces at low energies and are of less concern to the magnet operation.

Scatter from residual gases in the vacuum vessel will also produce secondaries. Lewin [63] has estimated the number of secondaries produced in a vacuum of p torr from the source strength given by:

$$125\, Np \cdot n/r \quad \text{particles/cm s}$$

with N the number of incident protons, r the radial distance, and n the number of secondaries.

One of the most dangerous but fortunately rare incidents is the accidental exposure of magnets to the full beam due to mis-steering as result of some failure.

At the slow pulse rates, a failure in this form is immediately apparent and the machine can be shut off after one pulse. The beam, however, is unfocused and it is hoped that the particle flux colliding with the magnet will be only a small fraction of the total flux ($\sim$ 1%).

4.9.3 Irradiation Effects on Type II Superconductors

It is known that few physical properties of superconductors type II are changed when exposed to nuclear irradiation. The critical current density, the critical temperature T_c, the lower and upper critical fields, and hysteretic losses are primarily affected when irradiated.

In NbTi and Nb_3Sn alloys, the critical temperature is reduced, so is the upper critical field. The hysteretic losses are increased, in general, but J_c does not change uniquely as a function of dose rate and must be studied more in detail. In composite conductors, due to the fact that the resistivity of the matrix material is substantially increased, the magnet becomes more unstable and fluxjump sensitive.

The mechanical properties of superconductors are also altered. As in the case of normal metals, superconductors undergo a transition from ductile to brittle condition (specifically NbTi), which in the case of multifilament composite conductors can yield to filament breakage.

The situation for Nb_3Sn is not yet clear; measurements on the mechanical properties of multifilament Nb_3Sn wires have not been performed.

Nb_3Sn irradiated [64] with fast neutrons ($E > 0.1$ MeV) have exhibited higher hysteretic losses with dose rates (0.5—1.5×10^8 n cm^{-2}).

The critical current density of Nb_3Sn (10...100 μm) thin tapes was enhanced 50% when irradiated at 50 °C with 10^{18} cm² fast neutrons [65]. Evaporated Nb_3Sn surfaces have shown the same general trend [66].

The current density of diffusion layers of Nb_3Sn irradiated up to a critical dose rate of 10^{17} p/cm² with protons and deuterons (1—1.3 MeV) was enhanced by a factor of six [67]. Above this critical value, the critical current density decreases. The enhancement in current carrying capacity is due to generation of clusters with dimensions of 10—100 Å, which are comparable to the coherence length of Nb_3Sn (~ 50 Å). The clusters are effective pinning centers.

Assuming the formation energy of clusters is 20 keV, the maximum value of J_c is obtained at a cluster expansion of 75 Å and an optimum mean distance between clusters of 240—310 Å. Each cluster consists of about 200—500 Frenkel defects. The reduction of the critical current density at higher dose rates is probably due to cluster overlapping. The current enhancement is annealed at 700—800 °C, indicating that cluster formation is the prime source of the generation of pinning centers.

Coffey et al. [58] have irradiated Nb_3Sn evaporated layers with 15 MeV deuterons up to a dose rate of 10^{17} d/cm². Nb_3Sn probes with initially low critical currents exhibit an enhancement at 5.7 K, and a reduction of J_c at 10.9 K. Nb_3Sn samples with initially high critical current always showed a reduction in J_c at any temperature and dose rate.

The critical temperature of Nb_3Sn was reduced by 1 K (~ 5%) and the upper critical field H_{c2} was lowered by about 15%.

Cold-worked NbTi probes were irradiated with deuterons at temperatures below 30 K. It was found that J_c was, in general, reduced. Annealing to 300 K yielded a recovery of J_c to about 95%.

Hassenzahl et al. [68] irradiated NbTi composite conductors with 13 to 15 MeV protons. At dose rates of 10^{18} p/cm² and temperature of 400 K, no appreciable reduction in J_c was noticed. At 77 K, the reduction was 2—5%. At 30 K, the reduction in critical current was stronger at low fields (4 T) than at higher fields (10 T).

Irradiation tests with 51.5 MeV deuterons at room temperature by Maurer et al. [69] have been performed on stabilized NbTi multicore conductors on the following geometries:

Conductors Type A: Copper matrix, 1 mm diameter; 361 filaments with 26 μm individual filaments diameters.

Conductors Type B: Copper matrix, 0.4 mm diameter; 61 filaments with 35 μm individual filaments diameters.

Results of Type A: At a dose rate of $1.1—1.24 \times 10^{11}$ rads, the take-off current was reduced 10—14% at transverse fields of 2.5—4 T. The critical current was reduced 15—22%. After a single irradiation period, no recovery to the original values was measured after 14 days exposition of the sample at room temperature. A second probe which was irradiated several times showed a reduction of the critical current by ~ 10% after the irradiation dose of 0.35×10^{11} rads. After the sample was warmed up and exposed to room temperature for 10 days, it recovered 97% of its initial J_c value.

Further irradiation up to 1.24×10^{11} rads reduced the J_c (seen from Fig. 4.9.2) to values of about 78% of its original number.

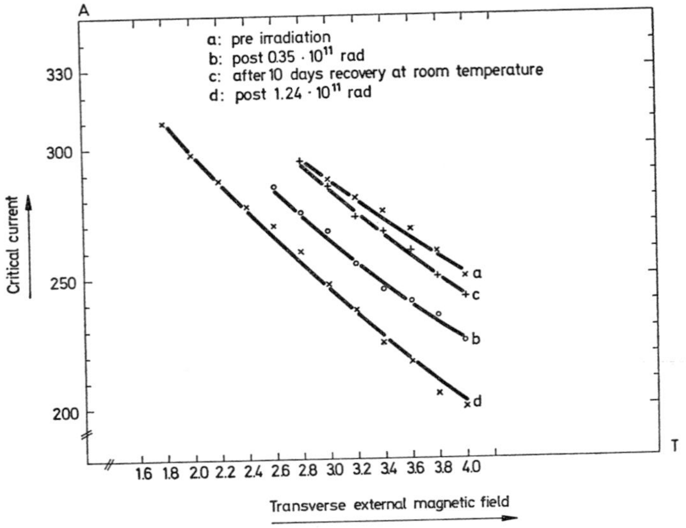

Fig. 4.9.2. Critical current-field characteristics prior and after irradiation at room temperature

Results of Type B: At dose rates of 10^{11} rads, only a reduction of J_c and the take-off current (~ 5%) was measured. Fig. 4.9.3 shows the general trend of the $(U - I)$ curves. The general trend of this curve did not change. However, samples having a $(U - I)$ characteristic (Fig. 4.9.3) exhibited, at 1.6×10^{10} rads, a reduction in the take-off and criti-

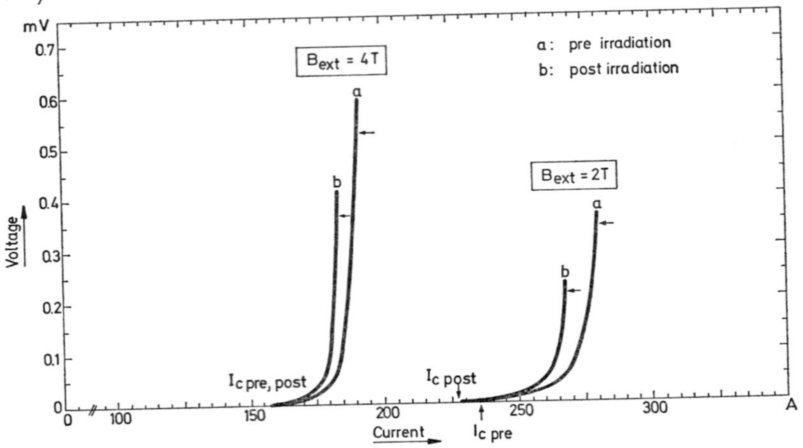

Fig. 4.9.3. Voltage-current characteristics of NbTi-Cu composites prior and after 51.5 MeV deuteron irradiation

cal currents of 5%. The stabilization characteristic of these samples is changed considerably. At the same B value, the irradiated probes showed a degradation of $\sim$ 20%.

4.9.4 Irradiation Effects on Normal Metals

Copper, aluminum, and cupronickel are widely used as matrix materials in conjunction with superconductors in magnets. In pulsed magnets, the use of In-Sn and Ag-Sn alloys as impregnants are becoming increasingly attractive. It is interesting to note that the electrical properties of these solders are not appreciably affected by irradiation.

Mechanical properties of coppers are known to change when exposed to nuclear irradiation. At 10^{20} n/cm^2, the yield strength of OFHC copper has increased 5 fold at 4.2 K, its tensile strength about 1.9 fold. The high material yield strength which indicates embrittlement is due to strain hardening. The change in electrical properties are somewhat related to the change in mechanical properties. The resistivity of copper and aluminum are changed considerably when exposed to nuclear irradiation. The increase in resistivity is temperature dependent.

Room temperature irradiated high purity copper and aluminum show that the resistivity is increased by a factor 2. Irradiated samples at 77 K exhibit an increase in resistivity of a factor 5. Samples irradiation at temperatures below 10 K exhibited an increase of resistivity by a factor of at least 40 ... 50.

New measurements by Sassin [60] on high purity copper and aluminum tapes ($r_{\text{Cu}} = 2800$; $r_{\text{Al}} = 1600$) at 9 K with 2.8 MeV electrons show that the resistivity of copper was increased from $\varrho_0 = 15.7 \times 10^{-9}$ Ohm · cm to $\varrho_{\text{irr}} = 590 \times 10^{-9}$ Ohm · cm.

This means that, at a rate of 1.15×10^{20} e/cm^2, the copper resistivity was increased by a factor of 37.5.

At the same dose rate, the resistivity of aluminum was changed from $\varrho_0 = 14.6 \times 10^{-9}$ Ohm · cm to $\varrho_{\text{irr}} = 578 \times 10^{-9}$ Ohm · cm (an increase of 39.6).

The increase in resistivity per electron per unit area is thus:

$$\frac{\Delta \varrho}{e^-/\text{cm}^2} \approx 5 \times 10^{-27} \text{ Ohm} \cdot \text{cm}/e^-/\text{cm}^2.$$

Newer measurements by Böning et al. [61] show even larger increases in resistivity. Copper ($r = 950$) irradiated with fast neutrons ($E > 0.1$ MeV) shows that the resistivity was increased from 0.178×10^{-8} Ohm · cm to 13.56×10^{-8} Ohm · cm at a dose rate of 2.52×10^{18} n/cm^2 (an increase of 76).

Aluminum ($r = 2160$) irradiated showed that the resistivity changed from 2.08×10^{-7} Ohm · cm to 310×10^{-7} Ohm · cm at a dose rate of 1.05×10^{11} n/cm^2 (an increase in resistivity of 149).

The increase of resistivity in copper and aluminum can be explained by the generation of new Frenkel pairs which change the scattering properties of conduction electrons. The mean free path of the electrons is generally reduced. The fact that the resistivity of copper and aluminum changes when irradiated has a profound effect on the stabilization of superconducting magnets. However, the increase in temperature (up to 300 K) anneals the effect about 95%. The irradiation imposed strain on the matrix material in the magnet section exposed to the highest dose rates produces a material which is fluxjump sensitive. As the superconductor has also changed its current carrying capability, a region of instability is produced which may endanger the safe performance of the magnet at high dose rates.

In Fig. 4.9.3, the short sample characteristic of 51.5 MeV deuterons irradiated (at room temperature) and not irradiated NbTi-copper composite is illustrated. After a dose of 1.6×10^{10} rads, we have found a decrease of stability at transverse external fields of 4 T resp. 2 T. At 4 T, I_c is unchanged, but the $I_{\text{take off}}$ decreases about 4% at 2 T, the I_c and take-off current decrease about 4%.

The irradiation effects on cupronickel, which is widely used as a matrix material, can be neglected. Irradiation effect on the metallic bond between superconductor and normal metal is also of concern. In the intermetallic layer between the superconductor and normal metal matrix, which is composed of ternaries of Nb, Ti, and Cu, a high thermal and electrical resistivity is generally encountered. The bond is responsible for a somewhat lower thermal diffusivity from the superconductor to the matrix. The intermetallic diffusion layer between the superconductor and the matrix is increased when the composite conductor is exposed to nuclear irradiation. Thus, at an increased magnetic diffusivity, the thermal diffusivity is further reduced. At present, measurements on thermal diffusivity over the intermetallic barrier are nonexistent due to complexity of the problem.

The effect of neutron irradiation on nonmagnetic steels have not been reported recently. However, it is expected that at dose rates of $\geqq 10^{17}\, n$ per cm^2 at 4.2 K, no significant changes in yield and tensile strength is to be expected.

4.9.5 Irradiation Effects on Magnet Insulations and Reinforcements

From a variety of thermoplastic and thermosetting materials, only a few insulation materials are being considered for superconducting high energy magnets: glass fiber tapes with or without mica, and impregnated or cast in epoxies.

If the coils are not impregnated in thermosettings, glass-fiber tapes are recommended as interturn and interlayer insulations. Glass fibers have a radiation damage threshold which is higher than 10^{11} rads. They exhibit high compressive strength ($> 3 \times 10^4$ kp/cm^2) and is adequate to prevent shorts between turns and layers.

If the coils are to be impregnated in suitable thermosettings, in order to prevent conductor movements and with it coil degradation, a variety of epoxies and epoxy Novalacs have been tested at low temperature irradiation environment. Glass-fiber impregnated thermosettings have shown no apparent change in mechanical and electrical properties at $5 \times 10^{18}\, n/\text{cm}^2$ at 4.2 K. Slight embrittlement was encountered (rupture elongation changed from 8.1 to 7.3% at 4.2 K) when the samples were irradiated up to 10^{17} n/cm^2.

The mechanical properties of fiberglass reinforced thermosettings can be improved by an order of magnitude (in irradiation dose) if proper epoxy functional materials are used in combination with the thermosettings to make the glass fiber compatible to the resin; the glass fiber tapes are to be heat-cleaned and chemically treated prior to impregnation.

4.9.6 Irradiation Effects on Helium

Irradiation of liquid helium at 4.2 K yields helium contamination and production of radioactive isotopes based on nuclear reactions [59]. Ionization may lead to a higher boil-off rate of the liquid. In Table 4.9.1, we have written the important reactions with their Q values:

Irradiation leads to the production of new elements such as ${}^3_2He$, ${}^3_1He$, ${}^5_2He$, ${}^5_3Li$, ${}^6_3Li$, etc.

Table 4.9.1
Production of radioactive helium isotopes due to irradiation

γ-irradiation:				
	${}^4_2He + \gamma$	$\rightarrow$	${}^3_2He + n$	— 20.577 MeV
		$\rightarrow$	${}^3_1H + p$	— 19.813 MeV
		$\rightarrow$	$d + d$	— 23.845 MeV
		$\rightarrow$	$d + n + p$	— 26.070 MeV
n-irradiation:				
	${}^4_2He + n$	$\rightarrow$	${}^5_2He + \gamma$	— 0.957 MeV
		$\rightarrow$	${}^3_1H + d$	— 17.588 MeV
		$\rightarrow$	${}^4_2He + n$	± 0
		$\rightarrow$	${}^3_2He + n + n$	— 20.577 MeV
		$\rightarrow$	${}^3_1H + n + p$	— 19.813 MeV
p-irradiation:				
	${}^4_2He + p$	$\rightarrow$	${}^5_3Li + \gamma$	— 1.967 MeV
		$\rightarrow$	${}^3_2He + d$	— 18.352 MeV
		$\rightarrow$	${}^4_2He + p$	± 0
		$\rightarrow$	${}^3_2He + n + p$	— 20.577 MeV
		$\rightarrow$	${}^3_1H + p + p$	— 19.813 MeV
d-irradiation:				
	${}^4_2He + d$	$\rightarrow$	${}^4_2He + d$	± 0
		$\rightarrow$	${}^5_2He + p$	— 3.182 MeV
		$\rightarrow$	${}^5_3Li + n$	— 4.192 MeV
		$\rightarrow$	${}^3_2He + {}^3_1H$	— 14.320 MeV
		$\rightarrow$	${}^6_3Li + \gamma$	— 1.471 MeV
		$\rightarrow$	${}^4_3He + n + p$	— 2.225 MeV

As these elements are partially unstable and do decay, they can contaminate the helium, which in a refrigerator system may lead to blocking of small passages. Fortunately, the amount of contaminants are small and no adverse effects have been reported to date.

References Chapter 4

[1] Livingston, J. D., Schadler, H. W.: Prog. in Materials Science **12,** 183 (1964).
[2] Labusch, R.: Crystal Lattice Defects **1,** 1 (1969).
[3] Beasley, M. R., Labusch, R., Webb, W. W.: Phys. Rev. **43,** 682 (1969).
[4] Friedel, J., De Gennes, P. G., Matricon, J.: Appl. Phys. Letters **2,** 119 (1963).
[5] Bean, C. P.: Rev. Mod. Phys. **36,** 31 (1964).
[6] Kim, Y. B., Hempstead, C. F., Strnad, A. R.: Phys. Rev. **131,** 2486 (1963).
[7] Morgan, G. H.: BNL int. Rep. AADD. 146 (1968).
[8] Hancox, R.: Proc. IEE **113,** 1221 (1966).
[9] Wilson, M. N.: Rutherford Laboratory Rep. RPP/A 89 (1972).
[10] Morgan, G. H.: J. Appl. Phys. **41,** 3673 (1970).
[11] McInturff, A. D., et al.: BNL int. rep. AADD. 179 (1971).
[12] Brechna, H., Ries, G.: IEEE Trans. Nucl. Sci. NS **18,** No. 3. 639 (1971).
[13] Ries, G., Brechna, H.: KFF Rep. 1371 (1972).
[14] Salmon, D. R., Catterall, J. A.: J. Phys. D: Appl. Phys. **3,** 1023 (1970).
[15] Eastham, A. R., Rhodes, R. G.: AC losses in superconducting wires, in: Proc. Int. Inst. of Refrig. Commission I. Tokyo 1970.
[16] Eastham, A. R., Rhodes, R. G.: Cryophysics and Cryoengineering, p. 147. Tokyo 1970.
[17] Pouillange, J. P., Prost, G.: GATS/71-24 (1971).
[18] Smith, P. F., et al.: Proc. 1968 Summer Study on Supercond. Devices, Part III, BNL 50155 (C 55), p. 913. Brookhaven 1968.
[19] Critchlow, P. R., Zeitlin, B.: J. Appl. Phys. **41,** 4860 (1970).
[20] Spurway, A. H., Lewin, J. D., Smith, P. F.: J. Phys. D: Appl. Phys. **3,** 1517 (1970).
[21] Smith, P. F.: Proc. 8th Int. Conf. on High Energy Acc. CERN 35 (1971).
[22] Brechna, H., Green, M. A.: Appl. Supercond. Conf. 1972, in: J. Appl. Phys. (to be published).
[23] McInturff, A. D., et al.: Brookhaven Natl. Lab. Report AADD. 179 CRISP 71-13.
[24] Green, M. A.: IEEE Trans. Nucl. Sci. NS **18,** 664 (1971).
[25] Thomas, D. B.: Proc. 8th Int. Conf. on High Energy Acc. CERN 190 (1971).
[26] McInturff, A. D., Dahl, P. F., Sampson, W. B.: Brookhaven Natl. Lab. Rep. BNL 16516 (1972).
[27] Jüngst, K. P., Krafft, G., Ries, G.: KFK 1217 (1970).
[28] Bogner, G., Salzburger, H., Franksen, H.: Proc. 3rd Int. Conf. Magnet Technology, Hamburg 1970.
[29] Bronca, G., et al.: GATS/70-37 (1970).
[30] Gilbert, W. S., Hintz, R. E., Voelker, F.: UCRL 18176 (1968).
[31] Hlasnik, C., Lefrancois, J. P., Pouillange, J. P.: SEDAP 69-308, SUP 84 (1969); SEDAP 18/11 (1969).
[32] Fietz, W. A.: Rev. Sci. Instr. **36,** 1621 (1965).
[33] McInturff, A. D.: Proc. 1968 BNL Summer Study Part II, BäL 50155 (C 55), 465 (1968).
[34] Brechna, H., Allen, M. A., Cobb, J. K.: J. Appl. Phys. **42,** 103 (1971).
[35] Sekula, S. T.: J. Appl. Phys. **42,** 16 (1971).
[36] Pech, T.: Rev. Phys. Appl. Tome **6,** 357 (1970).
[37] Wipf, S. L.: Phys. Rev. **161,** 404 (1967).

[38] Hart, H. R.: Proc. 1968 BNL Summer Study Part II, 571 (1968).
[39] Chester, P. F.: Rep. Progr. Phys. XXX/II, 561 (1967).
[40] Wilson, M. N., et al.: Brit. J. Phys. D: Appl. Phys. **3,** 1517 (1970).
[41] Hancox, R.: Proc. Intermag. Conf. Washington, D.C. (1968).
[42] Brechna, H., Garwin, E.: SLA proposal (1967).
[43] Hancox, R.: Proc. 3rd Int. Conf. Magnet Technology .Hamburg 1970.
[44] Brechna, H.: Bulletin S.E.V. **20,** 893 (1967).
[45] Steckly, Z. J. J., Zar, J. L.: IEEE Trans. NS **12,** 367 (1965).
[46] Coffey, D. L., Gauster, W. F.: Proc. 1968 BNL Summer Study Part III, 929 (1968).
[47] Gauster, W. F.: J. Appl. Phys. (1969).
[48] Williams, J. E. C.: Cryogenics. 234, Dec. (1963).
[49] Lock, J. M.: Cryogenics. 438, Dec. (1969).
[50] Rauh, M.: Diss. No. 4656. ETH Zürich, Juris (1971).
[51] Bogner, G.: Siemens Publ. (1970).
[52] Salzburger, H.: Siemens Akten-Notiz FL 52 (1968).
[53] Blume, L. F., Boyajian, A.: Trans. AIEE **38,** 577 (1919).
[54] Brechna, H.: Bul Oelikon No. 328—329, 89 (1958).
[55] Bewley, L. V.: Travelling Waves on Transmission Systems. London: Chapman and Hall 1951.
[56] Citron, A., et al.: Nucl. Instr. and Meth. **32,** 48 (1965).
[57] Goebel, K., Ranft, J.: CERN 70-16 (1970).
[58] Coeffey, H. T., et al.: Phys. Rev. **155,** 355 (1967).
[59] Brechna, H.: SLAC Publ. 469 (1968).
[60] Sassin, W.: Jul.-586-FN (1969).
[61] Böning, K., et al.: Int. Conf. on Vacancies and Interstitials in Metals 405 (1968).
[62] Brechna, H.: SLAC 40 (1965).
[63] Lewin, J. D.: RHEL/R 119 (1969).
[64] Swartz, H. R., Hart, H. R., Fleischer, R. L.: Appl. Phys. Letters **4,** 71 (1964).
[65] McEvoy, J. P., Decell, R. F., Novak, R. L.: Appl. Phys. Letters **4,** 43 (1964).
[66] Cullen, G. W., Novak, R. L.: J. of Appl. Phys. **37,** 3348 (1966).
[67] Wohlleben, K.: Z. f. angew. Physik **27,** 92 (1969).
[68] Hassenzahl, W. V., Rogers, J. D., Armstrong, W. C.: IEEE Trans. NS **18,** 683 (1971).
[69] Maurer, W., Brechna, H.: KFK Rep. 1469 (1972).

5. Cryogenics

*H. Brechna and H. M. Long**

Superconductivity has been and is regarded widely as low temperature phenomena**, and as such requires the use of cryogenic technology to achieve and maintain the needed environment. While transition temperatures for the onset of superconductivity have been observed to be as high as 21 K, the practical high field, high current density superconductors used in magnets and electrical machines are operated in the temperature range of 2 K to 8 K. This requires that helium be used as a cryogenic fluid; liquid from 2.1 K to 5.2 K and compressed as supercritical gas from 2.1 K to the desired temperature.

In the proceeding sections, a few properties, availability, handling, refrigeration and liquefaction of helium will be given. Some properties of other cryogens pertinent for the use of superconducting magnets will be discussed in brief, such as those of liquid nitrogen and hydrogen. Thermal, mechanical, and electrical properties of solid materials used in combination with superconductors will be presented.

5.1 General Properties of Cryogenic Fluids

All cryogens are colorless, odorless, and tasteless, both as liquids and gases. They are non-magnetic and do not conduct the electric current under normal conditions. The useful ranges of the cryogens are shown in Fig. 5.1.1. The critical, normal boiling, and triple points are indicated schematically at temperatures at which they occur.

Nitrogen, neon, and hydrogen exhibit liquid and melting ranges that are essentially normal. Nitrogen and neon can be related to each other reasonably well by the law of corresponding states. Hydrogen shows same large deviations, but agree in principle. Not so with helium. In the case of helium, there is no triple point and the liquid exists to indefinitely low temperatures. In Fig. 5.1.1., both the common isotope of helium, with mass 4, and the stable much less common isotope of mass 3 are shown. With these cryogens, fluid baths may be provided to the very lowest of temperatures. While there is no triple point for the helium,

* Oak Ridge Natl. Lab., Thermonuclear Division

** Superconductivity at room and higher temperatures is being studied by a number of investigators. But technical superconductors developed to date have the highest $T_c \leqq 21$ K.

there is in the case of helium 4 a change in properties at temperatures below the normal boiling point, the transition to the "superfluid state" at 2.19 K (He II). In this case, the liquid helium behaves in an anomalous fashion, exhibiting essentially zero viscosity and very high thermal conductivity. These unusual properties are akin to superconductivity, and therefore employed in some superconducting systems to further enhance the overall system operation.

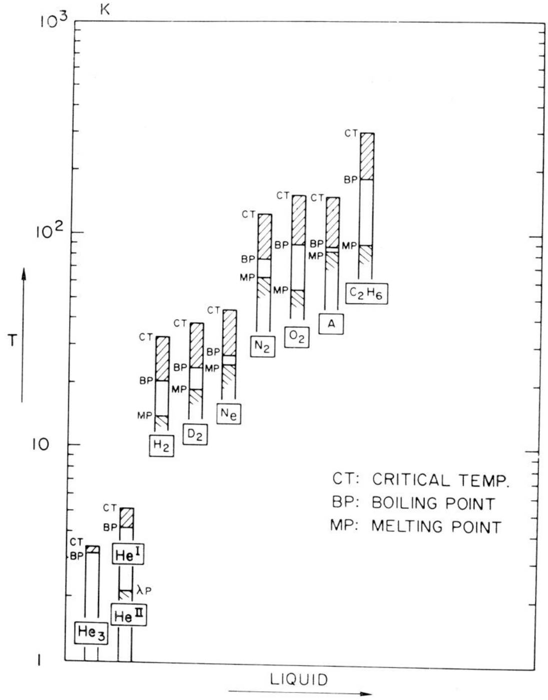

Fig. 5.1.1. Useful temperature range of cryogenic liquids

Although helium does not solidify in the usual manner, it can be solidified under pressure. The pressures, however, are higher than those usually encountered in superconducting apparatus, so we do not have to concern ourselves with problems due to solid helium. All other cryogens solidify at liquid helium temperatures. This must be considered when designing and operating superconducting magnets which may be precooled with liquid nitrogen or liquid neon.

A few physical properties of some cryogens are given in Table 5.1.1. Although properties of cryogenic liquids and materials required for equipment design are given later, more detailed information should be obtained from standard references [1, 2, 3, 4, 5, 6].

Table 5.1.1. *Physical properties of some cryogenic fluids*

Property	He II[a]	He 4	liquid H_2	liquid Ne	liquid N_2
Normal Boiling Point (K)	—[b]	4.215	20.39	27.2	77.32
Triple Point (K)	—[b]	—	13.96	44.5	126.2
Triple Point Pressure ($kp \cdot cm^{-2}$)	—[b]	—	0.0711	0.4265	0.1236
Critical Temperature (K)	—[b]	5.25	33.25	44.5	126.2
Critical Pressure ($kp \cdot cm^{-2}$)	—[b]	2.26	12.8	25.9	33.5
Specific Heat (Ws/g K)	7.51	4.48	6.185	1.86	8.582
Heat of Vaporization (Ws/g)	20.9	22.8	4.41	86.2	198
Heat of Fusion (Ws/g)	—	—	58.1	16.65	25.8
Density ($g \cdot cm^{-3}$)	0.147	0.125	0.07	1.204	0.8082
Latent Heat (Ws/g)	—	21	93	90	398
Dielectric Constant	1.057	1.049	1.23	1	1.44

[a] The λ-point of He II is at 2.17 K.
[b] He II behaves in an anomalous fashion. It has zero viscosity and high thermal conductivity.

5.1.1 Availability and Production

Gases such as helium, hydrogen, neon, and nitrogen are available worldwide from commercial sources as compressed gases in cylinders, the standard cylinder containing approximately six cubic meters (STP) compressed to about 150 kp/cm² or in form of liquids in suitable containers, dewars, or cryostats.

All cryogenic fluids are also commercially available in the US, Canada, and Western Europe in quantities from 25 to 30 liters up to many thousands of liters. Liquid nitrogen is available in most industrial areas around the world.

Nitrogen and neon are obtained as by-products of the oxygen production from distillation of liquid air. While many schemes are imployed for the production plants, they may be subdivided into two general kinds:

(a) Those which generate liquid oxygen and liquid nitrogen.
(b) Those which produce gaseous oxygen and gaseous nitrogen.

The *first kind* (a) are high pressure plants in which the air is compressed to 150 to 200 kp/cm² in order to provide the refrigeration which is withdrawn from the plant in the liquid products in addition to that required by the process.

The *second kind* (b) operates at low pressures, in the order of 6 to 8 kp/cm², since only the heat leak into the plant, the pressure losses and the separation energy need to be supplied by the compression energy.

Most modern air separation plants are of low pressure types (b) and thus the nitrogen must be liquefied in a separate system if liquid is desired.

Since the temperature at which neon liquefies are below those encouns tered in air destillation columns, the neon in the "processed air" collect- as a *non-condensable* gas in the nitrogen portion of the column along with two other components of the atmosphere, helium and nitrogen. These non-condensing gases must be removed; otherwise they would interfere with the separation process. In most plants they are returned to the atmosphere in a vent stream containing about 2.1% neon, 0.8% helium and 0.1% hydrogen with the remainder nitrogen. When neon is to be recovered, this vent stream is passed through a small condenser-rectifier in the plant where most of the nitrogen is liquefied and returned to the destillation column. The neon, helium and hydrogen rich stream from the condenser-rectifier is then passed through a carbon trap which further reduces the nitrogen impurity, yielding a crude neon-helium mixture containing about 95% neon and helium. This crude gas is compressed into cylinders for storage and further purification to produce pure neon and if desired pure helium. The pure products are available as a compressed gas, or in the case of neon as a liquid. The neon and helium concentration in the air are about 18.2 and 5.2 ppm (parts per million) respectively by volume. Enormous quantities of air must be processed in order to obtain reasonable quantities of these rare gases. The largest oxygen plant produce about 1.1×10^6 kg of oxygen per day and in so doing process about 3.9×10^6 m^3 of air, which, if 100% recovered will yield about 71 m^3 of gaseous neon and 20 m^3 of gaseous helium. These quantities correspond to 50 liters of liquid helium per day!

There are also some large oxygen plant complexes, which produce up to 5×10^6 kg of oxygen per day! If full recovery of neon and helium were possible the yield per day of these plants would be 250 liters per day of liquid neon and 135 liters of liquid helium. The total US production of oxygen in 1968 was 11×10^9 kg, this could have yielded a maximum of 500,000 liters of liquid neon and 270,000 liters of liquid helium.

Helium, however is not obtained for commercial use from the air but rather from certain natural gas streams, where it is present at concentrations ranging from fractions of a percent up to several per cent. The helium is recovered by a low temperature process in which the natural gas, predominantly methane is liquefied and the helium along with some nitrogen impurity is removed as non-condensable vapors. This crude helium is further refined by liquefying the nitrogen and stripping the helium with final purification to 99,995% (generally called grade A helium) by cold adsorption beds.

The pure helium is warmed up to room temperature and compressed to 150 to 200 kp/cm^2 for storage or shipment. In some modern commercial plants, the pure helium, while still cold is directed to a liquefier to produce liquid helium. The US production in 1969 of helium was about 24×10^6 m^3, in 1970 about 30×10^6 m^3. Of these at most 10% was employed for cryogenic purposes, yielding 35×10^6 liters of liquid.

Since 1963 a helium conservation program has been operated by the US government. This program was implemented to recover the helium otherwise lost through consumption of helium bearing natural gases as fuel. The natural gases, containing from 0.5 to 0.7% helium, are processed at several large plants which in aggregate remove about 100×10^6 m^3 of helium each year from the gas as "crude helium", this contains 50 to 85% helium. The remainder is principally nitrogen. This crude gas is pumped into depleted helium bearing gas formation which are controlled by the US Government Bureau of Mines as a natural reserve.

Helium activities accounted for almost half of the bureau's entire 1969 budget of $ 117 millions. The largest current use for helium is a purging and pressurizing agent in liquid fueled rockets. Expanding helium provides the pressure needed to push the rocket fuel to the engines, and helium keeps the propellant mixtures at proper temperatures. Helium's lightness and non-flamability make it the safest lifting gas, its small molecular size and rarity in the atmosphere make it superb for leak detection; its non-toxicity and lightness make it valuable as a breathing mixture for underwater work; and its ability to reach low temperature make it necessary for most kinds of super-cold applications. Helium is used extensively in shielded-arc welding and gas-chromatography. It may play an important role in the development of nuclear reactors, lasers and masers, magnetohydrodynamics, superconducting magnets, and superconducting generators, motors, and cables to transmit electrical power.

The chief existing sources of helium in the United States are primarily in Kansas, Oklahoma and Texas. The threat to this supply of helium lies in the fact that the natural gas supplies are used as a domestic household fuel. Unless the helium is extracted before, the gas is delivered to the customer, it is passed to the atmosphere when natural gas is burned. To save this disappearing resource the helium conservation program was launched in 1960. The US government has operated its own helium extraction plants but also buys helium from private companies. It is transported through pipelines to underground reservoirs near Amarillo, Texas, where it is stored.

At the end of 1969 the reservoirs contained about 0.8×10^9 m^3 of helium (STP), roughly a 30 years supply at current usage.

Hydrogen gas for liquid production is obtained from a variety of sources, most frequently from hydrocarbon gases which are cracked to obtain the hydrogen. Liquefaction occurs in relatively large mass plants in the US as well as in some much smaller laboratory installations.

5.2 Low Temperature Processes

5.2.1 Handling Cryogenic Fluids

Special procedures are required for safe handling of all cryogenic fluids because of their low temperatures, their volability, and the fact that the vaporized cryogens will not support life. Additional special precautions are required for the lower temperature boiling cryogens neon, hydrogen, and helium since they will condens air from the atmosphere and must *always* be stored in closed containers with controlled venting to prevent diffusion of air into the liquid compartment. The hydrogen vapor can form explosive mixtures with the atmosphere and must always be handled with great care. The relatively high cost of neon and helium dictate careful handling to minimize losses. The mandatory requirement of vacuum insulated equipment introduces potential hazards, as does the inevitable change in physical properties of solids as they are cooled by the cryogens.

Bulletins covering the precautions and safe practices for handling liquefied atmospheric gases, liquid hydrogen and liquid helium are available from commercial suppliers of the cryogens. These bulletins should be reviewed periodically to insure safe practices by laboratory personnel. Zabetakis [7] discusses the physiological, physical and chemical hazards of cryogens as well as presenting commentary on laboratory, plant and test site safety.

5.2.1.1 Safety Precautions

The following general precautions, while phrased specifically in terms of liquid helium, cover the main consideration which should be reviewed for each intended operation with cryogens:

Cover Eyes and Exposed Skin

Accidental contact with liquid helium or cold issuing gas with the skin or eyes may cause a freezing injury similar to a burn. Protect eyes and cover the skin, where the possibility of contact with cold fluid exists.

Keep Air and Gases Away from Liquid Helium

Liquid or cold gaseous helium can solidify any other gas. Solidified gases and liquids allowed to form and collect can plug pressure-relief passages and close relief valves. Plugged passages are hazardous because of continual need to relieve excess pressure produced when heat leaks into the cold helium. Therefore, one has always to store and handle liquid helium under positive pressure and in closed systems to prevent the infiltration and solidification of air or other gases.

Keep Exterior Surfaces Clean to Prevent Combustion

Atmospheric air will condens on exposed helium-cooled piping. Nitrogen, having a lower boiling point than oxygen, will evaporate first from condensed air, leaving an oxygen enriched liquid. This liquid may drip or flow to nearby surfaces. To prevent the possible ignition of grease, oil or other combustible materials which may be in contact with the air condensing surfaces, such areas must be cleaned to "oxygen clean" standards. Any combustible foam-type organic polymer insulation should be avoided.

Pressure-Relief Devices Must be Adequately Sized

Most cryogenic fluids require considerable heat for evaporation (5.5). Liquid helium, however, has a very low latent heat of vaporization.

Consequently, it evaporates rapidly when heat is introduced, or when liquid helium is first transferred into warm or partially cooled equipments and containers, where vacuum leaks and some damage may occur. Adding appreciable heat to the liquid, pressure relief devices for liquid helium equipment must therefore be of adequate capacity to release helium vapor resulting from such heat inputs, and thus prevent excessive pressure hazards. A quantitative method for determining the helium losses due to evaporation is discussed in 5.2.3.

Keep Equipment Area Well Ventilated

Although helium is non-toxic, it can cause asphyxiation in a confined area without adaquate ventilation. An atmospheric environment which does not contain enough oxygen for breathing can cause dizziness, unconsciousness, or even death. Helium, being colorless, odorless, and tasteless cannot be detected by the human senses*, and will be inhaled normally as if it were air. Without adequate ventilation, the vented helium will displace the normal air and give no warning that a non-life-supporting atmosphere is present. Thus one must store liquid containers in well ventilated areas.

5.2.2 Transferring Cryogenic Fluids

Transfer of cryogenic liquids with low-boiling-points is accomplished by means of suitable transfer lines from one vessel to another, or from a liquefier to the load without appreciable loss of the liquid by evaporation. The liquid transfer is usually indicated and sustained by an overpressure of the gas of the same nature as the liquid.

The design of transfer tubes is governed primarily by considerations of the liquid lost during transfer. The heat of vaporization of the liquid being transferred must be large compared to the heat capacity of the inner wall of the transferline in contact with the liquid, and the heat inflow to the stream of liquid.

Liquid Nitrogen

Liquid nitrogen can be transferred from dispensing containers by an overpressure of gaseous nitrogen through insulated or bare metal tubes, or if the dispenser is small enough, even by pouring. If the liquid must be transferred to another container which is located a few meters from the storage container, the transfer tube can be a thick walled rubber tube. The tube freezes rapidly and becomes rigid and fragile during transfer. Some caution should be taken not to move either vessel.

When transferring liquid nitrogen by pressure, the overpressure in the dispensing container should be kept to a minimum to reduce "flash loss", a loss occurring through vaporization upon depressurization of a saturated liquid. Discharge from a transfer line into an open receiver

* Helium in the air will cause an increase in the pitch of the human voice, but this is not a reliable indicator of either safe or hazardous helium concentrations.

usually causes considerable splashing and loss of liquid unless some care is taken such as the use of a phase separator shown in Fig. 5.2.1.

When filling a warm container with liquid nitrogen, the initial boiling is violent. Splashing in open vessels or rapid pressure rise in closed containers is evident. As the filling proceeds, the boiling becomes less violent, the vessel surface area, however, in this stage will be covered by

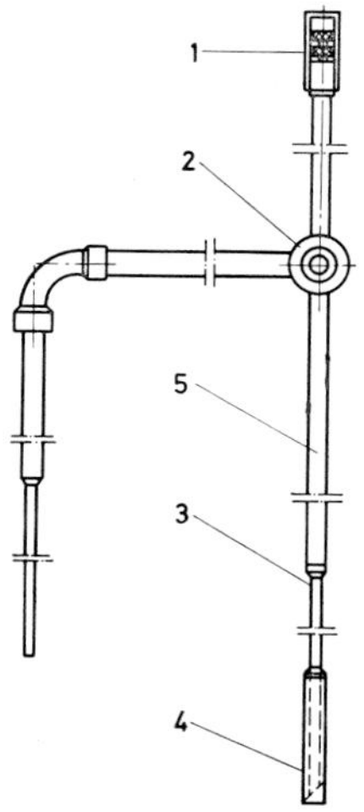

Fig. 5.2.1. Liquid He transfer line with liquid and gas phase separator to prevent splashing and liquid evaporation during transfer of liquid helium from a storage dewar to another dewar or to an experimental setup
1 Spindle for close up valve, *2* Safety valve, *3* He-tube, *4* Sinterbody (Liquid-gas-separator), *5* Vacuum jacket

a gas film. As the transfer proceeds, the gas film and the surface beneath it becomes progressively colder until the film breaks down and allows bulk liquid to contact the vessel surface. When this contact occurs, a second period of rapid boiling ensues. Splashing and a rapid pressure rise may occur again.

White clouds which frequently accompany nitrogen transfer is moisture condensed from the atmosphere not gaseous nitrogen which is colorless.

Hydrogen and Helium

The transfer of other cryogenic liquids, such as neon, hydrogen and helium, must be accomplished through insulated transfer lines to reduce heat leak. For these liquids the much lower latent heat of vaporization is an eminent factor in the efficient transfer. The design of transfer tubes requires more care than for nitrogen transfer. All the liquid helium in a container may evaporate in badly designed and not sufficiently thermally insulated transfer tubes.

The transfer line must be purged of air before liquid enters it. The inner wall of the transfer line must be precooled with the precooling vapor discharged to the atmosphere. The ball valve system (Fig. 5.4.6), greatly facilitates this purging precooling. The heat capacity of the evaporated gas also cools the tube wall. Each cm^3 of liquid helium produces gas with a heat capacity at constant pressure of 0.628 J/K, representing an enthalpy at room temperature of 167.4 J/cm^3 of liquid equivalent. To give an example we wish to transfer liquid through a stainless-steel transfer tube over a distance of 10 m. Stainless-steel has a density of $7.8\ g \cdot cm^{-3}$ and the atomic weight of 55.8. The tube has an inner diameter of 0.5 cm and a wall thickness of 0.05 cm. The amount of helium needed to cool down the tube from 300 to 4.2 K can be obtained from Fig. 6.2.24 which in this case is about 620 g or roughly 5 liters. Liquid helium requires the highest quality vacuum insulated transfer lines, because of the low latent heat of vaporization. As the viscosity of liquid helium (see Fig. 5.5.6) is very low, viscous flow losses are negligible. Since the density of the vapor is only $\frac{1}{8}$ of that of the liquid (at atmospheric pressure), the two phase flow losses in the transfer lines are also low. Nevertheless the transfer overpressure must be kept to an absolute minimum and consistent with adequate flow rate in order to minimize the flash losses. In most instances, the pressurization of some 1.33×10^3 to $2.67 \times 10^3\ N/m^2$ needed for transfer must be supplied by external means, since the heat leak into the helium vessels is kept as low as possible. A high pressure cylinder of pure gaseous helium can be used to supply warm gas at a regulated pressure or, alternatively, heat may be introduced by pulsing the vapor space with the air of a flexible bladder connected with the vapor space. Because the heat leak into transfer lines is usually much higher than in any other of the helium handling apparatus, the transfer times should be short, e.g. in a matter of minutes rather than hours. Where it is desirable to transfer liquid of a slow rate, the lines should be designed to have a very low heat leak.

In all cases, the vacuum jacket on the transfer lines should extend to the end of the line, otherwise the heat transfer from the surrounding vapor will reduce or eliminate liquid transport. This is especially important for transfer into warm apparatus, since the rapidly expanding helium vapors will have a considerable velocity and thus high local heat transfer values.

The rapid expansion of helium upon warming and the consequent high velocity will also cause high entrainment of the incoming liquid unless precautions are taken to aid in liquid collection. One of these is to extend the transfer line to the bottom of the receiving vessel and position it so that the liquid stream does not impinge on a massive object which may be warm and thus have a high heat capacity.

In addition to proper location of the liquid discharge, the apparatus should be designed to allow cold helium vapors to pass over the surface of the most massive parts, as the helium vapors contain some 75 times more total refrigeration than can be obtained by vaporization of the liquid. In order to achieve proper transfer, the vaporized helium must be vented from the apparatus at a sufficiently high rate to avoid back pressure which will reduce the liquid inflow. Frequently, this makes it desirable to have a large vent line for initial fill than for subsequent "topping" of the liquid where there is much less vaporization. For this "topping" operation, the transfer line should not extend into the liquid already present in the apparatus because the inflowing liquid-vapor-mixture will tend to splash it around, causing much liquid loss. A method of "initial filling" and topping a cryostat containing a superconducting magnet is illustrated in Fig. 5.2.2. Helium flows at the initial cool down through the tube to the bottom of the magnet. The evaporated cold gas passes through the internal helium passages in the coil and over the coil surface. The latent heat of the gas is utilized as described in section 5.5.1 and as shown in Fig. 5.5.3. When the liquid helium has reached its final

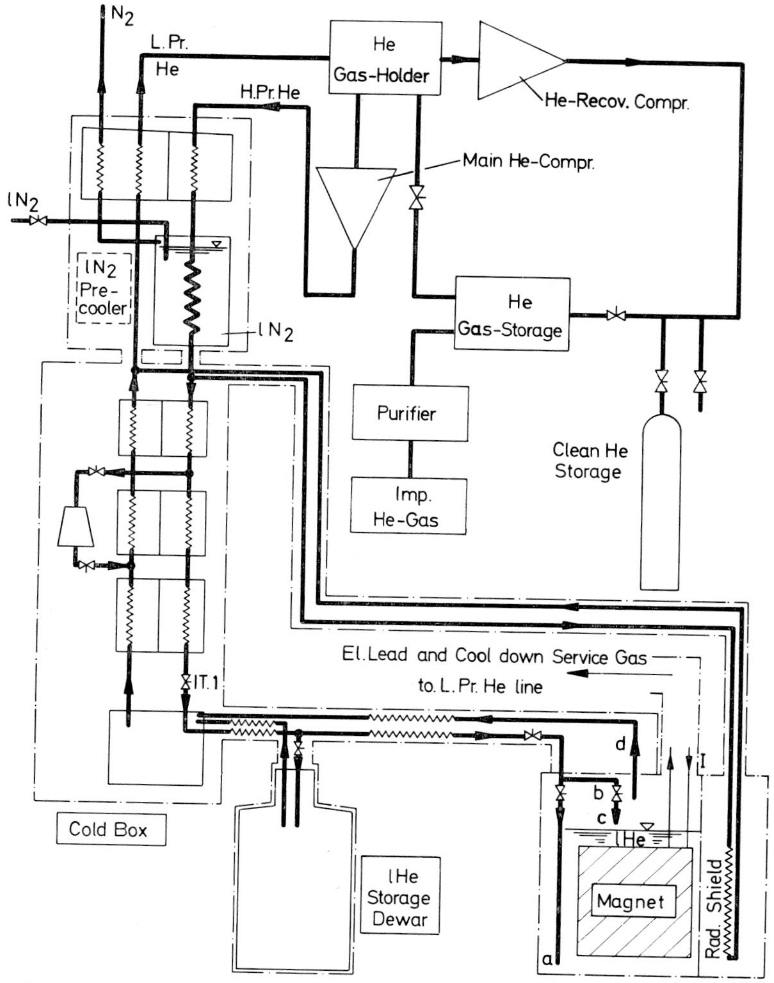

Fig. 5.2.2. Magnet and He refrigerator assembly. *a* Bottom He fill line, *b* Top He fill line, *c* Phase separator, *d* He gas vent line

level, indicated by a level gage, a valve closes this tube and the transfer line. Subsequent toppings happen through line ending on the top of the magnet. In order not to evaporate liquid helium through flashing, a phase separator or a gas breaker (Fig. 5.2.1) is mounted at the end of this tube.

Two level gages indicating maximum and minimum tolerable helium levels are generally used for automatic topping of the helium.

Transfer lines with flexible sections as shown in Fig. 5.2.3 make adjustments for bottom and top fill relatively easy. These transfer lines, however, are only suitable for small helium volumes. Flexible transfer lines are also more vulnerable than rigid ones and more "lossy". Loss values encountered for helium transfer lines are in the order of 0.2 to 1 W per m length.

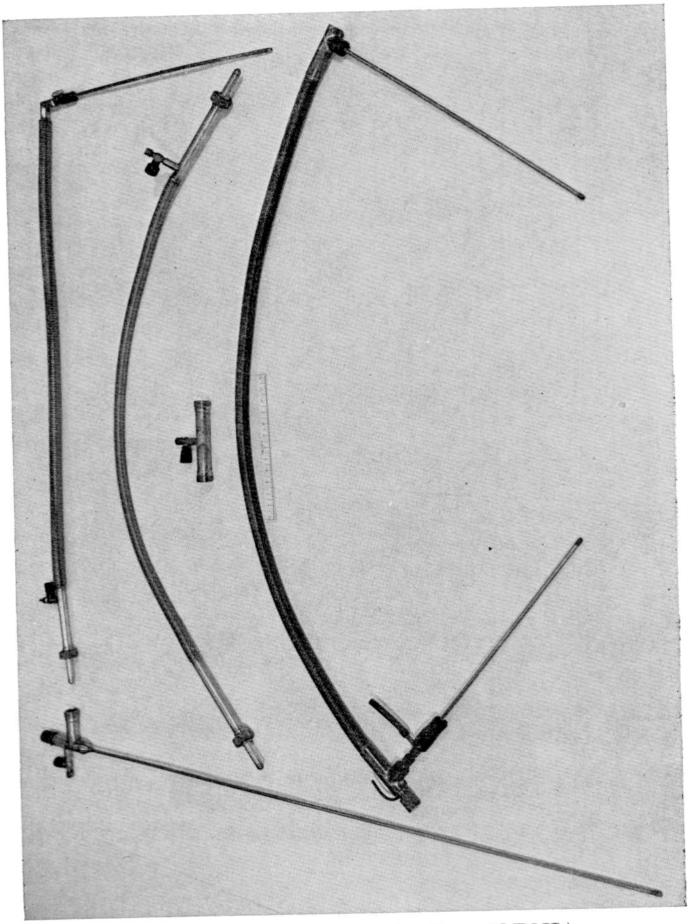

Fig. 5.2.3. Elastic transfer lines (ORNL)

Another method used with rigid transfer lines is to have the dispensing vessel on an adjustable platform. This method is cumbersome and only useful for small experiments. Modern transfer lines utilize superinsulation and a high vacuum around the inner filling tube. Suction valves must be provided to evacuate the room between the inner and outer tube. Nitrogen shields are used only in very long and large diameter transfer tubes ($\geqq$ 5 cm inner tube diameter) for permanent installations. Needle or ball valves are installed in practically all transfer lines and are used for initial purging, automatic helium transfer, etc. Reinforcing elbows and sections prevent cracking or damage of the rather delicate transfer lines.

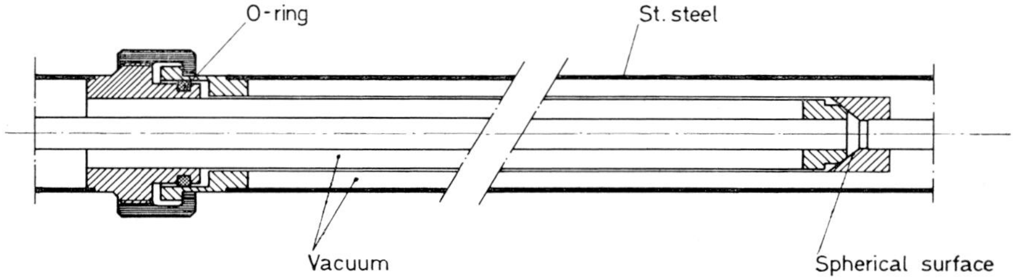

Fig. 5.2.4. Coupling unit between transfer line sections for liquid He

Sometimes, specifically in laboratories, two or more transfer lines must be coupled. Junctions are used for permanent or temporary connection of two transfer lines. The junctions, such as shown schematically in Fig. 5.2.4, are unfortunately lossy (in average about 1 Watt at 4.2 K per junction) and not useful for permanent installations. More detailed designs of transfer tubes are given by White [8]. Shipment and transfer of liquid helium is described in: "Technology of Liquid Helium" [9].

5.2.3 Liquid Level Measurement

5.2.3.1 Introduction

Measurement and control of the liquid level in a container or dewar have several reasons. In experimental dewars, the magnet is immersed in liquid. If the helium in the container is too low, upper portions of the superconducting magnet may not be covered by the liquid, or as due to liquid evaporation during transfer the level is even more lowered. The superconductor may be exposed to either somewhat higher temperatures in the dense liquid gas layer on top of the liquid, or the heat generated in pulsed superconducting magnets is not absorbed by the dense gas due to poorer specific heat. In the first case, the current carrying capacity of the magnet is reduced at higher temperatures (Fig. 1.6). In the second case the superconductor is not cooled down properly and may degrade.

In storage vessels the knowledge of the amount of helium in a dewar is self-evident. Although "dip sticks" are used in laboratories to check the amount of helium available in storage vessels prior to helium transfer, the method is cumbersome and some helium is evaporated each time the level is measured.

For continuously operating magnets, or superconducting devices, the automatic level control is a must. The method applied has to be absolutely reliable, as the information transmitted to the cryosystem to transfer helium to the experimental dewar must be given when a certain preset minimum level is reached. Helium transfer must stop when an upper level is attained.

5.2.3.2 Methods of Level Measurement

Many methods of level measurement are described in literature [10, 11, 12, 13]. In storage vessels the differential pressure due to the hydrostatic head of liquid in the container is measured. A non-volatile manometric fluid such as vacuum oil or dibutyl phtalate is required.

Another method of level indication is based on the considerable difference between the dielectric constants of liquids and vapors (Fig. 5.18). Two concentric open ended cylinders or parallel plates having a varying capacity with the liquid level are placed in the dewar. This method is not sensitive enough if heavy boiling occurs or during liquid transfer.

Heat transfer from hot solid surfaces to the liquid is much greater than to a gas or vapor. Thus, a vertical wire carrying a current has a higher temperature in the gasphase than in the liquid. For measuring continuous changes of the liquid level the total resistance of the heated wire carrying a small electric current is measured. To measure liquid helium levels superconducting wires are imployed. The resistance of the immersed section of the superconductor is zero. Metal wires used as level gages are tantalum, lead and niobium. The sensitivity of heated metal wires are somewhat to be desired and thus a chain of carbon resistors, capacitors, or the use of superconductors Type II is preferred.

In large storage dewars float and displacement type instruments are used as level gauges.

Transducers which respond to the change in one physical property in the liquid vapor interface to provide a varying output signal are becoming common as liquid helium detectors. In the following we list a few:

Resistive type sensors: The difference in the heat transfer rate which occurs between the sensor when immersed in liquid or when in vapor produce a temperature and corresponding resistance change in the sensing element. The resistance change is sudden and when reaching a threshold value it activates an output signal.

Capacitive type sensors: The difference in dielectric constant between vapor and liquid produces a change in the sensor capacitance. This change is detected and amplified to provide a usable signal.

Optical type sensors: The change in refractive index between liquid and vapor around an optical prism provides a variation in internally reflected light intensity. The light intensity is measured by a sensor cell whose varying output is detected and amplified.

Piezoelectric type sensors: The change in acoustic damping which occurs when the medium changes from liquid to vapor causes a change in energy dissipation in the resistive component of a crystal's equivalent impedance. The Q-value of the circuit decreases with the crystal in liquid, oscillations are damped. Circuit oscillation is detected to provide an output signal.

Magnetostrictive type sensors: These sensors also use the acoustic damping difference between liquid and vapor. A driving coil produces an oscillating magnetic field around a tubular magnetostrictive element. The element elongates and contracts at ultrasonic frequencies. A separate coil senses the motion of the element and provides a positive feedback signal to sustain circuit oscillation. When the element is restrained (e.g. when in liquid), the feedback signal is lost and oscillation stops. A detector circuit rectifies the oscillations to drive an output device.

Vibratory type sensors: The difference in viscous damping between liquid and vapor provides a variation in oscillation of a mechanical paddle which is driven by an oscillating solenoid slug. A similar slug mechanically linked to the paddle and oscillating in the magnetic field of a pick up coil produces a varying voltage that is converted to a usable output.

In order for a sensor to be useful as a liquid level detector the following conditions are to be met:

1. The average input power to the sensor must be low enough so that the temperature rise in the fluid is negligible.

2. The difference in the sensor response between liquid and gas phase must be detected as soon as possible.

3. The power pulse repetition rate must be as fast as possible in order to detect the liquid level in a dynamically changing situation.

4. When the sensor is exposed from a liquid to a gas phase, a small amount of liquid adhering to the sensor must be removed as fast as possible either by surface tensions or by evaporation with heat from the sensor power input.

All conditions are not compatible and compromises must be made. Condition (1) is met at a peak power level of 0.1 W/cm^2 at the film surface. Condition (2) is achieved in a few tenths of a second at a pulse repetition rate of 1 Hz. Condition (4) is realized by vaporizing the liquid in a maximum time of 10 seconds in normal gravity. This means that the real response time of the level gage is as high as 10 sec going from liquid to gas and about 1 sec going from gas to liquid. If faster pulse repetion rates of 10 to 20 Hz or so are utilized, the power level can be increased to 1 W/cm^2. In this case condition (4) is satisfied by vaporizing the liquid in less than 1.5 sec in normal gravity and about 3 sec in low gravity.

5.3 Liquefaction and Refrigeration

As in all chemical processes, low temperature processes are based on thermodynamic principles. Low temperature processes are, however, unique in the following respects:

(a) Mixtures of simple chemical species are involved. Thus estimates of thermodynamic properties are quite reliable.

(b) No chemical reactions are usually involved.

(c) Cryogenic processes can be designed precisely. Quantitative optimization is possible.

(d) Cryogenic designs require the development and maintenance of refrigeration units, that is, heat sinks at reduced temperatures.

(e) Successful optimization can be based on fundamentally calculated minimization of lost work in various processing steps.

(f) Safe, continuous operation requires the removal of impurities to extremely minute concentration levels.

(g) Construction materials must be carefully chosen to operate reliably between ambient temperature and the cryogenic temperature at which the cryogenic process occurs.

In this section we deal with refrigeration processes, units, and minimization of lost work, taking into account restrictions imposed by the properties of liquids. However, as superconducting magnets are generally operating at liquid helium levels we restrict the following sections to helium refrigeration and helium liquefaction.

There are three methods available for cooling a gas and thus developing refrigeration:

(i) Isobaric Cooling

The cooling of a gas by exchange of heat with a colder gas or fluid. This is a feature of all low temperature processes. However, no net refrigeration is produced.

(ii) Isenthalpic Expansion

The throttling of a fluid from high pressure to lower values with no work produced, is utilized. The expansion occurs at constant enthalpy, resulting in no temperature change for an ideal gas. A "real" gas may cool down or heat up during such a process depending on the sign of the Joule-Thomson coefficient:

$$\alpha_{JT} = \left(\frac{\partial T}{\partial p}\right)_H = -\left(\frac{\partial H}{\partial p}\right)_T \cdot \frac{1}{c_p} = \frac{1}{c_p}\left[T\left(\frac{\partial v}{\partial T}\right)_p - v\right]. \qquad (5.3.1)$$

(iii) Isentropic Expansion

Expansion of the fluid through some device (engine) to produce work, ideally occurs at constant entropy. This is an effective method of producing refrigeration.

These three methods may be represented on a thermodynamic diagram, such as a pressure enthalpy (p-H) diagram (Fig. 5.3.1.).

Constant pressure cooling is shown by the path AB, constant enthalpy expansion (valve) by either line AC or line BD. Constant entropy expansion (engine, turbine) by a line AE.

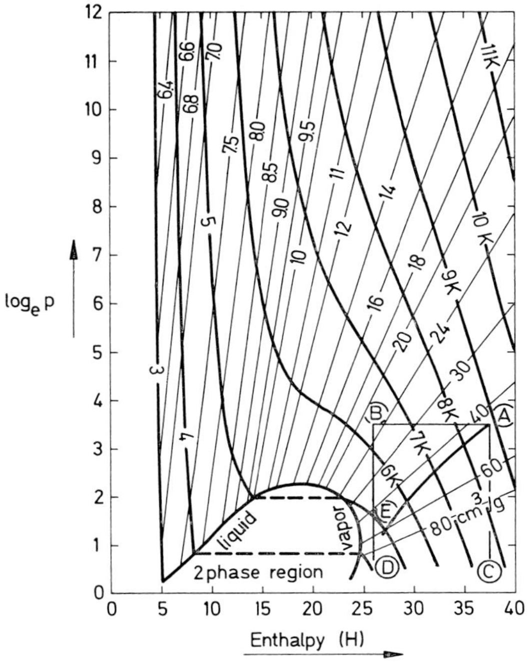

Fig. 5.3.1. Schematic representation of producing refrigeration, using a p-H-diagram:

$\overline{AB}$	Constant pressure cooling
$\overline{AC}$ or $\overline{BD}$	Constant enthalpy (valve) expansion
$\overline{AE}$	Constant entropy expansion (engine or turbine)

The Joule-Thomson expansion can produce either a temperature drop or a temperature rise, whereas the constant entropy expansion always produces a temperature drop.

A number of machines are commercially available in a range of sizes for single purpose duty as liquefiers or as refrigerators, or for combined services to provide liquefaction and refrigeration simultaneously. In large laboratories or large single installations the refrigeration plant may provide cooling at several temperature levels. In addition to providing refrigeration at various temperatures it can also liquefy helium.

Descriptions are given in literature and by manufacturers for machines suitable for superconducting magnet service to provide refrigeration from a few watts to several kilowatts at temperatures which range from 1.8 K up to 6 K. While the thermodynamic and mechanical designs of these machines become quite involved, particularly when multiple service is required, the basic principles are as seen above, straight foreward. Nearly all cryogenic processes in these machines use two or three of the above mentioned refrigeration sources. All practical machines, used in conjunction with superconducting magnets, employ helium gas as the working fluid, sometimes in combination with other fluids, such as nitrogen, even if for precooling purposes.

Table 5.3.1. *Minimum reversible work to liquefy N_2, H_2, and He at 1 kp cm^{-2}*

Gas	Pressure (kp cm^{-2})	T_2 (K)	T_1 (K)	ΔS (J g^{-1} K^{-1})
He 4	1	300	4.214	27.9
Hydrogen (normal)	1	300	20.39	53.3
Hydrogen (para)	1	300	20.268	62.5
Nitrogen	1	300	77.3	4.02

		ΔH (J g^{-1})	W_{liq} (J g^{-1})	W_{liq} (kJ/l^{-1})
He 4	1	1565	6,805	855
Hydrogen (normal)	1	3990	12,000	840
Hydrogen (para)	1	4530	14,220	995
Nitrogen	1	435	771	623

5.3.1 Basic Principles, Reversible Cycles

Refrigeration with a gaseous working fluid is achieved by successive operation of compression, heat transfer and expansion of the gas. These operations are collectively known as the operating cycle of the refrigerator. Work is done on the gas during compression, with the heat being rejected to the surroundings. Work is then performed by the compressed gas during expansion with heat being received from the refrigerated space or object. The work required to accomplish the refrigeration is given by the summation of the work done on and by the gas. The *specific work*, the ratio of the work required to accomplish the refrigeration, to the refrigerative effect is given by:

$$w = \frac{\Sigma W}{Q} = \frac{W_{comp} + W_{exp}}{Q}. \tag{5.3.2}$$

The specific work is used to evaluate a refrigerator or a refrigeration cycle. According to the second law of thermodynamics and the Carnot principle, the minimum possible value of the specific work required by an ideal cycle comprised of reversible, loss free operation is

$$\boxed{w_{rev} = \frac{T_0 - T_c}{T_c},} \tag{5.3.3}$$

with T_0 the temperature of heat rejection or compression and T_c the temperature of the heat load or refrigeration.

Fig. 5.3.2 is a plot of values of w_{rev} vs. T_c values ranging from 1 to 300 K, with T_0 chosen suitably to be 300 K. For the region of our particular interest for superconducting magnets (1.85 to 6 K) the ideal refrigerator would require from 161 to 49 Watts for each watt of refrigeration.

The ideal minimum specific work, while derived usually for the Carnot cycle consisting of alternative isothermal and isentropic steps, applies also to other ideal reversible cycles. Two of these, the Ericsson cycle

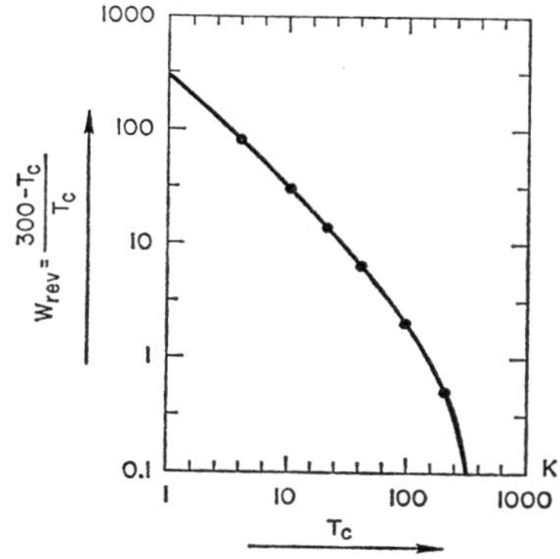

Fig. 5.3.2. Plot of reversible work versus temperature of a test load

and the Stirling cycle are shown along with the Carnot cycle on p-V and T-S diagrams in Fig. 5.3.3. All of these ideal cycles are alike in that they alternate isothermal (constant temperature) steps the difference being in the other step; isentropic (constant entropy) for Carnot cycle, isobaric (constant pressure) for the Ericsson cycle, and isochoric (constant volume) for the Stirling cycle.

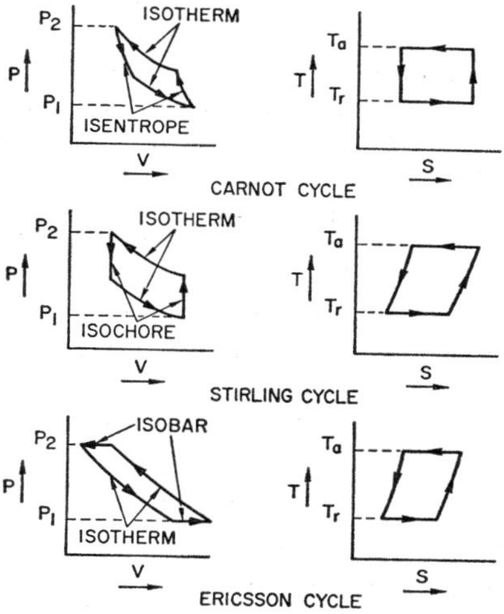

Fig. 5.3.3. Idealized p-V and T-S diagrams of a Carnot, Stirling, and Ericson cycles

To analyze the Ericsson and Stirling cycles, we first note that the isobaric and isochoric steps are heat transfer steps which balance each other and are assumed to be ideal with no losses. Therefore, the analysis need be concerned only with the isothermal compression steps.

If we consider an ideal machine with isothermal compression at temperature T_a of m moles per second of an ideal gas from pressure p_1 to p_2, we have

$$w_{\text{comp}} = \dot{m} \int_{p_1}^{p_2} p\, dv = \dot{m}\, k T_a \ln\left(\frac{p_2}{p_1}\right) \tag{5.3.4}$$

and for isothermal expansion at temperature T_r of the same quantity of gas

$$w_{\text{exp}} = \dot{m} \int_{p_1}^{p_2} p\, dv = \dot{m}\, k T_r \ln\left(\frac{p_2}{p_1}\right). \tag{5.3.5}$$

The refrigeration available is the heat required to maintain the expanding gas isothermal at temperature T_r, so that

$$Q = -\dot{m} \int_{p_1}^{p_2} p\, dv = \dot{m}\, k T_r \ln\left(\frac{p_2}{p_1}\right). \tag{5.3.6}$$

We have for the specific work of this ideal machine,

$$w_{\text{ideal}} = \frac{W_{\text{comp}} + W_{\text{exp}}}{Q} = \frac{\dot{m}\, k T_a \ln\left(\frac{p_2}{p_1}\right) - \dot{m}\, k T_r \ln\left(\frac{p_2}{p_1}\right)}{\dot{m}\, k T_r \ln\left(\frac{p_2}{p_1}\right)} = \frac{T_a - T_r}{T_r}. \tag{5.3.7}$$

This expression for the ideal work is the same as that presented for the ideal Carnot machine by Eq. (5.3.3). Real machines, will depart from the ideal and require greater specific work.

We have so far considered isothermal refrigeration only, but in many instances the refrigeration will be received over a temperature range T_1 to T_2. The evaluation of non-isothermal gas refrigeration in terms of the Carnot specific work has been discussed by Jacobs [14]. He assumes that an ideal non-isothermal refrigerator can be replaced by a series of ideal Carnot machines and thus for one incremental machine,

$$\left.\frac{dW}{dQ}\right)_T = \frac{T_0 - T}{T}. \tag{5.3.8}$$

Since the refrigeration dQ is supplied by gas flow $\dot{m}$ being warmed over a temperature interval dT, we have

$$dQ = \dot{m}\, c_p dT, \tag{5.3.9}$$

where c_p is the mean value of the specific heat over the interval dT.

From Eqs. (5.3.8) and (5.3.9) we have

$$dW = \dot{m}\, c_p \frac{T_0 - T}{T} dT. \tag{5.3.10}$$

If we assume c_p to be constant over the operating range T_1 to T_2 of the refrigerator, then we have

$$W = \dot{m}\, c_p \int_{T_1}^{T_2} \frac{T_0 - T}{T}\, dT = \dot{m}\, c_p \left[T_0 (\ln T_2 - \ln T_1) - (T_2 - T_1)\right]. \quad (5.3.11)$$

From Eq. (5.3.9), we have for the heat supplied

$$Q = \dot{m}\, c_p \int_{T_1}^{T_2} dT = \dot{m}\, c_p\, (T_2 - T_1) \quad (5.3.12)$$

and combining Eqs. (5.3.11) and (5.3.12), the specific work is

$$w_{\text{ideal}} = \frac{W}{Q} = \frac{T_0 (\ln T_2 - \ln T_1)}{T_2 - T_1} - 1 = \frac{T_0}{T_2 - T_1} \cdot \frac{1}{\ln\left(\frac{T_2}{T_1}\right)} - 1 . \quad (5.3.13)$$

Since the denominator of Eq. (5.3.13) is the logarithmic mean temperature (LMT) between T_1 and T_2, the expression may be written

$$\boxed{w_{\text{ideal}} = \frac{(T_0 - LMT_c)}{LMT_c},} \quad (5.3.14)$$

which is the same form as the Carnot specific work but with the log mean temperatures of the refrigeration load substituted for the refrigeration temperature. As pointed out by Jacobs, Eq. (5.3.14) reduces to Eq. (5.3.3) when the temperature interval approaches zero.

Non-isothermal refrigeration can be employed to meet most refrigeration requirements, provided that the highest temperature of the refrigerant T_2 is less than the maximum temperature allowed for the intended use. Thus, for superconducting systems T_2 must certainly be less than the highest permissible transition temperature of the superconductor and, therefore, the mean temperature of a non-isothermal refrigerator must be less than the temperature of an isothermal refrigerator to accomplish the same refrigeration task. By comparing Eq. (5.3.14) with Eq. (5.1.3), the non-isothermal specific work must therefore be higher than for the isothermal machine.

The specific work term has been adopted for use in the evaluation of cryogenic processes, since it is a direct measure of the effort required to achieve the low temperatures. The product of the specific work and thermal dissipation or heat inleakage gives the direct energy cost of the process involved for comparison with similar processes at other temperatures. The refrigeration industry has adopted the inverse of the specific work term, the coefficient of performance,

$$COP = \frac{Q}{\Sigma W}, \quad (5.3.15)$$

since this expression is in a convenient form for assessing the refrigeration capability of different refrigerators in terms of the power input.

Another very simple process is the Linde cycle, originally used to liquefy air. Fig. 5.3.4 shows this process schematically. The process is also based on a p-H diagram. In this process refrigeration is produced in the heat exchanger (steps 7 to 8) and in the Joule-Thomson expander (step 8 to 9). An enthalpy balance around the heat exchanger and liquid re-

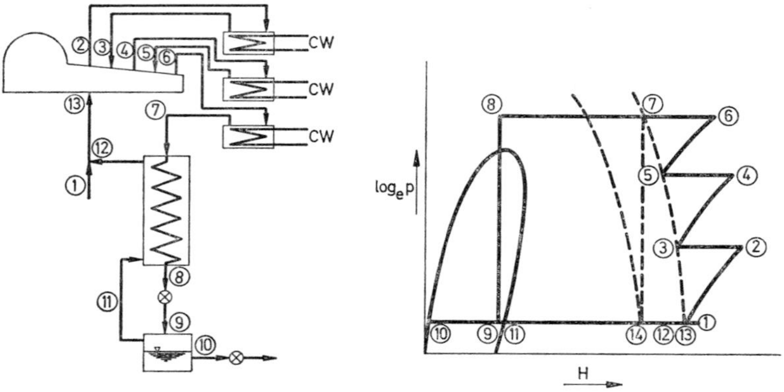

Fig. 5.3.4. Process of Linde Liquefaction; Refrigeration produced in steps 7—8; Joule-Thomson expansion in steps 8—9

ceiver can be used to determine the fraction (x) of gas liquefied in this process:

$$H_{10} \cdot x = H_7 - H_{12}(1 - x) \,, \tag{5.3.16}$$

with H = total enthalpy of the designated stream,
x = weight of the gas liquefied to weight of the gas compressed

$$x = \frac{H_{12} - H_7}{H_{12} - H_{10}} \,. \tag{5.3.17}$$

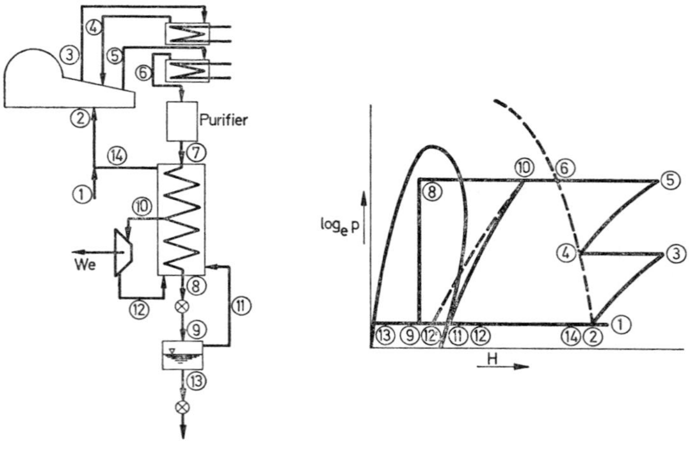

Fig. 5.3.5. Claude liquefaction cycle

For x to be positive H_{12} must be greater than H_7, even if the heat exchanger stipulations require H_{12} to be less than H_7. This process will produce liquid, only if the Joule-Thomson-effect at the *warm* end of the exchanger is positive, and then only if the exchanger is large enough to operate with a warm end ΔT less than $(T_{13} - T_{14})$. This warm end restriction is characteristic of the process, which generates refrigeration solely from internal heat transfer and Joule-Thomson expansion.

The standard compression-refrigeration cycle also uses Joule-Thomson expansion and heat exchange at constant pressure to generate refrigeration. Modified for low temperature refrigeration, it is identical to the Linde process except that all liquid formed is vaporized by an external warm stream, instead of being withdrawn.

The *Claude cycle* uses an expansion engine to remove energy as work and thus develope refrigeration. The schematic Claude diagram is shown in Fig. 5.3.5 along with the process path on p-H coordinates. Ideally the expansion (step 10 to 12) would be isentropic. The expander inlet temperature must be chosen such that the outlet condition has an acceptable quality, or fraction vapor. Since the expansion is adiabatic we have:

$$-W_{\text{exp}} = H_{12} - H_{10}\,. \tag{5.3.18}$$

The expander efficiency is given by:

$$\eta_{\text{exp}} = \frac{W_{\text{exp}}}{W_{\text{exp, ideal}}} = \frac{H_{12} - H_{10}}{H'_{12} - H_{10}}, \tag{5.3.19}$$

where H'_{12} = enthalpy at final pressure and expander inlet entropy.

The fraction of gas liquefied can be obtained from an enthalpy balance around the exchanger and liquid collector:

$$H_7 - y \cdot H_{12} = (1 - x)\, H_{14} + y \cdot H_{10} + x \cdot H_{13} \tag{5.3.20}$$

$$x = \frac{(H_{14} - H_7) + y\,(H_{10} - H_{12})}{H_{14} - H_{13}}, \tag{5.3.21}$$

with x = fraction of compressed gas that is liquefied,
y = fraction of compressed gas that is withdrawn to the expander.

The first term of Eq. (5.3.21) is the same as Eq. (5.3.17) and is generally negative.

The restriction of the Joule-Thomson-Effect is at the warm end of the exchanger.

The fraction of gas that can be withdrawn to the heat exchanger is limited by the ΔT of the heat exchanger based on an analysis of the heating and cooling paths through the exchanger. Fig. 5.3.6 shows cooling curves typical of Linde and Claude liquefaction cycles. For either process the returning low pressure gas would follow a path such as EF. Since this is a constant pressure process we have:

$$dH = \delta Q = -\dot{m}\, c_p \delta\, dT \tag{5.3.22}$$

and

$$\frac{dT}{\delta Q} = -\frac{1}{\dot{m}\, c_p \delta}. \tag{5.3.23}$$

Since the massflow $\dot{m}$ is constant and we can assume that the heat capacity $(c_p \delta)$ is nearly unchanged, the slope of the line EF is approximately constant. Line AC is typical of the high pressure gas cooling curve for a Linde liquefier. Lines ABD show the effect of removing high pressure gas to feed the turbine of a Claude cycle machine. The discontinuity

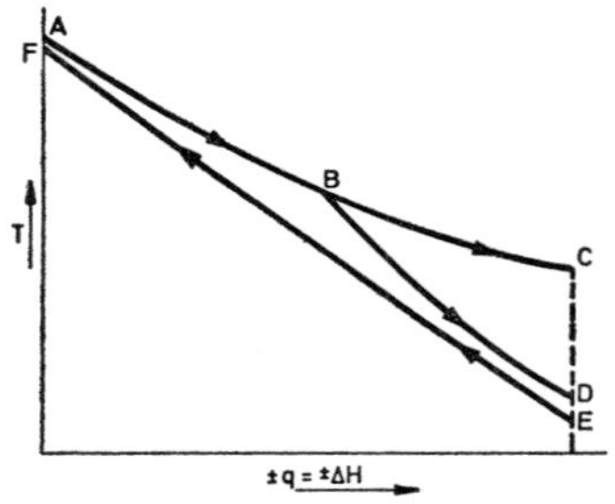

Fig. 5.3.6. Cooling curves typical of Linde and Claude liquefaction cycles

at B results from the reduction in the mass of the gas flowing through the exchanger below the turbine withdrawal point. The more gas that is withdrawn, the steeper BD becomes. The vertical line CE represent the temperature change ΔT, which is the driving force for the heat transfer at the appropriate point in the heat exchanger.

The work lost in a heat exchanger is also obtained from the cooling curves. Calculating the total entropy generation

$$\Delta S_T = \Delta S_H + \Delta S_C = \int\limits_H \frac{\delta Q}{T} + \int\limits_C \frac{d Q}{T} = \int\limits_H \frac{dH}{T} + \int\limits_C \frac{dH}{T}$$

$$\Delta S_T = \int \frac{T_H - T_c}{T_H \cdot T_c}\, dH\,. \tag{5.3.24}$$

Assuming $T_H - T_c = \Delta T$ and $T_H \approx T_c$ which is only valid if ΔT is small, than the work lost is:

$$W_{\text{lost}} = T_0 \int \frac{\Delta T}{T^2}\, dH\,. \tag{5.3.25}$$

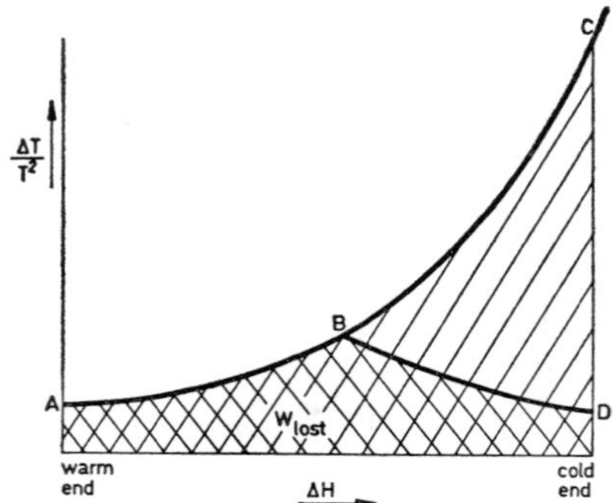

Fig. 5.3.7. Lost work in a heat exchanger

By using cooling curves similar to lines AC and EF of Fig. 5.3.6 and plotting $\Delta T/T^2$ versus H we obtain Fig. 5.3.7. The large temperature change ΔT at the end of the heat exchanger is the major contributor to the work lost. Methods to reduce W_{lost} are illustrated schematically in Fig. 5.3.8.

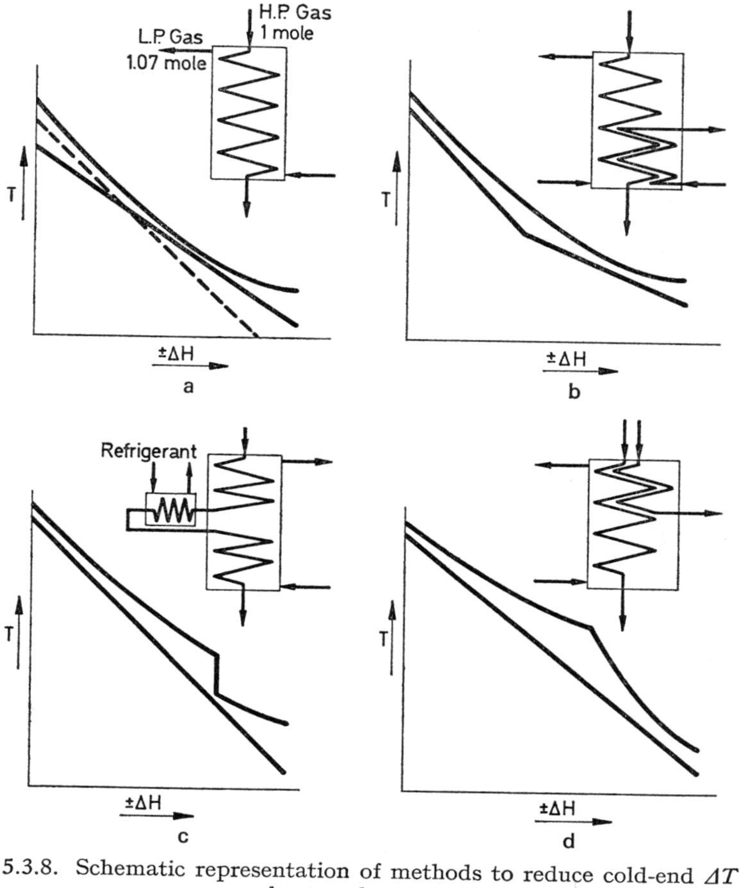

Fig. 5.3.8. Schematic representation of methods to reduce cold-end ΔT in heat exchangers
a The quantity of low pressure return gas is increased. The reduction in cold-end ΔT is partly off-set by the increase in warm-end ΔT, *b* Additional cold stream is fed to the exchanger and is withdrawn; it is partly reheated, *c* External refrigeration part is added in small quantities, *d* High pressure unbalance is added to the system

5.3.2 Efficiency of Real Cycles

Neither the specific work nor the COP term can properly be called an efficiency, if we adhere to the convention that efficiencies cannot exceed 100%. Even so, in some cases the specific work has been referred to as the "refrigeration efficiency". A more proper assessment, which cannot exceed 100%, is to compare actual machines to the Carnot cycle or to a completely reversible cycle employing the process steps of the cycle. The comparison with the Carnot cycle yields a figure of merit termed the "Percent Carnot" given by the expression,

$$\text{Percent Carnot} = [w_{\text{rev}}/w_{\text{act}}] \cdot 100 \tag{5.3.26}$$

$$\boxed{\text{Percent Carnot} = \frac{(T_0 - T_c)}{T_c} \cdot \frac{1}{w_{\text{act}}} \cdot 100\,.} \tag{5.3.27}$$

A review of the operating power requirements of a large number of refrigerators has been made by Strobridge [15]. From this review, he obtained values of Percent Carnot which he found to be strongly related to the refrigeration capacity and relatively independent of the operating temperature level. His results were summarized in Fig. 6.2.3, where the Percent Carnot for refrigerators and liquefiers operating in three temperature ranges is plotted versus the capacity in watts. A median curve is drawn for use in a general assessment of power requirements for refrigeration.

Thus, the power requirement for a given amount of refrigeration Q, at temperature T_c with heat rejection at temperature T_0 is obtained by finding the Percent Carnot η_Q for this value of Q from Fig. 6.2.3 and employing Eq. (5.3.27) in the form,

$$w_{\text{act}} = 100\,(T_0 - T_c)\,T_c\,\eta_Q \tag{5.3.28}$$

to obtain the specific work, The power is then

$$W_{\text{act}} = w_{\text{act}} \cdot Q\,. \tag{5.3.29}$$

The value of w_{act} obtained from Eq. (5.3.28) may be lower than the actual total specific work for some refrigerators in Strobridge's compilation since the work required to produce the liquid pre-coolants used in these machines has not been included in the calculation of the actual specific work. For a more accurate assessment, this work should be evaluated as discussed in Section 6.2.3 and then added to the compressor work used in Eq. (5.3.27) to determine the percent Carnot. Such an evaluation may indeed reduce the scatter in Fig. 6.2.3 and yield a more definitive curve for assessment of refrigerator power requirement.

The liquefiers included in Strobridge's compilation have been converted to equivalent refrigerators by the method presented above. While the compilation by Strobridge is valuable as a guide to power requirements it does not let one to assess the effect of component inefficiencies on the performance of a refrigerator. A detailed discussion of this assessment is beyond the scope of this chapter but a valuable expression, due to Grassman [16] for the effect of compressor and expander efficiency on the overall efficiency of ideal Brayton cycle (non-isothermal) refrigerators will be developed.

The Brayton cycle is shown in Fig. 5.3.9 with the ideal cycle on the left, the components in the middle and the real cycle with inefficiencies on the right. The ideal isothermal compressor supplies compressed gas to the heat exchangers, which have ideally zero transverse temperature and longitudinal pressure differentials and thence to an ideal reversible adiabatic (isentropic) expander which delivers cold low pressure gas to the refrigeration load where it is warmed and reintroduced to the heat exchangers for further warming before compression.

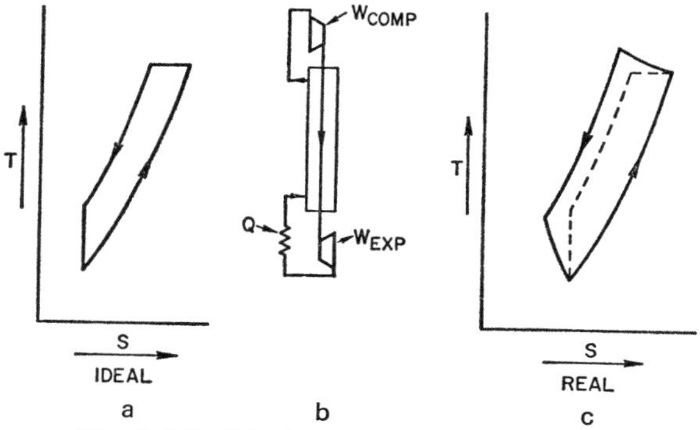

Fig. 5.3.9. Ideal and Real Brayton cycles
a Ideal, *b* Components, *c* Real cycle

For the analysis, we assume that the working fluid is an ideal gas and that there are no heat exchanger losses or heat inleakage into the system other than the refrigerated load.

In the ideal Brayton cycle, we have the compressor work given by the isothermal work from p_1 to p_2 with heat rejection to the atmosphere at T_a,

$$W_{\text{comp}} = \dot{m}\, kT_a \ln\left(\frac{p_2}{p_1}\right) \tag{5.3.4}$$

and if an actual compressor has an isothermal efficiency η_{is} then the real work of compression is

$$\boxed{W^*_{\text{comp}} = \dot{m}\, kT_a \cdot \ln\left(\frac{p_2}{p_1}\right) \cdot \frac{1}{\eta_{\text{is}}}\,.} \tag{5.3.30}$$

The refrigeration is produced by an expander whose ideal work is given by the reversible adiabatic (isentropic) relationship,

$$W_{\text{exp}} = \dot{m} \Delta H = \dot{m}\, c_p \cdot (T_1 - T_2)\,. \tag{5.3.31}$$

We have seen that this refrigeration can be considered to be supplied at a mean temperature given by the logarithmic mean value between T_1 and T_2,

$$T_m = \frac{T_2 - T_1}{\ln\left(\frac{T_2}{T_1}\right)}, \tag{5.3.32}$$

so that we have, upon substituting for $(T_1 - T_2)$,

$$W_{\text{exp}} = -\dot{m}\, c_p T_m \ln\left(\frac{T_2}{T_1}\right). \tag{5.3.33}$$

But, for isentropic expansion

$$\frac{T_2}{T_1} = \left(\frac{p_2}{p_1}\right)^{\frac{c_p - c_v}{c_p}}, \tag{5.3.34}$$

so that

$$W_{\text{exp}} = -\dot{m} c_p T_m \frac{c_p - c_v}{c_p} \cdot \ln\left(\frac{p_2}{p_1}\right) \tag{5.3.35}$$

$$= -\dot{m} k T_m \ln\left(\frac{p_2}{p_1}\right) \tag{5.3.36}$$

and if the machine has an adiabatic efficiency of η_{ad}, then the work output is

$$\boxed{W^*_{\text{exp}} = -\eta_{\text{ad}} \cdot \dot{m} k T_m \ln\left(\frac{p_2}{p_1}\right),} \tag{5.3.37}$$

and for the total work we have

$$W^*_{\text{tot}} = W^*_{\text{comp}} + W^*_{\text{exp}}$$

$$\boxed{W^*_{\text{tot}} = \dot{m} k \left[\frac{T_a}{\eta_{\text{is}}} - \eta_{\text{ad}} T_m\right] \ln\left(\frac{p_2}{p_1}\right).} \tag{5.3.38}$$

We now define an efficiency as the ratio of the ideal work to the actual work,

$$\eta = \frac{W_{\text{id}}}{W^*_{\text{tot}}}, \tag{5.3.39}$$

where the ideal work is that required by a reversible machine to produce the same output as the actual machine. Since the expander of the actual machine has an efficiency of η_{ad} its output is

$$\varphi^*_{\text{act}} = \eta_{\text{ad}} \dot{m}\, c_p (T_2 - T_1) \tag{5.3.40}$$

and the work of the ideal machine to produce this refrigeration is, from Eqs. (5.3.2), (5.3.14) and (5.1.40)

$$W_{\text{id}} = Q^*_{\text{act}} w_{\text{id}} \tag{5.3.41}$$

$$= Q^*_{\text{act}} \frac{T_a - T_m}{T_m} \tag{5.3.42}$$

$$= \eta_{\text{ad}} \cdot \dot{m}\, c_p \frac{(T_2 - T_1)\,(T_a - T_m)}{T_m}. \tag{5.3.43}$$

We have for the efficiency of the refrigeration cycle

$$\eta_{cy} = \frac{\eta_{\text{ad}} \cdot \dot{m} c_p \cdot \dfrac{(T_2 - T_1)\,(T_a - T_m)}{T_m}}{\dot{m}\, c_p \left(\dfrac{T_a}{\eta_{\text{is}}}\ \eta_{\text{ad}} \cdot T_m\right) \ln\left(\dfrac{T_2}{T_1}\right)}, \tag{5.3.44}$$

which can be reduced to

$$\eta_{cy} = \eta_{\text{ad}} \frac{T_a - T_m}{\dfrac{T_a}{\eta_{\text{is}}} - \eta_{\text{ad}} \cdot T_m} \tag{5.3.45}$$

$$\boxed{\eta_{cy} = \eta_{\text{ad}} \cdot \eta_{\text{is}} \frac{1 - \dfrac{T_m}{T_a}}{1 - \eta_{\text{ad}} \cdot \eta_{\text{is}} \cdot \dfrac{T_m}{T_a}}.} \tag{5.3.46}$$

With these expressions the effect of improving the efficiency of the machinery can be assessed. Note that the efficiencies occur as a product so that an improvement in either component is equivalent to the same percentage improvement in the other. Thus one should not pay excessively for high efficiency in one or the other if, by so doing, he neglects the other.

Note that as the ratio of the mean temperature to ambient gets small, as is the case for helium level refrigerators, then

$$\eta_{cy} = \eta_{\mathrm{ad}}\,\eta_{\mathrm{is}} \tag{5.3.47}$$

and one can immediately profit by the enhanced efficiency increase in either component. Trepp [17] points out that the product in Eq. (5.3.47) is frequently less than 50% since at best both expander and compressor have efficiencies in the 70 to 80% range. Actual machines have other losses in the heat exchanger and in heat leak which further reduce the cycle efficiency below the product of Eq. (5.3.47). The net result is that the overall cycle efficiency is well below 50%, as shown by Strobridge's compilation.

The very low efficiencies of small refrigerators show the compounding effect of the high surface to volume ratio of small machines, which causes high losses in all the components, added to the difficulty in achieving close tolerances for the small mechanical components.

5.3.3 Non-Isothermal Refrigeration

To meet a refrigeration requirement at temperature T_2 with a non-isothermal refrigerator operating from temperature T_1 to T_2, we have seen that the specific work is given by Eq. (5.3.14).

$$w_{\mathrm{ideal}} = \frac{W}{Q} = \frac{T_0 - LMT_c}{LMT_c}, \tag{5.3.14}$$

where LMT_c is the log mean temperature between T_1 and T_2,

$$\boxed{LMT_c = \frac{T_2 - T_1}{\ln\left(\dfrac{T_2}{T_1}\right)}.} \tag{5.3.32}$$

The refrigeration is supplied by a mass flow $\dot{m}$ of cold gas with constant specific heat c_p which is warmed from temperature T_1 to T_2 by the load,

$$Q = \dot{m}\,c_p(T_2 - T_1)\,. \tag{5.3.12}$$

From Eq. (5.3.12) we have that

$$T_2 - T_1 = \frac{Q}{\dot{m}\,c_p} \tag{5.3.48}$$

and

$$\frac{T_2}{T_1} = \frac{\dot{m}\,c_p \cdot T_2}{\dot{m}\,c_p \cdot T_2 - Q_2}, \tag{5.3.49}$$

so that upon substitution of Eqs. (5.1.41) and (5.1.42) in Eq. (5.1.28) we obtain,

$$LMT_c = \frac{Q}{\dot{m}\,c_p} \ln\left(\frac{\dot{m}\,c_p \cdot T_2}{\dot{m}\,c_p T_2 - Q}\right), \tag{5.3.50}$$

using this expression for LMT_e in Eq. (5.3.32) gives for the specific work,

$$w_{\text{ideal}} = \frac{\dot{m}\, c_p T_0}{Q} \left[\ln \left(\frac{\dot{m}\, c_p T_2}{\dot{m}\, c_p T_2 - Q} \right) \right]^{-1} - 1 . \tag{5.3.51}$$

To evaluate w_{ideal}, we need to know the values of the mass flow rate and properties of the working fluid in addition to the refrigeration load and the temperatures T_0 and T_2. For the expansion of an ideal gas, the outlet temperature T_2 is obtained as a function of inlet temperature T_0, efficiency η, pressure ratio p_2/p_0, the ratio of specific heats $\varkappa$ and heat inleakage Q from the relation:

$$T_2 = T_0 \left[1 - \eta + \eta \frac{p_2}{p_0} \cdot \frac{\varkappa - 1}{\varkappa} \right] + \frac{Q}{\dot{m} \cdot c_p} . \tag{5.3.52}$$

The refrigeration capacity Q, the figure of merrit η/η_c, and the power input to the compressor are plotted against various temperatures in Fig. 5.3.10. The design point of the refrigeration plant is somewhat to the right of the maximum efficiency state.

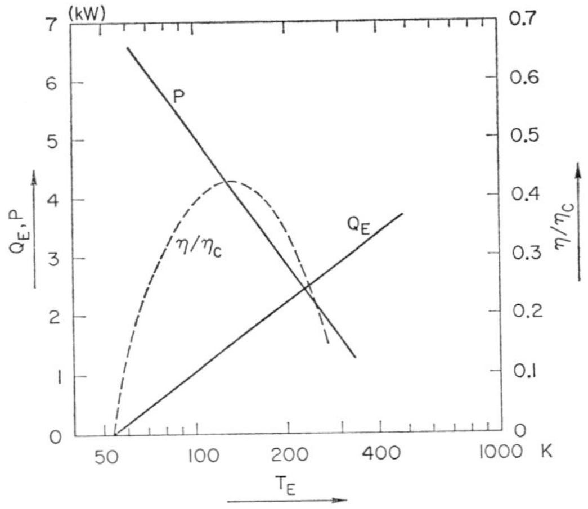

Fig. 5.3.10. Refrigeration capacity Q_E power consumption P and the figure of merit η/η_c of air liquefier at different temperatures

Fig. 5.3.11 illustrates a closed circuit helium refrigerator suitable to produce refrigeration in the temperature range of 10 to 50 K. The circuit consists of a compressor C which compresses the working medium in two pressure stages from 1 to 10 and 20 kpcm^{-2} isothermally, the counterflow heat exchangers X_1 to X_3, and the expansion engines E_1 and E_2, in which the part y of the high pressure stream is relaxed from a pressure p to the pressure p_0, while work is performed. The utilized cryocapacity $Q_{67} = \dot{m}\,(1 - y) \cdot (H_7 - H_6)$ is extracted at a gliding temperature, while the mechanical power $W = W_c - (W_{E1} + W_{E2})$ is required.

The high pressure gas flows through the additional heat exchanger X_4 (see Fig. 5.3.11) and is relaxed isenthalpically through the Joule-

Thompson valve V, where it is partially liquefied and reaches the helium liquefier, in which the liquefied gas is collected in the container B, and can be extracted.

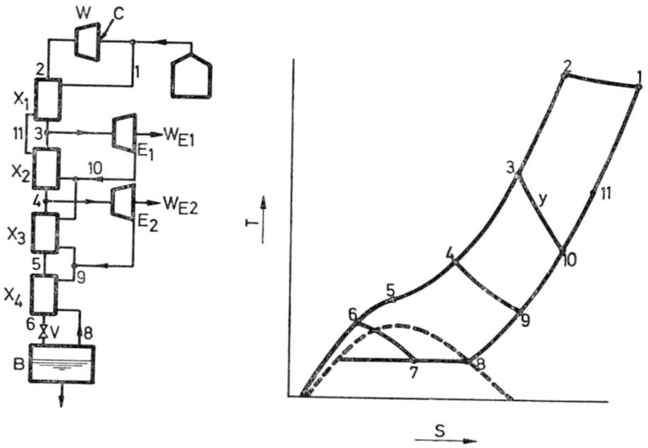

Fig. 5.3.11. Closed circuit refrigerator for production of refrigeration in the range of 10—50 K

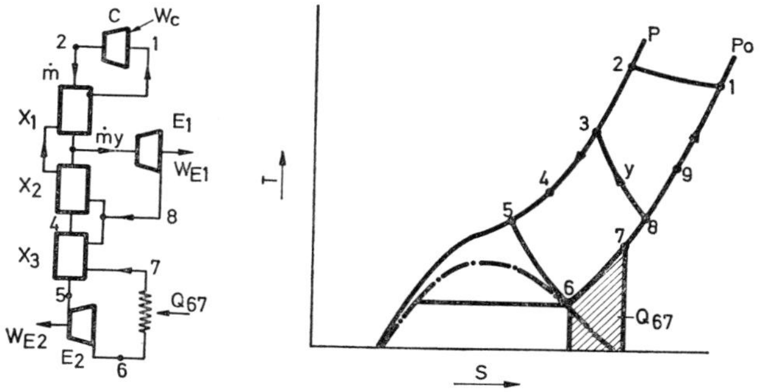

Fig. 5.3.12. Closed circuit refrigerator. *left* circuit diagram, *right* performance diagram

Fig. 5.3.12 shows the circuit diagram in more detail. Precooling with liquid nitrogen may eliminate the expansion machine E_1, which operates at a higher temperature. In this discussion of non-isothermal refrigerators, Jacobs [14] considered the case of equal lower temperatures for isothermal and non-isothermal refrigerators and showed, by an expression similar to Eq. (5.3.51), that the non-isothermal specific work is less than the Carnot work.

This apparent violation of the Carnot principle is, of course, due to the fact that in Jacobs' comparison, the non-isothermal refrigerator is at all times delivering refrigeration at a higher temperature than the Carnot machine. In either case, the analyses show that as the flow rate in gets large with consequent decrease in the temperature range of the refriger-ant, then the specific work tends to the Carnot value.

This analysis does not include the energy expended in or pressure drop required for circulating the cryogen, nor the heat in leakage due to the circulating system. These factors must be accounted for in a full analysis.

5.3.4 Practical Refrigerators

Of the three ideal cycles, only the Stirling cycle has been even closely approximated by practical gas refrigerators. In the Stirling cycle re-frigerators [18], the process steps are performed in a single integrated reciprocating machine on a closed volume of working fluid to produce a fixed quantity of refrigeration dependent upon the temperature. While this leads to a close approach to the thermodynamic ideal, it also leads to a rather inflexible system since the complete integrated mechanical and heat transfer system must be designed for a given refrigeration load at a given temperature. Comprehensive discussions of the Stirling cycle principle and machines are given in several papers and reviews [19, 20]. The performance of commercial Stirling machines for service in the three temperature ranges given in Strobridge's compilation are listed in Table 6.2.3 and shown in Fig. 6.2.3.

A modified Stirling-process proposed and realized by Gifford and Mc-Mahon [21] eliminates the use of the generated mechanical work in the expansion engine. This leads to very simple systems. Such refrigerators are obtained commercially in the range of 10 W at 15 K. In a three stage unit, utilizing special regenerators, even 7 K can be obtained. Due to their small cryo-output power and their poor overall efficiency, these refrigerators are not quite adequate to cool superconducting magnets.

Other practical low temperature refrigerators operate according to cycles in which inherently irreversible steps are performed by separate pieces of equipment, and even if the steps are perfectly accomplished, these cycles cannot achieve the low specific work of the ideal reversible cycles. The most frequently encountered practical cycles shown in Fig. 5.3.9 are the Brayton and Claude cycles.

The Brayton cycle employs a compressor, heat exchangers, and ex-pansion engine to produce a cold gas stream for non-isothermal refriger-ation, while the Claude cycle employs these components along with an expansion valve (Joule-Thompson or J-T valve) to produce liquid for iso-thermal refrigeration or for liquid withdrawal. The discrete components of these cycles render them less efficient than the integrated Stirling cycle, although in the larger machines the difference is not too great. The versatility of the Brayton and Claude cycles are their principle ad-vantage [18]. In contrast to the Stirling cycle machines, the gas stream can be directed for long distances between the components, and addition-al stages can be added with ease, all driven from a central main compres-sor. The components can be used in series or parallel to increase the efficiency by more closely approaching the ideal thermodynamic path, to cope with refrigeration requirements at various temperature levels, to provide variable capacity with time, or simply to increase the capacity over that available with single components.

Most of the refrigerators in Strobridge's compilation for Fig. 6.2.3 were, in fact, Brayton or Claude machines.

Both reciprocating and rotary machinery have been used in Brayton-Claude refrigerators, but the trend seems to be to reciprocating compressors and turbo expanders for large refrigeration loads. The reciprocating compressors are favored because the low molecular weight and consequent low density of gaseous helium leads to a very low head per stage in rotary machines, and thus to a large number of stages even for moderate overall pressure ratios. Modern non-lubricated compressors employing either the labyrinth sealing principle or fluoro carbon rings are capable of quite high efficiency (70 to 75%) and very high reliability while providing contaminent free high pressure helium gas. These compressors are available in a variety of sizes from a few cubic meters per hour to hundreds of cubic meters per second. In some installations, oil lubricated compressors have been employed with success, although the oil clean-up system must be carefully designed and close attention must be kept to the total operating time between flushing or regeneration of the clean-up system.

For cryogenic output capacities of more than 50 to 100 W at 4.2 K, turbines are used as expanders. In these turbines, the high pressure gas enters the rotor radially from outside and leaves the rotor axially through a diffusor. In contrary to reciprocating engines, the utilized pressure drop is limited to about 5 : 1. Thus, two turbines have to be connected in series. The rotors of small units have gas bearings and transmit their power at a speed of 10^5 to 2×10^5 r.p.m. over a charger whell located on the turbines to a closed He-refrigerator. The turbines designed by Sixsmith et al. [22, 23] have an 8 mm wheel diameter, reach a speed of 500,000 r.p.m., and have an efficiency of 50%. Oil lubricated bearings with a much larger wheel diameter have quite high efficiencies, up to 80%, but are useful only for large He-refrigerator units due to a high bearing torque.

The operating reliability of turbo-expanders has been generally very good, several systems having operated for many months unattended.

Reciprocating engines have the advantage in that their throughput can be changed in a wide range between 20 and 300 Nm^3/h by varying the speed of rotation. Their efficiencies are at a high expansion ratio of 30 : 1, about 60 to 80%. The most commonly used type is according to Collin's [24] design. In Collin's engine, a plastic coated piston hangs on a thin shaft and transmits its power over a balance wheel at relative small revolutions of 80 to 200 r.p.m. to an electrical generator. Gas-intake and gas-output are controlled with cams and mechanical valves. A second fundamentally different design is according to Doll and Eder [25, 26]. It is manufactured by Linde AG for displacement volumes of 20 and 50 cm^3 resp. Special processing of the piston surface enables a contact free movement of the piston within the cylinder, where a gap of only 2 μm is maintained between the two. Thus, high rotational speeds are achieved. A second feature of this engine is the control of the gas-intake and output by the piston itself through ring channels. The achieved efficiencies are between 65 and 75% at a compression ratio of 30 : 1 and a speed of 1200 r.p.m.

The remaining major process components of the refrigerators, the heat exchangers, has been supplied in a number of configurations. In most cases, these exchangers are designed to have high thermal efficiencies (98 to 99%) to insure very close temperature approaches between the streams. In some cases, however, the need for low pressure drop has dictated a sacrifice in overall thermal efficiency while in other cases the cost of sufficient heat exchanger surface to achieve high efficiency has not been warranted. For the larger machines the compact plate and fin exchangers are preferred in most installations although finned tubing has been used in several configurations.

For medium size refrigerators finned tubing seems preferable in most instances; for small refrigerators special matrix exchangers have been developed [18].

5.3.5 Liquefiers

The process for liquefying a gas by an ideal isentropic expansion is shown in the temperature entropy diagram of Fig. 5.3.13 (left) where steps a to b represent ideal isothermal compression, b to c represent isentropic expansion to the liquefaction point c where complete liquefaction occurs. The cycle is completed by the isothermal vaporization from c to d followed by isobaric warming from d to the starting point a. The cycle components for such a machine are shown in Fig. 5.3.13 (right) although for real gases of cryogenic interest the process is not workable, even in principle, since the pressure at point b is far beyond that achievable with process equipment as shown by the entries in Table 5.3.1 under p_b for liquefaction with c the normal boiling point.

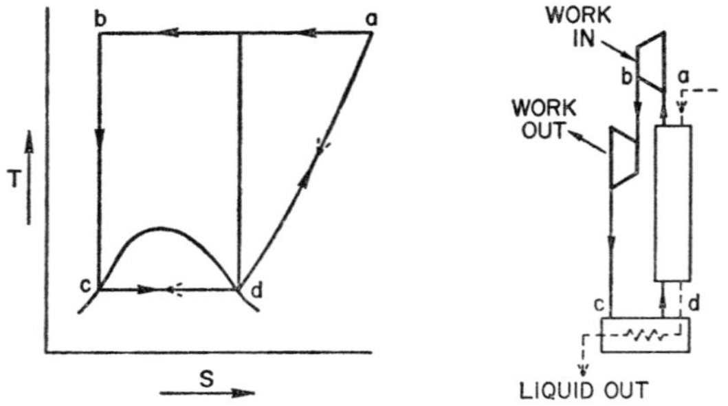

Fig. 5.3.13. Performance diagram and schematic of a Stirling refrigerator

In traversing the cycle of Fig. 5.3.13 we find that isothermal compression from a to b requires the heat extraction,

$$Q_{ab} = \dot{m}\, T_a (s_a - s_b)\,. \tag{5.3.53}$$

Isentropic expansion from b to c is adiabatic with no heat transfer while steps c to d and d to a require heat additions of

$$Q_{cd} = \dot{m}\lambda = \dot{m}\,(H_d - H_c) \qquad (5.3.54)$$

$$Q_{da} = \dot{m}\,(H_a - H_d)\,. \qquad (5.3.55)$$

The summation around the cycle yields the net work done in liquefying in moles per second of gas and returning it to its initial state,

$$w_{\text{liq}} = Q_{ab} - (Q_{cd} + Q_{da}) \qquad (5.3.56)$$

$$w_{\text{liq}} = \dot{m}\,[T_a(S_a - S_b) - (H_a - H_c)] \qquad (5.3.57)$$

$$w_{\text{liq}} = \dot{m}\,[T\Delta S - \Delta H]\,. \qquad (5.3.58)$$

In the cycle as described, the working fluid is vaporized and heated to room temperature with no net liquid production, Eq. (5.3.5); nevertheless, it gives the minimum work to liquefy $\dot{m}$ moles per second of gas. Since we can consider that the warming of the working fluid from d to a is accomplished by cooling an equal amount of gas from a to d, then the vaporization from c to d is accomplished by liquefying this gas to produce $\dot{m}$ in moles per second of liquid.

Thus, the net work to liquefy one mole per second of gas is given by

$$\boxed{W_{\text{liq}} = w_{\text{liq}}/\dot{m} = T\Delta S - \Delta H\,.} \qquad (5.3.59)$$

Since the steps taken in the liquefaction process are the ideal reversible steps, this work represents the minimum needed to achieve liquefaction. The values of the terms in Eq. (5.3.56) for several gases are given in Table 5.3.1 in terms of (J/g^{-1}) and (J/l). These values are the power to liquefy at the rates of one (gs^{-1}) and (ls^{-1}), respectively.

The enthalpy changes for the gas cooling step (Eq. (5.3.55)) and the condensation step (Eq. (5.3.54)) are shown separately in Table 5.3.2 to indicate the relative magnitude of the gas cooling and the condensation terms for the various gases.

If liquid is used solely in an isothermal refrigerator without employing the enthalpy (sensible heat or sensible refrigeration) of the vapor, this refrigeration is lost and, as shown in Table 5.3.1, represents a loss of 50% of the total available refrigeration in the case of nitrogen and 98.5% in case of helium. Of course, this lost refrigeration is not available at the boiling temperature, but it should be employed at higher temperatures to reduce heat leak into the apparatus or to provide the "gas cooling" refrigeration required for additional liquefaction, with the condensation refrigeration provided by other means, for example, an expansion engine.

If liquid is used to provide isothermal refrigeration, we can define an ideal specific work as the ratio of the minimum work of liquefaction to the latent heat of vaporization λ,

$$w'_{\min} = \frac{w_{\text{liq}}}{\lambda}, \qquad (5.3.60)$$

which can be compared with the reversible specific work required by a refrigerator w_{rev}. The values of $w'_{\min}$ and w_{rev} are given in Table 5.3.3. The

Table 5.3.2. *Minimum specific work to produce refrigeration by vaporization of N_2, H_2, and He at 1 kp cm^{-2}*

Gas	Pressure (kp cm^{-2})	T_2 (K)	T_1 (K)	$\Delta H_{cooling}$ (J g^{-1})	$\Delta H_{cooling}$ (kJ/l^{-1})	ΔH_{cond} (J g^{-1})	ΔH_{cond} (kJ/l^{-1})
He 4	1	300	4.214	1565	196	20.9	2.61
Hydrogen (normal)	1	300	20.39	3990	279	450	31.5
Hydrogen (para)	1	300	20.268	4530	317	448	31.4
Nitrogen	1	300	77.3	435	351	199	161

Table 5.3.3. *Comparison between reversible refrigeration work and minimum specific work to produce refrigeration by vaporization of N_2, H_2, and He at 1 kp cm^{-2}*

Gas	Pressure (kp cm^{-2})	T_2 (K)	T_1 (K)	W'_{min}	W_{rev}	W'_{min}/W_{rev}
He 4	1	300	4.214	326	70.4	4.63
Hydrogen (normal)	1	300	20.39	26.7	13.7	1.95
Hydrogen (para)	1	300	20.268	31.9	13.8	2.31
Nitrogen	1	300	77.3	3.89	2.89	1.34

ratio of these specific work terms evaluated for the same temperatures gives the ratio of the power costs for refrigeration by using boil-off liquid.

For helium level refrigeration at 4.2 K with heat rejection at $T_a =$ 300 K, we have

$$w'_{\text{min}} = 326 \left(\frac{\text{W}}{\text{W}}\right) \tag{5.3.61}$$

and

$$w_{\text{rev}} = 70.4 \left(\frac{\text{W}}{\text{W}}\right), \tag{5.3.62}$$

so that liquid refrigerant requires

$$\frac{w'_{\text{min}}}{w_{\text{rev}}} = \frac{326}{70.4} = 4.63 \tag{5.3.63}$$

times more power than a refrigerator of comparable efficiency. As we have seen that the real efficiency of refrigeration machines is sensitive to size, we should state that for the same power input to the equipment and thus to the same flow rate and efficiency, a refrigerator will produce 4.63 times the cooling achievable with a liquefier. This is the relationship used by Strobridge [15] to correlate the relative refrigeration capabilities of liquefiers and refrigerators. The ratio is however only an approximation for real machines, particularly for dual purpose liquefiers/refrigerators since the design of a unit may favor the liquefaction function over the refrigeration function or vice versa.

The work required for liquid pre-coolant for a liquefier or refrigerator can be determined by use of η_Q from Fig. 6.2.3 and the values of W_{liq} from Table 5.3.1 in the expression

$$w_{\text{pre}} = \frac{1}{\eta_Q} \cdot \dot{m}'_{\text{pre}} \cdot W_{\text{liq}}, \tag{5.3.64}$$

where $\dot{m}_{\text{pre}}$ is the rate of use of precoolant per unit of refrigeration or per unit of liquefaction of the machine.

This work must be added to the other specific work to determine the total specific work for the system. We have two choices for η_Q, either use the value corresponding to the actual power input to the source liquefier, in which case η_Q will be near the maximum of Fig. 6.2.3 or use the value corresponding to the precoolant use rate which will yield a much lower η_{Q}. The lower percent Carnot is probably the more realistic, since it will tend to account for the handling losses which effectively cause a higher cost for the liquid.

5.3.6 Real Liquefiers

Similar to refrigerators, the theoretical efficiency of helium liquefiers is not achieved in praxis. This is specifically true for the seldom utilized Joule-Thompson processes, where, by imploying precooling with liquid N_2 and H_2, the high pressure helium is cooled down below the inversion curve and is relaxed isentropically through a valve.

All commercially obtainable liquefiers are based on a process shown in Fig. 5.3.11. They use one or two expansion engines to produce refrigeration. The liquefied portion of the gas is either extracted immediately from the gas stream as shown in Fig. 5.3.11, which has the advantage of simplicity in design, or as illustrated in Fig. 5.3.14, is delivered to a second compressor C_2. The helium gas is cooled in a heat exchange process and is liquefied finally through the valve V_2 at a low pressure and relaxed to normal pressure.

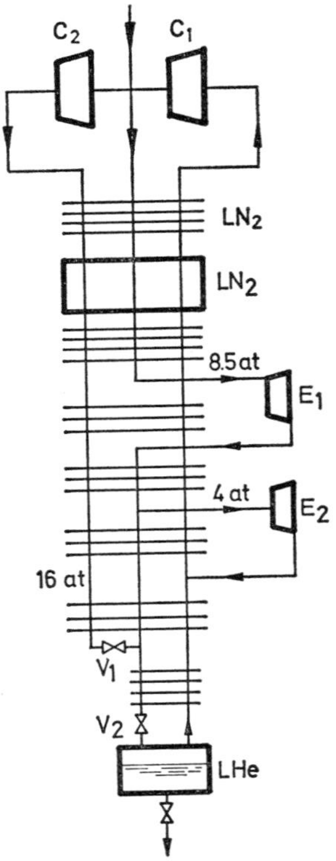

Fig. 5.3.14. Schematic representation of a He-liquefier

The refrigeration circuit, operated by the compressor C_1, is closed. It displays a 4.2 K refrigerator, whose refrigeration capacity is used to condense the He-gas delivered from C_2. These helium plants are highly reliable, as only He is used in the circuit and the liquefied helium is passed once through the liquefier. Recently Johnson, Collins, and Smith [27] have used a second reciprocating engine instead of the expansion valve V (see Fig. 5.3.12). In this engine, the helium gas having a temperature of about 7 K is relaxed isentropically and leaves the liquefier as a nearly saturated liquid. This procedure delivers about 33% more liquid at the same throughput. Smaller liquefiers for about 10 l/h liquid He are obtained commercially from Philips. The liquefier imploys two two-step cryogenerators, in which the gas is cooled over heat exchangers in steps of 90, 60, 40 and 25 K through a separate liquefaction circuit until the high pressure gas is relaxed in an ejector and partially liquefied. The specific power input of these liquefiers is about 2.5 kW/h per liter liquid helium.

5.4 Handling and Storage of Cryogenic Fluids

Liquid cryogens are shipped in thermally insulated containers ranging in size from a few tens of liters to several thousands of liters. They are stored in those places where they are used in similar containers. For users of small quantities, the same containers are frequently employed for shipping and storage, while for delivery to users where large quantities of cryogen is required, transport containers are special purpose vessels built into automotive or railway chassis and sometimes contain pumps to facilitate the transfer into storage. Factory built storage vessels for large users range in size from 1,500 liters to 40,000 liters, while vessels constructed on site may range up to a million liters or more.

The equipment to handle each of the cryogens, nitrogen neon, hydrogen, and helium is different because of their physical properties, their value, and their cost. However, it is desirable to use the same equipment for all cryogens in cases of emergency.

A frequently used small container for shipping, storing, and dispensing *liquid nitrogen* is a 100 to 160 liter vessel which employs vacuum superinsulation to achieve a low evaporation rate (1.5 Vol.-% per day). It is a "light weight construction" (product to vessel weight ratio of 1.3), while being small enough to be conveniently handled by one man (0.5 m diameter $\times$ 1.5 m high). Specification of these containers, as well as the 100 liter dewar is given in Table 5.4.1.

Table 5.4.1. *Specification of liquid nitrogen storage containers*

		Type		
		LS-175 PB	LS-600	LS-1900
Capacity (lit)		169	600	1900
Overall dimensions	Height (m)	1.54	1.2	1.6
	Outer dia. (m)	0.58	0.9	1.4
	Length (m)	—	2.2	2.6
Weight	Empty (kg)	104	345	890
	Full (kg)	224	830	2420
Evaporation rate per day ((Vol.-%)		1.5	1	0.5
Construction	Inner vessel	St. steel	Aluminum	Aluminum
	Outer vessel	Carbon steel	Carbon steel	Carbon steel
	Insulation	super-insulation	super-insulation	super-insulation

Liquid neon [5] is shipped and stored in small containers which have nitrogen cooled radiation shields to reduce the heat leak to the neon so that no liquid is lost. The density of liquid neon ($1.2 \text{ g} \cdot \text{cm}^{-3}$) poses a problem in using containers designed for liquid helium ($0.125 \text{ g} \cdot \text{cm}^{-3}$) and liquid hydrogen ($0.07 \text{ g} \cdot \text{cm}^{-3}$), unless the containers are only partially filled. On the other hand, the high volumetric heat of vaporization of neon, 29 Wh/liquid liters is about 40 times that of helium and 33 times that of hydrogen. This means that the containers will still perform reasonably well with low volume percent boil off. Containers used commonly for neon shipment are those described below in connection with helium shipment. Shipping containers for *liquid hydrogen* are similar in appearance, size and general construction to those used for liquid nitrogen, but are designed to have lower heat leak in order to maintain evaporation rates in the order of a few percent per day range. Because of the hazardous nature of hydrogen, special precautions are required for venting the gas. Permits are needed for the transport of containers.

The low volumetric heat of vaporization of *helium*, 0.78 Wh/liq.lit., results in a low permissible heat leak if evaporation rates are to be held to the few percent per day rate desired for most transport and storage containers. Early investigators have shown, that by using the high vacuum insulated containers developed for liquid nitrogen, with very thin neck tubes and liquid nitrogen as a "shielding" refrigerant, satisfactory

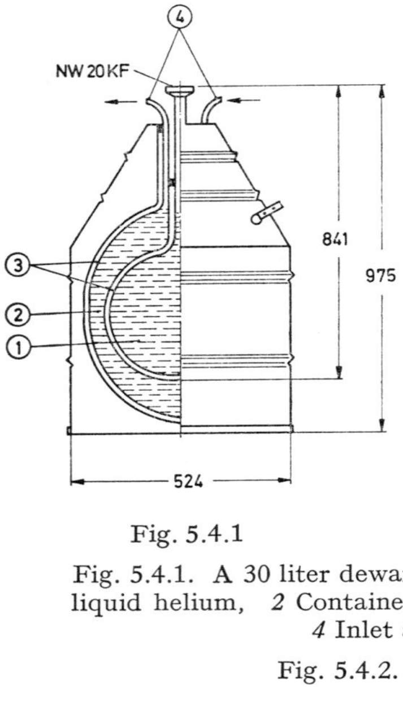

Fig. 5.4.1

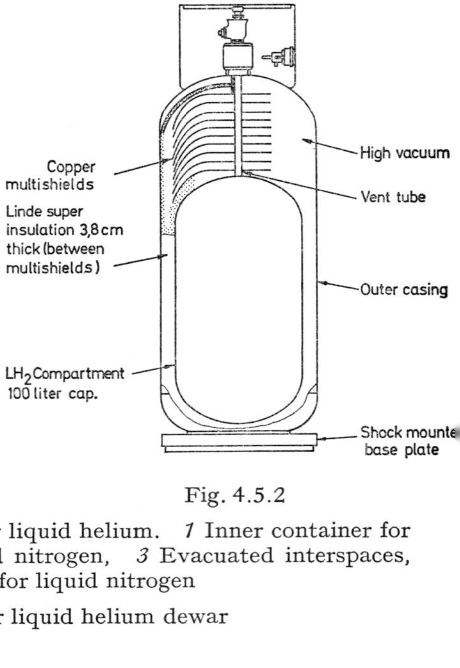

Fig. 4.5.2

Fig. 5.4.1. A 30 liter dewar vessel for liquid helium. *1* Inner container for liquid helium, *2* Container for liquid nitrogen, *3* Evacuated interspaces, *4* Inlet and outlet for liquid nitrogen

Fig. 5.4.2. A 100 liter liquid helium dewar

performance could be achieved. Containers of this general type (Fig. 5.4.1) are still available. Their requirement of providing liquid nitrogen for thermal shielding and precooling, and their fragility makes them cumbersome to use in the laboratory. Nevertheless, the first commercial shipment of liquid helium in the US and the liquid helium pool in England were accomplished by using the type of containers in sizes of 25, 50 and 100 liters.

More rugged containers have been developed based on the use of "superinsulation", which employ only the refrigeration in the boil-off vapor to shield off the liquid helium contents [28]. These containers in the 100 lit. size shown in Figs. 5.4.2 and 5.4.3 are similar in size and appearance to the 160 lit. superinsulated liquid nitrogen servicing containers. Specifications of some liquid helium containers are given in Table 5.4.2.

These transport and storage containers for liquid helium do not require any liquid nitrogen shielding and are fully equipped with safety relief valves and bursting discs.

Table 5.4.2. *Specification of liquid helium storage containers*

		Type	
	LSHe-30	LSHe-102	LSHe-500
Capacity (lit.)	30	102	500
Overall dimensions: Height (m)	1.12	1.59	1.7
Overall dimensions: Outer dia. (m)	0.475	0.51	1.06
Overall dimensions: Length (m)			
Weight: Empty (kg)	37.7	99.5	33.8
Weight: Full (kg)	41.45	112.0	400.5
Evaporation rate per day (Vol.-%)	0.80	1.00	1.00
Construction: Inner vessel	St. steel	St. steel	St. steel
Construction: Outer vessel	Aluminum	St. steel	Aluminum
Construction: Insulation	superinsul. and multishields	superinsul. and multishields	superinsul. and multishields

All liquid helium transport and storage containers must be maintained with positive internal pressure to prevent air inleakage in the container, which may subsequently freeze within the container and plug up the liquid withdrawal line. The positive pressure is assured by using a pressure relief valve to close the container, the valve opening when the heat leak produces some pre-set internal pressure in the container. Care must be taken lest these pressure relief valves freeze open when the containers vent at high rates, leaving a channel for air leakage. Many containers are built with a single thin wall neck of access tube which must be opened for insertion and removal of a transfer tube, at which time the possibility of air inleakage and subsequent plugging is also great.

The general method of circumventing the neck-plug hazards due to the malfunction of a pressure relief valve is to provide dual pressure relief passages as shown in Fig. 5.4.4. Containers with single neck tubes, which cannot otherwise be altered, should be fitted with an insert into the neck to provide two discharge paths. Such a device is shown in Fig. 5.4.5. The outer annular path should be permanently connected to

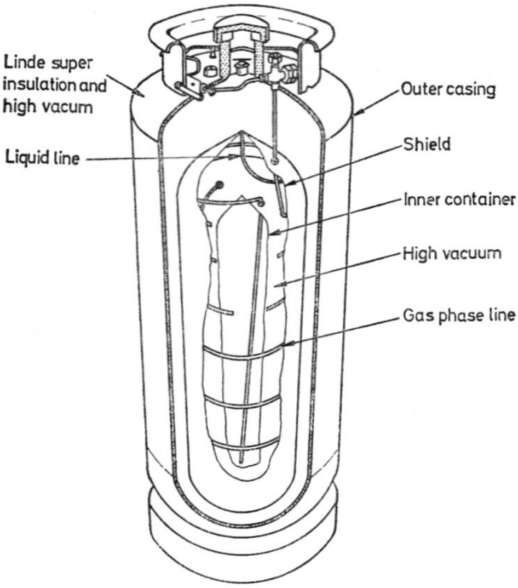

Fig. 5.4.3.
A 100 liter liquid helium dewar

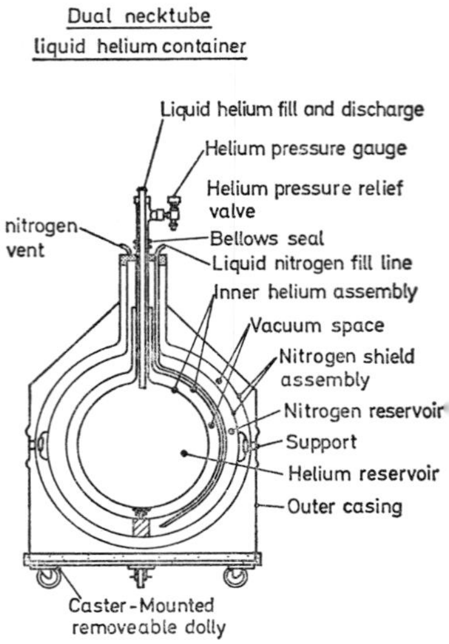

Fig. 5.4.4.
Dual neck tube liquid helium container

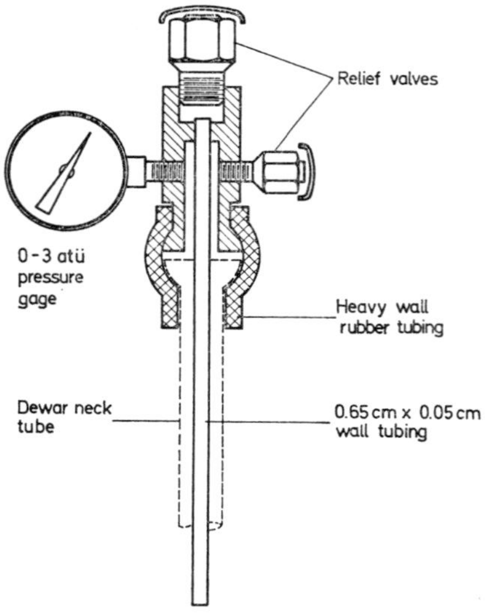

Fig. 5.4.5.
Dewar insert which provides two discharge paths circumventing helium dewar neck plug hazards

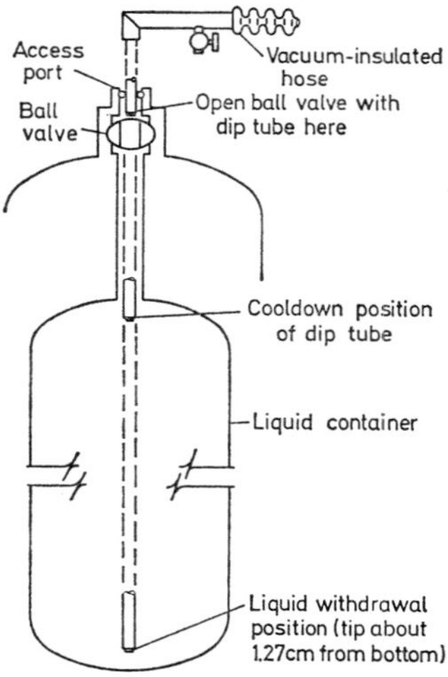

Fig. 5.4.6.
Use of a ball valve to reduce plugging hazards in liquid helium dewars

a pressure relief valve. A removable pressure relief valve should be used to stop the withdrawal access. The better alternative, of course, is a permanent dual relief passage.

A positive way to reduce the hazard of plugging the withdrawal passage during insertion and withdrawal of the transfer line is to use a ball valve to close the passage as shown schematically in Fig. 5.4.6. By inserting the transfer tube into the 0-ring seal with the ball valve closed and then cracking the valve to purge the transfer tube, the possibility of air inleakage is minimized. Similar protection can be achieved by a permanently installed withdrawal line terminated by a valve suitable for liquid helium.

5.5 Physical Properties of Cryogenic Fluids

The cryogens being of interest to cool down superconducting magnets and to maintain the operational temperature are liquid or gaseous helium, liquid nitrogen, and liquid hydrogen. The physical properties of these three cryogens are given in literature extensively [1, 2, 3]. Tables and sets of curves of material properties are compiled by the U.S. National Bureau of Standards [6]. It is thus redundant repeating data at this place. In this section, we have compiled only those physical properties which are required in designing superconducting magnet systems.

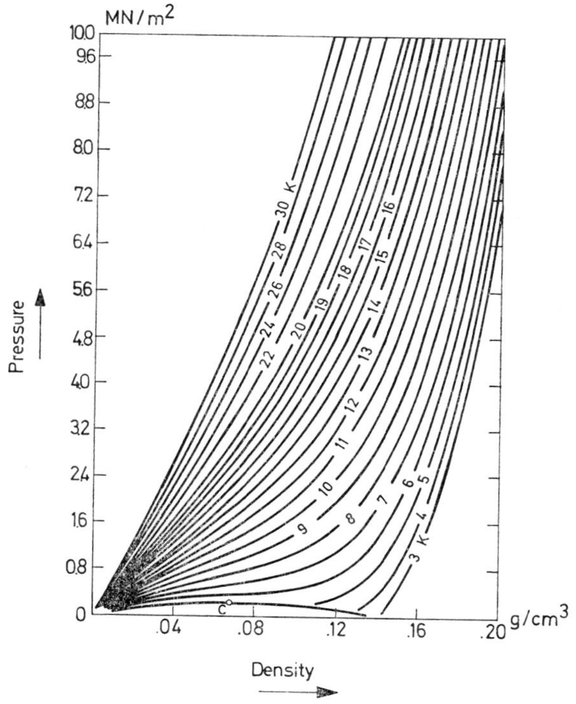

Fig. 5.5.1. Pressure-density diagrams of helium (1 MN/m² = 10 kp/cm²)

5.5.1 Helium

Major data on helium have been compiled by Keesom [29]. Newer data are presented by Lounasmaa [30]. Mann and Stewart [31] have correlated the data given by Keesom and Lounasmaa. The data presented in this chapter are the results of these correlations, recent measurements, and extension of existing data in the range of 3 to 30 K.

(*a*) *Pressure Density Diagram*

Fig. 5.5.1 is a graph of density versus temperature for helium at various pressures ranging from atmospheric pressure to 10 MN/m². The density changes most rapidly in this temperature range.

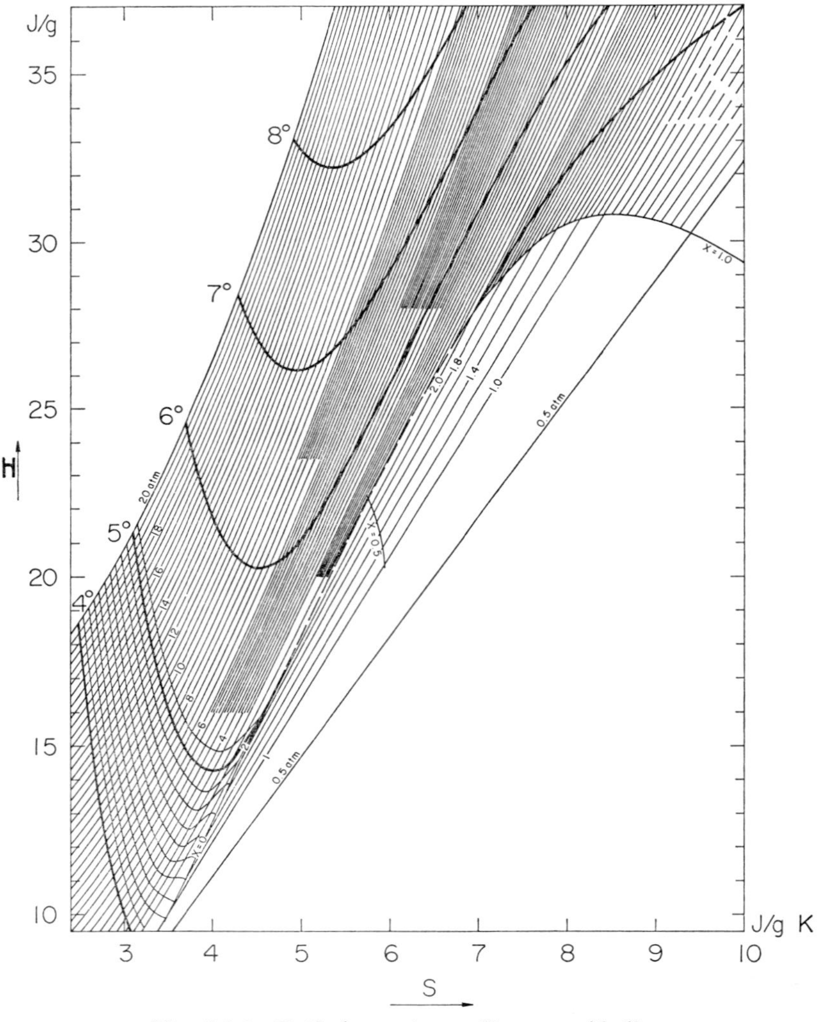

Fig. 5.5.2. Enthalpy-entropy diagram of helium

(*b*) *Enthalpy-Entropy Diagram*

The data of Fig. 5.5.2 are suitable for the designs of cryogenic equipment operating in the temperature range of 4 to 8 K.

(*c*) *Enthalpy and Heat of Vaporization*

The curves given in Figs. 5.5.3 and 5.5.4 are useful in designing magnets with internal coolant passages, where the latent heat of helium can be utilized in a heat exchange process.

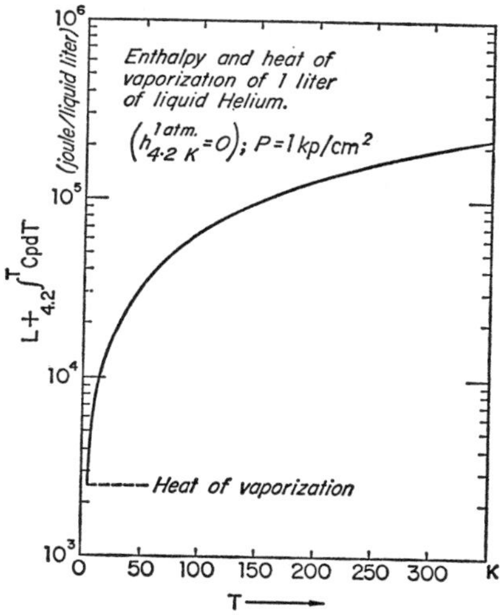

Fig. 5.5.3. Enthalpy-temperature diagram of helium at 1 kp/cm²

(*d*) *Specific Heat Temperature Diagram*

The diagram given in Fig. 5.5.5 gives the specific heat related to pressure and temperature changes near the critical point.

(*e*) *Viscosity*

As seen in Fig. 5.5.6, the viscosity of helium is a rather weak function of pressure at room temperature. At temperatures near the liquid-vapor region, the viscosity is a strong function of pressure.

(*f*) *Thermal Conductivity*

Fig. 5.5.7 gives thermal conductivity data as a function of temperature. The thermal conductivity has a similar temperature characteristic as the viscosity; they are strong functions of pressure at temperatures around 5 K.

(*g*) *Velocity of Sound*

The knowledge of the velocity of sound is of interest in the designs of safety valves and rupture discs. If helium is evaporated rapidly due to a full or partial magnet quench, the speed of gaseous helium passing

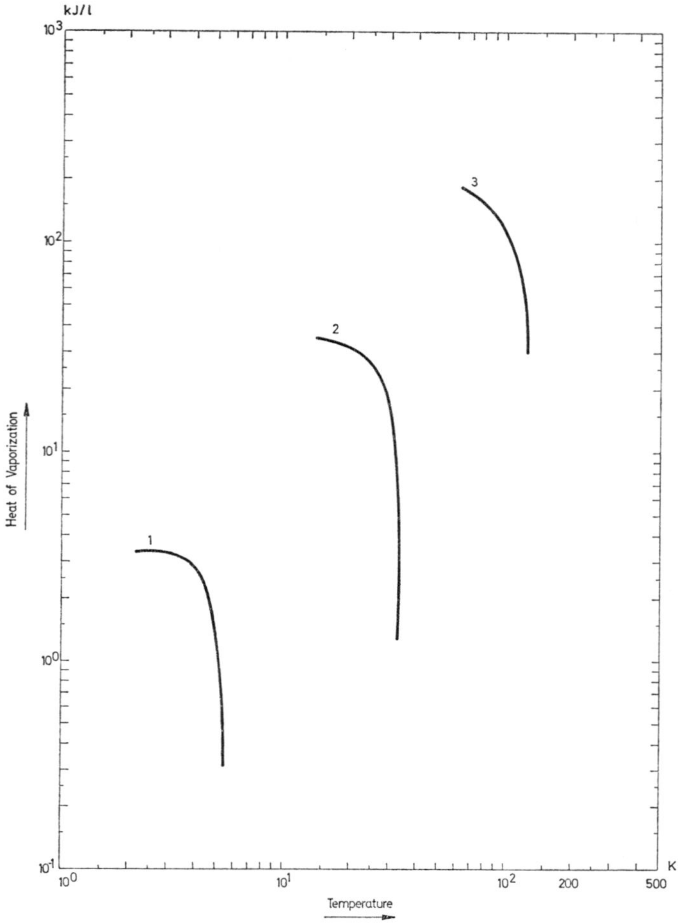

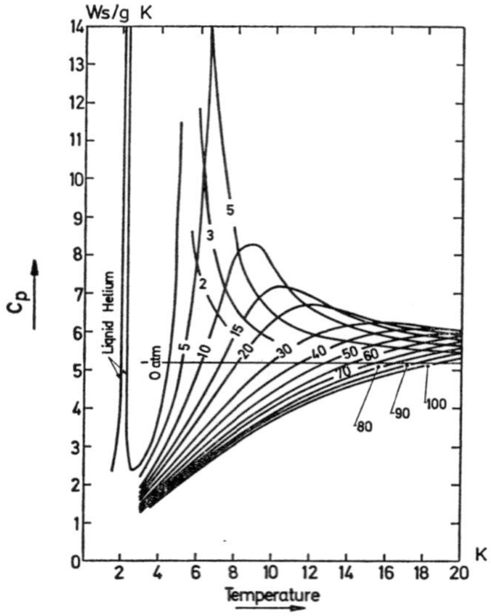

▲ Fig. 5.5.4.
Heat of vaporization-temperature diagram of cryogenic liquids
1 liquid helium,
2 liquid hydrogen,
3 liquid nitrogen

◀ Fig. 5.5.5.
Specific heat-temperature diagram of helium

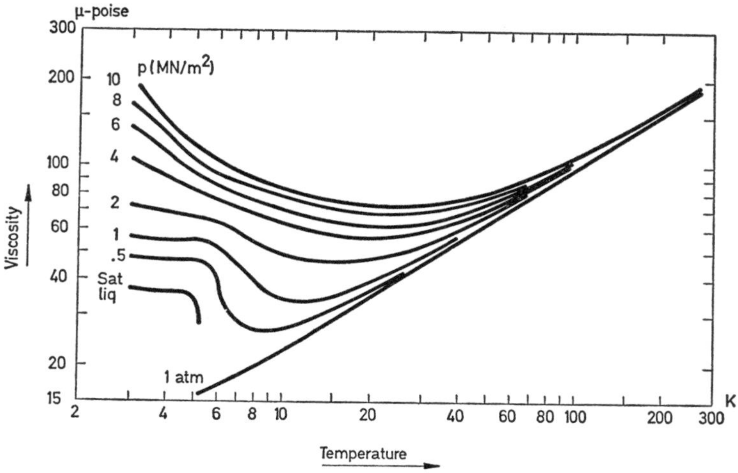

Fig. 5.5.6. Viscosity vs. temperature of helium

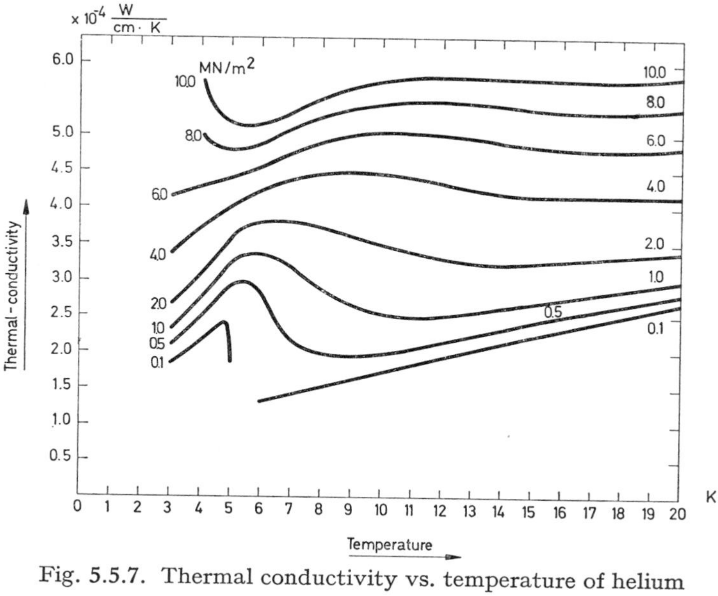

Fig. 5.5.7. Thermal conductivity vs. temperature of helium

through necks and through safety valves must be only a fraction of the sound velocity, otherwise the pressure and the vibrations developed may lead to the destruction of the necks and safety valves. The velocity of sound is also needed to design J-T-valves.

The velocity of sound for helium is given in Fig. 5.5.8 as a function of temperature. The data are compared to the speed of sound for nitrogen and hydrogen.

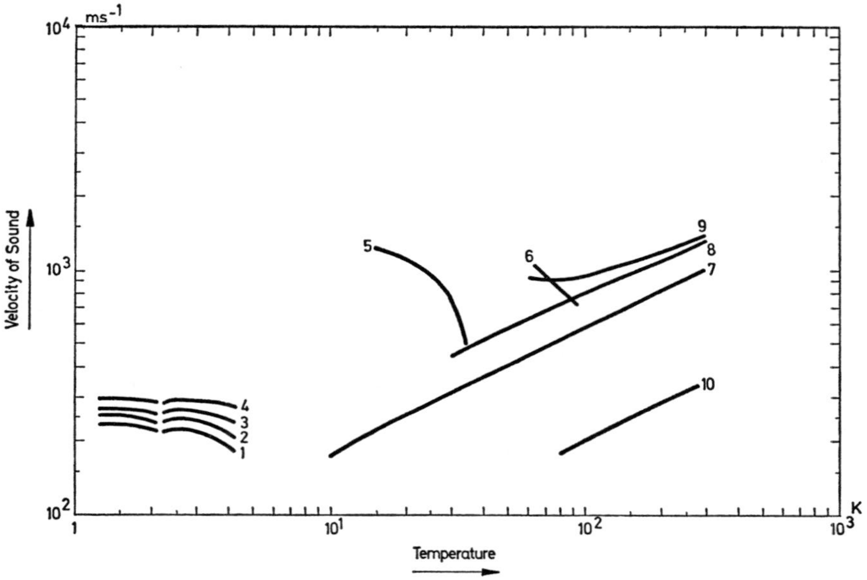

Fig. 5.5.8. Speed of sound of cryogenic liquids and gases
1 Liquid helium: saturated liquid, *2* Liquid He at 2.5 kp/cm², *3* Liquid He at 5 kp/cm², *4* Liquid He at 10 kp/cm², *5* Liquid hydrogen: saturated liquid, *6* Liquid nitrogen: saturated liquid, *7* Gaseous He: 1 kp/cm², *8* Gaseous H_2: 1 kp/cm², *9* Gaseous H_2: 100 kp/cm², *10* Gaseous N_2: 1 kp/cm²

(*h*) *Dielectric Strength*

The voltage breakdowns of grade A helium measured in gaps of 0.01 to 0.7 cm height are:

a.c. voltage	250 kV/cm	saturated liquid	(1 kp/cm²)
	320 kV/cm	supercritical helium	(4 kp/cm²)
d.c. voltage:	470 kV/cm	saturated liquid	(1 kp/cm²)

For comparison we give values of voltage breakdown for polyethylen tape (0.1 cm thick) = 450 kV/cm (a.c.) and 900 kV/cm (d.c.).

5.5.2 Hydrogen

Liquid hydrogen is seldom used in superconducting systems. However in hydrogen bubble chamber magnets, liquid or gaseous hydrogen is used for cooling heat shields. Hydrogen is not used for precooling superconducting systems. However after the discovery of the (Nb, Al, Ge) systems, which have a critical temperature of $\sim$21 K, hydrogen as a cryogen may become attractive. In this section physical property data, which seem to be of interest in conjunction with superconducting systems are compiled for liquid hydrogen. Few properties of liquid hydrogen were given in Table 5.1.1.

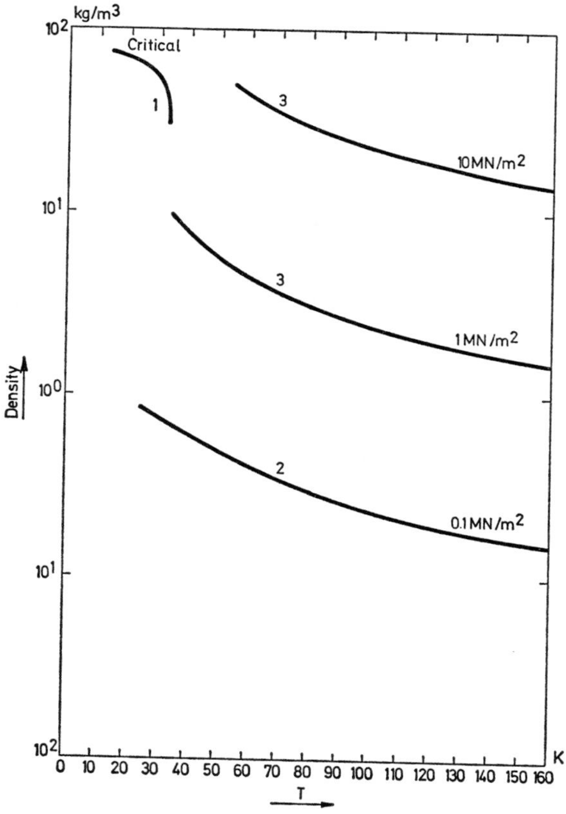

Fig. 5.5.9. Density-temperature diagram for n.hydrogen

1 Saturated liquid, *2, 3, 4* n.hydrogen gas at different pressures

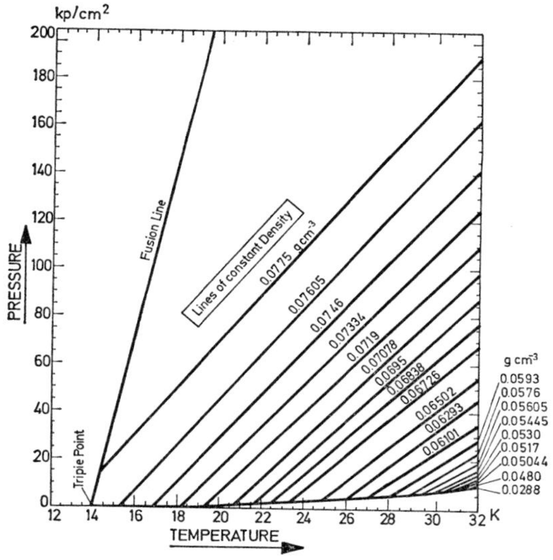

Fig. 5.5.10. Pressure, density-temperature chart for n.hydrogen

(*a*) *Density-Temperature Diagram*

Figs. 5.5.9 and 5.5.10 illustrate the density of liquid and gaseous hydrogen versus temperature at pressures of 0.1 ... 10 MN/m².

(*b*) *Heat of Vaporization*

The heat of vaporization versus temperature of liquid hydrogen is illustrated in Fig. 5.5.4.

(*c*) *Specific Heat Temperature Diagram*

Curves of specific heat at constant volume, pressure, and for saturated liquid are given in Fig. 5.5.11.

(*d*) *Viscosity* [32]

The viscosity of hydrogen as a function of temperature is presented in Fig. 5.5.12.

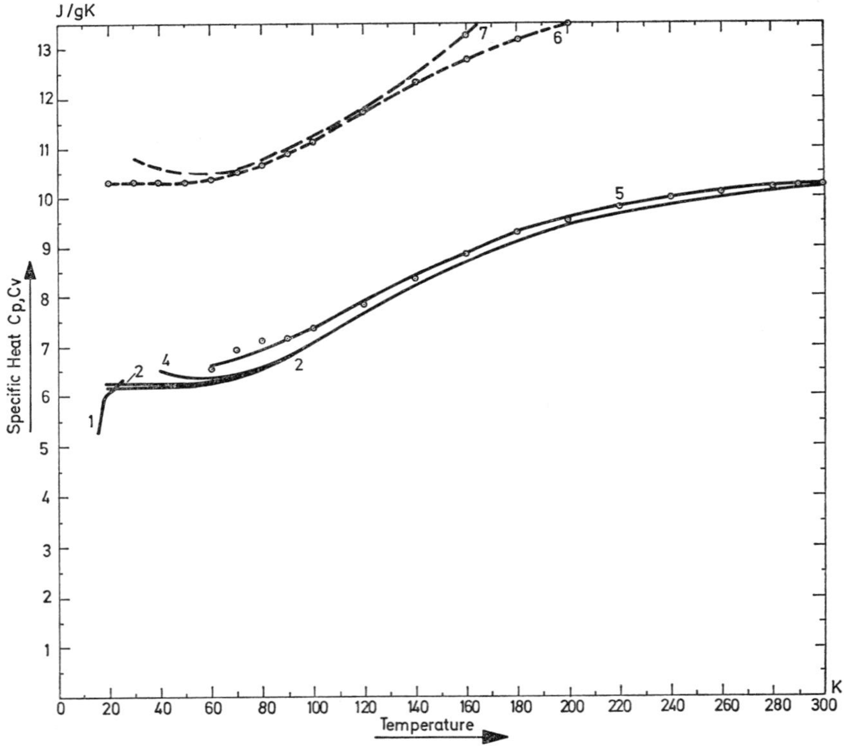

Fig. 5.5.11. Specific heat (c_p, c_v) vs. temperature for n.hydrogen

1—5: c_v
- *1* Liquid hydrogen, gaseous hydrogen at 0 ... 100 kp/cm²
- *2* (0 kp/cm²)
- *3* 1 kp/cm²
- *4* 10 kp/cm²
- *5* 100 kp/cm²

6—7 · c_p of gaseous hydrogen at 0 and 1 kp/cm²

(*e*) *Thermal Conductivity* [32]

The thermal conductivity of liquid and gaseous hydrogen versus temperature is shown in Fig. 5.5.13.

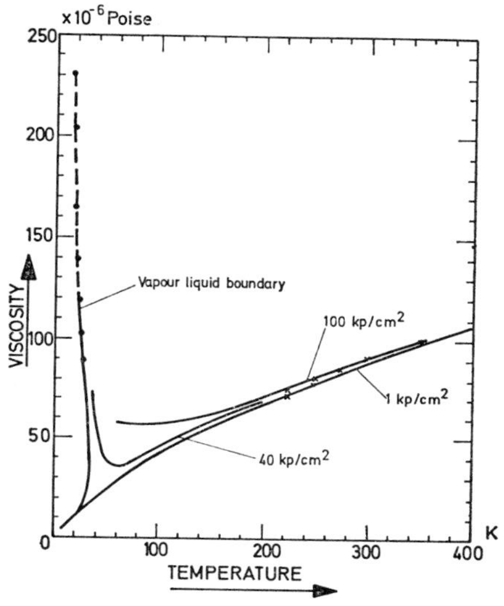

Fig. 5.5.12.
Viscosity vs. temperature of n.hydrogen

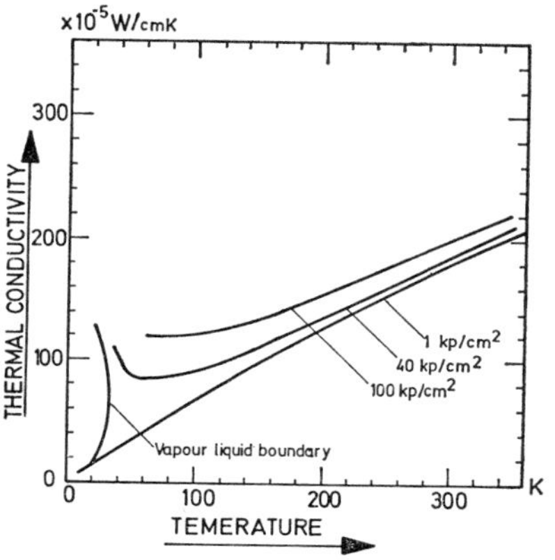

Fig. 5.5.13.
Thermal conductivity vs. temperature of n.hydrogen

(*f*) *Speed of Sound*

The speed of sound of liquid and gaseous hydrogen is given in Fig. 5.5.8.

5.5.3 Nitrogen

Liquid nitrogen is commonly used to precool magnet systems and to cool down and hold the temperature of heat shields at about 78 K. It is a safe refrigerant, is inactive, and is neither explosive nor toxic. It is widely used commercially and is prepared by the fractionation of liquid air.

(*a*) *Density Temperature Diagram*

Fig. 5.5.14 illustrates the density of saturated liquid, gaseous nitrogen, and of nitrogen vapor.

(*b*) *Heat of Vaporization*

Heat of vaporization versus temperature is given in Fig. 5.5.4. These data are of considerable interest to design nitrogen, precooled helium cryostats, and to operate helium and hydrogen liquefiers and refrigerators with nitrogen precooling.

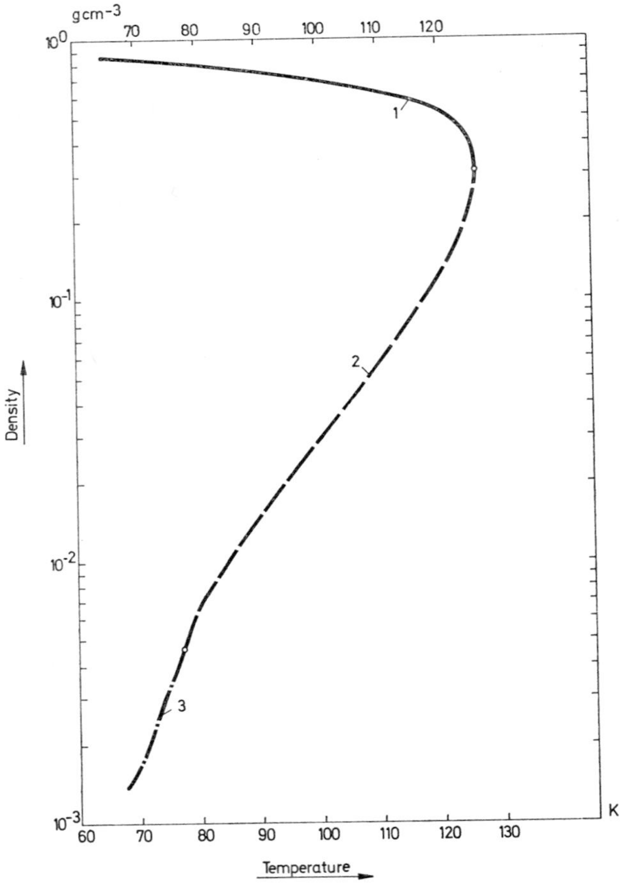

Fig. 5.5.14. Density-temperature diagram of nitrogen. *1* Liquid nitrogen, *2* Gaseous nitrogen, *3* Saturated nitrogen vapor

(*c*) *Specific Heat Temperature Diagram*

The specific heat versus temperature is shown in Fig. 5.5.15.

(*d*) *Speed of Sound*

The speed of sound in liquid and gaseous nitrogen is illustrated in Fig. 5.5.8.

(*e*) *Dielectric Constant*

The dielectric constant of nitrogen at saturation pressure is given in Fig. 5.5.18.

(*f*) *Viscosity* [32]

The viscosity of N_2 is shown in Fig. 5.5.16.

(g) *Thermal Conductivity* [32]

The thermal conductivity of nitrogen is illustrated in Fig. 5.5.17.

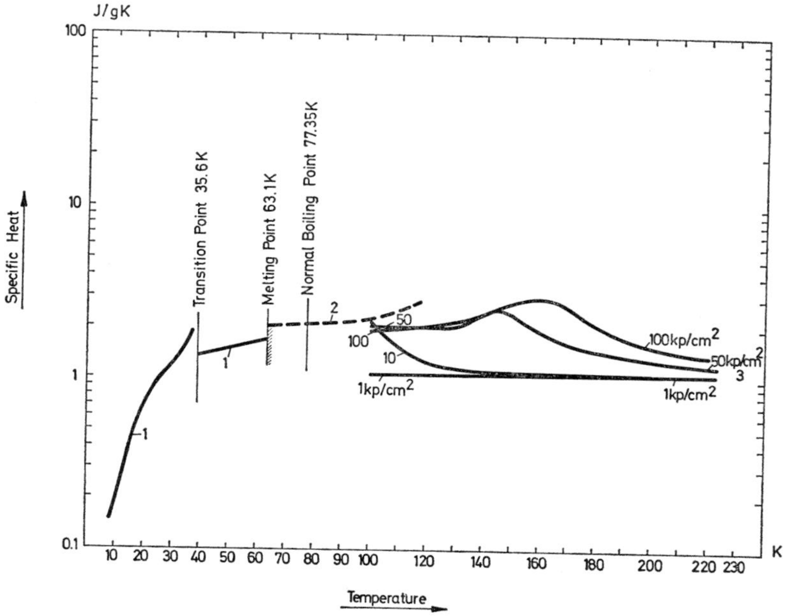

Fig. 5.5.15. Specific heat, c_p, versus temperature diagram of nitrogen. *1* Solid-nitrogen, *2* Liquid nitrogen, *3* Gaseous nitrogen at different pressures

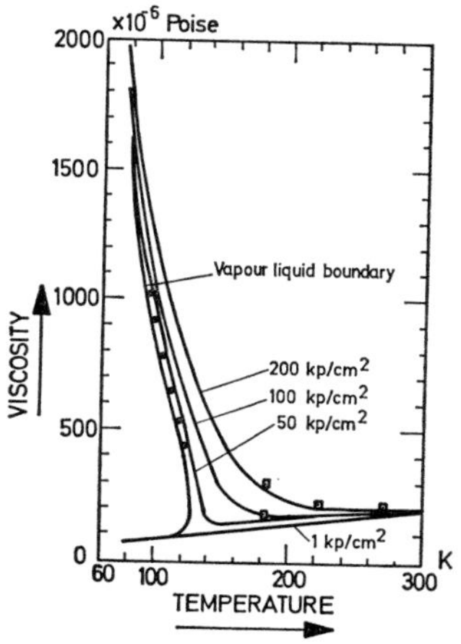

Fig. 5.5.16. Viscosity-temperature diagram of nitrogen

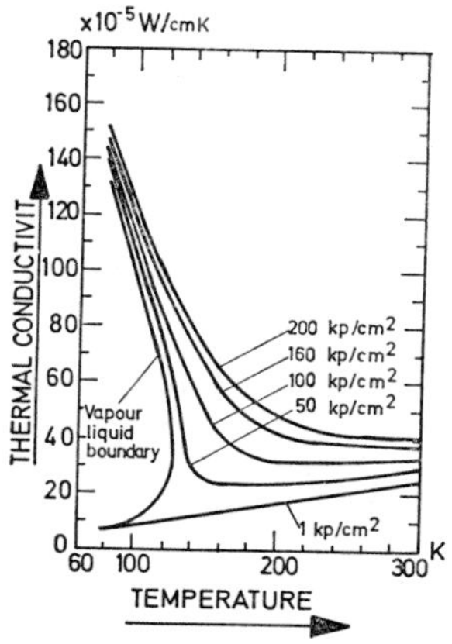

Fig. 5.5.17. Thermal conductivity-temperature diagram of nitrogen

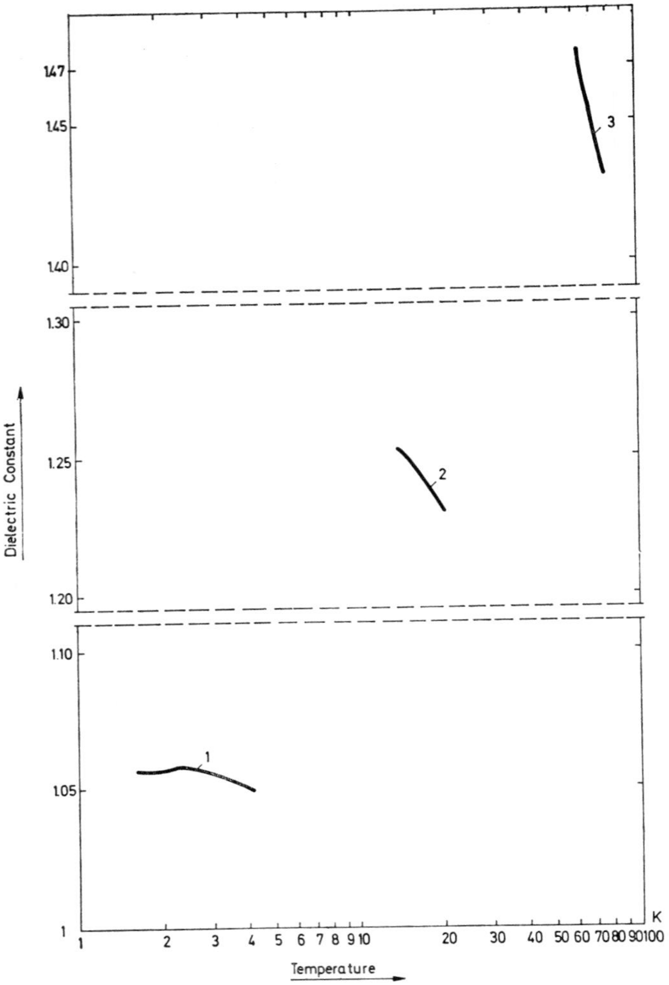

Fig. 5.5.18. Dielectric constant of cryogenic liquids
1 Liquid He; *2* Liquid H_2; *3* Liquid N_2

5.6 Physical Properties of Solids

5.6.1 Introduction

Composite or multifilament conductors have a transport current density of at least one order of magnitude higher than watercooled normal conductors, they are exposed to magnetic fields which are higher by at least a factor of 2 to 3 than in watercooled magnets and to unbalanced stress environments. Considering these, the effect of magnetomechnical stresses on conductors and the problem of coil support, prevention of wire movements in coils are of great importance. In addition, the effect to thermo-mechanical stresses on the conductor, the insulation and coil auxiliary parts, while the magnet is cooled down to temperatures below the critical temperature of the superconductor and thermally and magnetically cycled, have to be considered. The coil must be supported adequately without damage to the active conductor material, the conductors must be wound rigidly to prevent degradation.

Superconducting pulsed magnets must be reliable over the lifetime of the accelerator, in which they are a part. If we assume that the pulse duration of such a magnet is six seconds (see Chapter 6) and the magnet is in operation over 8,000 hours per year, than over a 10 year operation (which may concur with the lifetime of the synchrotron) the coils will be pulsed about 5×10^7 times. As the coil active parts of the conductor, such as the superconductor, the matrix materials and the inactive parts such as the insulations are stressed periodically near the limit of their yield strength, the problem of fatigue must be investigated thoroughly. Unfortunately only limited information in this area is presently available. No solid is perfectly pure or homogeneous. There are always impurity atoms present which cause irregularities in the crystal lattice. Among various types of impurities we name a few (see also Fig. 4.1.2).

Point defects: Two types of point defects are widely encountered: Interstitials (an extra atom is resting in the interstices between regular crystal sites) and vacancies (atoms missing from normally occupied positions).

Imperfections are dislocations (edge and screw dislocations). Part of the crystal is shifted or slipped relative to its adjacent part.

Impurities: Impurity atoms are found at lattice sites, solid solutions and foreign particles in the crystal.

Grain boundaries are obstacles, as they interrupt the orientation of neighbouring crystals.

Both, point defects and dislocations are always present to some degree in metallic crystals. Few mechanical properties of solids can be explained only by their presence, e.g. the interaction of dislocations accounts for the yield stress and ductility of materials. The modulus of elasticity depends mostly on the regular arrangement of atoms in the crystal.

Material ductility (plastic deformation) depends upon the mobility of dislocations within the material. The movement of dislocations are impurity atoms, which lock dislocations by their distortion of the lattice. Vacancies and interstitials provide obstacles to the movement of dislocations. Hard foreign particles in the crystal increase the necessary stress to move dislocations. Grain boundaries are discontinuities in the orientation of neighbouring crystals and thus we may conclude that metals with a small grain size have a higher yield stress than metals with a large grain size (annealing increases grain size and work hardening reduces the grain size). After sufficient plastic deformation has occurred, further local dislocation motion is blocked by larger numbers of dislocations and finally rupture of atomic bonds will eventuate.

The deformation of thermosettings, elastomers and plastics takes place by uncoiling their chainlike long molecules. Van der Waals bonds which are weak will rupture first.

A sliding of long molecules may come afterwards. Finally, the stronger chemical bond is ruptured. In this final stage, the material exhibits more resistance to stress. Cross-linked thermosettings are brittle, their structures and stress resistance are not unlike glass at cryogenic temperatures.

Thin plastic films and glass-fiber tapes are quite common as insulation in superconducting magnets. They display at low temperatures, a remarkable degree of flexibility and high strength. Among the various materials used, we may mention polyamid (H-Film, Kepton), Polycarbonate (Lexan), polyethylene tetraphalate (Mylar) and polyvinyl chlorides of the rigid types (Nylon). The yield strength of a Mylarfilm (10^{-2} cm thick), is increased from 10^3 kp/cm² (300 K) to 3×10^3 kp/cm² (4.2 K) without loosing flexibility. Extruded Nylon (type 101) has a room temperature (300 K) tensile strength of 0.7×10^3 kp/cm². The value is increased approximatively four fold at 4.2 K. The impact strength (Izod) of Nylon (type 101, with 2.5% water) is changed slightly from room temperature to 77.3 K. Measured values were 0.276 mkp/cm² at 300 K, 0.13 mkp/cm² at 77.3 K and the extrapolated value to 4.2 K was 0.08 mkp/cm².

Structural materials used for coil support such as bandages and reinforcements are various types of stainless steels (304, 304 L, 314, 314 L etc.), unidirectional glass filaments preimpregnated but in a semicured stage, glass-fiber tapes, titanium, beryllium copper (2 ... 2.5% Be) and others. In addition to their high tensile properties it is important that either their thermal contractions are matched to that of the conductor or elastic members must be utilized, which allow the support structures to retain a nearly constant stress on the coils.

Material strength is increased at cryogenic temperatures, compared to room- or elevated temperatures, but nevertheless the gain in material strength is not nearly sufficient to overcome or counteract the strain imposed by the combined forces acting upon the conductor.

Stresses on the composite conductor are manifold, their effects on the material behaviour and properties complex and thus simplified assumptions unavoidable. We sum them up as follows:

(i) The conductor is prestressed during winding (at room temperature) (see Fig. 6.2.9) to prevent the coil from loosing tension while cooled down. Loose turns lead to wire movements and to premature quenching and training of the conductor. Prestraining the wire, if performed only in a controlled manner will yield an uniform material stressing, and increase low temperature material resistivity (Fig. 5.6.1).

If prestraining is not performed in a controlled fashion the stress distribution during and after cool down depends on the position of the conductor in the magnet and may lead to regions of instability.

(ii) If the coil is wound on a high tensile strength former, with a different thermal contraction coefficient than the composite conductor, such as stainless steel (see Fig. 6.2.8), the cooling down of the magnet results in additional strains induced by the different thermal contraction between coil support and conductor, which again yield cold working of the conductor substrate.

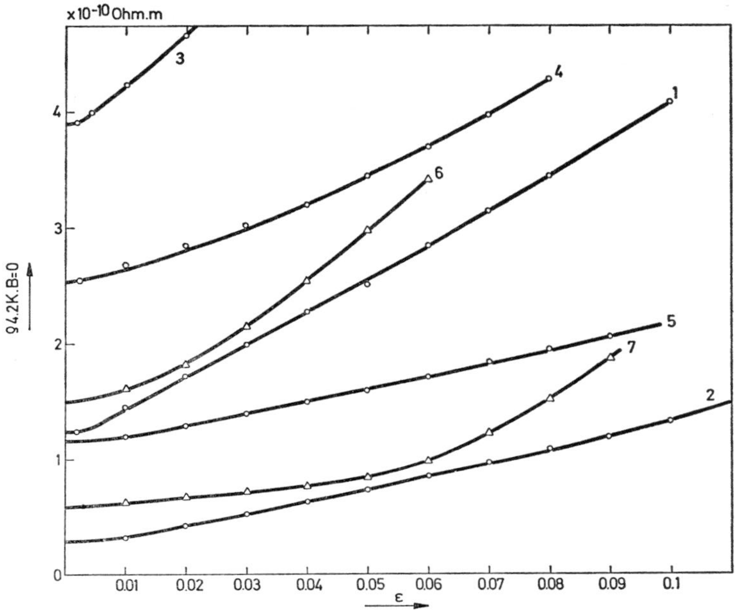

Fig. 5.6.1. Resistivity vs. strain of coppers and aluminums

1 ETP-Cu: $\frac{1}{4}$ hard ($E = 1.24 \times 10^6$ kp/cm^2)
2 ETP-Cu: annealed at 600 °C, for 1 h
3 OFHC-Cu: hard ($E = 1.23 \times 10^6$ kp/cm^2)
4 OFHC-Cu: $\frac{1}{2}$hard ($E = 1.23 \times 10^6$ kp/cm^2)
5 OFHC-Cu: annealed at 600 °C, for 1 h
6 99.995% pure al. as received ($E = 0.79 \times 10^6$ kp/cm^2)
7 99.995% pure al. annealed at 300 °C for 2 h

(iii) If the coil is self supporting, but restrained from free movements, stress distribution and the azimuthal contraction may not be uniform over the magnet and may lead to local changes in material characteristics.

(iv) If the coil is self supporting but not restrained from free movements or free contraction, the aperture shape and the useful magnet bore volume may change uncontrolled leading to a new and perhaps undesirable field distribution.

(v) When the superconducting coil is energized, magnetomechanical forces (Lorentz forces) act on the conductor in such a manner that circumferential strains are superimposed to the strains mentioned above. The conductor is either cold worked further or moves under the combined effects of the stress environment.

Cold work increases primarily the residual resistivity of the substrate and thus effects the cryogenic (or cryostatic) stability of the magnet and yield material embrittlement. If the conductor strain exceeds the matterial yield limit (e.g. 0.15% yield), plastic deformation will occur. Under this condiction the material will creep. Fig. 5.6.2 illustrates the stress strain properties of copper when magnetomechanically cycled, such as in case of pulsed synchrotron magnets. After a number of cycles the material plastic creep may lead to a loosely wound coil and may exhibit training effects and degradation.

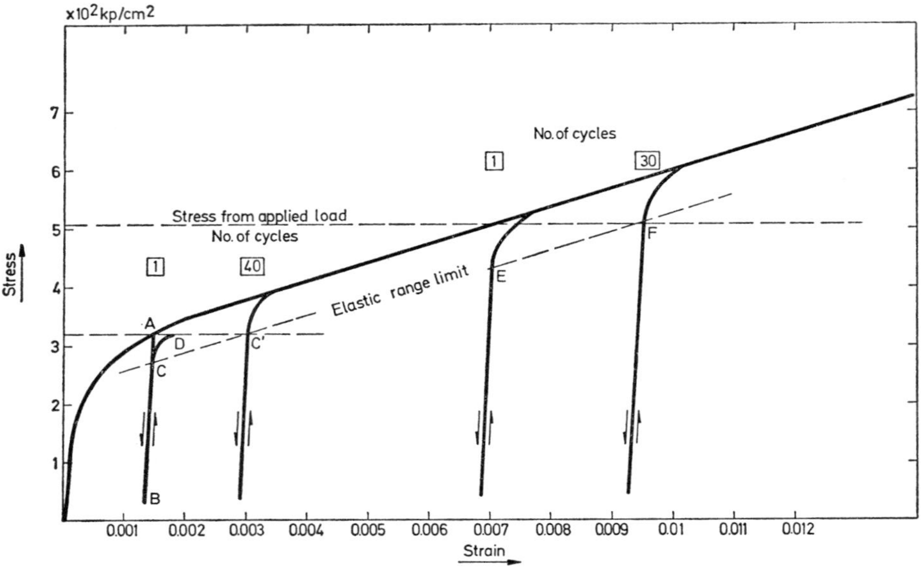

Fig. 5.6.2. Magneto-mechanically-stressed-strained OFHC-copper at 4.2 K

The interturn and interlayer insulation is also affected by the thermal and magnetomechanical stresses. The insulation may rupture and turn to turn insulation can break, leading to local short circuits.

The result of interturn and interlayer short circuits are, prolonged charging time and in case of a quench, the release of magnetic field energy. The energy discharge will occur primarily into the short circuit which may drive the conductor and consequently the coil into a normal state. In case of many shorts distributed over the coil (not lacalized shorts), only stabilized coils are able to withstand the energy discharge without being damaged. The above complication is not intended to paint a gloomy picture of the problems facing the magnet designer, but to encourage a more thorough study of thermomechanical properties of material being utilized in superconducting magnets.

5.6.2 Mechanical Properties of Solids

We briefly treat stress-strain, impact strength, flexibility, thermal contraction, and elastic and plastic behaviour of a number of solids commonly used in superconducting magnets.

5.6.2.1 Stress-Creep Relation

In general, materials used in superconducting magnets are seldom stressed beyond their elastic limits. However it may happen that the materials used may be stressed to their plastic limit. In this case plastic creep is observed.

In straining the material we observe three regions:

1. Strain in elastic region ε_1 (time independent). In this region Hooke's law is valid

$$\varepsilon_1 = \frac{1}{E} \cdot \sigma . \tag{5.6.1}$$

2. Region of viscous creep ε_2. The strain is roughly proportional to stress and time

$$\varepsilon_2 = C_2 \cdot \sigma \cdot t . \tag{5.6.2}$$

3. Strain due to elastic creep ε_3. This strain has at least one, but usually several relaxation times

$$= \sum_{i=1}^{n} \varepsilon_{3,i} (1 - e^{-j\tau}) . \tag{5.6.3}$$

with the relaxation time expressed as:

$$\tau = C_3 \cdot \exp(-C_4/T) .$$

τ is a function of temperature.

$C_2 \ldots C_4$ are constants, and are determined experimentally. The test procedure to obtain ε_2 and ε_3 is as follows:

The stress on the material is reduced to zero and the retarded creep phenomenon is observed optically. Fig. 5.6.3 illustrates the stress-strain behaviour of a thermoset schematically.

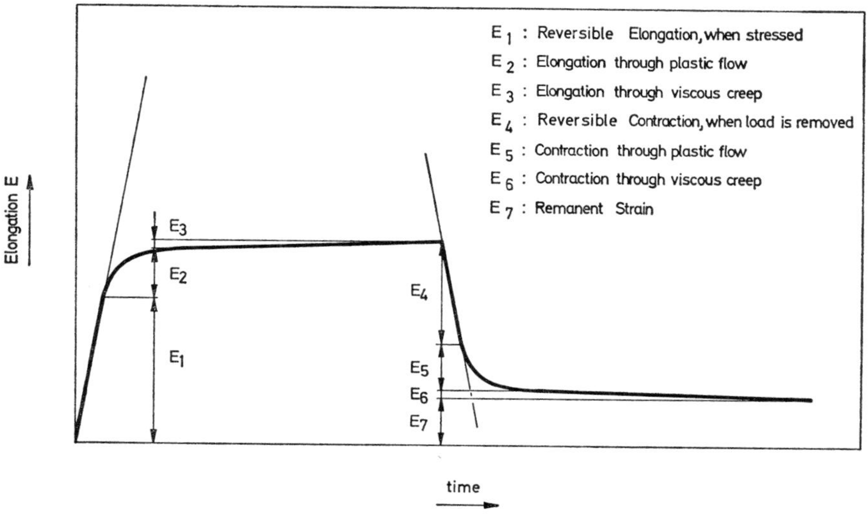

Fig. 5.6.3. Schematic representation of strain-time-relation in elastic and plastic region

If the material is stressed beyond its elastic limit (e.g. 0.1% yield) even if the material is relieved of any preloading afterwards a small preloading history will prevail.

Effect of stress on polycrystalline metals are elastic and plastic strain, recrystallization, crystal defects, dislocations (edge, screw), embrittlement, change in residual resistivity, and change in magnetoresistance behaviour.

When materials are cooled, the thermal vibration of atoms is less effective in assisting dislocation motion at cryogenic temperatures. Ductility is decreased at lower temperatures and material embrittlement is observed.

Embrittlement in materials (introduction of impurities with dislocation) is observed in metals, alloys, thermosettings, and elastomers. Only Polytetrafluoroethylene is an exception to the rule. It can be deformed plastically down to 1 K.

The mechanism by which thermosettings and elastomers respond to stress is not the same as in metals at low temperatures. The decreased thermal energy at cryogenic temperatures allows the attractive intermolecular forces to become more effective, the material resists deformation more strongly. The transition from ductile to brittle condition occurs at a temperature range, where the thermal contraction of materials are large, indicating that the increased rigidity of the materials results from a decrease in effective distance between adjacent molecules.

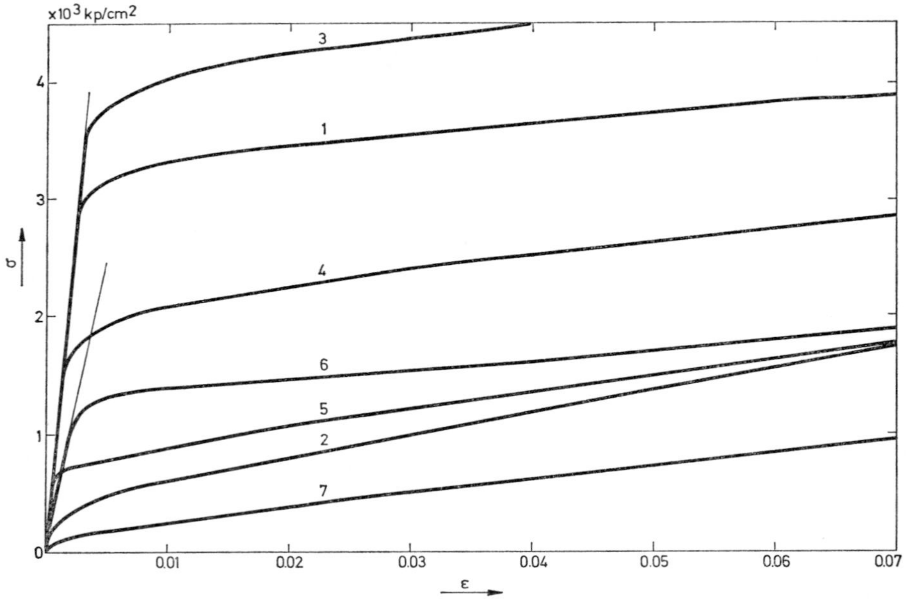

Fig. 5.6.4. Stress-strain diagrams of coppers and aluminums
1—7 See caption Fig. 5.6.1

5.6.2.2 Stress-Strain Relation

The most common test to measure the strength of solids are tension tests yielding elongation. Data obtained are yield and tensile strength and the elastic modulus. Compression and impact strength are important material properties needed to predict conductor and insulation behaviour.

With few exceptions the tensile strength of solids is greater at low temperatures than at room- or elevated temperatures. For polycrystalline metals the strength at 4.2 K may be two to five times that at room temperature. For plastics the strength at 77.2 K may be two to eight times greater than the room temperature value. Glasses show less increase in strength at low temperatures ($\sim$ factor 2) compared to values at 300 K. The same result is obtained in filled or glass-fiber reinforced thermosettings.

The classical stress-strain curves (shown in Fig. 5.6.4) for copper and aluminum have 2 sections:

(i) An initial rise representing the approximate linear reversible relationship between stress and elongation (the elastic characteristic of the material).

(ii) The section representing the plastic behaviour. The figures represent typical examples of OFHC and ETP coppers widely used as a substrate in composite conductors and pure aluminum, annealed and cold worked. The stress-strain relationship of Nb-Ti and OFHC copper composite and annealed ETP copper is illustrated in Fig. 5.6.5. Stress-strain relation for Nb-Ti wire (0.025 cm diameter) is given in Fig. 5.6.6 at 300 K and 4.2 K.

Cyclic stress-strain characteristic of ETP copper was given in Fig. 5.6.2. At the first cycle the sample is strained to a point A. The stress is now reduced, until B is reached. Reapplying the same initial load, the material is elastic until point C. The material is in a plastic range and the curve moves to D. Repeating the stress-strain cycles, the material keeps extending until C and D coincide. The extension is indicated by the line C—C'. The stress on the material is further repeated to the point E—F etc.

Few types of stainless-steels when strained at cryogenic temperature, become slightly ferromagnetic. The proper choice of stainless-steel, which retain its nonmagnetic behaviour is important, specifically for high homogeneity magnets, or if the problem with residual fields become disturbing.

In few cases, beryllium copper is selected as a reinforcement, even though the price of this material is higher than stainless-steel. In Fig. 5.6.6 the stress-strain curve of 2% Be Copper is given for comparison.

Insulation in the form of tapes, films, or spacers is used in superconducting magnets. Glass-reinforced epoxies and polyesters are utilized frequently. In addition glass fiber tapes impregnated in epoxies as interturn and interlayer insulation, nylon, dacron, thermosettings filled with Al_2O_3 or quartz, and unfilled are common in magnet design to achieve conductor rigidity, to prevent wire motion, and to define coil geometry.

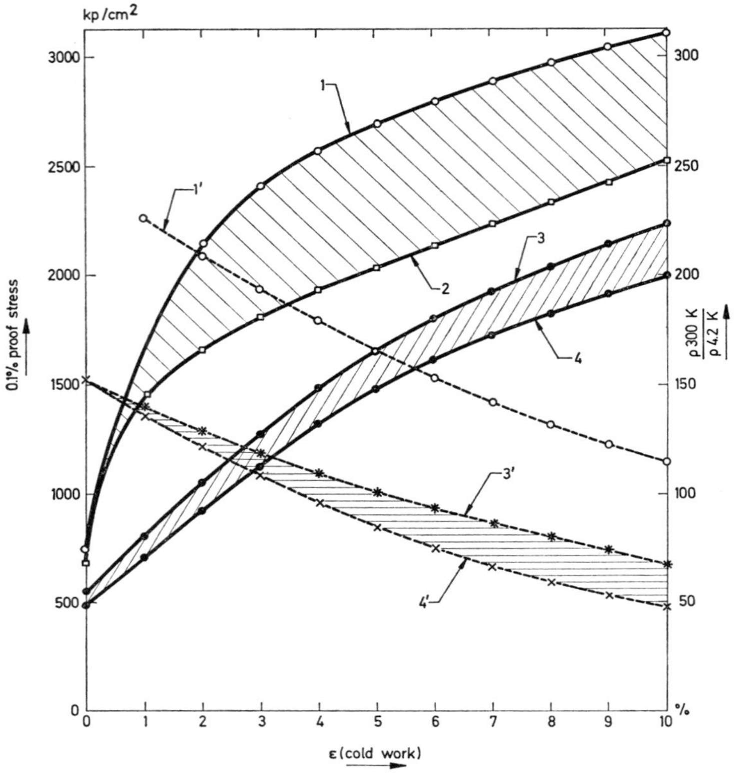

Fig. 5.6.5. Stress-strain and resistivity-strain relationship for coppers and composite conductors (IMI)

$\sigma - \varepsilon$
- *1* Nb(60%)Ti and Cu: $\frac{3}{1}$ Cu to superconductor ratio
- *2* Nb(60%)Ti and Cu: $\frac{5}{1}$ Cu to superconductor ratio
- *3* 99.995% pure OFHC copper, as received
- *4* 99.995% pure OFHC copper annealed

r
- *1'* 99.995% pure OFHC copper annealed
- *2'* Nb(60%)Ti and Cu: $\frac{5}{1}$ to superconductor ratio
- *3'* Nb(60%)Ti and Cu: $\frac{3}{1}$ to superconductor ratio

Simple static test methods on glass fiber reinforced thermosettings (such as tensile, flexural and shear) do not divert from those on metals and reinforced plastics. While the tensile specimen with expanded ends in metals are specified by ASTM, NEMA or DIN norms, there are no standards for probes to study mechanical properties of reinforced plastics at low temperatures. Clamping the test specimen and inserting appropriate pressure at low temperatures have lead to more complicated clamp designs.

Glass fiber is brittle; it does not flow or work harden to any appreciable extent. However measurements indicate that the tensile strength of glass is dependent upon the length of time and the load which is applied. It

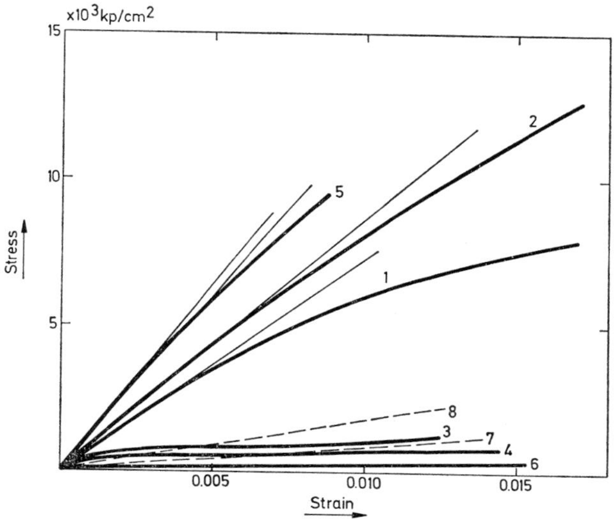

Fig. 5.6.6. Stress-strain diagram of materials:
1 Nb(60%)Ti-wire, 0.025 cm diameter at 300 K ($E = 7.15 \times 10^5$ kp/cm²)
2 Nb(60%)Ti-wire, 0.025 cm diameter at 4.2 K ($E = 8.14 \times 10^5$ kp/cm²)
3 Composite conductor: Cu/NbTi $= \frac{10}{1}$
4 OFHC-copper annealed
5 Be(2%)Cu: ($E = 1.23 \times 10^6$ kp/cm²) at 4.2 K
6 Glassfiber epoxy structureν glassfiber direction to stress direction $\alpha = 55°$
7 As (6): $\alpha = 45°$
8 As (6): $\alpha = 30°$

has been shown that "static" fatigue of glass is due to attack of water vapour and carbon dioxide on the surface. With cyclic loading, the glass fiber should not suffer except from its environment. This has been confirmed experimentally.

The glass fiber woven in a tape or mat exhibits more pronounced fatigue properties than bulk glass. There are two reasons:

(a) To make the glass fiber compatible to the thermosetting, the glass fiber is heat cleaned and a surface finish is applied.

(b) The type of weave also has a small effect.

The mechanism of interaction between the resin, the filler, and the glass fiber reinforcement is a simple one. When the impregnation is performed, the thermosetting must wet the surface of the fiber or filler thoroughly. During curing, it polymerizes or condenses around the fiber (filler) to form a rigid solid. During this curing process, not only does the resin stick to the fiber but, also shrinks. Thus, the mechanism of adhesion is partially mechanical. However, it is known that adsorption and diffusion are equally important. Finally, chemical and physical effects have to be considered. The mechanical adsorption and the diffusion processes involve intermolecular forces, the physical effects involve electrostatic forces, and the chemical effects involve the chemical bond.

In order to improve the bond strength of the filled impregnant, wetting of the inorganic surfaces, by means of organic polymers (silicone-chemicals) such as Silan coupling agents, has been very successful. When treating the E-glass-cloth or the inorganic filler (quartz, Al_2O_3) with the organic polymer and impregnating it with epoxy, the flexural strength of the system is increased by a factor of about two. This improvement in overall mechanical strength permits loading of epoxies to about 60% by weight of the total mixture. As seen in Fig. 6.2.8, such an epoxy filler system has practically the same thermal contraction characteristics as a composite superconductor.

Because of the complexity of reinforced thermosettings, the assumptions of homogeneous isotropic materials to obtain some average elastic modulus, average tensile, or yield data may lead to errors. Samples must be tested in two or three directions. Test data also vary with reference to the direction of glass filament to the stress direction, respectively. Thus, even in simple geometries, it is not an easy matter to predict low temperature performance of complex systems.

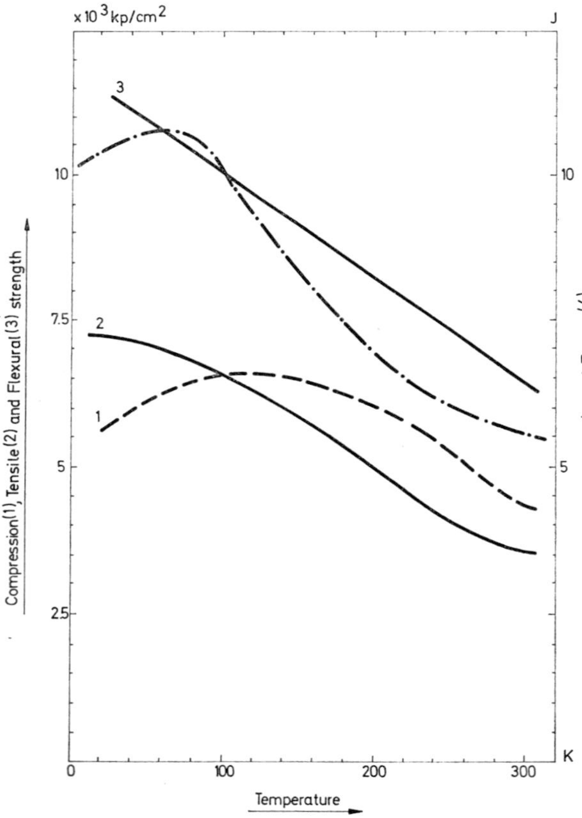

Fig. 5.6.7. Tensile, compressive strength and impact energy of glassfiber reinforced epoxy resins vs. temperature. Glasscontent 35% by the total weight

The stress-strain relationship may yield valuable test data when comparing properties of different thermosettings, the methods of preparation, pre and past curing, method of loading them with suitable fillers, glass fibers (chopped or continuous), etc. However, data on impact energy and the knowledge of the work to initiate fracture is of equal importance.

Data on tensile strength of glass fiber reinforced epoxy resin and its impact energy and compressive strength are given in Fig. 5.6.7. To find a relationship between strain energy and strain behaviour, the integration of the area under the σ-ε-curve at different temperatures and plotting the integrated data versus temperature is employed.

5.6.2.3 Fatigue

Transient stresses are induced in a solid during the cooling process, thermal cycling or magneto-mechanical cycling. If the thermal contraction of the various magnet components are different, observable effects are produced. In complex systems such as glass fiber reinforced thermosettings bonded to copper two effects are the most obvious:

Due to the different stress-strain properties of the various components shear stresses are produced at the surface layers between the components. During cool down the temperature of the thermosetting (bond) lags behind that of the metal because of its poorer thermal conductivity. Due to the fact that there will be differences in thermal contraction between the metal and the adhesive, thermal stresses are induced in the interface layer. If the thermal stress is greater than the bond strength, the bond may shear and fracture.

Stresses are also encountered if the materials are subjected to transient magnetomechanical strain, however there will be no differential expansion and contraction effects between the various magnetcomponents. The material is ruptured because of tensile fatigue properties. In

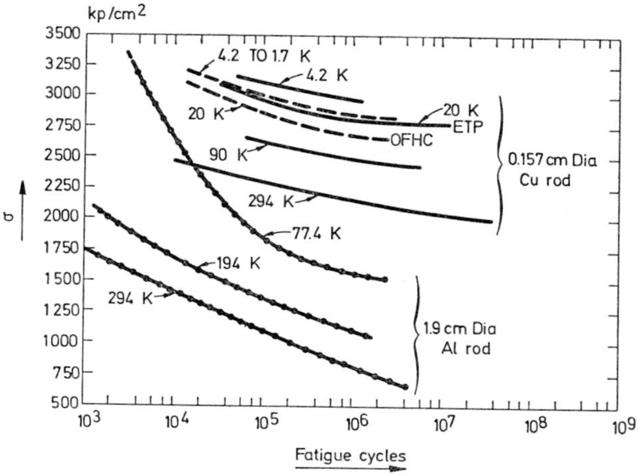

Fig. 5.6.8. Fatigue-strength of copper and aluminum rods at different temperatures

Fig. 5.6.8 fatigue properties of aluminium and copper and in Fig. 5.6.9, fatigue curves of glass fiber reinforced epoxies are illustrated.

Table 5.6.1 gives a few mechanical properties of epoxy fiber glass-laminates.

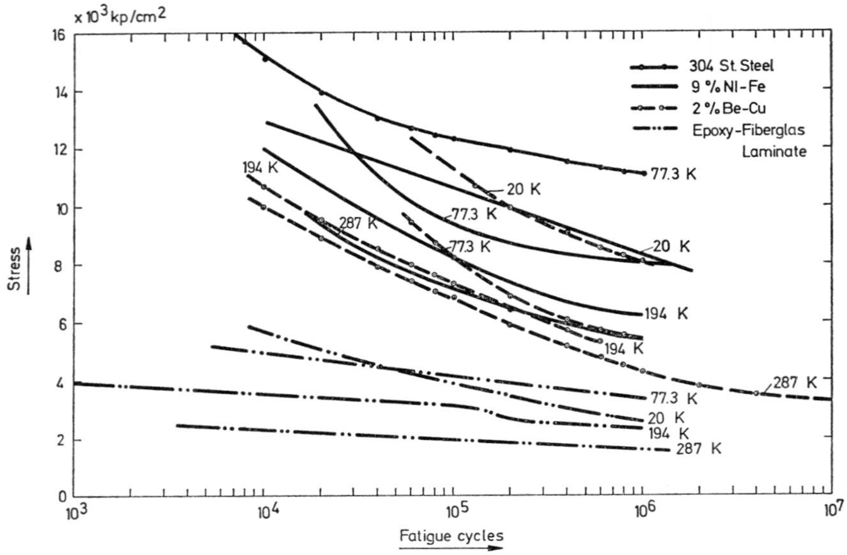

Fig. 5.6.9. Fatigue-strength of some structural materials at different temperatures

As mentioned, fiber glass is not affected by mechanical stress-cycling. The fatigue behaviour of glass-reinforced epoxies depends predominantly upon the behaviour of the resin. The fatigue behaviour of various resins have been studied extensively at room and elevant temperatures. Only scattered data are available at cryogenic temperatures.

From the available data fatigue curves on unfilled resins at room temperature and glass-reinforced resins at different temperatures are illustrated in Fig. 5.6.9.

Table 5.6.1. *Mechanical properties of epoxy-fiberglass-laminate* (FMNA with 35% glass-content by weight)

		Temperature		
		300 K	77.3 K	4.2 K
Tensile strength	(kp/cm²)	3,000	6,450	7,150
Compressive strength	(kp/cm²)	3,000	7,700	7,500
Flexural strength	(kp/cm²)	5,600	10,500	10,000
Fatigue strength	(kp/cm²)			
10^3 cycles		3,000	5,700	7,100
10^4 cycles		2,360	4,940	5,650
10^5 cycles		1,930	4,150	3,940
10^6 cycles		1,640	3,350	2,500

5.6.3 The Work of Fracture

Impact tests on specimen are often made on metals to obtain an estimate of their toughness. In particular the Charpy-V-notch test is widely used. It is well accepted that a high breaking stress or yield strength is not the only criterion when making a choice of a material for a particular application, because it is often found that when the material fractures a high breaking strength is accompanied by only a small amount of energy absorption. This means that the material is brittle, and its failure is sensitive to cracks or flaws that develop during service.

As an example we illustrate the tensile and impact properties of low carbon steel and soft magnetic 9% nickeliron as a function of temperature in Fig. 5.6.10. Low carbon steel is brittle at 4.2 K and fracture sensitive. 9% Ni-Fe has good mechanical properties. Its magnetic property is comparable to that of low carbon steels.

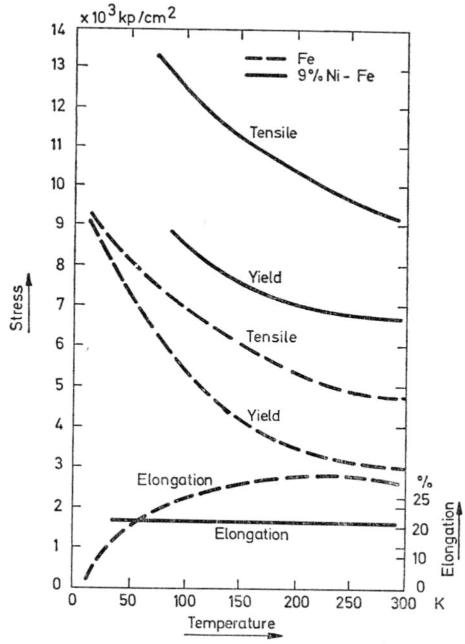

Fig. 5.6.10. Stress-temperature diagrams and elongation-temperature diagrams of stainless-steel (304) and 9% Ni-Fe (soft-magnetic)

The criterion for crack growth is given by the Griffith-relation:

$$-\frac{\partial U}{\partial A} = \gamma\,, \quad (\mathrm{J/cm^2}) \tag{5.6.4}$$

where U is the elastic strain energy stored in the structure, A is the area of the fractured face, and γ is the surface energy. When this criterion is fulfilled, it is energetically possible for the crack to grow. If it should do so, its subsequent behaviour depends upon the change of $-\partial U/\partial A$, as the crack increases in size. If $-\partial^2 U/\partial A^2$ is positive, the crack will accelerate because the energy being released is more than sufficient to create the new surface area. If $-\partial^2 U/\partial A^2$ is negative, there may come a point during the propagation of the crack when $-\partial U/\partial A$ becomes less than γ and external work must be done to keep the crack moving. In this case the growth of the crack can be controlled.

To obtain the amount of energy which produces a crack and to make it grow, a hard machine must be used. At the instance when the fracture begins, further loading must be stopped. The crack initiation in thermosetting is of considerable interest and its measurement quite important, but is difficult at cryogenic temperatures.

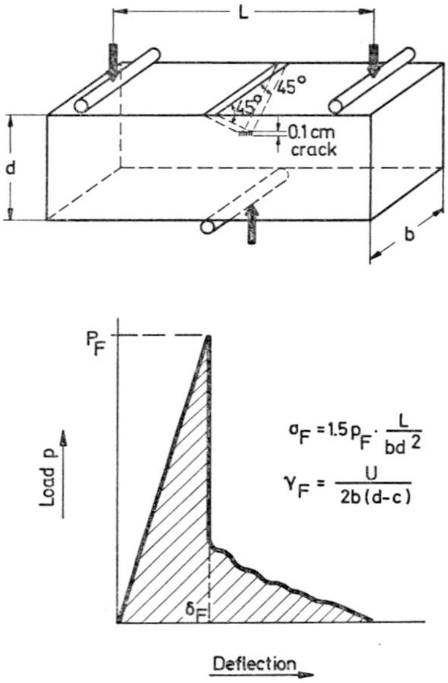

Fig. 5.6.11. Test set up (schematic) to measure strain energy at low temperatures. c is the depth of the notch

(i) Geometries of specimen and a method to determine γ is given by Tattersall and Tappin [33], and modifications of the notch form is proposed by RHEL. The data published for epoxies scatter and in order to obtain reproducable data, the specimen are precracked at room temperature, with a controlled crack depth of approx. 0.1 cm at the top of the triangular notch. The precracking is accomplished by mechanically cycling the specimen in a vibrator. However present data scatter. The load deflection curve is illustrated schematically in Fig. 5.6.11. When the notch depth is small compared to the depth of the sample, then the surface energy γ is given by:

$$\gamma = -\frac{\partial U}{\partial A} = \frac{(1 - m^2)\, \pi\, (\sigma_F^2)\, c}{2E}, \tag{5.6.5}$$

where m is Poisson's ratio, E is Young's modulus, σ_F the fracture stress and c the depth of the triangular cut. In Eq. (5.6.5) plain strain condition ($3 P_f l/2 bd^2$) is assumed. The work of fracture is given by:

$$\gamma_F = \frac{U}{2b(d-c)}. \tag{5.6.6}$$

Data on work of fracture of several materials at room temperature is given in Table 5.6.2.

Table 5.6.2. *Work of fracture at 300 K and 4.2 K*

Material	Work of fracture (J/cm²)
Copper	5
Brass	3
Cast iron	0.4
Poly styrene (tough)	0.4
Graphite	22×10^{-3}
Alumina	4×10^{-3}
Glass	10^{-3}
Glassfilament-Epoxy 300 K (20 to 30% glass cont. by volume)	$(3–5)\cdot 10^{-2}$
Epoxy at 300 K ($E = 6.5 \times 10^5$ kp/cm²)	$1.3\cdot 10^{-2}$
Epoxy at 4.2 K ($E = 7.15\times 10^5$ kp/cm²)	$(2–5)\cdot 10^{-2}$

5.6.4 Thermal and Transport Properties of Solids

In this section we treat the specific heat, thermal conductivity and thermal diffusivity, the thermal and electrical conductivity of solids pertinent to the behaviour of superconducting magnets. Thermal contraction (expansion) of solids will be treated briefly, to sum up the various properties of solids in the temperature range from 4 to 300 K.

(a) Specific Heat and Thermal Capacity

The heat capacity of a substance is defined as the energy gradient with respect to temperature, while the volume is kept constant. Thus:

$$C_v = \left(\frac{\partial E}{\partial T}\right)_v, \tag{5.6.7}$$

with E = internal energy of the substance and T = its temperature. C_v is the amount by which the internal energy changes due to a rise in temperature of one degree when heat is applied to the substance. According to the rule of Dulong and Petit the thermal capacity is a constant and independent of temperature:

$$C_v = \frac{dE}{dT} = 3 N_0 k = 25.7 \quad \left(\frac{\text{J}}{\text{mole}\cdot\text{K}}\right).$$

However as illustrated in Fig. 5.6.12 the thermal capacity is strongly dependent on temperature and is zero at 0 K. The theory of heat capacity was advanced by Debye who treated a crystalline solid as an infinite elastic continuum and visualized the excitement of all the possible elastic standing waves in the material. The lattice heat capacity per mole according to Debye is expressed by:

$$D = 3\,nR\left[3\left(\frac{T}{\theta_D}\right)^3 \int\limits_0^{\theta_D/T} \frac{x^4 e^x}{(e^x - 1)^2}\,dx\right], \qquad (5.6.8)$$

with n = number of atoms per molecule, R the universal gas constant per mole $= 8.3 \times 10^{-14}$ J/mole · K, θ_D = Debye temperature defined by

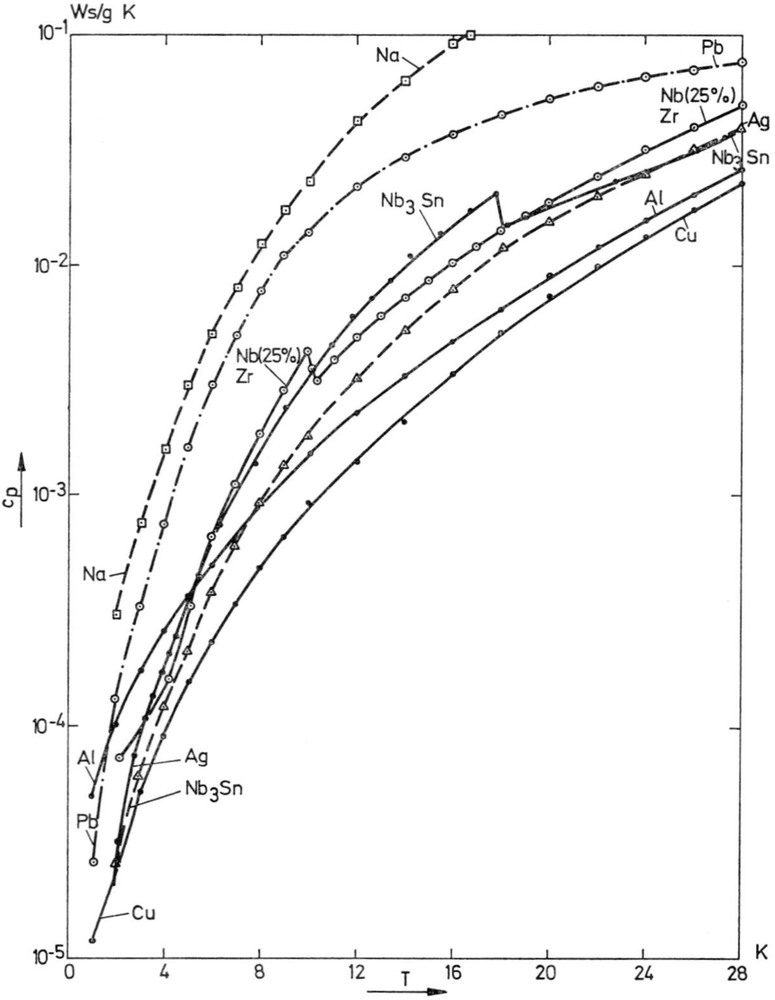

Fig. 5.6.12. Specific heat vs. temperature of normal metals and Type II superconductors

$(h\omega_D/k_B)$ while the value of $\omega_D = (6\pi^2 N \cdot v_s/V)^{\frac{1}{3}}$, v_s = average velocity, V/N the volume per atom, k_B = the Boltzmann constant $= 1.38\times10^{-23}$ (J/K), x is a dimensionless variable defined by $x = h_v/k_B T$ and h = Planck's constant $= 6.626\times10^{-34}$ Joule/s.

The heat capacity according to Eq. (5.6.8) is dependent only on (T/θ_D) and the mathematical function of x.

The agreement between Debye theory and the experimental results is remarkably good for many solids, specifically at $T \ll \theta_D$. This is even more surprising if we consider that Debye assumed a parabolic form of the vibration spectrum for every solid without regard for individual differences.

Debyes theory was later amended by considering interatomic forces in the lattice. Born, Blackman and Kármán calculated the frequency spectrums in more detail.

While normal materials such as Ag, Al, Cu or others behave with respect to C_v as predicted by the theory, superconducting materials illustrate an anomalous behaviour. At the critical temperature, the specific heat exhibits a discontinuity which is in concurrence to the transition from superconducting to non-superconducting state.

The specific heat in the normal state of Nb_3Sn can be expressed with good approximation by:

$$\boxed{C_n = \alpha T + \beta T^3,} \qquad (5.6.9)$$

where α and β are constants proportional to the density of states, at the Fermi surface and the Debye-temperature.

The specific heat jump at $\Delta C/T_c = 31$ m J/(gr at) $\cdot$ K² is reported by Vieland [34]. The normalized jump $\Delta C/\alpha T_c$ is close to the BCS prediction of 1.43 with ($\alpha = 21$), which is obtained by linear extrapolation of the normal state data.

At the low temperature range $3.2 < T_c/T < 16$, in addition to the predicted T^3 term, an additional term T is present. For Nb_3Sn the Debye-temperature is found experimentally to be $\theta_D = 228$ K.

Data on the specific heat of common metals, few alloys such as constantan, monel metal, soft solders, supporting materials such as silica glass, Pyrex, Teflon, Nylon, casting and bonding materials such as epoxies, Bakelite varnish are published in literature [8] in the temperature range of 1 to 100 K.

Correlation equations for the specific heat of materials used in magnets are given below:

(i) Stainless-steel (17% Cr; 10,1% Ni; 0.86% Mn):

$$c_p = 0.464\ \mathrm{T} + 3.8\times10^{-4}\ \mathrm{T}^3 \qquad \text{(J/kg K)}$$
for T < 10 K

(ii) Nylon and cold setting epoxies (CIBA):

$$c_p = 1.8\times10^{-2}\ \mathrm{T}^3 \qquad \text{(J/kg K)}$$
for T < 4 K

(iii) Carbon resistors (10 Ohm, 0.1 W) (Allen Bradley):

$$c_p = 6.4\times10^{-3}\ \mathrm{T} + 1.3\times10^{-3}\ \mathrm{T}^3 \ \text{(J/kg K)}$$
for 1 K $\leqq$ T $\leqq$ 4 K.

It may be noted that the specific heat of thermosettings, varnishes and bonding agents are greater then that for copper at temperatures between 1 K and 20 K. Their contribution to the specific heat measurements in calorimeters can not be neglected.

(b) Thermal Conductivity

Considering a system which is not in thermal equilibrium, it is of practical interest to study a non-equilibrium steady state, such as placeing the opposite ends of a sample in thermal contact to systems of two different temperatures. The energy transport from the higher to a lower temperature system is accomplished through the thermal conductivity of the contact link between the two parts. The quantity transported through the system is the energy U. The flux of the energy is defined by:

$$\dot{q}_u = \frac{Q}{A} = -k \operatorname{grad} T . \tag{5.6.10}$$

If we consider the solid as an aggregation of interconnected atoms, vibration of any atom in the solid affects the others. The propagation takes place through interatomic forces in the material. Actually collective vibration of an entire array of atoms occurs rather than vibration of individual atoms. In a crystalline solid these vibrations may be considered as displacement waves travelling through the solid. With increase in temperature, thermal vibrations increase in amplitude. The transmission of a disturbance through a solid is governed by the density of the material and the modulus of elasticity. In an isotropic solid, the transmission speed $\bar{v}$ of a longitudinal wave is given by:

$$\bar{v} = (E/\delta)^{\frac{1}{2}} \tag{5.6.11}$$

with E the modulus of elasticity and δ the density of the material. E relates the restoring force to the displacement and δ the mass of material displaced by the vibration. The greater change in an elastic modulus during the vibration (if T is increased) leads to more effective scattering of thermal waves. In 1929 Peierls postulated the term of directional cooperative quantized thermal vibrations (phonons) of different frequencies in the crystal. Phonon is a "particle" of thermal energy. A thermally excited solid is treated theoretically as a gas of phonons.

A second mechanism for the low temperature transport of heat is the electronic thermal conduction (the transport of thermal energy) by the motion of conduction electrons.

There are two main scattering processes that limit the electronic thermal conductivity. The first is the scattering of conduction electrons by thermal vibrations of the lattice, as represented by the electron-phonon resistivity ϱ_i, a characteristic property of a metal. The second process is the scattering of conduction electrons by imperfections (impurity atoms and lattice defects), as represented by the imperfections or electron-defect resistivity ϱ_0. This scattering is most important at low temperatures. The total reciprocal electronic thermal conductivity k_e is usually assumed to be the sum of the two resistivities (Matthiessen's rule):

$$\boxed{\frac{1}{k_e} = \varrho_i + \varrho_0 .} \tag{5.6.12}$$

The absolute magnitude and temperature dependences of the two resistivity terms have been extensively studied both experimentally and theoretically.

The two terms of the thermal resistivity equation can be given explicitly by:

$$\varrho_i = A \cdot T^n \quad n \approx 2\text{—}3; \quad T < 40\text{ K}$$

$$\varrho_0 = B \cdot T^{-1} \quad \text{at all temperatures.}$$

The constant A in the electron-phonon resistivity term is related to the characteristic properties of the material. B in the imperfection resistivity term is related to imperfections and residual electrical resistivity of the particular sample. The electron-phonon resistivity of a metal reaches a constant value above 40 K.

Equation (5.6.12) is also amended by a deviation term $\varrho_{i,0}$ given by the expression [35]

$$\varrho_{i,0} = \frac{\alpha \varrho_i \varrho_0}{\beta \varrho_i + \gamma \varrho_0},$$

α, β and γ are constants of order unity and are determined experimentally.

In pure annealed metals the phonon-electron resistivity has approximately the same temperature dependence as the phonon dislocation resistivity. Separation of the two scattering mechanisms may not be unambiguous. For unannealed specimen the dislocation resistivity will greatly outweigh the other.

In an actual crystal which contain imperfections, phonons are scattered. From the concept of phonon-scattering the lattice thermal conductivity was derived as:

$$\boxed{k_L = \frac{1}{3} C_v \cdot \bar{v} \cdot l .} \tag{5.6.13}$$

The thermal conductivity increases directly with the mean free path length l, the heat capacity and the speed of the particles $\bar{v}$. The thermal conductivity c_v of a solid can be expressed in terms of the conductivity k_0 at the Debye temperature.

$$k_L = k_0 \cdot \left(\frac{\theta_D}{T}\right). \tag{5.6.14}$$

This expression holds when $T \geqq \theta_D$.

At cryogenic temperatures where $T \ll \theta_D$ then

$$\boxed{k_L = k_0 \cdot \left(\frac{\theta_D}{T}\right)^3 \exp\left(\frac{\theta_D}{bT}\right).} \tag{5.6.15}$$

With sufficient reduction in temperature the mean free path length increase until it reaches the dimensions of the crystal, then phonons are reflected from the crystal surfaces. With l and $\bar{v}$ being constants, k_L depend on the dimensions of the crystal and according to Eq. (5.6.12) on the thermal capacity, which vanishes at zero temperature.

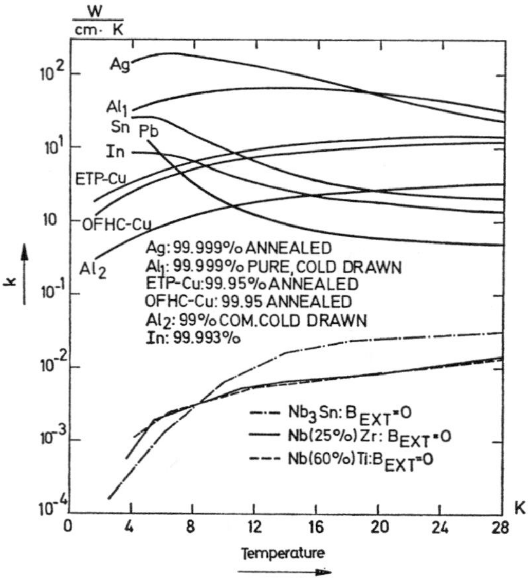

Fig. 5.6.13. Thermal conductivity of normal metals and Type II superconductors

In Fig. 5.6.13 the thermal conductivity of a few metals and Type II superconductors are illustrated as a function of temperature. The thermal conductivity of common superconducting alloys is at least two orders of magnitude smaller than normal metals used in composite conductors. For Nb_3Sn the thermal conductivity in the temperature range of 18.3 to 6.5 K is dominated by an electronic component that is impurity limited [36]. Below 6.5 K the number of normal electrons has been so reduced that the phonon contribution rises only to be limited by boundary scattering with a constant mean free path and a T^3 temperature dependence characteristic of the low temperature specific heat.

The thermal conductivity curves for coppers plotted in Fig. 5.6.13 are based on commercially available samples and not of the high purity types. These coppers are more readily obtained commercially and used in conjunction with superconductors. The two aluminum samples represent cold drawn hard and annealed wires.

The theory of thermal conductivity is idealized. No real crystal is free of defects and these provide additional scattering sources for phonons. Impurity and interstitial atoms in the lattice, vacancies and dislocations provide irregularities capable of scattering phonons. A quantitative treatment of thermal conductivity is still not available. Nonmetallic systems, complex insulations, thermosettings etc. can be treated in the same way as crystalline dielectrics, even if the phonon-theory is based on structural regularity of the solid. In Fig. 5.6.14 the thermal conductivity of an epoxy system (F-MNA) is given. To match the thermal contraction of the epoxy to that of a composite superconductor the epoxy has been filled with quartz. The thermal conductivity is improved by about a factor of four (at 4.2 K) when loaded with inorganic fillers. The filler concentration has however two disadvantages: The mechanical strength of the epoxy system is reduced and the viscosity of the system is greatly increased. The first disadvantage can be remedied partially by adding about 1 to 2% epoxy functional materials (e.g. Z 6040) to the epoxy or heat cleaning and chemically treating the filler with Silans or Garans. To be able to impregnate glass fibers with these highly filled viscous epoxies, it is recommended to use three component epoxies (epoxy-hardner and accelerator). By reducing the amount of the accelerator in the mixture the impregnation temperature can be raised, thus lowering viscosity.

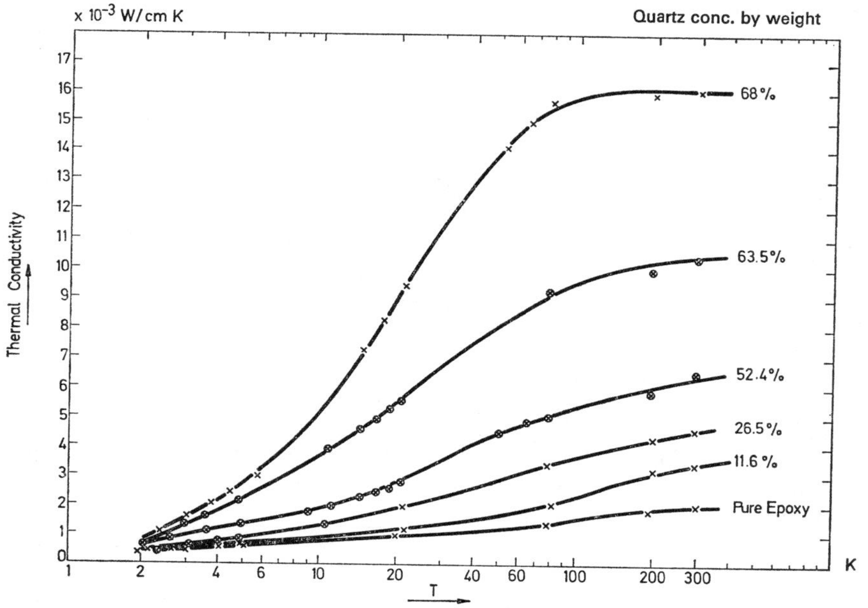

Fig. 5.6.14. Thermal conductivity of a mineral filled epoxy. The mineral content is given in percentage of the total weight of the mixture

The effect of the magnetic field (transverse or longitudinal) on thermal conductivity of metals is not given in detail. In superconductors such as NbTi the effect of phonon contribution below 5 K is somewhat more pronounced. The $\partial k/\partial T$ curve is steeper. However above 5.2 K no changes in k has been observed as a function of B.

In normal metals the effect of transverse magnetic fields is generally a reduction of the thermal conductivity. However the effect is not in agreement with the Wiedemann-Franz-law as shown in Section 4.8.2.1. At temperatures below 10 K, the Lorentz number can be assumed to be constant and independent of temperature ($L = 2.45 \times 10^{-8}\ \mathrm{V^2\ K^{-2}}$ for $T \ll \theta_D$).

(c) Thermal Diffusion

Thermal diffusivity plays quite an important role in the stability criteria of superconducting magnets. Enthalpy stabilization requires that the thermal diffusivity must be higher than the magnetic diffusivity in order to confine the region of normality, or to cool down the normal area and restore superconducting condition. The rate of field rise $\dot{B}$ in superconducting magnets, which is proportional to the magnetic diffusivity, is also limited to some extent by the thermal diffusivity.

If a system is in diffuse contact on both ends with two reservoirs of various potentials, the direction of particle flow will increase the entropy of the system. The driving force of diffusion is taken as the gradient of particle concentration, and the flux is the number of particles passing through the sample connecting the two systems per unit area in unit time:

$$\dot{q}_D = - D \cdot \mathrm{grad}\, n\,. \tag{5.6.16}$$

D, the diffusivity, is given by

$$D = \frac{1}{3} \cdot \bar{v} \cdot l\,. \tag{5.6.17}$$

Combining Eqs. (5.6.17) and (5.6.13), we may also write:

$$\boxed{D = \frac{k}{C_v} = \frac{k}{C \cdot \delta},} \tag{5.6.18}$$

as a ratio of thermal conductivity and the heat capacity of the material.

In Fig. 5.6.15, we have compiled data for thermal diffusivity of a number of materials, including a Type II superconductor, in function of temperature. The vast difference between the thermal diffusivities of normal metals and superconductors explains the reason why copper or aluminium is desired as a matrix.

(d) Electrical Conductivity

Several lengthy papers or books on the subject of electrical conductivity or its reciprocal value the electrical resistivity have been published recently [37, 38]. Therefore only a brief discussion will be given here.

When an electric field is applied to a metal, its quasi-free conduction electrons are accelerated, resulting in the transport of electrical energy. The transport process by electrons is again limited by the collision process. From Ohm's law, we may write for the electrical resistivity:

$$\varrho = \frac{2\, m \bar{v}}{(n e^2 l)}, \tag{5.6.19}$$

with e the charge of the electron, m its mass, n is the number of electrons capable of exchanging energy with the applied field per unit volume, $\bar{v}$ is the randome mean thermal velocity and l the mean free path.

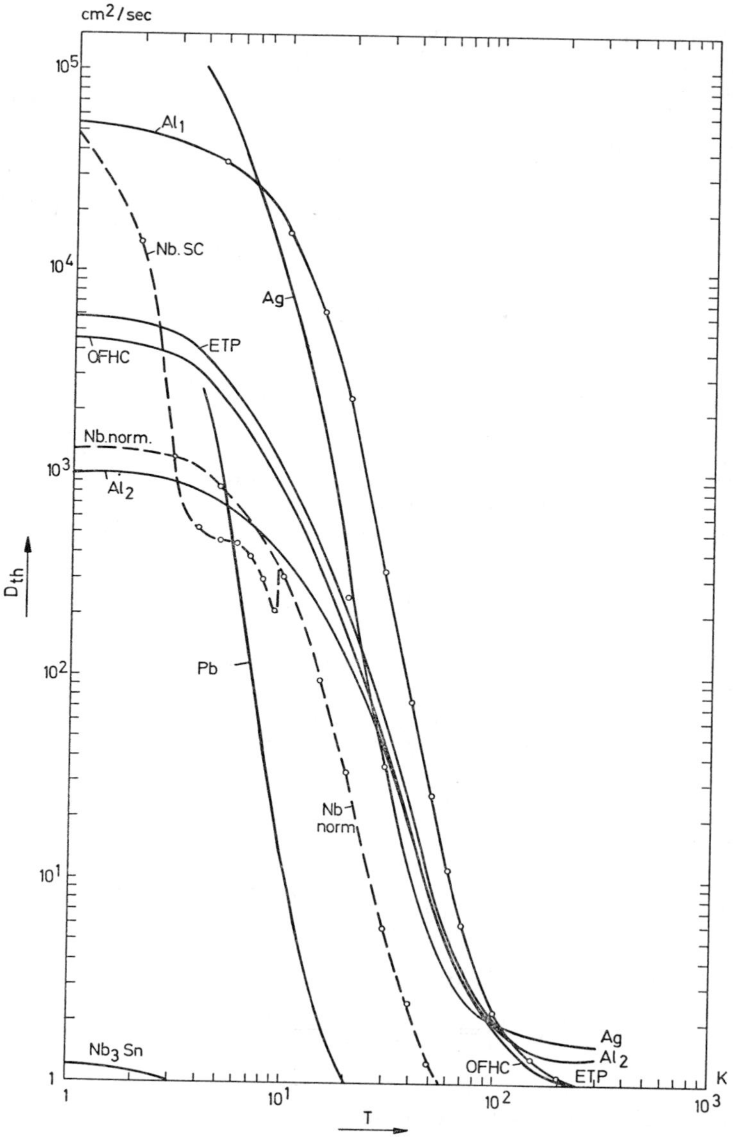

Fig. 5.6.15. Thermal diffusivity of normal metals and Type II superconductors. (D_{th} of Nb, Ti at 4.2 K is 1.2 cm²/s)

In Eq. (5.6.19), m and e are constants, n (for a given metal) is also constant, and $\bar{v}$ is only slightly dependent on temperature because of the degeneracy of the conduction electrons. Thus, the problem of electrical resistivity reduces to a mechanism limiting the mean free path.

In Section 5.6.4 (b) on thermal conductivity in metals we noted that the mean free path of electrons was limited by collision with phonons and with imperfections in the crystal lattice. The first is a dynamic scattering, the second a static one. The scattering mechanism can be considered independently to a good approximation. The total electrical resistivity ϱ, is thus assumed to be the sum of two separate resistivities, the residual or imperfection resistivity ϱ_0 and the intrinsic resistivity ϱ_i. This means that Matthiessen's rule for electrical resistivity is obeyed:

$$\varrho = \varrho_i + \varrho_0 . \tag{5.6.20}$$

The expression for the magnitudes and temperature dependence of the two components are approximated by:

$$\varrho_i = \alpha \cdot T^n \quad n \cong 4 \dots 5 \qquad T < 40 \text{ K}$$

$$\varrho_0 = \beta = \text{constant.}$$

Deviations from additivity for electrical resistivity have been observed experimentally for copper and its dilute alloys by Grüneisen [39] and others. The electrical resistivity of dilute copper alloys is further complicated by the existance of a minimum near 10 K.

To examine electrical resistivity data is to compare experimental values for some forms of the Bloch-Grüneisen equation, which is written by MacDonald [40] as:

$$\varrho = \frac{C}{\theta_D M}\left(\frac{T}{\theta_D}\right)^5 \cdot J_s\left(\frac{\theta_D}{T}\right), \tag{5.6.21}$$

where M is the atomic weight, C is a constant, and $J_s\,(\theta_D/T)$ is the Debye integral given by:

$$J_n(z) = \int_0^z \frac{z^n\,dz}{(e^z - 1)\,(1 - e^{-z})}.$$

To examine the temperature dependence, Laubitz [41] writes for the resistivity:

$$\varrho = C'T\left[1 - \frac{1}{18}\left(\frac{\theta_D}{T}\right)^2\right]\left[1 + 6\gamma \int_{\theta_D}^{T} (T)\,dT\right]\left(1 - \frac{\Delta\sigma}{\sigma_0}\right), \tag{5.6.22}$$

where the Debye integral is approximated by

$$J_s\left(\frac{\theta_D}{T}\right) = \frac{1}{4}\left(\frac{\theta_D}{T}\right)^4\left[1 - \frac{1}{18}\left(\frac{\theta_D}{T}\right)^2\right]$$

and $\Delta\sigma/\sigma_0$ is the correction to the Bloch-Grüneisen equation.

Eq. (5.6.22) is also useful in the simpler form:

$$\varrho = C'T\left[1 - \frac{1}{18}\left(\frac{\theta_D}{T}\right)^2\right]\left[1 + 6\gamma \int \alpha\,(T)\,dT\right]. \tag{5.6.23}$$

This equation agrees with Eq. (5.6.22) within 0.05% at $T = \theta_D$ and -2.3% at $T = 0.5\,\theta_D$.

To obtain the temperature dependence of ϱ we assume:

$$\alpha(T) = C_1 + C_2 T$$

and obtain:

$$\varrho = -\frac{6\gamma \cdot C'\theta_D^2 \cdot C_1}{18} + C'\left[1 - \frac{\theta_D^2 \cdot 6\gamma \cdot C_2}{36}\right]T + 6\,C'\gamma C_1 T^2 + \frac{1}{2}C'C_2\,6\,\gamma T^3 - \frac{C'\theta_D^2}{18}\left(\frac{1}{T}\right). \tag{5.6.24}$$

The constant term is less than 1% and can be ignored.

The term with T^3 goes to zero if α is assumed constant.

Thus:

$$\varrho = C'T + 6\,C'C_1\gamma T^2 - \frac{C'\theta_D^2}{18} \cdot \frac{1}{T}. \tag{5.6.25}$$

The experimental data can be least square fitted to an equation:

$$\varrho = AT + BT^2 + C/T \tag{5.6.26}$$

and one obtains a maximum standard deviation of $\sim 1.3\%$ for temperatures above 100 K for copper.

For $T/\theta_D > 0.5$, the Bloch-Grüneisen equation can be simplified as:

$$\varrho = \frac{A' \cdot T}{\theta_D^2}, \tag{5.6.27}$$

which gives the linear relation ship at high temperatures.

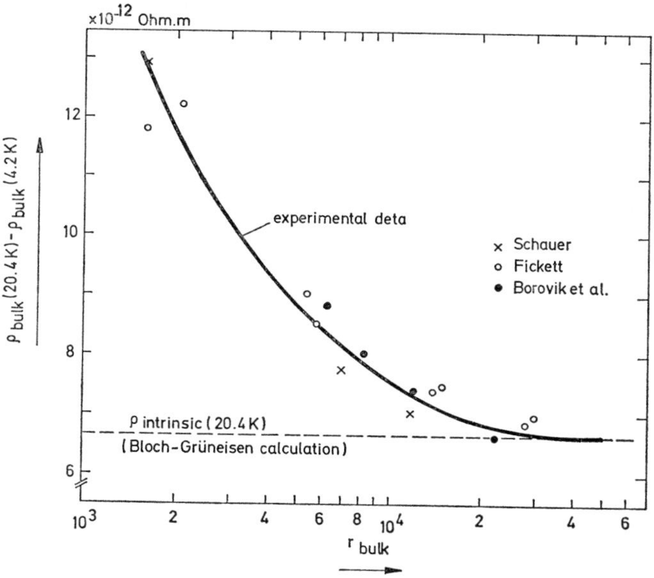

Fig. 5.6.16. Calculated and measured resistivity ratios ($B = 0$) of aluminum tapes vs. bulk resistivity ratio

For $T/\theta_D < 0.1$, the Eq. (5.6.21) is approximated by

$$\varrho = \frac{A'' \cdot T^5}{\theta_D^6} \tag{5.6.28}$$

and predicts that the resistance will have a T^5 dependence at low temperatures. The validity of the Bloch-Grüneisen equation is illustrated in Fig. 5.6.16 for aluminum of various purities in the temperature range of 4.2 to 20.4 K. The data show that Matthiessen's rule is violated and the two components of Eq. (5.6.20) are not independent. The interaction between dispersion phenomena is increased with defect concentration.

The effect of elastic and plastic deformation and cold-work on the increase of resistivity in copper and aluminum was given in (5.6.1). The increase in resistivity can be summarized by the relation:

$$\Delta\varrho = C \cdot \varepsilon^n , \tag{5.6.29}$$

where ε is the plastic strain. In Fig. 5.6.1, the resistivity of copper and aluminum was shown as a function of strain at 4.2 K.

It may be of interest to present the electric resistivity of a few engineering alloys for comparison. Fig. 5.6.17 gives resistivity curves for stainless steels.

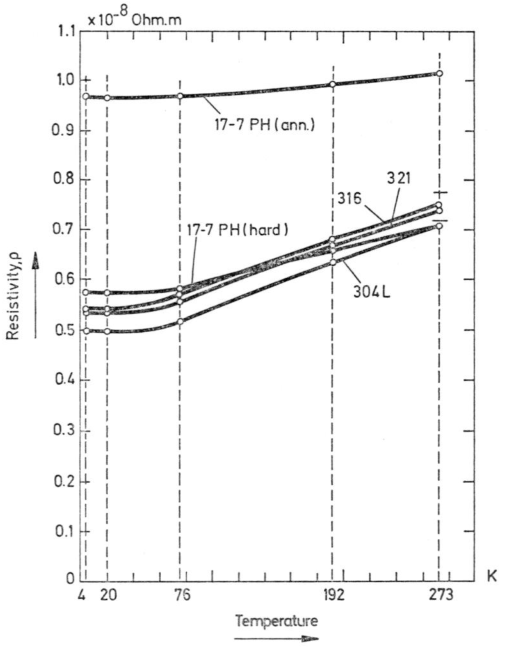

Fig. 5.6.17. Resistivity-temperature dependence of stainless-steels

Magneto-Resistance Effects

The resistance of both metals and semi-conductors change when exposed to magnetic fields either transverse or longitudinal to the direction of the current flow. Notable early measurements on metals are those by Kapitza [42] who used magnetic fields up to 30 T.

The statistical link between macroscopic electrical resistivity and the parameters describing the microscopic electron collision process is the Boltzmann equation. The presence of the magnetic field enters the Boltzmann equation in such a way as to affect all the electron collision processes. Hence, the change in resistivity of a sample as a function of applied field is more generally a multiplicative factor rather than additive as suggested by Matthiessen's rule.

The general Boltzmann equation is not adequate for simple analysis. However with certain approximations one can obtain Kohler's rule [43] for metals:

$$\frac{\varrho\,(B,T)}{\varrho\,(0,T)} = f\left(\frac{B}{\varrho\,(0,T)}\,,\,\varphi\right), \tag{5.6.30}$$

where $\varrho\,(B,T)$ denotes the resistivity as a function of the field B and temperature T; $\varrho\,(0,T)$ is the resistivity at no field, and φ is the angle between the magnetic field B and the current direction. The function f is unique for a given polycrystalline metal, which means that it is independent of small variations in purity in normally pure metals. The magneto-resistance data are presented in the general form:

$$\frac{\Delta\varrho\,(B,T)}{\varrho\,(0,T)} = \frac{C_1 \cdot B^2}{1 + C_2\,B^2}\,, \tag{5.6.31}$$

where C_1 and C_2 are material constants and $\Delta\varrho$ is the change in resistivity.

For aluminum the transverse magneto-resistance, fitting most experimental data within 20%, is given by [44]:

$$\frac{\Delta\varrho}{\varrho\,(0,T)} = \frac{\varrho\,(B,T) - \varrho\,(0,T)}{\varrho\,(0,T)} = b^2\,\frac{1 + 1.77 \cdot 10^{-3}\,b}{1.8 + 1.6\,b + 0.53\,b^2}\,, \tag{5.6.32}$$

with
$$b = B\,(\text{Tesla}) \cdot \frac{\varrho_{\text{reference}}}{\varrho\,(0)} \cdot 10^{+1}$$

and
$$\varrho_{\text{reference}} = 2.75 \times 10^{-8}\ \text{Ohm} \cdot \text{m}\,.$$

Values of $\Delta\varrho/\varrho$ are plotted in Figs. 6.2.11 and 6.2.12 for aluminum.

It may be pointed out that newer experimental data by Schauer [45], Borovik et al. [46] and Fickett [47] are not in agreement with Eq. (5.6.32). The qualitative agreement between Kohler's rule and measurements are satisfactory showing that with increasing purity the magneto-resistance of Al is increased markedly. Saturation effects are observed at fields higher than 3 T.

The temperature variation of the magneto-resistance and the effective resistivity ratio $r_{\text{tape}}\,(B,T)$ for different purities are given in Fig. 6.2.12 and here we observe at transverse fields some non-saturation behaviour up to 4 T, which were the limit of measurements. Probably several phenomena are acting together, such as small angle electron-phonon scattering, magnetic breakdown in specific crystal directions, or the generation of necklike channels to neighbouring Brillouin zones [48], which remove conductivity electrons.

The anomalous resistivity behaviour of aluminum is disturbing, as pure aluminum may be utilized for low temperature magnets at liquid hydrogen temperatures or as a matrix in composite conductors.

Also of interest are the curves presented in Fig. 6.2.11, which are Kohler plots of various Al-types with various residual resistivities. While Kohler's diagram indicates a much smaller value of $\Delta\varrho/\varrho$, measured data indicate a substantial deviation from this rule. For $r = 5\times10^4$ even at $B = 2$ T no saturation effect is observed.

The measured magneto-resistance data for polycrystalline coppers comply with Kohler's rule quite satisfactory, as illustrated in Fig. 6.2.13.

Based on Pippard's [49] model of a Fermi surface of copper, Ziman [50] has derived a model in which a transverse field has two boundary regions.

Low field region:

$$\frac{\Delta\varrho_\perp}{\varrho(0,T)} \cong (r\cdot B)^2, \qquad (5.6.33)$$

with $\varrho(0,T)$ as the resistivity at 4.2 K and no field and r as the residual resistivity ratio.

High field region:

$$\frac{\varrho_\perp}{\varrho(0,T)} \cong 0.23\times10^{-2}\,(r\cdot B)\,, \qquad (5.6.34)$$

with B in (Tesla). According to Ziman the major variable is $(r\cdot B)$, which indicates that with increasing r the quadratic dependence of $\varrho(B)$ is limited to decreasing B values.

Magneto-resistance data for high strength copper alloys are given by Stevenson and Bolton [51]. The Kohler plot of commercially available coppers, Zr-Cu (0.12% Zr) and Cr-Coppers (0.3% Cr) are given in Fig. 5.6.18.

The copper alloys have a tensile strength of 4,280 kp/cm² at 300 K and thus are of technical interest. Specifically, Cr-Cu may be used as a substrate in composite conductors. The resistivity of the alloys after being annealed and aged at 290.8 K is for Cr-Cu about 1.93×10^{-8} Ohm · m and for Zr-Cu alloy 1.63×10^{-8} Ohm · m.

When samples (wires with a diameter of 0.005 cm to 0.0127 cm) were drawn, the resistivity ratio was affected quite severely. Cr-Cu showed a resistivity ratio of 2.4 at 4.2 K and Zr-Cu a resistivity ratio of 5.6 at 4.2 K compared to room temperature values.

Annealing for 11 hours at 500 °C in argon atmosphere, followed by a water quench, improved the condition substantially. The ratios given in Fig. 5.6.18 were obtained at 4.2 K.

The resistivity of NbTi, NbZr and NbSn vs. temperature is illustrated in Fig. 5.6.19.

Lorentz Number

An electric field or a temperature gradient causes a drift in the transport phenomenon in a metal. This drift is restricted only by collisions of electrons with imperfections in the lattice, whether they are static defects or lattice vibrations. When the electron distribution function is disturbed

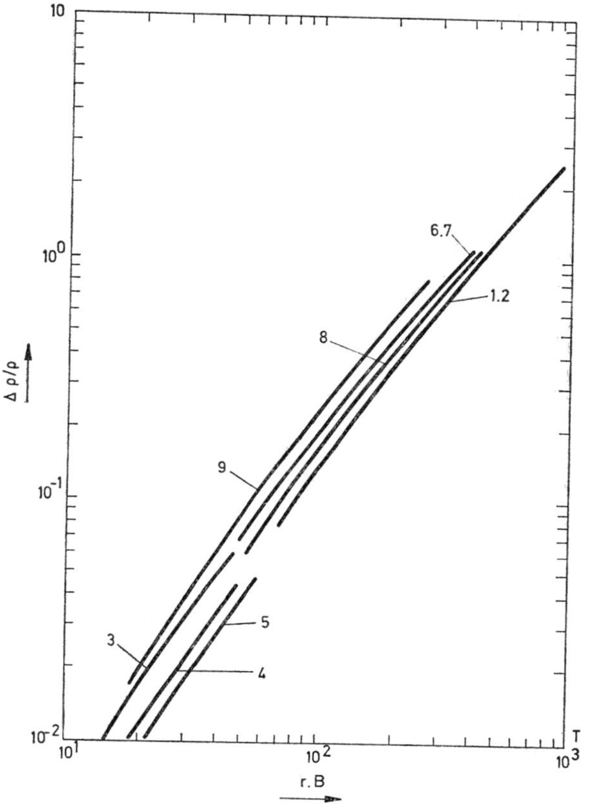

Fig. 5.6.18. Kohler diagrams of copper and copper alloys:

1 Copper wire $r = 88.5$	*4* Cr-Cu wire $r = 4.37$	*7* Zr-Cu wire $r = 55.1$
2 Copper wire $r = 89.1$	*5* Cr-Cu wire $r = 4.7$	*8* Zr-Cu wire $r = 43.0$
3 Cr-Cu wire $r = 11.7$	*6* Zr-Cu wire $r = 51.5$	*9* Zr-Cu wire $r = 20.0$

from its equilibrium value, the rate of return to equilibrium (by collision process) may be expressed in terms of a relaxation time τ. The Boltzmann equation expresses the equilibrium situation. A solution of the Boltzmann equation is given by Klemens [52]:

$$\frac{1}{\varrho} = \frac{e^2}{12\,\pi^3} \int \frac{\tau v^2 dS}{|\mathrm{grad}_k E|} \tag{5.6.35}$$

$$k = \frac{k_B^2\, T}{36\,\pi} \int \frac{\tau v^2 dS}{|\mathrm{grad}_k E|}, \tag{5.6.36}$$

where dS is an element of the Fermi surface in k space, v is the electron velocity, and k_B is the Boltzmann constant. If the relaxation time is the same for the electric and thermal transport, then

$$\boxed{\frac{1}{\varrho k T} = \frac{\pi^2}{3} \left(\frac{k_B}{e}\right)^2 = L\,.} \tag{5.6.37}$$

The value of L is due to Sommerfeld and is independent of relaxation time and band structure. The Lorentz number L is a constant and its numerical value is:

$$L = \frac{\pi^2}{3}\left(\frac{k_B}{e}\right)^2 \cong 2.45\times10^{-8}\,(\mathrm{V}^2\,\mathrm{K}^{-2})$$

for $T > \theta_D$ and for $T \ll \theta_D$.

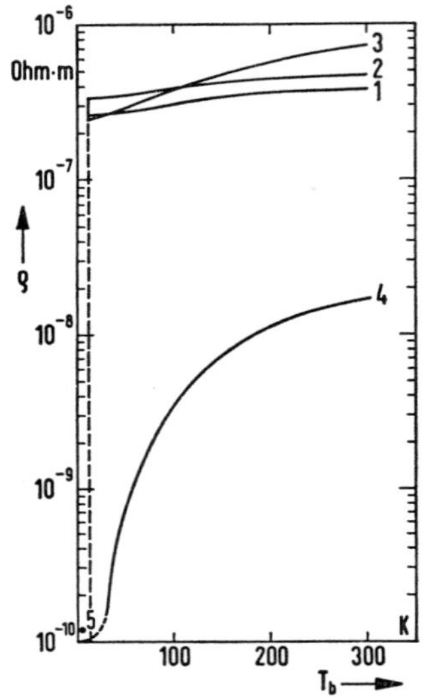

Fig. 5.6.19. Electrical resistivity of superconducting Type II wires and of OFHC copper

1 Nb_3Sn
2 Nb(25%)Zr
3 Nb(60%)Ti
4 99.999% pure annealed copper
5 99.995% pure OFHC copper as received, not annealed (4.2 K, $B = 0$) $= 1.2\times10^{-8}$ Ohm · cm

The Lorentz number is a constant if the conduction electrons are scattered elastically. This condition is approximately true at high and very low temperatures, where the residual resistivity is predominant. At intermediate temperatures the condition of elasticity no longer holds and the Lorentz number is considerably smaller than the Sommerfeld value.

Lorentz numbers for Al are given by Arp [53], for pure coppers by Powell [54] and White [55]. For copper the maximum deviation from Sommerfeld numbers occurs at approximately 25 K ($L = 0.7\times10^{-8}\,\mathrm{V}^2\,\mathrm{K}^{-2}$) and for single crystal-aluminum at 100 K ($L = 1.2\times10^{-8}\,\mathrm{V}^2\,\mathrm{K}^{-2}$), for commercial aluminum (1100—0) at 100 K ($L = 1.55\times10^{-8}\,\mathrm{V}^2\,\mathrm{K}^{-2}$).

(e) Thermal Contraction

The atoms of a solid have been assumed to vibrate symmetrically about some equilibrium position in the solid. The detailed character of the vibrations was assumed to depend upon temperature independent forces of attraction and repulsion between atoms.

Building on the formulation of lattice dynamics originally put foreward by Debye, Grüneisen developed an approximate thermodynamic equation of state for solids. He was able to show that the dimensionless ratio:

$$\gamma = V \cdot \beta\,(\varkappa \cdot C_v) \tag{5.6.38}$$

should be constant for all solids.

V is the volume of the solid, β is the linear coefficient of thermal expansion (contraction), $\varkappa$ is the isothermal compressibility and C_v the thermal capacity.

The experiment shows that γ is not an universal constant and changes with respect to the material. The reason that γ is not constant is due to $\varkappa$, which decreases with increasing atomic number. The other factors in Eq. (5.6.38) remain relatively constant. The coefficient β of the thermal expansivity is defined by:

$$\beta = \frac{1}{V}\left(\frac{\partial V}{\partial T}\right)_f \tag{5.6.39}$$

and the Grüneisen's relation is rearranged to:

$$\beta = \frac{C_v \cdot \gamma \cdot \varkappa}{V}. \tag{5.6.40}$$

Neither the volume nor the compressibility varies greatly with temperature, but with decreasing temperature they change in the same direction. The ratio $\varkappa/V$ is quite insensitive to temperature. If γ were also insensitive to temperature for a specific material, than the expansivity would follow the same trend as C_v.

In Fig. 6.2.8 the thermal contraction of several metals, including epoxies (filled and unfilled) versus temperature is illustrated.

As in case of the thermal capacity the thermal expansivity exhibits also anomalies, such as seen for stainless steel (304) and "Pyrex" glass.

As in case of thermal capacity, there should be a contribution to the expansivity of metals from their conduction electrons. This contribution should be linear with temperature and should be distinguishable from the lattice contribution at very low temperatures, where lattice expansivity varies with T^3. These dependencies however have not been verified experimentally.

5.7 Heat Losses

Heat losses occur in cryogenic apparatus as a result of three modes of heat transport:

Heat conduction: The transport of heat between continuous parts of a solid at different temperatures.

Heat convection: The transport of heat on boundary layers between different media.

Heat radiation: The transport of heat by electromagnetic flux.

The design of cryogenic apparatus is usually directed toward either maximizing or minimizing the heat transport with the device or parts of it, or between the apparatus and its environment.

5.7.1 Heat Conduction

The magnitude of one dimensional heat conduction is calculated by the Fourier equation

$$Q = \frac{-k A \, dT}{dl} \quad (\mathrm{W}), \qquad (5.7.1)$$

where k is the thermal conductivity, A is the area normal to the heat flow path and dT/dl is the thermal gradient along the flow path. The thermal conductivity in general, is a function of temperature, as shown in Fig. 5.7.1, although for small temperature intervals can be considered a constant. For most cryogenic apparatus, however, the temperature intervals are sufficiently large enough that the temperature dependence of k must be taken into account. This is sometimes accomplished by assigning an average value to the conductivity denoted by $\bar{k}$, and termed the apparent conductivity over the temperature range. Where this is not possible or desirable, the Fourier equation is integrated to yield, for the case of constant cross section area,

$$\boxed{\frac{QL}{A} = -\int_{T_1}^{T_2} k(T)\, dT .} \qquad (5.7.2)$$

The values of the integral in Eq. (5.7.2), the conductivity integral, are tabulated for many metals and alloys in literature [6]. Several of thermal conductivity measurements are reproduced in Section 5.6 of this chapter.

For temperature ranges that do not extend to 4 K, the value to 4 K from the lower temperature should be subtracted from the measured value for transport between the upper temperature and 4 K, thus

$$\frac{-QL}{A} = \int_{T_1}^{T_2} k(T)\, dT = \int_{T_1}^{4K} k(T)\, dT - \int_{T_2}^{4K} k(T)\, dT . \qquad (5.7.3)$$

The presented data or the Eqs. (5.7.2) or (5.7.3) may be used to calculate the heat leak for most solid structural members and control and instrumentation connections. If the member is a composite such as insulated electrical wire, then the insulation sheath and the wire are calculated as being in parallel, this is an approximation, since the differing temperature dependence of the thermal conductivities will cause lateral heat flow between the two bodies and the heat flow is not purely one dimensional as assumed by Eq. (5.7.1).

For the case where there is a change in area of the thermal conductor, Garwin [56] has shown that the conductivity integrals may still be used. Thus, if a cylinder or bar of the length l_1 and area A_1 is connected thermally in series with a cylinder of length l_2 and area A_2 between temperatures T_1 and T_2, then the thermal flux will be given by

$$Q\left(\frac{l_1}{A_1}+\frac{l_2}{A_2}\right)=\int_{T_1}^{T_2} k(T)\,dT$$

$$Q=\left(\int_{T_0}^{T_2} k(T)\,dT-\int_{T_0}^{T_1} k(T)\,dT\right)\frac{A_1\cdot A_2}{A_1 l_1+A_2 l_2}\,. \quad (5.7.4)$$

The same result occurs if the shorter and longer sections are interchanged, thus, a reduction in area at the hot or cold end are equivalent, regardless of the actual value of the thermal conductivity at the ends.

Since the heat flux is determined by the area under the $k(T)$ curve, a small region of T, with $k(T)$ very small, will reduce the flux only slightly, while a small region with very large $k(T)$ will contribute greatly to the heat flux. Thus, the high conductivity peaks occurring in pure metals, as shown in Fig. 5.7.1 should be avoided if at all possible.

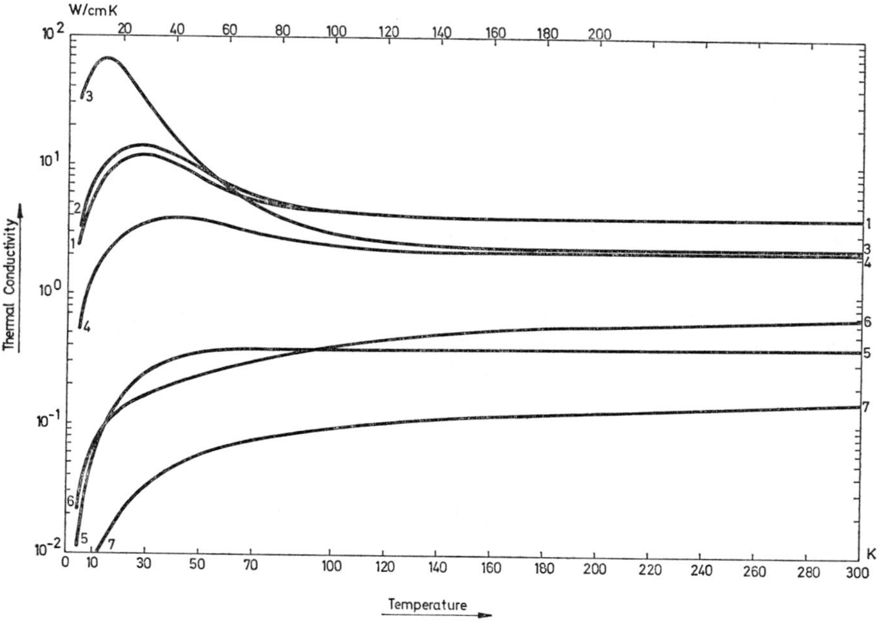

Fig. 5.7.1. Thermal conductivity of various metallic materials as a function of temperature

1 OFHC-copper: 99.95% pure, annealed
2 ETP-copper: 99.95% pure, annealed
3 Aluminum: 99.996% pure single crystal
4 Aluminum: 99.99% cold drawn
5 Cupro-nickel: 90% copper, 10% Ni — annealed
6 Silver solder: 50% silver, 15% copper, 16.5% Zinc, 18% cadmium
7 Stainless-steels: 304, 305

The Fourier equation is employed to calculate the heat leak through stagnant gases and liquids as well as through amorphous insulating materials. Frequently an average value of conductivity $\bar{k}$ (sometimes a directly measured quantity) for a given temperature difference will be used rather than the conductivity integral. In many problems the heat flux desired is the radial conduction between concentric cylinders and for the case of cylinders of length l and radii r_1 and r_2 we have

$$d\,Q = kl \cdot 2\,\pi r \frac{dT}{dr} \tag{5.7.5}$$

$$Q = kl \cdot 2\,\pi \frac{\Delta T}{\ln\left(\frac{r_1}{r_2}\right)}. \tag{5.7.6}$$

Expression for Q for other geometries can be found in the standard treatise on heat transfer and heat conduction [57].

Finally, the heat conduction which occurs in a gas at low temperature and low pressure is of interest in designs of many apparatus. Corruccini [58] gives the expression

$$\boxed{Q = C a_0\, p_{mm} (T_2 - T_1)} \tag{5.7.7}$$

for the heat transfer between surfaces at temperatures T_1 and T_2 by a gas at pressure p_{mm} (measured by a gauge at room temperature) with an accomodation coefficient of a_0. The value of C is approximately 0.028, 0.053 and 0.016 for helium, hydrogen and nitrogen respectively. The accomodation coefficient will at most be equal to 1, so that an upper limit to the heat transfer will be obtained.

Superconducting magnets are composed of materials with different thermal properties. In case heat is generated in a certain area in the coil effective volume, this heat is conducted along the metallic wire, but also across the conductor to the coolant, or from the conductor to the reinforcements, insulation or other materials adjacent to the conductor. As seen from Fig. 5.6.13 the thermal conductivity of copper and aluminum in the temperature range 4 to 8 K is in the order of 0.6 to 5 W/cmK, the thermal conductivity of insulation materials are much worse ($\sim 10^{-3}$ W per cmK). Superconductors Type II also have very poor thermal conductivity ($\sim 10^{-3}$ W/cmK), but this is overcome by reducing their sizes to values where the ratio of surface to cross sectional area become large and place them in substrates of high thermal conductivity. The heat generated in the superconductor diffuse into the substrate and is convected by the coolant (if direct cooling is provided).

If insulation barriers are foreseen between conductors and layers the thermal conductance of the structure is reduced. To avoid coil heating the generation and distribution of heat in the coil must be studied in detail.

Superconducting pulsed coils are sometimes impregnated in thermosettings, waxes and thermoplasts or other suitable materials, to prevent the strands from movements when the current and field are changed, and to eliminate friction, which may lead to local heating. In this case the conductor is not in direct contact with the coolant and the heat must be conducted through layers of low thermal conductivity materials to heat drains (copper screen soldered to tubes carrying helium) or directly to the coolant.

Various methods of cooling composite coils are common in superconducting magnets:

(a) *Open coil cooling* (Fig. 7.4.2)
The conductor is wrapped spirally with a fiber glass tape. However a substantial conductor area must be still exposed to the coolant by direct contact.

(b) *Edge cooling* (Fig. 6.2.2b)
Common in double pancake coils. Only edges of tapes or rectangular shaped conductors are cooled.

(c) *Layer cooling* (Fig. 6.2.2b)
Spacers between layers enable the coolant to be in contact with at least two conductor surfaces.

(d) *Heat drains* (Fig. 2.5.1f)
Used in potted or impregnated coils, where direct access of the coolant to the conductor may be difficult. The heat drain is a metallic screen or sheet soldered to metal tubes, where helium is pressurized through them, or the screen extends beyond the coil sections into the bath (see Chapter 4).

(e) *Indirect cooling* (Fig. 2.5.1b)
The coil is subdivided into a number of potted layers or pancakes with coolant channels provided between individual sections.

(f) *Pure conduction* (Fig. 2.5.1a)
Tightly wound coils with coolant passages provided only to the external surfaces of the coil.

In a coil with internal heat sources, the temperature distribution is given by the Poisson's equation:

$$k_x \frac{\partial^2 T}{\partial x^2} + k_y \frac{\partial^2 T}{\partial y^2} + k_z \frac{\partial^2 T}{\partial z^2} + w_v = 0\,. \tag{5.7.8}$$

Solution of this equation for one and two dimensional cases is given in Section 2.5.

The thermal conductivity of the composite coil is expressed by:

$$\boxed{k_{\text{res}} = \frac{\sum_i (d_i)\,(\prod_i k_i)}{\sum_i (d_i \prod_{k \neq i} k_k)}}$$

$$i = 1, 2, 3, \ldots \qquad k = 1, 2, 3, \ldots \tag{5.7.9}$$

If the coil consists of a metallic conductor ($i = 1$) with a thickness d_1, an insulation ($i = 2$) with thickness d_2, and is impregnated in a thermoset ($i = 3$) of thickness d_3, the overall thermal conductivity of this composition is expressed according to Eq. (5.7.9) by:

$$k_{\text{res}} = \frac{(d_1 + d_2 + d_3) \cdot k_1 \cdot k_2 \cdot k_3}{d_1 \cdot k_2 \cdot k_3 + d_2 \cdot k_1 \cdot k_3 + d_3 \cdot k_1 \cdot k_2}.$$

For coils wound with cylindrical conductors as shown in Fig. 5.7.2a, b we study two cases:

(a) Coils not impregnated. The area between conductors is filled by liquid or gaseous helium.

(b) Coils impregnated in a thermoset, a thermoplast, or by some liquid which solidifies at liquid helium temperature.

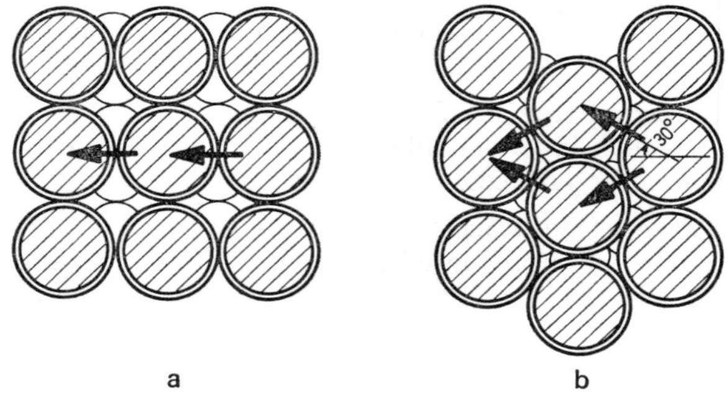

Fig. 5.7.2. Coil cross section with cylindrical conductors. *a* Schematic representation of the heat flux between cylindrical conductors, *b* Schematic representation of the heat flux between displaced cylindrical conductors

In both cases we assume that there are no vacuum pockets. If the heat conduction path is interrupted by cracks or fissures, the thermal conduction is jeopardized. In the equation for the overall thermal con-

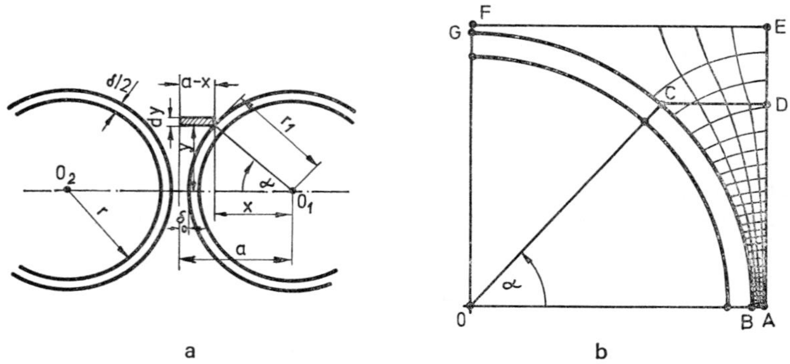

Fig. 5.7.3. Schematic representation of the heat flux between adjucant conductors. *a* Conductor positioning to calculate overall heat conductivity, *b* The thermal conductivity of the conductor coating is assumed 10 times better than that of air. Practically no heat is conducted through the area CDEFG

duction of the coil Eq. (5.7.9) an estimated thickness of the air gap and the thermal conductivity of gaseous helium must be taken into consideration.

From Fig. 5.7.3a, b, the thermal conductivity of a strip dy and a thermal length of $2\,(a - x)$ per unit conductor length is given by [59].

$$dk_1 = \frac{k_{He} \cdot dy}{2\,(a - x)} = k_{He} \cdot \frac{r_1 \cdot \cos(\alpha)\, d\alpha}{2\,[a - r_1 \cos(\alpha)]}, \tag{5.7.10}$$

with k_{He} the thermal conductivity of helium either in liquid or gaseous form.

The thermal conductivity of the area between two adjacent conductors from the surface of the insulations per unit conductor length is obtained by integrating Eq. (5.7.10):

$$k_1 = 2 \int_0^a dk_1 = k_{He} \cdot \left[\frac{2a}{\sqrt{a^2 - r_1^2}} \tan^{-1}\left(\sqrt{\frac{a + r_1}{a - r_1}} \cdot \tan\left(\frac{\alpha}{2}\right)\right) - \alpha\right]. \tag{5.7.11}$$

In general the ratio $(2\,r_1/\delta_0)$ is much larger than unity. Thus we may simplify the expression for k_1 without lossing accuracy as:

$$k_1 = k_{He} \left[\sqrt{\frac{2r_1}{\delta_0}} \cdot \tan^{-1}\left(\sqrt{\frac{2r_1}{\delta_0}} \cdot \tan\left(\frac{\alpha}{2}\right)\right) - \alpha\right].$$

With:

$$\sqrt{\frac{2r_1}{\delta_0}} = \xi$$

$$\tan^{-1}\left(\sqrt{\frac{2r_1}{\delta_0}} \tan\left(\frac{\alpha}{2}\right)\right) = \varrho$$

we get

$$k_1 = k_{He} \cdot (\xi \cdot \varrho - \alpha)\,.$$

If we neglect α, compared to ξ and ϱ, we obtain the simple expression:

$$k_1 \approx k_{He} \cdot \xi \cdot \varrho\,. \tag{5.7.12}$$

As the conductor insulation is subjected to the same heat flux as the helium and also the heat distribution is identical, we may derive the thermal conductivity values for parallel conductor location (Fig. 5.7.2a).

$$k_p = 2 \cdot k_{He} \cdot k_{isol} \cdot \frac{\varrho_1 \cdot \xi \cdot x}{k_{He} \cdot \xi^2 + 2\,k_{isol} \cdot x}, \tag{5.7.13}$$

and for conductors in shifted or displaced location (Fig. 5.7.2b)

$$k_s = 4 \cdot k_{He} \cdot k_{isol} \cdot \frac{\varrho_2 \cdot \xi \cdot x}{k_{He} \cdot \xi^2 + 2\,k_{isol} \cdot x}, \tag{5.7.14}$$

where:

$$\varrho_1 = \tan^{-1}\left(\xi \tan\frac{\alpha}{2}\right) \quad \text{for} \quad \alpha = \frac{\pi}{3}$$

$$\varrho_2 = \tan^{-1}\left(\xi \tan\frac{\alpha}{2}\right) \quad \text{for} \quad \alpha = \frac{\pi}{6}$$

$$x = \frac{2r}{\varrho}.$$

If the coil is impregnated with a thermoset which has a thermal conductivity k_{th}, then Eq. (5.7.11) is still valid:

$$k_1 = k_{th} \cdot \left[\frac{2a}{\sqrt{a^2 - r_1^2}} \cdot \tan^{-1}\left(\sqrt{\frac{a + r_1}{a - r_1}} \cdot \tan\frac{\alpha}{2} \right) - \alpha \right]. \tag{5.7.15}$$

If the composite conductor consists of square or rectangular shaped conductors, as illustrated in Fig. 5.7.4a, b, the overall thermal conductivity is expressed by [60].

$$k_{cond.} = k_m \cdot \left[\frac{\sqrt{1 + r} - 1}{\sqrt{1 + r}} + \frac{k_{sc}}{(\sqrt{1 + r} - 1)\, k_{sc} + k_m} \right], \tag{5.7.16}$$

where r = volume of matrix material/volume of superconductor,
k_m = thermal conductivity of the matrix,
k_{sc} = thermal conductivity of the superconductor.

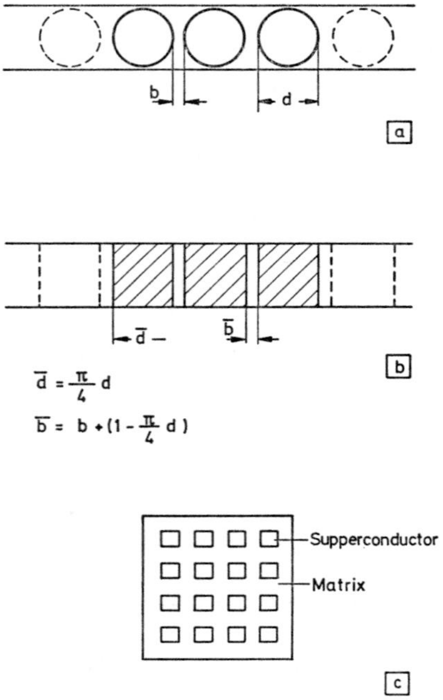

Fig. 5.7.4. Coil cross section with square conductors

The thermal conductivity of an impregnated coil with square conductors is thus obtained from:

$$k_{coil} = \frac{(d_{is} + d_{cond})\, k_{is} \cdot k_{cond}}{d_c \cdot k_{is} + d_{is} \cdot k_{cond}}. \tag{5.7.17}$$

Again if one has to account for vacuum pockets this equation must be corrected accordingly.

5.7.2 Convection

Heat transfer by natural convection is frequently ignored in superficial treatment of cryogenic apparatus, while forced convection is dealt with in some detail, since it is the mode of heat transfer in heat exchangers .The fundamental equation for heat transfer by convection is

$$\boxed{Q = hA\,(T_s - T_f)\,,} \tag{5.7.18}$$

where h is the local coefficient of heat transfer between a surface of the area A at temperature T_s and the fluid at temperature T_f. In many cases natural convection is suppressed by baffles which prevent motion of the fluid. The heat transport is then calculated by the Fourier equation as discussed above. Where natural convection is not suppressed the value of h may be calculated from a knowledge of the geometry of the system. Lorentz derived an equation for h by considering the gravitationally induced flow up a vertical wall of infinite length but finite height l,

$$h = 0.548 \cdot \left(\frac{g\,\delta^2 c_p k^3 \theta_s}{\eta\, l\, T_e} \right)^{0.25}, \tag{5.7.19}$$

where $\theta_s = T_s - T_e$, the temperature of the surface above the environment T_e, k the thermal conductivity, c_p the specific heat, δ the density, η the viscosity of the fluid and g the gravitational constant.

By the method of dimensional analysis, the heat transfer by natural convection in a vertical tube of diameter d is given by

$$N_{\mathrm{Nu}} = 0.55\;[N_{\mathrm{Gr}} \cdot N_{\mathrm{Pr}}]^{0.25} \tag{5.7.20}$$

or

$$\frac{hd}{k} = 0.55 \left(\frac{g\beta d^3 \theta_s}{\nu^2} \cdot \frac{\nu}{D_{\mathrm{fl}}} \right)^{0.25}, \tag{5.7.20a}$$

where N_{Nu}, N_{Gr} and N_{Pr} are the dimensionless Nusselt, Grashof, and Prandtl numbers and β is the coefficient of cubic thermal expansion, ν is the kinematic viscosity and D_{fl} is the thermal diffusivity of the fluid. This expression can be used to calculate the heat transfer to the wall of tubes which penetrate the low temperature region and do not have large gas flows in them.

For vent tubes and other connections with large gas flows, the force convection equations must be employed. For turbulent flow the Dittus-Boelter equation [61]

$$\left(\frac{hd}{k} \right) = 0.023 \left(\frac{dG}{\eta} \right)^{0.8} \left(\frac{c_p \eta}{k} \right)^{0.4} \tag{5.7.21}$$

can be used to calculate the heat transfer coefficient. White [8] has reduced this expression to the approximate value

$$h \approx \frac{(\dot{m})^{0.8}}{(d)^{1.6}}, \tag{5.7.22}$$

where $\dot{m}$ is the mass flow rate and d is the diameter of a circular tube.

For forced laminar flow the heat transfer is given by the equations

$$N_{\mathrm{Nu}} = 1.86 \left(N_{\mathrm{pe}}\right)^{0.33} \left(\frac{d}{l}\right)^{0.33} \left(\frac{\eta_a}{\eta_s}\right)^{0.14}, \tag{5.7.23}$$

$$\frac{hd}{k} = 1.86 \left(\frac{v\,d}{D_{\mathrm{fl}}}\right)^{0.33} \left(\frac{d}{l}\right)^{0.33} \left(\frac{\eta_a}{\eta_s}\right)^{0.4} \tag{5.7.24}$$

where N_{pe} is the dimensionless Péclet number, v is the velocity, D_{fl} the thermal diffusivity and the subscripts a is for the fluid mean temperature from entrance to exit and s is for the surface temperature.

So far we have given correlation equations for forced flow. It is important however to study heat transfer properties of cryogenic fluids. With reference to Fig. 5.7.5, which illustrates the temperature-entropy diagram we see that small pressure and temperature ranges enclose a given fluid phase or condition. For pressure and temperature ranges required by some cooling systems one may encounter the helium fluid in a number of phases or states. It may be difficult in the design of cryogenic apparatus to avoid working in regions around the critical point, where the helium has some undesired properties and where the heat transfer phenomenon is not well understood. In dealing with liquid helium one faces the heat transfer problems which involve the liquid, the supercritical fluid and a two phase fluid. The two phase and supercritical fluid regions are among the more difficult heat transfer areas to study for any

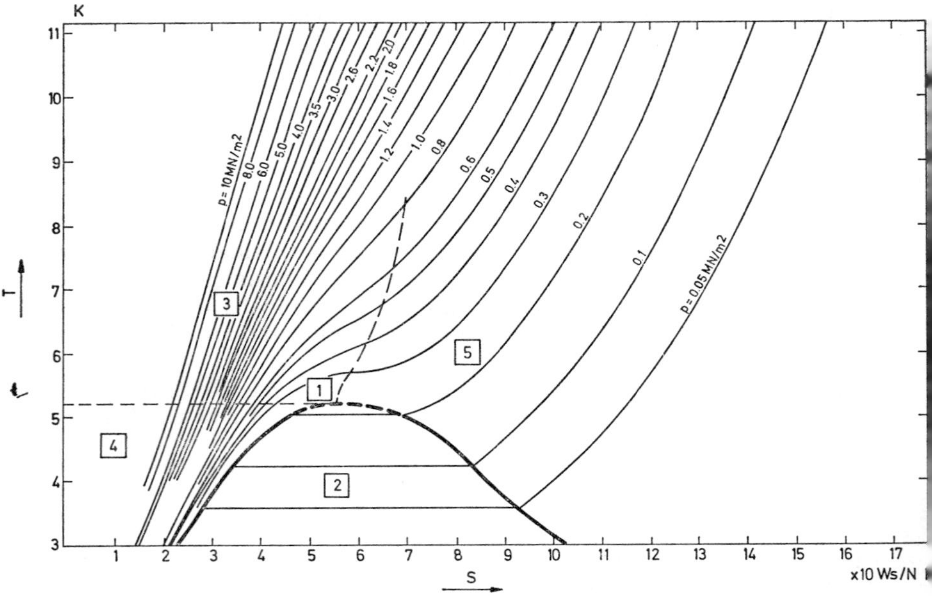

Fig. 5.7.5. Temperature entropy diagram for helium with schematic representation of the five areas in the T-S diagram, where helium behaves differently. *1* Helium near critical temperature, *2* Two phase helium, *3* Supercritical helium, *4* Compressed liquid, *5* Gaseous helium

cryogen. Difficulties in heat transfer studies with liquid helium is encountered on examing the viscosity (Fig. 5.5.6), and the thermal conductivity (Fig. 5.5.7) of liquid helium. We find that the data available to date have a poor degree of reliability.

Comparing conventional fluids (such as water) to helium we find that their physical properties are quite diverging. At normal boiling point the viscosity of water is about 100 times higher than that of helium, the specific heat about equal to that of helium and the density about eight times as great. Therefore in considering the conventional heat transfer Eq. (5.7.21) which involve Reynolds $(d \cdot G/\eta)$ and Prandtl $(c_p \cdot \eta/k)$ numbers we see that with helium quite different combinations of values of these dimensionless groups are obtained. As the correlations of Eq. (5.7.21) are based on empirical relationship of dimensionless groups, the extension of these expressions into regions where fluid properties differ by such a margin cannot be justified without experimental verifications. NBS has endeavoured since 1969 to study transport properties of superfluid, supercritical and two phase helium. A review paper giving the progress of the work may be obtained [62], but the termination of the experimental work is expected by 1973.

Regions of Phase and Transport Property Behaviour

Transport phenomena are influenced by two sets of conditions:

(i) Physical properties.
(ii) Flow structure.

With respect to the flow structure, one is concerned with the fluid behaviour in the boundary layer (Laminar or turbulent). To study heat transfer properties of helium, the T-S-diagram is divided into five regions [63], with respect of transport properties and phase behaviour (Fig. 5.7.5).

Region 1

Phase separation is either not distinct or not possible.

Region 2

Phase separation with equilibrium is distinct.

Region 3

Pressurized liquid helium region.

Region 4

Supercritical helium region.

Region 5

Gaseous region.

The line of maximum specific heat separates region 1 from 5. On the left side of the line, one experiences pressure and density oscillation. In region 3 (fluid), the physical properties of helium differ from conventional liquids, but conventional correlations still hold.

For regions 1 and 2, only recently few data have become available. We summarize briefly some of the properties of helium in these five regions.

1. Helium 4, region 1

In this region, reliable property data are lacking. Recently viscosity, thermal conductivity [64], and heat transfer data [65] have been measured.

It is interesting to note ,that the classical Dittus-Boelter Eq. (5.7.21) agrees reasonably well with experimental heat transfer data measured in the range of 0.3 to 2 MN/m². In this equation, the Nusselt coefficient must be 0.022 instead of 0.023. The standard deviation for this correlation is 14.8%. Modification from the Dittus-Boelter equation includes another term as a ratio of the wall to bulk temperature, which varies between 1.02 ... 3. The correlation equation for pressurized helium with a standard deviation of about 8.5% is given by:

$$\mathrm{Nu} = 0.0259 \left(N_{\mathrm{Re}}\right)^{0.8} \cdot \left(N_{\mathrm{Pr}}\right)^{0.4} \cdot \left(\frac{T_w}{T_b}\right)^{-0.716}$$

or explicitely:

$$\boxed{\frac{h \cdot d}{k} = 0.0259 \left(\frac{dG}{\eta}\right)^{0.8} \cdot \left(\frac{c_p \cdot \eta}{k}\right)^{0.4} \cdot \left(\frac{T_w}{T_b}\right)^{-0.716}} \quad . \qquad (5.7.25)$$

Various modes of heat transfer correlations are presented in Fig. 5.7.6. Comparing forced supercritical helium versus pool boiling or superfluid helium cooling system, we see that at low Reynold numbers, the heat transfer values are in the same range. Pressurized helium is more attractive only at higher Reynold numbers ($\geqq 10^6$).

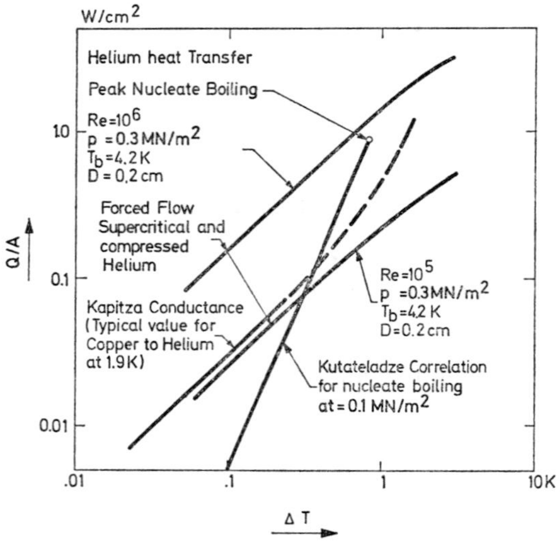

Fig. 5.7.6. Heat transfer vs. temperature rise at the boundary between wall and bulk liquid helium

During heat transfer near the critical pressure ($p_c = 0.23$ MN/m²) oscillations were observed by a number of investigators. A possible criterion for the onset of oscillations is the presence of vastly different density states of the fluid within the system. Near the critical region large density changes occur with small temperature changes (see Fig. 5.5.1). Oscillations may also occur in the region far from the critical region, if the wall to bulk temperature is high. The frequency of the oscillations are less than 0.1 Hz.

Two disadvantages of pressurized helium should be mentioned. When the helium is passing through a coolant passage, helium is warmed up in the process.

Compressed liquid or supercritical helium passing at a speed v through a hydraulic passage will generate frictional losses at the interface on the passage walls. These losses can be calculated from shearing stresses on the helium boundary layer [66]

$$\tau_0 = \frac{1}{8}\lambda \cdot \frac{\delta_f}{g}\,\bar{v}^2. \tag{5.7.26}$$

λ is a frictional resistance coefficient, explicitely written as:

$$\lambda = 0.3164\left(\frac{\bar{v}\cdot d_h}{\nu}\right)^{-\frac{1}{4}}, \tag{5.7.27}$$

with ν the dynamic viscosity of the fluid.

Inserting λ into Eq. (5.7.26) we get:

$$\tau_0 = 3.955\times 10^{-2}\left(\frac{\delta_f}{g}\right)\left(\bar{v}\right)^{\frac{7}{4}}\left(\nu\right)^{\frac{1}{4}}\left(d_h\right)^{-\frac{1}{4}}. \tag{5.7.28}$$

We see that τ_0 changes with $(\bar{v})^{7/4}$ rather than proportional to the average velocity as for laminar flow, known from Stokes' law. The frictional losses per unit area is obtained from:

$$P_{fr} = \tau_0\cdot\bar{v} = 3.88\times 10^{-6}\left(\frac{\delta_f}{g}\right)\cdot\left(\bar{v}\right)^{\frac{11}{4}}\cdot\left(\nu\right)^{\frac{1}{4}}\cdot\left(d_h\right)^{-\frac{1}{4}}\quad\left(\frac{\mathrm{W}}{\mathrm{cm}^2}\right). \tag{5.7.29}$$

A hydraulic passage in the coil may have a length in excess of 50 m. Several passages can be connected in parallel to keep the pressure drop low.

For compressed helium at a pressure of 5 kp/cm² and 4.5 K we have the following data:

$$\begin{aligned} \delta &= 0.145\ \mathrm{g/cm^{-3}}, \\ \frac{\delta}{g} &= 1.477\times 10^{-4}\ \mathrm{g/s^2/cm^{-4}}, \\ \eta &= 4.7\ \ \times 10^{-5}\ \mathrm{g/cm^{-1}/s^{-1}}, \\ \nu &= 3.24\ \times 10^{-4}\ \mathrm{cm^2/s^{-1}}. \end{aligned}$$

If we assume a speed of helium of $v = 10^2$ cm/s⁻¹ through a circular coolant passage of $d_h = 0.5$ cm, we get from Eq. (5.7.29) the losses per unit coolant surface area of:

$$P_{fr} = 2.76\times 10^{-5}\ \mathrm{W/cm^{-2}}.$$

For the 50 m long coolant passage, the frictional losses would be about 0.2 watts.

Referring to Fig. 5.5.2, which illustrates the pressure-enthalpy characteristic of helium, and operating at lines of constant enthalpy, the change in pressure over a hydraulic passage leads to a temperature increase. If the refrigerator delivers the compressed liquid helium at $T =$

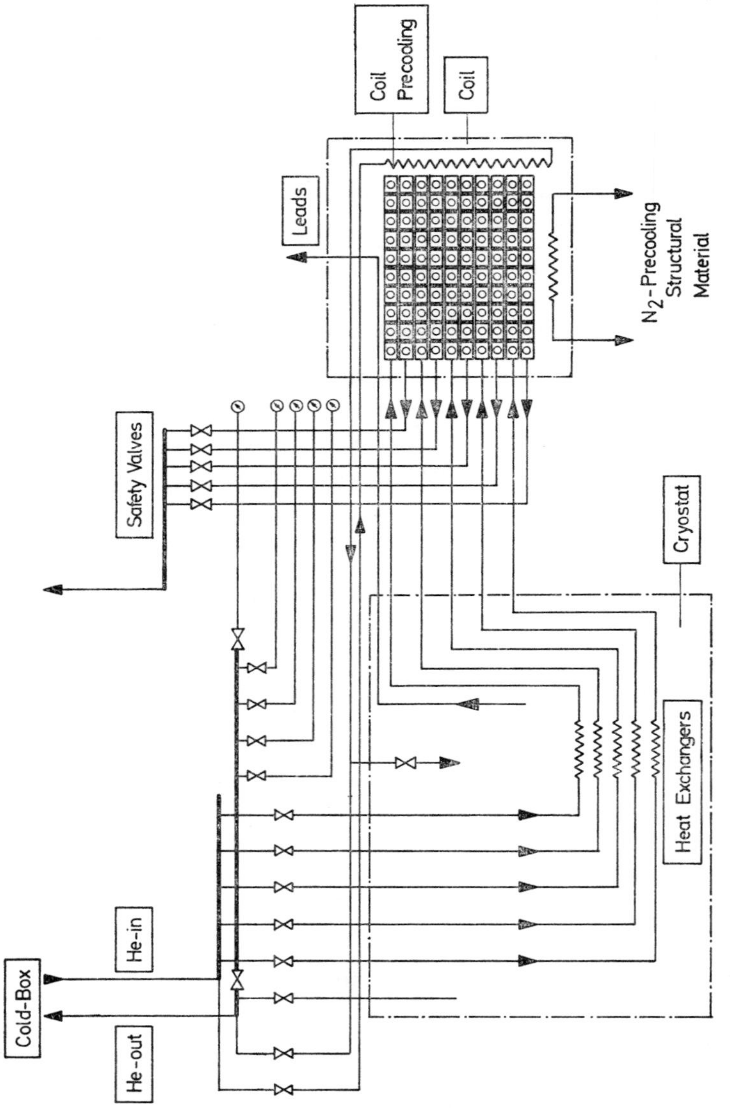

Fig. 5.7.7. Flow diagram of pressurized helium through a coil consisting of hollow composite conductors (CERN)

5 K to the magnet with the entrance pressure of 1 MN/m², a pressure reduction over one hydraulic passage of 0.5 MN/m² will yield a temperature rise of 0.15 K. But if the pressure over a hydraulic passage is reduced from 1.5 to 1 MN/m², the temperature rise may be 0.5 K, which, for Nb_xTi being exposed to fields of 5 T or higher, may mean a reduction in current density of ~35 to 50%, as seen from Fig. 1.6. However, intermediate pressure reductions can be taken where the pressure is reduced in steps rather than gradually over the entire hydraulic passage, or the temperature of the entering helium can be reduced using a heat exchanger.

The major advantage of pressurized helium is the simplicity of magnet operation, which adds to the magnet reliability. The volume of helium required to maintain the magnet operation is exceedingly smaller than when pool boiling is utilized. If the magnet is operated under normal conditions only a fraction of the normal amount of helium must be forced through the hydraulic passages of the coil (Fig. 5.7.7). The refrigerator

Table 5.7.1. *Some properties of compressed liquid and supercritical helium*

	Saturated liquid He at 4.2 K $p = 1$ kp/cm²	Pressurized He $p = 4$ kp/cm² $T_b = 5$ K	Supercritical He $p = 4$ kp/cm² $T = 15$ K
Density (g/cm^{-3})	0.125	0.180	0.12
Specific heat c_p (Ws/gK)	4.48	2.2	6.0
Thermal conductivity (W/cm K)	2.71×10^{-4}	3.5×10^{-4}	5×10^{-4}
Viscosity (g/cm^{-1} s^{-1})	3.56×10^{-5}	$\sim4.5\times10^{-5}$	7.5×10^{-5}
Reynold number factor[a] δ/η	3.52×10^{3}	4×10^{3}	1.6×10^{3}
Prandtl number[b]	0.590	~0.28	~0.9
Grashof number factor[c]	3×10^{9}	3×10^{9}	2×10^{8}

[a] Reynolds number: $v \cdot d_h \cdot \delta/\eta$ (d_h = hydraulic diameter $= 4 \cdot A/0$)
[b] Prandtl number: $c_p \cdot \eta/k$
[c] Grashof number: $g \cdot 1/v \cdot (\delta v/\delta T)\, p \cdot d_h^3 \cdot (T_w - T_{fl})/\nu^2$

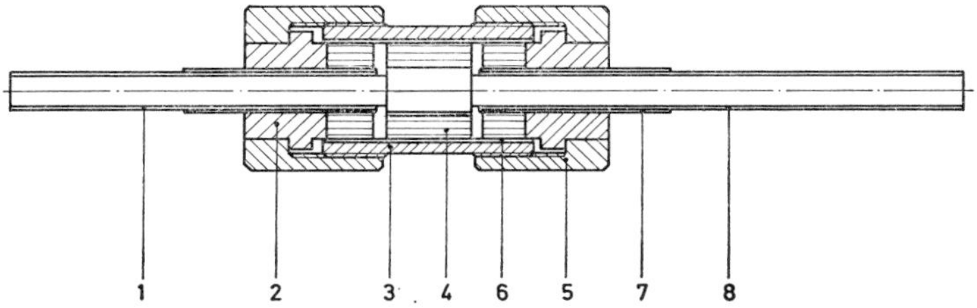

Fig. 5.7.8. Vacuum tight insulating joints made of oxidized aluminum bushings used in the circuit shown in Fig. 5.7.7 (Morpurgo). *1* and *8* St.steel he-tube, *2* Compression piece (anticorodal), *3* and *4* tube and cylinder (anticorodal), *5* Collar (anticorodal), *6* and *7* Teflon tubes

plant is small. Adequate quantities of helium can be stored in high pressure cryostats and, in cases of emergency, pumped through the magnet, or to cool down the magnet after a quench.

The design of insulating sections, between hollow conductor and helium manifold may prove to be somewhat complicated at the high pressure section. A possible design is illustrated in Fig. 5.7.8.

2. Helium 4, Region 2

This is the region of boiling heat transfer and we have to make a distinction in this region between pool boiling and forced cooling.

Pool Boiling

The general boiling curve shown in Fig. 5.7.9 may be divided into the following sections:

(i) Initiation of nucleate boiling (*OA*).
(ii) Nucleate boiling curve (*BC*).
(iii) Maximum nucleate boiling flux (*C*).
(iv) Film boiling curve (*DE*).
(v) Minimum film boiling flux (*F*).

The following observations are of interest. The first temperature rise (*OA*) occurs at very low heat flux values until inception of bubbles on the wall surface. This first temperature rise may effect the performance of superconducting coils, when fluxjumps, or in case of pulsed superconducting magnets where localized heating at low heat fluxes may occur.

Measurements by Lyon [67], Smith [68], and Johannes [69] show the following interesting results:

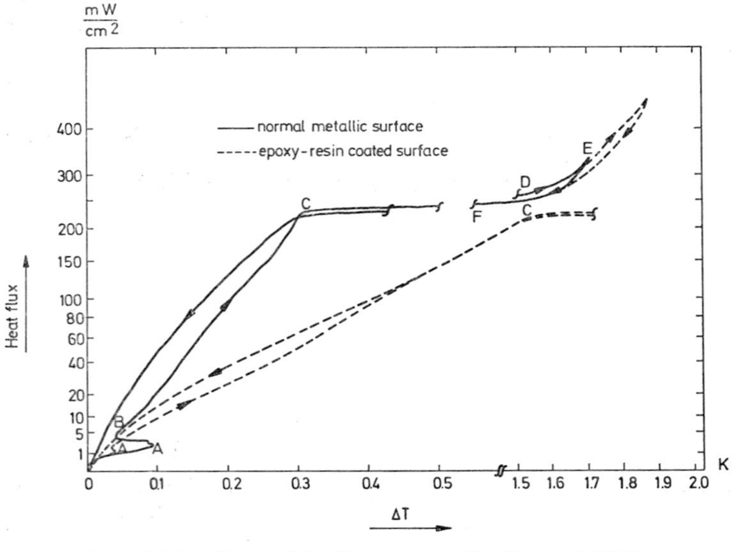

Fig. 5.7.9. General boiling curve of helium (IEKP)

Rough surfaces exhibit a hysteretic behaviour in the heat flux temperature rise characteristics as seen in Fig. 5.7.9. Polished and smooth surfaces illustrate little hysteretic behaviour. The hysteresis is due to the difference in the number of nucleation sites while increasing and decreasing the heat flux. On increasing the heat flux few sites are activated, where on decreasing the heat flux more sites remain active. The value of the peak nucleate boiling flux (P.N.B.F.) varies with the nature of the heated surface. Table 5.7.2 gives a summary of P.N.B.F. values at 1 kp cm^{-2} for horizontal and vertical surfaces reported in literature.

Table 5.7.2. *Influence of the surface nature on the* $(q/A)_{max}$ [69]

Heated Surface	$(q/A)_{max}$ Horizontal Surface (W/cm²)	Vertical Surface (W/cm²)
Mica	—	0.5
Aluminum	0.8	0.6
Copper-tin	0.84	—
Stainless-steel	0.75	0.375
Platinum	0.8	0.58
OFHC-Copper	1.0	0.58
Copper (99.999%)	1.2	—

The effect of surface coating is also reported. According to Cummings [70] a coating of ice enhances the P.N.B.F. by about 55%. Surfaces coated with epoxy resins (~100 μm thick) enhanced the P.N.B.F. values to 35% for larger gaps and 20% for narrow channels.

Lyon [71] reported that dirty surfaces do not change the peak nucleate boiling flux but change the boiling curve to more unstable regions. Very thin epoxy surfaces (~7 μm) on composite conductors did not change the nucleate boiling curve but lead to a more stable magnet performance. The effect of the channel width on horizontal coolant passages in a magnet is reported by Whetstone [72] and is given in Fig. 5.7.10. About 85% of the conductor surface is directly exposed to the helium.

The effect of heated surface orientation on the peak nucleate boiling flux is studied by Lyon [67] and shown in Fig. 5.7.11. The lowest flux is obtained with a heated surface pointing downwards, the highest with the surface pointing upwards (ratio ~4 : 1). The height of the vertical heated wall has a direct consequence on P.N.B.F.: The higher the heated wall the smaller peak heat fluxes are obtained as shown in Fig. 5.7.12. Wilson's data for large gaps (0.2 cm height) are also indicated in this figure [73].

Typical nucleate boiling data can be obtained from Fig. 5.7.11. The lower portion of the curves are defined by the inception of bubble formation. The upper portion of the curves indicate the initiation of film boiling. The maximum heat flux is plotted against the temperature difference between the bulk of the fluid and the heated surface. The incep-

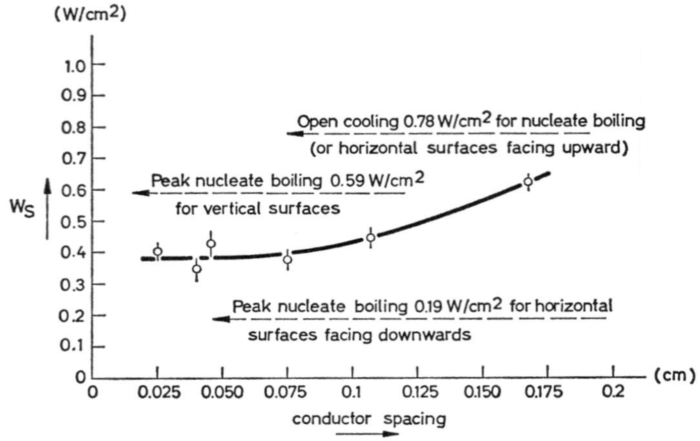

Fig. 5.7.10. Heat flux versus temperature rise in superconducting solenoids

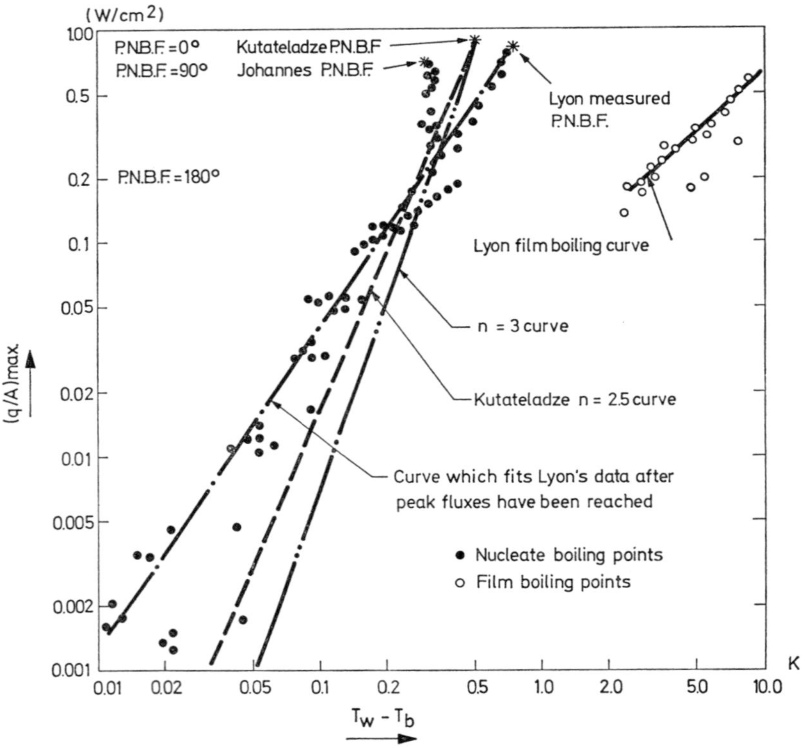

Fig. 5.7.11. Peak nucleate boiling flux data versus temperature rise

tion of bubbles causes a sharp rise in the slope of the curve. This slope remains constant and is proportional to about the *third* power ($n = 3$) of the temperature difference. The analytical expression which represents nucleate boiling phenomena has the general form:

$$\left(\frac{q}{A}\right) = f\,(\text{Properties})\,(T_{\text{wall}} - T_b)^n. \tag{5.7.30}$$

The multiplier of the ΔT value is a function of the heated surface as well as a function of fluid. Among a number of correlation equations the equation derived by Kutateladze [74] is reasonably adequate for cryogenic fluids:

$$\left(\frac{q}{A}\right) = C_1 \cdot \left[\frac{c_{p,\text{fl}} \cdot k_{\text{fl}}}{h_{\text{fg}} \cdot \delta_v}\right] \cdot \frac{(\delta_{\text{fl}})^{1.282} \cdot (p_b)^{1.75}}{(\sigma_{\text{fl}})^{0.96} \cdot (\eta_{\text{fl}})^{0.626}} (T_w - T_b)^{2.5}. \tag{5.7.31}$$

C_1 is a constant. One may also note the exponent of ΔT which is $n = 2.5$. The maximum nucleate boiling heat flux is defined as the flux, at which the nucleate boiling curve becomes discontinuous. The boiling then enters a "film boiling" region, associated with rapid temperature increase. The expression for peak nucleate pool boiling is given by:

$$\left(\frac{q}{A}\right)_{\max} = C_2 \cdot h_{\text{fg}} \cdot (\delta_v)^{0.5}[\sigma_{\text{fl}}(\delta_{\text{fl}} - \delta_v)]^{0.25}. \tag{5.7.32}$$

In this expression C_2 indicates a numerical factor, depending on the system of units used to characterize the fluid properties. σ_{fl} is the surface tension of the fluid to its own vapor.

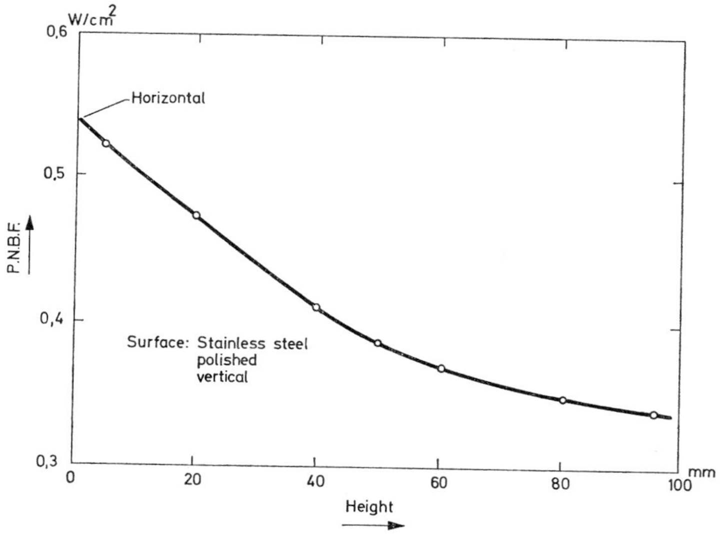

Fig. 5.7.12. Peak nucleate boiling flux versus channel height

By reducing the value of $\Delta T = T_w - T_h$ on the film boiling curve, a point is reached where the film boiling flux is a minimum. This point is of considerable interest and is derived for cylinders with diameters of $d_h \geqq 1.0$ cm:

$$\boxed{\left.\frac{q}{A}\right)_{\min} = C_3 \cdot (h_{fg} \cdot \delta_v) \left[\frac{\sigma_{fl}(\delta_{fl} - \delta_v)}{(\delta_{fl} + \delta_v)^2}\right]^{0.25}.} \tag{5.7.33}$$

Finally for the designer, the P.N.B.F. values versus channel geometry given in Fig. 5.7.13 may be of interest. Superconducting magnets with various internal channel widths are included and show that direct cooled conductor behave similar to well defined geometries.

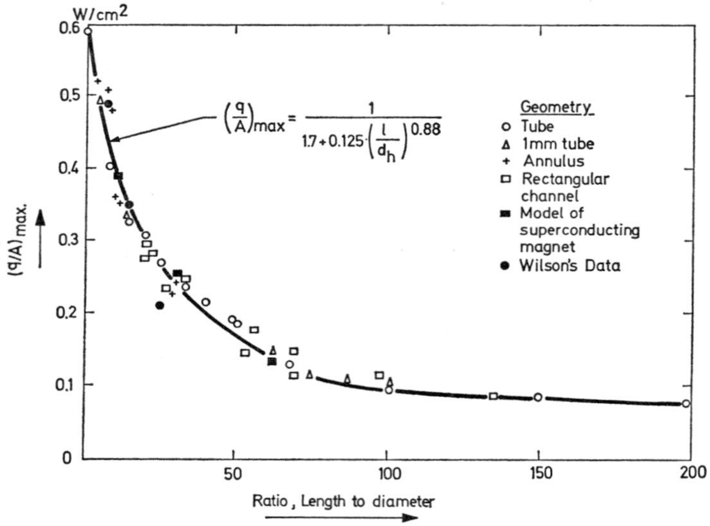

Fig. 5.7.13. Peak heat flux vs. ratio of length to diameter for various channel geometries (Wilson's data are from [73])

Forced Convection Boiling

The process of nucleate boiling of liquid helium is of considerable interest from the point of view of the design and construction of superconducting d.c. and a.c. magnets. In pulsed magnets, the overall energy density and overall heat dissipation may not be large (< 100 mW/cm³) but in particular areas, due to flux motion, flux jumps and flux sweeping or due to material characteristics at high fields higher heat fluxes (> 100 mW/cm²) may occur.

An improvement in the cooling process leads to an enhancement in the magnet performance.

Usually superconducting magnets are placed in a pool of liquid helium. Circulating the helium has been considered only in a few isolated cases, because of design complications and the requirement of a pump. The following reasons have led to a more careful study and adaptation of forced liquid helium cooling.

(a) Immersion of a magnet in a liquid helium pool requires large quantities of helium, which is undesirable in large magnet systems. In case of an overall quench vast amounts of helium may evaporate. In systems, where forced helium cooling is applied, the amount of helium required to cool the magnets can greatly be reduced.

(b) Forced liquid helium cooling improves the heat transfer to the coolant. Coil dimensions, and thus overall size of the system, can be reduced.

To maintain forced cooling, suitable pumps and provisions to force and guide the refrigerant through internal coil passages have to be imployed. The control system used in providing each coolant passage with an adequate amount of fluid and to guard the helium velocity through each passage is a complicated problem. While in the case of heat removal with supercritical helium the velocities of the fluid through passages may be high in a two phase fluid, where relatively low pressures can be imployed, the velocity of helium through coolant passages may be less than 1 m per second. The designer is advised to study the helium distribution in internal channels of a magnet much more carefully in this case than in the case of pool boiling, where correlation data are quite adequate.

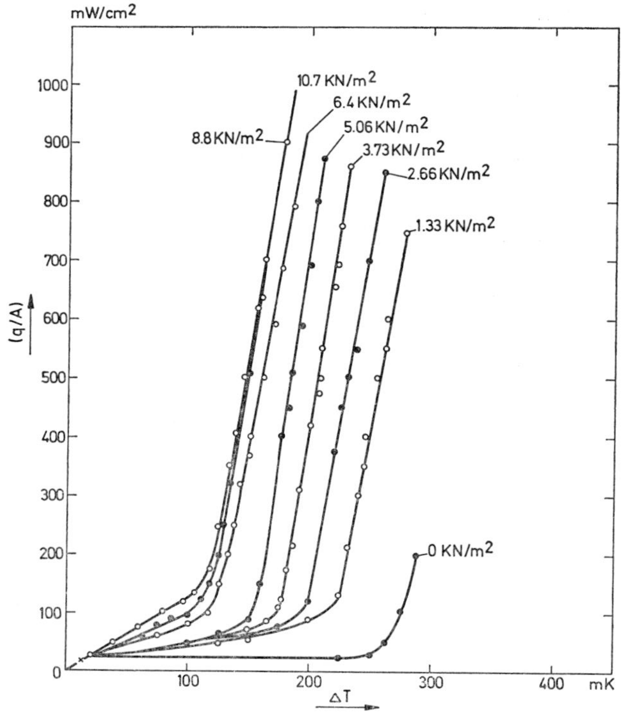

Fig. 5.7.14. Heat flux versus temperature rise for a two phase helium flow at various bath pressures (1 kN/cm² = 10^{-2} kp/cm²)

Jergel et al. [75] have studied the effect of forced helium circulation on heat transfer in narrow channels. Some of their data given for 0.3 and 0.6 mm high vertical channels, 50 cm long, 10 mm wide are plotted in Fig. 5.7.14.

For *laminar* flow the results obtained so far were correlated by

$$\left.\frac{q}{A}\right)_{\max,\,\mathrm{lam}} = 1.95\,(d_h)^{0.5} + 1.1\,\bar{v}^2. \qquad (5.7.34)$$

If the hydraulic diameter ($d_h = 4 \cdot A/0$) is expressed in (cm) the fluid velocity $\bar{v}$ in cm/s^{-1}, the maximum heat flux for nucleate boiling $q/A)_{\max}$ is expressed in W · cm^{-2}.

In the turbulent flow region the maximum heat flux for nucleate boiling is proportional to $\bar{v}^2$:

$$\left.\frac{q}{A}\right)_{\max,\mathrm{turb}} = C \cdot 5.7\, v^2 + \text{additional terms}. \qquad (5.7.35)$$

According to these results, the heat transfer improvement is more impressive in the turbulent flow region. The results suggest, that there is a $q/A)_{\max}$ value associated with each d_h with no flow. The Eq. (5.7.34) gives for the 1, 0.6 and 0.3 mm channels the values of $q/A)_{\max,\,\mathrm{lam}} =$ 0.87; 0.67; 0.48 W/cm^2. The measured values were 0.16; 0.2 and 0.04 to 0.07 W/cm^2!

There are of course two discrepancies between the usual correlation equations and the experimental data:

(i) The Eq. (5.7.34) gives optimistic values for $v = 0$. It is recommanded to use correlations given in the previous sections for nucleate pool boiling.

(ii) The wide spread in maximum heat flux values (0.04 ... 0.07 W per cm^2) for the small channel of 0.3 mm height indicates, that the heat transfer is affected by bubble entrapment, a result which was also found by other investigators mentioned above.

Comparing the modified Dittus-Boelter correlation Eq. (5.7.25) with the new experimental data obtained for a 0.6 mm channel ($d_h = 0.12$ cm) for turbulent flow ($v = 10^2$ cm · s^{-1}), using $\Delta T = 0.4$ K, we obtain:

$$\left.\frac{q}{A}\right)_{\max} \cong 130\ \mathrm{mW/cm^2}, \qquad \text{rom Eq. (5.7.25)}$$

$$\left.\frac{q}{A}\right)_{\max} \cong 800\ \mathrm{mW/cm^2}. \qquad \text{from Eq. (5.7.34)}$$

The discrepancy is even more significant, as the correlation equation obtained for compressed liquid or supercritical helium does not seem to be applicable for forced two phase flow. It is known that the boundary layer in a channel between heated wall and bulk fluid is much smaller than in laminar flow. The velocity gradient is much larger. This means that the bubble size prior to detachment from the wall is much smaller under turbulent flow conditions, being about the size of the boundary layer. However, it is clear that any sweeping of the bubbles away from the surface improves the heat transfer process.

3. Helium 4, region 3 and 4

Region 3 and 4, devoted to the single phase helium fluid, has recently become attractive. Heat transfer measurements at high pressure up to 2 MN/m² are reported by Dorey [76] and recently by Girratano [77]. The classical Dittus-Boelter Eq. (5.7.25) given earlier is generally satisfactory.

Jacob [78] recommended the two expressions given below for flat channels:

$$\frac{h \cdot d_h}{k} = 0.664\ (N_{\mathrm{Re}})^{0.5} \cdot (N_{\mathrm{Pr}})^{0.5} \tag{5.7.36}$$

for laminar flow and

$$\frac{h \cdot d_h}{k} = 0.036\ (N_{\mathrm{Re}})^{0.8} \cdot (N_{\mathrm{Pr}})^{1.0}\,, \tag{5.7.37}$$

for turbulent flow. Using this correlation equation and compare it to the heat flux numbers reported by Jergel we obtain:

$$\frac{q}{A} = h\Delta T = 152\ \mathrm{mW/cm^2} \qquad \text{from Eq. (5.7.37)}$$

and

$$\frac{q}{A} = h\Delta T \cong 800\ \mathrm{mW/cm^2}, \qquad \text{from Eq. (5.7.34)}$$

we had again chosen $\Delta T = 0.4$ K.

From this comparison, we see that the forced convection in region 2 is more attractive than heat transfer in the supercritical region.

4. Helium 4, region 5

Helium behaves in this region as in the three regions discussed above. The correlations derived sofar are satisfactory. Data for viscosity, thermal conductivity specific heat and density in this region are given in Figs. 5.5.6, 5.5.7, 5.5.5 and 5.5.1.

With these data the heat transfer values can be calculated.

5.7.3 Radiation

The energy emitted by radiation from a black body of area A at a temperature T is proportional to the fourth power of T.

$$Q_b = \sigma_s \cdot A \cdot T^4 \quad (\mathrm{W})\,, \tag{5.7.38}$$

where

$$\sigma_s = 5.67 \times 10^{-8}\ \mathrm{W/m^{-2}\ K^{-4}}.$$

The Eq. (5.7.38) represents Stefan-Boltzmann law and σ_s a fundamental constant.

For practical application one would like to determine the energy which is lost through radiation. The reduction of the energy is obtained from:

$$Q_b = \sigma_s A\ (T^4 - T_{\mathrm{a}}^4)\,, \tag{5.7.39}$$

with T_a indicating the ambient temperature where the radiator is located.

The wave length λ_{max}, for which the ratio of radiation intensity per wave length intervall reaches a maximum is inversely proportional to the temperature T. This law, known as *W*. Wien's law is written explicitely:

$$\lambda_{max} \cdot T = 2.88 \times 10^{-3} \quad (\text{m K}) . \tag{5.7.40}$$

In other words, the peak intensity of radiation at any two temperatures occurs at a wave length λ such, that Wien's displacement law hold:

$$\lambda_1 T_1 = \lambda_2 T_2 = 2.88 \times 10^{-3} \quad (\text{m K}) . \tag{5.7.41}$$

Infrared radiation occurs in the range of 0.8×10^{-6} m to 400×10^{-6} m and visible light in the range 0.4×10^{-6} m to 0.8×10^{-6} m. The corresponding temperature ranges are from Wien's law 7 to 3600 K and 3600 to 7000 K resp. Thus, the characteristic radiation of the surface encountered in cryogenic apparatus is infrared radiation and the wave length ranges from about 10×10^{-6} m at room temperature to nearly 10^{-2} m at liquid helium temperature.

Perfe t black body surfaces do not exist in nature. Materials employed in practical systems only approximate black bodies in that they absorb and emit only a fraction of the energy given by Eq. (5.7.38). Thus

$$Q_g = \varepsilon \cdot \sigma_s \cdot A \cdot T^4 \quad (\text{W}) , \tag{5.7.42}$$

where ε is the *emissivity*, a value less than one.

These surfaces are known as gray bodies since they do approximately respond to the wave length distribution, given by Planck's law.*

The values of ε are found in practice in the range from > 0.9 for black anodized aluminum to 0.02 for clean aluminum foil. Fig. 6.2.20 illustrate total emissivity values measured on different material surfaces.

The incident energy that is not absorbed, is transmitted through the surface or reflected from it. The reflection from a surface may be *specular*, ray like (such as in a mirror), or *diffuse* where a ray is scattered in many directions (such as by a ground glass). The energy transport due to radiation between facing surfaces in some geometrical arrangements depends upon the specular or diffuse nature of the surfaces.

The heat transport between two plane parallel surfaces, whether diffuse of specular, with emissivities ε_1 and ε_2 at temperatures T_1 and T_2 is given by

$$Q = \sigma_s A \left(T_1^4 - T_2^4\right) \frac{\varepsilon_1 \varepsilon_2}{\varepsilon_1 + \varepsilon_2 - \varepsilon_1 \varepsilon_2} , \tag{5.7.43}$$

when $\varepsilon_1 \simeq \varepsilon_2 \simeq 1$,

$$Q_{\varepsilon=1} = \sigma_s A \left(T_1^4 - T_2^4\right) , \tag{5.7.44}$$

when $\varepsilon_1 \gg \varepsilon_2$

$$Q_{\varepsilon_2} = \sigma_s A \left(T_1^4 - T_2^4\right) \varepsilon_2 \tag{5.7.45}$$

and when $\varepsilon_1 = \varepsilon_2 = \varepsilon \ll 1$

$$Q_\varepsilon = \sigma_s A \left(T_1^4 - T_2^4\right) \varepsilon/2 . \tag{5.7.46}$$

---

* M. Planck's law postulates that the radiation intensity is distributed over wave lengths.

For long coaxial cylinders or concentric spheres with areas A_1 and A_2 and diffuse reflecting surfaces the heat transport is given by

$$Q = \sigma_s A \, (T_1^4 - T_2^4) \frac{\varepsilon_1 \, \varepsilon_2}{\varepsilon_1 + (A_1/A_2)\,(1 - \varepsilon_2)\,\varepsilon_1}. \tag{5.7.47}$$

For specular reflection the term $A_1/A_2 = 1$ and the resulting heat transport is the same as Eq. (5.7.43).

The overall radiant heat transport can be reduced by introducing a thermally isolated surface between two surfaces at temperatures T_1 and T_2. This will reflect part of the radiation, absorb another part and then emit the absorbed part at an intermediate temperature between T_1 and T_2, especially if the surface has a low emissivity.

For the heat transfer with n shields between plane parallel bounding surfaces, we have

$$Q = \sigma_s A \, (T_1^4 - T_2^4) \frac{\varepsilon_i \, \varepsilon_k}{(n - 1)\,\varepsilon_i + 2\,\varepsilon_k}, \tag{5.7.48}$$

where

$$\varepsilon_i = \frac{\varepsilon_0 \, \varepsilon_s}{\varepsilon_s + \varepsilon_0 - \varepsilon_0 \, \varepsilon_s} \quad \text{and} \quad \varepsilon_k = \frac{\varepsilon_s}{2 - \varepsilon_s}.$$

ε_0 being the emissivity of the bounding surfaces and ε_s the emissivity of the shields.

For the case where $\varepsilon_0 = \varepsilon_s = \varepsilon$

$$Q = \sigma_s A \, (T_1^4 - T_2^4) \frac{\varepsilon}{(n + 1)\,(2 - \varepsilon)}. \tag{5.7.49}$$

If ε is small as is the case for bright shields, then the heat transport is greatly reduced, if a large number of shields can be installed in a thermally isolated manner. The modern multilayer superinsulations [79] (Fig. 5.7.15) achieve this to a good approximation by using either a fibrous separator between shields of metallic foils, or coated surfaces, or by using point contacts between the coated surfaces. In all these cases, convective heat transfer is effectively eliminated by a high vacuum between the bounding surfaces.

For the case where the boundaries are highly reflective $\varepsilon_0 \ll 1$ but the shields are nearly black, $\varepsilon_s \simeq 1$, the heat transport is expressed by

$$Q_{nb} = \sigma_s A \, (T_1^4 - T_2^4) \frac{\varepsilon_0}{(n - 1) \cdot \varepsilon_0 + 2}. \tag{5.7.50}$$

For small values of n the Eq. (5.7.50) reduces to Eq. (5.7.46) which indicates that the shields are of little value. For very large n numbers, however, the result approaches Eq. (5.7.49) and the heat transport is greatly reduced. The effectiveness of fine powder insulations is explained by Eq. (5.7.50), since they approximate a large number of highly absorbing isothermal bodies or shields.

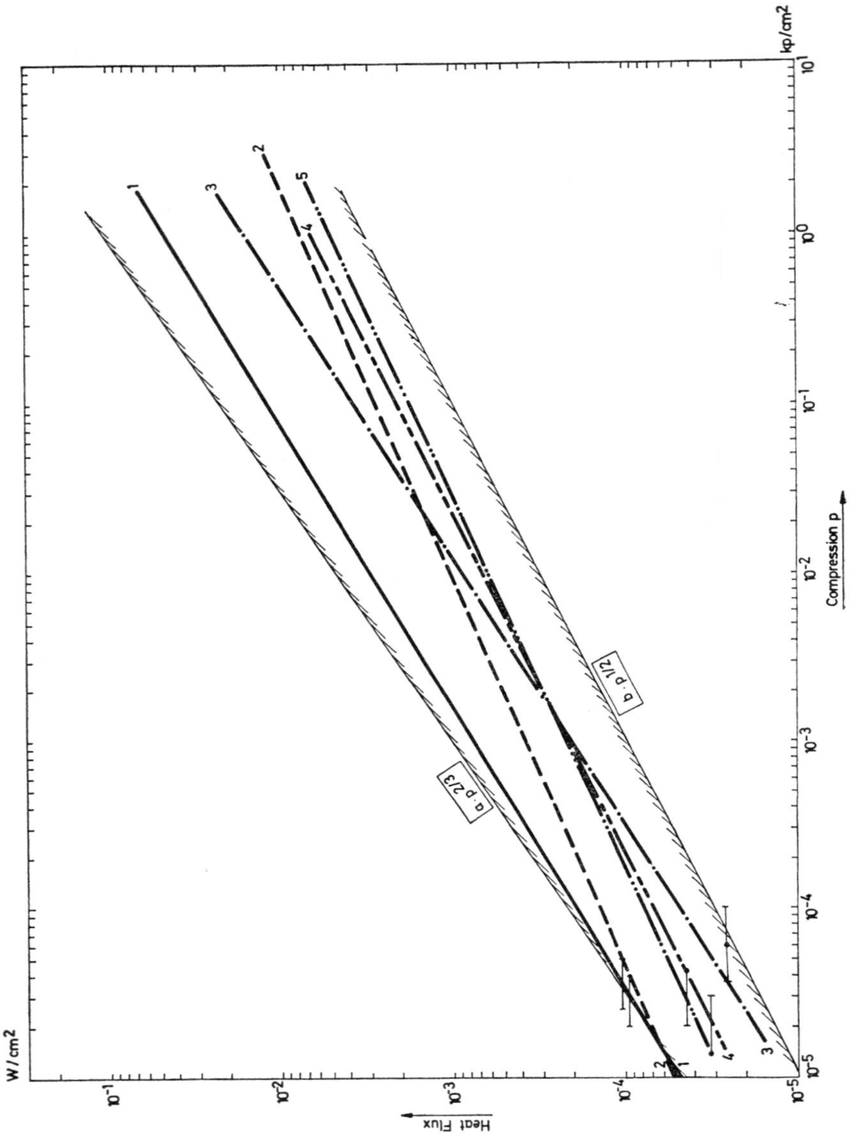

Fig. 5.7.15. Effect of external pressure on the heat flux through multilayer superinsulations:

1 10 layers of 6.35×10^{-4} cm aluminized polyester (both sides) with 22 layers of 2.5×10^{-3} cm glass fabric

2 10 layers of 6.35×10^{-4} cm aluminized polyester (both sides) with 10 layers of 7.6×10^{-3} cm silk netting with $10^{-2} \times 1.2$ cm² stripes of glass netting

3 10 layers of 5.08×10^{-3} cm tempered aluminum with 11 layers of 17.8×10^{-3} cm nylon netting

4 10 layers of 6.35×10^{-4} cm aluminized polyester (both sides) with layers of 7.6×10^{-3} cm silk netting with $2 \times 10^{-3} \times 0.6$ cm² stripes of glass netting

5 10 layers of 6.35×10^{-4} cm aluminized polyester (both sides) with 22 layers of 7.6×10^{-3} cm silk netting

5.7.4 Methods to Minimize Thermal Losses

The equations presented in the previous sections contain the necessary information to minimize the thermal losses to the environment. For the case of conduction through solids of given thermal conductivity, it is obvious that the cross-sectional area should be reduced and the length of the solid (being exposed to a thermal gradient) increased. The results of Garwin's analysis [56] show that little can be done by adjusting the areas unless the area is reduced at the position where the thermal conductivity has a peak value. Reduced areas at either end of the solid are not useful.

For convective heat transfer low pressures are the remedy or, if that is not possible, then small diameters are desirable as seen from Eq. (5.7.25) Where possible, of course, the convective heat transfer from venting gases should be employed to conserve the sensible refrigeration in the venting gas. In the case of neck tubes, this can be accomplished by inserting a loose fitting plug of a low conductive material such as a foam plastic which will cause the vapor to flow in a small area close to the wall and thus enhance the heat transfer.

Energy transport by radiation exists in all low temperature systems and the use of reflective walls as in the original Dewar vessel, is the first step, along with the required high vacuum of less than $1.33 \times 10^{-3}\ \mathrm{Nm}^{-2}$ to eliminate convection. Radiation shields are obviously the next step, the best practical system being the multishielded insulations. The fiber matt supported shield insulations, the superinsulations, are convenient to use and maintain high performance. The layer density is controlled to a great degree by the fiber matt and the results are consistent with theory for simple cylindrical shapes, if there are not too many penetrations of the insulation which act to short circuit the shields. When aluminum foil is used as the reflective shields, short circuits are particularly bad, as the thermal conductivity of aluminum is very high.

The point contact separation suffers from poorly controlled layer density but has an overall lower density which is attractive in some applications. The easy handling is deceptive, since the thermal convection is dependent upon layer density as shown in Fig. 5.7.15. Fiber matt separated reflective shields are commercially available. Optimally about 25 layers of superinsulation can be applied in a one cm wide gap.

Further improvement over simple vacuum or multilayer insulation can be achieved through using a refrigerated shield in the insulation space, with the refrigeration being obtained from the vapor vented from the container. In this case, the vapor is vented in such a way, that the sensible heat is exchanged with the shield. Three methods have been developed:

(i) Employing a discrete shield with the vapor vented through a tube heat exchanger distributed over the shield.

(ii) Using a shield supported from the neck tube and deriving its refrigeration from the gas vented from the neck tube.

(iii) Employing several shields supported by multilayer insulation and connected for heat transfer to the neck tube.

Finally, the radiation down the neck tubes or other vents and openings can be minimized by placing several reflective shields along the length or by blackening the inside surface of the tube, so that the radiation is absorbed along the wall and re-emitted at a lower temperature, therefore in effect producing a large number of shields.

5.7.5 Application

One of the interesting applications of the first and second law of thermodynamics, and the use of the correlations obtained in section 5.7.2, is the calculation of liquid helium losses by the conduction and radiation of a magnet cryostat. We apply to a magnet cryostat indicated by an open system, noted by a boundary b_0, illustrated in Fig. 5.7.16, the first law of thermodynamics

$$Q)_{b0} = Q_c + Q_r \tag{5.7.51}$$

$$Q)_{b0} = \dot{m}\,(u + pV)_{v,T_1} = \left.\frac{dU}{dt}\right)_{b0}. \tag{5.7.52}$$

In these equations Q_c is the heat conducted down the cryostat neck, Q_r is the heat radiated to the flask from a temperature T_3, and $\dot{m}$ is the time rate of mass flow.

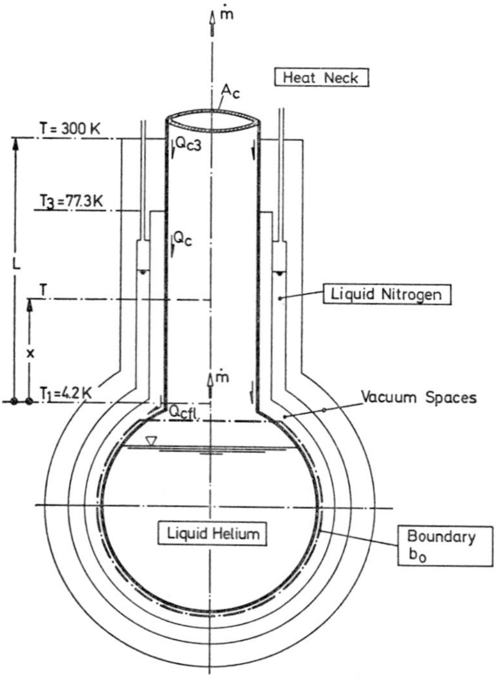

Fig. 5.7.16. Schematic representation of a liquid helium dewar with liquid nitrogen precooling and vacuum spaces

And:
$$\left.\frac{dU}{dt}\right)_{b0} = -\dot{m} \cdot u_{f,T1} .$$

The subscripts f refer to the saturated liquid and v to the vapor states.

Eq. (5.2.52) is rewritten as

$$Q_c + Q_r = \dot{m}\,[(u + pV)_v - u_f]_{T_1} . \tag{5.7.53}$$

Simplifying,

$$Q_c + Q_r = \dot{m} \cdot G$$

and expressing the heat radiated, by

$$Q_r = \sigma_s \cdot \varepsilon \cdot A_1 (T_3^4 - T_1^4) , \tag{5.7.54}$$

we can combine the equations in a form:

$$\frac{\dot{m}}{\varepsilon \cdot A_1} = \frac{\sigma_s (T_3^4 - T_1^4)}{G - Q_c/\dot{m}} . \tag{5.7.55a}$$

The equation is of the form:

$$\boxed{\frac{m}{\varepsilon \cdot A_1} = f_1(Q_c/\dot{m})} . \tag{5.7.55 b}$$

The heat conducted through a cryostat neck is given by:

$$Q_{\text{net}} = H_{\text{out}} - H_{\text{in}} = Q_c - Q_{c,\text{f1}} , \tag{5.7.56}$$

where we assumed that the temperatures of the gas and the flask wall are identical.

For a constant pressure process:

$$H_{\text{out}} - H_{\text{in}} = \dot{m} \int_{T_{\text{f1}}}^{T} c_p(T)\, dT .$$

If c_p is assumed to be constant over a small temperature range, we may write:

$$H_{\text{out}} - H_{\text{in}} = \dot{m}\, c_p (T - T_{\text{f1}}) . \tag{5.7.57}$$

Thus

$$Q_c - Q_{c,\text{f1}} = \dot{m}\, c_p (T - T_{\text{f1}}) . \tag{5.7.58}$$

The heat conduction along the wall is given by:

$$Q_c = k \cdot A_c \cdot \frac{dT}{dx} . \tag{5.7.59}$$

For simplicity let us assume that the coefficient of thermal conduction is a linear function of temperature, then

$$k = aT - B ,$$

and we define:

$$k_{\text{f1}} = aT_{\text{f1}} - b ,$$

then:

$$k = a\,(T - T_{\text{f1}}) + k_{\text{f1}} . \tag{5.7.60}$$

Combining these conduction equations we obtain:

$$Q_c = Q_{c,\mathrm{fl}} + \dot{m}\, c_p (T - T_{\mathrm{fl}})$$

$$k A_c \frac{dT}{dx} = Q_{c,\mathrm{fl}} + \dot{m}\, c_p (T - T_{\mathrm{fl}})$$

$$\left(a \cdot (T - T_{\mathrm{fl}}) + k_{\mathrm{fl}}\right) \cdot A_c \cdot \frac{dT}{dx} = Q_{c,\mathrm{fl}} + \dot{m}\, c_p (T - T_{\mathrm{fl}}) \,,$$

which yields:

$$\frac{dx}{A_c} = \left[\frac{k_{\mathrm{fl}} + a\,(T - T_{\mathrm{fl}})}{Q_{c,\mathrm{fl}} + \dot{m}\, c_p (T - T_{\mathrm{fl}})}\right] dT \,. \tag{5.7.61}$$

If we define:

$$T - T_{\mathrm{fl}} = \tau$$

and integrate Eq. (5.7.61) we get:

$$\frac{x}{A_c} = \frac{k_{\mathrm{fl}}}{\dot{m}\, c_p} \left[\ln\left(Q_{c,\mathrm{fl}} + \dot{m}\, c_p \cdot \tau\right)\right]_0^\tau$$

$$+ \frac{a}{(\dot{m}\, c_p)^2} \left[Q_{c,\mathrm{fl}} + \dot{m}\, c_p \tau - Q_{c,\mathrm{fl}} \cdot \ln\left(Q_{c,\mathrm{fl}} + \dot{m}\, c_p \cdot \tau\right)\right]_0^\tau$$

at $x = L$, $T = T_H$, and $\tau = \tau_H$, then:

$$\frac{L}{A_c} = \frac{1}{(\dot{m}\, c_p)} \left\{a \tau_H + \left[k_{\mathrm{fl}} - Q_{c,\mathrm{fl}} \left(\frac{a}{\dot{m}\, c_p}\right)\right] \ln\left(\frac{Q_{c,\mathrm{fl}} + \dot{m}\, c_p \tau_H}{Q_{c,\mathrm{fl}}}\right)\right\}. \tag{5.7.62}$$

This equation is of the form:

$$\boxed{\dot{m} \frac{L}{A_c} = f_2 (Q_{c,\mathrm{fl}}/\dot{m}) \,.} \,. \tag{5.7.63}$$

Combining Eqs. (5.7.55) and (5.7.63) we may write:

$$\boxed{\frac{1}{\left(\frac{L}{A_c}\right) \cdot \varepsilon A_1} = f_3 \left(\frac{Q_c}{\dot{m}}\right).} \tag{5.7.64}$$

One may plot $\dot{m}/\varepsilon A_1$ as a function of $[(L/A_c)\, \varepsilon A_1]^{-1}$, which is shown in Fig. 5.7.17. As $[(L/A_c)\, \varepsilon A_1]^{-1}$ is known from the cryostat geometry the corresponding value of maximum flow can be obtained directly from the graphs. The data of Fig. 5.7.17 are in agreement with measured data given by Wechsler [80, 81].

Using stainless steel as a cryostat material and approximating its thermal conductivity in the temperature range of 4 to 100 K by

$$k = 1.18\,\mathrm{T} - 2.36 \,,$$

we obtain the helium loss rate.

It is interesting to note that below a certain value of

$$\left(\frac{L}{A_c}\, \varepsilon A_1\right)^{-1} = 4.72 \times 10^4 \quad (\mathrm{cm}^{-1}) \,,$$

the helium loss rate is not reduced, and remains at 6.8×10^{-3} lit per (day · cm^2). As further reduction of the ratio $(L/A_c\, \varepsilon A_1)^{-1}$ does not reduce the helium loss rate, we may conclude that the loss rate is essentially due to radiation.

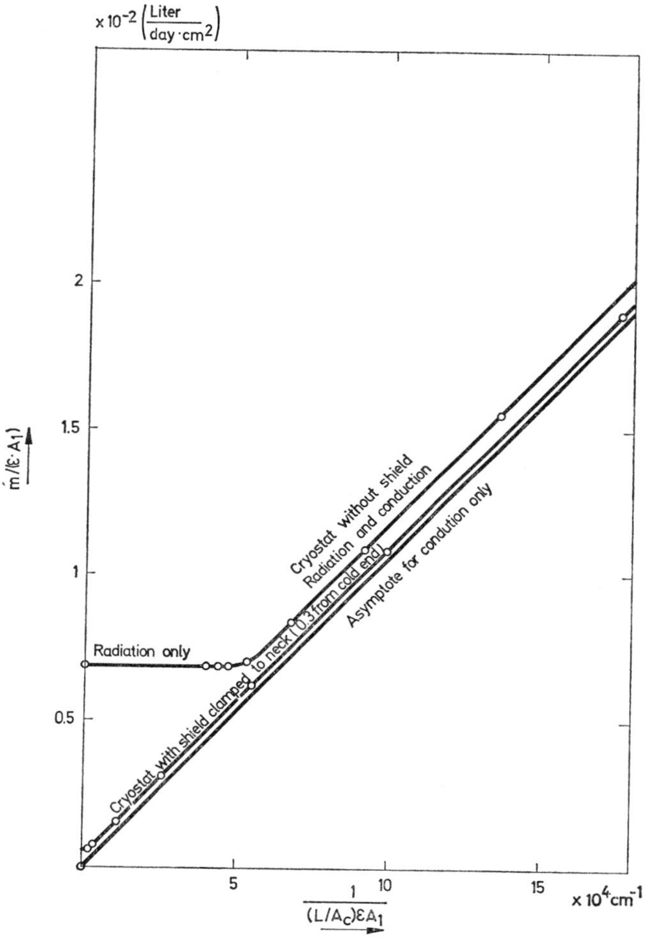

Fig. 5.7.17. Specific values to calculate helium boil off of simple helium dewars

References Chapter 5

[1] Strowbridge, T. R.: The thermodynamic properties of nitrogen from 64 K to 300 K between 0.1 and 200 atm. NBS Tech. Note 129 (1962).

[2] Dean, J. W.: A tabulation of the thermodynamic properties of normal hydrogen from low temperatures to 300 K and from 1 to 100 atm. NBS Tech. Note 120 (1961).

[3] Roder, H. M., Weber, L. A., Goodwin, R. D.: Thermodynamic and related properties of parahydrogen. NBS Monograph 94 (1965).

[4] Mann, D. B.: The thermodynamic properties of helium from 3 to 300 K and from 0.5 atm to 100 atm. NBS Tech. Note 154 (1962).

[5] McCarty, R. D., Stewart, R. B.: Advances in Cryogenic Engineering 9, p. 161. New York: Plenum Press 1964.

[6] Johnson, V. J.: Properties of Materials at Low Temperatures. A Compendium. New York: Pergamon Press 1961.

[7] Zabatakis, M. G.: Safety with Cryogenic Fluids. New York: Plenum Press 1967.

[8] White, G. K.: Experimental Techniques in Low Temperature Physics, 2nd Ed., Oxford 1968.

[9] Kropschot, R. H., Birmingham, B. W., Mann, D. B.: Technology of Liquid Helium. NBS Monograph 111 (1968).

[10] Zimmerman, I. E., Root, R. C.: Rev. Sci. Instr. **32,** 853 (1961).

[11] Sequin, H. J., Leonard, R. W.: Rev. Sci. Instr. **37,** 1743 (1966).

[12] Burgeson, D. A., Pestalozzi, W. G., Richards, R. J.: Advances in Cryogenic Engineering 9, p. 416. New York: Plenum Press 1969.

[13] Efferson, K. R.: Adv. Cryogenic Eng. **15,** 124. New York: Plenum-Press (1970).

[14] Jacobs, R. B., Collins, S. C.: J. Appl. Phys. **11,** 491 (1940).

[15] Strobridge, T. R.: IEEE Trans. NS **16,** 1104 (1969).

[16] Grassmann, P.: Schweiz. Bauzeitung **79,** 797 (1961).

[17] Trepp, Ch.: Proc. First Int. Cryogenic Engineering Conf. 277. Tokyo 1967.

[18] Long, H. M.: Refrigeration at Cryogenic Temperatures (ASME Meeting, 1963).

[19] Fleming, R. B.: Advances in Cryogenic Engineering 14, Paper E 6, p. 197. New York: Plenum Press 1969.

[20] Hogan, W. H., Stuart, R. W.: Design considerations for cryogenic refrigeration (ASME meeting, 1963).

[21] McMahon, H. O., Gifford, W. E.: Advances in Cryogenic Engineering 5, p. 354. New York: Plenum Press 1960.

[22] Sixsmith, H.: Proc. First Int. Symp. on Gas-Lubricated Bearings, ACR 49 (1959).

[23] Birmingham, B. W., Sixsmith, H., Wilson, W. A.: Advances in Cryogenic Engineering 7, p. 30. New York: Plenum Press 1962.

[24] Collins, S. C.: Advances in Cryogenic Engineering 2, p. 8. New York: Plenum Press 1956.

[25] Doll, R., Eder, F. X.: Kältetechnik **16, 5** (1964).

[26] Doll, R., Eder, F. X.: Advances in Cryogenic Engineering 9, p. 561. New York: Plenum Press 1964.

[27] Johnson, R. W., Collins, S. C., Smith, J. L.: Advances in Cryogenic Engineering 16, p. 171. New York: Plenum Press 1971.

[28] Paivanas, J. A., Robert, O. P., Wang, D. I. J.: Advances in Cryogenic Engineering 10 (Sections A—L), p. 197. New York: Plenum Press 1964.

[29] Keesom, W. H.: Helium. Amsterdam: Elsevier 1942.

[30] Lounasmaa, O. V.: ASME Trans. **72,** 751 (1950).

[31] Mann, D. B., Stewart, R. B.: NBS Tech. Note 8 (1959).

[32] Diller, D. E., Hanley, H. J. M., Roder, H. M.: Cryogenics **10,** 286 (1970).

[33] Tattersall, H. G., Tapin, G.: J. of Math. Science **1,** 296 (1966).

[34] Vieland, L. J., Wicklund, A. W.: Phys. Rev. **166,** 424 (1968).

[35] Powell, R. L., Roder, H. M., Hall, W.: J. Phys. Rev. **115,** 314 (1959).

[36] Cody, G. D., Cohen, R. W.: Rev. Mod. Phys. **120** (1964).

[37] Jones, H.: Handbuch der Physik 19. Berlin: Springer 1956.

[38] Gerritsen, A. N.: Handbuch der Physik 19. Berlin: Springer 1956.

[39] Grüneisen, E.: Ann. Phys. **16,** 530 (1933).
[40] MacDonald, D. K. C.: Handbuch der Physik 14. Berlin: Springer 1956.
[41] Laubitz, M. J.: Can. J. Phys. **45,** 3677 (1967).
[42] Kapitza, P.: Proc. Roy. Soc. London, Ser. A, 123, 292, 342 (1929).
[43] Kohler, M.: Ann. Physik **40,** 601 (1941).
[44] Corruccini, R. J.: NBS Tech. Note 218 (1964).
[45] Schauer, W., Specking, W., Turowski, P.: KFK Rep. 1219 (1970).
[46] Borovik, E. S., Volotskaya, V. G., Fogel, N. Y.: Sov. Phys. JETP **18,** 34 (1964).
[47] Fickett, F. R.: Phys. Rev. B, **3,** 1941 (1971).
[48] Lück, R.: Aluminium **12,** 733 (1969).
[49] Pippard, A. B.: Phil. Trans. Roy. Soc. London, Ser. A, **250,** 325 (1957).
[50] Ziman, J. M.: Proc. Roy. Soc. London, Ser. A, **226,** 436 (1954).
[51] Stevenson, R., Bolton, R.: Can. J. Phys. **41,** 985 (1963).
[52] Klemens, P. G.: Australian J. Phys. **7,** 70 (1954).
[53] Arp, V.: Proc. 1968 Summer Study on Supercond. Devices, Part III, p. 1095, Brookhaven 1968.
[54] Powell, R. L.: Ph. D. Thesis, Univ. of Cambridge (1966).
[55] White, G. K.: Australian J. Phys. **6,** 397 (1953).
[56] Garwin, R. L.: Rev. Sci. Instr. **27,** 826 (1956).
[57] Rohsenow, W., Choi, H.: Heat, Mass and Momentum Transfer. New York: Prentice-Hall 1961.
[58] Corruccini, R. J.: Vacuum **7—8,** 19 (1957—1958).
[59] Unger, F.: Archiv f. Elektr. **41,** 357 (1955).
[60] Kafft, G.: KFK Rep. 1347 (1970).
[61] Dittus, F. W., Boelter, L. M. K.: Publ. in Engineering **2,** 443, Berkeley: Univ. of Calif. 1930.
[62] Arp, V., et al.: NBS Rep. 9763 (1970).
[63] Smith, R. V.: Proc. 1968 Summer Study on Supercond. Devices, Part I, p. 249. Brookhaven 1968.
[64] Hanley, H. J. M., Childs, G. E.: Cryogenics **9,** 106 (1969).
[65] Boissin, J. C., et al.: Advances in Cryogenic Engineering 8, p. 607. New York: Plenum Press 1968.
[66] Kestin, J.: Boundary Layer Theory, 4th ed. New York: McGraw-Hill 1960.
[67] Lyon, D. N.: Int. J. of Heat and Mass Transfer **7,** 1097 (1964).
[68] Smith, R. V.: Cryogenics **9,** 11 (1969).
[69] Johannès, C.: Paper B 3, Cryogenic Engineering Conf. Boulder, Col. 1970.
[70] Cummings, R. D., Smith, J. L.: Bull. ITR Annex 1966-5 (1966).
[71] Lyon, D. N.: Advances in Cryogenic Engineering 10, p. 371. New York: Plenum Press 1965.
[72] Whetstone, C. W., Boom, R. W.: Advances in Cryogenic Engineering 13, p. 68. New York: Plenum Press 1968.
[73] Wilson, M. N.: Bull. de l'Inst. Int. de froid **46,** annex 5 (1966).
[74] Kutateladze, S. S.: AEC Translation 3770, 1949 (1952).
[75] Jergel, M., Hechler, L., Stevenson, R.: Cryogenics **10,** 413 (1970).
[76] Dorey, A. P.: Cryogenics **5,** 146 (1965).
[77] Girratano, P. J., Arp, V. D., Smith, R. V.: Cryogenics **11,** 385 (1971).
[78] Jacob, M.: Heat Transfer. New York: John Wiley 1957.
[79] Black, I. A., Glaser, P. E.: Advances in Cryogenic Engineering 11, p. 26. New York: Plenum Press 1966.
[80] Wechsler, A., Jacket, H. S.: Rev. Sci. Instr. **22,** 282 (1951).
[81] Wechsler, A.: J. Appl. Phys. **22,** 1463 (1951).

6. Economic Consideration in the Design of Water-cooled, Cryogenic, and Superconducting Magnets

6.1 Introduction

Experimental physics generally requires a great number and variety of magnets for analyzing, bending, focusing, and guiding electrically charged particles. Magnetic fields, having specified magnitudes and spatial distribution, are being used in high energy physics (e.g., magnetic moment of hyperons) in conjunction with accelerators, for magneto optics (e.g., Zeeman Effect), for solid-state physics (e.g., susceptibility, Larmor precession), for magneto-acoustic effects (e.g., Fermi surfaces), and many other areas, such as in antiferromagnetism, in biological applications, effect of magnetic fields on ecology, and recently application of magnetic fields for medical purposes.

We have already mentioned that for particle energies in the GeV range and requirements of high-precision measurements in all areas of physics, it becomes increasingly difficult to comply and meet the requirements economically when using classical techniques in producing magnetic fields. In Section 1, we discussed four methods of magnetic field generation and in this section, we compare economical aspects of these methods. However, we will be brief in the economical studies of pulsed magnets of short duration and low duty cycle, simply because their area of utilization is much more limited than continuous operation magnets. To compare water-cooled, cryogenic, and superconducting magnets from an economical point-of-view, giving some overall data for capital and operation costs, one may consider technical feasibility, safety of operation, reliability, and field reproducibility.

Compared to water-cooled magnets, access to cryogenic and superconducting magnets is somewhat more cumbersome and magnet repair, in case of insulation breakdown, more difficult. Magnet designs with room-temperature accessible bores are preferred over closed designs, even if the refrigeration losses are higher due to thermal radiation and heat conduction, we will base our considerations on magnets having identical accessible working bores.

The great asset of Type II superconducting magnets is in achieving high fields. It is tempting to take advantage of the high-field properties of Type II alloys by using the relation:

$$\boxed{\int B\,dl = p \cdot \theta/3\times 10^{-1}} \quad \text{(Vs/m)} \tag{6.1.1}$$

with p the particle momentum in GeV/c, θ the deflection angle in radians. The magnetic field is expressed in Vs/m^2, and $\int_{-\infty}^{+\infty} dl$, the effective magnet length, in m.

Eq. (6.1.1) indicates that for the same particle energy, the effective bending radius of the particle track is inversely proportional to the field strength. However, for many reasons, a minimum length is desirable. We summarize a few:

(a) Charged particles have a finite decay time. The decay may occur in small-bore magnets outside the useful bore volume and not accessible to observation or photography.

(b) Measurement accuracy may suffer, due to loss of chamber resolution. The so-called vertical dip angle of particles is independent of the magnetic field. If the experimental chamber is too small, its resolution may be so poor that it may be inadequate.

(c) Control of fringing field in high-field magnets of short axial length may be difficult. This will affect the beam entry and exit, specifically if the end field is an appreciable portion of the total useful field length. Control of $\int B\,dl$ or $\int \frac{dB}{\partial r}\,dl$ may be difficult, as was shown in Chapter 2.

(d) Achievement of a specified homogeneity within the useful bore volume may become difficult.

(e) At high fields, magnetomechanical forces may become exorbitant and mechanical material strength may force the choice of an upper field limit upon the designer.

There is no doubt that reducing the useful bore or gap volume by using higher fields is more desirable for most experiments. For instance, if the chamber diameter can be reduced by a factor of two by increasing the transverse field by a factor of four to keep the same measurement accuracy, there are several advantages besides the saving in the cost of buildings and auxiliary parts (ease in photography of interactions, scanning simplification). Photographic reduction of smaller track fields (i.e., from a 1 m diameter to 35 mm film size, instead of from 2 m diameter to 70 mm film size) into smaller film sizes is easier without losing resolution and thus saving cost of film.

However, a generalization of the beneficial properties of high-field superconductors may prove to be erroneous. Close co-operation between physicist and magnet designer is essential to determine the magnet size within the frame of possibility and practicability, and to determine the optimum size of the experimental area and field strength. There are several ways to compare magnets of various types (water-cooled, superconducting, etc.):

(a) *Cost comparison for specific kinds of magnets*

This comparison is the most accurate and enables one to study in detail the properties of the magnet system. For large and expansive experimental setups, this is the only correct method of comparison.

(b) *General cost comparison*

Cost formulas for all kinds of magnets may be derived, based on optimized magnet systems. Depending on the classification of magnets, one may derive equations, curves and tables which give a rough estimate of costs.

Method (b) can be adapted to obtain costs for the following magnet types:

b_1. d.c. superconducting magnets,
b_2. d.c. superconducting magnets with iron return yokes,
b_3. conventional d.c. air-core magnets,
b_4. conventional d.c. iron-core or iron-yoke magnets,
b_5. cryogenic d.c. air-core magnets,
b_6. cryogenic d.c. iron-core magnets

The major drawback in the general cost formula is the continuous change in the price of material and labor; however, it may be adapted for specific cases and circumstances to fit the cost comparison formula.

6.2 Cost Comparison for Specific Magnet Systems

Only a few magnet systems are considered in this section. In the first case, where solenoids are discussed, we compare the important family of pulsed magnets to room, cryogenic, and superconducting magnets. For all other types, such as dipole (bending) magnets, magnetic lenses (quadrupoles), and exponential-type magnets, we compare the water-cooled magnets to cryogenic and superconducting magnets.

6.2.1 Solenoids and Split Coils

For a solenoid with an axial length of $2\,b$, bore diameter of $2\,a_1$, and outer diameter of $2\,a_2$, one calculates the field at the center from the well-known Fabry relation:

$$\boxed{B_{0,0} = \lambda \cdot J \cdot a_1 \cdot \mathfrak{J}(\alpha,\beta)}\,, \qquad (6.2.1)$$

with $\mathfrak{J}(\alpha,\beta)$ the current density factor introduced by Gauster [1]:

$$\boxed{\mathfrak{J}(\alpha,\beta) = \mu_0 \beta \ln \frac{\alpha + (\alpha^2 + \beta^2)^{\frac{1}{2}}}{1 + (1 + \beta^2)^{\frac{1}{2}}}}\,, \qquad (6.2.2)$$

with $\alpha = a_2/a_1$,
$\beta = b/a_1$,
λ = space factor = ratio of active conductor area to the cross-sectional solenoid area.

Eq. (6.2.2) is valid for solenoids with square cross-sectional area and uniform current density distribution.

Given the central field, the overall current density, and the bore diameter $2\,a_1$, the value of $\mathfrak{J}(\alpha,\beta)$ is fixed, but the magnitude of α and β is not. Since the magnet cost depends upon the length of the conductor in superconducting magnets, and upon the magnet shape as well as the power requirement in water-cooled and cryogenic temperature magnets, we study each system separately.

6.2.2 Water-Cooled Solenoids with Uniform Current Density Distribution

In water-cooled solenoids, the main concern is the power requirement to generate a given magnetic field. Fabry's relation for a power-optimized coil, relating the central field to the required power, is written:

$$\boxed{B_{0,0} = G(\alpha,\beta) \cdot \sqrt{\frac{P \cdot \lambda}{a_1 \varrho}},} \tag{6.2.3}$$

with P the required power, and ϱ the material resistivity; $G(\alpha,\beta)$ is the geometry factor. The relation between $G(\alpha,\beta)$ and $\mathfrak{J}(\alpha,\beta)$ is easily found by expressing the power in terms of the ohmic resistance of the conductor and the coil geometry:

$$P = \lambda J^2 \varrho a_1^3 \cdot 2\pi\beta(\alpha^2 - 1), \tag{6.2.4}$$

which yields

$$G(\alpha,\beta) = \frac{\mu_0}{\sqrt{2\pi}} \left(\frac{\beta}{\alpha^2 - 1}\right)^{\frac{1}{2}} \cdot \ln \frac{\alpha + (\alpha^2 + \beta^2)^{\frac{1}{2}}}{1 + (1 + \beta^2)^{\frac{1}{2}}}. \tag{6.2.5}$$

In Eq. (3.5), the power is expressed in Watts, the bore radius in m, and the resistivity in Ohm · m. Curves of $\mathfrak{J}(\alpha,\beta)$ and $G(\alpha,\beta)$ are published by Gauster [1], Fabry [2], Bitter [3] and Montgomery [4].

The coil volume is expressed in the form:

$$V = a_1^3 \cdot 2\pi\beta(\alpha^2 - 1) = a_1^3 \cdot v, \tag{6.2.6}$$

with v the volume factor.

The capital cost of the magnet system depends on the weight of the coil, the conductor material, the coil configuration and conductor shape, and the cost of the power supply.

In considering the weight and volume of the solenoid, one must make a choice between the various materials suitable. For low and medium-field magnets, two metals are of interest: copper and aluminum. Copper has a density of 8.89×10^3 kg/m³ at 300 K, aluminum 2.699×10^3 kg/m³. Copper is 3.3 times as dense as aluminum, which makes the aluminum magnet lighter. However, the resistivity of aluminum (see Table 1.6) is about 60% that of copper which, for the same magnet size and central field, requires 60% more power.

Based on present values, Fig. 6.2.1 gives the price of power supplies per kW for various power ranges [5]. We see that at high power levels, the power supply cost is appreciable, and the magnet optimization for aluminum conductor must be performed accordingly.

For large solenoids, or Helmholtz-type magnets, three-dimensional coil geometries (beam transport, plasma experiments) with useful field volumes exceeding 0.1 m³, water-cooled hollow square conductors are preferred. Coils in solenoids are composed of double pancakes with one or more hydraulic passages connected in parallel; electrical passages, however, connected in series (Fig. 6.2.2). It is easily seen that hollow conductors with continuous lengths of 100 m or so are required for each hydraulic passage. Due to manufacturing limitations, the length of a hollow conductor depends on the overall weight of a single billet, which can be handled in a press. With the present systems, conductors having cross sections of $(5\times5)\times10^{-4}\,m^2$ and a cooling passage diameter of 3×10^{-2} m can be manufactured in a continuous length of ~ 25 m.* Thus several waterproof joints must be produced by means of brazing, hard soldering, electronic beam welding, or other techniques which are time-consuming. Thus coil-winding becomes more expensive for copper solenoids than for aluminum coils, for which the aluminum conductor may have a continuous length of several hundred meters.

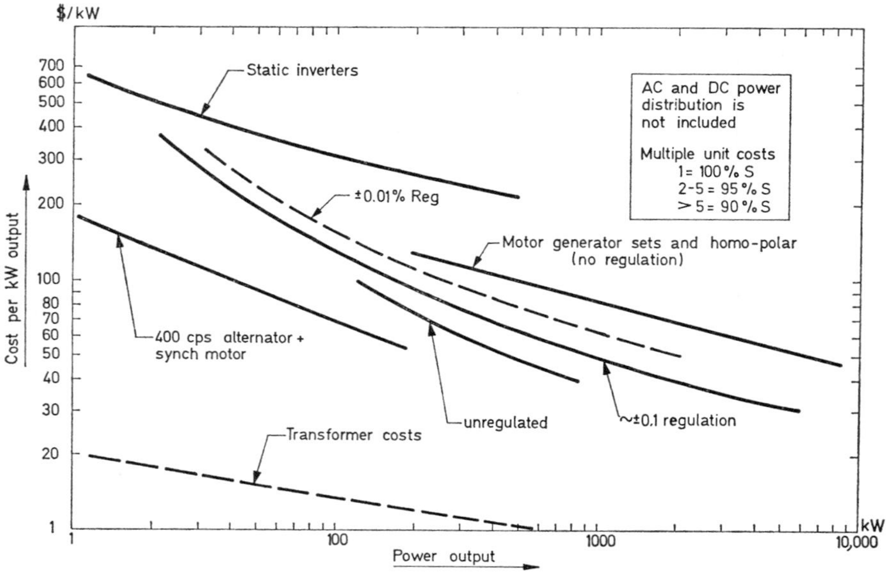

Fig. 6.2.1. Cost of power supplies including switches and voltage control to feed electromagnets with the following criteria:

Load resistance: 0.025 ... 0.4 Ohms
Load voltage: $\leqq$ 600 volts
Load current: $\geqq$ 500 Amp
Primary voltage: 470 V, 4,160 V and 12,470 V.

Newer joining techniques, such as explosive methods, electron beam welding, and cold weld techniques, are being explored but have not been successful in large-size conductors.

As an example, let us consider a split-coil magnet with parameters given in Table 6.2.1. We calculate four possibilities:

(a) Copper coil with no iron return yoke.
(b) Copper coil with iron return yoke.
(c) Aluminum coil with no iron return yoke.
(d) Aluminum coil with iron return yoke.

* Continuous copper casting process yields much greater length; however, overall material uniformity (density variation) and the surface condition of the cooling passage are presently unacceptable.

The central magnetic field has been fixed to 2.6 T, as well as the magnet inner accessible diameter and the accessible axial separation between the two coil halves. The major requirement is radial access to the central field section and axial access from the top for optical purposes. A magnet of this type is useful for bubble- and spark-chamber experiments. The field homogeneity over $r = \pm\, 0.5$ m radius (useful bore diameter), as well as over $z = \pm\, 0.25$ m, is supposed to be better than 2%. The magnetic field of a pair of coils of equal rectangular cross-sectional area, along the symmetry axis $r = 0$, and uniform current density distribution, is given by:

$$B_r = 0$$

$$B_z = \mu_0 \cdot \lambda J a_1 \left\{ (\gamma + \beta) \ln \frac{\alpha + [\alpha^2 + (\gamma + \beta)^2]^{\frac{1}{2}}}{1 + [1 + (\gamma + \beta)^2]^{\frac{1}{2}}} - (\gamma - \beta) \ln \frac{\alpha + [\alpha^2 + (\gamma - \beta)^2]^{\frac{1}{2}}}{1 + [1 + (\gamma - \beta)^2]^{\frac{1}{2}}} \right\}. \tag{6.2.7}$$

With $\gamma = \frac{b + g}{a_1}$ and $2\,g$ the axial coil separation.

For a pair of solenoids with unequal rectangular cross-sections, we may write:

$$B_r = 0$$

$$\begin{aligned} B_z = \mu_0 \lambda J a_1 \Big\{ &(\gamma' + \beta') \ln \frac{\alpha + [\alpha^2 + (\gamma' + \beta')^2]^{\frac{1}{2}}}{1 + [1 + (\gamma' + \beta')^2]^{\frac{1}{2}}} \\ &- (\gamma' - \beta') \ln \frac{\alpha + [\alpha^2 + (\gamma_1' - \beta_1')^2]^{\frac{1}{2}}}{1 + [1 + (\gamma' - \beta')^2]^{\frac{1}{2}}} \\ &+ (\gamma'' + \beta'') \ln \frac{\alpha + [\alpha^2 + (\gamma'' + \beta'')^2]^{\frac{1}{2}}}{1 + [1 + (\gamma'' + \beta'')^2]^{\frac{1}{2}}} \\ &- (\gamma'' - \beta'') \ln \frac{\alpha + [\alpha^2 + (\gamma'' - \beta'')^2]^{\frac{1}{2}}}{1 + [1 + (\gamma'' - \beta'')^2]^{\frac{1}{2}}} \Big\}. \end{aligned} \tag{6.2.8}$$

Or in simplified form:

$$B_z = \mu_0 \lambda J a_1 \cdot F\,(\alpha, \beta', \beta'', \gamma', \gamma'') \,. \tag{6.2.9}$$

The power requirement of the coil may be expressed as:

$$P = 2\,\pi a_1^3 J^2 \cdot \lambda \cdot \varrho\,(\alpha^2 - 1)\,(\beta' + \beta'') \,. \tag{6.2.10}$$

Eliminating the current density from Eqs. (6.2.9) and (6.2.10), we get:

$$\boxed{B_z(0,0) = \frac{\mu_0}{\sqrt{2\,\pi}} \cdot \frac{F\,(\alpha, \beta)}{[(\alpha^2 - 1)\,(\beta' + \beta'')]^{\frac{1}{2}}} \cdot \left(\frac{P \cdot \lambda}{a_1 \cdot \varrho}\right)^{\frac{1}{2}}}\,, \tag{6.2.11}$$

with F being the geometry factor for a pair of coils. If P is expressed in Watts, a_1 in m, the resistivity ϱ in Ohm · m, we get B_z in Teslas.

For our particular magnet, we have chosen $B_z(0,0) = 2.6$ T and $a_1 = 0.7$ m, for all four magnets studied. Table 6.2.1 gives the optimized coils and power requirement. In Table 6.2.2. we have calculated the prices of these magnets based on the current values for coil manufacturing.

Table 6.2.1. *Technical data on a pair of water-cooled coils, with and without iron return yoke**

	Copper coils no iron	Aluminum coils no iron	Copper coils with iron	Aluminum coils with iron
Central Field (T)	2.6	2.6	2.6	2.6
a_1 (m)	0.7	0.7	0.7	0.7
α	2.2	3.2	1.95	2.5
$\beta' = \beta''$	0.6	0.6	0.45	0.55
γ	0.871	0.871	0.721	0.821
$F(\alpha, \beta, \gamma)$	0.609	1.017	0.4239	0.697
λJ (A/m^2)	5.68×10^6	2.90×10^6	5.30×10^6	3.06×10^6
J_{cond} (A/m^2)	8.74×10^6	4.46×10^6	8.15×10^6	4.72×10^6
P (MW)	7.2	10.3	4.8	6
V_{coil} (m^3)	6.5	15.4	3.6	12.44
W_{coil} (kg)	57.8×10^3	41.6×10^3	32×10^3	33.6×10^3
W_{Fe} (kg)	—	—	135×10^3	199×10^3
B_{coil} (T)	2.6	2.6	1.9	1.88
B_{Fe} (T)	—	—	0.7	0.72

* All magnets are optimized. The space factor is estimated at $\lambda = 0.65$.

In Table 6.2.2, we observe one interesting feature of iron-bound magnets. The power requirements and the overall cost of the magnet are less than for magnets without ferromagnetic flux return path, a fact which is encountered even in high-field magnets, where the iron is saturated. This interesting feature is observed, as we see in Chapter 6.3, also for high-field superconducting magnets.

Iron in combination with solenoids has been treated extensively in Chapter 2. At present, we have calculated optimized iron-bound magnets for comparison. To optimize the iron-bound magnet, several computer codes solving non-linear quasi-Poisson equations are being utilized for two-dimensional and axially symmetric magnets.

We also see, from Tables 6.2.1 and 6.2.2, that aluminum coils are more economical than copper coils; however, the lower capital cost is counteracted by the higher operating cost. It is entirely a matter of operational time of such a magnet if, over a period of several years (magnet lifetime), an aluminum magnet is to be preferred over a copper coil. If, for example, the magnets outlined in Table 6.2.1 are to operate 8000 hours per year at a power cost of \$ 10^{-2}/kWh, we would definitely choose the iron-bound copper coil. If the operating time of the magnet is less than 1000 hours per year, which is usual in some experimental magnets, the aluminum conductor, with ferromagnetic return yokes, is to be adopted.

Other problems facing the coil designer are the coil layout, the method of cooling, the support of magnetomechanical and thermal stresses, and the choice of adequate insulation, which can withstand electrical breakdown under severe environmental conditions, e.g. water vapor, dust, and nuclear irradiation.

Table 6.2.2. *Cost comparison between various coil configurations with and without iron return yokes*

Cost of (U.S. Dollars)	Copper coil no iron	Aluminum coil no iron	Copper coil with iron	Aluminum coil with iron
Coil	3.82×10^5 [a]	1.87×10^5 [b]	2.11×10^5	1.51×10^5
Support structure	0.3×10^5	0.3×10^5	0.3×10^5	0.35×10^5
Iron[c]	—	—	1.49×10^5	2.19×10^5
Power supply	3.6×10^5	4.84×10^5	2.4×10^5	3.0×10^5
Safety interlocks, Instrumentation	0.5×10^5	0.5×10^5	0.5×10^5	0.5×10^5
Design, Engineering, Installation	0.7×10^5	0.7×10^5	0.7×10^5	0.7×10^5
Miscellaneous	0.6×10^5	0.6×10^5	0.5×10^5	0.6×10^5
Total (dollars)	9.52×10^5	8.81×10^5	8.00×10^5	8.85×10^5

[a] The present price of a copper coil wound into double pancakes and insulated is approximately \$ 6.60/kg. The price of extruded copper conductor is currently \$ 2.20/kg.

[b] The present price of an aluminum coil wound into double pancakes and insulated is \$ 4.50/kg. The price of extruded aluminum conductor is currently \$ 1.65/kg.

[c] The price of low-carbon steels already machined at matching surfaces is estimated at \$ 1.10/kg.

We will discuss the method of cooling utilized in such solenoids, which is also common in high-energy beam transport and experimental magnets, and types and methods of insulation.

The direct water-cooling of the current-carrying conductor is by far the most efficient. In magnets outlined in Table 6.2.1, square or rectangular conductor with circular, rectangular, or some other shape of the coolant passage is used most extensively. Each hydraulic passage has a length of several meters, depending on the power density and the surface heat flux of the conductor. By using demineralized water, a heat flux in the order of 4×10^4 W/m²/°C has been achieved. For more compact coil designs, edge-cooling is preferred, where the surface heat flux of $(8 \ldots 9) \times10^4$ W/m²/°C may be obtainable. However, insulation of edge-cooled conductors may prove to be more difficult than for hollow conductors. Fig. 6.2.2 shows methods of coil design for uniform current density distribution over the entire cross section.

In order to be able to see if the economic comparison given in Table 6.2.2 is also technically feasible, we give the method of calculating the cooling of water-cooled coils.

The power density in coils with uniform current density is calculated readily from [6]

$$w_v = \frac{P}{2\,\pi a_1^3\,(\alpha^2 - 1)\,\beta\cdot\lambda_i\cdot\lambda_c\cdot\lambda_r\ldots}\ (\mathrm{W/m^3})\,. \qquad (6.2.12)$$

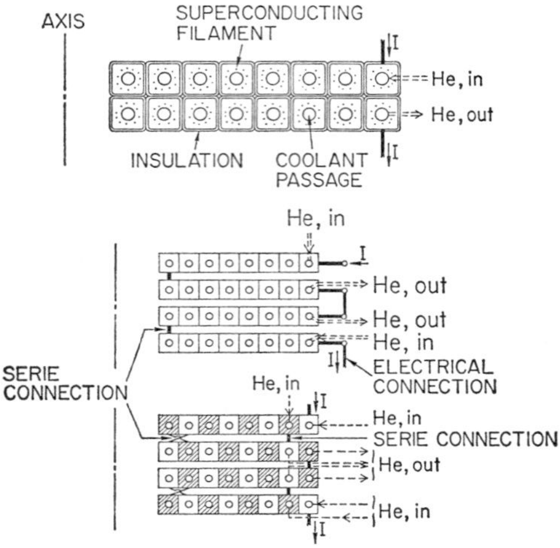

Fig. 6.2.2a

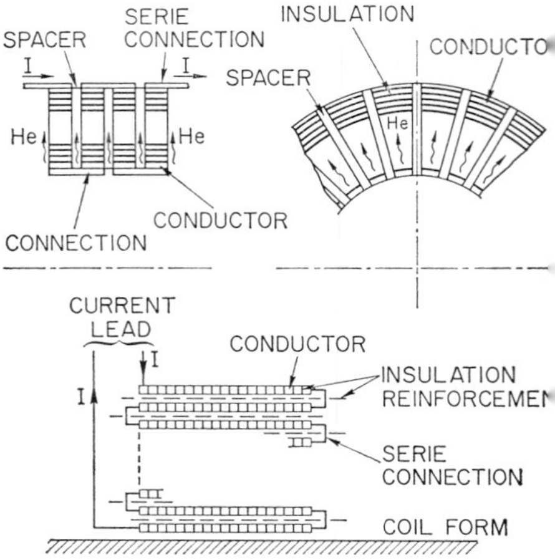

Fig. 6.2.2b

Fig. 6.2.2a. Internally cooled superconducting double pancake configurations. Top: Conventional pancake arrangement; middle: Conventional coil arrangement with the turns electrically connected in series, the helium passages in parallel; bottom: Interleaved double pancake arrangement. The coil is wound such that no surge voltage or voltage oscillations will occur (power supply failures, magnet quench etc.) due to the interleaving which increases the conductor series capacitance in relation to the ground capacitance

Fig. 6.2.2b. Edge cooled arrangements. Top: Pancakes wound of composite tapes are spaced by glassfiber epoxy structures to allow free helium edge cooling; bottom: Layer type coil winding

In this equation, the space factor is subdivided and specified as due to insulation (λ_i), cooling passage (λ_c), reinforcements (λ_r), etc.

The relation between power density w_v and heat flux w_s is given by the simple expression:

$$\boxed{\frac{w_v}{w_s} = \frac{\lambda_c}{1 - \lambda_c} \cdot \frac{4}{d_h},} \tag{6.2.13}$$

with d_h being the hydraulic diameter of the individual conductor. Combining these two equations, we get the expression for the heat flux:

$$w_s = \frac{P \cdot d_h}{8\,\pi a_1^3 \cdot (\alpha^2 - 1)\,\beta \cdot \lambda_i} \cdot \frac{1 - \lambda_c}{\lambda_c^2} \quad (\mathrm{W/m^2})\,. \tag{6.2.14}$$

The heat transfer from the conductor cooling surface to the coolant is a function of the velocity of water, and a function of the surface roughness of the coolant passage walls.

The heat transfer coefficient, h, is given by:

$$h = \frac{w_s}{\Delta T_{w,b}},$$

where $\Delta T_{w,b}$ indicates the temperature drop from coolant passage wall to the bulk liquid.

The heat transfer coefficient for single-phase flow* is given by the relation [7]:

$$\boxed{h = C \cdot \frac{k_f}{d_h} \cdot \left(\frac{d_h v \cdot \delta}{\eta}\right)_f^{0.8} \cdot \left(\frac{c_p \cdot \eta}{k}\right)_f^{0.4}} \quad \left(\frac{\mathrm{W}}{\mathrm{m^2\,K}}\right), \qquad (6.2.15)$$

with k_f = heat conductivity of the fluid in (W/m K),
δ = density (kg/m³),
c_p = specific heat in (Ws/kg K),
η = viscosity (kg/m · s),
d_h = hydraulic diameter (m),
v = coolant velocity (m/s),
C = Nusselt number $= 2.3 \times 10^{-2}$.

Empirically, the heat transfer coefficient for turbulent flow through cylindrical tubes is written for single-phase flow of water by:

$$h = 2 \times 10^3\,(1 + 1.5 \times 10^{-2} \cdot T_m) \cdot \frac{v_l^{0.87}}{d_h^{0.13}} \quad (\mathrm{W/m^2\,K})$$
$$T_m = 0.9\,T_b + 0.1\,T_{\text{wall}}\,. \qquad (6.2.16)$$

The velocity of water through a pipe is a function of the pressure drop along the pipe, which is given by the empiric equation of Darcy [8]:

$$\boxed{\Delta p = 5.1 \times 10^2\, v_l^2 \left[f \cdot \frac{l}{d_h} + C_c + C_e\right]} \quad (\mathrm{N/m^2})\,. \qquad (6.2.17)$$

f indicates the friction factor, l and d_h are the length and hydraulic diameter of the coolant passage in m, v_l is the velocity of the liquid (coolant) in m/sec, and C_c and C_e are the loss coefficients at the exit and entry of the liquid in the coolant passage. According to Kays [9], the maximum values for exit and entry are (for square entry and exit):

$$C_c = 0.5 \quad \text{and} \quad C_e = 1.0\,.$$

For long pipes, C_c and C_e can be neglected, when compared to the term $f \cdot l/d_h$ in Eq. (6.2.17).

The friction factor is a function of the absolute surface roughness and the Reynolds number. A first-order crude estimate of the pressure drop across the pipe is given by the correlation:

$$\Delta p = 5\,l \cdot \frac{v^{1.75}}{d_h^{1.25}}, \quad (\mathrm{N/m^2}) \qquad (6.2.18)$$

with l and d_h expressed in m, v in m/sec.

The limits of the surface heat flux are dictated by the burn-out heat flux when the magnet is operated under conditions of surface boiling. In many applications, such as the cases in Table 6.2.1, the heat flux (power per unit cooling surface) is dictated by the optimum power conditions or, in large high-energy laboratories, by the availability of power for the

* Correlation equations for laminar and turbulent flow are given in Chapter 5.7.

particular magnet. The conductor insulation dictates the upper limit of the operational temperature. In general, the exit temperature of the cooling water is kept below the boiling point in order to prevent difficulties due to two-phase flows.

The coolant velocity through the passage is given by the maximum available pressure in the hydraulic system in the laboratory installations and the geometry and surface condition of the hydraulic passage.

As an example, we calculate the temperature rise in the two coils without iron, specified in Table 6.2.1.

We assume for the hollow conductor configuration the space factors of $\lambda_i = 0.9$ and $\lambda_c = 0.72$, due to insulation and presence of internal coolant passages. The coil is composed of a number of double pancakes, illustrated in Fig. 6.2.2a, insulated by means of glass fiber tapes impregnated by a suitable thermosetting. No additional reinforcements or support structure are assumed; thus, the space factor will be $\lambda_r = 1$.

The coil MMF is proportional to the coil geometry and the overall current density in Eq. (6.2.1):

$$NI = 2\,\lambda J a_1^2 (\alpha - 1)\,\beta\,, \tag{6.2.19}$$

which is, for the copper coil (we calculate only one module):

$$(NI)_{\mathrm{Cu}} \cong 4.2 \times 10^6 \ \mathrm{A}\,,$$

and for the aluminum coil:

$$(NI)_{\mathrm{Al}} = 3.76 \times 10^6 \ \mathrm{A}\,.$$

For both coils we chose a current of 10^4 A, which gives the number of turns per module: $N_{\mathrm{Cu}} = 420$ and $N_{\mathrm{Al}} = 376$. The effective conductor area for each conductor is thus: $A_{\mathrm{eff,Cu}} = 11.44 \times 10^{-4}\ \mathrm{m}^2$ and $A_{\mathrm{eff,Al}} = 22.4 \times 10^{-4}\ \mathrm{m}^2$. The copper conductor has the external dimension of $(3.8 \times 4) \times 10^{-4}\ \mathrm{m}^2$ and a coolant passage with the hydraulic diameter of 2.3×10^{-2} m, which gives $a\lambda_c = 0.72$. The conductor corners are rounded to prevent insulation damage and eliminate potential buildup due to sharp points [10]. In general, if the corner radius is equal to half the thickness of insulation between conductors, the dielectric stress enhancement at the corner is small and may be ignored. For large conductor sizes, as in our example, the ratio $r/g \gg 1$, and thus the electrical stress is simply the voltage across the insulation. The number of turns per pancake is 21; the number of pancakes in each module is 20. The insulation thickness in the radial and axial directions is chosen to be 10^{-3} m on each side of individual turns.

We assume the available pressure in the laboratory is such that we may utilize a pressure drop of approximately $10^6\ \mathrm{N/m}^2$ across each hydraulic passage. The length of the conductor in each double pancake is obviously

$$l = N_{dp} \cdot \pi a_1 (\alpha + 1) = 42 \cdot \pi \cdot 0.7 \cdot 3.2 \cong 296 \ \mathrm{m}\,.$$

From Eq. (6.2.19), assuming $v = 2.9$ m/sec, we get:

$$\Delta p = 9.1 \times 10^5 \ \mathrm{N/m}^2.$$

At the velocity of $v = 2.9$ m/sec, the Reynold's number for water at an assumed average temperature of $T = 50\,^{\circ}\mathrm{C}$ is $N_{\mathrm{Re}} = 1.2 \times 10^5$, corresponding to the friction coefficient $f = 1.7 \times 10^{-2}$. Thus, from Eq. (6.2.17) we have as a comparison:

$$\Delta p = 9.35 \times 10^5 \text{ N/m}^2,$$

which is slightly higher than the value obtained from Eq. (6.2.18).

We assume that each double pancake consists of one hydraulic passage. The bulk liquid transports the power of 3.6×10^6 W in each module. The temperature rise of water through the passage is given by:

$$\Delta T = \frac{P}{N_{hp} \cdot U \cdot c_p}, \tag{6.2.20}$$

with N_{hp} = the number of hydraulic passages,
U = mass flow $= v \cdot \pi d_h^2/4$ (m³/sec),
$c_p \delta$ = heat capacity in Ws/m³ K.

For our copper coil:

$$\Delta T = \frac{3.6 \times 10^6}{10 \cdot 12 \times 10^{-4} \cdot 4.182 \times 10^6} = 72\,°\text{C} .$$

The heat transfer coefficient is calculated from Eq. (6.2.15):

$$h = 1.25 \times 10^4 \text{ W/m}^2 \text{ K} ,$$

which yields a surface temperature drop of $\Delta T_w = 0.85\,°\text{C}$ from the conductor wall to the bulk coolant.

Since the temperature rise in the copper cross section is calculated to be $\Delta T_m = 0.3\,°\text{C}$, and, assuming an inlet water temperature of 20 °C, the maximum temperature in the conductor is:

$$T_{\max} = T_{\text{in}} + \Delta T + \Delta T_W + \Delta T_{\text{Cu}} = 93.15\,°\text{C} .$$

If T_m would exceed the boiling temperature of water, each double pancake would have to be subdivided into a number of hydraulic passages in parallel, such that the maximum water temperature is less than the boiling temperature of water.

For the aluminum coil, we obtain the following data: Conductor outer dimensions $(5.5 \times 5.7) \times 10^{-4}$ m², with a coolant passage with $d = 3.38 \times 10^{-2}$ m diameter. Assuming 10^{-3} m insulation on each side of the conductor, we obtain 27 turns in the radial direction and 7 double pancakes for each module. Each double pancake has an additional face insulation of 10^{-3} m thickness. The space factors are thus: $\lambda_i = 0.91$ and $\lambda_c = 0.714$. Assuming one hydraulic passage per double pancake, we have: $l_{hp} = 262$ m. With $v = 3.8$ m/sec, we get: $N_{\text{Re}} = 1.95 \times 10^5$, $f = 1.6 \times 10^{-2}$, and $\Delta p = 9.1 \times 10^5$ N/m². The temperature rise of water, according to Eq. (6.2.21) is $\Delta T = 51.5\,°\text{C}$. The heat transfer coefficient calculated from Eq. (6.2.16) at an average temperature of 40° is: $h = 0.71 \times 10^4$ W/m²/°C. Thus with $w_s = 2.645 \times 10^4$ W/m² a temperature drop of $\Delta T_W = 3.7\,°\text{C}$ is obtained. With $\Delta T_{\text{Cu}} = 0.5\,°\text{C}$, assuming the water inlet temperature is 20 °C, we obtain the maximum conductor temperature of $T_{\max} = 75.7\,°\text{C}$. Thus both magnets are useful and no temperature problems may be encountered.

It can also be shown that conductor strain due to magnetomechanical forces calculated in Chapter 2 are within the material strength of the copper or aluminum conductor chosen.

6.2.3 Cryogenic Magnets with Uniform Current Density Distribution

The optimum design of a cryogenic magnet system depends on the proper choice of temperature at which the magnet operates, the choice of conductor material and shape, and the optimum coil configuration. The temperature and conductor choice depends on the total losses (Joules heating and refrigeration) and the current price of the conductor, as well as coil manufacturing cost.

Obviously the cost optimum corresponds in essence to the minimum conductor losses for a selected current density. These losses are the sum of the actual electrical losses and the power requirement of the refrigeration system. The so-called "loss parameter, w" [11] of a magnet gives a good indication for the choice of temperature and is expressed in the simplified form:*

$$\boxed{w = \frac{\varrho\,(T,B)}{\varrho_{300\,\mathrm{K}}} \cdot \left[1 + \frac{1}{R\,(T) \cdot \eta_{\mathrm{isoth}} \cdot \eta}\left(\frac{T_0}{T} - 1\right)\right]}, \qquad (6.2.21)$$

with

$\frac{T_0}{T} - 1 = \frac{W_n}{Q_{\mathrm{Carnot}}}$ = Ratio of the amount of power required by a Carnot machine to the refrigeration produced, with W_n the net power and Q the refrigeration produced,

R = Performance efficiency for the actual refrigeration cycle,

η_{isoth} = Isothermal compressor efficiency,

η = A product of all efficiencies encountered, such as for the heat exchanger (η_{H}), gas transport (η_{GT}), heat conduction and radiation (η_{CR}), etc.

In Eq. (6.2.21), T_0 is the temperature at which heat is being ejected by the refrigerator. We chose $T_0 = 300$ K, which is common.

The loss parameter is indicative of the extent to which real refrigerators deviate from ideal machines. An ideal refrigerator would require 70.4 W of power to produce 1 W of refrigeration at 4.2 K (liquid helium temperature). A real refrigerator, however, requires more power to produce one Watt of refrigeration, depending on the size of the refrigeration plant. For a 100 Watt refrigerator, the required input power would be 740 W to produce one Watt at 4.2 K. The ratio of the Carnot efficiencies to the actual efficiency is called percent Carnot and is expressed as:

$$\text{Percent Carnot} = \frac{\left.\frac{W_c}{Q}\right)_{\mathrm{Carnot}}}{\left.\frac{W_c}{Q}\right)_{\mathrm{actual}}} \cdot 100 = \frac{\frac{T_0}{T} - 1}{\left.\frac{W_c}{Q}\right)_{\mathrm{actual}}} \cdot 100\,, \qquad (6.2.22)$$

where W_c is the net input power required, and Q the refrigeration produced.

---

* $\varrho\,(T, B)$ is the conductor resistivity as a function of temperature and magnetic field. The effect of strain, magnetic field, thermal and mechanical cycling is dealt with in detail in the sections below.

Curves for percent Carnot are published by Strobridge [12]. Fig. 6.2.3 illustrates percent Carnot values based on actual plants. The curve fitts the data and is used as an averaged percent Carnot as a function of net refrigeration.

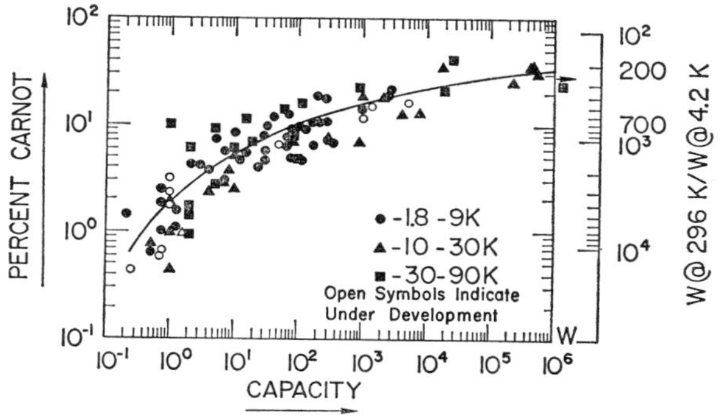

Fig. 6.2.3. Percent Carnot of helium refrigerators either in operation or under construction (after Strowbridge)

In Table 6.2.3, we give the power requirements for an "ideal" and an actual 100 W refrigerator:*

Table 6.2.3. *Power requirements for a 100-Watt refrigerator*

Fluid	T (K)	Refrigeration (W/W) ideal	Refrigeration (W/W) actual	Percent Carnot
Helium	4.2	70.4	740	9.5
Hydrogen	20.4	13.7	196	7
Neon	27.2	10	111	9
Nitrogen	77.4	2.88	18	16

The loss parameter is now rewritten in terms of percent Carnot in the form:

$$w = \frac{\varrho\,(T,B)}{\varrho_{300\,\mathrm{K}}} \cdot \left[1 + \frac{T_0/T - 1}{\%\ \mathrm{Carnot}\,(T)}\right]. \tag{6.2.23}$$

It is recommended to use for $\varrho_{300\,\mathrm{K}}$ the resistivity of a known or commonly utilized metal (pure copper or aluminum), and to compare the change of resistivity as a function of temperature and percent Carnot. Approximate data are obtained by using the curve given in Fig. 6.2.3. In Fig. 6.2.4, we have calculated w values for pure copper, aluminum, silver, and tin, and for electrical grade aluminum, as well as for high-

* See Chapter 5.3.

conductivity, commercially available, copper. Loss factors are compared to the room temperature resistivity of pure copper ($\varrho_{300\,\mathrm{K}} = 1.75 \times 10^{-8}$ Ohm · m). For pure copper and aluminum, the minimum value of w is reached at temperatures in the vicinity of liquid hydrogen temperature. Hard-drawn copper, with $r = \varrho_{300\,\mathrm{K}}/\varrho_T = 1.65 \times 10^2$, has its minimum loss parameter close to liquid neon temperature, and with smaller r values, the minimum w value shifts towards higher temperatures.

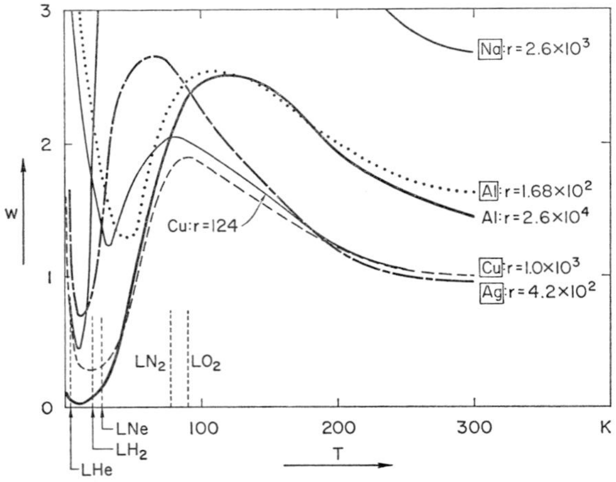

Fig. 6.2.4. Loss parameters w for various pure metals as a function of operating temperature

It is understood that percent Carnot values given in Table 6.2.3 are based on present data obtained from a large number of refrigerators. These values may change, but dramatic improvements are certainly not to be expected in the near future.

Fig. 6.2.4 indicates that high-purity aluminum has a minimum w value of around 10 K, and w values are slightly higher at 4.2 and 21 K, which indicates we may operate the magnet with liquid hydrogen. However, the magneto-mechanical forces in a coil of the size given in Table 6.2.1 are high, and beyond the 0.15% yield strength of ultrapure aluminum. Thus we must direct our attention first to the Lorentz forces and magneto-mechanical stresses on the conductor before we attempt to make a final choice of the best conductor. Another advantage of aluminum as a conductor is its low magneto-resistance value transverse to the magnetic field, as compared to copper.

For a cryogenic magnet, we must use a cryostat containing the cryogen and the coil, a vacuum tank, and ample space for superinsulation and metallic low temperature shields (if necessary), with cold gas cooling at some appropriate temperature between the room temperature wall in the exterior and the cryogenic temperature in the interior of the vacuum vessel or low temperature shields with liquid nitrogen cooling. It is assumed that the ferromagnetic return path of the magnet (being at room temperature) is shaped so as to contain an optimized cryogenically cooled magnet.

The coil inner diameter must be larger in order to accommodate the cryostat. We take a spacing of 2.5×10^{-2} m on each side of the coil, keeping the accessible radial opening unchanged ($2 g_1 = 0.38$ m). The coil separation is accordingly larger. But since, for most experiments, radial accessibility is only partially required, we assume that only 50% of the radial opening is used.

For both coils, we use a common vacuum vessel and cryogen container. Thus we are able to keep actual coil separation to about $2 g_2 = 0.44$ m. The coil inner diameter is now 1.5 m. For a field of $B(0,0,0) = 2.6$ T, we optimize the magnet by using the resistivity of pure multicrystal aluminum at 21 K, which is 2.6×10^{-11} Ohm · m, corresponding to $r = 1000$. This is compared to a magnet using pure copper with $r = 600$, which has, at 21 K, a resistivity of $\varrho = 2.9 \times 10^{-11}$ Ohm · m. The magnet parameters are given in Table 6.2.4. The space factors of the coils are estimated to be 0.75 for the aluminum coil and 0.8 for the copper coil.

From Tables 6.2.4 and 6.2.5, we make the observation that the major expense is in the refrigerator and in the coils. Both types of magnets are more expensive than the room-temperature magnets given in Tables 6.2.1 and 6.2.2. The operational cost, however, is much less. For the aluminum coil with a total net refrigeration power of ~ 7 kW, the total installed power for a refrigerator providing 21 K is about 500 kW (at 300 K). The copper coil requires the total installed power of 550 kW for the 7.5 kW refrigeration at 21 K. This is only 10% of the power required for the room-temperature copper magnet.

The power requirement changes considerably, especially for a pure aluminum coil, if the effects of mechanical strain and magneto-resistance are entered into the material resistivity.

At present, aluminum conductors with $r > 10^4$ at 4.2 K are available only in strip form. Hollow conductors with supercritical helium pumped through the coolant passage are attractive but may pose some manu-

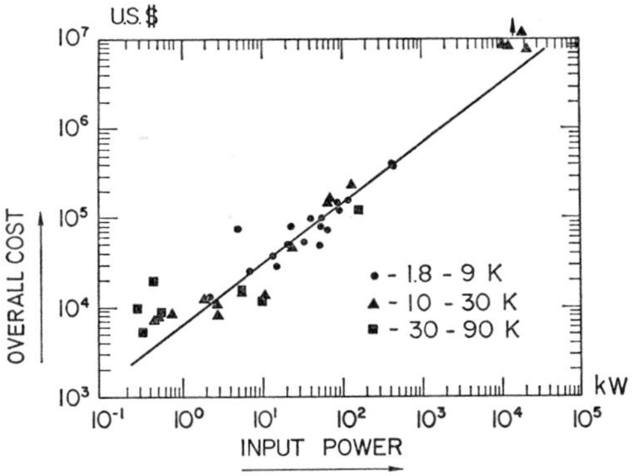

Fig. 6.2.5. Cost of refrigerators for various temperature ranges as a function of input power

facturing problems. To obtain the conductor configuration, we calculate the ampere-turns per coil:

$$NI)_{\text{coil}} = 2\,\lambda J \cdot a_1^2\,(\alpha - 1)\,\beta\,. \tag{6.2.19}$$

The coil overall current density is calculated from Eq. (6.2.9):

$$\lambda J = \frac{B_{z,\text{coil}}}{\mu_0 \cdot a_1 \cdot F\,(\alpha,\beta',\beta'')}\,.$$

Table 6.2.4. *Technical data on a pair of cryogenic coils with iron return yokes*

	Aluminum coils with iron $r = 10^3$ [b]	Copper coils with iron $r = 600$ [b]	Aluminum coils[a] with iron $r = 400$ [c]	Copper coils[a] with iron $r = 176$ [c]
Central field (T)	2.6	2.6	2.6	2.6
a_1 (m)	0.75	0.75	0.75	0.75
α	1.8	1.8	1.8	2
$\beta_1 = \beta_2$	0.5	0.5	0.5	0.5
γ	0.793	0.793	0.793	0.793
$F\,(\alpha,\beta,\gamma)$	0.38	0.38	0.38	0.465
λJ (A/m^2)	5.90×10^6	5.91×10^6	5.90×10^6	4.80×10^6
J_{cond} (A/m^2)	7.87×10^6	7.87×10^6	7.87×10^6	6.00×10^6
P (kW)	~7	7.5	~18	22
V_{coil} (m^3)	6	6	6	8
W_{coil} (kg)	15.4×10^3	53.3×10^3	15.4×10^3	70×10^3
W_{Fe} (kg)	190×10^3	190×10^3	190×10^3	210×10^3
B_{coil} (T)	2.1	2.1	2.1	2.1
B_{Fe} (T)	0.5	0.5	0.5	0.5

[a] The magnetoresistance and strain effects on the conductor resistivity have been taken into account.

[b] $r = \varrho_{300\,\text{K}}/\varrho_{21\,\text{K}}$; usually r indicates the residual resistivity or the ratio $\varrho_{300\,\text{K}}/\varrho_{4.2\,\text{K}}$.

[c] The thermal contraction forces, thermomechanical forces, and magnetoresistance, as well as influence of thermal and magnetic cycling, have been taken into account.

In this equation, only the contribution of the solenoid to the total field is accounted for. For our particular example, $B_{z,\text{coil}} = 2.1$ T.
With the data of Table 6.2.4 for aluminum coils:

$$NI)_{\text{coil}} = 2 \cdot 5.90\times10^6 \cdot (0.75)^2 \cdot 0.8 \cdot 0.5 = 2.68\times10^6.$$

Choosing $N = 3000$, $I = 893$ A, we get the overall cross section of the conductor, including insulation, reinforcements, and coolant passages:

$$A_t = \frac{893}{5.90\times10^6} = 1.5135\times10^{-4}\ \text{m}^2$$

and the actual conductor cross section:

$$A_c = \frac{893}{7.87\times10^6} = 1.134\times10^{-4}\ \text{m}^2.$$

Table 6.2.5. *Cost comparison of various cryogenic coils with iron yokes*

Cost of (U.S. dollars)	Aluminum coil $r = 10^3$ [a]	Copper coil $r = 600$	Aluminum coil $r = 400$	Copper coil $r = 176$
Coil	$3.6\ \times 10^5$ [b]	3.5×10^5	3.6×10^5	4.6×10^5
Coil support, structure, etc.	$0.5\ \times 10^5$	0.4×10^5	0.5×10^5	0.4×10^5
Cryostat	0.25×10^5	0.3×10^5	0.3×10^5	0.3×10^5
Vacuum tank	0.45×10^5	0.5×10^5	0.5×10^5	0.5×10^5
Iron	$2.1\ \times 10^5$	2.1×10^5	2.1×10^5	2.3×10^5
Power supply [c]	—	—	—	—
Refrigerator	$4\ \times 10^5$ [d]	4.4×10^5	6.5×10^5	$7\ \times 10^5$
Design, Engineering, Installation	$1.4\ \times 10^5$	1.3×10^5	1.4×10^5	1.4×10^5
Miscellaneous	$1\ \times 10^5$	$1\ \times 10^5$	$1\ \times 10^5$	$1\ \times 10^5$
Total (dollars)	$13.3\ \times 10^5$	13.6×10^5	15.9×10^5	17.5×10^5

[a] r is expressed in this table as $\varrho_{300\,\mathrm{K}}/\varrho_{21\,\mathrm{K}}$.
[b] The price of pure aluminum strip is estimated at \$ 25/kg wound into a coil.
[c] Power supply cost is about \$ 1000 to \$ 2000 and is negligible, compared to the other expenses.
[d] Additional heat losses have been taken into account due to heat conduction and radiation. For overall cost, see Fig. 6.2.5. Refrigerator operating at 21 K.

The aluminum strip conductor has a width of 2.3×10^{-2} m and a thickness of 4.93×10^{-3} m, while the insulation and reinforcing strips of the same width have a thickness of 1.06×10^{-3} m. The coolant passages are 2×10^{-3} m high.

Each coil is composed of 30 pancakes, with 100 turns per pancake radially. The conductor is edge-cooled. Radial spacers are located between pancakes in spoke form. The maximum circumferential distance between spacers is given by the strain due to axial magnetomechanical forces. In order to divert the strain from the aluminum strip to a supporting band, a 2% beryllium-copper or stainless-steel strip is wound bifilar under pre-tension directly with the aluminum. The pre-tension may be constant or it may be varied. A thin insulation tape is required to insulate adjacent radially-spaced turns.

The resistivity of pure (99.999%) aluminum is changed when it is subjected to mechanical strain. Exploratory measurements by Brooks and Purcell [13], Arp [14], Brown [15] and others, giving the change in resistivity as a function of plastic strain and applied stress, are shown in Fig. 6.2.6. Newer published data by ANL, NBS and NASA have been included in Fig. 6.2.7. The data scatter a great deal and more systematic experimental work in this area is required.

Stresses on the conductor are composed of:
— Winding stresses, including pre-tension.
— Thermal stresses, due to different expansion coefficients of conductor, coil form, and reinforcing (Fig. 6.2.8).
— Magnetomechanical stresses, due to Lorentz forces.

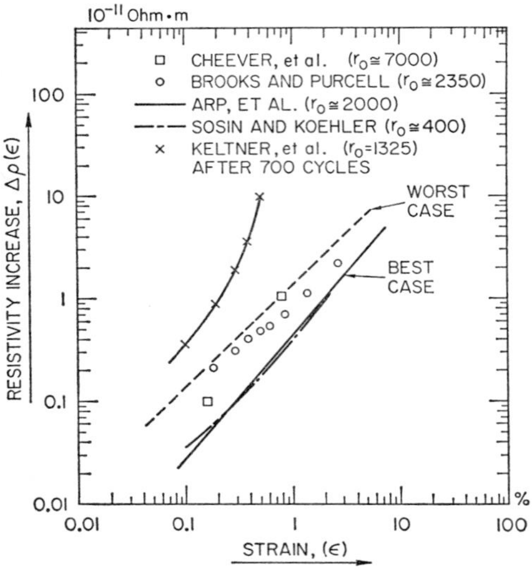

Fig. 6.2.6. Resistivity increase as a function of material strain for pure aluminum (NASA)

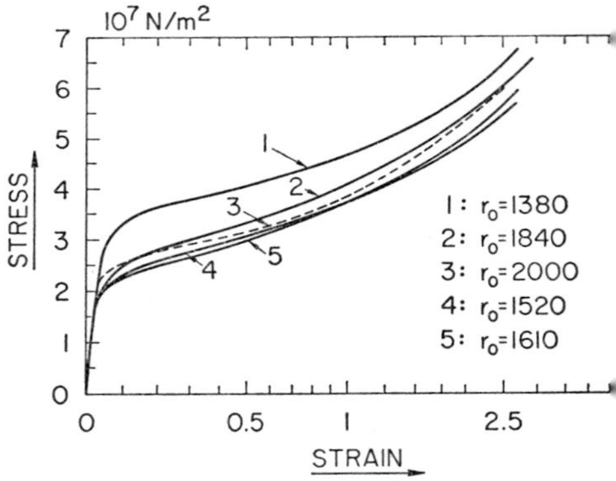

Fig. 6.2.7. Stress-strain diagrams of aluminum with various purities (NBS

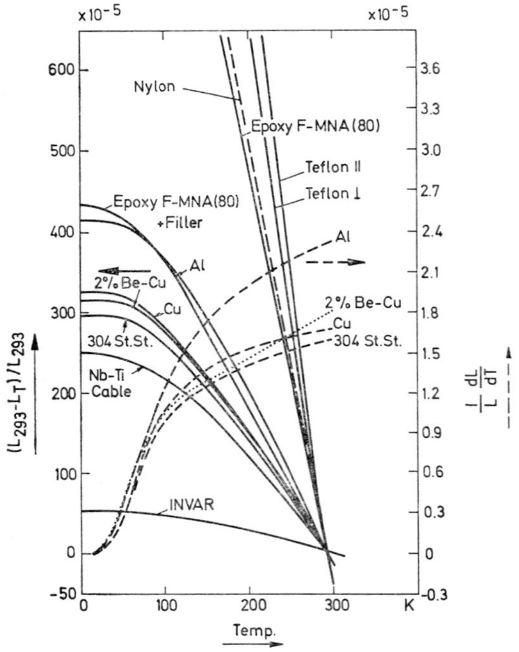

Fig. 6.2.8. Thermal contraction integra and differential thermal contraction of various materials being used as conducto and structural materials in superconduc ing magnets

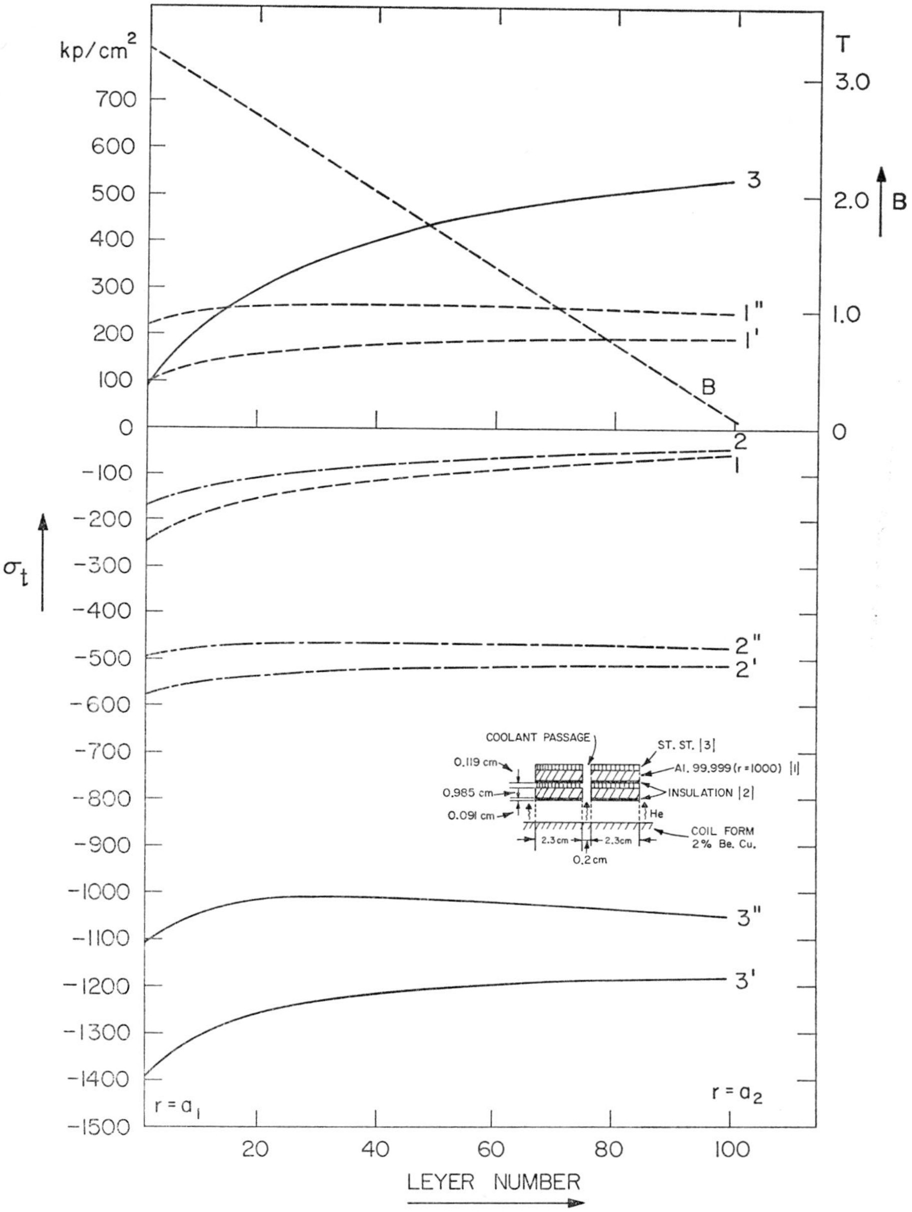

Fig. 6.2.9. Hoop-stress distribution over axially symmetrical coil configurations assuming linear field distribution over the coil radius. *1, 1′* and *1″* Aluminum conductor, *2, 2′* and *2″* Insulation, *3, 3′* and *3″* Stainless steel reinforcing strip. With an applied pretension on the stainless steel band (*3′*) it is possible to limit the maximum stress on the aluminum to values which are less than its yield strength (*1′*)

By varying the stress on the reinforcing strip (see Fig. 6.2.9), as well as inserting external pressure, we are able to keep the maximum resulting stress on the conductor over the whole coil to less than 300 kp/cm², or to any desired value corresponding to the strength of material used as conductor. For the magnet we are considering, the change in resistivity at this stress is due to winding tension and thermal contraction ($\sigma_{\max} \cong 200$ kp/cm²), yielding $\Delta\varrho_0 \cong 2 \times 10^{-12}$ Ohm · m. When the coil is energized ($\sigma_{\max} = 247$ kp/cm²), the change in resistivity is increased to $\Delta\varrho_{\max} = 4.5 \times 10^{-12}$ Ohm · m. Thus the resistivity is increased to $\varrho_0 \cong 2.8 \times 10^{-11}$ Ohm · m over the coil. When the coil is cycled thermally say 1000 times, which is not unusual over the lifetime of a magnet, the resistivity is increased at the above load gradually (see Fig. 6.2.10), and we may assume that ϱ_0 reaches finally a value of 3×10^{-11} Ohm · m.

The increase in resistivity due to the magnetic field, averaged over the whole coil, is $\varrho_m \cong 1.3 \cdot \varrho_0 = 1.3 \times 3 \times 10^{-11} = 3.9 \times 10^{-11}$ Ohm · m. Thus the average resistivity at 21 K is increased from the original value of $\varrho_{0,0} = 2.6 \times 10^{-11}$ Ohm · m to a new value of $\varrho = 6.5 \times 10^{-11}$ Ohm · m, which yields the ratio: $r = 400$. For an optimized coil, calculated in Table 6.2.4, the power requirement would be 18 kW at 21 K. The installed power to produce 18 kW at 21 K is, according to Fig. 6.2.3, approximately 1.3 MW.

Aluminum is a material which sofar has shown best cryogenic performance with respect to magnet applications. Using processes such as "organo electrolysis" and "zone refining" a residual resistivity ratio of $r = r\,[B = 0;\, T = 4.2\ \text{K}] = \varrho\,(293\ \text{K})/\varrho\,(0;\, 4.2\ \text{K})$ of several ten thousand can be achieved in bulk material.

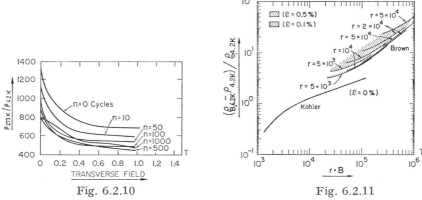

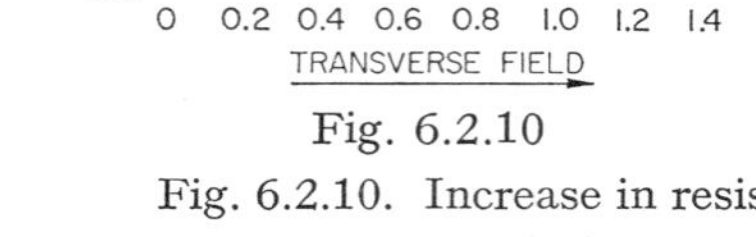

Fig. 6.2.10

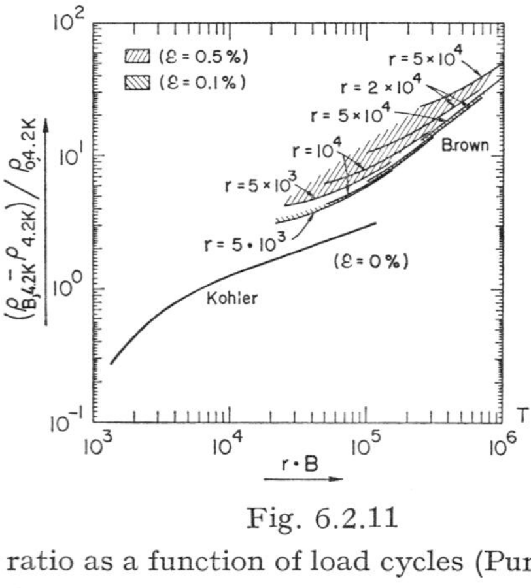

Fig. 6.2.11

Fig. 6.2.10. Increase in resistivity ratio as a function of load cycles (Purcell)

Fig. 6.2.11. Kohler's diagram of aluminum tapes, strained and unstrained. The high purity aluminum tapes exhibit stronger deviation from Kohler's rule

Aluminum worked to wires and tapes in order to be useful in coil design, looses as we have seen part of this very important low temperature property. The resistivity of aluminum increases with decreasing d/l, where d is the tape or wire smaller dimension and l is the mean free path of the electron. This effect called "size effect" sets an lower boundary value to the thickness of the tape. The resistivity also changes, when mechanically strained. Due to very low mechanical strength stainless steel as a backing tape may become imperative in high field coils.

Finally the resistivity of aluminum is increased if a magnetic field is applied to the conductor. In Fig. 6.2.11 the resistivity change is plotted versus $B \cdot r$ for aluminum of various residual resistivity ratios. The values of $\Delta\varrho/\varrho$ are measured at 4.2 K.

The temperature and field dependence of aluminum is illustrated in Fig. 6.2.12. While at liquid helium temperature aluminum follows Kohler's rule, it deviates at higher temperatures.

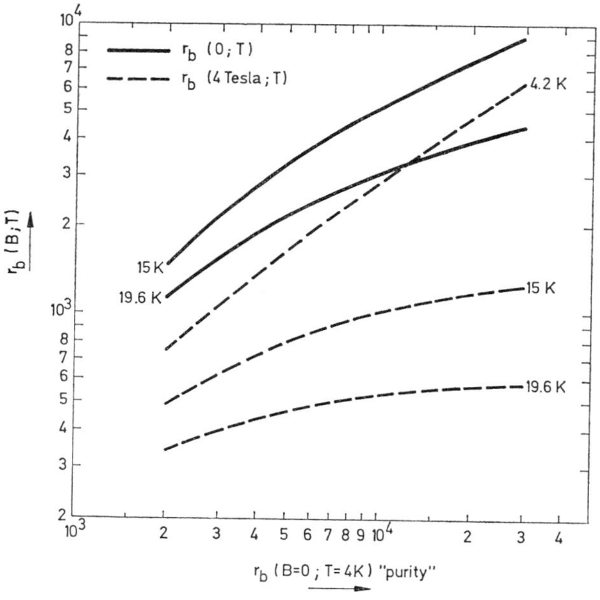

Fig. 6.2.12. Temperature and field dependence of the resistivity ratio of aluminum

Another disadvantage of aluminum is the lack of good bond to the superconductor. Only in isolated instances at an exorbitant price the bond between aluminum and $Nb_{(x)}Ti$ was satisfactory. Attempts to produce multifilamentary conductors with aluminum substrates has been unsuccessful sofar. Only if the $Nb_{(x)}Ti$ is bonded first to copper and then to aluminum good stabilizing results have been achieved.

Nevertheless aluminum has been used for stabilizing Nb_3Sn tapes. As we see in Chapter 7, an aluminum tape wound bifilar with Nb_3Sn has enhanced the performance of the coil more than 12%. The original fear that aluminum may flow under the influence of magnetomechanical stresses have not been justified.

Aluminum has a much higher thermal diffusivity than copper, although the thermal conductivity of aluminum is comparable to that of copper.

As in the case of aluminum, similar calculations performed for a copper coil give the following values of resistivity: The combined magneto-mechanical and thermal contraction strains result in an average stress value on the copper conductor of $\sim$110 kp/cm^2. It was assumed that the copper strip ($2.3 \times 10^{-2} \times 0.65 \times 10^{-2}$ m) was prestressed with a uniform load of 21 kp/cm^2. No reinforcing strips are necessary. The copper resistivity is slightly increased (see Chapter 5.6) giving a value of $r = 545$. The average field over the coil is nearly 1.8 T. The increase in resistivity of copper due to the magnetoresistance is, according to Fig. 6.2.13, approximately $\varrho_m = 2.1 \cdot \varrho_0$. Thus the combined results of mechanical strain and magnetoresistance yield a final value of $\varrho = 9.56 \times 10^{-11}$ Ohm $\cdot$ m, or $r = 176$. With this new value, we obtain a power of 22 kW at 21 K for the copper coil. The power requirement at 300 K is 1.8 MW.

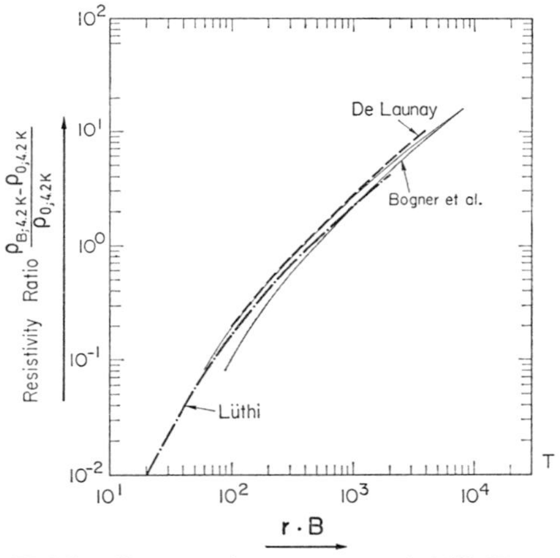

Fig. 6.2.13. Kohler diagram for copper at 4.2 K (Bogner Siemens)

If we compare Tables 6.2.2 and 6.2.5, we see clearly that the room temperature d.c. magnet is less expensive. However, the operational cost of the cryogenic aluminum coil is attractive. Depending on the operating hours per year, we may estimate the operational cost of a cryogenic magnet.

Let us assume the magnet operates 2000 hours per year. According to Table 6.2.1, the magnet power for the optimized copper coil (iron bound) at room temperature was 4.8×10^6 Watts. Estimating \$ 10^{-2} per kWh, the cost of power alone for this magnet will be \$ 9.6×10^4 per year.

We compare the operational cost of the aluminum coil with the iron return yoke with the cryogenic magnet in more detail. The operational cost of the cryogenic magnet originates from two major sources:

(a) Cool-down of magnet and container.
(b) Production of refrigeration at operational temperature.

Two ways are open to produce 21 K, which was the chosen operational temperature: gaseous helium or liquid hydrogen. As we concern ourselves with liquid helium cooling for superconducting magnets later, and large liquid hydrogen systems have been in operation for some time without incident, and handling, storage, and protection of hydrogen-cooled systems are well known, we discuss the refrigeration using liquid hydrogen.

The refrigerator has essentially three functions:
— To cool down the apparatus to the correct operating temperature.
— To produce the necessary cryogenic fluid for filling the cryostat.
— To remove static and dynamic loads which arise during operation, while maintaining the relevant temperature within specific limits.

Reliability in all cryogenic and superconducting magnets is of utmost importance, since the operation depends on the presence of the cryogen. Although large equipment has remained cold for a year or more at a time, one may also consider the warm-up and cool-down of the magnet several times during the year especially if we assume that the magnet is 25% of the year in operation.

Two possibilities are open: One may warm up the magnet to a temperature where the enthalpy of the coil is still low, say 60 K or so. Or one may warm up the magnet to any desired temperature between liquid nitrogen and room temperature. In the first case, the refrigerator is operated only at partial power, while in the second case, one may require additional temporary facilities for storage of liquid hydrogen or additional refrigeration.

The copper coil has a weight of 70×10^3 kg and the aluminum coil 15.4×10^3 kg. A total amount of 5.6×10^9 J (Cu coil) and 1.8×10^9 J (Al coil) must be removed from 300 to 21 K. The coolant requirements, assuming we cool the aluminum magnet with hydrogen only, at different temperatures, are as follows: To cool down 1 kg Al in the range of 300 to 140 K, we require (0.4 to 0.1) kg of H_2 maximum (3100 to 770 liters), and 0.08 to 0.04 kg H_2 minimum (616 to 308 liters, as seen from Fig. 6.2.14. Thus, to cool down the aluminum coil from 300 to 140 K, we require the impressive amount of 47.7×10^6 liters maximum or 47.4×10^5 liters minimum, depending on whether or not we can utilize the sensible heat of hydrogen. It certainly pays off to wind the coil such that the gas can penetrate the turns and have direct contact with the cryogen.

In the same manner, we may calculate the amount of hydrogen required to cool down the stainless steel coil support, the stainless steel hydrogen container, and the insulation.

To cool down the coil to 80 K, we require the average amount of 2.3×10^6 liters of H_2 and finally to cool it down to 21 K, the average amount of 70×10^3 liters of H_2 must be provided. All values are calculated at atmospheric pressure.

Our refrigerator (18 kW at 21 K) must provide the equivalent of 8.2×10^5 J at (21 K to 80 K) temperature, 7.38×10^7 J at (80 K to 140 K) temperature and 2.2×10^8 J in the temperature range of 140 K to 300 K in the form of hydrogen gas. Referring to Fig. 6.2.3, we estimate the Carnot efficiencies at these temperature ranges and calculate the power output of the refrigerator. With the refrigerators accounted for, the cooldown can be accomplished in 14 hours for the Cu coil and 5.6 hours for the Al coil. However, cooldown can also be accomplished by supplemental liquid hydrogen or a combination of liquid nitrogen and liquid hydrogen.

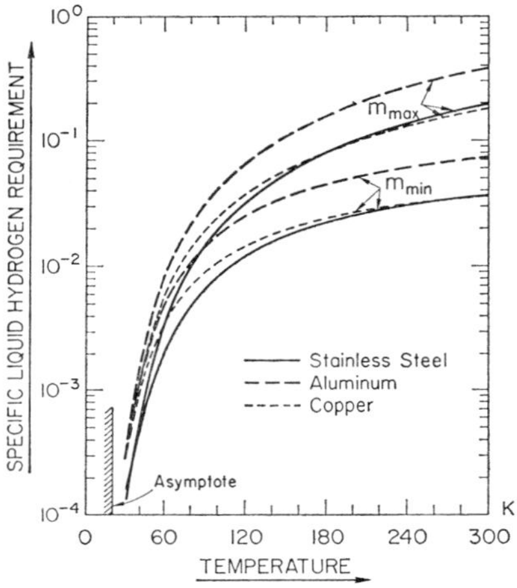

Fig. 6.2.14. Specific liquid hydrogen requirement (mass of liquid to mass of solid) to cool down various materials to liquid hydrogen temperature: $m_{\max}$ No enthalpy cooling, $m_{\min}$ enthalpy cooling

For our particular application, a closed circuit refrigerator discussed in 5.3 is chosen to perform the cooldown operation. As the ohmic losses at 21 K are predominant, the cooldown requirement does not impose additional power buildup. In addition to the ohmic losses, one may consider for all cryogenic magnets losses due to electrical leads, heat radiation and convection. Vapor-cooled shields at a temperature between 21 and 300 K (say $\sim$ 60 K), incorporation of superinsulation in the vacuum container, losses through transferlines (evacuated with superinsulation) are important.

The major part of losses in d.c. cryogenic magnets are ohmic. Static and dynamic losses due to heat radiation and thermal conduction are small. It is shown in the latter part of this chapter that heat losses due to radiation and convection are dominant in superconducting magnets, while ohmic losses are contributed only by non-persistent joints and normal-to-superconductor connections in electrical leads.

The operating cost of the magnet (we choose the aluminum coil) is derived as follows:

Power cost =

$$\begin{pmatrix}\text{power at}\\ \text{300 K in kW}\end{pmatrix}\cdot\left(\frac{\text{power cost}}{\text{kWh}}\right)\cdot(\text{time}) = 500\times 10^{-2}\times 2\times 10^{3} = \$\,10^{4}$$

Liquid nitrogen for refrigeration ($\sim$ 2 times power cost) $= 2\times\$\,10^{4}$

Make-up hydrogen for liquefier ($\sim$ equal to power cost) $= \$\,10^{4}$

Labor (1 man/shift) $\cdot$ (4.3 shifts) $\cdot \dfrac{\$\,13{,}000}{\text{man-year}} \cong 6\times\$\,10^{3}$

Total: $\$\,4.6\times 10^{4}$ per year.

This number can be compared to the power cost of the room temperature aluminum coil with iron from Table 6.2.1, which will be:

$$(\text{power at 300 K in kW})\cdot\left(\frac{\text{power cost}}{\text{kWh}}\right)\cdot(\text{operating time})$$

$$= 6\times 10^{3}\times 10^{-2}\times 2\times 10^{3} = \$\,1.2\times 10^{5}$$

Labor cost $= \$\,6\times 10^{3}$

Total: $\$\,1.26\times 10^{5}$ per year.

If we compare the ten-year cost of the aluminum coil with iron return yoke to the cryogenic magnet, we obtain the following cost breakdown:

Ten-year cost comparison between a 300 K and a 21 K aluminum coil with iron return yoke

	Aluminum coil with iron, 300 K	Aluminum coil with iron, 21 K
Capital cost	8.85×10^{5}	15.9×10^{5}
10-year interest cost at 8%	7.1×10^{5}	12.8×10^{5}
10-year operating cost at 2000 hours of operation/year	12.6×10^{5}	4.6×10^{5}
Total ten-year cost	$\$\,28.55\times 10^{5}$	$\$\,33.3\times 10^{5}$

It is evident that continuous-duty cryogenic magnets of large volume cannot compete presently with room-temperature magnets. However, magnets with long or short pulse duration, specifically for low duty cycle accelerators, may very well be cooled cryogenically.

6.2.4 Superconducting Magnets

To compare superconducting magnets economically to the two above cases of water-cooled and cryogenic magnets, we no longer concern ourselves with power requirements due to ohmic losses in the magnet, which we may neglect. Losses due to current leads and cables are small, and the refrigerator must provide for the steady-state operation of the refrigeration duce to heat convection through support structures and current leads, and due to thermal radiation through the container surface. The size of the refrigerator is thus small, compared to those estimated for cryogenic magnets. However, capital cost of the magnet itself, due to the high prices of conductor, is of concern, and coil optimization is more rewarding.

As seen in Chapters 1 and 2, there is a relationship between overall current density in a superconductor as a function of the magnetic field at the conductor, of the mechanical stresses, and the conductor temperature.

To optimize the coil, we have to find for a certain field and corresponding current density in the conductor, the minimum volume of the coil envelope, which will yield the minimum length of conductor in the coil. To find the minimum coil volume, we refer to the relation

$$B_z = \mu_0 \lambda J \cdot a_1 F(\alpha, \beta_1, \beta_2) . \tag{6.2.9}$$

To express F in simpler terms, let us assume that the coil halves are identical. We choose a split-coil configuration, as we have done in Section 6.2.2. For a given axial separation between the coil halves, required by the experimental accessibility, we rewrite Eq. (6.2.8) in the form:

$$F = \beta_1 \ln \frac{\alpha + [\alpha^2 + \beta_1^2]^{\frac{1}{2}}}{1 + [1 + \beta_1^2]^{\frac{1}{2}}} - \beta_2 \ln \frac{\alpha + [\alpha^2 + \beta_2^2]^{\frac{1}{2}}}{1 + [1 + \beta_2^2]^{\frac{1}{2}}},$$

with $\beta_1 = (2b + g)/a_1$ and $\beta_2 = g/a_1$, $2g$ being the axial coil separation. The envelope of the coil volume is given by:

$$V = 2\pi a_1^3 \cdot (\alpha^2 - 1) \cdot (\beta_1 - \beta_2) = a_1^3 \cdot v(\alpha, \beta_1, \beta_2) . \tag{6.2.24}$$

The coil geometry factor was expressed in the form:

$$G(\alpha, \beta_1, \beta_2) = \frac{\mu_0}{(2\pi)^{\frac{1}{2}}} \cdot \frac{F(\alpha, \beta_1, \beta_2)}{[(\alpha^2 - 1)(\beta_1 - \beta_2)]^{\frac{1}{2}}},$$

$$G(\alpha, \beta_1, \beta_2) = \mu_0 \cdot \frac{F(\alpha, \beta_1, \beta_2)}{[v(\alpha, \beta_1, \beta_2]^{\frac{1}{2}}} . \tag{6.2.25}$$

For any value of B_z and corresponding λJ, we will have a constant value of $F(\alpha, \beta_1, \beta_2)$. The minimum volume is thus found by maximizing the geometry factor $G(\alpha, \beta_1, \beta_2)$.

To find the comparative superconducting coil to the solenoids calculated in Section 6.2.2, we recall that $\beta_2 = g/a_1 = 22/75 = 0.2933$. For this value we have calculated the distribution of $F(\alpha, \beta)$ and from Eq. (6.2.22), the maximum value of G in Fig. 6.2.15. A general plot of F values for a split coil arrangement is given in Fig. 6.2.16.

The conductor volume within the coil envelope is simply given by

$$V_{\text{cond}} = \lambda \cdot V_{\text{coil}} .$$

The space factor is determined by the size of the coolant passages, the size and shape of reinforcements (if necessary), the conductor insulation, etc. Thus:

$$V_{\text{cond}} = \lambda \cdot a_1^3 \cdot 2\pi (\alpha^2 - 1)(\beta_1 - \beta_2) .$$

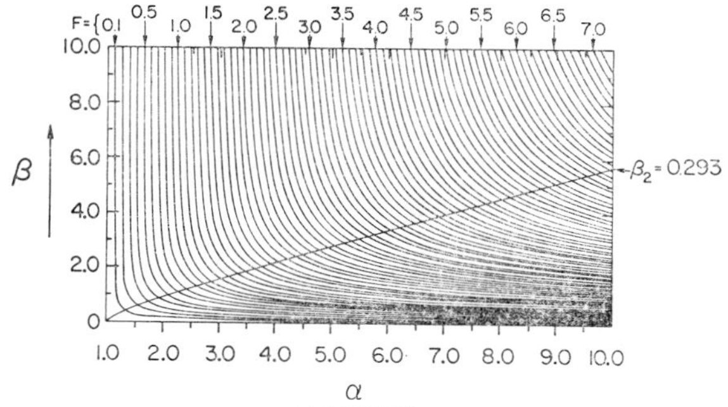

Fig. 6.2.15. Volume minimized split coil arrangement. For a $\beta_2 = 0.2933$, F is calculated as a function of α and β_1

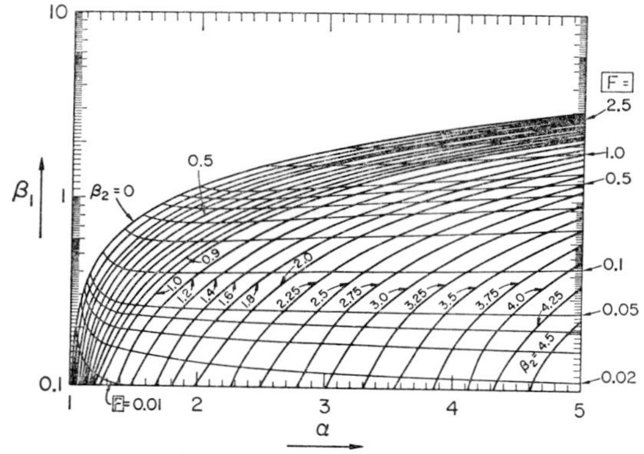

Fig. 6.2.16. Volume minimized split coil arrangements

The conductor length is found, using the number of turns per unit area:

$$l = N \cdot \pi a_1 (\alpha + 1) ,$$

with N the total number of turns. With the number of turns per unit area expressed in the form

$$n = N/2\, a_1^2\, (\alpha - 1)\, (\beta_1 - \beta_2)$$

we obtain the conductor length utilized in a split coil in the form:

$$l = n \cdot 2\, \pi a\, {}_1^3 (\alpha^2 - 1)\, (\beta_1 - \beta_2) = n\, a_1^3 \cdot v\, . \tag{6.2.26}$$

We consider now two different ways of optimizing the superconducting coil:

(a) Superconducting Coil with Constant Current Density Distribution

For a central field of 2.6 T, the field at the conductor is 4.5 T maximum (Fig. 6.2.17). At this field region, Nb_xTi is the most interesting alloy, which exhibits, at a relatively low price, the highest current density (Fig. 6.2.18).

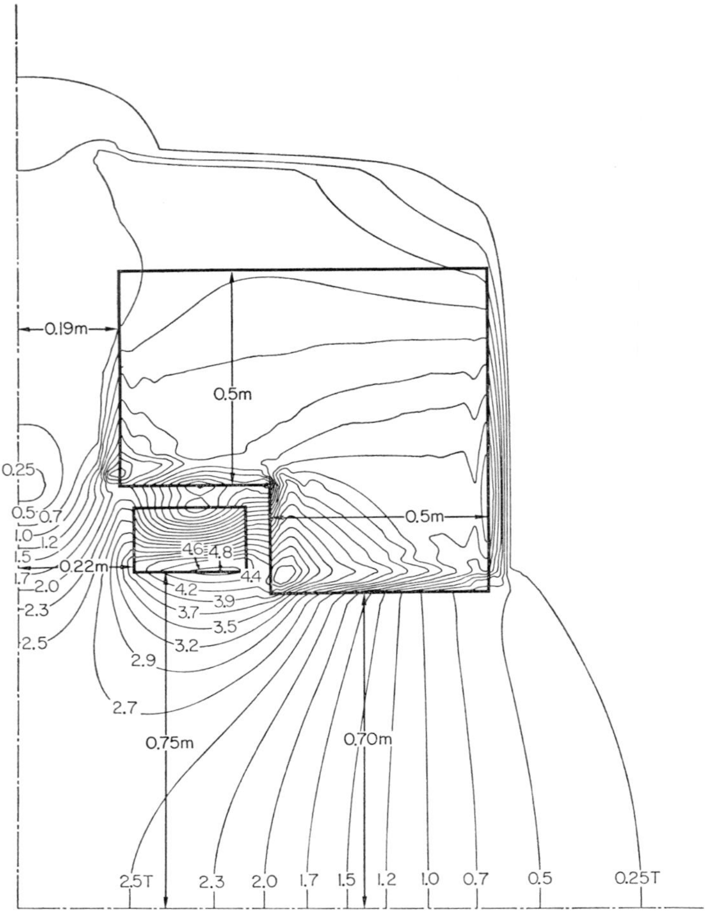

Fig. 6.2.17. Magnet configuration (only $\frac{1}{4}$ of the magnet iron core and coil is shown) used for comparison of various modes of cooling

However, as seen from Sections 1 and 4, the superconductor is imbedded in a normal conductor for cryostatic or dynamic stability purposes. It is more important to build a magnet with predictable performance, well protected and not self-destructive in case the magnetic field collapses due to internal or external disturbances (quench). The amount of substrate to be paralleled with the superconductor is calculated as follows. The cryostatic stability, according to Stekly [16] is given by:*

$$\boxed{I_c^2 \cdot \varrho \leq h\,(T_c - T_b) \cdot A_n \cdot S_n}\,. \tag{6.2.27}$$

* See also 4.7.3.4 Eq. (4.7.61).

The derivation assumes a linear variation of current-carrying capacity of the superconductor with the temperature.

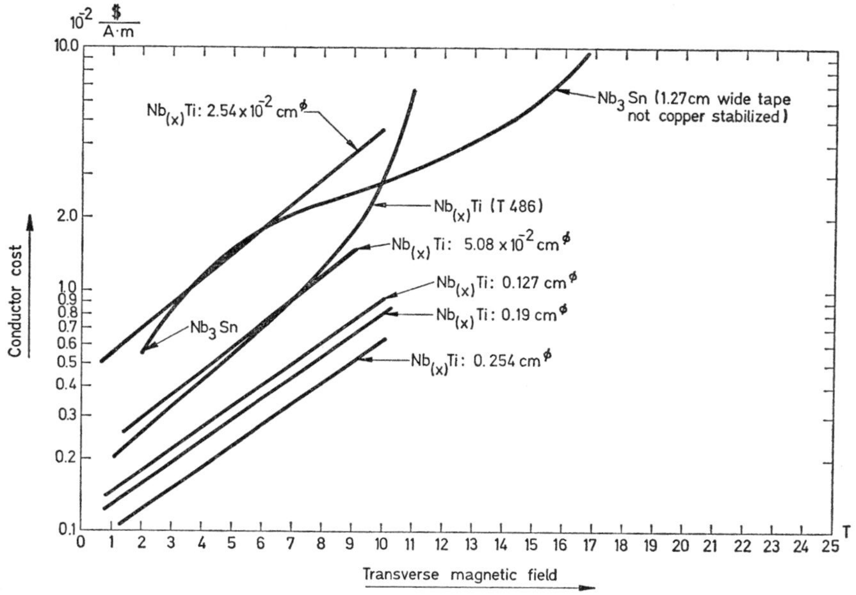

Fig. 6.2.18. Estimated prices of superconducting composite conductors. Prices quoted per Amp · m are only for simple multifilamentary conductors (twisted and untwisted): with 100 filaments; *Ratio:* Cu : Sc 2:1 for 2.54×10^{-2} cm/diam., 3 : 1 for 0.254 cm/diam., and Nb_3Sn not Cu stabilized tapes. The prices can be used with some restrictions. Cable prices used for pulsed superconducting magnets are not given

The design heat flux $w_s = h\,(T_c - T_b)$ for superconducting coils may be obtained from Fig. 5.7.6. For most large coils, where helium can cool at least 50% of the conductor surface by direct contact, the assumption of a heat flux of $\sim 4 \times 10^3$ W/m² is justifiable. If a hollow composite conductur is used, where supercritical helium is pumped through internal coolant passages, the heat flux of $\sim 5 \times 10^3$ W/m² may be assumed.

For the conductor perimeter per unit length S_n, we can write the expression $f \cdot (A)^{\frac{1}{2}}$, where f accounts for the conductor shape.

Assuming all the current flows through the substrate and the superconductor is in a normal state, then we may write for the current density in the substrate:

$$\boxed{J \leqq (h\Delta T \cdot f/\varrho)^{\frac{2}{3}} \cdot I^{-\frac{1}{3}}}\,, \qquad (6.2.28)$$

where I is the nominal coil current.

The resistivities of various metals being utilized or considered for use in superconducting magnets are given in Table 1.6. A conductor cross-section selected according to Eq. (6.2.25) is fully stabilized. The overall coil current densities achieved this way are low and the magnet coil will be large in size and accordingly expensive. We may consider, however, the following way to reduce the size and cost of the magnet:

No matter how stable a superconductor is, there is always a chance that some kind of a failure may occur which could lead to a quench: The heat may propagate along the conductor, or persist at the spot of disturbance. At any rate, serious damage may occur by overheating the coil. Failures such as blockage of coolant passages within the coil, loss of vacuum, breakdown of power supply or of the refrigerator can cause the coil to convert to a normal state, heat up, and lead to insulation and conductor damage.

Protection of the coil against overheating, excessive loss of helium, increase in pressure in the container which may damage it, destruction of the superconductor, and voltage breakdown of insulation when the coil quenches, has been proposed and executed by several authors [17, 18] in various laboratories. The classical method is the dissipation of the electromagnetic energy into external resistors, discussed in more detail in Chapter 7.

Here we briefly determine the extent of the protective circuit and the optimum current density in the composite conductor, which determines an upper temperature limit when a region of normality has appeared in the coil. If the coil becomes resistive, the magnet is disconnected by means of fast-acting switches from the power supply. The current flows through an external resistor and the field decay is dictated by the time constant $\tau = L/R$.

The essential requirement for protection is that the temperature at any part of the magnet should not exceed a certain specified value, dictated by the characteristics of the superconductor or the insulation, whichever is more sensitive.

The energy dissipation takes place in the substrate in a composite conductor. The contribution of the superconductor to the enthalpy stabilization is small and, in our calculation, negligible. Assuming all the helium surrounding the conductor is evaporated and the helium enthalpy is neglected, we may write the energy balance per unit volume of substrate:

$$[j\,(t)]^2 \varrho\, dt = c_p \cdot \delta \cdot dT\,,$$

with $c_p \delta$ the heat capacity of the substrate. Integrating this expression, we get:

$$\int_0^\infty [j\,(t)]^2 dt = \int_{T_b}^{T_m} \left(\frac{c_p \delta}{\varrho}\right) dT = \int_{T_b}^{T_m} f\,(T)\, dT\,. \tag{6.2.29}$$

The right-hand side of the equation depends on the characteristics of the substrate. Curves of $\int_{T_b}^{T_m} f\,(T)\, dT$ are plotted for several substrates in Fig. 6.2.19.

Referring now to the circuit given in Fig. 7.4, the decay of current depends on the total inductance of the system, mainly the coil and lead inductance, and the resistance of the normal conductor region, the leads, and the external resistor. The current decay is exponential with a time constant:

$$\tau = \frac{\sum_i L_i}{\sum_i R_i},$$

$$i = 1, 2, \ldots$$

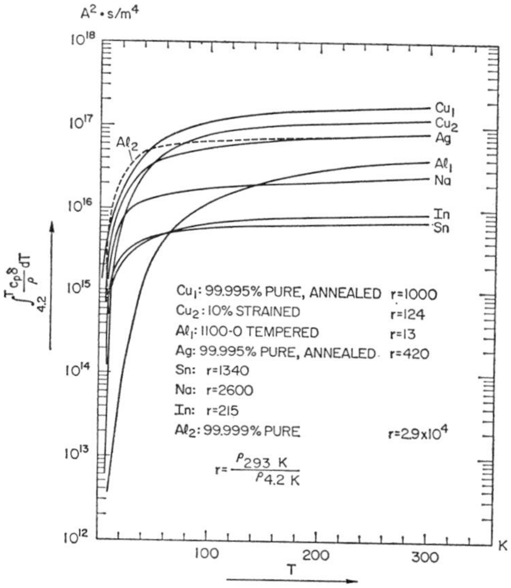

Fig. 6.2.19. Curves of $\int_{4.2K}^{T'} \frac{c_p \delta}{\varrho} dT$ as a function of temperature for various metals

Hence, Eq. (6.2.29) becomes:

$$\frac{1}{2} J^2 \tau = \int_{T_b}^{T_m} f(T) \cdot dT = g(T_m), \tag{6.2.30}$$

where J is the initial current density in the substrate and T_m the maximum allowable temperature. By inserting the expressions

$$E = \frac{1}{2} L I^2 \quad \text{and} \quad U_m = I \cdot R,$$

in Eq. (6.2.30), with E the stored magnetic field energy and U_m the maximum voltage when the field decays, we get the relation between the initial current density in the substrate and the coil parameter:

$$J \leqq \left[g(T_m) \cdot U_m \cdot \frac{I}{E} \right]^{\frac{1}{2}}. \tag{6.2.31}$$

With the expressions of current density for cryostatic stabilization (Eq. (6.2.28)) and for coil protection (Eq. (6.2.31)), we may eliminate the current I as an independent variable from these two equations, and get

$$\boxed{J_{\text{opt}} = [g\,(T_m) \cdot U_m/E]^{0.2} \cdot [h\Delta T \cdot f/\varrho]^{0.4}} \,. \tag{6.2.32}$$

For our particular magnet, with the central field of $B_{0,0} = 2.6$ T, the field energy is approximately 12.5 MJ. We have used annealed OFHC copper as substrate and limited the maximum temperature to $T_m = 30$ K. From Fig. 6.2.19, we get: $g\,(30\text{ K}) \cong 10^{16}$ A² s/m⁴. With the values,

$$\begin{aligned} h\Delta T &= 4\times10^3 \text{ W/m}^2,\\ U_m &= 10^2 \text{ Volts},\\ f &= S_n/(\text{A})^{\frac{1}{2}} = 3,\\ \varrho_n &\cong 1.7\times10^{-10} \text{ Ohm}\cdot\text{m}, \end{aligned}$$

we obtain the optimum current density in the conductor substrate:

$$\begin{aligned} J_{\text{opt}} &= \left(\frac{10^{16}\times10^2}{12.5\times10^6}\right)^{0.2} \cdot \left(\frac{4\times10^3\times3}{1.7\times10^{-10}}\right)^{0.4}\\ &= 5.25\times10^7 \text{ A/m}^2. \end{aligned}$$

Assuming a space factor of $\lambda \cong 0.8$, the overall current density in the magnet is:

$$\lambda J = 4.2\times10^7 \text{ A/m}^2.$$

From Eq. (6.2.9), we get:

$$F = \frac{20\times10^3}{0.4\,\pi\times4.2\times10^3\times75} = 0.051\,.$$

The corresponding minimum volume is obtained from Fig. 6.2.16, with

$$\alpha = 1.20\,, \quad \beta_1 - \beta_2 = 0.37\,, \quad v_{\min} \cong 1.03$$

for the Helmholtz pair.

The maximum field at the conductor is calculated from Fig. 2.1.19 to be 4.8 T, which, from measurements with small- and medium-size coils, corresponds to a critical current density of $J_{sc} = 6\times10^8$ A/m² for heat-treated Nb(50%)Ti conductors.

The ratio of substrate area to superconductor area is thus:

$$\frac{A_{\text{Cu}}}{A_{\text{sc}}} = \frac{J_{\text{sc}}}{J_{\text{Cu}}} = \frac{6\times10^8}{5.25\times10^7} = 11:1\,.$$

We choose a current of $I = 3000$ A and obtain the conductor cross sections:

$$A_{\text{Cu}} = 0.572\times10^{-4}\text{ m}^2 \quad \text{and} \quad A_{\text{sc}} = 0.05\times10^{-4}\text{ m}^2.$$

Thus the conductor has the dimensions of:

$$0.78\times0.8\times10^{-4}\text{ m}^2.$$

Assuming coolant passages of 10^{-3} m width between turns and layers, we get from the number of ampere-turns per coil:

$$NI)_{\text{coil}} = 2\,a_1^2\,(\alpha - 1)\,\beta\cdot(\lambda J) = \underline{1.8\times10^6\text{ A}}\,,$$

the number of turns per coil:

$$N_{\text{coil}} \cong 600\,,$$

which can be divided into the number of turns per layer of 34 and the number of layers per coil of 18.

The conductor length per coil is therefore: $l \simeq 6200$ m, including some additional material to be used for the leads and joints.

A detailed breakdown of the cost of this magnet is given in Table 6.2.6. The approximate cost of the superconducting wire with respect to current and field is given in Fig. 6.2.18. The cost refers to simple composite conductors and is subject to change. It may be used simply as a reference. More exact and up-to-date values can be obtained from the manufacturer of the specific conductor.

(*b*) *Current Optimized Superconducting Coils*

As seen in Section 2, the field distribution over the coil in radial direction decays practically linearly. With the decrease in field intensity in radial direction, the critical current density in the superconductor can be increased accordingly, while the cross section of the normal material follows Eq. (6.2.32). At lower fields, the magnetoresistance effect becomes less pronounced and the cross section of the substrate is reduced. As the overall conductor cross section becomes smaller, the overall coil cross-sectional area can be reduced, compared to a constant current density coil configuration. Assuming β is fixed, i.e. the coil length does not change with the coil radius in order to simplify coil manufacturing and support, then we may approach the optimization of coils in two ways:

1. Reduce the conductor cross-sectional area according to the field decay as a function of coil radius from layer to layer. This task, from a manufacturing point-of-view, is uninteresting and expensive.

2. Divide the entire coil cross-sectional area into several concentric modules. Each section will have a constant overall current density. However, the current density in each module will be different. We may energize each module from separate current sources, or, by choosing the conductor cross sections according to the optimum current density given by the properties of the composite conductor and the field distribution, energize all modules from one current source by connecting them in series.

As an example, we again choose the split coil described above. We divide each coil section arbitrarily into three modules. The current chosen is 3000 A. The axial coil separation between coil sections will be $2\,g = 0.44$ m.

Module I:

$$a_{1,\text{I}} = 0.75 \text{ m}\,, \quad g = 0.22 \text{ m}\,, \quad B_{0,\text{I}} = 0.9 \text{ T}\,, \quad B_{\max} = 4.5 \text{ T}\,.$$

From Eq. (6.2.32), we calculate: $J_{\text{opt}} = 4.65 \times 10^7$ A/m².
Fig. 1.3 indicates that at $B_{\max} = 4.5$ T, we may choose the current density through the $Nb_{(x)}Ti$ superconductor in the coils:

$$J_{\text{sc}} = 6 \times 10^8 \text{ A/m}^2.$$

The current densities correspond to the cross sections:

$$A_{\text{Cu}} = 6.45 \times 10^{-5}\ \text{m}^2, \quad A_{\text{sc}} = 5 \times 10^{-6}\ \text{m}^2.$$

With a space factor of $\lambda \cong 0.8$, we obtain the overall current density in module I:

$$(\lambda J)_{\text{I}} = 0.8 \times \frac{3 \times 10^3}{6.95 \times 10^{-5}} = 3.45 \times 10^7\ \text{A/m}^2 .$$

The optimized coil parameters are obtained according to Fig. 2.1.(16—21) of Chapter 2:

$$D_{\text{I}} = \frac{B_{\text{I}}(0,0,0)}{\mu_0 \lambda J \cdot g} = 0.0944; \quad \beta_{2,\text{I}} = \frac{g}{a_1} = 0.2933; \quad \beta_{\text{I}} = \left.\frac{2b+g}{a_1}\right)_{\text{I}} \cong 0.68 .$$

(Fig. 2.1.18)

$$\beta_{1,\text{I}} = \frac{b}{a_1} \cong 0.19; \quad \alpha = \left.\frac{a_2}{a_1}\right)_{\text{I}} = 1.1; \quad v_{\min} = 0.26; \quad l_{\text{cond}} = 2435\ \text{m} .$$

(Fig. 2.1.17)

Module II:

$$a_{1,\text{II}} \cong 0.825\ \text{m}, \quad g = 0.22\ \text{m}, \quad B_{0,\text{II}} = 0.6\ \text{T}, \quad B_{\max} \cong 2.3\ \text{T} .$$

As in module I, we obtain the current densities:

$$J_{\text{opt II}} = 5.5 \times 10^7\ \text{A/m}^2; \quad J_{\text{sc II}} = 9 \times 10^8\ \text{A/m}^2.$$

The corresponding copper and superconductor areas are:

$$A_{\text{Cu}} = 5.46 \times 10^{-5}\ \text{m}^2; \quad A_{\text{sc}} = 3.34 \times 10^{-6}\ \text{m}^2; \quad (\lambda J)_{\text{II}} \cong 4.4 \times 10^7\ \text{A/m}^2,$$

and, therefore: $D_{\text{II}} = \dfrac{0.6}{0.4\,\pi \times 10^{-6} \times 4.4 \times 10^7 \times 0.22} \cong 0.05$.

$$\beta_{2,\text{II}} = \frac{0.22}{0.825} = 0.267; \quad \beta_{1,\text{II}} = 0.615; \quad \beta_{\text{II}} = 0.175; \quad \alpha = 1.05;$$

$$v_{\min} = 0.2254; \quad l_{\text{cond}} = 1750\ \text{m};$$

Module III:

$$a_{1,\text{III}} = 0.866\ \text{m}, \quad g = 0.22\ \text{m}, \quad B_{0,\text{III}} = 0.5\ \text{T}, \quad B_{\max} = 1.4\ \text{T} .$$

$$J_{\text{opt, III}} = 6 \times 10^7\ \text{A/m}^2 \therefore A_{\text{Cu}} = 5 \times 10^{-5}\ \text{m}^2 ,$$

$$J_{\text{sc}} = 10 \times 10^8\ \text{A/m}^2 \therefore A_{\text{sc}} = 3 \times 10^{-6}\ \text{m}^2 ,$$

$$(\lambda J)_{\text{III}} = 4.6 \times 10^7\ \text{A/m}^2 .$$

$$D_{\text{III}} = 0.041; \quad \beta_{2,\text{III}} = 0.254; \quad \beta_{\text{I, III}} = 0.588; \quad \beta_{\text{III}} = 0.167;$$

$$\alpha = 1.045; \quad v_{\min} = 0.193; \quad l_{\text{cond}} = 1900\ \text{m} .$$

The conductors in each of the three sections are connected in series. Methods of producing low-resistance joints are common and described in Chapter 4.8. The helium cryostat is similar to a container for liquid hydrogen, but more expensive. We have ignored at present the price of the power supply and have shown the magnet price in Table 6.2.6. These prices are now directly comparable to those of the room temperature and the cryogenic magnet.

Assuming we have perfected the electromechanical bond between superconductor and high-purity aluminum (which at present is not the case), we can exercise the same calculation as for the copper stabilized conductor. Due to the much smaller ϱ_n value of aluminum ($\varrho_n \cong 5\times10^{-11}$ Ohm · m, including magnetoresistance, mechanical strain due to Lorentz forces, and thermal contraction), we achieve a higher J_{opt} than calculated for copper.

For our example, the optimum current density, according to Eq. (6.2.32) would be: $J_{Al} = 8.8\times10^7$ A/m². The overall current density $\lambda J = 6.6\times10^7$ A/m². The coil would have the following parameters: $\alpha = 1.15$; $\beta_1 - \beta_2 = 0.31$. The total amount of conductor used would be: $l_c = 5900$ m for a current of $I = 3000$ A. The price breakdown for this magnet is given in Table 6.2.6.

Table 6.2.6. *Cost estimate of a B* (0,0,0) = 2.6 T *superconducting Helmholtz coil with iron return yoke*

Cost of (in U.S. dollars)	Superconducting coil with uniform current density (Cu substrate)	Superconducting coil current optimized (Cu substrate)	Superconducting coil uniform current density (Al substrate)
Conductor	110,000	70,000	95,000
Coil-winding, support structures, insulation, current leads, safety	200,000	190,000	200,000
Helium tank	25,000	25,000	25,000
Power supply and control	20,000	20,000	20,000
Safety	6,000	6,000	6,000
Instrumentation, power	8,000	8,000	8,000
Vacuum tank (stainless steel)	30,000	30,000	30,000
Instrumentation, vacuum	9,000	9,000	9,000
Multiple layer insulation	5,000	5,000	5,000
Energy vent shut-off	7,000	7,000	7,000
Refrigerator (100 W), including transferlines, purifier and compressors	130,000	130,000	130,000
Helium surge tank	10,000	10,000	10,000
Helium storage tank	15,000	15,000	15,000
Iron return yoke	115,000	110,000	105,000
Design, Engineering, Installation	180,000	180,000	180,000
Miscellaneous	30,000	30,000	30,000
Total $	900,000	845,000	875,000

6.2.5 Operating Cost of Superconducting Coils

When operating superconducting magnets, two distinct refrigeration problems exist: The cool-down and the steady-state operation.

Cool-down of the coil structure will require considerable refrigeration, since the cold components of a typical magnet may weigh several hundred to several thousand kilograms. Various refrigeration methods may be applied to cool down and operate the magnet system, such as precooling the magnet with liquid nitrogen to about 80 K, then supplying liquid helium and venting the evaporated gas to the atmosphere. This method, although simple, would be prohibitively expensive and we do not consider it further. The method of utilizing an individual refrigerator or liquefier with the magnet is more practical and has been considered here in detail. A suitable quantity of liquid helium may be stored for filling the coil enclosure with helium, or for cases of emergency, such as refrigerator breakdown.

The type of thermal insulation chosen for the superconducting magnet enclosure determines the temperature at which the evaporated helium enters the refrigeration circuit. If the gaseous helium enters the magnet system at a temperature near ambient ($\sim$ 300 K) temperature, the cryogenic process cycle will employ the principle of a helium liquefier. If the gaseous helium return temperature is near 4.2 K, then a refrigeration cycle will be employed. The choice between the two methods is not trivial and must be studied for each case specifically. It may be mentioned that it requires approximately four times more power to produce a unit of liquid helium using a liquefier as it does using a refrigerator which recondenses cold gas!

The magnet had the following parameters, which are of interest for this section: Field energy, 12.5×10^6 J, conductor length, 12,400 m (both coils); conductor cross section, 0.622×10^{-4} m^2 (copper and superconductor). Thus the conductor weight is 6.8×10^3 kg. The weight of the coil form (stainless steel) is estimated at 1300 kg. Insulation and support structure weigh another 1000 kg. The helium container has a weight of 800 kg. We must cool-down a weight of $\sim$9900 kg from 300 to 4.2 K. A total of 7.5×10^8 J must be removed from the magnet and helium cryostat. As mentioned above for the cryogenic magnet, the cool-down may be accomplished by using a closed circuit refrigerator, by using liquid helium, or by using a combination of liquid nitrogen and liquid helium.

We assumed that the magnet operates over 2000 hours per year intermittently, and we may anticipate that there will be periods of several weeks when the magnet is not operated. Let us assume that the magnet must be cooled-down 10 times over a year. The refrigerator, which supplies coolant to the magnet, is capable of cooling-down the magnet at a certain rate. We may also use liquid nitrogen to cool-down the magnet to 80 K, further by means of gaseous helium to about 20 K, and by bulk liquid to 4.2 K, where the specific heat of the materials is small. Whatever scheme is used, let us first calculate the size of the refrigerator required to provide coolant for steady-state operation and then answer the method of cool-down, as well as the magnet temperature, in the intermediate periods, while the magnet is not energized.

Steady-State Heat Load

The heat load imposed on the refrigeration system originates from heat conduction through support structures, electrical leads and thermal insulation, heat radiation through thermal insulation and the radiation surface of the helium container, and ohmic loss heating through convection (non-superconductive) and electrical leads. The heat load is transferred from the higher temperature environment to the low temperature parts, and its magnitude is given by the method of refrigeration and the method and degree of thermal shielding.

Superinsulation

A multi-layer insulation (so-called superinsulation) consisting of alternating layers of highly reflective materials (aluminum foil, aluminized mylar) and low-conductivity spacer material (fiberglass, nylon net, crinkled nylon) is applied properly on the helium container. The space containing the superinsulation is evacuated to at least 10^{-4} mm Hg for the insulation to be effective. For such an insulation, we may achieve thermal conductivity of the order of 10^{-4} W/m K. Utilizing shields at temperatures between 4.2 and 300 K (say 78 K with liquid nitrogen cooling), we may reduce this number. Heat is transferred primarily by radiation and conduction through the spacer material. The thermal conductivity is expressed as:

$$k = \frac{1}{n} \cdot \left\{ h_c + \frac{\sigma \varepsilon \cdot T_2^3}{2 - \varepsilon} \left[1 + \left(\frac{T_1}{T_2} \right)^2 \right] \left[1 + \frac{T_1}{T_2} \right] \right\}. \quad (6.2.33)$$

In this equation

n = number of layers per unit thickness,
h_c = heat transfer coefficient through the spacer material (W/m² K),
σ = Steffan Bolzman constant $= 5.672 \times 10^{-8}$ W/m² (K)⁴,
ε = effective total emissivity of the solid material (Fig. 6.2.20a, b),
T_2, T_1 = absolute temperatures at the warm and cold sides, respectiv.

The thermal conductivity of the superinsulation is sensitive to n. If the superinsulation is packed too tightly, k increases. In practice, about 2×10^3 layers/m may be applied. However, the optimum n may be found for any temperature range from Eq. (6.2.33) and the emissivity curves of Fig. 6.2.20.

For our particular magnet, with a surface area of $A = 6$ m² and an insulation thickness of 10^{-2} m, we get the heat loss through the superinsulation from 300 to 4.2 K of 18 watts, which is a large number. Using a nitrogen-cooled shield, the heat loss is reduced to 5 watts, which is reasonable. As the work required to provide a unit of refrigeration at 4.2 K is approximately 20 times that required at 80 K, in the case of a superconducting magnet, we shield the 4.2 K environment either with liquid nitrogen or with a shield (or shields) cooled with effluent helium gas at some temperature between 20 and 80 K.

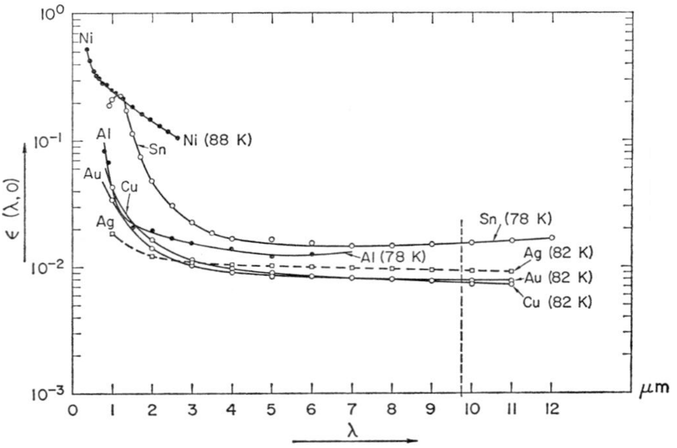

Fig. 6.2.20a. Total spectral emittance of various metal surface as a function of wave length ("total" applies to the emittance obtained by integration over all wave lengths)

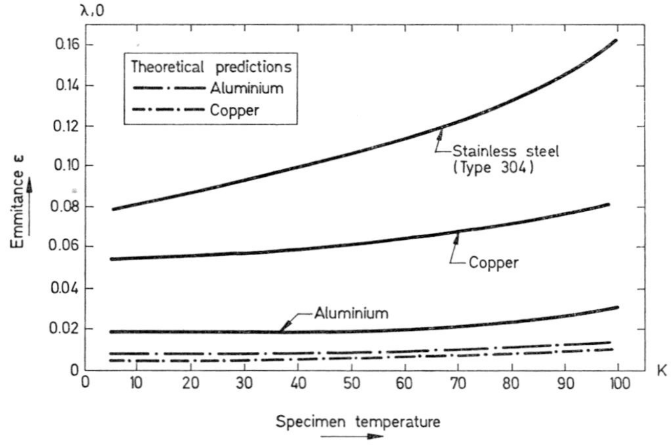

Fig. 6.2.20b. Emittance curves for various metals

Coil Support

There are two kinds of support with which we are concerned: Support to prevent structural loading of the multi-layer insulation, and supports to adequately locate the coil and its helium container within the room temperature vacuum vessel (Fig. 6.2.21).

The size of the support structure is determined mainly by the expected magnetic stresses on the coil due to unbalanced forces, resulting from the relative location of individual coils with respect to each other and with respect to the iron yokes. Another reason for introducing support structures is the coil weight. Another support method is shown in Fig. 6.2.22 which represent an NAL design.

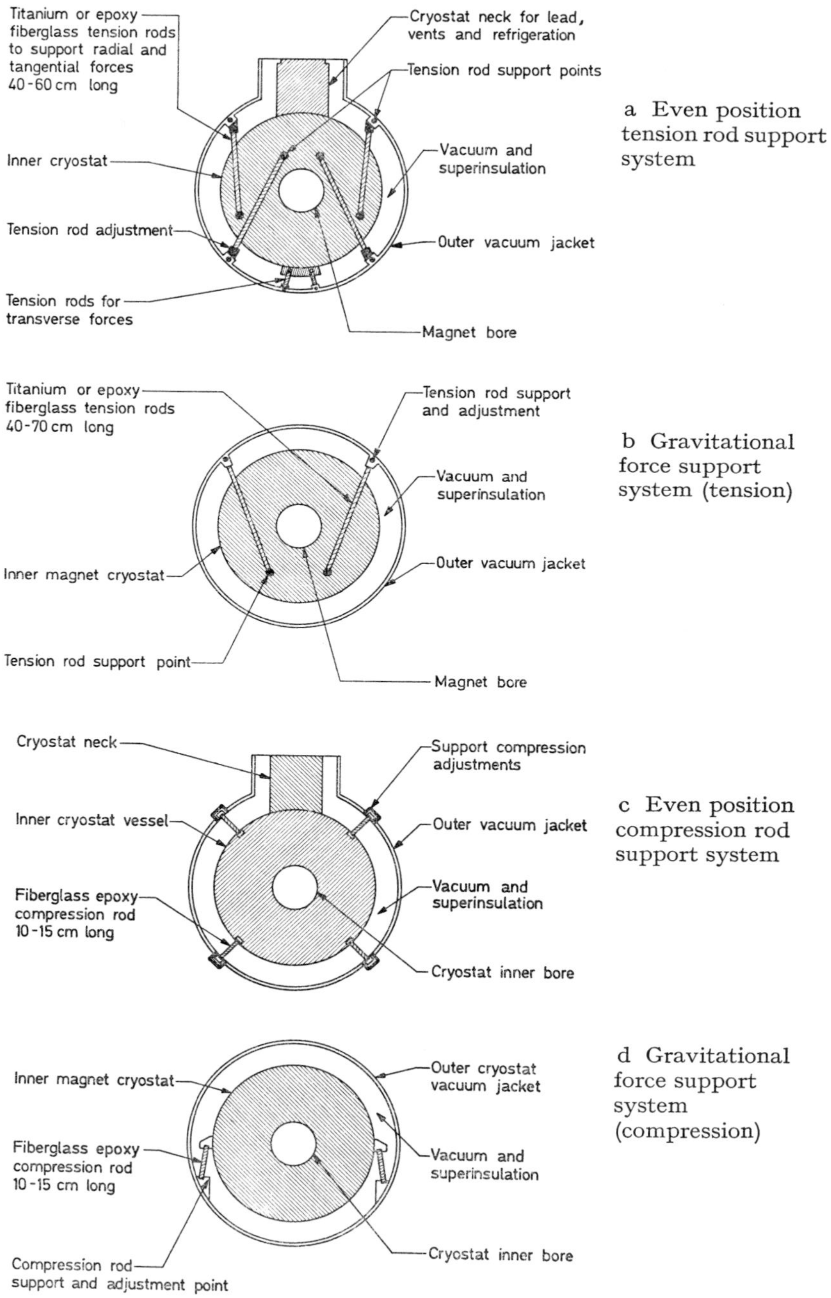

Titanium or epoxy fiberglass tension rods to support radial and tangential forces 40-60 cm long
Cryostat neck for lead, vents and refrigeration
Tension rod support points
Inner cryostat
Vacuum and superinsulation
Tension rod adjustment
Outer vacuum jacket
Tension rods for transverse forces
Magnet bore
a Even position tension rod support system
Titanium or epoxy fiberglass tension rods 40-70 cm long
Tension rod support and adjustment
Vacuum and superinsulation
Inner magnet cryostat
Outer vacuum jacket
Tension rod support point
Magnet bore
b Gravitational force support system (tension)
Cryostat neck
Support compression adjustments
Inner cryostat vessel
Outer vacuum jacket
Fiberglass epoxy compression rod 10-15 cm long
Vacuum and superinsulation
Cryostat inner bore
c Even position compression rod support system
Inner magnet cryostat
Outer cryostat vacuum jacket
Fiberglass epoxy compression rod 10-15 cm long
Vacuum and superinsulation
Cryostat inner bore
Compression rod support and adjustment point
d Gravitational force support system (compression)

Fig. 6.2.21. Tension and compression rod support systems to support the magnet within the cryostat

As a support material, low-thermal conductivity high-tensile strength materials are used. Small cross sections and long lengths of supports are sought to reduce heat leak. Stainless steel bolts or high tensile wires are utilized. To support the coil alone, assuming no unbalanced forces between iron and coil, the heat leak introduced by support structure is 2.2 W for a 1000 kg coil structure from 300 K. If an 80 K temperature

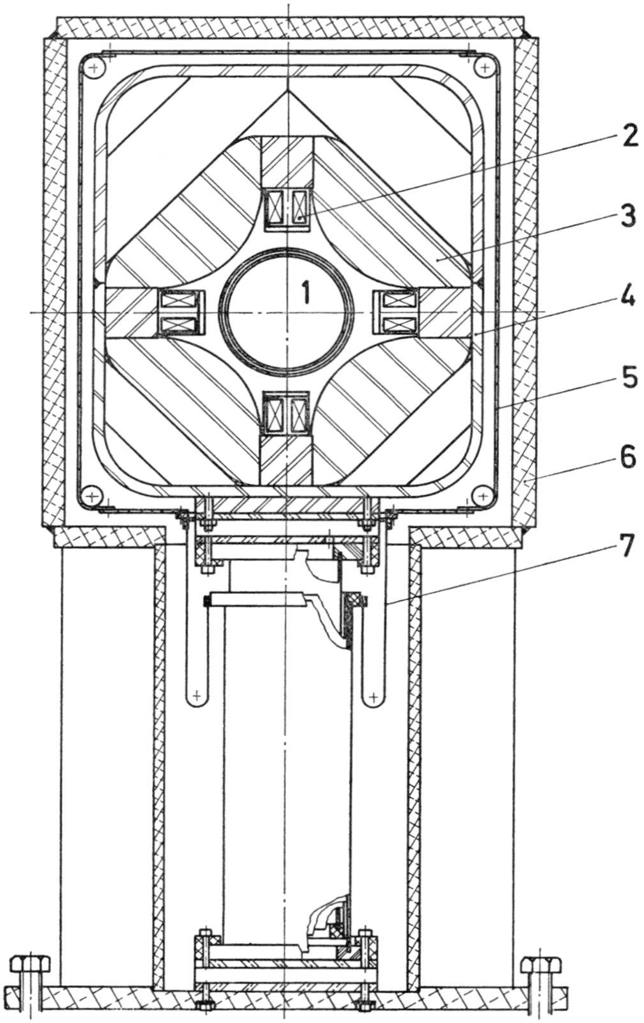

Fig. 6.2.22. Magnet and cryostat assembly. The superconducting quadrupole is supported by means of a central structure (NAL-design). *1* Aperture, *2* Coil, *3* Iron pole, *4* Helium container, *5* Nitrogen shield, *6* Room temperature part, *7* Nitrogen connection to support

shield is provided (the support is thermally in contact with the heat shield), the heat leak is reduced to 1.1 W for a coil weighing 1000 kg. For our coil, the heat leak would be approximately 11 watts. Assuming some unbalanced forces, mainly if the coil is not perfectly centered within the iron yoke, the increased cross-sectional area leads to an additional heat leak of 9 watts.

*Electrical Leads**

Heat leak due to electrical leads has been studied by several authors [19, 20, 21]. Modern leads are cooled in a counterflow manner, passing the effluent gas through a number of copper layers or wires. If the sensible heat ($m \cdot c_p dT$) of the helium gas is utilized, the optimum heat load at design current should correspond to 2 watts per 1000 amperes per pair of leads. However, the cold gas passing through the lead would not be available to the refrigerator and additional coolant capacity must be provided. A new method for small magnets is to operate the magnet after it has been charged in persistent mode and disconnect the leads from the magnet. The leads are withdrawn above the liquid helium level. As for large superconducting magnets, this design is still in a preliminary stage. The heat load in our magnet, due to leads, will be 6 watts (3000-A leads).

Electrical Joints

Manufacturing limitations force us to provide several joints within the coil. There are also copper-to-superconductor connections from the electrical leads to the conductor. Presently it is possible to manufacture 3000-A composite conductors in 500-m continuous lengths. Methods of producing low-resistance joints, such as simple overlapping of conductors, electron beam welding, joints by means of explosive methods, etc., are widely known and it is possible to produce electrical joints having a resistance of 10^{-8} to 10^{-9} Ohms. Although the heat load due to electrical joints is small compared to the other losses discussed, we assume a total loss to be one watt.

Summary

The total heat load to the superconducting magnet studied is 32 watts. Some additional loss must be encountered through transfer tubes from the refrigerator to the magnet. Modern transfer tubes with no liquid nitrogen shields but utilizing superinsulation have a loss of $\sim$0.5 ... 0.8 W per linear m for a 2-cm inner diameter transfer line.

In addition, we have to concern ourselves with losses when the helium is intermittently replenished into the helium reservoir. Controlling the helium flow to maintain the helium level in the magnet reservoir is difficult. Continuous flow to the reservoir would require sophisticated devices for flow regulators. The simplest method is to start filling when a minimum helium level in the reservoir is reached. Level gages are used to indicate maximum-minimum helium levels in the reservoir. If we assume

* See also section 4.8.2.1.

that the refrigerator is located 10 m from the magnet, the total heat load of the magnet (including transferring and handling losses) will be 40 watts.

We chose a refrigerator having a capacity of 100 watts, which provides ample capacity even to liquefy additional helium during steady-state operation.

Cool-Down

Cool-down of the coil structure will require considerable refrigeration, since a substantial material weight must be cooled-down to 4.2 K. We found that to cool-down the magnet from 300 to 4.2 K, approximately 7.5×10^8 J must be removed. To cool-down the magnet system from 80 to 4.2 K, the amount of 5.3×10^7 J must be removed.

The curves in Figs. 6.2.23 and 6.2.24 indicate the specific liquid requirement for cool-down [22]. The maximum and minimum requirements refer to the degree to which the sensible heat $\left(\int_T c_p dT\right)$ of the coolant is utilized. To cool-down 1 kg of copper from 80 to 4.2 K approximately 0.27 kg of helium would be required. If the sensible heat of helium is fully utilized, only 0.022 kg of helium may be required. For a practical magnet, sufficient cooling surface must be provided in order to reduce helium requirement. As we will not be able to utilize all the sensible heat of helium, we require some number between the maximum and minimum curves. We will need roughly 0.15 kg of helium per kg of copper, and 0.1 kg of helium per kg of stainless steel, i.e., a total of 1.1×10^3 liters of helium to cool the coil-down from 80 to 4.2 K.

To calculate the cool-down time, enthalpy values for helium as function of temperature and pressure may be consulted. The temperature interval is subdivided. From Fig. 6.2.24, we determine the amount of

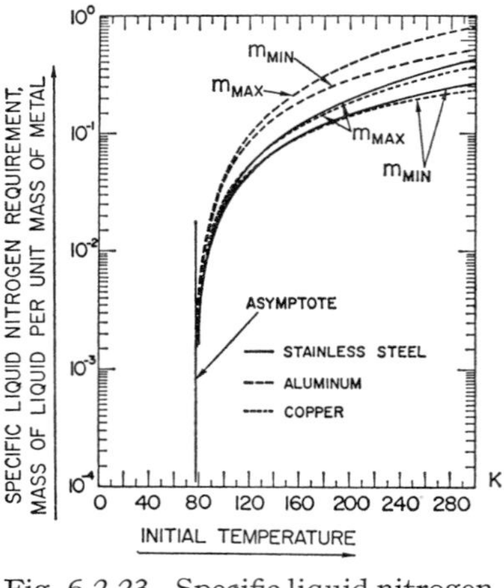

Fig. 6.2.23. Specific liquid nitrogen requirement to cool down various metals down to 78 K:

$m_{\min}$ Enthalpy cooling,
$m_{\max}$ No enthalpy cooling

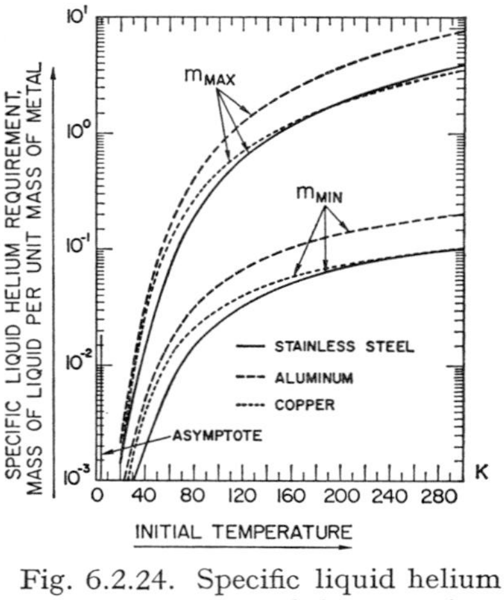

Fig. 6.2.24. Specific liquid helium requirement to cool down various metals down to 4.2 K:

$m_{\min}$ Enthalpy cooling,
$m_{\max}$ No enthalpy cooling

helium to be used during the various temperature intervals, and obtain the cool-down storage energy. Using (% Carnot) efficiencies at various temperature intervals, we are able to obtain the time, which will elapse until the magnet is cooled from 80 to 4.2 K. For our particular 100-W refrigerator, we may be able to utilize about 80 watts at 4.2 K for magnet cool-down. At higher temperatures, the losses through transfer, conduction and radiation are smaller, and we may utilize better refrigeration values due to higher values of percent Carnot.

For our particular magnet, the cool-down from 80 to 4.2 K will require approximately 100 hours if the 100-watt refrigerator is utilized. To accelerate cool-down, one may use auxiliary refrigeration capacity, if available.

As the cool-down time is fairly long and cumbersome, one may consider the possibility of warming up the magnet between experiments to about 20 K and operating the refrigerator at a fraction of its capacity. The cool-down from 20 to 4.2 K can be accomplished in 15 hours with the available refrigerator.

A 100-watt refrigerator (at 4.2 K) requires a total installation power of 75 kW. We may now calculate the operating cost of the magnet:

Power cost (steady state)	$75\times2\times10^3\times10^{-2}$	=	\$ 1,500
Power cost (10 cool-downs per year)	$75\times10^3\times10^{-2}$	=	750
Liquid nitrogen for refrigerator	$5\times$power cost	=	11,250
Make-up helium for refrigerator	$2\times$power cost	=	4,500
Helium liquefaction in the time between runs		=	2,000
Labor (1 man/shift) · (2 shifts) · \$ 13,000/man-year		=	26,000
	Total cost per year	=	\$ 46,000

The ten-year cost of the superconducting coil with iron return yoke is summarized in the table below:

Ten-year cost of a superconducting split-coil magnet with iron return yoke, and a central field of 2.6 T

Capital cost	\$ 900,000
10-year interest cost at 8%	720,000
10-year operating cost at 2000 hours operation per year	460,000
Total for ten years	\$ 2,080,000

Compared to room temperature and cryogenic magnets, the superconducting coil is clearly more economical.

It may be pointed out that there is a possibility to further reduce operating costs, cut down helium losses through leaks, etc., but these improvements are not the subject of discussion in this section.

6.2.6 Long Solenoids

With reference to the original Fabry relation [2], we wrote the field equation in the center of an axially symmetric solenoid:

$$B\,(0,0) = \mu_0(\lambda J)\, a_1\beta \cdot \ln\left[\frac{\alpha + (\alpha^2+\beta^2)^{\frac{1}{2}}}{1+(1+\beta^2)^{\frac{1}{2}}}\right]. \qquad (6.2.1)$$

If $a_1\beta \gg a_1$ and $a_1\beta \gg a_2$, i.e. if the coil length compared to the radial dimensions of the coil is large, we may rewrite Eq. (6.2.1) in the form:

$$\boxed{B\,(0,0) = \mu_0(\lambda J)\cdot a_1 \cdot (\alpha - 1)}\,. \qquad (6.2.34)$$

The coil volume was given by:

$$V = 2\,\pi a_1^3 \cdot (\alpha^2 - 1)\,\beta\,. \qquad (6.2.6)$$

With: $$\alpha - 1 = \frac{B\,(0,0)}{\mu_0(\lambda J)\,a_1}; \quad \alpha + 1 = 2 + \frac{B\,(0,0)}{\mu_0(\lambda J)\,a_1},$$

we obtain:

$$\boxed{V = 2\,\pi a_1 \frac{B\,(0,0)}{\mu_0(\lambda J)}\left[2\,a_1 + \frac{\beta}{\mu_0(\lambda J)}\cdot\beta\right]}\,. \qquad (6.2.35)$$

The coil surface area is of interest for cooling, however for specific magnet types, the method of cooling and the chosen internal coil structure, must be considered. Few possible coil designs were shown in Figs. 6.2.2a, b. With the new types of conductors it is possible to wind coils (conventional, cryogenic and superconducting) with inner cooling Fig. 6.2.2a, edge cooling Fig. 6.2.2b or through thermal convection by potting the coils in suitable thermosets and having double pancakes cooled by immersing them in liquid He or forcing liquid helium through channels.

If the solenoid is not internally cooled (no internal coolant channels and no provisions for spacers, ducts etc.), the coil surface is given by:

$$\boxed{S = 2\,\pi\left[2\,a_1 + \frac{B\,(0,0)}{\mu_0\lambda J}\right]\left[2\,a_1\beta + \gamma\cdot\frac{B\,(0,0)}{\mu_0\lambda J}\right]}\,. \qquad (6.2.36)$$

γ is the fraction of the cooled endsurface of the solenoid perpendicular to the coil axis. In Eq. (6.2.36) both the inner and outer coil surfaces are cooled. If these surfaces are again partially cooled due to reinforcements, etc. only the fraction of the cooled surface must be taken into account.

Instead of calculating specific solenoids, we give equations for the power consumption of conventional and low temperature solenoids. The coolant requirement can be obtained from Section 6.2.5.

The power required for a long conventional or cryogenic solenoid can be obtained from:

$$\boxed{P = \lambda J^2 \cdot \varrho \cdot a_1^3 \cdot 2\,\pi\beta\,(\alpha^2 - 1)}\,, \qquad (6.2.4)$$

or in terms of the central field:

$$P = 2\pi \cdot \varrho \cdot \frac{(B\,(0,0))^2}{\lambda \mu_0^2} \cdot a_1 \beta \cdot \frac{\alpha + 1}{\alpha - 1} \tag{6.2.37}$$

$$= 2\pi \cdot \varrho \cdot \left(\frac{B\,(0,0)}{\mu_0}\right)^2 \frac{a_1 \beta}{\lambda} \cdot \left(1 + \frac{2\,\mu_0 \lambda J\, a_1}{B\,(0,0)}\right). \tag{6.2.38}$$

We see from Eq. (6.2.37) that the power required for a long solenoid of a given length decreases with α, but the fabrication cost increases with α. The minimum general cost occurs at values of $\alpha \gg 1$. For solenoids having a bore diameter of $2\,a > 1$ m, the value of α chosen will seldom exceed two.

6.2.7 Magnets for Energy Storage

Electrical energy may be stored in a number of ways: As electric charge in a capacitor, as chemical energy in accumulators and explosives, as nuclear energy in a reactor as kinetic energy in mechanical systems and as potential energy in compressed gases.

Energy is extracted from natural resources such as coal, oil, natural gases, hydraulic and powerplants, fusion of atomic nuclei etc.

Since it is not feasible to generalize the most perfect method of energy storage, regions in which particular forms of energy storage are suitable must be defined.

Methods of field implosion (see Chapter 1.3) by means of detonators are suitable for times up to 1 µs. Energy densities reached are about 10^5 J/cm³. Rotating machines and batteries are used for slow energy extraction in the range of minutes to hours and may reach energy-densities up to ~200 J/cm³.

Energy can be stored in the magnetic field of a coil. A reasonable field generated by a superconducting coil (~15 T) gives an energy density of ~ 90 J/cm³. Energies in the order of MJ to GJ can be discharged suitably in ms to several seconds depending on the choice of the load, the switching mechanism and the superconductor used in the storage coil. A combination of inductive storage coils and rectifier inverters is suitable for energy pumping in electrical networks.

Among a variety of different coil geometries we have selected three types which are discussed briefly. Details in the choice of switches, method of energy compensation, magnetomechanical forces and coil reinforcement are given in literature [23].

(*a*) *Solenoids*

Equations relating the stored energy to the coil volume, the current density and a method optimizing the coil volume is given in Table 6.2.7. A special case in which energy stored is a maximum at a given amount of superconductor is the Brookscoil with $\alpha = 2$ and $\beta = 0.5$. Factors to calculate the coil inductance as a function of α an β are presented in Fig. 6.2.25.

Table 6.2.7. *Energy stored in solenoids*

Solenoid

$$L = \Lambda\, a_1 N^2$$
$$E = 2\,\Lambda\, a_1^5\,(\lambda J)^2 (\alpha - 1)^2\, \beta^2$$
$$V_c = 2\,\pi\, a_1^3\,(\alpha^2 - 1)\,\beta$$
$$NI = 2\,a^2 (\alpha - 1)\,\beta\,(\lambda J)$$
$$E = \frac{\Lambda}{2^{2/3}\pi^{5/3}}(\lambda J)^2 \left[\frac{\beta}{(\alpha - 1)^2(\alpha + 1)^5}\right]^{\frac{1}{3}} V_c^{5/3}$$
$$B\,(0,0,0) = G\,\frac{NI}{a_1}\left[\frac{\pi\,(\alpha + 1)}{2\,\beta\,(\alpha - 1)}\right]^{\frac{1}{2}}$$

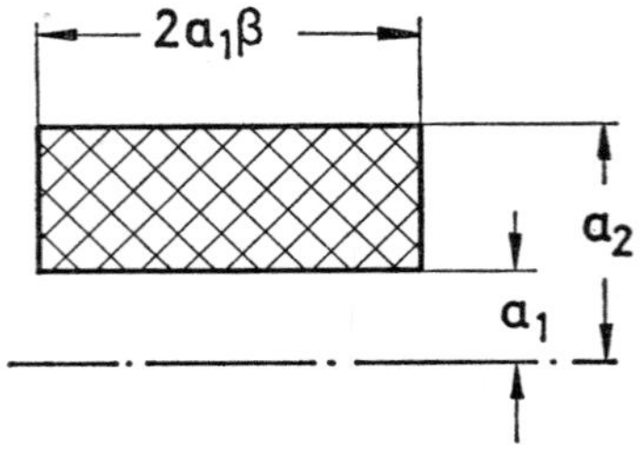

Brooks Coil

$$\alpha = 2$$
$$L = 25.485\times 10^{-7}\, a_1 N^2 \qquad \text{(H)}$$
$$E = 3.03\times 10^{-8}\, V^{5/3}\,(\lambda J)^2 \qquad \text{(J)}$$

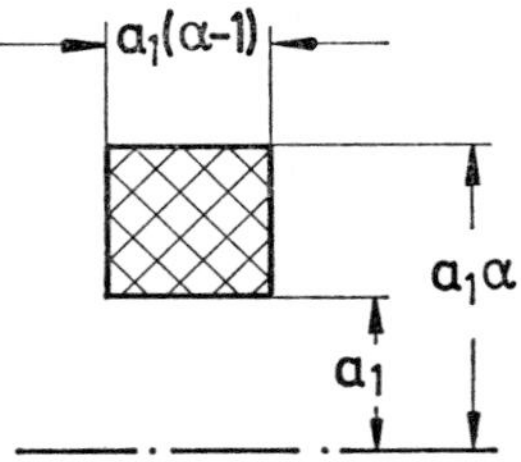

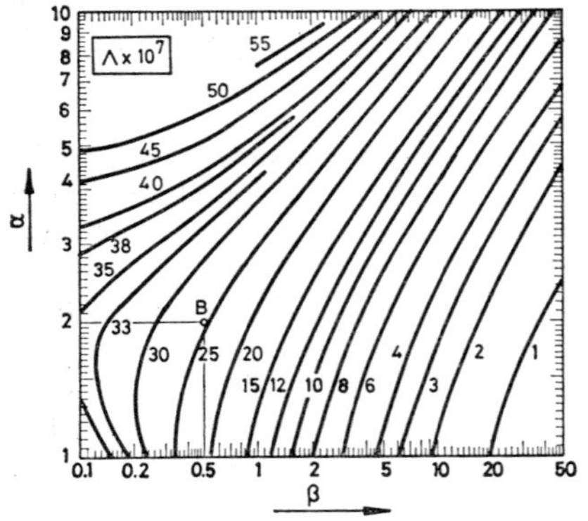

Fig. 6.2.25. Lines of constant $\Lambda = L/n\,a_1^2$, L (Hy) and $a_1\,(m)$, B = Brooks coil

(b) Spheres

Fig. 6.2.26 illustrates a possible spherical coil geometry. The current distribution in the coil follows a sin (θ) distribution. Table 6.2.8 presents pertinent equations relating the stored energy to the effective amount of superconductor. Geometry factors $f\,(\alpha)$ for optimization of the spherical coil can be calculated simply from the Table 6.2.8. Screening the field of a coil from the environment in order to prevent the fringing field to influence experimental set up is imperative for solenoids and spheres. The field of a sphere is screened effectively by a second sphere with the currents in the screen flowing in the opposite direction to the main coil. The superconductor is utilized optimally if the ratio of the radii of both the screening and the main coil is 1.59, and the ratio of current densities is 0.25.

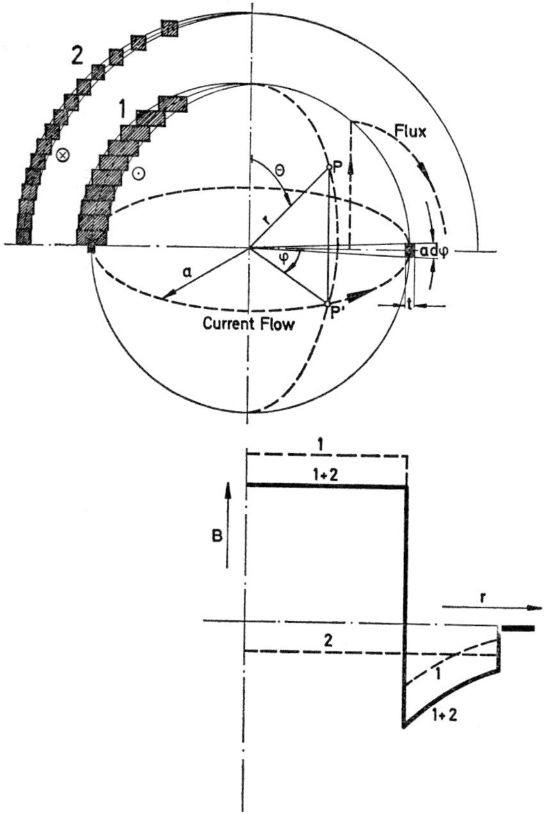

Fig. 6.2.26. Coil arrangement for spherical storage magnets.
1 Main coil, *2* Field screening coil.
Field distribution is shown in the bottom

Thin wall sphere

$$NI = 2a_1^2(\alpha-1)(\lambda J_0)$$

$r < a_1$

$$B_r(r,\Theta) = \frac{2}{3}\mu_0 a_1(\alpha-1)(\lambda J_0)\cos\Theta$$

$$B_\Theta(r,\Theta) = -\frac{2}{3}\mu_0 a_1(\alpha-1)(\lambda J_0)\sin\Theta$$

$r > a_1$

$$B_r(r,\Theta) = \frac{2}{3}\mu_0\left(\frac{a_1}{r}\right)^3 a_1(\alpha-1)(\lambda J_0)\cos\Theta$$

$$B_\Theta(r,\Theta) = \frac{1}{3}\mu_0\left(\frac{a_1}{r}\right)^3 a_1(\alpha-1)(\lambda J_0)\sin\Theta$$

$$E_{tot} = \frac{4\pi}{9}\mu_0 a_1^5(\alpha-1)^2(\lambda J_0)^2$$

$$E_{tot} = \frac{1}{3\left(\frac{4\pi}{3}\right)^{2/3}}\mu_0\cdot\frac{(\alpha-1)^2}{(\alpha^3-1)^{5/3}}(\lambda J_0)^2 V^{5/3}$$

$$Q_{SC} = \pi^2 a_1^3(\alpha-1)(\lambda J_0)$$

$$E_{tot} = \frac{4}{9\pi^3}\mu_0\frac{Q_{SC}^2}{a_1}$$

Thick wall sphere

$$NI = a_1^2(\alpha^2-1)(\lambda J_0)$$

$r < a_1$

$$B_r(r,\Theta) = \frac{2}{3}\mu_0 a_1(\alpha-1)(\lambda J_0)\cos\Theta$$

$$B_\Theta(r,\Theta) = -\frac{2}{3}\mu_0 a_1(\alpha-1)(\lambda J_0)\sin\Theta$$

$r > a_2$

$$B_r(r,\Theta) = \frac{\mu_0}{6}a_1(\alpha^4-1)(\lambda J_0)\left(\frac{a_1}{r}\right)^3\cos\Theta$$

$$B_\Theta(r,\Theta) = \frac{\mu_0}{12}a_1(\alpha^4-1)(\lambda J_0)\left(\frac{a_1}{r}\right)^3\sin\Theta$$

$a_1 \leq r \leq a_2$

$$B_r(r,\Theta) = \frac{2}{3}\mu_0\left[a_1\left[\alpha-2\left(\frac{a_1}{2r}\right)^3\right]-\frac{3}{4}r\right](\lambda J_0)\cos\Theta$$

$$B_\Theta(r,\Theta) = \frac{2}{3}\mu_0\left[\frac{9}{8}r-a_1\left(\frac{a_1}{2r}\right)^3-a_1\alpha\right](\lambda J_0)\sin\Theta$$

$$E_{tot} = \frac{2\pi}{45}\mu_0 a_1^5(\lambda J_0)^2 f(\alpha)$$

$$E_{tot} = \frac{\mu_0}{30\left(\frac{4\pi}{3}\right)^{2/3}}(\lambda J_0)^2 V^{5/3} g(\alpha)$$

$$Q_{SC} = \frac{\pi^2}{3}(\lambda J_0)a_1^3(\alpha-1)(\alpha^2+\alpha+1)$$

$$E_{tot} = \frac{2\mu_0}{5\pi^3}\cdot\frac{Q_{SC}^2}{a_1}h(\alpha)$$

$$f(\alpha) = (\alpha-1)^2(\alpha^3+2\alpha^2+3\alpha+4)$$

$$g(\alpha) = \frac{(\alpha-1)^2(\alpha^3+2\alpha^2+3\alpha+4)}{(\alpha^3-1)^{5/3}}$$

$$h(\alpha) = \frac{\alpha^3+2\alpha^2+3\alpha+4}{(\alpha^2+\alpha+1)^2}$$

Table 6.2.8. *Energy stored in spherical coils*

(c) Toroidal Coils

Table 6.2.9 gives pertinent data relating the coil volume to the energy being stored. The toroidal coil does not require a screen, to shield the fringing flux. Fig. 6.2.27 illustrates geometry factors $f(\alpha, \beta)$ for optimization of toroidal coils.

To give an example, we have calculated three types of magnets for a stored energy of 10^{10} J. The Brooks coil seems to be most economical. The choice of α and β depends not only on E, and the amount of superconductor alone, but also on the maximum tolerable stress-limit of the conductors being used. Since the stresses, E and Q_{sc} are functions of α, and β, by selecting tolerable values for the hoop stress, one may calculate the optimum Q_{sc} for a required energy. Table 6.2.10 gives a comparison between the three types of inductive storage coils.

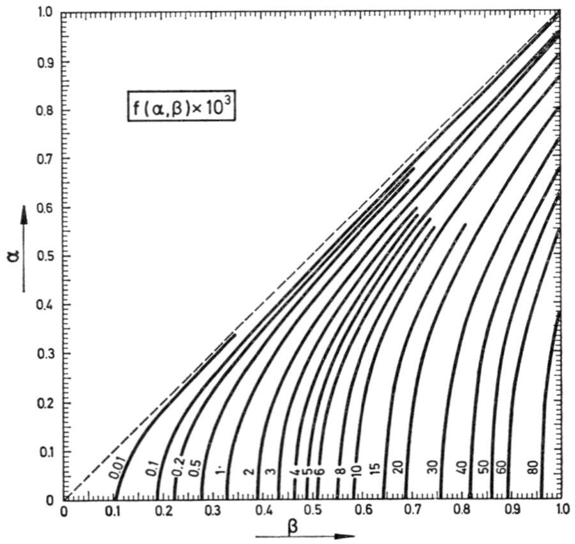

Fig. 6.2.27. Geometry factors for toroidal coils

The effective weight of storage coils and their costs depend on the overall current density being selected, the price of the superconductor and method of coil reinforcement. Since the current-density in large storage coils is quite small, cryastatic stabilization will be not difficult to achieve and high strength materials such as Be-Cu or non-magnetic stainless-steel can be selected as a matrix. To illustrate the effect of current density on the effective cost per energy, Fig. 6.2.28 illustrates the specific weight and price of Brooks coils as a function of stored energy.

Table 6.2.9. *Energy stored in toroidal coils*

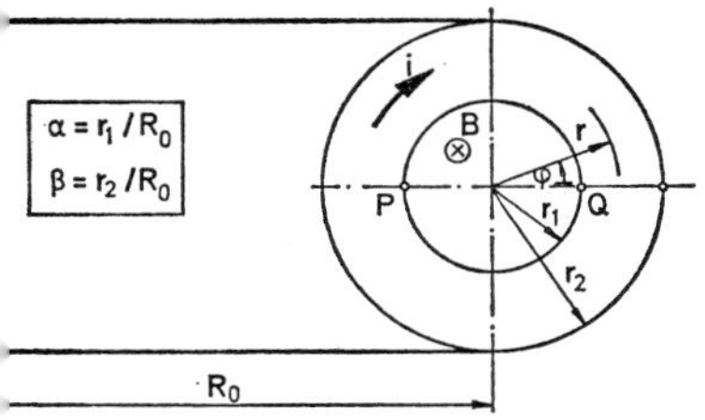

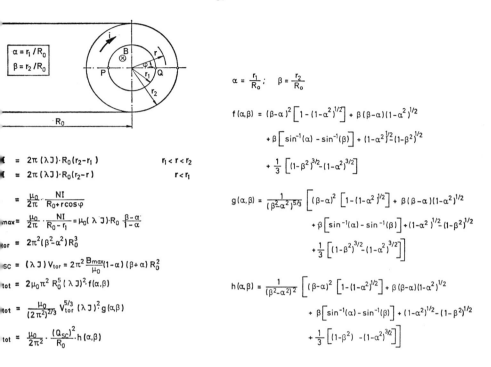

$$\alpha = \frac{r_1}{R_0}; \quad \beta = \frac{r_2}{R_0}$$

$= 2\pi(\lambda J)\cdot R_0(r_2 - r_1) \qquad r_1 < r < r_2$

$= 2\pi(\lambda J)\cdot R_0(r_2 - r) \qquad r < r_1$

$= \frac{\mu_0}{2\pi}\cdot\frac{NI}{R_0 + r\cos\varphi}$

$_{max} = \frac{\mu_0}{2\pi}\cdot\frac{NI}{R_0 - r_1} = \mu_0(\lambda J)\cdot R_0\,\frac{\beta-\alpha}{1-\alpha}$

$_{tor} = 2\pi^2(\beta^2-\alpha^2)R_0^3$

$_{SC} = (\lambda J)\,V_{tor} = 2\pi^2\frac{B_{max}}{\mu_0}(1-\alpha)(\beta+\alpha)R_0^2$

$_{tot} = 2\mu_0\pi^2 R_0^5(\lambda J)^2 \cdot f(\alpha,\beta)$

$_{tot} = \frac{\mu_0}{(2\pi^2)^{2/3}}\,V_{tor}^{5/3}(\lambda J)^2 \cdot g(\alpha,\beta)$

$_{tot} = \frac{\mu_0}{2\pi^2}\cdot\frac{(Q_{SC})^2}{R_0}\cdot h(\alpha,\beta)$

$$f(\alpha,\beta) = (\beta-\alpha)^2\left[1-(1-\alpha^2)^{1/2}\right] + \beta(\beta-\alpha)(1-\alpha^2)^{1/2} + \beta\left[\sin^{-1}(\alpha) - \sin^{-1}(\beta)\right] + (1-\alpha^2)^{1/2} - (1-\beta^2)^{1/2} + \frac{1}{3}\left[(1-\beta^2)^{3/2} - (1-\alpha^2)^{3/2}\right]$$

$$g(\alpha,\beta) = \frac{1}{(\beta^2-\alpha^2)^{5/3}}\left[(\beta-\alpha)^2\left[1-(1-\alpha^2)^{1/2}\right] + \beta(\beta-\alpha)(1-\alpha^2)^{1/2} + \beta\left[\sin^{-1}(\alpha) - \sin^{-1}(\beta)\right] + (1-\alpha^2)^{1/2} - (1-\beta^2)^{1/2} + \frac{1}{3}\left[(1-\beta^2)^{3/2} - (1-\alpha^2)^{3/2}\right]\right]$$

$$h(\alpha,\beta) = \frac{1}{(\beta^2-\alpha^2)^2}\left[(\beta-\alpha)^2\left[1-(1-\alpha^2)^{1/2}\right] + \beta(\beta-\alpha)(1-\alpha^2)^{1/2} + \beta\left[\sin^{-1}(\alpha) - \sin^{-1}(\beta)\right] + (1-\alpha^2)^{1/2} - (1-\beta^2)^{1/2} + \frac{1}{3}\left[(1-\beta^2) - (1-\alpha^2)^{3/2}\right]\right]$$

Table 6.2.10. *Comparison between inductive energy storage system*

Stored energy 10^{10} J		Central field 10 T		Maximum current 10^4 A	
Brooks coil		Spherical coil		Toroidal coil	
$_1$	= 2.44 m	a_1	= 2.36 m	R_0	= 6.4 m
J_0	= 9.6×10^2 A cm^{-2}	α	= 1.6	r_1	= 0.96 m
Q_{sc1}[a]	= 1.3×10^9 A · m	$\lambda J_{0,1}$	= 1.1×10^3 A cm^{-2}	r_2	= 1.8 m
NI	= 56.9×10^6 A	$\lambda J_{0,2}$	= 0.28×10^3 A cm^{-2}	λJ_0	= 1.1×10^3 A cm^{-2}
		Q_{sc1}	= 1.5×10^9 A · m	Q_{sc}	= 3.1×10^9 A · m
$_1$	= 13×10^4 m	l_1	= 15×10^4 m	l	= 3.0×10^4 m
$_\theta$	= 3.2×10^3 kp cm^{-2}	σ_θ	= 3.3×10^3 kp cm^{-2}	$\sigma_{\theta,max}$	= 3.4×10^3 kp cm^{-2}
Q_{sc2}[b]	= 1.2×10^9 A · m	Q_{sc2}	= 1.5×10^9 A · m	$\sigma_{r,max}$	= 1.4×10^3 kp cm^{-2}

(1) refers to the main coil

(2) refers to the screen

6.2.8 Beam Transport and Accelerator Magnets

In this category we study three types of magnets:

Dipolemagnets, used for bending charged particles or momentum analysis, quadrupole magnets for beam focusing and solenoids, which form an image of the point source of particles. Solenoids have been studied above in Sections 6.2.5 and 6.2.6. Below we compare only focusing properties of solenoids with magnetic lenses, specifically quadrupoles. Multipole magnets are not treated.

(*a*) *Dipole Magnets*

Considerable progress has been made in recent years in all aspects of the design of superconducting beam transport magnets. The design of beam transport magnets has changed from the original C, H, and window frame type magnets usually chosen in conventional iron core or iron bound magnets, to new types with iron around the coils, where the circumferential current distribution or the coil shape is designed such to generate a perfect dipole field within a large fraction of the magnet aperture. In Chapter 2 detailed dipole designs with the iron yoke in the proximity to coils, or around the dewar were given. For the economic consideration we select only two dipole designs (the intersecting circles and the $\frac{2}{3}$ rule coils) and calculate the pertinent magnet parameters. Comparison of other types of magnets will be similar and is not repeated.

Superconducting Type II composite conductors are developed for low duty cycle pulsed magnets (for synchrotron type accelerators), which have low losses, and for d.c. applications. Presently, magnetic fields in the order of 4 to 6 T are chosen as peak values for accelerator magnets as well as for beam transport magnets. The choice of the maximum field value is based on economical consideration specifically in conjunction with high energy accelerators.

Intrinsically stable multicore conductors, cables or braids are utilized for a variety of magnets. The advances in conductor technology has been such that composite cables can be used economically for d.c. applications. $Nb_{(x)}Ti$ filaments with diameters of 7 μm are already commercially available. The fine superconducting filaments are embedded in suitable matrix materials such as copper, copper alloys (cupro-nickel has been used in few cases and pure aluminum). The individual conductor is drawn to its final size and the filaments in the conductor twisted to a degree that they are magnetically decoupled. A mixture of copper and cupronickel has also been employed with various degrees of success.

The strands are either insulated with suitable thermosettings, or impregnated with silver-tin (7%/93% ratio) and indium-tin solder, then annealed during ~ 20 hours at temperatures of about 350 °C in order to form an oxyde (bronze) surface. The conductors are transposed into a cable, pressed and formed into a specified shape (square or rectangular) and to required tolerances.

Often these multiconductor-cables are impregnated with a solder, such as the above mentioned silver-tin alloy and poisoned to increase its low temperature resistivity by adding about 10 to 20% b.w. bismuth to the alloy in order to reduce cross-coupling between strands. The cable is wrapped in a tape (fiber-glass wound openly) or fully insulated. In other cases the cable is wound into a coil and then impregnated with suitable thermosettings, such as epoxies filled with inorganic fillers to match the thermal contraction coefficient of the conductor material or other materials such as stearin etc. The filamentary superconductors have very high critical current densities. Some of the recently developed conductors reach current densities of 2×10^5 A/cm² at 5 T and at 4.2 K. With ratios of normal to superconducting materials in the order of 1 : 1 to 2 : 1 the current density in a single conductor will be about 6×10^4 to 10×10^4 A per cm². The cable has a packing factor of about 50%. Thus one may expect overall current densities in a dipole coil (including spacers for cooling, reinforcements, additional cable insulation) in the order of 2.5×10^4 to 3.0×10^4 A/cm² at 5 T.

The following properties are desired in a twisted, transposed multifilamentary conductor or cable:

(i) Small diameter filaments to reduce hysteretic losses (see Chapter 4).

(ii) High current density to achieve high magnet efficiency.

(iii) High critical temperature to minimize temperature problems or allow magnet operation at temperatures above 4.2 K (see Table 6.2.11).

(iv) Adequate matrix resistivity to permit a reasonable twist rate (for pulsed magnets about 4 twists per cm). For a copper matrix this corresponds to a resistivity ratio of $r = 40 \ldots 100$.

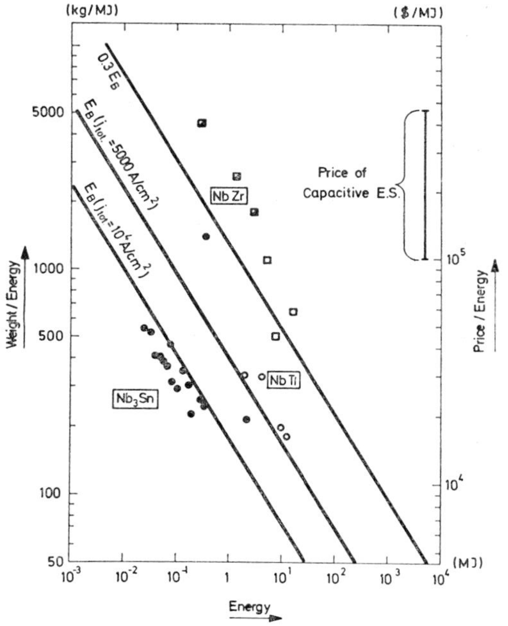

Fig. 6.2.28. Price and weight of superconducting Brooks coils v. energy

Very fine filaments have the disadvantage of breakage during the manufacturing, twisting, cabling and coil winding process. As each conductor contains hundreds of filaments (> 500) the breakage of filaments are generally undetected until the coil is wound and tested. An addition of Tantalum provides ductility to $Nb_{(x)}Ti$. Also better cabling methods may prevent breakage of conductors within the cable. The problem however is serious enough to be of concern.

Table 6.2.11. *Some properties of technical high temperature Type II superconductors*

Material	Critical temp. (K)	Critical field (T) at		Critical current density (A/cm²) at 5 T	
		4.2 K	8 K	4.2 K	8 K
$Nb_{(x)}Ti$	10.6	11.8	5	$\sim 3 \times 10^5$	$< 10^4$
Nb_3Sn	18.05	20	16.5	10^6 [a]	6×10^5
V_3Ga	14.5	21	17	$\sim 10^5$	$\sim 4 \times 10^4$
Nb_3Bl	18.67	30.8	27.5	—	—
$Nb_3[Al_{0.8}Ge_{0.4}]$	20.7	40.5	35	$> 10^5$	2×10^4
$Nb_2Hf\,Zr$	10.5	22	?	5×10^5	—
NbN	16	> 21	~18	2.5×10^6 [b]	—

[a] The thickness of the Nb_3Sn layer is assumed to be $\sim 1\ \mu m$. The layer thickness of Nb_3Sn/Nb is about 2.5×10^{-3} cm. The sample width is 0.5 cm.
[b] The sample, tested by Gavaler [24] has a cross-sectional area of $1.2 \times 10^{-9}\ cm^2$.

Higher critical temperatures is achieved in V_3Ga and Nb_3Sn than for $Nb_{(x)}Ti$. Already V_3Ga multifilament conductors are commercially available, which allow the coil to operate at 8 K. Thus the refrigeration cost can be reduced by at least 50%! Some work is being done to produce composite Nb_3Sn wires. But due to the complexity of the problems and the brittleness of the Nb_3Sn progress is still slow. Only short length of conductor (< 300 m) with 300 filaments have been drawn and tested. A new set of specifications pertinent to composite conductors are now in effect for most magnets:

Table 6.2.12. *Composite wire specifications*

Individual wire diameter (conductor)	$\leqq 0.04$ cm
Number of filaments in a wire	> 300
Filament size	$\leqq 7\ \mu m$
Current density in wire	$> 5 \times 10^4\ A/cm^2$ at 6 T
Superconductor content in wire	40—50%
Matrix resistivity ratio r	20—100
Ratio J_c, 3 T/J_c, 6 T	$\leqq 2$
Effective composite resistivity at J_c	$\leqq 10^{-12}\ Ohm \cdot cm$

The main difficulty encountered in producing the material specified in Table 6.2.12 was a premature onset of resistivity caused by narrow spots in filaments [25]. This behaviour is illustrated in Fig. 6.2.29, where the effective resistivity of the composite is plotted against the current density. The onset of resistivity begins at a lower current than I_c for smaller filaments made from the same billet. However this effect does not depend on the absolute size of the filaments, but on the degree of reduction of the billet. A small initial billet will produce more uniform filaments because less overall reduction is required.

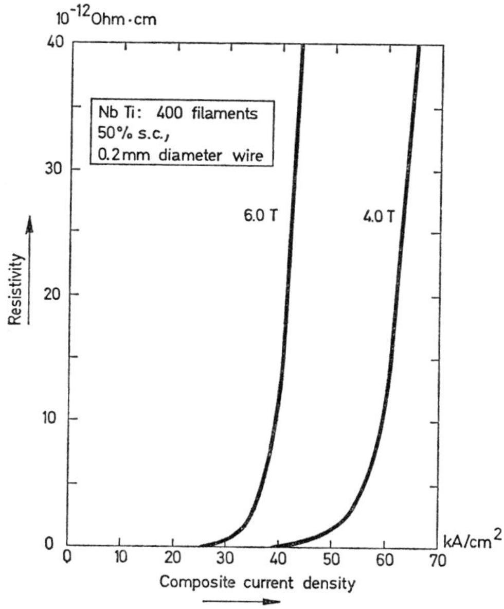

Fig. 6.2.29. Effective composite resistivity as a function of current density

The composite effective resistivity is of considerable importance in pulsed magnets, since it leads to power dissipation during the flat top operation if the magnet is energized near to the critical current. As a practical value, the effective resistivity of the composite conductor at critical current was chosen not to exceed 10^{-12} Ohm · cm. Since only a small fraction of the conductor being wound into the coil is exposed to high fields and operates near its critical current value, the overall effective coil resistance will be small. Since the effective wire resistance is small, it can be shown that Joules losses are smaller than hysteretic losses (Chapter 4.1).

Two possible magnet designs are illustrated in Figs. 6.2.30 and 6.2.31. In Fig. 6.2.30, the iron is the proximity to the coil and is cooled by liquid helium. The iron core is composed either of laminated insulated silicon-steel sheets (each sheet about 0.13 to 2.0 mm thick) or of insulated iron wire wound directly over the coil. The flux return passage is separated from the coil by means of insulated spacers.

The iron flux return path is impregnated with high strength epoxies to obtain rigidity. Although for the pulse magnets considering a flux density change of $dB/dt = 2$ T/sec, fairly thick individual laminations may be utilized (> 2 mm), damage of the lamination edge (lattice distortion) due to stamping and shearing may set an upper limit to the lamination thickness.

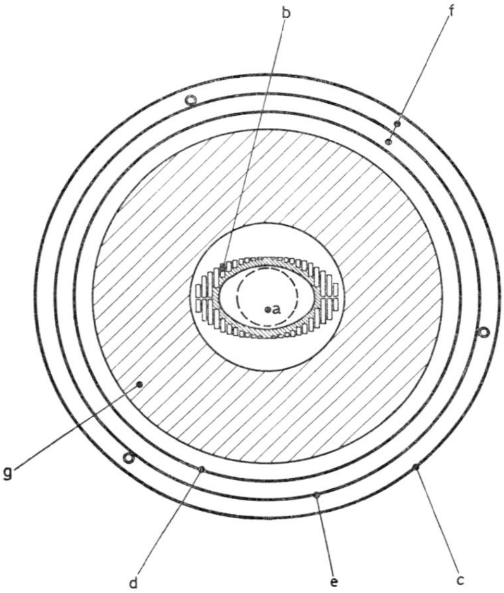

Fig. 6.2.30

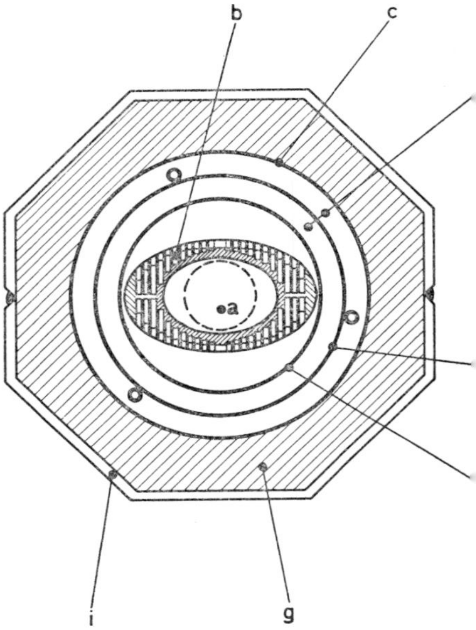

Fig. 6.2.31

Fig. 6.2.30. Coil and iron core configuration. The iron shield is located around the coil. *a* Magnet Aperture, *b* Coil, *c* Cryostat, *d* Helium Container, *e* Nitrogen Shield, *f* Vacuum and Superinsulation, *g* Iron Shield

Fig. 6.2.31. Coil iron and cryostat arrangement with iron core placed around the cryostat and forms the room temperature part of the cryostat. *a* Magnet Aperture, *b* Coil, *c* Room temperature Vessel, *d* Helium Container, *e* Nitrogen Shield, *f* Vacuum and Superinsulation, *g* Iron Shield

In Fig. 6.2.31 the iron shield is placed at some radial distance from the coil, and is at room temperature. The cryostat is located between the iron shell and the coil. In this case, the iron core is not saturated, while in the case of Fig. 6.2.30 portions of the iron are saturated if the aperture field exceeds 4 T.

The placement of the iron core adjacent to the coil (iron being cooled to liquid He temperature) has the following advantages:

(i) The iron confines the field and supports the magnetomechanical forces and the coil weight, and prevents the coils from sagging.

(ii) The total magnetic field energy is reduced. This feature allows magnets to be placed closer to each other along the beam. The iron yoke is generally extended over the coil in the axial direction. Reduction of energy also reduces the peak voltages to ground and reduces the cost of the energy storage.

(iii) The iron contribution to the coil field is considerable, shown in Chapter 2 and Fig. 2.2.34. Thus, the field enhancement due to the iron saves valuable conductor material.

(iv) The magnetomechanical forces between the iron and coil are not transmitted over the dewar due to possible asymmetries resulting from assembly, which may force the designer to choose thick-walled cryostats or complicated reinforcing structures. It may be pointed out that a 0.1 mm axial shift between a coil and iron yoke may produce a force of approximation 10^3 kp per unit length of magnet at a central field of 5 T. If the coil is supported by means of tension or compension rods, heat leak is increased.

(v) The iron yoke shields the magnetic field of the coil from the surrounding areas and shield external fields of being effective in changing field distribution in the aperture.

The apparent disadvantages of the iron shield placed adjacent to the coil are:

(a) The required amount of the coolant to cool-down the iron to the operating temperature of the superconductor may be substantial in large magnets. However, in laminated or coiled iron shields, enthalpy-cooling is possible. The iron yoke can be provided with a large number of coolant passages in axial and radial directions.

(b) The generated heat by eddy currents and hysteretic losses in the iron must be removed by the coolant. It can be shown that these losses are small at low frequencies (pulse duration $>$ 10 sec), and of the same order as the eddy current losses in the composite conductor.

(c) The iron shield affects the field distribution over the dipole aperture and along the magnet axis. Specifically, sextupole fields are disturbing. As we saw in Chapter 2.2, it is possible to eliminate the sextupole component by changing the radial thickness of the iron shield for *one* particular flux or to shape the coil accordingly. For pulsed magnets when the field is raised, sextupole and dodecapole components occur with variable amplitudes.

The sextupole component can be reduced and eliminated at the highest aperture field. At lower fields, the iron should not be saturated, if possible. The radial gap may still be fairly large, but the iron should be placed in the coolant merely to support the coil weight and the magnetomechanical forces. In the axial direction, the field distortion of the magnet ends due to different values of relative permeability in iron is not negligible. Yoke end corrections may help somewhat, but in highly saturated iron shields, iron shaping does not have the desired effect as expected in the unsaturated iron.

Lines of constant permeability in the xy-plane of a dipole iron configuration is shown in Fig. 6.2.32. The maximum operating field in the magnet aperture is 4.2 T. The field distribution will change when the

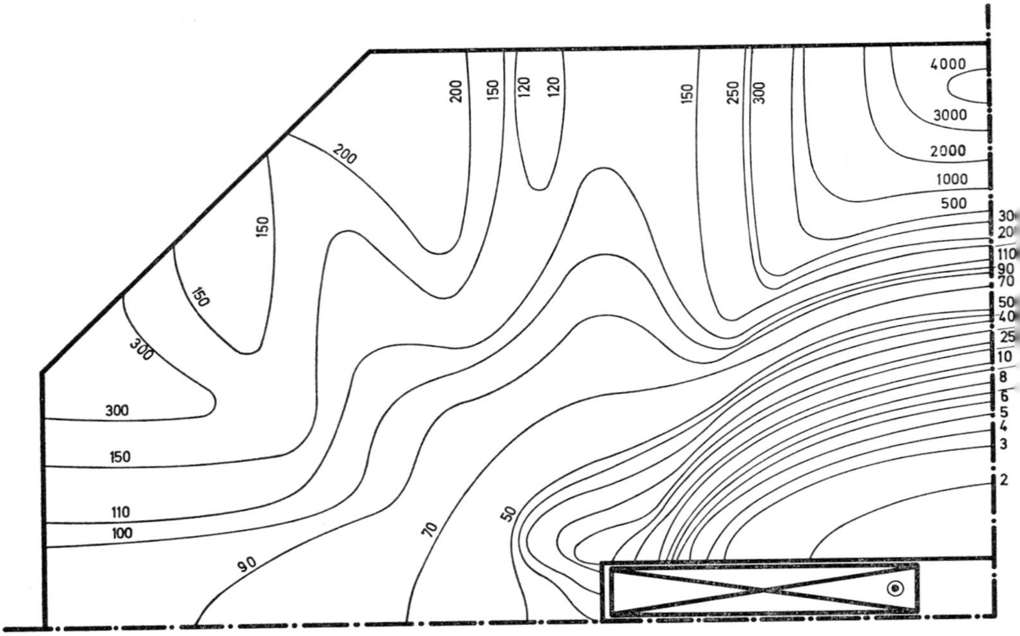

Fig. 6.2.32. Lines of constant permeability in the iron yoke of a picture frame magnet. The magnet central field is 4.2 T

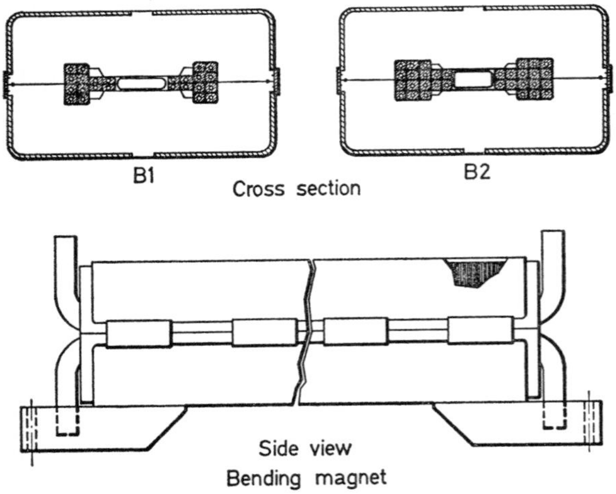

Fig. 6.2.33

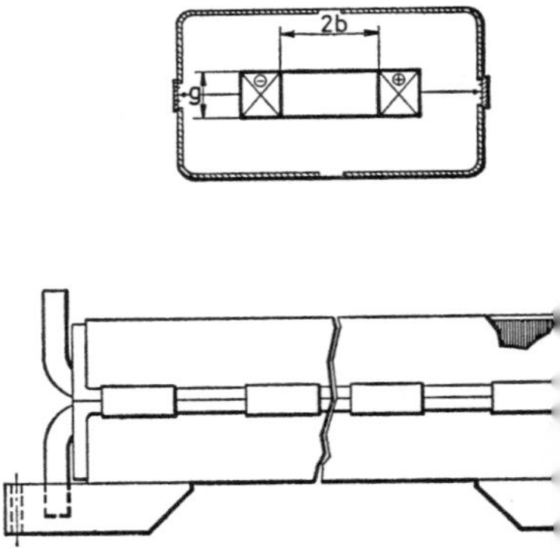

Fig. 6.2.34

Fig. 6.2.33. Dipole magnet configurations of the NAL main accelerator. The magnet can be energized up to 2.4 T corresponding to a maximum proton energy of 500 GeV in the main ring

Fig. 6.2.34. Picture frame magnet configuration

field is raised from low values to the maximum flat top value in synchrotron type magnets, and may affect field distribution during the beam injection period. The field disturbance during the change in field is of considerable importance; it is being studied for static and dynamic cases by a number of magnet designers.

In d.c. type magnets, field errors can be corrected by means of auxiliary coils.

Low temperature properties of pure iron is measured by Bozorth [27] and Brechna [28]. Magnetization of silicon-iron is measured by McInturff [29]. While the saturation value in pure iron increases at cryogenic temperatures compared to 300 K values, it is practically unchanged for silicon-steels. The permeability increases at low temperatures considerably at low and medium field regions, as was shown in Chapter 2.2.2. The area under the hysteresis curve increases at 4.2 K by a factor of about 2 compared to 300 K values.

The choice of the radial iron thickness and the selection of the radial gap between the coil and the iron shield cannot be generalized. Most appropriate would be to compute the field, over the iron using one of the computercodes mentioned in Chapter 2 and determine the field distribution over the iron.

Dipole Configurations

A common dipole configuration consists of two or four current blocks shown in Fig. 6.2.33. Iron return yokes and iron mirrors retain the flux.

For a coil configuration according to Fig. 6.2.34, with two current blocks we gave in Eq. (2.1.107) the field equations for constant current density. If the gap height corresponds to the height of the coil, g; the gap width is $2\,b$ and the coil width is $2\,h$, then the field in the center of the gap is given by:

$$B_y = \frac{\mu_0(\lambda J)}{\pi} \cdot g \ln \left[\frac{1 + \left(\frac{2\,h + 4\,b}{g}\right)^2}{1 + \left(\frac{2\,h}{g}\right)^2} \right]. \tag{6.2.39}$$

The cross sectional area of the coil is

$$A = g \cdot 2\,b$$

and its volume: $V = 2\,bg \cdot l_m = 4\,bg \cdot (l + l_e)\,.$

For a dipole with a cos (θ) current density distribution (Fig. 6.2.35), we get the two field components for $r < a$:

$$\begin{aligned} B_r &= -\frac{\sqrt{3}}{\pi} \mu_0 \lambda J\, a_1 \cdot (\alpha - 1) \sin\theta \\ B_\theta &= -\frac{\sqrt{3}}{\pi} \mu_0 \lambda J\, a_1 \cdot (\alpha - 1) \cos\theta\,. \end{aligned} \tag{6.2.40}$$

For $r = x$ and $\theta = 0$ which is on the median plane:

$$B_x = 0; \quad B_y = -\frac{\sqrt{3}}{2} \mu_0 \lambda J\, a_1 (\alpha - 1)\,. \tag{6.2.41}$$

We gave the ampereturns of such a coil in Section 2:

$$NI = \lambda J\, a_1^2\,(\alpha^2 - 1)\,\frac{2\pi}{3}.$$

The outer coil radius

$$a_2 = a_1 + \frac{\pi}{\sqrt{3}\,\mu_0}\cdot\frac{B_y(0,0)}{(\lambda J)}. \tag{6.2.42}$$

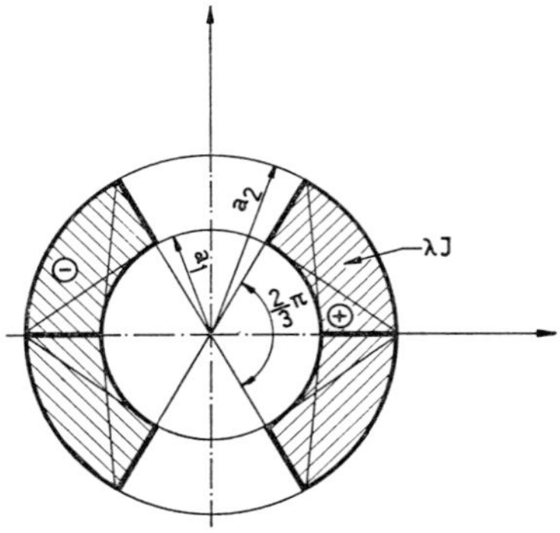

Fig. 6.2.35. Dipole coil configuration ($\frac{2}{3}$ rule design)

The coil volume is given by

$$V = \frac{4\pi^2}{3}\,a_1^2\,(\alpha^2 + 1)\cdot l_m\,, \tag{6.2.43}$$

and the external and internal cooling surfaces:

$$S = \frac{4\pi}{3}\cdot a_1\,(\alpha + 1)\cdot l_m\,, \tag{6.2.44}$$

with l_m the mean length of the coil.

A current distribution according to two overlapping circles given in Eq. (2.1.27) yields a dipole field:

$$B\,(0,0) = \frac{\mu_0}{2}\cdot s\cdot(\lambda J)\,. \tag{6.2.45}$$

With

$$2\,a = s + d\,; \quad \cos\varphi = \frac{s}{s+d},$$

we obtain the area of the current density distribution

$$A = \pi a^2\cdot\left[1 - \frac{2\varphi - \sin 2\varphi}{\pi}\right]. \tag{6.2.46}$$

Assuming the coil cross section is unchanged at the coil ends the coil volume is given by:

$$V = A\cdot l_m = 2\,A\,(l + l_e)\,. \tag{6.2.47}$$

Assuming the surfaces adjacent to the aperture are not cooled due to the presence of a coil support the coil surface area is obtained from:

$$S = 4\,\pi a\left(1 - \frac{\varphi}{\pi}\right)\cdot(l + l_e)\,. \tag{6.2.48}$$

(*b*) *Magnetic Focusing Lenses*

Magnetic quadrupole lenses provide stronger focusing than solenoids or inclined magnetic surfaces. Since their introduction in 1952, magnetic quadrupoles have become an essential part of any beam transport system.

The longitudinal magnetic field produced by a solenoid has been used for focusing charged particles, but compared to a pair of quadrupole lenses (doublet) or a set of three quadrupoles (triplets) of the same performance, the power dissipation of a solenoid is much higher. For conventional or cryogenic type solenoids, their use has been restricted to low momentum beams, to situations where azimuthal symmetry is important, or if the solid angle of a system has to be increased. With superconducting solenoids, the use of solenoids for focusing purposes could be extended to high momentum beams.

To make an economical comparison one has to study several types of lenses and in addition compare conventional, cryogenic, pulsed and superconducting magnets.

A *magnetic quadrupole lens* is characterized by two quantities: $l = l_{\mathrm{eff}}$ the axial effective length, and

$$\boxed{K = k^2 l = G \frac{l}{BR} = G \cdot l \cdot \frac{300\,Z}{[E\,(E + 2\,E_0)]^{\frac{1}{2}}}}, \qquad (6.2.49)$$

the lens strength. G is the field gradient, Z the atomic number, E the kinetic energy of the particle and E_0 its rest energy.

The expression:

$$\boxed{BR = \frac{[E\,(E + 2\,E_0)]^{\frac{1}{2}}}{3 \times 10^2}} \qquad (6.2.50)$$

is called the rigidity of the particles.

Here E and E_0 are expressed in MeV; B in (T) and R in (m).

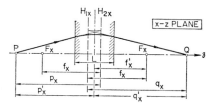

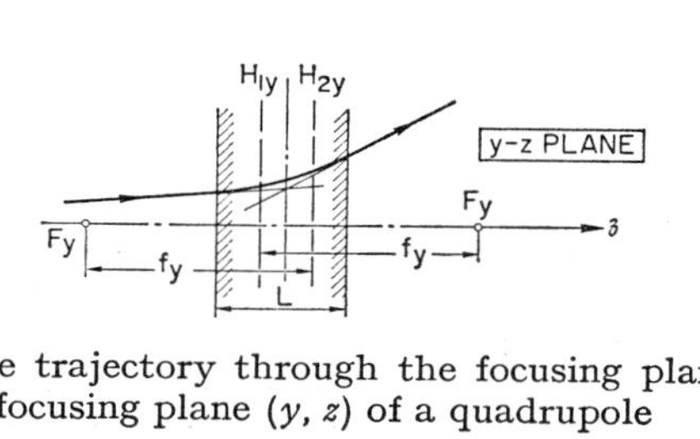

Fig. 6.2.36. Particle trajectory through the focusing plane (x, z) and defocusing plane (y, z) of a quadrupole

The direction of fields and forces and the motion of particles in a quadrupole lens have been studied by several authors [30, 31]; they are not repeated. The quadrupole has a focusing and defocusing plane, as seen in Fig. 6.2.36. For each plane one can derive one equation of motions:

$$m\frac{d^2x}{dz^2} = -k^2x \tag{6.2.51}$$

in the focusing plane and

$$m\frac{d^2y}{dz^2} = +k^2y \tag{6.2.52}$$

in the defocusing plane.

With

$$k^2 = \frac{G}{BR}$$

and (BR) the magnetic rigidity of the particles. The solution of the two equations in matrix notation is given by:

$$\begin{pmatrix} x \\ x' \end{pmatrix} = \begin{pmatrix} \cos(kl) & \frac{1}{k}\sin(kl) \\ -k\sin(kl) & \cos(kl) \end{pmatrix} \begin{pmatrix} x_0 \\ x_0' \end{pmatrix} \tag{6.2.53}$$

in the focusing plane and

$$\begin{pmatrix} y \\ y' \end{pmatrix} = \begin{pmatrix} \cosh(kl) & \frac{1}{k}\sin(kl) \\ k\sinh(kl) & \cosh(kl) \end{pmatrix} \begin{pmatrix} y_0 \\ y_0' \end{pmatrix}. \tag{6.2.54}$$

x_0, x_0', y_0, y_0' are the source coordinates, x, x', y, y' the image coordinates as illustrated in Fig. 6.2.36. For $z = l$ we also used the effective magnetic length.

From Eqs. (6.2.53) and (6.2.54) the following lens parameters are obtainable:

Focusing plane	Defocusing plane
Focusing power:	
$P_F = -k\sin(kl)$	$P_D = +k\sinh(kl)$
Focal distance (from principal plane):	
$F_F = \dfrac{1}{k\sin(kl)}$	$F_D = \dfrac{1}{k\sinh(kl)}$
Focal distance (from edge of magnet):	
$f_F = \dfrac{1}{k\,\mathrm{tg}(kl)}$	$f_D = \dfrac{1}{k\,\mathrm{tgh}(kl)}$
Image distance:	
$q_F = \dfrac{p\cos(kl) + \frac{1}{k}\sin(kl)}{pk_1\sin(kl) - \cos(kl)}$	$q_D = -\dfrac{p\cosh(kl) + \frac{1}{k}\sinh(kl)}{pk\sinh(kl) + \cosh(kl)}$
Quadrupole magnification:	
$m_F = -pk\sin(kl) + \cos(kl)$	$m_D = pk\sinh(kl) + \cosh(kl)$

The ampere turns of a quadrupole lens (per pole) with iron return yoke are given by:

$$(NI)_p = \frac{1}{\mu_0} (G) \frac{R^2}{2}, \quad \text{(A)} \tag{6.2.55}$$

the main flux (per pole):

$$\phi_n = R^2 (G) \operatorname{cotg} (2\varphi^*), \quad \text{(Vs)} \tag{6.2.56}$$

its effective length:

$$l_{\text{eff}} = l + C \cdot R, \quad \text{(m)} \tag{6.2.57}$$

with $0.9 \leqq C \leqq 1.1$. The constant, C, relates the physical length of the quadrupole to the effective length. R is the aperture radius.

The power requirement (per pole), is expressed by:

$$P_p = \frac{1}{2\mu_0} \cdot G \cdot R^2 \cdot \varrho \cdot l_m (\lambda J), \quad \text{(W)} \tag{6.2.58}$$

The coil volume (per pole):

$$V_p = \frac{1}{2\mu_0 (\lambda J)} \cdot G \cdot R^2 l_m, \quad (\text{m}^3) \tag{6.2.59}$$

and the coil surface area (per pole):

$$S_p = \frac{f \cdot R}{\sqrt{\lambda J}} \cdot \sqrt{\frac{1}{2\mu_0} G} \cdot l_m. \quad (\text{m}^2) \tag{6.2.60}$$

In the above equations, l_m is the mean length of the coil, f is a factor linking the coil cross sectional area to the coolant surface: $f\sqrt{A} = S$; and G is the field gradient (Vs/m³). p is the object distance from the entrance magnet surface, q the image distance from the magnet exit surface.

Lens-Combinations

Except for a few applications, quadrupoles are found in beam transport systems in doublets, triplets or in periodic multiplet combinations. The main reason is that lens combinations must be focusing in both planes. In doublets, there is a lack in symmetry in both planes. Triplets are more expensive and require more space, but have the advantage of independent control over the focusing action in both planes.

Doublets

The image distance of a doublet arranged according to Fig. 6.2.37 is given by:

$$q = -\frac{\left[d - \dfrac{p\cos(kl) + \dfrac{1}{k}\sin(kl)}{pk\sin(kl) - \cos(kl)}\right] \cdot \cosh(kl) + \dfrac{1}{k}\sinh(kl)}{\left[d - \dfrac{p\cos(kl) + \dfrac{1}{k}\sin(kl)}{pk\sin(kl) - \cos(kl)}\right] \cdot k\sinh(kl) + \cosh(kl)}. \tag{6.2.61}$$

The focal distance from the magnet exit plane:

$$f_{fd} = \frac{1 - dk\,\mathrm{tg}\,(kl) - \mathrm{tgh}\,(kl)\,\mathrm{tg}\,(kl)}{k\,\mathrm{tg}\,(kl) + dk^2\,\mathrm{tg}\,(kl)\,\mathrm{tgh}\,(kl) - k\,\mathrm{tgh}\,(kl)}, \tag{6.2.62}$$

and the magnification:

$$m_{FD} = [-pk\sinh(kl) + \cos(kl)]\left[d - \frac{p\cdot\cos(kl + \frac{1}{k}\sin(kl)}{pk\sin(kl) - \cos(kl)}\sinh(kl) + \cos(kl)\right]. \tag{6.2.63}$$

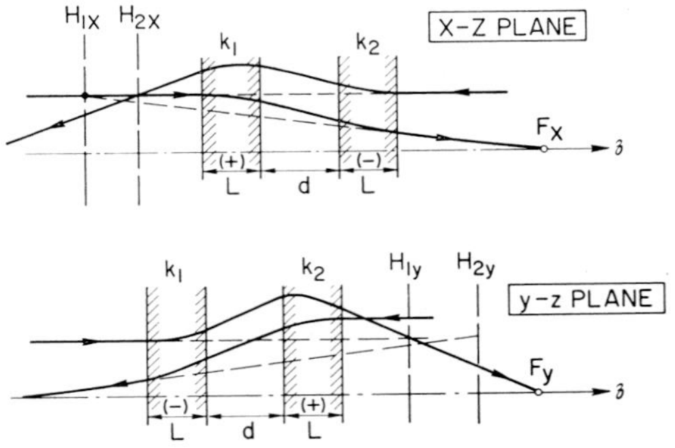

Fig. 6.2.37. Particle trajectory through the focusing plane (x, z) and defocusing plane (y, z) of a quadrupole doublet

Similar equations can be derived if the beam passes first through the defocusing plane and then through the focusing plane. The magnification in this case is given by:

$$m_{DF} = [pk\sinh(kl) + \cosh(kl)]\left[d + \frac{-p\cosh(kl) + \frac{1}{k}\sin(kl)}{pk\sinh(kl) + \cosh(kl)}k\sin(kl) - \cos(kl)\right]. \tag{6.2.64}$$

The focal distance from the down-beam magnet exit surface is given by:

$$f_{df} = \frac{1 + \mathrm{tgh}\,(kl)\,[dk + \mathrm{tg}\,(kl)]}{k\,\mathrm{tg}\,(kl)\cdot[1 + dk\,\mathrm{tgh}\,(kl) - k\,\mathrm{tgh}\,(kl)]}. \tag{6.2.65}$$

Comparing the two focal distance, we see that: $f_{df} \neq f_{fd}$.

With respect to the principal plane the focal distance is given by:

$$\frac{1}{F_{fd}} = k\cdot\sin(kl)\cdot\cosh(kl) + dk\sinh(kl)\cdot\sin(kl) - \cos(kl)\cdot\sinh(kl) = \frac{1}{F_{df}}. \tag{6.2.66}$$

For required object and image distances King [31] has derived the following equations:

First Element

$$K_1^2 = (k_1^2 l)^2 = \frac{\left(\frac{d}{l} + \frac{p}{l} + \frac{q}{l} + 2\right)\left(\frac{d}{l} + \frac{p}{l} + \frac{3}{2}\right)}{\left(\frac{d}{l} + 1\right)\left(\frac{p}{d} + \frac{1}{2}\right)^2\left(\frac{d}{l} + \frac{q}{l} + \frac{3}{2}\right)}. \tag{6.2.67}$$

Second Element

$$K_2^2 = (k_2^2 l)^2 = \frac{\left(\frac{d}{l} + \frac{p}{l} + \frac{q}{l} + 2\right)\left(\frac{d}{l} + \frac{q}{l} + \frac{3}{2}\right)}{\left(\frac{d}{l} + 1\right)\left(\frac{q}{l} + \frac{1}{2}\right)^2\left(\frac{d}{l} + \frac{p}{l} + \frac{3}{2}\right)}. \tag{6.2.68}$$

If we allow the image distance to approach infinity we obtain the focal strength:

$$K_1^2 = (k_1^2 l)^2 = \frac{\frac{d}{l} + \frac{p}{l} + \frac{3}{2}}{\left(\frac{d}{l} + 1\right)\left(\frac{p}{l} + \frac{1}{2}\right)^2}, \tag{6.2.69}$$

$$K_2^2 = (k_2^2 l)^2 = \frac{1}{\left(\frac{d}{l} + 1\right)\left(\frac{d}{l} + \frac{p}{l} + \frac{3}{2}\right)}. \tag{6.2.70}$$

Another set of useful parameters for doublet are the quantities K_{01} and K_{02} expressed as a function of $p/l = q/l$ as:

$$K_{01} = \Omega \cdot \left(\frac{l}{R}\right)^2; \quad K_{02} = \frac{G \cdot l^2}{p} = \frac{BR}{p}\left(\frac{l}{R}\right)^2. \tag{6.2.71}$$

From these two equations we obtain the figure of merrit:

$$\boxed{p \cdot \Omega_{\max} = \frac{K_{01}}{K_{02}}(B \cdot R)}. \tag{6.2.72}$$

$\Omega_{\max}$ is the maximum solid angle of a doublet. p is the source distance from the entrance surface of the first quadrupole, R is the aperture radius and B the field in the aperture.

The two parameters K_{01} and K_{02} are given in Fig. 6.2.38 for the practical case of $p = q$ i.e. object distance equal to image distance, as a function of the normalized object distance.

Triplets

The triplet is a combination of three quadrupoles of alternating polarity. Most triplets used in praxis are symmetrical, i.e. the two outer quadrupoles are of equal strength and length and are equally separated from the central element (Fig. 6.2.39).

We state the following cases:

(a) The two outer elements are identical and focusing.

(a1) All three quadrupoles are identical, the focal length is given by:

$$\frac{1}{F} = -\frac{1}{f}\left(1 - \frac{d^2}{f^2}\right). \tag{6.2.73}$$

(a2) The inner quadrupole has the double strength of the outer quadrupoles, then:

$$\frac{1}{f_i} = \frac{2}{f_0} \quad \text{and} \quad \frac{1}{F} = -\frac{2\,d}{f_0^2}\left(1 - \frac{d}{f_0}\right). \tag{6.2.74}$$

(a3) The inner quadrupole has half the strength of the outer quadrupole:

$$\frac{2}{f_i} = \frac{1}{f_0}, \qquad \frac{1}{F} = -\frac{1}{2\,f_0}\left(3 - \frac{d^2}{f_0^2}\right). \tag{6.2.75}$$

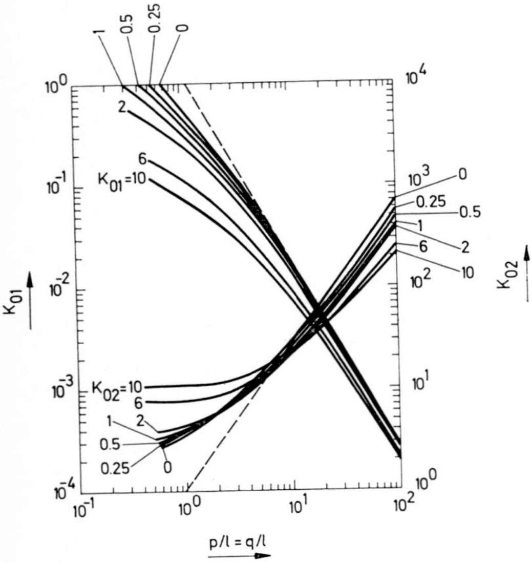

Fig. 6.2.38

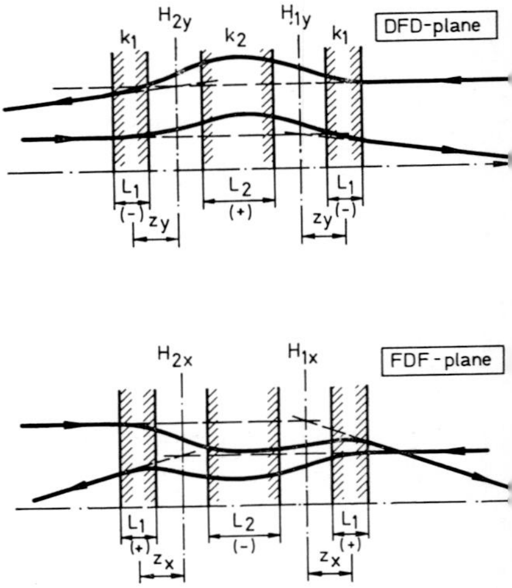

Fig. 6.2.39

Fig. 6.2.38. Strength of a doublet as a function of the normalized object distance $K_{01} = \Omega\,(l/R)^2$; $K_{02} = G \cdot l^2/p$; K_{02} is the reduced gradient of the second quadrupole

Fig. 6.2.39. Particle trajectory through the planes (x, z) and (y, z) of a quadrupole triplet. H_{1x} and H_{2x} resp. H_{1y} and H_{2y} are location of principal planes. DFD is Defocusing, Focusing-Defocusing arrangement. FDF is the focusing, defocusing-focusing arrangement

The focal distance with respect to the principal planes is obtained from:

$$\frac{d}{F_{FDF}} = -\left(1 - \frac{d}{f_0}\right)\left[2 - \left(1 - \frac{d}{f_0}\right)\left(2 + \frac{d}{f_i}\right)\right], \tag{6.2.76}$$

(b) The outer elements are identical and defocusing.

(b1) All three quadrupoles are identical:

$$f_i = f_0 = f$$

$$\frac{1}{F} = \frac{1}{f}\left(1 - \frac{d^2}{f^2}\right). \tag{6.2.77}$$

(b2) The inner quadrupole has the double strength of the outer quadrupole:

$$\frac{1}{f_i} = \frac{2}{f_0}, \qquad \frac{1}{F} = -\frac{2\,d}{f_0^2}\left(1 + \frac{d}{f_0}\right). \tag{6.2.78}$$

The focal distance, with respect to the principal plane is given by:

$$\frac{d}{F_{DFD}} = -\left(1 + \frac{d}{f_0}\right)\left[2 - \left(1 + \frac{d}{f_0}\right)\left(2 - \frac{d}{f_i}\right)\right]. \tag{6.2.79}$$

If we require, that the focal distances of the two planes must be the same, then from:

$$\frac{1}{F_{FDF}} = \frac{1}{F_{DFD}}, \tag{6.2.80}$$

$$\frac{1}{f_i} = \frac{2}{f_0\left(1 + \frac{d^2}{f_0^2}\right)}. \tag{6.2.81}$$

Only if $d \ll f_0$:

$$\frac{1}{f_i} \approx \frac{2}{f_0}. \tag{6.2.81}$$

Solenoids

The longitudinal magnetic field produced by a solenoid has been used for ion and electron focusing. As compared to a conventional doublet or triplet of the same performance, the power dissipation in a solenoid is much higher, doublets or triplets are preferred for general beam transport purposes. Superconducting solenoids however are still being used with good success for focusing purposes in electron microscopes.

The focal distance of a solenoid is expressed by [33]

$$\boxed{\frac{1}{F} = \frac{B^2 l}{4 \cdot (BR)^2}}, \tag{6.2.82}$$

with (BR) again the rigidity of the particle given in Eq. (6.2.50).

The comparison of the Eqs. (6.2.76) and (6.2.82) based on the focal length of a beam of charged particles of a certain rigidity allows the calculation of all lens parameters. When the choice is made, detailed calculation will show the optimum final design based on conventional, cryogenic, or superconducting magnets.

6.3 Cost Comparison in General

Economic comparison of magnet system for accelerators, experimental magnets, and beam transport systems has been made by Green [34], Strobridge [35], the LRL [36] and BNL [37] accelerator study groups, Haskell and Smith [38], and others. The general cost comparison has the advantage of obtaining an order of magnitude type cost figure, but lacks the elegance that a final item-by-item cost figure must be obtained. General cost comparisons sometimes tend to show optimistic figures, as it is not possible to work out details. Prices of materials, labor, and design are continuously changing; thus, final cost figures may be misleading.

The cost of the system given here is only an outline and a possible guideline. For more accurate calculations, we refer to the above chapters.

There are three basis of cost comparisons:

The "capital" cost of an equipment, C_c.

The "time based" cost, C_t, which is the capital cost multiplied by the interest factor and the number of years.

The "operational" or "running" cost.

The superconducting system is in operation over a certain fraction of a year. In electrical machines, this fraction is about 8000 hours per year. The running time of an accelerator may vary, but in general, it is shorter.

In *superconducting magnets*, the main item is the cost of the coil, which is proportional to the coil volume.

The capital cost of the coil is composed of the cost of the superconductor, the coil structure, and the coil winding.

The cost of the cryostat (dewar) is proportional to the coil surface S. The capital cost of the cryostat is made up of the helium container, vacuum enclosure, supports, superinsulation, and or nitrogen shield.

The capital cost of the refrigerator is a function of the heat input into the cryostat due to magnet losses (e.g. in pulsed magnets), heat leak through the cryostat surface, heat radiation and heat inleak along current leads. It may be useful to determine whether or not a separate refrigerator would be required to cool-down the nitrogen shield, if bulk liquid nitrogen is available, or if heliumgas cooling at the same intermediate temperature may be utilized.

Another factor which may influence the capacity of the refrigerator is the cool-down time.

The running cost of a helium refrigerator is given by the efficiency of the system, the helium losses through leaks (even in a close circuit system, which hopefully is less than 1% of the total helium capacity per month), and the type of operation (either if the system is operating continuously (d.c.), intermittant, or with interruptions of longer durations).

The capital cost of power supplies is not insignificant for pulsed or synchrotron type magnets, but is rather small in d.c. type magnets.

A very significant part in the cost estimate are safety and monitoring equipments in a magnet system. It is rather difficult to estimate the cost of auxiliary parts such as heliumtransfer lines, flow meters, cold valves, helium monitoring system, pressure and vacuum measuring equipment, helium level gages, magnet quench protection circuits, field screening and field correcting elements short and long range field measuring equipment, and feed back circuits to reduce the current or to shut off the magnet in case the field changes abnormally in discrete points of a beam transport system where continuous measurements are performed.

In the optimization of a circuit one may also study the possibility of using one refrigerator to cool-down and maintain the temperature of a number of magnets, or using small refrigeration units for only one or a limited number of magnets. One important problem to be solved is the distribution of helium from the refrigerator to the magnets in order to maintain reliable magnet operation.

In the design of a large system, to give an example, we consider the CERN II synchrotron with an ultimate energy of 1000 GeV, a ring diameter of 2.2×10^3 m, and a magnetic length of $\sim$5000 m. If we assume a pulse duration of 6 sec in which the magnetic field is raised during injection to 5 T in 2 sec, a flat top duration of 2 sec for ejection and finally a field reduction time of 2 sec, modern pulsed magnets can be designed having a total Joule loss value of $\sim$15 watts per magnet meter length. As shown in Chapter 4, these losses are comprised of eddy current losses in the matrix material, hysteretic losses in the superconductor, self field losses due to the pulsed transport currents flowing through the superconductor, auxiliary losses due to the non-homogeneous field over the coil and some additional eddy current losses, iron losses (eddy current and hysteretic), static and dynamic losses of the cryostat, and the support structure. In addition we may assume static losses of about 2 W per m of transferlines. If we operate the system to 4.5 K, the total amount of refrigeration required will be less than 100 kW at this temperature. The problem of optimization will be to select a suitable number of refrigerators (minimum one unit), and to design a suitable helium distribution, monitoring and control system. The input power of such a refrigeration system, if we select 1 kW units at 4.5 K would be 30 MW (see Fig. 6.2.3). If the magnets could be wound with multifilament V_3Ga conductors the total installed power would drop to about 20 MW, and also the capital cost of the refrigerator system.

Such a system requires a very elaborate safety and protection circuit, optimum energy storage, and distribution units which can also be superconducting, field correction elements such as multipoles (mainly sextupoles), or field decoupling superconducting screens or shields. The useful aperture of the magnet system may have a cold vacuum which leads to a substantial saving. Quick disconnects must allow the removal and replacement of a magnet in a short time without the necessary warming up of the unit to be removed. It may also be advisable to keep a number of reserve dipoles at 20 K for emergency cases.

Magnets are required in the main ring for the beam transport to the experimental areas, beam extraction, and in experimental areas. Practically all these magnets can be superconducting.

To boost up the machine energy to 500 GeV, superconducting dipoles may be inserted in the circumferential space between conventional magnets (missing magnet scheme). The conventional water-cooled dipoles can be replaced at a later date by superconducting dipoles to achieve a final energy in excess of 1000 GeV.

This scheme, though attractive, has a few drawbacks, specifically from an economical point of view. In addition, the accelerator may not be very flexible in its operation. Two other schemes are thus under study. The first scheme sees the placing of superconducting magnets on top or concentric to the conventional ring, the second scheme proposes an additional concentric ring in a separate tunnel. The last scheme has the added advantage that the experiments do not have to be interrupted during the construction of the second ring. It is also feasible to add more concentric rings and boost up the final energy to several 1000 GeV.

To give an example we compute the cost of a high energy synchrotron based on current commercially available superconducting Type II composite materials, current technological practices and present day materials available. The prices quoted are, in a few cases, somewhat different from the values given in previous tables of this chapter.

They are corrected to account for additional requirements and to include new features.*

We base the price of the synchrotron on the following numbers:

(a) The cost of the superconductor is \$ 3×10^{-3} per Amp · m for Nb-Ti at 4 T. The overall current density of the coil is 3×10^{2} A/cm² at 4 T.**

(b) The cost of magnet procurement is assumed to be 30% of the total cost of the Nb-Ti composite. (The iron shield is in the proximity to the coil and cooled to liquid helium temperature, but placed such that the iron is not saturated.)

(c) The power supply cost is assumed to be \$ 35/kW peak power (Fig. 6.2.1). The cost figure includes installation and housing.

(d) The cost of the magnet cryostat is \$ 1,500/m.

(e) Refrigeration costs are based on data from Fig. 6.2.5. Recent advances in the design of transferlines are included [39]. The a.c. losses of the magnet (static and dynamic) are assumed to be 20 watt/m of magnet including static losses of transferlines.

(f) The cost of the tunnel ($3\times3{,}5$ m²) is assumed to be at \$ 3,500/m, including penetration, shielding, enclosure and foundation.

(g) Cost of utilities is assumed to be \$ 100/kW of power handled.

(h) The cost of electrical power is assumed to be \$ 0.01/kW hr.

Table 6.3.1 gives crude cost estimates of a number of accelerators under construction, proposed or estimated [34]. The number quoted may have undergone some changes since they have been reported first.

For calculation purposes, the following data have been used:

Coils

Dipoles with circular aperture have a $\cos\theta$ current density distribution.

Quadrupoles with circular aperture have a $\cos 2\theta$ current density distribution.

The iron cylindrical shell is spaced at a distance around the coil where the maximum flux density in the iron does not exceed 2 T. The cost of

* The cost figures are taken from a study by M. A. Green for GESSS.

** The "advanced" superconductors are assumed to have a current density higher by a factor of 2 compared to presently available materials. The assumption is justified by the improvements of new superconductors.

The basic machine parameters used for cost comparison are the following:

Fixed Parameters

Maximum energy	1000 GeV
Injection energy	25 GeV
Intensity	10^3 pp pulse
Ratio machine radius to magnetic radius	1.38

Variable Parameters

Superconductor[a]	$Nb_{(x)}Ti$
Dipole peak field[b]	3 ... 7 T
Cycle time[c]	6, 12 and 24 sec
Magnet aperture[d]	2.5, 7.5, 10, 12.5 and 15 cm
Magnet field homogeneity	10^{-3}

[a] The superconducting filament size is 5 μm for $Nb_{(x)}Ti$.

[b] We assume that in the near future V_3Ga and Nb_3Sn multifilament conductors will become available. For these materials, which we call advanced or futuristic conductors one can assume flux densities in the aperture of 3 ... 8 T.

[c] The cycle time is subdivided as follows: $\frac{1}{3}$ cycle time for the field rise time (injection) and for field decay of fall time. The pause between two pulses is 0.5 sec. Flat top time (ejection) is $\frac{1}{3}$ pulse duration with a minimum 0.5 sec.

[d] The aperture is assumed circular.

superconductors can be obtained from Fig. 6.2.18 the price of laminated silicon steel from Table 6.2.2.

Correction coils

The field correction is lumped together in the form of one or several dipoles and multipoles. The cost of correcting coils does not excceed 5% of the total magnet cost.

Power supply

If E_{max} denotes the peak stored energy, E_{min} the minimum stored energy, then the peak power is given by:

$$P_{max} = 2 \cdot (E_{max} - E_{min})/T_{rise}$$

and the r.m.s. power by:

$$P_{rms} \cong 0.65\, P_{max}\,.$$

The cost of the generator (\$ 0.010 r.m.s. power (W) + 0.02 d.c. power (W)), the cost of the motor (\$ 0.002 r.m.s. power areas (W) + 0.02 d.c. power (W)), the rectifier cost (\$ 0.005 peak power (W) + 0.015 r.m.s. power (W)), the cost of the flywheel (0.0003 E_{max} (Joule)), the cost of leads and the cost of installations must be included.

Cryostats

The cost of the magnet cryostat is given by the weight of the stainless-steel cylinders, the flanges, heat shields, the superinsulation, the procurement and installation.

Table 6.3.1. *Comparison of cost estimates and parameters of a number of accelerators, proposed or under construction* [34]

	Conventional	Superconducting		
Accelerator	NAL [40] 500 GeV Energy (1968)	LRL [36] 10 GeV Energy (1970)	Green Est. [41] 1000 GeV Energy (1969)	BNL [37] 2000 GeV Energy (1970)
Injection Energy	8 GeV	50 MeV	25 GeV	30 GeV
Intensity	1.5×10^{12} ppp	2×10^{12} ppp	1.5×10^{14} ppp	10^{13} ppp
Dipole field	2.25 T	5 T	5 T	(4 ... 6) T
Aperture	5.07×10.14 cm²	10.14 cm ⌀	10.15 cm ⌀	~3.83 cm ⌀
Pulse duration	~10 s	10 s	15 s	~4 ... 6 s
Acc. cost estim. 10^6 US $/GeV				
Magnets incl. cryostats	0.044	0.050	0.054	0.020
Power supply	0.012	0.003	0.024	0.008
Refrigerator	—	0.047	0.012	0.018
Radio frequency and electronics	0.005	0.011	0.029	0.009
Vacuum	0.005	0.003	0.004	0.002
Control, injection, extraction	0.010	0.009	0.012	0.005
Enclosure and plant facilities	0.055	0.038	0.037	0.015
Total capital cost	0.131	0.161	0.173	0.077
10 year operating cost	0.140	0.060	0.055	0.035
Capital and operating cost	0.271	0.221	0.228	0.112

Refrigeration

To estimate the amount of coolant required to operate a magnet system, one has to estimate:

(*a*) *Heat load*
(a1) Superinsulation: 1.1 W/m (length)
(a2) Cryostat static losses: 2.5 W/m
(a3) Counter flow current leads: 2 mW per A per lead
(a4) Magnet static losses: 5.5 W/m
(a5) Transfer lines: 2 W/m incl. couplers, connections

(*b*) *Beam heating*

(*c*) *Beam losses in circulation*

(*d*) *A.C. losses in superconducting magnets*

Hysteretic losses in superconductor and iron lamination.
Self field losses.
Eddy current losses (in normal substrate and iron lamination).
Auxiliary losses in coil, cryostat and structural material.

Radio frequency

The r.f. power is obtained from:

$$(E_{ej} - E_i) \cdot I_{\text{beam}} \cdot 1.6 \times 10^{-19} / T_{\text{rise}}$$

r.f. peak power: 2 r.f. power,
cavity tuning,
tuning of structural parts and energy distribution,
wave guides.

Vacuum system

Vacuum pipes between magnets which will have
superinsulated jackets,
vacuum joints,
roughing pumps,
high vacuum pumps.

Power related facilities

As a lump sum about \$ 100/kW.

Tunnel

\$ 3500/m.

Based on the above price, and cycle assumptions, machine and magnet parameters Fig. 6.3.1 shows the cost of a synchrotron with a cycle times of 6 sec and a magnet aperture of 7.5 cm diameter. The cost of the assumed futuristic superconducting magnets is lower. The minimum of the cost curves are fairly flat, such that the dipole field range (4 ... 6) T can be selected without appreciable price variation. At higher fields the price of the accelerator increases rapidly, which is not surprising due to the critical current behaviour of the superconductor, the losses and the complicated controles in current regulations etc. Fig. 6.3.2 illustrates the cost of machines as a function of the dipole aperture assumed circular. The effect of the cycle time on the machine cost is quite substantial. However, cycle times longer than about 15 sec do not have an appreciable effect on the cost.

The magnet conductor chosen is $Nb_{(x)}Ti$ multicore conductor with filament sizes of 5 μm embedded in a copper matrix. For this conductor the capital cost and the total cost of the accelerator have a minimum at 3.5 ... 4.5 T peak magnetic field.

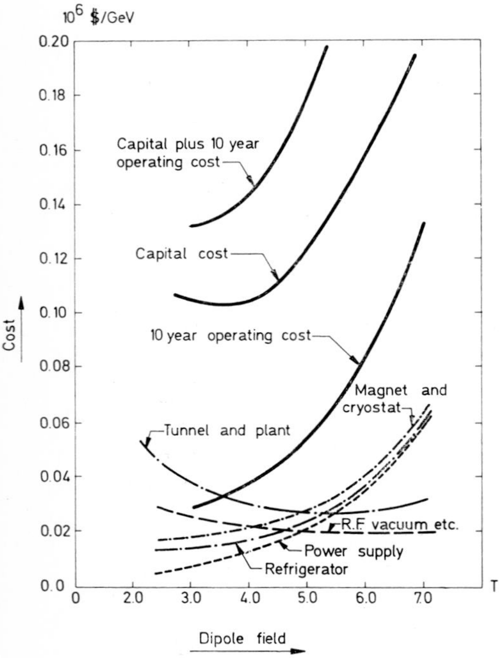

Fig. 6.3.1

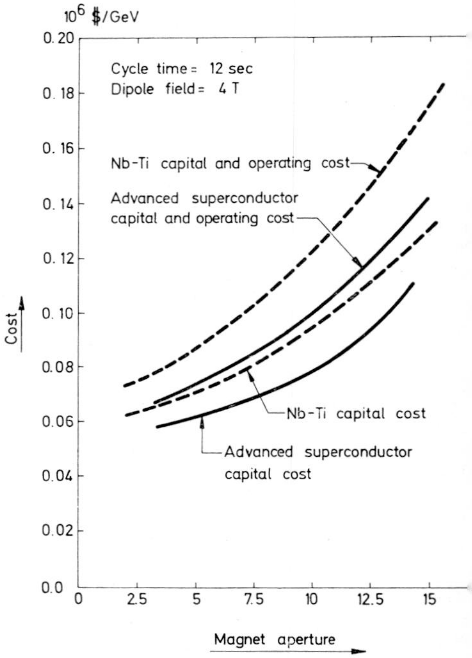

Fig. 6.3.2

Fig. 6.3.1. Accelerator component costs per GeV/c as a function of the dipole field. The aperture of the dipole has a diameter of 7.6 cm. Pulse duration is 6 sec. Superconductor selected as the coil winding is Nb-Ti copper composite

Fig. 6.3.2. The effect of the aperture size on the cost of a synchrotron accelerator

We may conclude from the curves in Fig. 6.3.2 and the Table 6.3.1, which gives the cost figures of the NAL accelerator, that the price of a superconducting synchroton will be about a factor of 2 less, than of a conventional synchrotron of the same beam energy. As the power cost is expected to increase, due to newer more stringent requirements because of environmental pollution (thermal pollution), the difference between normal and superconducting machines may become even more pronounced.

A superconducting synchrotron will have a higher energy by about a factor of 2 ... 3 compared to a conventional synchrotron of the same size.

Cycle times in the order of 10 sec seem realistic for high energy accelerators, whether superconducting or conventional. Very high injection energies seem technically more practical, specifically as the field tolerance requirements in the magnet aperture can be relaxed to about 10^{-3}; this has already been achieved for superconducting dipole magnets. It may also be pointed out that if the iron return yoke is placed around the coil within the cryostat and is cooled by a cryogen, the iron losses at a cycle time of 10 sec will be less than 8% of the total static and dynamic losses of the magnet coil and cryostat, which with new designs, have a static loss of 2 W per meter magnet length.

Cryogenic magnets with coils wound of high purity aluminum or sodium, are specifically attractive for pulsed operation in d.c. hybrid type magnets, or small high field d.c. solenoids. The initial hope that aluminum coils may have low running cost at 21 K at magnetic fields in excess of 4 T has been dimmed somewhat due to the discovery that Kohler's rule is not applicable at high temperatures. With increased purity, the electric resistivity of aluminum due to magnetic fields (magneto-resistance effect) is much higher than expected. The magneto-resistance effect, treated in more detail in Chapter 5 and 6.2, is responsible for the unexpected higher running cost.

Only at liquid helium temperatures, magneto-resistance saturation effects at magnetic fields exceeding 3 T could be observed in aluminum tapes with $r > 2{,}000$.

The new high purity tapes are also exceedingly expensive compared to copper or electric grade aluminum used for room temperature applications, and are comparable to the price of superconductors. The cost of cryogenic magnets may be somewhat higher than room temperature magnets. The ohmic losses at 4.2 K of cryogenic pulsed magnets based on aluminum coils under the same conditions are about a factor 5 ... 8 higher than in superconducting pulsed magnets. Including the capital cost of the refrigeration system, cryostats, etc. to the cost of magnets, it can be seen that cryogenic low duty cycle, long time pulsed magnets or d.c. magnets can hardly compete with superconducting magnets. Compared to room temperature systems, the running cost of cryogenic systems over 10 years should be a factor of 2.5 ... 5 lower.

Conventional magnets

Direct watercooled copper or aluminum coils (electric grade aluminum is often chosen) with or without iron return yokes (iron bound and iron core magnets) have been an important part in practically all branches of physics in the past decades. They are encountered in a great variety, but their aperture fields seldom exceed 2 T in large magnets due to excessive ohmic losses in the conductor and thus high operating costs.

Computer programs optimizing iron core and iron bound magnets for high energy application are available but it is more common to match new magnets to already existing power supplies.

Also convenient designs, such as accessibility to the high field region, field uniformity, and space requirements, are preferred over optimized designs. With the availability of fast computers and the existance of two-dimensional computer programs the optimum shape of the iron return yoke can be determined. The requirement in the field shape, field distribution determines the shape of the coil. Whence the coil shape, its volume and weight are calculated, the magnet capital cost is obtained from current prices of material and labour.

In conventional magnets, the cost of the power supply is an important parameter. Power supplies with stringent requirements on current stability are naturally expensive. The cost of power supplies per kW output is higher for low output than for high output power. For large systems of magnets it is thus common to energize several magnets from one power supply and provide means to shunt the current adequately in individual magnets, a method which is also adapted in superconducting magnet systems.

In this general cost comparison it was not attempted to give a simplified equation but to present a background for a rough estimate. If single magnets of certain types are compared, the equations linking beam optical performance of magnets are given. The magnet parameters, its volume, weight and cooling surface are calculated, and using prices for material, labour, including depreciation, escalation, etc., the exact price of the magnet can be obtained.

For larger systems optimization programs may give a rough preliminary estimate. These programs are certainly adequate for cost estimates and for proposals as a basis for comparing different systems.

References Chapter 6

[1] Gauster, W. F.: Comm. Electronics **52,** 822 (1961).
[2] Fabry, C.: Eclairage Electrique **17,** 133 (1898).
[3] Bitter, F.: Part II: Rev. Sci. Instr. **7,** 482 (1936).
[4] Montgomery, D. B., Terrel, J.: NML Publ. 1515 (1961).
[5] Harris, C. A.: SLAC TN-66-39 (1966).
[6] Brechna, H., Montgomery, D. B.: NML Publ. 62-1 (1962).
[7] McAdams, W. H.: "Heat Transmission," 228. New York: McGraw Hill 1954.
[8] Jones, P. J.: Petroleum Production. New York (1946).
[9] Kays, W. M., London, A. L.: Trans. ASME **74,** 1179 (1952).
[10] Brechna, H., Oster, E.: Proc. Int. Symp. Magnet Techn., 313 (1965).
[11] Burnier, P., Laurenceau, P.: Rev. Général de l'Electricité **74,** 542 (1965).
[12] Strowbridge, T. R.: I.E.E.E. Trans. NS **16,** 3 (1969).
[13] Purcell, J., Jacobs, R. B.: Cryogenics June (1963).
[14] Arp, V.: Proc. 1968 Summer Study. BNL, p. 1095 (1968).
[15] Steckly, Z. J. J., Zar, J. L.: I.E.E.E. Trans. Nucl. Sci. NS **12,** 367 (1965).

[16] Brown, G.V.: NASA, Tech. Mem. TMX-52571 (1969).
[17] Smith, P. F.: Rev. Sci. Instr. **34,** 368 (1963).
[18] Maddock, B. J., James, G. B.: Proc. I.E.E. **115,** 543 (1968).
[19] Williams, J. E. C.: Cryogenics, Dec. (1963).
[20] Lock, J. M.: Cryogenics, 438, Dec. (1969).
[21] Keilin, V. E., Klimenko, E. Yu.: Cryogenics Aug. (1966).
[22] Jacob, R. B.: Paper J. 6 Adv. Cryog. Eng. **7,** 529 (1963).
[23] Brechna, H., Arendt, F.: KFK Report 1964 (1972).
[24] Gavaler, J. R., et al.: JAP **42,** 54 (1971).
[25] Sampson, W. B., et al.: Particle Acc. **1,** 173 (1970).
[26] Britton, R. B.: Proc. 1968 Summer Study. BNL, p. 893 (1968).
[27] Bozorth, R. M.: "Ferromagnetism." D. van Norstrand (1964).
[28] Brechna, H.: SLAC TN-65-87 (1965).
[29] McInturff, A. D., Claus, J.: BNL Rep. AADD. 162 (1970).
[30] Banford, A. P.: The Transport of Charged Particle Beams E. & F. N. Spon Ltd. 1966.
[31] Brown, K. L.: SLAC Rep. 91 (1970).
[32] King, M. N.: RHEL Int. Rep. MPS/EP 22 (1962).
[33] Edwards, D. N., Rose, B.: Nucl. Instr. and Methods **7,** 135 (1965).
[34] Green, M. A.: UCRL 20299 (1971).
[35] Strowbridge, T. R., et al.: NBS Rep. 9259 (1966).
[36] LRL 200 GeV Acc. Design Study. UCRL 16000 (1968).
[37] BNL-Adv. Synchrotron Study Group. BNL 15430 (1970).
[38] Haskell, G. P., Smith, P. F.: Nucl. Instr. and Methods **30,** 277 (1964).
[39] White, G. K.: "Exp. Techniques in Low Temp. Phys." 2nd Ed. Oxford: Clarendon Press 1968.
[40] NAL-Design: Rep. (1968).
[41] Green, M. A.: I.E.E.E. Trans. NS **16,** 1082 (1969).

7. Examples of Superconducting Magnet Systems

In this chapter, we describe in detail a few magnets either built and tested or being constructed. The choice, although arbitrary, has been such that even though the magnets described have a few common marked features, their designs are basically different. Magnets built and indicated in Table 1.8 are similar to those given in this chapter. For comparison, we have selected a few high field solenoids (Fig. 7.1), wound with Nb_3Sn tapes without iron yoke. The ANL magnet, presently the largest magnet in operation, generates a field of 1.8 T at an overall current density of 1700 A/cm². This magnet, with an iron return yoke, can also be built from conventional conductors; but there are two reasons for the choice of building a superconducting magnet: To show that a large multi-megajoule superconducting magnet can be used for high-energy experimental purposes even though the capital cost of this magnet would be comparable to a water-cooled magnet, and to illustrate the importance of a superconductor from the point-of-view of power saving in areas where overall power densities in large laboratories becomes problematic.

Fig. 7.1. High field superconducting solenoids (I.G.C.)

The MIT hybrid magnet gives a new approach in the utilization of superconducting magnets with water-cooled solenoids, a design which is of interest in achieving high magnetic fields with moderate power requirement. The ORNL magnet for plasma research, a combination of various magnets constructed of Nb_3Sn tapes and $Nb_{(x)}Ti$ composite conductors, is interesting from the point-of-view of magnet design; but if water-cooled magnets would be utilized, it would necessitate an exorbitant amount of power to generate the same field.

The SLAC magnet, a typical utility magnet, has a conservative design using already old-fashioned monofilamentary cables, but illustrates experimental difficulties with cables constructed with mono-filament multistrand cables. Reasons why portions of this magnet were rewound in order to make the coil more useful as an experimental magnet were safety of operation. energizing of the three module magnet from one or two current sources, and increasing charging speed.

7.1 The Argonne National Laboratory 3.7-m Hydrogen Bubble-Chamber Magnet

Following the alphabetical sequence, the ANL Bubble-Chamber Magnet is currently the largest operating superconducting magnet by far. It is interesting to note that in the area of high-energy physics, only the bubble-chamber physicist has endeavored to use multi-megajoule superconducting magnets instead of water-cooled copper solenoids with ferromagnetic return yokes. ANL has already pioneered in this respect, by utilizing superconducting magnets for a 3.7-m diameter [1], and earlier for a 0.25 m diameter bubble chamber [2]. The 0.25 m bubble chamber system, completed in 1965, is still being used sporadically. The ANL 4.78-m diameter bubble-chamber magnet, in double Helmholtz configuration, was first energized in April 1969, and reached 1.8 T in the bore center. Fig. 7.1.1 illustrates the cross-section through the magnet and chamber system. The basic features of this magnet and a few system parameters are given in Table 7.1.1. The coils wound from Nb(48%) Ti and OFHC composite conductor strip are immersed in liquid helium contained in a toroidally-shaped helium vessel. However, iron return yokes, used to confine the fringing field and moderately boost up the field generated by the coil MMF, were utilized for the first time in large superconducting magnets. The composite strip, in 1967 a novelty [3], consists of six superconducting filaments which are in good thermomechanical bond to the copper matrix. The new technology in composite conductor manufacturing started about 1966 to replace monofilamentary multistrand cables.

The coil consists of four sections with a total number of 30 pancakes, each having 84 turns. The transport current through the conductor to produce 1.8 T is 2200 A. The magnet inductance is 40 H which, by using a 10-Volt power supply, would permit to charge the magnet in 8000 s. However, the first time the magnet was energized, the current was cycled through the coil in steps raised and reduced intermittently until, after

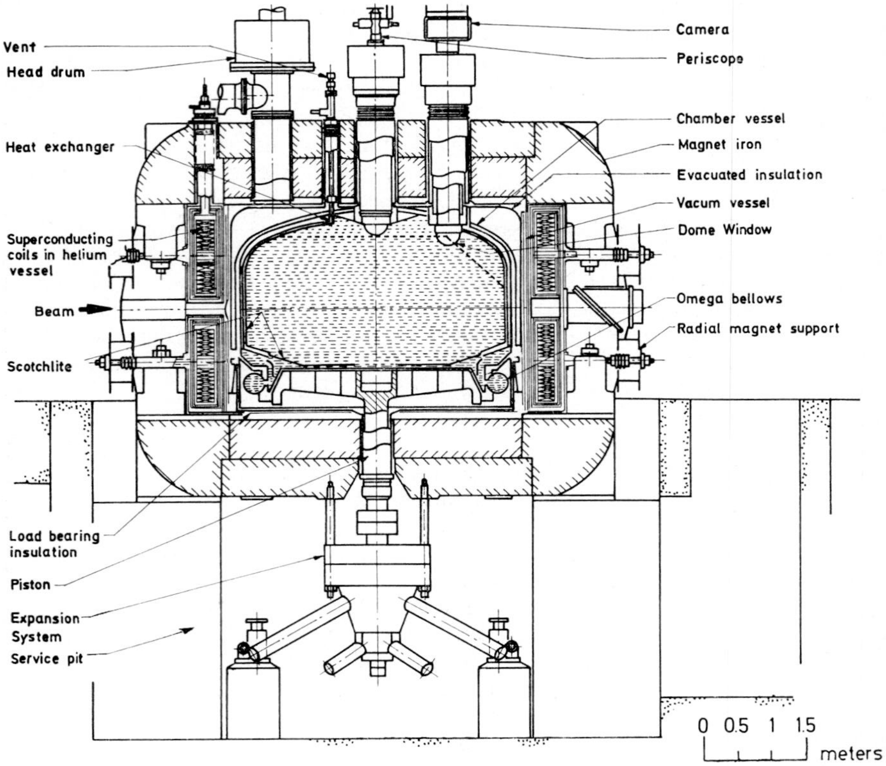

Fig. 7.1.1. Cross-section through the ANL 3.7 hydrogen bubble-chamber system

approximately 26 hours, the design current of 2200 A, corresponding to the field of 1.8 T, was reached. The magnet is in operation since 1969.

The coil and cryostat weigh a total of 90×10^3 kg and are hung from the magnet yoke by pairs of stainless steel pipes (Fig. 7.1.2).

The heat leak through the support structure to the helium bath could be limited to 50 W by means of helium/gas-cooled tubings (50 K).

The vacuum vessel is fabricated from 2.5 cm thick rolled aluminum plates to withstand 1 kp cm^{-2} of external and 1.36 kp cm^{-2} of internal pressure. An emergency vent of 25 cm diameter opens at a pressure of 1.36 kp cm^{-2}. The vacuum tank acts also as a secondary shorted turn with a time constant of 8 s.

To energize the magnet, a three-phase full-wave Thyristor-controlled power supply provides a 10-Volt maximum voltage and a current of 2200 A. The power supply controle unit is equipped with necessary safety interlocks and fault-indicating lamps.

The magnet windings are protected against a sudden release of the magnetic stored energy (quench) by means of parallel shunts (shown in

Fig. 7.1.3). When a resistive region is developed in the coil, both the power supply and the stabilizer resistor are disconnected from the circuit, and the coil current is forced to flow through the so-called dump resistor (0.1 Ohm). The energy stored in the magnet (80×10^6 J) is dissipated at an initial rate of 0.44 MW per second (2200 A, 200 V, design condition). It is anticipated that 60% of the field energy is dissipated in the external resistors. The field decay time constant is 2380 s.

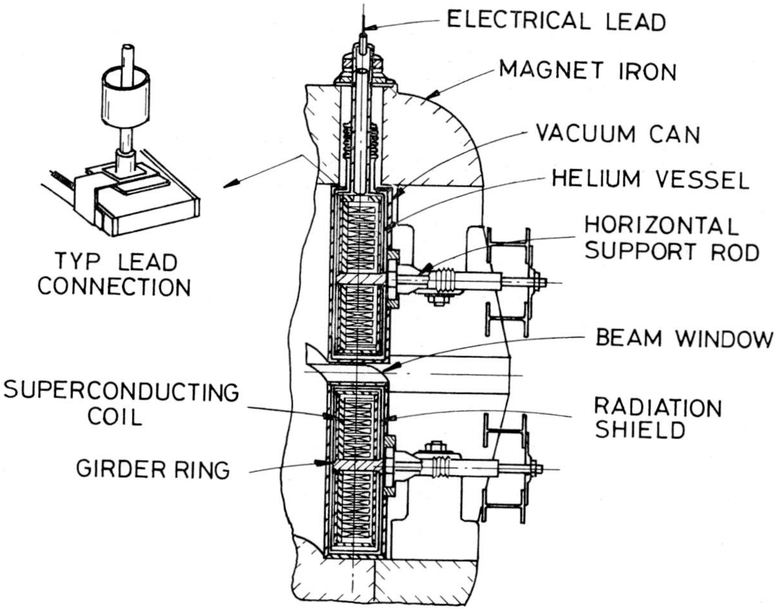

Fig. 7.1.2. Schematic representation of the ANL magnet cross-section showing the location of the coil in the helium vessel and vacuum enclosure, their respective locations, and the support structure

Each pancake is edge-cooled. Between adjacent turns, teflon tapes and epoxy insulation are utilized. The conductor is "cryostatically" stabilized for a heat flux of 0.13 W/cm² corresponding to a temperature rise of 0.1 K in the conductor over the bulk helium temperature. Between adjacent pancakes, aluminum blocks (5×1.25 cm), epoxy-coated, are spaced at 6° intervals.

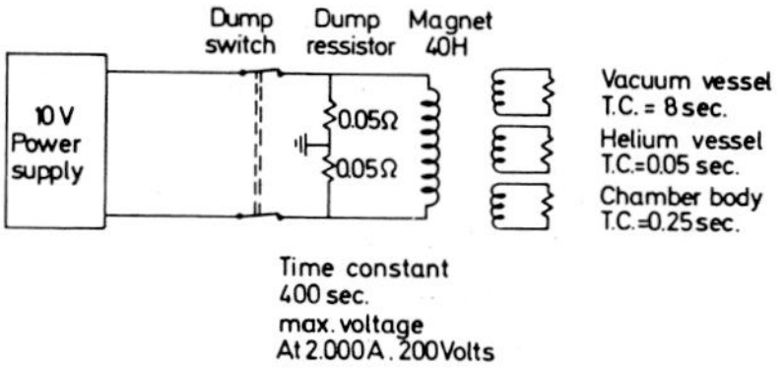

Fig. 7.1.3. Simplified diagram of the energyzing and protective system of the ANL magnet

These blocks are connected to copper braids to keep the temperature gradient over the coil to less than 10 K during the cool-down period.

Practically all the data given in Table 7.1 are calculated. There were no measured data available on the actual performance of the coil after the first magnet operation.

During the magnet-charging, two quenches were encountered, roughly at 900 and 1350 A. Reasons for the quenches, as well as pertinent experience with the coil, are not given by ANL. It is hoped that by gaining experience with the magnet, more data will become available.

The conductor length per pancake is 1325 m. The conductor supplied by the manufacturer had a continuous length of 220 m and required that six individual pieces be connected by means of lap joints using 30-cm long copper bars, lead-tin soldered, and riveted to the conductor.

The cross-sectional area of each superconducting filament (a total of 6) is 0.0835 cm², which yields a current density in the superconductor of 4.4×10^4 A/cm². The filament cross-section is flattened, due to the manufacturing process, with an approximate ratio of the major to the minor diameter of 20 : 1 or more, which lead to the anisotropy effect of the conductor, parallel and perpendicular to the main field, and is reflected in a J_c distribution given in Fig. 7.1.4.

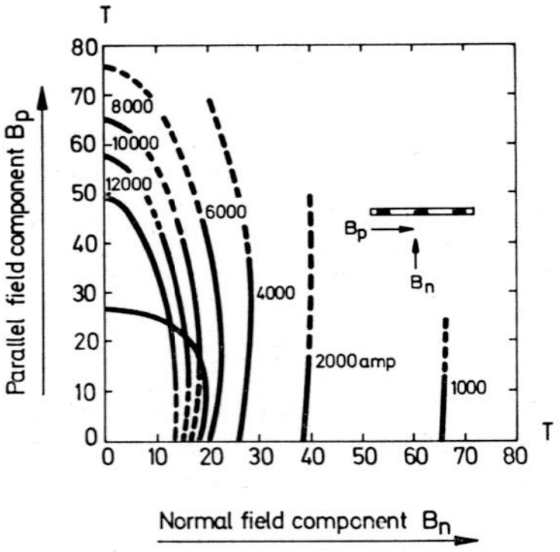

Fig. 7.1.4. Critical current versus parallel and normal field for the ANL magnet conductor

Another interesting aspect of the ANL magnet complex is the cryogenics system. The weight of the various components to be cooled are given by:

Cryostat (stainless steel, welded construction)	33×10^3 kg
Coils (mainly copper)	46×10^3 kg
Aluminum tie rods and spacers	1.5×10^3 kg

To cool-down this weight from ambient to 4.5 K, a total of 6.7×10^9 Joules must be removed. The cooling system is designed such that over a time period of 400 s, power at a rate of 2×10^5 W can be released. The helium gas will be removed from the cryostat such that the safe operating pressure is not exceeded in the cryostat and the transfer lines.

The refrigerator is designed for the following conditions:

(i) Continuous magnet operation.

(ii) Providing 500 W of refrigeration at 50 K.

(iii) To shield thermally the magnet system.

(iv) To provide 400 W at 4.45 K to maintain liquid helium quantities both in the storage dewar and in the liquid helium reservoir, when precooled with hydrogen.

Approximately 25 liters of helium are evaporated per hour through the electric leads and the coil support structure. The gas is warmed up to ambient temperature and is returned to the suction side of the compressors.

If the liquid hydrogen precooling is replaced by liquid nitrogen, the refrigerator performance is altered to 320 W at 4.45 K and 500 W at 80 K. Basically, the refrigerator will furnish cooled helium for magnet cool-down, provides liquid helium for filling and maintaining constant helium level over the magnet, and finally, it will keep a 10^4 liter helium dewar continuously filled. The refrigerator is an expansion type device with a reciprocating engine and has a thermodynamic efficiency of ~55%.

Fig. 7.1.5. One of the two 4.8 m i.d. coil halves of the 1.8 T ANL magnet. A section of the stainless-steel helium vessel is shown suspended above the assembly

To cool-down the magnet from ambient temperature to 4.45 K, about 140 hours were required. During this time, 30,000 liters of liquid nitrogen and 45,000 liters of liquid hydrogen were evaporated into the air.

The helium refrigeration system provides an adequate helium storage to permit eight hours of magnet operation without refrigerator, which seem appropriate for emergency repair and maintenance work.

We may note that the conductor resistance at 4.2 K and 1.8 T is 0.031 Ohm. To produce 1.8 T at 2200 A cryogenically (without superconductor), a power of 180 kW would be necessary!

The impressive size of the coil during assembly of the upper section is shown in Fig. 7.1.5.

Table 7.1.1. *The ANL 3.7-m, 1.8 T, bubble-chamber magnet characteristics*

Coils	
Operating current at 1.8 T field (A)	2200
Field energy (J)	80×10^6
Coil inductance (H)	40
Weight of copper in windings (kg)	45×10^3
Weight of superconductor (kg)	300—400
Coil axial compressive force (kp)	6.8×10^5
Compressive stress on coil separators (kp/cm²)	140
Average current density in conductor (A/cm²)	1700
Total ampereturns (30 pancakes, 84 turns each)	5.5×10^6
Coil inner diameter (m)	4.78
Coil outer diameter (m)	5.28
Axial coil length (m)	3.04
Iron yoke	
Weight of iron (kg)	1.45×10^6
Conductor	
Conductor cross-section (cm²)	5.0×0.25
Cu:superconductor ratio	24 : 1
$\varrho_{273\,K}/\varrho_{4.2\,K}$	200
Recovery current (calculated) (A)	3300 at 2.5 T
Stable current (calculated) (A)	4300 at 2.5 T
Design stress (kp/cm²)	840
Stabilized for maximum heat flux (W/cm²)	0.13
Power supply	
Power supply voltage (V)	10
Current range (A)	0—3000
Current regulation (long term in %)	0.1
Helium refrigerator	
Normal operation with liquid hydrogen precooling	
Cooling of insulation shield around magnet reservoir	500 W at 50 K
Liquid helium maintenance and storage	400 W at 4.45 K
Counterflow lead cooling and support	25 liters/hour

7.2 The CERN Liquid Hydrogen Bubble-Chamber Magnet

In July 1967, the European Organization for Nuclear Research, the Federal Republic of Germany, and the French Commissariat à l'Energy Atomique agreed to jointly build a hydrogen bubble-chamber of about 33.5 m³ volume and a visible volume of 22 m³ at CERN. The chamber is intended for experiments using particle beams from the intersecting storage rings and the European 300 GeV synchrotron.

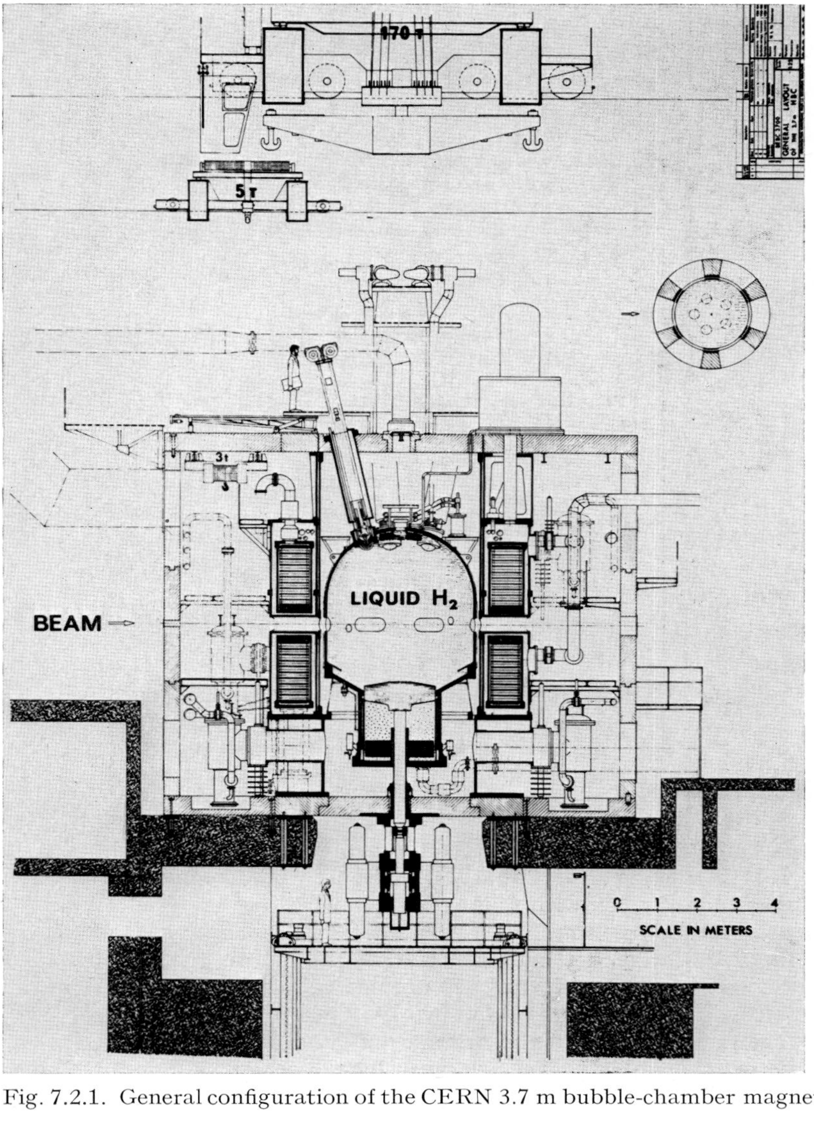

Fig. 7.2.1. General configuration of the CERN 3.7 m bubble-chamber magnet

The 3.7 m diameter bubble-chamber is surrounded by a superconducting magnet designed to produce a magnetic field of 3.5 T. The field homogenity within the cylindrical chamber volume is better than ± 3%. As illustrated in Fig. 7.2.1, the coil consists of two sections placed in two anular helium cryostats and separated by metallic spacers. A shield of low carbon steel surrounds the bubble-chamber magnet at a distance of about 3 m from the coils. The shield is primarily intended for safety reasons and to shield the magnetic field such that at 1 m distance around the iron shield, the flux density is less than 0.1 T.

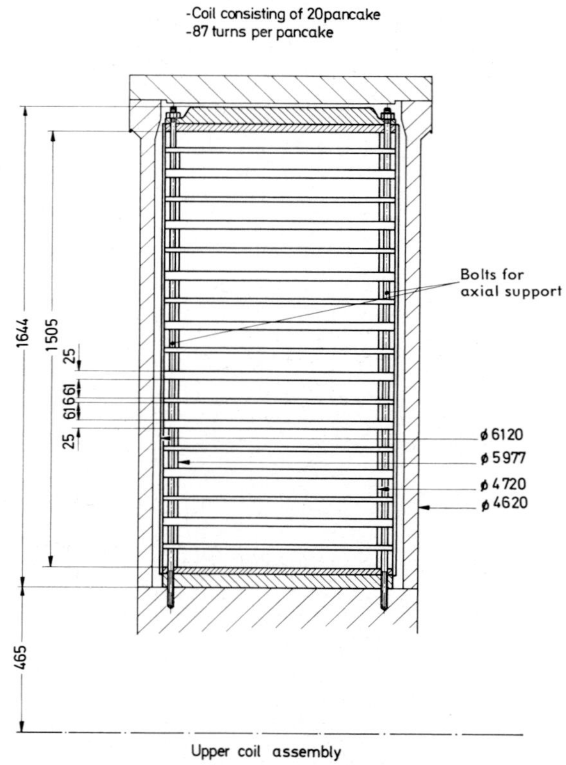

Fig. 7.2.2. Coil structure of the CERN bubble-chamber magnet

It is interesting to note that if the magnet would be built with OFHC copper and operated with water-cooling, a total power of 60 MW would be necessary to produce a field of 3.5 T in a split coil arrangement (shown in Fig. 7.2.2), where the coil inner diameter is 4.72 m. The total installed power input of the refrigerator to provide refrigeration of 900 W in the superconducting system is about 360 kW.

The main features of the magnet are given in Table 7.2.1. The coil structure shown in Fig. 7.2.2 is given by the coolant requirements and the electromagnetic forces.

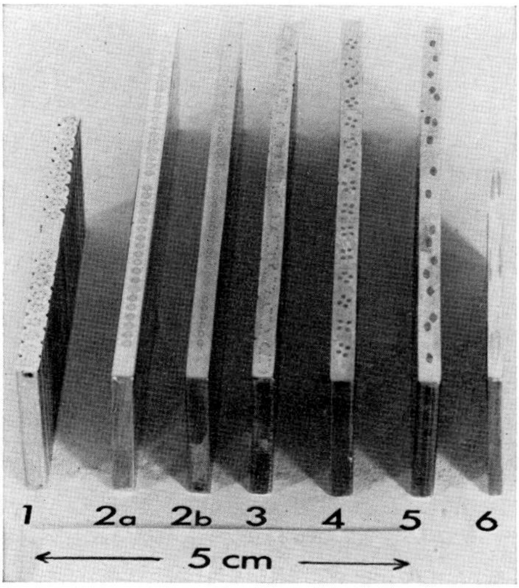

Fig. 7.2.3. Cross-section of Cu-NbTi composite conductors tested prior to the choice of the final configuration

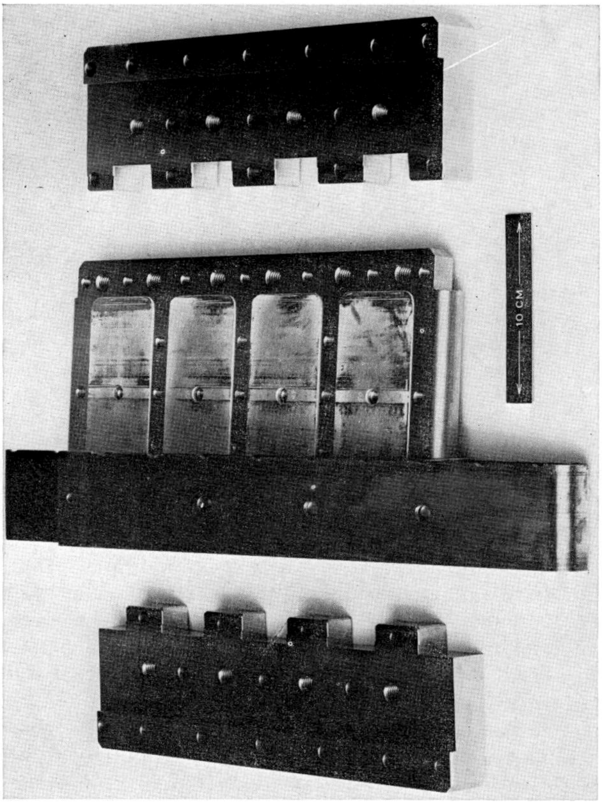

Fig. 7.2.4. Conductor joints used in the CERN bubble-chamber magnet

Each coil section (there are two) is composed of 20 pancakes with 87 turns each. The pancakes are separated by glass-reinforced epoxy plates, which withstand the compressive force and guarantee the cooling of each double layer. The conductor is reinforced by means of none-ferromagnetic stainless-steel and is, in addition to the edge cooling, provided with a coolant tape made of a copper band with milled-out spacers. In addition, each turn is provided with a heating tape and an insulation tape. The composite conductor consists of 200/224 NbTi filaments with individual diameters of 0.196 to 0.23 mm. The matrix material chosen is annealed OFHC copper. The ratio of copper to superconductor is about 26 : 1. Fig. 7.2.3 shows the cross-section of the composite conductors used for winding test coils, from which two types of conductor marked by 1 and 3 were finally selected for the main winding.

Fig. 7.2.5. Section of the large CERN bubble-chamber magnet

The conductor length in each pancake is 1500 m, which, to avoid joints within the pancakes, required manufactur ng composite conductors with a continuous length for at least one pancake. In order to comply with this condition, one manufacturer used electron beam welding to splice longitudinally conductor sections. Pancakes and double pancakes are joined between polished stainless steel blocks. Fig. 7.2.4 shows such a joint. The composite tapes at the beginning and ends of two adjacent pancakes are placed side by side and sandwiched between indium sheets covering both tapes, and stainless-steel plates, and compressed by means of strong bolts. The resistance of each joint at liquid helium temperature is less than 10^{-10} Ohms. Fig. 7.2.5 shows a coil section, where the joints are clearly visible.

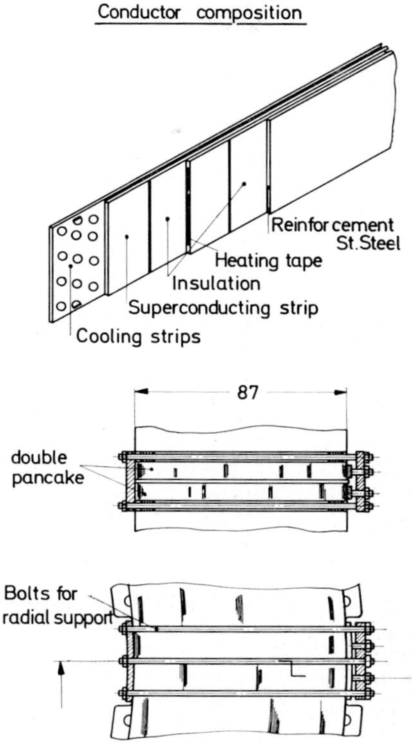

Fig. 7.2.6.
Conductor composition and pancake arrangements

The conductor composition and the pancake arrangement are shown in Fig. 7.2.6. The composite conductor various insulation tapes, heating and reinforcement strips are fed from separate storage coils into the main coil as shown in Fig. 7.2.7. The coil winding plant is illustrated in Fig. 7.2.8, where the main coil is wound (1) and the pancakes after winding, compacting and covering with protective tapes to ward them from dust, metalchips and water are stored (11) until needed for final assembly.

The function of the cooling strip and the reinforcing tape are obvious: The axial forces between the two coil sections are 9×10^6 kp, radial stress 1.23×10^3 kp/cm^2. The conductor reinforcement is prestressed during winding with 2×10^3 kp/cm^2. The heating strip consists of an aluminum tape sandwiched between two layers of insulations. If the heating strip is switched on, it heats up each pancake and enforces the transit on from superconducting to normal condition in a short time, dissipating 250 kJ in helium. The heating strip is especially useful if the magnet current is changed according to experimental requirement. When the magnet is fully energized from a 17 Volt powersupply, its field energy will be about 800 MJ. The magnet has been energized once in 1972.

1
2
3
4
5
6
7
8
9
10
11

Fig. 7.2.7. Winding detail of one pancake. The various tapes sandwiched to gether are from left to right, the cooling copper tape with milled out buttons, insulation tape (Polyamide) a stainless-steel reinforcement tape (6×0.2 cm^2), insulation ribbon (Polyester) an aluminum heating tape ($5.6 \times 0{,}1$ cm^2), insulation tape and the composite conductor ($6{,}1 \times 0.3$ cm^2)

Table 7.2.1. *The CERN 3,7 m, 3.5 T bubble-chamber magnet characteristics*

Coils	
Nominal current for 3.5 T field in the center (A)	5700
Ampere turns (40 pancakes, 87 turns) (A)	20×10^6
Coil inductance (H)	48
Stored field energy (J)	800×10^6
Theoretical value of power dissipated in the "normal" coil at 4.2 K, $B = 0$, (W)	1.1×10^6
Weight of copper in winding (kg)	92×10^3
Weight of superconductor (kg)	3×10^3
Weight of stainless steel incorporated in coil (kg)	60×10^3
Weight of insulation (kg)	11×10^3
Weight of stainless steel in container (kg)	200×10^3
Coil axial compressive force (kp)	9×10^6
Conductor constraining stress (kp/cm^2)	10^3
Constraining stress in reinforcement (kp/cm^2)	2×10^3
Coil inner diameter (m)	4.72
Coil outer diameter (m)	5.976
Iron yoke	
Inner diameter (m)	12
Outer diameter (m)	12.7
Weight of top and bottom plates (0.5 m thick) (kg)	$\sim 10^6$
Conductor (NbTi, Copper stabilized)	
Conductor cross-section (cm^2)	6.1×0.3
Copper to superconductor ratio	$\sim 26:1$
$\varrho_{273\,K}/\varrho_{4\cdot 2\,K}$	200
Maximum field at conductor (T)	5.1
Current density in the superconductor (A/cm^2)	80×10^3
Overall current density in conductor (A/cm^2)	1350
Number of superconducting filaments	200/224
Diameter of each filament (cm)	$3 \times 10^{-3}/4 \times 10^{-3}$
Continuous length of each conductor (m)	1560
Recovery current at 5.1 T (A)	7000
Conductor resistivity at 5.1 T and $\sigma = 1.2 \times 10^3$ kp/cm^2 (Ohm · cm)	3.5×10^{-8}
Power supply	
Voltage (V)	17
Current range (A)	0—6000 A
Magnet charging time (h)	6
Long term current regulation (%)	0.1

Fig. 7.2.8. The CERN (BEBC) coil winding arrangement
1 Coil winding table, *2* Composite conductor, *3* Heating type, *4* Reinforcement tape, *5* Stainless-steel tape, *7* Bath to clean the conductor, *8* Bath to clean the reinforcement tape, *9* Guidance arm with microswitch, *10* Conductor tension device, *11* Complete pancake

Reinforcing strip (Stainless-steel 316 L)

Cross-section (cm^2)	6×0.2
Continuous length (m)	250
Magnetic permeability at 4.5 K and law field	<1.05

Cooling strip (E.T.P. Copper)

Cross-section prior to machining (cm^2)	6×0.2
Unit length (m)	80

Insulating tape (Polyester and Polyimid film)

Thickness (μ)	$2 \times (150 + 50)$
Breakdown voltage at 4.5 K (V)	30×10^3

Heating tape (Aluminum 98.3%)

Cross-section (cm^2)	5.6×0.1

Refrigerator

Cool-down (specification) (Magnet coolded-down with He Gas)

Gas recovery temp. (K)	Flow (gs^{-1})	Refrigerator power (kW)
300	220	92.5
100	390	18.5
40	540	36

Normal operation (specification) Magnet

1.5 kW at $4.4 \leqq T \leqq 4.5$ K saturated vapor and
25 gs^{-1} at $45 \leqq T \leqq 300$ K

Estimated losses

Liquid Nitrogen shield for magnet cryostat	2 kW/80 K
Liquid Helium Dewar (25,000 lit) has a boil off of 0.5% per day	1 W head load or 30 W for refrigerator
Heat conduction through supports (6 feet)	50 W
Other heat conduction	100 W
Radiation to lN_2 shield/cryostat	150 W
Radiation to unshielded parts	100 W
Joule heating	100 W
Current leads (4.4 K)	150 W
Instrumentation wires (Sensing leads)	30 W
Transfer lines	200 W
25,000 lit He Dewar	20 W

7.3 Composite Magnet System, the McGill and MIT Hybrid Magnets

Generation of magnetic fields higher than 15.0 T is of considerable interest in many areas of physics. If the required useful experimental region is large (> 100 cm³), the cost of a single magnet and power supply system which can produce fields of, say, 20 T or more for continuous operation would be exorbitant.

We have seen in the previous chapters that very high fields can be generated by pulsing magnets, or to cool-down the coil conductor material to liquid hydrogen, or liquid helium temperatures. But we have also seen that many experiments cannot be performed with the short pulse time of a few milliseconds, specifically if the experimental data must be taken during the flat top of the pulse. Cryogenic magnets for continuous operation require large amounts of refrigerants and the cost of the refrigeration system is high, as we saw in Chapter 6.

Stevenson [5] pointed out that cryogenic magnets (for intermittent operation during a few minutes) can be built economically if the magnet is divided into a number of concentric coils. The subsections (modules) must be self-supporting and each one optimized for the required central field. For a cryogenic magnet, the input power into the coil P, and the central field, $B_{0,0}$, are linked together by the well-known Fabry relation:

$$B_{0,0} = G \cdot \left(\frac{P \lambda}{a_1 \varrho} \right)^{\frac{1}{2}}, \tag{7.3.1}$$

or in terms of a uniform distribution of current density:

$$H_0 = \mathfrak{J} \cdot a_1 \cdot (\lambda J) , \tag{7.3.2}$$

with G and $\mathfrak{J}$ being geometry and current density factors, λ the space factor, a_1 the solenoid bore radius, ϱ the conductor resistivity, and J the current density.

For cryogenic magnets, considerable power is saved if pure metals are utilized (Chapter 6). Specifically, pure aluminum is of much interest, where the resistive ratios (r) from room temperature to liquid helium temperature of about 10^4 in tape can be utilized.

Stevenson proposed the construction of a two-module magnet to generate a field of 22 T during 1.4 minutes. He uses two Al coils. Helium gas at 13 K inlet temperature and 18 K outlet temperature circulates through coolant passages under a pressure of 25 kp/cm². The conductor is about 2 K warmer than the coolant. At the design field, the resistance ratio for the chosen aluminum from room temperature of 4.2 K is about 1000 with no stress effects, and about 330 under magnetomechanical stress. At the operational temperature, $\varrho \cong 8.04 \times 10^{-9}$ Ohm · cm. The power requirement and the expected maximum stress levels are calculated briefly as follows:

For a solenoid, we may write Fabry's relation as follows:

$$P = \left(\frac{B}{G} \right)^2 \cdot \frac{\varrho a_1}{\lambda}, \tag{7.3.3}$$

which is expanded for two concentric modules, a and b, as:

$$\begin{aligned} P = P_a + P_b &= \left(\frac{B}{G} \right)_a^2 \cdot \left(\frac{\varrho a_1}{\lambda} \right)_a + \left(\frac{B}{G} \right)_b^2 \cdot \left(\frac{\varrho a_1}{\lambda} \right)_b \\ &= k_a B_a^2 + k_b B_b^2 \\ &= k \cdot B^2. \end{aligned}$$

The minimum power required is calculated to be:

$$P = \frac{k_a k_b}{k_a + k_b} \cdot B^2 . \tag{7.3.4}$$

If both coils are optimized and $a_{1,a} = 2$ cm, $a_{1,b} = 11$ cm, then with $\alpha_a = 5.3$, $\alpha_b = 2.4$, $\beta_a = 5$, $\beta_b = 2.4$, we have the geometry factors $G_a = 0.162 \times 10^{-4}$, $G_b = 0.147 \times 10^{-4}$ for uniform current distribution [6]. Space factors given are $\lambda_a = 0.747$ and $\lambda_b = 0.877$. If one assumes that ϱ is the same for both modules, which is not the case due to the lower field at the outer coil, then we get for $B_a = B_b = 11$ T, the power:

$$P_a = 10 \text{ kW}; \quad P_b = 57 \text{ kW} .$$

Thus, to generate a field of 22 T, a power of 67 kW would be required

A single solenoid to generate the same field optimized with a radius of $a_1 = 2$ cm would require a power of 73 kW with $\alpha = 12.8$, $\beta = 10$, and $G = 0.12 \times 10^{-4}$. As illustrated in Fig. 6.2.3, a refrigerator to remove a continuous load of 67 kW will require an installed power of 2 MW and will cost (according to Fig. 6.2.5) about \$ 1.5×10^6. A cryogenic system of this magnitude is not competitive with a room temperature setup. On the other hand, if the magnet is operated for the time of a few seconds to one minute (as proposed by Stevenson), a cryogenic system handling a load of 6×10^6 Joules at 13 to 18 K with helium gas with a mass of 3.6×10^5 g becomes quite attractive. The field is pulsed in a triangular shape with a duration of 1.4 minutes or so, for experiments where field homogeneity is not required. During this time, the refrigerant is pumped into the magnet. After the pulse, the coolant is stored until the mass of 3.6×10^5 g of helium at 13 K is compressed in a storage vessel and the magnet is ready for the next pulse.

Another interesting feature of the composite solenoid is the stress distribution over the coil, specifically on the conductor. Recalling that the force $\boldsymbol{F}$ produced by the magnetic field $\boldsymbol{B}$ and the current density $\boldsymbol{J}$ are expressed in the form

$$d\boldsymbol{F} = (\boldsymbol{J} \cdot \boldsymbol{B})\, dv . \tag{7.3.5}$$

The tangential stress in an unsupported conductor to balance the magnetic force is given in simplified form:

$$\sigma = J \cdot B a_1 , \tag{7.3.6}$$

since J and B are perpendicular.

Eliminating J from Eqs. (7.3.2) and (7.3.6), we get:

$$\sigma = \frac{B^2}{\mathfrak{J} \cdot \lambda} . \tag{7.3.7}$$

The stress depends only on the field and the coil configuration, and not on any particular coil dimension. In Eq. (7.3.7),

$$\mathfrak{J} = \mu_0 \beta \ln \frac{\alpha + (\alpha^2 + \beta^2)^{\frac{1}{2}}}{1 + (1 + \beta^2)^{\frac{1}{2}}} . \tag{7.3.8}$$

For a composite coil, we may write from Eq. (7.3.7)

$$\sigma = \frac{B_i \cdot (B_i + B_0)}{\mathfrak{J}_i \cdot \lambda_i} . \tag{7.3.9}$$

With $\mathfrak{J}_i$ the current density factor of the inner coil and B_i the field generated by the inner coil alone.

Eq. (7.3.1) shows the effect of two-module coils, where the stress at the inner bore diameter is proportional to $B_i \cdot (B_i + B_0)$, rather than $(B_i + B_0)^2$ for a single-module magnet. For the above particular case of a two-module coil:

$$\mathfrak{J}_i = 3.76 \times 10^{-6}$$
$$B_i = B_0 = 11 \text{ T},$$

at 15,000 A, and thus:

$$\sigma = 865 \text{ kp/cm}^2.$$

This is a rather high stress level for a pure aluminum conductor at 13 K and will require special conductor reinforcements. It should be pointed out, however, that Eq. 7.3.9 is valid only for an unsupported conductor without the counteraction of neighboring turns.

The magnet system under construction at *MIT Francis Bitter Magnet Laboratory* [7] consists of water-cooled and superconducting solenoids (shown in Fig. 7.3.1). This magnet system deviates from the cryogenic magnet proposed by Stevenson in several ways, which we describe below.

Part of the field is generated by superconductors and part by the water-cooled copper conductor. Such a system has the advantage that higher fields can be produced economically, than can be generated by

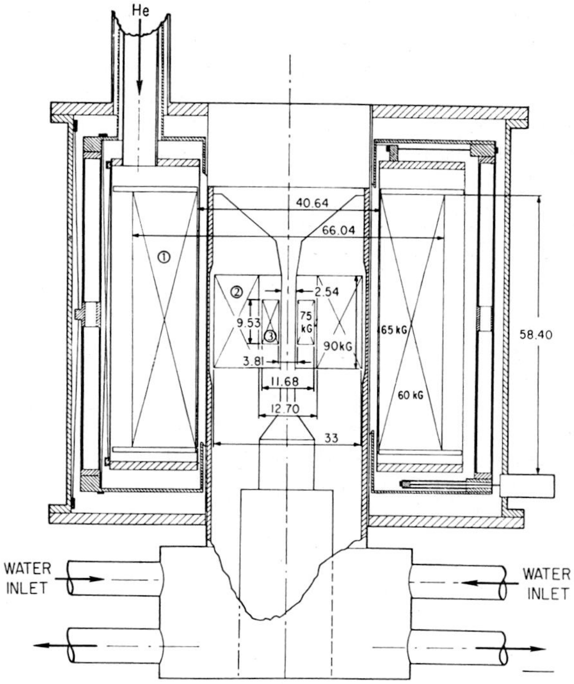

Fig. 7.3.1. MIT-Hybrid magnet. *1* Superconducting coil, *2* Copper coil, *3* Copper coil insert

either of the parts alone. Presently, one of the obvious reasons for building composite-type magnets is the fact that the upper critical field of commercially available Type II superconductors is limited. Thus, the superconducting portion of the magnet generates the lower field; the high field portion is generated by the water-cooled conductor. The two parts are separated by an appropriate cryostat and are energized from various current sources.

Let us concentrate on the superconducting portion first: Close to the critical field, the B—J curve has a "knee" and the current-carrying capacity in the conductor is drastically reduced if the field is increased above this somewhat characteristic value. The choice of the superconducting material is linked closely to the field at the conductor. Referring to Fig. 6.2.18, we see that Nb_xTi* composites are less expensive than other Type II superconductors at lower fields. A cross-over point is the field of ~ 8 T.

The superconducting MIT hybrid magnet generates 6 T in the center with a field of ~ 6.6 T at the conductor. The conductor material chosen is Nb_xTi composite conductor with 60 fine superconducting filaments imbedded in a copper matrix. The room temperature bore has a diameter of ~ 31 cm.

The inner magnet part (non-superconducting) was chosen to be water-cooled copper, the main reason being that the MIT magnet laboratory is equipped with sufficient power (8 MW for continuous operation). Given now the external field of 6 T from the outer superconducting magnet, a field of 9 T can be produced with an intermediate coil (uniform current density distribution) having a bore of 12.7 cm with a power of ~ 4.25 MW. An inner coil with a bore of 3.8 cm diameter finally generates a field of 7.5 T with a power consumption of 0.75 MW (radial current density distribution), the total field thus being 22.5 T with a power level of ~ 5 MW at a working bore of 2.54 cm.

The designers claim that by replacing presently-built inserts with new ones at a power level of 10 ... 16 MW, fields close to 30 T can be generated.

Although the state of the art in superconducting technology permits the building of superconducting coils producing even 10 to 12 T, they have taken the prudent approach of building, as a first step, the superconducting magnet generating 6 T at a field energy level of 2 Mjoules. The magnet is energized from an 8-kW power supply, the helium is provided from an external 250-liter dewar. Between successive runs, the coil is warmed up to 20 K and kept at this temperature by means of an regenerative refrigerator. The superconducting magnet is charged in 15 minutes. Field-sweeping operation is accomplished by the two water-cooled inner solenoids. An energy dissipation system, similar to that described in Section 7.1 ,with a shunt resistance of 0.1 Ohm and a two-pole breaker, is foreseen to disconnect the magnet from the power supply if a quench occurs.

---

* The subscript x denotes the ratio of Nb to Ti, such as Nb (50%) Ti or Nb (60%) Ti, etc., which differs according to the manufacturer.

We summarize some of the features of this magnet system: All water connections are located at the bottom of the coil. The water-cooled inserts can be removed from the magnet without disturbing the cryostat.

The innermost coil at 22.5 T has a Bitter disc-type design with radial or axial cooling, described in papers by Montgomery [8] and Brechna [9]. The principal design problem in the magnet cryostat is the support of the magnetic interaction forces between the inserts and the superconducting coils. The interaction force results in an axial spring constant of 7200 kp/cm and a radial displacement of 1070 kp/cm. The axial force is a restoring force and proper coil alignment to superpose the magnetic centers may eliminate it. However, if asymmetric failures prevail in the inserts, accelerating forces act during a 300-ms period, and peak forces of 24 g may result.

The radial force is unstable and necessitates provisions for restoring radial springs. To counteract both forces, spring action is provided.

Table 7.3.1. *MIT hybrid magnet — main characteristics*

Superconducting coil	
Central field generated (T)	6
Inside coil diameter (cm)	40.7
Number of double pancakes	24
Turns of pancake	52
Design current (A)	1500
Total conductor length (m)	4500
Coil weight (kg)	680
Stored energy (J)	2×10^6
Conductor parameters	
Cross-section (cm^2)	1×0.2
Copper-to-superconductor ratio	8 : 1
Number of superconducting strands	60
Filament diameter (cm)	0.023
Maximum heat flux (W/cm^2)	0.25
Coolant passages (cm^2)	0.0435×0.76
Edge-cooling of each pancake face	80%
Heat losses	
Thermal conduction and radiation losses (W)	0.52
Lead losses at 1500 A (W)	3
Cool-down from 70 K (liters)	140
Cool-down from 20 K (liters)	5
Intermediate water-cooled coil	
Central field generated (T)	9
Inside coil diameter (cm)	12.7
Power consumption (W)	4.35×10^6
Inner water-cooled coil	
Central field generated (T)	7.5
Inside coil diameter (cm)	3.8
Working bore diameter (cm)	2.54
Power consumption (W)	0.75×10^6

The main features of this magnet system are summarized in Table 7.3.1.

The turns per pancake are separated by 0.254 cm long by 0.038 cm high polyester glass spacers cemented on both sides of the conductor.

The pancake faces are exposed to the coolant (80%) by means of Micarta strips, which separate the pancakes and support axial forces of 136,000 kp which are at the median plane, resulting in a compressive force of 385 kp/cm². The peak conductor stress is 760 kp/cm² and thus moderate, which does not require the use of reinforcing tapes wound parallel to the conductor.

The superconducting outer coil was tested successfully in 1972. The completed coil assemply is anticipated to be operational in 1973.

7.4 The Oak-Ridge-IMP*-Superconducting Coil System

A superconducting mirror quadrupole system has been constructed at ORNL's Thermonuclear Division [10] with the aim to create a multi-keV proton plasma by injecting an energetic H° beam into a target plasma whose ionic component is cold protons. Design work for this system started in the later part of 1966. The NbTi mirror coils were completed and became operational in July 1969. The Nb_3Sn quadrupoles were completed and tested in 1971. The entire system shown in Fig. 7.4.1 was scheduled to become operational in the later part of 1971.

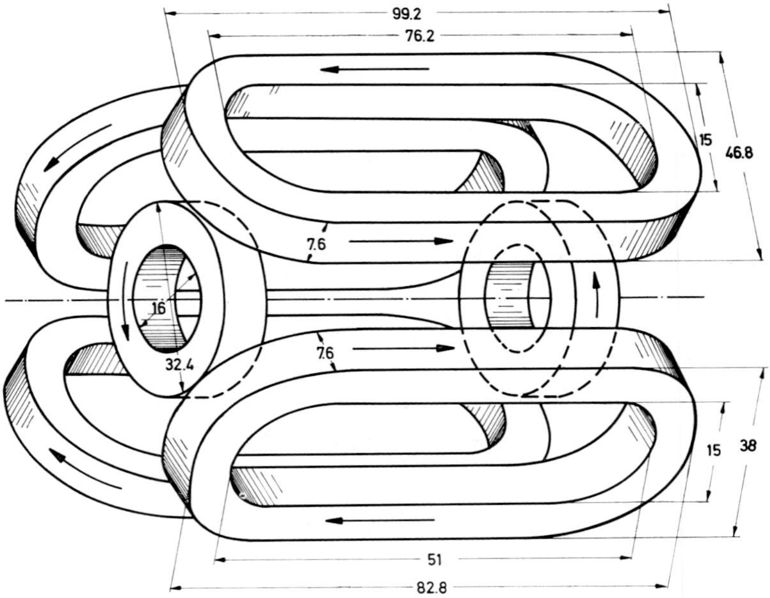

Fig. 7.4.1. The ORNL-IMP superconducting coil system

* Injection into microwave plasma.

Criteria for the IMP magnetic fields are briefly summarized:

1. The field configuration should be d.c. and of "magnetic well" geometry to provide stabilization against MHD modes of plasma instability.

2. The central field value and the well depth should be separately controllable in order to permit wide variation in position of the field contour for microwave resonance relative to both B_0 and the last closed magnetic contour.

3. Operation with resonant heating at 35.7 GHz ($B_{res} \cong 1.25$ T). A fairly deep magnetic well (depth 1.5 ... 2.0 B_0) was desired for the experiments.

To comply with the requirements, the central field of 2 T and a throat field of 4 T (to give a mirror ratio of 2 : 1) were produced by mirror coils having a clear bore of 14.7 cm and a coil case separation of 14.6 cm. The mirror coils are surrounded by a quadrupole magnet system. With all six of the coils energized to their design values, a maximum field of about 8.5 T occurs in the windings of the quadrupoles; a maximum field of 5.9 T occurs in the windings of the mirror coils.

Table 7.4.1 gives most pertinent data of the system.

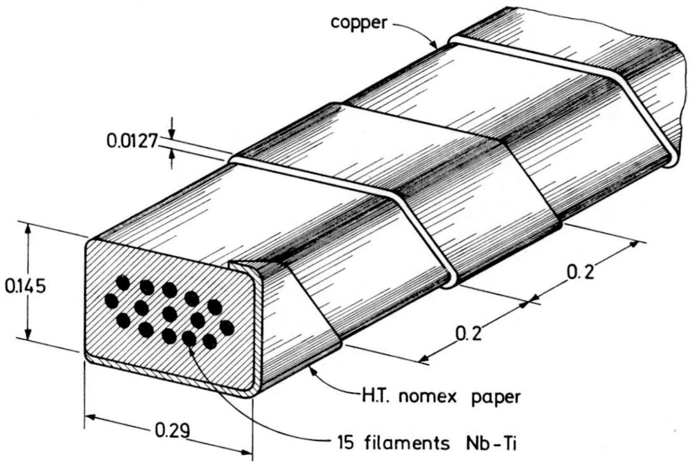

Fig. 7.4.2. The IMP-Mirror Coil composite conductor

An optimized design using water-cooled copper coils was studied. It was found that about 10 to 15 MW power would be necessary to generate the above fields! A hybrid system with conventional mirror coils and superconducting coils was studied, but abandoned because of excessive forces between the two systems. A completely superconducting system appeared to be a reasonable solution.

The conductor for the mirror coil consists of stabilized NbTi multifilament composite conductor, the quadrupole conductor of aluminum stabilized Nb_3Sn tapes. As the choice of the appropriate conductor met with a variety of problems, we briefly describe tests performed prior to the conductor selection.

NbTi Conductor for Mirror Coils

The composite conductor shown in Fig. 7.4.2 was selected. To facilitate winding and to withstand the electromagnetic forces, a rectangular cross section was chosen. Insulation was provided by a 50% coverage of Nomex* paper. Nomex paper allows less motion of the conductor under electromagnetic stresses than mylar; however, it has a very low shear strength.

Fig. 7.4.3. ORNL-Cusp Coil testing arrangement

ORNL has employed two standard test methods, the short sample test and a coil test. While the short sample test is quite common in various laboratories, the coil test utilizes about 300 m of conductor wound into two identical coils, which are operated in a series cusp mode (Fig. 7.4.3). They are tested with the horizontal axis in the vertical field of a 6 T copper coil energized from a 6 MW power supply. As indicated, the peak forces are axially outward at the bottom of the coil and axially inward at the top. The maximum field is in the winding section near the

* Dupont trade name.

median plane. The cusp test reproduces the condition found in the mirror-quadrupole system best. Test results for both short sample and cusp coil are given in Fig. 7.4.4. It may be pointed out that at the time of the tests twisted filaments in composite conductors were not commercially available. When the conclusions of the tests were to select for the mirror coil, the conductor with 15 filaments having individual diameters of 0.0284 cm, and a copper to superconductor ratio of 3.5 : 1 was chosen. The conductor did not exhibit fluxjumps at low fields but only small fluxjumps at high fields. Although the short sample tests showed a conductor performance of 2×10^4 A/cm² at 6 T, the design of the mirror coil is more conservative by choosing an overall current density of 6560 A per cm² at 5.9 T. In tests of the individual mirror coils, each reached short sample performance at 6.7 T with 10^4 A/cm² with 0.2 A/sec charging rate. With faster charging rates, the critical current was degraded in quantitative agreement with cusp coil tests.

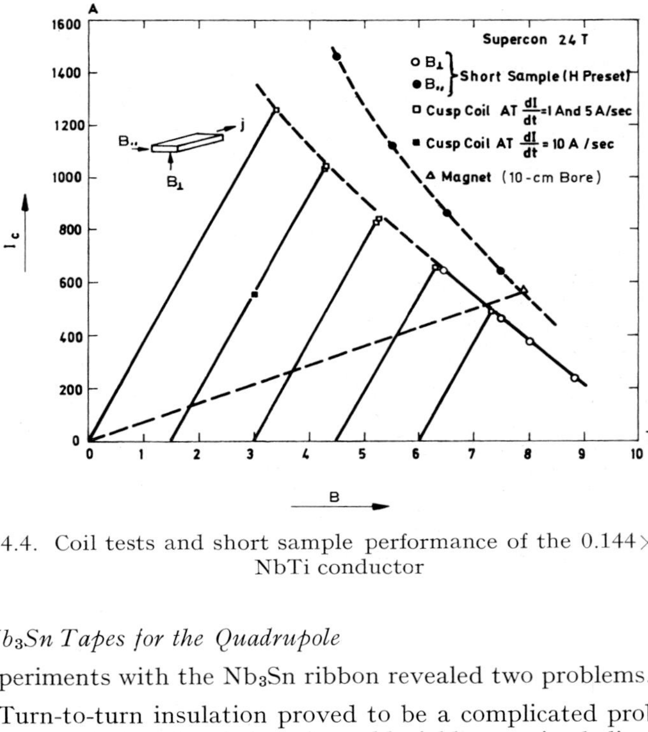

Fig. 7.4.4. Coil tests and short sample performance of the 0.144×0.29 cm² NbTi conductor

The Nb_3Sn Tapes for the Quadrupole

Experiments with the Nb_3Sn ribbon revealed two problems.

1. Turn-to-turn insulation proved to be a complicated problem. No interturn insulation (local shorts) would yield excessive helium boil off when the magnet was charged or discharged.

2. The Nb_3Sn tape developed an instability when the magnetic field perpendicular to the face of the ribbon ($B_\perp$) reached the values of 2.5 to 3 T. Frequently, this instability caused a superconducting to normal transition.

As an interturn insulation, a thin coating (1.27×10^{-4} cm) of Al_2O_3 of filled colloidal graphite worked best. It had a resistivity of > 500 Ohm · cm at 4.2 K when a pressure of 55 kp/cm² was applied.

The reason for the occurrence of the instability in Nb_3Sn tapes when exposed to $B_\perp$ fields is not clear. However, it may be due to heating effects. To eliminate this effect, two methods were applied and tested in the cusp coil experiment: The use of small high purity Al-shunts and the co-winding of an aluminum tape with the superconducting ribbon. The first method was difficult to construct, as the shunts were damaged easily. The Al tape wound with the superconductor (1.5×10^{-3} cm thick with $r = 2000$) was much more successful. It provided eddy current damping, which prevents sudden changes in the B component, it acted as a heat sink, a thermal path from inside the pancake to its edge, and as a current shunt.

Diffusion processed Nb_3Sn tape, which was 1.27 cm wide and 2×10^{-2} cm thick, was sandwiched with Al ribbon and insulation tape. The design current density of 13,500 A/cm² at 8 T (field at the winding) could be achieved. The distance between pancakes was 0.31 cm; thus, an overall space factor of less than 0.7 was obtained.

Table 7.4.1. *ORNL-IMP-coil specifications*

Mirror coils (NbTi-Copper composite)

Design current density (A/cm²)	6558	
Current to give the necessary current density (A)	383.2	
Maximum field at winding (T)	5.88	
Number of turns (2 sections)	2498	2489
Coil area (cm²)	146	
Inductance (H)	2	
Conductor cooling surface (%)	50	
Weight of completed coil (kg)	272	
Conductor length (m)	1905	1897
Field energy (J)	147×10^3	

Quadrupole coils

Design current density (A/cm²)	13,500			
Current to give the necessary current density (A)	800.6	804.7	823.6	821.4
Maximum field at the winding	7.2	7.2	8.5	8.5
Maximum $B_\perp$ (T)	3.9	3.9	6.2	6.2
Number of turns (4 sections)	1196	1190	1820	1825
Coil area (cm²)	70.93	70.93	111	111
Conductor cooling (edge cooling) (%)	50			
Weight of completed coil (kg)	145	145	336	336
Conductor length (m)	2000	1980	5560	5580
Field energy (J)	2.24×10^6			

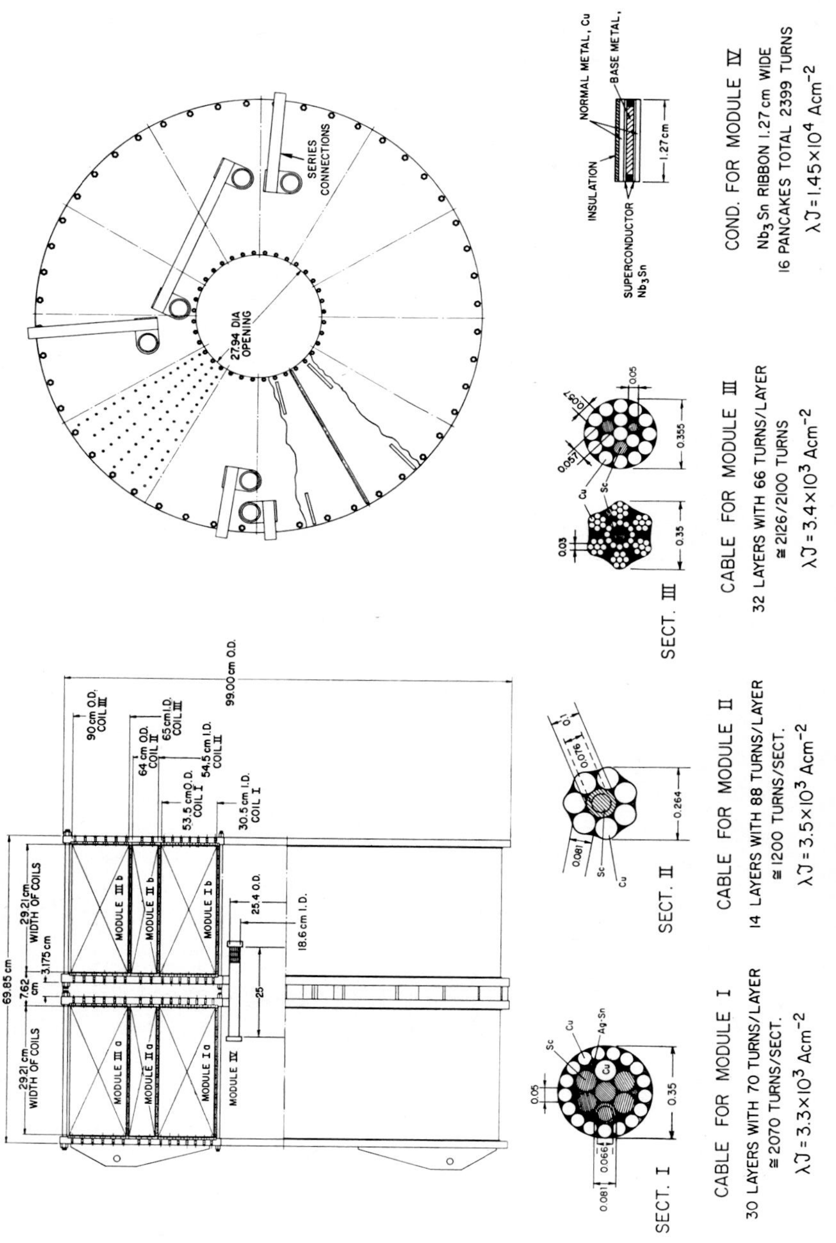

Fig. 7.5.1. General layout of the SLAC 7 T magnet with a Nb_3Sn tape wound coil to generate 10.3 T

7.5 The SLAC 7 T, 30-cm Bore, Helmholtz Magnet

The SLAC magnet, originally built as a model for a large bubble- or spark-chamber magnet system, is now being used as a test facility. The design was started early in 1966 and the magnet completed in October 1967. It has since been in intermittent operation.

The basic magnet configuration is seen in Figs. 7.5.1 and 7.5.2. Each coil section consists of three modules. Coils I and II are wound with a Nb(22 at %) Ti and copper cable, impregnated with a silver-tin alloy (7% Ag) for rigidity and better thermal conduction to the bath. Coil module No. III consists of Nb(25%)Zr and copper cable. The coil subdivision had three reasons: To current-optimize the conductor, to limit the axial and radial magnetomechanical forces, and to operate each part of the coil separately (if required). Thus, each coil module is mechanically self-suppourting and can slip axially on each other. The axial coil separation can be chosen according to the experimental requirements. Each coil may be energized from a separate current source, or modules I and III can be operated in series from one power supply.

Fig. 7.5.2a. SLAC coil arrangement

The cable, a predecessor of the composite multi-strand conductor, is spirally wrapped with a multi-filament cylindrical nylon braid (Nomex) in order to achieve free helium passage between adjacent turns (Fig. 7.5.2b). The coolant gap between adjacent turns is approximately 0.5 mm. Heat transfer measurements and calculations, reported in Chapter 4 on small coil configurations with various widths of coolant passages, indicate that the conductor is stabilized to a maximum heat flux of 0.4 W/cm² in the coil. The magnet current in the inner coil section I is 500 A for stable coil performance at 7 T design field. Using the data in Fig. 6.2.13 and assuming that theoretically 25% of the cooled perimeter is covered by the insulation braid, the maximum heat flux (for cryostatic stability) per unit length is calculated from Eq. (6.2.27) with the maximum field at the conductor being 7.5 T, the copper resistivity is 4.5×10^{-8} Ohm · cm.

ig

one gets

$$h\Delta T = \frac{25 \times 10^4 \times 4.5 \times 10^{-8}}{3.85 \times 10^{-2} \times 0.824} = 0.355 \text{ W/cm}^2.$$

However, after repeated operations, several quenches, thermal cycles, and indications of short-circuited turns, it is unlikely that the original gap width has been still retained. Calculated compressive radial stress on the conductor amount to 175 kp/cm², and the axial compressive stress on the conductor is 300 kp/cm². The combined compressive stresses on the insulation is $\sim 6.5 \times 10^3$ kp/cm², a value which is below measured values for braided polyamid, shown in Table 7.5.1.

The advantage of such an insulation scheme is somewhat hampered by the later observation of fatigue due to thermomechanical stresses. As indicated in Section 7.4, the conductor can move under the influence of electromagnetic forces yielding fluxjumps and premature quenches. The insulation was squashed in high stress areas, yielding a more axially compact coil and narrower coolant passages. In addition, sporadic

Table 7.5.1. *Mechanical and electrical properties of some insulation tapes at 300 K and 78 K*[a]

Properties	E-Glass fibers[b] multi-filament	Glass fibers impregnated (multifilament tapes) Epoxy	Polyester	Silicone	Polyester mono-filament	Polyamid[c] (multifilament braid) Heat treated	Silicone impregnat.
Tensile strength (kp/cm^2)	2.9×10^4	3.9×10^3	3.2×10^3	1.5×10^3	10^3	4×10^3	3×10^3
Compressive strength (kp/cm^2)	$>3\times10^4$	3.5×10^3	2.5×10^3	1.5×10^3	4×10^3	1.2×10^4	0.8×10^4
Breaking elongation (%)	<1	<1	<1	<1	11	2	2
Initial E modulus (kp/cm^2)	7.5×10^5	4×10^5	2.1×10^5	2×10^5	7×10^4	10^4	0.8×10^4
Density (g/cm^3)	2.55	1.6—2	1.9	1.5—1.8	1.1—1.4	1.3	1.3
Specific heat (Ws/g K)	0.8	0.8	0.8	0.8	0.6	0.7	0.7
Thermal expansion coefficient, K^{-1}	4.8×10^{-6}	$(1.3—1.8)\times10^{-5}$	1.6×10^{-5}	2×10^{-5}	7×10^{-4}	6×10^{-4}	8×10^{-4}
Thermal conductivity (W/cm K)	1.04×10^{-2}	8×10^{-3}	1.1×10^{-2}	10^{-2}	1.2×10^{-3}	1.3×10^{-3}	1.5×10^{-3}
Volume resistivity (Ohm · cm)	10^{13}—10^{-5}	10^{16}	10^{14}	4×10^{15}	10^{14}	10^{14}	10^{14}
Dielectric constant	6.3	—	—	—	6.6	—	—
Dielectric strength (V/cm)	37	2×10^5	4×10^4	10^5	10^5	0.8×10^5	10^5

[a] Average of 5 samples. Strength value ± 30%.

[b] Brittle, fractured after repeated thermal cycling and during wrapping around conductor.

[c] Also commercially available under trade name Nylon (Dupont); tapes available in various widths and thicknesses; retains plasticity even at 4.2 K.

interturn and interlayer shorts were observed, which adversely affecte the coil charging time.

The magnet is energized by means of helium counterflow current leads and is suspended by a stainless steel tube attached to the top flange. The vertical dewar, designed rather conservatively in classical fashion, has a liquid nitrogen jacket. To reduce helium evaporation by radiation and helium gas conduction, a vacuum vessel is placed inside the helium tank. When the experimental setup is placed inside the magnet bore, the cylindrical central bore is closed and the helium gas is forced through a spiral passage on the outer surface of the vacuum tank. With the coil not energized, the hourly helium evaporation is about 4 liters. When the coil is energized to full field, the helium evaporation is 12 liters per hour. Pertinent magnet parameters are shown in Table 7.5.2. At 7 T, the magnet field energy is approximately 5.2 MJ.

Cool-Down Method

The magnet is initially cooled-down by means of liquid nitrogen pumped to triangularly-shaped copper vessels attached to the middle stainless steel flanges. When a temperature of ~ 100 K is reached, the cool-down is performed by means of a helium gas transferred directly from a 7-watt helium liquefier operated as a refrigerator, with 25 W capacity at 4.5 K. Generally, it is attempted to use the latent heat of the cold gas, but the magnet system does not permit a very efficient cooling and most of the cold gas passes through the magnet bore and outside coil surface. The warm gas is collected from the dewar in gas bags and a recovery system of low- and high-pressure helium gas tanks by means of an automatically activated compressor. After repurification, the gas is re-refrigerated or reliquefied (Fig. 7.5.3), and either stored or transferred back to the magnet dewar.

Table 7.5.2. *SLAC 7 T, 30 cm bore, magnet parameters without high field insert module IV*

Coil parameters		
Maximum central field (T)		7.2
Field energy (J)		5.5×10^6
Number of turns:	Section I	4140
	Section II	2800
	Section III	4140
Operational currents, I/II/III (A)		475/600/500
Overall current density (A/cm²)		4000
Axial gap (cm)		8 min.
Charging time (hrs)		1
Nucleate boiling heat flux (W/cm²)		0.38
Magnet weight (kg)		1600
Coil inductance		34
Conductor hoop stress (kp/cm²)		1100
Radial stress (max) (kp/cm²)		178
Axial stress (kp/cm²)		400
He evaporation at 6.4 T level (l/hr)		12

When the temperature of 40 K is monitored by means of thermocouples and carbon resistors in the magnet, liquid helium is transferred to the dewar directly from the liquefier or helium storage vessels.

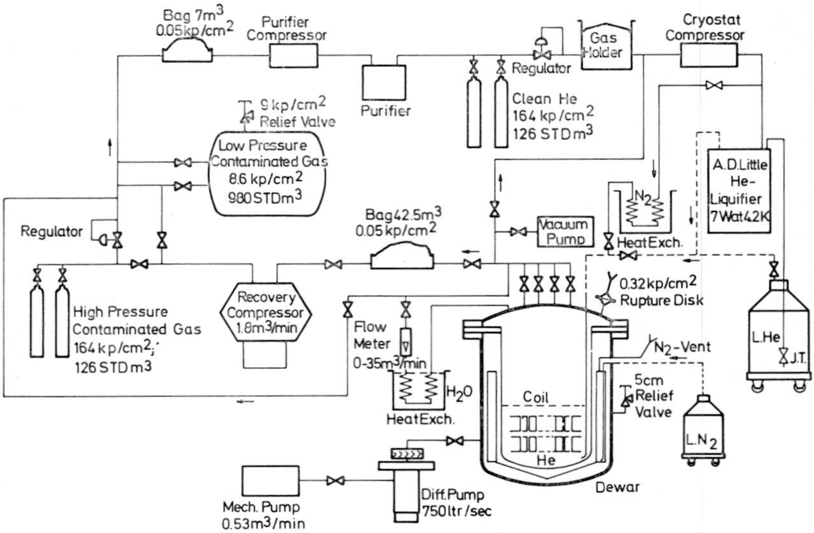

Fig. 7.5.3. SLAC-superconducting magnet and cryogenic system

The magnet has a total weight of 1600 kg. A total of 42,000 m of conductor is wound in the six modules. The cool-down of the magnet from 100 to 40 K requires 48 hours, and from 40 to 4.2 K, six hours. The total amount of liquid helium to cool-down the magnet to 4.2 K and fill the dewar until the coil is immersed is 650 liters.

The helium evaporation of the magnet corresponds to 12 liters of helium per hour, corresponding to ~ 8 W liquefaction. Thus the refrigerator can provide sufficient cooling for continuous magnet operation. This type of operation would require closed-loop refrigeration.

Theoretically, very little helium should be lost to ambient, but through the piping, joints, and material faults (mostly during and after experiments), about 15% of the total amount of helium stored per month is lost to ambient through leaks. Helium gas is purchased in large quantities and stored in high-pressure bottles, from which the gas is purified, reliquefied, and stored in dewars. (Total present capacity is ~ 1200 liters.)

The magnet dewar can withstand about 2 kp/cm² internal pressure. However, a 10-cm diameter rupture disc is provided which explodes when the internal pressure exceeds 1.4 kp/cm².

Operational Experience

When the magnet was first energized, it became evident that the conductor choice for module II was inadequate. A single superconducting filament with a diameter of 0.076 cm, incorporated each within seven strands of copper (0.081 cm diameter each), experienced fluxjumps which led to several catastrophic quenches at 360 A, corresponding to a current density of 6540 A/cm² over the conductor, and 7.9×10^4 A/cm² in the superconductor at a maximum field of ~ 4.7 T — a disappointing result!

There are two reasons why flux movements over the entire magnet and catastrophic fluxjumps in module II were encountered:

1. Due to Lorentz forces, the conductor experiences macroscopic wire movements. The conductor insulation is compressed. These wire movements generate flux motion and fluxjumps.

2. The superconductor in module II with its large diameter is unstable by itself. The heat generated by fluxjumps and flux motion is not carried away fast enough due to low thermal diffusivity; the conductor exhibits training and degradation.

To illustrate the unstable performance of the superconducting cable in module II, the following simple calculation is performed:

P. F. Smith [12], J. Stekly [13], and R. Hancox [14] have calculated the diameter of individual filaments in an "intrinsically stable" composite conductor:

$$d_f^2 < \left(T_0 \frac{k_s}{\varrho_n}\right) \left(\frac{1}{J_m \cdot J_c}\right), \qquad (7.5.1)$$

with $T_0 = J_c \Big/ \left(-\frac{dJ_c}{dT}\right) \cong \frac{T_c}{2}$ and for $Nb_xTi \cong 5$ K,

k_s = thermal conductivity of the superconductor = 1.2×10^{-3} W per cm K at 4.2 K,

J_m = overall current density in the composite conductor,
= 6540 A/cm² at 360 A,
= 9080 A/cm² at 500 A,
(corresponding to about 5 T at the conductors)

J_c = critical current density in the superconductor,
= 1.2×10^5 A/cm² at 5 T,

ϱ_n = resistivity of copper at 5 T = 3.25×10^{-8} Ohm · cm.

Thus: $d_f < 1.53 \times 10^{-2}$ cm (for 360 A),

and $d_f < 1.3 \times 10^{-2}$ cm (for 500 A).

The superconductor used has a diameter of 7.6×10^{-2} cm! The overall composite conductor current density is dependent upon the specific heat of the substrate and the overall ratio of superconductor to copper diameter,

$$J_{av} = \frac{1}{d_n} \left[\frac{3 \times 10^9}{4\pi} \cdot c_p \delta \cdot \frac{d_s}{d_n} \cdot T_0\right]^{\frac{1}{2}}, \qquad (7.5.2)$$

with $c_p \delta$ = thermal capacity of the copper matrix,
= 9×10^{-4} Ws/cm³ K,

d_s = diameter of the superconducting filament,
= 7.6×10^{-2} cm,

d_n = cable diameter, assuming the impregnation is faultless and the heat can be absorbed readily,
= 2.642×10^{-1} cm.

The expected average current density for enthalpy cooling would be

$$J_{av} = 2100\ \mathrm{A/cm^2} < 6540\ \mathrm{A/cm^2}\ !$$

This value corresponds to a conductor current of 115 amperes.

The coil intermediate section, which contributes only about 1.3 T to the total field, is unstable. Specifically, when the magnet field was raised above 6.5 T, the helium boil-off would exceed 20 liters/hour. In order to avoid repeated quenches, the magnet operation (for experimental purposes) was limited to 6 T.

For continuous and long-range experiments, the coil Section 2 had to be modified. It was replaced by a square composite conductor ($0.18 \times 0.18\ \mathrm{cm^2}$) with a take-off current of 600 A and 6 T. The copper-to-superconductor ratio is 4 : 1, and the average current density over the conductor is $15 \times 10^3\ \mathrm{A/cm^2}$. The magnet is now suitable for experimental work up to 7.2 T maximum central field.

Coil Protection

It was stated that with the exception of the new module, the coil is not cryostatically stabilized. Flux motion due to wire movements cannot be prevented in this design. To protect the magnet against damage in case of a quench, a circuit (given in Fig. 7.5.4), was used in order to dissipate more than 90% of the field energy externally. Compared to the first two quenches, at 6 T and 6.47 T, where approximately 300 liters of helium were evaporated each time within 15 sec, the helium evaporation (using the protection system) is limited to ~ 15 to 20 liters when the magnet quenches. However, the recharging of the magnet is cumbersome and thus variable shunt resistors R_1, R_2, R_3 are used, which, in case of

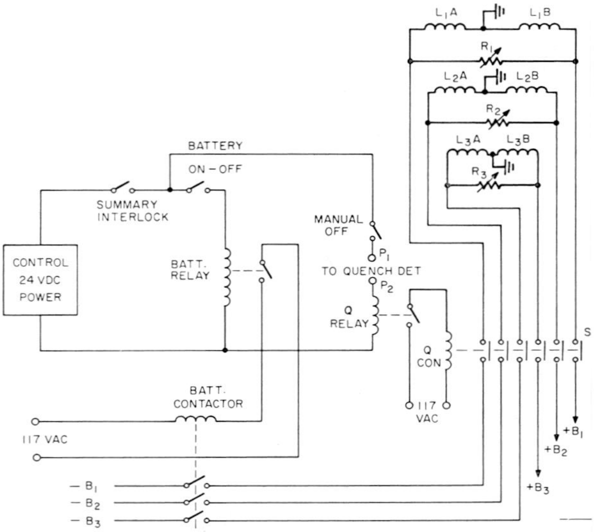

Fig. 7.5.4a. Schematic representation of the SLAC coil protection system

a magnet quench, yield with the coil inductance a field time constant of 5 seconds. The use of low-resistance shunts and diodes has the additional advantage that the power supply can be disconnected entirely; the magnet energy is dissipated in a programmed fashion over R_1–R_3.

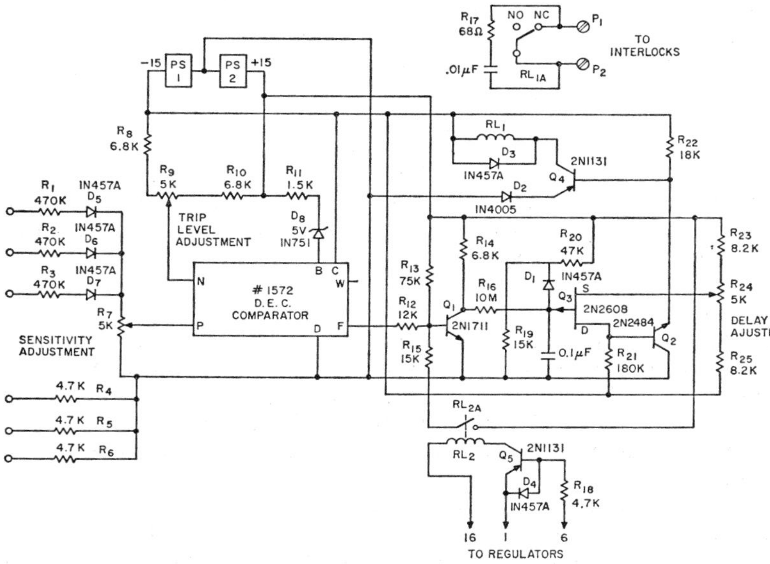

Fig. 7.5.4b. Detail of the SLAC coil protection system

In Fig. 7.5.4a, the magnet modules are indicated by ($L_{1A} \dots L_{3B}$) and are directly connected to external variable water resistors, $R_1 \dots R_3$. The coil is energized from stabilized current sources, $B_1 \dots B_3$. The main three-pole switch, S, is activated manually by means of a fast-acting relay Q, or automatically by means of the system (Fig. 7.5.4b) attached to $P_1 \dots P_2$.

The automatic switching system uses a voltage comparison method, where the coil terminal voltage is continuously compared to a pre-set constant voltage by means of a comparator. The following requirements, however, have to be met:

(a) In case of a fluxjump or appearance of a transient voltage, the switch S should not be activated immediately.

(b) When the magnet current is changed, resulting in a positive or negative inductive voltage, the switch S should not be activated.

The comparator in the "quench detector" is commercially available.* When the magnet is being charged, the input voltage to the diodes ($D_5 \dots D_7$) is set such that they are reverse biased. Thus the inductive charging voltage, due to the increase in current $\left(L \dfrac{di}{dt}\right)$, has no effect.

* Manufactured by Digital Equipment Corporation, Maynard, Massachusetts.

However, if the terminal voltage also indicates a resistive component (part of the magnet current flows through the conductor substrate), the diodes ($D_5 \ldots D_7$) start to conduct. When the voltage exceeds the trip level, the switch S is activated. Lowering the magnet current deliberately, again the voltage $\left(L \frac{di}{dt}\right)$ would activate S. However, a signal from the power supply control circuit prevents actuation by closing the switch RL_{2A}. The switch RL_{2A} is closed as long as the magnet current is reduced. After the magnet current has reached its lowest set value, RL_{2A} will open, allowing the voltage detector to assume control.

When terminal P of the comparator is more positive than terminal N, terminal F will be at — 3 Volts with respect to common terminal D. This voltage allows Q_1 to be cut off and C_1 to be charged through resistors R_{16} and R_{14} towards + 15 Volts. When the voltage at C_1 reaches approximately 5 Volts, diode D_1 starts to conduct, holding the voltage at C_1 to a constant value of 5 Volts. However, as the voltage across C_1 starts to rise from zero to + 5 Volts, transistor Q_3 (F.E.T.) is cut off, which turns off Q_2 and switches on Q_4. The relay RL_1 is energized, which activates RL_{2A}. Q is now de-energized, which trips off the main switch S. The cut-off voltage level at Q_3 is set by R_{22}, which controls the response time at any disturbance. The response time, in any case, is about 0.5 second.

The switching oscillograms of the magnet at a field of 6 T are given in Fig. 7.5.5. At $t = 0$, the switch S is activated. ($R_1 \ldots R_3$) are set at

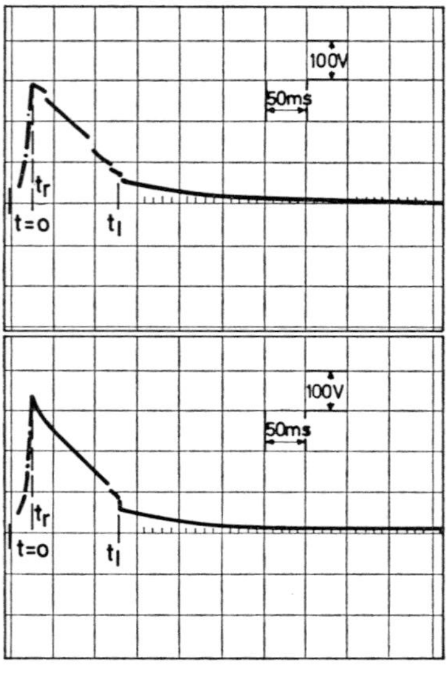

Fig. 7.5.5. Voltage-time-quench characteristics of the SLAC magnet

one Ohm each. The voltage is increased to the maximum allowable value of $I \cdot R = 350$ Volts. The total field energy is dissipated according to

$$L : \sum_{i=1}^{n} R_i = 120 \text{ ms} ,$$

until the recovery current value is reached at t_1. The magnet is superconducting and the rest of the magnet energy is dissipated according to the time constant $\tau = L/R_i$, with R_i being solely the shunt resistance and the resistance of the leads and cables.

If a partial or complete coil quench occurs and the current flows through the conductor shunt material (substrate or matrix), the conductor temperature will rise to a certain limit given by the conductor and protective circuit characteristic parameters. Generally, the maximum temperature seldom exceeds 70 K in stabilized magnets. In order to prevent thermal and thermomechanical damage to the conductor and insulation, it is necessary to limit the maximum temperature reached to some maximum value T_m. (See Chapter 6 for choice of J_{opt}.)

The normal region along the conductor propagates approximately 10 to 20 m/s.

Some quench time constants are as follows:

(i) Quench propagation time constant in the coil is 0.01 to 0.1 sec.

(ii) The time constant required to deposit the magnetic field energy as heat in the coil is 0.7 to 1.5 sec.

(iii) The time constant of the heat removal from the coil to the helium bath is 10 to 20 sec.

Magnet Cool-down and Helium Boil-Off

The weight of the whole system, including the insert coil and the support structure, is 1620 kg, of which 1060 kg is the weight of copper and superconductor (Nb_3Sn in the Max Planck coil module IV*, Nb(50%)Ti in the SLAC coil), 350 kg stainless steel, 140 kg aluminum, and 70 kg of insulation and miscellaneous metals and materials. To cool-down this mass from 300 to 4.2 K, a total of 1.11×10^8 Joules must be removed from the magnet. With the SLAC refrigeration system (7 watt liquefier operated as a refrigerator yields $\sim$ 25 watts at 4.5 K), this cool-down can be accomplished at a slow rate of an average of 3 K per hour. It was decided to cool-down the magnet system by using supplementary gaseous and liquid nitrogen, which was pumped through a built-in heat exchanger within the coil modules.

From 100 K down, the 7 W liquefier was operated as a refrigerator, with the helium gas circulated through the coil in a closed circuit. Part of the sensible heat of helium is now utilized.

An average of 2 ... 3 K per hour cooling was about the optimum condition achieved. The cool-down to 40 K was accomplished in 71 hours.

---

* The module IV (Max Planck Institute Munich) was placed within the SLAC coil to test its performance at a backing field of 6 T. The total coil assembly produced a central field of 9.9 T with a maximum field of 10.3 T at the conductor (see Fig. 7.5.1).

Assuming supplementary liquid helium would be available to cool-down the magnet from 100 to 4.2 K, about 160 kg of liquid helium would have been required (see Fig. 6.2.24). Instead, cooling the magnet down from 100 to 20 K by means of gaseous helium and from 20 to 4.2 K by using bulk liquid helium, about 190 liters of helium was used.

To fill the container with liquid helium, approximately 480 liters of helium were used. The static helium boil-off, measured with a laminar flowmeter, was about 4 liters per hour. The steady-state helium boil-off, when the magnet had reached a central field of 9.9 T with all 7 modules, was ~ 8 liters per hour. Two pairs of current leads, with counterflow cooling, were used to carry 500 amperes to modules I, III, and IV. These leads have exhibited total losses of 4.6 liters/hour at 500-A current level (static and dynamic values). As module II was energized only to ~ 60 A, total losses in these additional current leads encountered were 0.8 liter per hour.

Referring to earlier operations of the magnet system, 14 to 16 liters of helium were evaporated per hour when only modules I ... III were energized. Recent modifications and improvements of the system now allow the operation of the magnet system continuously from the 7 watt liquefier.

The heat load imposed on the magnet refrigeration system has three origins: *Heat conduction* through support and electrical leads, *heat radiation* through the thermal insulation, and *ohmic losses* through the electrical leads. During magnet charging, i.e. raising and lowering the magnet current, losses are produced due to generation of eddy currents in the substrate and fluxjump resistivity in the superconductor.

The losses during magnet charging depend on the speed with which the magnetic field is increased or decreased, the conductor cross-section, conductor dimensions parallel and perpendicular to the field, the dimensions of the superconductor, and the rate of twist. With the available current regulators in the cryogenics laboratory, the minimum rate at which we were able to change the magnet current was 5 A per minute. The module IV was energized at this speed, while the current in the SLAC coil (I ... III) was increased at a rate of about 7 ... 10 A/min. This dynamic heat load resulted in an additional 3 to 5 liters per hour boil-off during magnet charging, but as it was temporary, it did not affect the long-range steady-state operation of the magnet.

The steady-state heat load of 8 liters of helium per hour for a 6 Mjoule unit having a container diameter of 1 m, is achieved, thanks to thermal shielding, careful application of superinsulation, which yielded a thermal conductivity in the order of $(1 \dots 2) \times 10^{-6}$ W/cm K, and liquid-nitrogen-cooled heat shields. As indicated, the heat leak and ohmic losses due to the electric leads are the largest part of the total heat load. Three pairs of leads were utilized:

(a) A pair of leads designed and built at SLAC, utilizing the sensible heat of helium gas. This pair of leads has a static heat load of 1.3 liters per hour and an ohmic loss of 1.3×10^{-3} W/A lead at 500-A current level, as illustrated in Fig. 7.5.6, curves a_1 and a_2. These leads are essentially adequate for use at 1000 A, with 0.98×10^{-3} W/A/lead performance. They are built from two spirally-wound copper conductors, with one channel blocked to helium passage by means of glass wool, and only one surface cooled by the evaporated helium. These leads are efficient, but costly. They were used to energize Modules I and III.

(b) A pair of leads built at SLAC by means of cooling fins. The leads are simple in design, but have a high dynamic boil-off (curve b_2 in Fig. 7.5.6). Their static boil-off is considerably less than the other leads. These leads were used to energize Module II.

(c) The Max Planck module IV was energized through a pair of leads commercially acquired. These leads have a high static boil-off of 2.9 liters per hour, but an ohmic loss of 2.1×10^{-3} W/A lead at the 500-A level, which was the specified operation value. The lead performance is shown in Fig. 7.5.6, curves c_1 and c_2.

Additional losses are attributed to series connections between modules, to joints from current leads to the composite superconductor, and

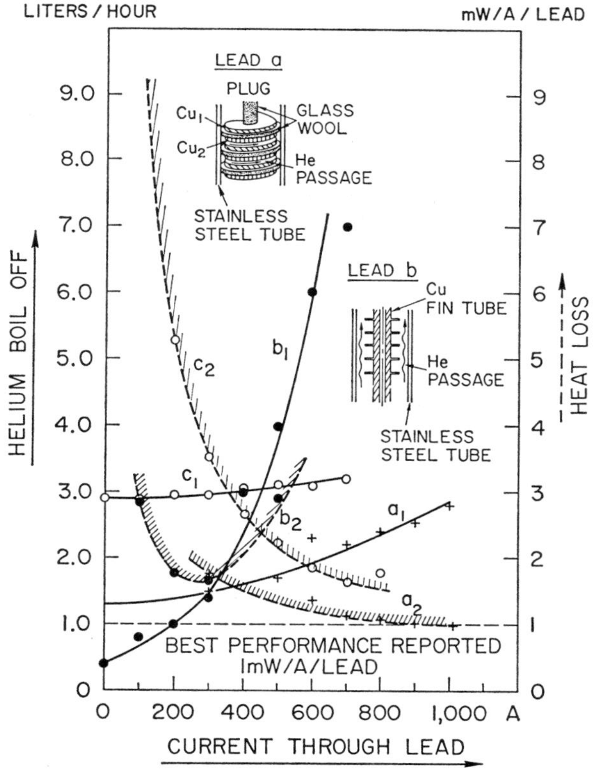

Fig. 7.5.6. Static and dynamic performance of current leads used at the SLAC magnet system

to internal joints within each module. The series connections may have a resistance of $\sim 10^{-7}$ to 10^{-8} Ohms per joint, but the total summation may add up for the balance of losses encountered.

A short note may be devoted to transfer and flashing losses when helium is transferred from 500-liter dewars to the helium container, including the magnet system. The transfer lines are rigid, evacuated, double-walled stainless steel tubes, with lengths of 2 to 3 m. They are home-built and as no superinsulation or nitrogen-cooled shields are used, losses are high and can be estimated at ~ 1 watt per m length of transfer line, compared to new, highly evacuated transfer lines having loss values of 0.3 W/m length transfer line. As transfer lines are in use only a fraction of the operation time, this loss in the laboratory is not too important.

Additional helium in the magnet container is evaporated when a jet of warm helium gas is blown through the transfer line into the magnet cryostat. This was unavoidable, as the transfer line had to be cooled from room temperature down to 4.2 K each time liquid helium is transferred from the storage dewar to the container and no liquid-gas separator was used. The flashing losses can be kept to a minimum if, at the end of the transfer line, a metallic gas breaker is installed which prevents the jet of helium gas from mixing with the liquid, or if continuous helium filling is introduced. In this case, approximately 2 to 5 liters of helium are additionally evaporated each time helium is transferred.

In order to be fully equipped for all possibilities of liquid losses, SLAC can store approximately 1200 liters of helium in 5 dewars prior to each large scale experiment. During the steady-state operation, the liquefier produces $\sim$ 10 liters of helium per hour, and thus the magnet steady-state operation can be extended ad libitum.

References

[1] Jones, R. E., et al.: Paper F 1. Advances in Cryogenic Engineering **15,** 141 (1970).

[2] Laverick, Ch., Lobell, G.: Rev. Sci. Instr. **36,** 825 (1965).

[3] Wong, J., et al.: J. of Appl. Phys. **39,** 2518 (1968).

[4] Wittgenstein, F.: Rev. Industries Atomique **5/6,** 23 (1970).

[5] Stevenson, R.: Canadian J. of Physics **41,** 2102 (1963).

[6] Stevenson, R.: Canadian J. of Physics **42,** 1343 (1964).

[7] Montgomery, D. B., et al.: Adv. in Cryog. Engng **14**, 88 (1969).

[8] Montgomery, D. B.: Solenoid Magnet Design. John Wiley Interscience (1969).

[9] Brechna, H.: MIT National Magnet Lab. Rep. NML 62-1 (1962).

[10] Efferson, K. R., et al.: 4th Symp. on Engng. Probl. of Fusion Research (1971).

[11] Brechna, H., et al.: Paper C 3 Adv. in Cryog. Engng. **13**, 116 (1968).

[12] Smith, P. F., et al.: Proc. 1968 Summer Study on Supercond. Devices, p. 839. Brookhaven 1968.

[13] Steckly, Z. J. J., Zar, J. L.: Trans. I.E.E. **12,** 367 (1965).

[14] Hancox, R.: Proc. I.E.E. (London) **113,** 1221 (1966).

Subject Index

101860

1-MONTH